BARRON'S

AP®

CHEMISTRY

5TH EDITION

Neil D. Jespersen, Ph.D.
Professor of Chemistry
St. John's University
Jamaica, New York

BARRON'S

About the author:

Professor Neil Jespersen earned his B.S. degree in chemistry at Washington and Lee University (Lexington, Va.) and his Ph.D. degree at The Pennsylvania State University. He specializes in bio-analytical research projects and has mentored more than 100 undergraduates and 25 graduate students. Dr. Jespersen teaches graduate and undergraduate analytical chemistry courses and regularly teaches sections of general chemistry. He currently serves on the board of the American Chemical Society.

Credits:
Art on pages 79 and 80 from SAT Subject Test in Chemistry, 9th Ed., by Joseph A. Mascetta. Barron's Educational Series, Inc., 2008.

All inquiries should be addressed to:
Barron's Educational Series, Inc.
250 Wireless Boulevard
Hauppauge, New York 11788
http://www.barronseduc.com

ISSN (print only): 2150-3362
ISSN (print w-CD-ROM): 1949-9205
Library of Congress Control Number (print only): 2009200366
Library of Congress Control Number (print w/CD-ROM): 2007203731

ISBN-13 (book only): 978-0-7641-4050-1
ISBN-10 (book only): 0-7641-4050-7
ISBN-13 (with CD-ROM): 978-0-7641-9523-5
ISBN-10 (with CD-ROM): 0-7641-9523-9

PRINTED IN THE UNITED STATES OF AMERICA

9 8 7 6 5 4 3 2

Contents

Preface

You are about to embark on one of the more intellectually challenging experiences of your life, the Advanced Placement Examination in Chemistry. Fewer than 1 percent of all high school students take this exam. Whatever the outcome, you are to be congratulated as one of a select group. As a conscientious student, you can use this review book to help you increase your score. A higher score can lead to college course credit and a head start in your selected career.

The AP Examination in Chemistry is different from other exams and tests that you have taken. *Explain*, *compare*, and *predict* are three important words often used on the AP Chemistry Exam. Remembered facts and calculation procedures are the basic groundwork of chemistry; however, high scores require a thorough understanding of chemical principles and relationships. Chemistry is rich in these relationships. The key to success on the exam is to think like a chemist and to apply your knowledge of one or more basic principles to provide a logical description of how chemicals behave.

This review book is designed with you, the student, in mind. It concentrates on the topics that are essential for a good score on the AP Chemistry Exam. In particular, the book is designed to provide insights into the use of basic principles to answer seemingly complex questions.

The discussion in each chapter is interspersed with exercises in which subject-matter problems are presented and solved. At the end of each chapter are questions to test your understanding of the topics discussed. These, together with the two diagnostic and three practice tests, provide hundreds of questions with a range of difficulty and complexity typical of an advanced placement exam. This review material will help you to pinpoint weak areas on which you need more preparation, and the explained answers can be used to identify sources of error or confusion.

Acknowledgments

First and foremost, a very special thank-you to my wife, Marilyn Zak Jespersen, who spent countless hours reading and correcting the manuscript and suggesting changes. Marilyn's contributions have made this book readable, understandable, and user-friendly. No other person could have been as dedicated to the work as she was.

I am grateful also to Professor James Brady for many fruitful discussions and ideas in the years we have been colleagues at St. John's. We share the idea that our job is not to teach chemistry but to excite students into learning it. In our recent projects, he has continued to hone my skills as a writer.

Finally, I thank the editors and reviewers for their suggestions and encouraging comments during the production of the book.

Introduction

IMPORTANT FACTS ABOUT THE ADVANCED PLACEMENT EXAMINATION IN CHEMISTRY

This examination is given in May each year at selected sites throughout the country. Exact dates, locations, and application forms are available in most guidance counselor offices. Information is also available from the following College Board Advanced Placement Program Offices.

National Office
Advanced Placement Program
45 Columbus Avenue
New York, NY 10023-6992
212 713-8066
E-mail: ap@collegeboard.org

Middle States Regional Office
Serving Delaware, District of Columbia, Maryland, New Jersey, New York, Pennsylvania, Puerto Rico, and the U.S. Virgin Islands
Two Bala Plaza, Suite 900
Bala Cynwyd, PA 19004-1501
866 392-3019
E-mail: msro@collegeboard.org

Midwestern Regional Office
Serving Illinois, Indiana, Iowa, Kansas, Michigan, Minnesota, Missouri, Nebraska, North Dakota, Ohio, South Dakota, West Virginia, and Wisconsin
6111 N. River Road, Suite 550
Rosemont, IL 60018-5158
866 392-4086
E-mail: mro@collegeboard.org

New England Regional Office

Serving Connecticut, Maine, Massachusetts, New Hampshire, Rhode Island, and Vermont

470 Totten Pond Road
Waltham, MA 02451-1982
866 392-4089
E-mail: nero@collegeboard.org

Southern Regional Office

Serving Alabama, Florida, Georgia, Kentucky, Louisiana, Mississippi, North Carolina, South Carolina, Tennessee, and Virginia

3700 Crestwood Parkway NW, Suite 700
Duluth, GA 30096-7155
866 392-4088
E-mail: sro@collegeboard.org

Southwestern Regional Office

Serving Arkansas, New Mexico, Oklahoma, and Texas

4330 South MoPac Expressway, Suite 200
Austin, TX 78735-6735
866 392-3017
E-mail: swro@collegeboard.org

Western Regional Office

Serving Alaska, Arizona, California, Colorado, Hawaii, Idaho, Montana, Nevada, Oregon, Utah, Washington, and Wyoming

2099 Gateway Place, Suite 550
San Jose, CA 95110-1051
866 392-4078
E-mail: wro@collegeboard.org

The Educational Testing Service has a web page at http://www.collegeboard.org. The web page for the advanced placement examination is http://www.collegeboard. org/ap/chemistry/html/index001.html/

Application forms and fees for the examination are usually due 1 month before the examination date. Late registration may not be accepted and a penalty fee is charged.

TEST CONTENT AND DISTRIBUTION

The material on the AP Chemistry Examination parallels the suggested course content for an AP course. The following lists topics and the approximate percentage of the exam devoted to each one. Note that this is not an exhaustive list of all possible topics, but it is a general guide.

STRUCTURE OF MATTER (20%)

Atomic Theory and the Structure of the Atom
  Evidence supporting the atomic theory
  Atomic masses, atomic numbers, mass number, isotopes
  Electronic structure of the atom
    Energy levels, atomic spectra
    Quantum numbers, atomic orbitals
  Periodic trends and relationships
    Ionization energies, electron affinities, electronegativity
    Atomic and ionic radii, oxidation states
Chemical Bonding
  Inter- and intramolecular binding forces
    Ionic and covalent bonding
    Hydrogen bonds, dipole-dipole and van der Waals forces
      (including London forces)
    Forces related to states, properties, and structure of matter
    Bond polarity, electronegativity
  Models of molecules
    Lewis structures, resonance
    VSEPR
    Valence bond theory, hybrid orbitals, sigma and pi bonds
  Molecular geometry, structural isomerism
    Geometry of simple molecules, organic and inorganic
    Coordination complexes
    Dipole moments, molecular polarity
    Relationship of properties to structure
Nuclear Chemistry
  Nuclear equations, radioactivity
  Half-lives and applications

STATES OF MATTER (20%)

Gases
    Laws of ideal gases
        Ideal gas law (equation of state)
        Partial pressures (Dalton's law)
    Kinetic-molecular theory
        Interpretation of gas laws
        Avogadro's hypothesis
        Relationship between kinetic energy and temperature
        Deviations from ideal gases (real gases)
Liquids and Solids
    Kinetic-molecular theory applied to liquids and solids
    Phase diagrams of a pure substance, triple point, critical point
    Changes of state
    Structure of solids and crystals; lattice energy
Solutions
    Types of solutions and solubility
    Concentration units (normality is not tested)
    Raoult's law
    Colligative properties, osmosis
    Nonideal behavior of solutions

REACTIONS (35–40%)

Reaction Types
  Acid-base reactions; Arrhenius, Brönsted-Lowry, and Lewis theories
  Coordination complexes and amphoterism
  Precipitation reactions
  Oxidation-reduction reactions
    Oxidation number
    Electron transfer in oxidation and reduction
    Electrochemistry including electrolytic cells and Faraday's laws, galvanic cells and standard reduction potentials, Nernst equation prediction of the direction of a reaction
Stoichiometry
  Ionic and molecular species in chemical systems, net ionic equations
  Balancing equations including redox equations
  Mass and volume relationships using the mole concept, empirical formulas, and limiting reactants
Equilibrium
  Dynamic equilibrium concept, Le Châtlier's principle and equilibrium constants
  Quantitative use of equilibrium
    Equilibrium constants for gas-phase reactions
    Equilibrium constants for reactions in solutions
      Acid-base equilibrium, and pH calculations
      Solubility product calculations including common ions
      Buffer and hydrolysis equilibria
Kinetics
  Rates of reaction, general concepts, and factors
  Determination of rates, rate laws, reaction order, and rate constants from experimental data including graphs
  Effect of temperature on rates
  Activation energy and catalysis
  Relationship of rate-determining step to rate laws and reaction mechanisms
Thermodynamics
  State functions
  First law of thermodynamics, enthalpy change, heats of formation and reaction, Hess's law, calorimetry
  Second law of thermodynamics including the concept of entropy, free energies of formation and reaction, the relationship between free energy, enthalpy, and entropy.
  Relationships between free energy change, equilibrium constants, and electrode potentials

DESCRIPTIVE CHEMISTRY (10–15%)

There are a large number of facts, principles, and concepts not listed above that will be needed to demonstrate a knowledge of chemistry on the AP Exam. In addition, the principles, concepts, and properties of chemicals are part of the real world outside the classroom. This should be a continuing part of the AP course. Some appropriate areas are

Chemical reactivity and reactions and a knowledge of chemical nomenclature

Relationships in the periodic table that allow prediction of chemical and physical properties. These can be horizontal, vertical, or diagonal relationships.

Organic chemistry, structures, functional groups, nomenclature, and chemical properties.

LABORATORY (5–10%)

College-level chemistry courses include laboratory exercises. These involve measurement, preparation of solutions, experimental setup, and synthesis and analysis of various compounds. The AP Exam will have questions involving the laboratory experience. Most of those questions involve general laboratory procedures, including

Making and recording appropriate observations of chemical substances and reactions

Making quantitative measurements and recording data appropriately

Calculating results from quantitative data and making appropriate interpretations of these results

Communicating the results of experiments in laboratory reports

FORMAT OF THE EXAMINATION

The Advanced Placement Examination in Chemistry consists of two sections. Section I and Section II are weighted equally (50 percent each) toward your overall score.

Section I is a 75-question, multiple-choice set of questions and is essentially the same as in past years in format and administration. You will have 90 minutes for this section, and calculators are NOT allowed. Any calculations will be simple mathematics that generally should not require a calculator. This section is designed to test the breadth of a student's knowledge of chemistry. Because each AP Chemistry course cannot possibly cover all possible topics, it is expected that there will be questions that you cannot possibly answer. No student is expected to know the answers to all of the multiple-choice questions.

Section II has two parts. Part A is allotted 55 minutes and Part B 45 minutes. You are allowed to use calculators for only Part A of Section II. Part A has three quantitative questions; one of these will focus on chemical equilibrium. There will be one laboratory-based question. If it appears in Part A, it will be a quantitative question. If it appears in Part B, it will be a qualitative question. Part B will also contain one question that asks you to write three balanced chemical equations from word descriptions of the reactants. You will need to determine the reaction products and write net ionic equations, if appropriate. There will also be a short question about each of these reactions.

CHEMICAL CALCULATIONS

Below is a list of problem types and calculations that the test taker should have mastered after taking an advanced placement course. The student should be able to assess calculation results in regard to reasonableness, significant figures (including the results of logarithmic and exponential operations), and precision of measurements.

1. Percentage composition
2. Empirical formulas from experimental data and molecular formulas from empirical formulas
3. Molar masses from physical and colligative properties, including gas density, freezing-point measurements, and boiling-point measurements
4. Gas laws, including the ideal gas law, Dalton's law of partial pressures, and Graham's law of effusion
5. Stoichiometric relations using the concept of the mole; titration calculations
6. Mole fractions; molar and molal solutions
7. Faraday's laws of electrolysis
8. Equilibrium constants and their applications, including their use for simultaneous equilibria
9. Standard electrode potentials and their use; Nernst equation
10. Thermodynamic and thermochemical calculations
11. Kinetics calculations

CALCULATOR POLICY

The policy toward calculators used on the AP Chemistry Examination has evolved with the power and capabilities of calculators themselves. The intent of the policy is to make the examination fair to all students. As noted above, calculators may be used only on Section II, Part A, of the exam. You will be allowed to use any calculator, including programmable and graphing calculators. You will not be able to use any type of computer, including laptops, PDAs, and devices with an alphanumeric keyboard or stylus/pen input. If there are any questions, you should consult the AP Chemistry website or one of the Educational Testing Service offices.

SCORING OF THE EXAMINATION

The multiple-choice section is machine-graded. Care must be taken to be sure that there are no stray marks on your answer sheet and that all erasures are clean and complete. Also be sure that each answer has only one response marked. There is a penalty for wrong answers. One quarter (0.25) is deducted for each incorrect multiple-choice answer. There is no penalty for leaving a question unanswered.

Highly trained high school and college chemistry teachers using a predefined scoring system grade the free-response section. Scoring is reviewed several ways to ensure consistent results. In addition, different readers grade each question to avoid carryover from one question to the next. Exams are chosen randomly for rescoring to be sure that scores do not drift during the scoring sessions.

Scores for Sections I and II are combined to obtain an overall score. These are then translated into the 1 to 5 rankings that students and colleges receive. The scoring varies little from year to year, as the table below indicates.[1]

AP Examination Score	Meaning	2006 Results (%)	2005 Results (%)	2004 Results (%)
5	Extremely well qualified	17.0	15.0	16.4
4	Well qualified	18.4	18.3	17.6
3	Qualified	22.4	22.7	22.9
2	Possibly qualified	17.7	19.7	18.9
1	No recommendation	24.5	24.3	24.3
Students taking the exam		87,465	78,453	71,070

[1] *Information provided by the AP Chemistry website.*

Approximately 59 percent of those taking the exam earn a score of 3 or better.

HOW TO USE THIS REVIEW BOOK

Every student taking the AP Chemistry Examination is an individual with different plans and needs. Used properly, this book can be individualized to help any student maximize their score on the AP Exam no matter what their previous preparation or success with learning chemistry has been. There are two important factors: (1) the time available for your preparation before the exam and (2) your current knowledge of chemistry. Look under the heading below that seems to fit you best.

I'm at the Top of My Class and Just Need a Tune-Up

This review book has the most problems and questions of any currently available. If you are at the top of your class, the first place to start is with a sample of the AP Exam. There are five exams available, three at the end of this book and two electronic versions. Take these exams to sharpen your responses and learn how to manage your time.

I Do Pretty Well, But I Can Do Better

You will need a little time to review all the topics. We suggest taking one of the five practice exams to see how you do. If you do well, study the topics that show up as weaknesses. If your score is less than desired, you might want to try the diagnostic tests.

Average Describes Me, But I Need to Excel

Start with a diagnostic test. Evaluate where the problems lie and read those parts of the book thoroughly. Try a practice exam and evaluate your results. If needed, take another diagnostic exam to locate and correct your major problems.

Chemistry Confuses Me, Help

Even the totally confused can be helped. Start with the diagnostic tests. Review the chapters on subjects that show up as problems in the analysis. Repeat the diagnostic tests as needed. The more time you have, the better. However, the diagnostics in this book help identify the areas that can produce the a better score with the time available.

ORGANIZING YOUR REVIEW

1. **Plan to start reviewing as soon as possible.** You don't go on a diet and lose 10 pounds in 1 day, you cannot exercise for a week and run a marathon, and you cannot read this book in one sitting and have a thorough review of chemistry. The sooner you start reviewing, the more leisurely pace you can use to digest information. Cramming does not allow time for concepts to gel and for relationships to become apparent.

2. **Set a schedule for your review.** Depending on the available time before the AP exam, divide your review into reasonable study blocks. Schedule more time than you need, plan on 2-hour study sessions, and reward yourself with time off for topics you know well.

3. **Actively review.** Don't just read this book; actively study it. Use a red pen to cancel units in problems. Write out the answers to all problems; don't solve them mentally. Make notes of topics that you find confusing.

4. **Ask questions.** AP teachers are dedicated educators who want you to succeed, but don't expect them to teach the whole course over again. Ask specific questions (write them down beforehand). If you need help with a problem, show your teacher that you've tried to solve it and indicate where you got stuck. Education is learning how to ask questions and get them answered.

5. **Assess your progress and review your weaknesses.** Use the diagnostic tests to help you concentrate on specific areas. Use the practice exams to become accustomed to test conditions.

The following schedule may be helpful in planning your review sessions. Each directive represents a 2-hour session. The schedule should be read horizontally, not vertically. For example, in the first week of review, you would take the first diagnostic test on Monday, review the first most needed chapter on Wednesday, and review the second most needed chapter on Friday.

The schedule requires 8 weeks to complete and leaves all weekends, as well as Tuesdays and Thursdays, free. You can change this to a Tuesday/Thursday/Saturday schedule or any other sequence that assures regular review. This schedule can be compressed into as few as 3 weeks if review is done on a daily basis.

Practice Exam 3 is included for additional practice.

	Monday	Tuesday	Wednesday	Thursday	Friday
Week 1	Take first diagnostic test.		Review first most-needed chapter.		Review second most-needed chapter.
Week 2	Take second diagnostic test.		Review first most-needed chapter.		Review second most-needed chapter.
Week 3	Chapter 1		Chapter 2		Chapter 3
Week 4	Chapter 4		Chapter 5		Chapter 6
Week 5	Chapter 7		Chapter 8		Chapter 9
Week 6	Chapter 10		Chapter 11		Chapter 12
Week 7	Chapter 13		Chapter 14		Take first practice exam.
Week 8	Review		Review weak areas.		Take second practice exam.

(If results indicate continued weakness after the second practice test, repeat the last week of review and take the third diagnostic test.)

WHAT TO EXPECT OF THE AP CHEMISTRY EXAMINATION

1. **Exam difficulty.** There is no question that the Advanced Placement Examination in Chemistry is difficult, and there are at least three reasons. First, there will be topics that you never covered in class. Because of the volume of material, even college-level courses do not include all of the topics presented on the exam. Don't waste time on questions dealing with unfamiliar material. Second, the examination is long. The multiple-choice section allows an average of slightly more than 1 minute per question. Third, many questions combine two or more concepts.

2. **Section I: Multiple-choice questions.** There are three types of questions. *Factual questions* require quick recall of important facts about chemicals and their reactions. *Conceptual questions* ask you to assess how a theory, law, or concept is applied in chemistry. *Estimation questions* require only simple mathematical solutions, using fundamental equations of chemistry.

Good scores are achieved by correctly answering as few as 50 percent of the 75 multiple-choice questions. On average, students who answer all the questions score 60 percent on the first 25 questions, about 50 percent on the next 25 questions, and only 30 percent on the last 25 questions. The increasing difficulty of the questions and a lack of time are the main reasons for the dramatic decrease in the scores for the last 25 questions.

3. **Section II: Free-response questions.** Part A requires calculated solutions to problems. Part B involves writing chemical reactions and essay answers. Part A allows the use of a calculator and tables of equations. Part B does not allow a calculator or tables of information. One question in Part B requires writing chemical equations from a descriptive sentence. The three remaining questions refer to general chemistry and laboratory principles. This is illustrated in the practice exams at the end of the book. Concise, well-thought-out answers are important. Explain what you are doing, and state all assumptions. Before starting calculations, you should show fundamental equations, with units. A knowledge of many principles of solubility, ionization, complexation, and oxidation-reduction is required in order to write chemical equations with correct reactants and products. Identify the type of reaction: combustion, double replacement, and so on. Essay questions ask you to explain fundamental principles as applied to specific substances. Questions about common household and environmental chemicals are also asked.

Readers are looking for key terms and concepts, along with their correct usage, in explanations. Define terms clearly, and avoid using any you are not sure are correct.

HOW TO MAXIMIZE YOUR SCORE

The Advanced Placement Examination in Chemistry is designed so that the average score will be approximately 50 percent. This is done by careful selection of the difficulty of the questions and of the length of the exam itself. There are well-known concepts and methods for ensuring that you will achieve the maximum score you deserve.

The multiple-choice section is designed to test your recall of fundamental chemical concepts and the use of these concepts to solve basic chemistry problems. The questions cover the entire AP course syllabus and are designed with various levels of difficulty. Each question has five choices, only one of which is the most appropriate answer. There is a penalty for incorrect answers, and the multiple-choice section is graded using the formula

$$\text{SCORE} = \left(\begin{array}{c}\text{CORRECT} \\ \text{RESPONSES}\end{array}\right) - 0.25\left(\begin{array}{c}\text{INCORRECT} \\ \text{RESPONSES}\end{array}\right)$$

The free-response section involves written answers to selected types of questions. In Part A of this section, calculation-type questions are asked. A calculator is allowed only in this section. In addition, a special table of chemical equations and symbols will be made available. Logical solutions to problems are sought by the graders. Proper use of numbers and significant figures is also important. Show all steps and fundamental equations before starting calculations. Part B of the free-response section has four questions. One question requires that the student generate the correct chemical equation from a written description of a reaction. These equations need not be balanced. However, the correct products should be predicted, and spectator ions should be eliminated. The physical state, such as gas (g), liquid (l), or solid (s), is also helpful, although not required, in illustrating knowledge of the topic. The remaining three questions in Part B are essay questions. Two are required questions, with a choice available for the third. One of the required questions will focus on laboratory work in AP courses. In the numerical and essay sections the answers are graded by well-trained teams of chemistry professionals, usually college and high school teachers. Intensive training and testing of the graders ensures that the tests are graded uniformly. The grading is reviewed by several people to confirm that grading has been done correctly.

Strategies for Multiple-Choice Questions

SHOULD YOU GUESS?

> **TIP**
>
> Always take a guess if you can definitely rule out two or more answers.

The probability of selecting the correct choice by random guessing is one in five (20 percent), and the deduction for an incorrect answer is one in four (25 percent). **Random guessing is never appropriate**. However, if it is possible to definitely eliminate any of the responses, the probability of success rises. Eliminating one choice raises the odds to 25 percent, which is the break-even point. If two or more choices can be definitely eliminated, guessing from the remaining ones is a good strategy. For this strategy to work, however, you must be certain that the eliminated choices are wrong.

If you are not totally certain that a response can be eliminated, assign it a fraction depending on your level of certainty. If you are 50 percent sure that two answers can be eliminated, that is the same as eliminating one certain response, and you may be justified in guessing between the remaining choices.

DISTRACTERS

In the design of multiple-choice questions, the writer constructs the responses so that one choice is correct and the others are "distracters." A distracter is a response that looks good at first glance but has a serious flaw that makes it incorrect. The better the design of the test, the more distracters will be found in each question.

One popular method for constructing distracters is to use subtle changes in the wording to make a response incorrect. For example:

1. **All** chemicals become more soluble as the temperature increases.
2. **Most** chemicals become more soluble as the temperature increases.

or

1. The reaction is **exo**thermic.
2. The reaction is **endo**thermic.

Careful reading of questions and understanding of terminology are very important. The distinctions between "most" and "all" in the first set above and between "exothermic" and "endothermic" in the second are obviously significant. To ensure selecting the best answer to a nonnumerical problem, be sure to read each response before selecting one. Often a good-sounding, but incorrect, response is listed before the correct one. Another approach is to read the responses in reverse order. Pay special attention to responses that are exactly the opposite of each other as in the "exothermic"/"endothermic" example above. One response must necessarily be wrong and may also provide a clue as to whether or not both are incorrect. Also pay special attention to responses that differ by only one word, as in the first set above. Once again, they may provide a clue as to the correct way to think about the problem.

For numerical problems, some distracters provide answers in which the data are simply used in the wrong manner. For instance, for the question "What is the value of 5 divided by 2?" the answer choices may be as follows:

(A) 2.50 $\left(\dfrac{5}{2}\right)$

(B) 0.40 $\left(\dfrac{2}{5}\right)$

(C) 3.00   (5.00 – 2.00)
(D) 7.00   (5.00 + 2.00)
(E) –3.00   (2.00 – 5.00)

The parentheses show the calculation method used to obtain the answers. The question hinges on understanding the term *divided by*, and keeping in mind that 2, not 5, is the divisor, before the proper calculation can be made to obtain answer A. You must understand the proper method for using the data.

REASONABLENESS

There are, however, some common methods for increasing the probability of choosing the correct answer to a numerical chemistry problem. It is important to remember the *principle of reasonableness*. This means that answers must reflect obedience to fundamental principles, such as the conservation of matter and energy. Your personal experiences in everyday life also may provide clues as to the reasonableness of answers.

For example, if 2 grams of one reactant are mixed with 5 grams of another, it is impossible to have any more than 7 grams of product even under the best conditions (law of conservation of matter). Therefore any response that is greater than 7 grams may be eliminated very quickly. As another example, if a hot solution is added to a colder one, the final temperature must be somewhere between the low of the cold solution and the high of the hot solution. Any other responses may be eliminated as incorrect without any calculations at all.

Pointers about the reasonableness of answers will be given throughout this book.

ESTIMATING ANSWERS

The new Advanced Placement Tests place minimal focus on the use of calculators. Instead concepts are tested, and most mathematical problems use simple numbers

TIP

Estimating answers to mathematical problems is an important skill.

that do not require calculators. In effect, these questions test your ability to set up problems rather than any ability to solve problems or do mathematical operations. In accord with this new approach, students must now understand how to quickly estimate rather than calculate answers. Below are the basic principles of estimation. Many of the problem solutions at the end of the chapters and for the sample tests indicate methods for estimating answers and also show detailed calculations.

The **first principle** of estimation is that all problems must still be set up in a rigorous and logical manner. This has not changed from the times when calculators were allowed. The **second principle** of all estimation is to round numbers to one, or at most two, significant figures. The **third principle** is to round in a manner that makes cancellation simple and to take every opportunity to cancel. The **fourth principle** is to add and subtract in groups. A few examples of these principles follow. Although the following are broken into steps, there are no rigorous steps to memorize.

EXAMPLE 1

Solve for X.

$X = 25.34 \times 1.890 \times 0.00318$ Step 1: round to one or two digits
       $\times\ 4.1689 \times 9.823$
   $= 25 \times 2 \times 0.003 \times 4 \times 10$ Step 2: note that $4 \times 2 \times 25 = 200$
   $= 200 \times 0.003 \times 10$ Step 3: simplify the decimal
   $= 0.6 \times 10 = 6$ Step 4: finish up

Notice (step 1) that you should try to round up as much as you round down. Also notice (step 2) that you DO NOT have to multiply numbers in sequence. The actual calculator answer is 6.24. This is off by 0.23, but when estimating, that is perfectly alright. You now know that the correct answer cannot be 13,248 and that it cannot be 5.2×10^{-5}.

EXAMPLE 2

Calculate Y.

Here we demonstrate how to estimate the high and low limits of a calculation.

$Y\ =\ 3.642 \times 1.102 \times 7.785$ Step 1: round to simple numbers,
       $\times\ 178.2 \times 51.65 \times 0.00219$ except 3.5
   $=\ 3.6 \times 1 \times 8 \times 200 \times 50$ Step 2: note that $200 \times 50 = 10,000$
       $\times\ 0.002$
   $=\ 10,000 \times 3.6 \times 8 \times 0.002$ Step 3: $10,000 \times 0.002 = 20$
   $=\ 20 \times 3.6 \times 8$ Step 4: 3.6 is between 3 and 4;
$Y_1\ =\ 20 \times 4 \times 8 = 640$ (maximum)   do two calculations
$Y_2\ =\ 20 \times 3 \times 8 = 480$ (minimum)

The calculated answer is 626.7.

In this problem we chose one number that could be rounded either up or down and rounded it up to find a maximum and down to find a minimum. This is a useful technique to set limits on your answers.

EXAMPLE 3

Calculate Z.

$Z = 21.47 \times \dfrac{0.000341}{0.06625} \times \dfrac{294.3}{771.9}$

Step 1: round numbers, paying attention to cancelling opportunities

$= 20 \times \dfrac{0.0003}{0.06} \times \dfrac{300}{800}$

Step 2: cancel 2 of 20 with 8 of 800 to get

$= 10 \times \dfrac{0.0003}{0.06} \times \dfrac{300}{400}$

Step 3: cancel zeros in 300 and 400

$= 10 \times \dfrac{0.0003}{0.06} \times \dfrac{3}{4}$

Step 4: divide center terms by 3 to obtain the following

$= 10 \times \dfrac{0.0001}{0.02} \times \dfrac{3}{4}$

Step 5: multiply out the numerator and denominator

$= \dfrac{0.003}{0.08}$

Step 6: multiply the numerator and denominator by 1000 each

$= 3/80$

Step 7: estimate the answer as about 4 parts in 100, or 0.04

The calculated answer is 0.0421. Once again the estimate is in the ballpark. It is important to notice how advantage was taken of simple math operations in cancelling. Good cancelling saves a lot of work and reduces errors. Finally, each separate step was written out above. In real examples, these steps are done on a single equation without rewriting it.

EXAMPLE 4

Calculate A.

$A = \dfrac{2.847 \times 10^{-3}}{9.113 \times 10^{6}} \times \dfrac{6.321 \times 10^{-8}}{14.34 \times 10^{-23}}$

Step 1: regroup all exponential terms together and all numbers together as shown

$= \dfrac{2.847 \times 6.321}{14.34 \times 9.113} \times \dfrac{10^{-3} \times 10^{-8}}{10^{6} \times 10^{-23}}$

Step 2: evaluate exponents

$= \dfrac{2.847 \times 6.321}{14.34 \times 9.113} \times 10^{6}$

Step 3: evaluate numbers; simplify first

$= \dfrac{3 \times 6}{14 \times 9} \times 10^{6}$

Step 4: divide numerator and denominator by 3

$= \dfrac{1 \times 6}{14 \times 3} \times 10^{6}$

Step 5: divide numerator and denominator by 3 again

$= \dfrac{1 \times 2}{14 \times 1} \times 10^{6}$

Step 6: divide numerator and denominator by 2

$= \dfrac{1 \times 1}{7 \times 1} \times 10^{6} = \dfrac{1}{7} \times 10^{6}$

or 0.14×10^{6}

The calculated answer is 0.1377×10^{6} or 1.377×10^{5}.

EXAMPLE 5

Calculate *B*.

$$B = 20.5 + 2.346 + 102.33 + 33.62 + 5.009$$

Round off to one significant figure and add:

$$B = 20 + 2 + 100 + 30 + 5$$
$$= 157$$

The calculated answer is 163.8.

EXAMPLE 6

Estimate pH from [H$^+$], where pH is defined as $-\log$ [H$^+$].

In almost all cases the hydrogen ion concentration is expressed exponentially, for example, 4.1×10^{-5} M. It turns out that the pH for a 4.1×10^{-5} M solution is 4.39. This is between pH 4 and 5. In fact, any hydrogen ion concentration with an exponent of 10^{-5} has a pH between 4 and 5. Extending this, we can estimate the pH to within one unit by simply looking at the power of 10 for the hydrogen ion concentration. The pH is ALWAYS a maximum of the positive value of the exponent and a minimum of one pH unit less than that. If [H$^+$] = 3.8×10^{-6}, we can quickly say that the pH is between 5 and 6. If [H$^+$] = 3.8×10^{-11}, the pH is between 10 and 11.

We can use exactly the same principle if [OH$^-$] is given and you want to know the pOH. For example, if [OH$^-$] = 1.3×10^{-2}, the pOH will be between 1 and 2. If [OH$^-$] = 3.3×10^{-6}, the pOH will be between 5 and 6.

In both cases the exponent tells us the maximum value for the pH or pOH. One less than this maximum is the minimum value that the pH can be.

EXAMPLE 7

Estimate [H$^+$] from pH.

If you are given a pH of 4.7, the corresponding [H$^+$] will be $10^{-4.7}$. The decimal exponent is unusual but does not violate any rules of mathematics. At times it is convenient to work with decimal exponents.

In cases where the decimal exponent is inconvenient, use a high and a low value to establish a range. For example, the [H$^+$] of $10^{-4.7}$ can also be a high of 10^{-4} to a low of 10^{-5}.

EXAMPLE 8

Determine additional logarithmic measurements.

If you want more precision in logarithms, it is necessary only to remember that $\log 2 = 0.3$ and $\log 3 = 0.5$.

For example, if the [OH$^-$] = 3.3×10^{-6}, we can calculate the pOH as $-\log$ (3.3×10^{-6}), which is $-\log 3.3$ and $-\log$ (10^{-6}). The former is -0.5 and the latter is $+6$. The two add up to 5.5 for the pOH.

Strategies for the Free-Response Section

NUMERICAL CALCULATIONS

Any scientific calculator, including a graphing calculator, can be used for Section II Part A of the exam. Be certain that the batteries are fresh and that you are familiar with the operation of your calculator. Write appropriate chemical reactions. Always write the fundamental equations or laws that the question requires. Identify variables. Use correct algebra in the solution and show as many algebraic steps as possible. Clearly state any assumptions you have used and verify that each assumption is valid before reporting the answer. Check also that you have used the correct number of significant figures in calculations.

For example, consider this problem: Calculate the pH of a 0.100 M solution of hydrofluoric acid, $K_a = 6.9 \times 10^{-4}$.

To logically solve the problem you need a chemical reaction, the equilibrium law, a simplification, and a solution, as shown in the following steps.

Reaction: $HF \rightleftharpoons H^+ + F^-$ or $HF + H_2O \rightarrow H_3O^+ + F^-$

Equilibrium law used: $K_a = \dfrac{[H^+][F^-]}{[HF]}$ or $\dfrac{[H_3O^+][F^-]}{[HF]}$

Simplified equation: $[H^+] = \sqrt{K_a C_a}$, where C_a is the initial HF concentration and the assumption is that $[H^+] \ll C_a$.

Solution:
$$[H^+] = \sqrt{(6.9 \times 10^{-4})(0.100)}$$
$$= \sqrt{(6.9 \times 10^{-5})}$$
$$= 8.3 \times 10^{-3} \text{ (This agrees with the assumption.)}$$
$$pH = -\log[H^+]$$
$$= -\log(8.3 \times 10^{-3})$$
$$= 2.08$$

CHEMICAL EQUATIONS

You will be asked to translate a written description of a chemical reaction into a chemical equation. This question draws on a storehouse of general reactions that you have encountered in your courses. You will have to write the balanced net ionic equations for three reactions. There will be a short question to answer about each reaction. Identify the states of the substances as (*aq*), (*g*), (*s*), or (*l*) to show more knowledge.

ESSAY QUESTIONS

Take time to think about your answer and organize your response. The graders are often looking for key words, such as the names of physical laws (give the equation if you don't remember the name), basic concepts, or theories. They are also looking for the proper usage of these words in the context of the problem. Incorrect statements detract from your response. Avoid using terms or concepts that you are not sure of. Write enough to fully explain your answer. Keep in mind, however, that chemists tend to be very concise and precise. Avoid complicated sentences, flowery language, and rambling, overlong responses.

FINAL PREPARATIONS FOR THE EXAM

Just as an athlete needs to prepare for the "big game," the student must prepare physically, mentally, and emotionally for the "big test." Here are some suggestions:

1. Eat well to have enough energy for the exam. A good dinner the night before and a relaxed breakfast on the day of the exam provide the energy essential to peak performance.
2. Get plenty of sleep. A full 8 hours of sleep is recommended for a rested body and a well-functioning mind. The night before the exam is no time to cram; in fact, such last-minute study may be detrimental.
3. The night before the exam, assemble the things you will need: plenty of #2 pencils with erasers, a scientific calculator with fresh batteries, a watch, and your admission card for the AP exam. You should also plan what you will wear to the test. Comfortable, loose-fitting clothes, including items such as sweaters that can be layered or removed to suit the room temperature, are best.
4. Be sure your transportation to the test center is reliable. Set your alarm so that you can leave early. Allow time to deal with the unexpected: a traffic jam, flat tire, or late-running bus.
5. Minimize distractions and worries. Leave all valuables at home so that you do not worry about them during the test. Put all unrelated matters firmly out of your mind.
6. Be confident of your ability. A positive attitude is very important in successful test taking.
7. Relax. This one test will not make or break your career. Enjoy the exam and show the world how well you can do.

DIAGNOSTIC TESTS

Diagnostic Test 1
ANSWER SHEET

1 Ⓐ Ⓑ Ⓒ Ⓓ Ⓔ	13 Ⓐ Ⓑ Ⓒ Ⓓ Ⓔ	24 Ⓐ Ⓑ Ⓒ Ⓓ Ⓔ	35 Ⓐ Ⓑ Ⓒ Ⓓ Ⓔ
2 Ⓐ Ⓑ Ⓒ Ⓓ Ⓔ	14 Ⓐ Ⓑ Ⓒ Ⓓ Ⓔ	25 Ⓐ Ⓑ Ⓒ Ⓓ Ⓔ	36 Ⓐ Ⓑ Ⓒ Ⓓ Ⓔ
3 Ⓐ Ⓑ Ⓒ Ⓓ Ⓔ	15 Ⓐ Ⓑ Ⓒ Ⓓ Ⓔ	26 Ⓐ Ⓑ Ⓒ Ⓓ Ⓔ	37 Ⓐ Ⓑ Ⓒ Ⓓ Ⓔ
4 Ⓐ Ⓑ Ⓒ Ⓓ Ⓔ	16 Ⓐ Ⓑ Ⓒ Ⓓ Ⓔ	27 Ⓐ Ⓑ Ⓒ Ⓓ Ⓔ	38 Ⓐ Ⓑ Ⓒ Ⓓ Ⓔ
5 Ⓐ Ⓑ Ⓒ Ⓓ Ⓔ	17 Ⓐ Ⓑ Ⓒ Ⓓ Ⓔ	28 Ⓐ Ⓑ Ⓒ Ⓓ Ⓔ	39 Ⓐ Ⓑ Ⓒ Ⓓ Ⓔ
6 Ⓐ Ⓑ Ⓒ Ⓓ Ⓔ	18 Ⓐ Ⓑ Ⓒ Ⓓ Ⓔ	29 Ⓐ Ⓑ Ⓒ Ⓓ Ⓔ	40 Ⓐ Ⓑ Ⓒ Ⓓ Ⓔ
7 Ⓐ Ⓑ Ⓒ Ⓓ Ⓔ	19 Ⓐ Ⓑ Ⓒ Ⓓ Ⓔ	30 Ⓐ Ⓑ Ⓒ Ⓓ Ⓔ	41 Ⓐ Ⓑ Ⓒ Ⓓ Ⓔ
8 Ⓐ Ⓑ Ⓒ Ⓓ Ⓔ	20 Ⓐ Ⓑ Ⓒ Ⓓ Ⓔ	31 Ⓐ Ⓑ Ⓒ Ⓓ Ⓔ	42 Ⓐ Ⓑ Ⓒ Ⓓ Ⓔ
9 Ⓐ Ⓑ Ⓒ Ⓓ Ⓔ	21 Ⓐ Ⓑ Ⓒ Ⓓ Ⓔ	32 Ⓐ Ⓑ Ⓒ Ⓓ Ⓔ	43 Ⓐ Ⓑ Ⓒ Ⓓ Ⓔ
10 Ⓐ Ⓑ Ⓒ Ⓓ Ⓔ	22 Ⓐ Ⓑ Ⓒ Ⓓ Ⓔ	33 Ⓐ Ⓑ Ⓒ Ⓓ Ⓔ	44 Ⓐ Ⓑ Ⓒ Ⓓ Ⓔ
11 Ⓐ Ⓑ Ⓒ Ⓓ Ⓔ	23 Ⓐ Ⓑ Ⓒ Ⓓ Ⓔ	34 Ⓐ Ⓑ Ⓒ Ⓓ Ⓔ	45 Ⓐ Ⓑ Ⓒ Ⓓ Ⓔ
12 Ⓐ Ⓑ Ⓒ Ⓓ Ⓔ			

Diagnostic Test 2
ANSWER SHEET

1 Ⓐ Ⓑ Ⓒ Ⓓ Ⓔ	13 Ⓐ Ⓑ Ⓒ Ⓓ Ⓔ	24 Ⓐ Ⓑ Ⓒ Ⓓ Ⓔ	35 Ⓐ Ⓑ Ⓒ Ⓓ Ⓔ
2 Ⓐ Ⓑ Ⓒ Ⓓ Ⓔ	14 Ⓐ Ⓑ Ⓒ Ⓓ Ⓔ	25 Ⓐ Ⓑ Ⓒ Ⓓ Ⓔ	36 Ⓐ Ⓑ Ⓒ Ⓓ Ⓔ
3 Ⓐ Ⓑ Ⓒ Ⓓ Ⓔ	15 Ⓐ Ⓑ Ⓒ Ⓓ Ⓔ	26 Ⓐ Ⓑ Ⓒ Ⓓ Ⓔ	37 Ⓐ Ⓑ Ⓒ Ⓓ Ⓔ
4 Ⓐ Ⓑ Ⓒ Ⓓ Ⓔ	16 Ⓐ Ⓑ Ⓒ Ⓓ Ⓔ	27 Ⓐ Ⓑ Ⓒ Ⓓ Ⓔ	38 Ⓐ Ⓑ Ⓒ Ⓓ Ⓔ
5 Ⓐ Ⓑ Ⓒ Ⓓ Ⓔ	17 Ⓐ Ⓑ Ⓒ Ⓓ Ⓔ	28 Ⓐ Ⓑ Ⓒ Ⓓ Ⓔ	39 Ⓐ Ⓑ Ⓒ Ⓓ Ⓔ
6 Ⓐ Ⓑ Ⓒ Ⓓ Ⓔ	18 Ⓐ Ⓑ Ⓒ Ⓓ Ⓔ	29 Ⓐ Ⓑ Ⓒ Ⓓ Ⓔ	40 Ⓐ Ⓑ Ⓒ Ⓓ Ⓔ
7 Ⓐ Ⓑ Ⓒ Ⓓ Ⓔ	19 Ⓐ Ⓑ Ⓒ Ⓓ Ⓔ	30 Ⓐ Ⓑ Ⓒ Ⓓ Ⓔ	41 Ⓐ Ⓑ Ⓒ Ⓓ Ⓔ
8 Ⓐ Ⓑ Ⓒ Ⓓ Ⓔ	20 Ⓐ Ⓑ Ⓒ Ⓓ Ⓔ	31 Ⓐ Ⓑ Ⓒ Ⓓ Ⓔ	42 Ⓐ Ⓑ Ⓒ Ⓓ Ⓔ
9 Ⓐ Ⓑ Ⓒ Ⓓ Ⓔ	21 Ⓐ Ⓑ Ⓒ Ⓓ Ⓔ	32 Ⓐ Ⓑ Ⓒ Ⓓ Ⓔ	43 Ⓐ Ⓑ Ⓒ Ⓓ Ⓔ
10 Ⓐ Ⓑ Ⓒ Ⓓ Ⓔ	22 Ⓐ Ⓑ Ⓒ Ⓓ Ⓔ	33 Ⓐ Ⓑ Ⓒ Ⓓ Ⓔ	44 Ⓐ Ⓑ Ⓒ Ⓓ Ⓔ
11 Ⓐ Ⓑ Ⓒ Ⓓ Ⓔ	23 Ⓐ Ⓑ Ⓒ Ⓓ Ⓔ	34 Ⓐ Ⓑ Ⓒ Ⓓ Ⓔ	45 Ⓐ Ⓑ Ⓒ Ⓓ Ⓔ
12 Ⓐ Ⓑ Ⓒ Ⓓ Ⓔ			

Diagnostic Test 3
ANSWER SHEET

1 Ⓐ Ⓑ Ⓒ Ⓓ Ⓔ	13 Ⓐ Ⓑ Ⓒ Ⓓ Ⓔ	24 Ⓐ Ⓑ Ⓒ Ⓓ Ⓔ	35 Ⓐ Ⓑ Ⓒ Ⓓ Ⓔ
2 Ⓐ Ⓑ Ⓒ Ⓓ Ⓔ	14 Ⓐ Ⓑ Ⓒ Ⓓ Ⓔ	25 Ⓐ Ⓑ Ⓒ Ⓓ Ⓔ	36 Ⓐ Ⓑ Ⓒ Ⓓ Ⓔ
3 Ⓐ Ⓑ Ⓒ Ⓓ Ⓔ	15 Ⓐ Ⓑ Ⓒ Ⓓ Ⓔ	26 Ⓐ Ⓑ Ⓒ Ⓓ Ⓔ	37 Ⓐ Ⓑ Ⓒ Ⓓ Ⓔ
4 Ⓐ Ⓑ Ⓒ Ⓓ Ⓔ	16 Ⓐ Ⓑ Ⓒ Ⓓ Ⓔ	27 Ⓐ Ⓑ Ⓒ Ⓓ Ⓔ	38 Ⓐ Ⓑ Ⓒ Ⓓ Ⓔ
5 Ⓐ Ⓑ Ⓒ Ⓓ Ⓔ	17 Ⓐ Ⓑ Ⓒ Ⓓ Ⓔ	28 Ⓐ Ⓑ Ⓒ Ⓓ Ⓔ	39 Ⓐ Ⓑ Ⓒ Ⓓ Ⓔ
6 Ⓐ Ⓑ Ⓒ Ⓓ Ⓔ	18 Ⓐ Ⓑ Ⓒ Ⓓ Ⓔ	29 Ⓐ Ⓑ Ⓒ Ⓓ Ⓔ	40 Ⓐ Ⓑ Ⓒ Ⓓ Ⓔ
7 Ⓐ Ⓑ Ⓒ Ⓓ Ⓔ	19 Ⓐ Ⓑ Ⓒ Ⓓ Ⓔ	30 Ⓐ Ⓑ Ⓒ Ⓓ Ⓔ	41 Ⓐ Ⓑ Ⓒ Ⓓ Ⓔ
8 Ⓐ Ⓑ Ⓒ Ⓓ Ⓔ	20 Ⓐ Ⓑ Ⓒ Ⓓ Ⓔ	31 Ⓐ Ⓑ Ⓒ Ⓓ Ⓔ	42 Ⓐ Ⓑ Ⓒ Ⓓ Ⓔ
9 Ⓐ Ⓑ Ⓒ Ⓓ Ⓔ	21 Ⓐ Ⓑ Ⓒ Ⓓ Ⓔ	32 Ⓐ Ⓑ Ⓒ Ⓓ Ⓔ	43 Ⓐ Ⓑ Ⓒ Ⓓ Ⓔ
10 Ⓐ Ⓑ Ⓒ Ⓓ Ⓔ	22 Ⓐ Ⓑ Ⓒ Ⓓ Ⓔ	33 Ⓐ Ⓑ Ⓒ Ⓓ Ⓔ	44 Ⓐ Ⓑ Ⓒ Ⓓ Ⓔ
11 Ⓐ Ⓑ Ⓒ Ⓓ Ⓔ	23 Ⓐ Ⓑ Ⓒ Ⓓ Ⓔ	34 Ⓐ Ⓑ Ⓒ Ⓓ Ⓔ	45 Ⓐ Ⓑ Ⓒ Ⓓ Ⓔ
12 Ⓐ Ⓑ Ⓒ Ⓓ Ⓔ			

Diagnostic Test 1

Directions: Answer the following multiple-choice questions. You may use a calculator and the periodic table on page 621, but no other information. Limit your time to 60 minutes. If you do not finish in 60 minutes, note the number of questions answered and then continue until all the remaining questions are answered. Record your total time. Score your test with the answer key at the end of the test. Also at the end of the test are tables to help diagnose your strengths and weaknesses. Review the topics that have the most errors and then continue with Diagnostic Test 2. For more of a challenge, do not use your calculator when answering these questions.

1. A certain beam of monochromatic light has a frequency of 3.0×10^{15} hertz. The wavelength of this radiation is

   (A) 300 nm in the ultraviolet region
   (B) 500 nm in the visible region
   (C) 2,000,000 meters in the radio wave region
   (D) 20 cm in the infrared region
   (E) 200 pm in the x-ray region

2. Of the following oxo acids, which is predicted to be the strongest acid?

   (A) HBrO
   (B) HClO
   (C) HIO
   (D) $HClO_3$
   (E) HIO_3

3. What is the equilibrium law for the reaction

   $$C_3H_8(l) + 5\,O_2(g) \rightarrow 3\,CO_2(g) + 4\,H_2O(l)$$

   (A) $K = \dfrac{[CO_2]^3}{[O_2]^5}$

   (B) $K = \dfrac{[C_3H_8][O_2]^5}{[CO_2]^3[H_2O]^4}$

   (C) $K = \dfrac{[C_3H_8]}{[H_2O]^4}$

   (D) $K = [CO_2]^3$

   (E) $K = \dfrac{[O_2]^5}{[CO_2]^3}$

4. The rate law may be written using stoichiometric coefficients for which of the following?

   (A) precipitation reactions
   (B) acid-base reactions
   (C) elementary processes
   (D) solubility reactions
   (E) oxidation-reduction reactions

5. A certain radioactive material has a half-life of 36 minutes. Starting with 10.00 grams of this material, how many grams will be left after 2 hours?

   (A) 1.00 grams
   (B) 1.5×10^{-4} grams
   (C) 1.2 kg
   (D) 0.25 gram
   (E) 3.33 grams

6. A solution is prepared by dissolving 30.0 grams of $Ni(NO_3)_2$ in enough water to make 250 mL of solution. What is the molarity of this solution?

   (A) 0.496 molar
   (B) 0.656 mol/L
   (C) 3.3 molar
   (D) 6.3×10^{-3} molar
   (E) 0.0205 mol/L

7. Hard materials such as silicon carbide, used for grinding wheels, are said to be examples of

   (A) ionic crystals
   (B) molecular crystals
   (C) metallic crystals
   (D) network crystals
   (E) amorphous solids

8. The equilibrium constant, K_c, for the dissociation of HI into hydrogen gas and iodine vapor is 21 at a certain temperature. What will be the molar concentration of iodine vapor if 15 grams of HI gas is introduced into a 12.0-L flask and allowed to come to equilibrium?

   (A) 4.58 mol/L
   (B) 0.00687 mol L^{-1}
   (C) 4.4×10^{-3} M
   (D) 9.76×10^{-3} M
   (E) 0.117 mol/L

9. What is the molarity of a sodium hydroxide solution that requires 42.6 mL of 0.108 M HCl to neutralize 40.0 mL of the base?

   (A) 0.0641 M
   (B) 1.64 M
   (C) 0.115 M
   (D) 0.400 mol/L
   (E) 0.203 mol/L

10. Changing which of the following will change the numerical value of the equilibrium constant:

    (A) the pressure of reactants
    (B) the pressure of products
    (C) the temperature
    (D) the total mass of the chemicals present
    (E) the units from partial pressure to moles per liter

11. The simplest alkene has

    (A) at least two pi bonds
    (B) at least four sigma bonds
    (C) a tetrahedral configuration
    (D) cis-trans isomers
    (E) optical isomers

12. Helium effuses through a pinhole 5.33 times faster than an unknown gas. That gas is most likely

    (A) CO_2
    (B) CH_4
    (C) C_5H_{12}
    (D) C_8H_{18}
    (E) C_6H_6

13. The Pauli exclusion principle states that

    (A) no two electrons can have the same energy
    (B) no two electrons with the same spin can occupy an orbital
    (C) no two electrons can occupy separate orbitals
    (D) no two electrons can pair up if there is an empty orbital available
    (E) no two electrons can have the same four quantum numbers

14. Which of the following can form hydrogen bonds?

    (A) $CH_3OCH_2CH_3$
    (B) HCN
    (C) CH_3OCH_2Br
    (D) CH_3NH_2
    (E) $(CH_3)_3N$

15. The subatomic particle with the least penetrating ability is

    (A) a beta particle
    (B) a neutron
    (C) a proton
    (D) an alpha particle
    (E) a gamma ray

16. Which of the following is expected to be a polar molecule?

    (A) PCl_4F
    (B) BF_3
    (C) CO_2
    (D) $Si(CH_3)_4$
    (E) SCl_6

17. All of the following ions have the same electron configuration except

    (A) Rb^+
    (B) Se^{2-}
    (C) As^{5+}
    (D) Sr^{2+}
    (E) Br^-

18. In which of the following is the negative end of the bond written last?

    (A) O—S
    (B) Br—N
    (C) N—C
    (D) P—Cl
    (E) Cl—I

19. What is the molar mass of a monoprotic weak acid that requires 26.3 mL of 0.122 *M* KOH to neutralize 0.682 gram of the acid?

    (A) 212 g/eq
    (B) 4.70 g/eq
    (C) 147 g/eq
    (D) 682 g/eq
    (E) 55.9 g/eq

20. Which of the following correctly lists the individual intermolecular attractive forces from the strongest to the weakest?

    (A) induced dipole < dipole-dipole
        < hydrogen bond
    (B) dipole-dipole < induced dipole
        < hydrogen bond
    (C) induced dipole < hydrogen bond
        < dipole-dipole
    (D) dipole-dipole < hydrogen bond
        < induced dipole
    (E) hydrogen bond < dipole-dipole
        < induced dipole

21. A mechanism is a sequence of elementary reactions that add up to the overall reaction stoichiometry. A substance that is produced in one elementary reaction and consumed in another is called

    (A) a catalyst
    (B) an intermediate
    (C) a reactant
    (D) a complex
    (E) an activated complex

22. How many moles of propane (C_3H_8) are there in 6.2 g of propane?

    (A) 14.1 mol
    (B) 1.4×10^{-1} mol
    (C) 71 mol
    (D) 7.1 mol
    (E) 1.4×10^2 mol

23. What is the work involved if a gas in a 2.0-liter container at 2.4 atmospheres pressure is allowed to expand against a pressure of 0.80 atmosphere?

    (A) –3.2 L atm
    (B) –9.62 L atm
    (C) –4.8 L atm
    (D) +14.4 L atm
    (E) +4.82 L atm

24. Monatomic ions are often

    (A) very soluble
    (B) radioactive
    (C) isoelectronic with a noble gas
    (D) highly colored
    (E) very electronegative

25. A first-order reaction has a half-life of 34 minutes. What is the rate constant for this reaction?

    (A) $3.4 \times 10^{-4} \ s^{-1}$
    (B) $2.04 \times 10^{-2} \ s^{-1}$
    (C) $2.9 \times 10^{-1} \ min^{-1}$
    (D) 34 min
    (E) $1.53 \ s^{-1}$

26. The electrolysis of molten magnesium bromide is expected to produce

    (A) magnesium at the anode and bromine at the cathode
    (B) magnesium at the cathode and bromine at the anode
    (C) magnesium at the cathode and oxygen at the anode
    (D) bromine at the anode and hydrogen at the cathode
    (E) hydrogen at the cathode and oxygen at the anode

27. The oxidation number of chlorine in the perchlorate ion, ClO_4^-, is

    (A) +2
    (B) −3
    (C) −1
    (D) +7
    (E) +5

28. A certain reaction has a $\Delta H° = -43.2$ kJ mol^{-1} and an entropy change $\Delta S°$ of +22.0 J $mol^{-1} \ K^{-1}$. What is the value of $\Delta G°$ at 800 °C?

    (A) −66.8 kJ mol^{-1}
    (B) +21.2 kJ mol^{-1}
    (C) −21.2 kJ/mol
    (D) −2365 kJ mol^{-1}
    (E) −1.96 kJ/mol

29. How many electrons, neutrons, and protons are in an atom of ^{52}Cr?

    (A) 24 electrons, 24 protons, 24 neutrons
    (B) 27 electrons, 27 protons, 24 neutrons
    (C) 24 electrons, 28 protons, 24 neutrons
    (D) 24 electrons, 24 protons, 28 neutrons
    (E) 24 electrons, 27 protons, 25 neutrons

30. The ideal gas law is successful for most gases because

    (A) room temperature is high
    (B) volumes are small
    (C) gas particles do not interact significantly
    (D) gases are dimers
    (E) gas molecules move slowly

31. A Lewis acid

    (A) donates protons
    (B) makes a solution acid by release of H_3O^+ ions
    (C) donates electron pairs
    (D) is an electron pair acceptor
    (E) is always an oxo acid

32. The electron configuration $1s^2 \ 2s^2 \ 2p^6 \ 3s^2 \ 3p^6 \ 4s^2$ signifies the ground state of the element

    (A) V
    (B) Ti
    (C) Co
    (D) Ca
    (E) Nb

33. Solvents A and B form an ideal solution. Solvent A has a vapor pressure of 345 torr at room temperature. At the same temperature, solvent B has a vapor pressure of 213 torr. What is the vapor pressure of a mixture where the mole fraction of A is 0.285?

    (A) 213 torr
    (B) 345 torr
    (C) 568 torr
    (D) 251 torr
    (E) 484 torr

34. Which of the following reactions is expected to have the greatest decrease in entropy?

    (A) $HCl(aq) + NaOH(aq) \rightarrow NaCl(aq) + H_2O(l)$
    (B) $CH_4(g) + 2\ O_2(g) \rightarrow CO_2(g) + 2\ H_2O(l)$
    (C) $C(s) + O_2(g) \rightarrow CO_2(g)$
    (D) $2\ SO_2(g) + O_2(g) \rightarrow 2\ SO_3(g)$
    (E) $CaCO_3(s) \rightarrow CaO(s) + CO_2(g)$

35. What is the conjugate acid of the $H_2PO_4^-$ ion?

    (A) HPO_4^{2-}
    (B) $H_2PO_4^-$
    (C) H_3PO_4
    (D) PO_4^{3-}
    (E) $H_2PO_4^-$ only has a conjugate acid

36. The name for $CH_3CHBrCHClCH_2CH_3$ is

    (A) dibromotrichloropropane
    (B) dibromotrichloropentane
    (C) 2-bromo, 3-chloropentane
    (D) 2,3-bromochloropentane
    (E) 2-chloro-3-bromopentane

37. Each resonance form of the nitrate ion, NO_3^- has how many sigma and how many pi bonds?

    (A) 1 sigma and 2 pi
    (B) 2 sigma and 1 pi
    (C) 1 sigma and 1 pi
    (D) 3 sigma and 2 pi
    (E) 3 sigma and 1 pi

38. Twenty-five milligrams of sucrose $(C_{12}H_{22}O_{11})$ is dissolved in enough water to make 1.00 liter of solution. What is the molality of the solution?

    (A) 7.3×10^{-5}
    (B) 7.31×10^{-2}
    (C) 73.1
    (D) 1.36
    (E) 1.36×10^{-5}

39. Which of the following is considered a metalloid?

    (A) Cr
    (B) Mn
    (C) Si
    (D) S
    (E) Bi

40. Approximately how many half-lives will it take to reduce 10.0 kg of a radioactive substance to 1.0 microgram of that substance?

    (A) 5
    (B) 8
    (C) 13
    (D) 29
    (E) 34

41. Which of the following is a reduction half-reaction?

    1. $Cu^{2+} + e^- \rightarrow Cu^+$
    2. $Cu^+ + e^- \rightarrow Cu^0$
    3. $Fe^{2+} \rightarrow Fe^{3+} + e^-$

    (A) 1 only because copper(II) ions are reduced
    (B) 3 only because the iron is reduced
    (C) 1 and 2 because they both reduce copper ions
    (D) 1 and 3 because they do not have insoluble ions
    (E) all are reduction half-reactions because they have electrons in the equation

42. Which of the following is expected to have the largest bond polarity?

    (A) S—O
    (B) P—F
    (C) C—B
    (D) C—N
    (E) N—O

43. How many electrons, neutrons, and protons are in an atom of Cr?

    (A) 24 electrons, 24 protons, 24 neutrons
    (B) 27 electrons, 27 protons, 24 neutrons
    (C) 24 electrons, 28 protons, 24 neutrons
    (D) more information is needed to answer this question
    (E) 24 electrons, 27 protons, 25 neutrons

44. Which of the following is incorrectly named?

    (A) $CaCl_2$ calcium chloride
    (B) $Fe(NO_3)_3$ iron(III) nitrate
    (C) $AlBr_3$ aluminum tribromide
    (D) $K_2Cr_2O_7$ potassium dichromate
    (E) NH_4Br ammonium bromide

45. The name for the following is

    $$CH_3CCl_2CH_2CH_3$$

    (A) butane
    (B) 2-dichloropentane
    (C) 2,2-dichlorobutane
    (D) 2,2-chloropropane
    (E) 3,3-chloropentane

Diagnostic Test 1
ANSWER KEY

1. A	10. C	19. A	28. A	37. E
2. D	11. B	20. E	29. D	38. A
3. A	12. D	21. B	30. C	39. C
4. C	13. E	22. B	31. D	40. E
5. A	14. D	23. A	32. D	41. C
6. B	15. A	24. C	33. D	42. B
7. B	16. A	25. A	34. B	43. D
8. C	17. C	26. B	35. C	44. C
9. C	18. D	27. D	36. C	45. C

EVALUATING YOUR RESULTS

Score your test using the answer key. Then complete the following tables. The first table is designed to find your general strengths and weaknesses based on four broad categories. The second table is a more specific diagnostic chart that will suggest which particular chapters you should concentrate your studies on. In combination, these two tables will help you focus your efforts on the material that needs the most study.

QUESTION CATEGORIES

QUESTION TYPE	QUESTIONS	NUMBER WRONG
Basic Facts	2, 3, 7, 10, 11, 13, 15, 16, 21, 24, 29, 35, 37, 39, 44, 45	
Basic Concepts	4, 12, 14, 18, 27, 28, 30, 31, 32, 36, 41, 42	
Calculations	1, 5, 6, 8, 9, 12, 19, 22, 23, 25, 27, 28, 33, 38, 40	
Mixed Concepts	9, 17, 20, 26, 32, 34, 35, 43	

BREAKDOWN BY TOPICS

CHAPTER	QUESTIONS	NUMBER WRONG
1 Structure of the Atom	1, 13, 32	
2. Periodic Table	29, 39, 43	
3. Nuclear Chemistry	5, 15, 40	
4. Ionic Compounds and Reactions	17, 24, 44	
5. Covalent Compounds	16, 18, 37, 42	
6. Stoichiometry	9, 19, 22	
7. Gases	12, 30	
8. Liquids and Solids	7, 14, 20	
9. Solutions	6, 33, 38	
10. Equilibrium	3, 8, 10	
11. Kinetics	4, 21, 25	
12. Thermodynamics	23, 28, 34	
13. Redox and Electrochemistry	26, 27, 41	
14. Acids and Bases	2, 31, 35	
15. Organic Chemistry	11, 36, 45	

ANSWERS EXPLAINED

1. **(A)** The basic equation is $\lambda \nu = c$, which reads: The wavelength times the frequency is equal to the speed of light, which is 3.0×10^8 m/s.

$$\lambda = (3.0 \times 10^8 \text{ m s}^{-1})/(1.0 \times 10^{15} \text{ s}^{-1})$$
$$= 3.0 \times 10^{-7} \text{ meters}$$

At this point we see that c is not correct and that several answers start with 300 or 500. Increasing 3.0 to 300 requires decreasing the exponent from –7 to –9 to get

$$= 300 \times 10^{-9} \text{ meters}$$

Substituting the metric prefix "n" for $\times 10^{-9}$ we get 300 nm, which is response (A).

2. **(D)** Among HBrO, HClO, and HIO, the more electronegative central atom indicates a stronger oxo acid. Therefore HClO is the strongest of these three. For the same reason, we choose $HClO_3$ over HIO_3 as the strongest acid in this pair. Comparing HClO and $HClO_3$, we select $HClO_3$ as the strongest because it has the larger number of unshared or double-bonded oxygen atoms that stabilize the anion, creating a stronger acid.

3. **(A)** By convention, liquids are not written as part of the equilibrium law. Liquids always have the same number of molecules per liter, which is a constant, combined into the value of the K. The only nonsolid substances in this reaction are $O_2(g)$ and $CO_2(g)$. In this reaction, CO_2 to the third power appears in the numerator because it is a product, and O_2 to the fifth power is in the denominator.

4. **(C)** Only reactions that are known to be elementary processes (i.e., actual collisions of molecules) can be used to write a rate law. Most other reactions proceed by multistep mechanisms that have to be deduced from experimental evidence.

5. **(A)** With a half-life of 36 minutes, there will be $120/36 = 3.33$ half-lives. Therefore there are more than 3 and less than 4 half-lives. Three half-lives leaves 1/8 of the material, and 4 half-lives leaves 1/16 of the material. 1/8 of the initial 10 grams is a little more than 1 gram, while 1/16 of 10 grams is a little more than 1/2 gram. The only answer that falls in that range is 1.00 gram.

Use the integrated rate equation: $\ln(g_0/g_t) = kt$, also $\ln(2) = kt_{1/2}$.

Therefore $k = (0.693)/36 \text{ min} = 0.0193 \text{ min}^{-1}$.

Then

$$\begin{aligned}
\ln(g_0) - \ln(g_t) &= (0.0193 \text{ min}^{-1})(120 \text{ min}) \\
\ln(10) - \ln(g_t) &= 2.31 \\
2.30 - \ln(g_t) &= 2.31 \\
-\ln(g_t) &= 0.01 \\
\ln(g_t) &= -0.01 \\
g_t &= 1.00
\end{aligned}$$

6. **(B)** Molarity is the moles of solute in each liter of solution; molarity = mol solute/L solution

$$\begin{aligned}
\text{mol Ni(NO}_3)_2 &= 30.0 \text{ g Ni(NO}_3)_2 \\
&= 0.1640 \text{ mol Ni(NO}_3)_2
\end{aligned}$$

$$\begin{aligned}
\text{molarity Ni(NO}_3)_2 &= (0.1640 \text{ mol Ni(NO}_3)_2)/0.250 \text{ L} \\
&= 0.656 \text{ mol Ni(NO}_3)_2/\text{L}
\end{aligned}$$

7. **(B)** Network crystals are held together with covalent bonds. These bonds make the crystal one large, rigid, molecule. As a result, the macroscopic substance is very hard.

8. **(C)** The reaction is $2 \text{ HI}(g) \leftrightarrow H_2(g) + I_2(g)$, and the equilibrium law is

$$K = \frac{[H_2][I_2]}{[HI]^2} = 21$$

$$? \text{ mol } \frac{HI}{L} = \frac{15 \text{ g HI}}{12 \text{ L}} \left[\frac{1 \text{ mol HI}}{128 \text{ g HI}} \right] = 9.77 \times 10^{-3} \text{ mol L}^{-1}$$

When HI reacts, $2x$ moles of HI form x moles of H_2 and x moles of I_2. The equilibrium law is

$$\frac{(x)\,(x)}{(9.77 \times 10^{-3} - 2x)^2} = 21 \quad \text{Take the square root of both sides of the equation.}$$

$$\frac{(x)}{9.77 \times 10^{-3} - 2x} = 4.58$$

$$x = 4.47 \times 10^{-2} - 9.16x$$
$$10.16x = 4.47 \times 10^{-2}$$
$$x = 4.4 \times 10^{-3} \text{ mol L}^{-1}$$

Because $x = [I_2]$, that is the concentration of I_2 vapor expected. We keep only two significant figures because K_c had only two significant figures.

9. **(C)** Molarity is a ratio of units (moles per liter); therefore, the setup of the calculation must start with a ratio, such as another molarity.

Setup:

$$?M_{\text{NaOH}} = 0.216\,M_{\text{HCl}}$$

Expand molarities to mol/L and insert conversion factors:

$$? \frac{\text{mol NaOH}}{\text{L NaOH}} = \frac{0.108 \text{ mol HCl}}{\text{L HCl}} \times \frac{1 \text{ mol NaOH}}{1 \text{ mol HCl}} \times \frac{0.0426 \text{ L HCl}}{0.0400 \text{ L NaOH}}$$
$$= 0.115\,M \text{ NaOH}$$

10. **(C)** ONLY temperature changes will result in changes in the numerical value of the equilibrium constant.

11. **(B)** Alkenes have a double bond. The double bond consists of one sigma and one pi bond connecting a pair of carbon atoms. The simplest alkene, ethane, has one pi bond and five sigma bonds. Geometrically, they are triangular planar around the carbon atoms (sp^2 hybrid). Because of that geometry, the simplest alkene molecule does not have isomers or optical activity.

12. **(D)** Graham's law of effusion is written as

$$\frac{\text{rate}_1}{\text{rate}_2} = \sqrt{\frac{\text{mass}_2}{\text{mass}_1}}$$

Defining helium as compound 1 in the equation above, we can substitute

$$5.33 = \sqrt{\frac{\text{mass}_2}{4}}$$

Square both sides:

$$28.4 = \frac{\text{mass}_2}{4}$$

$$\text{mass}_2 = 4 \times 28.4 = 114 \text{ g/mol}$$

The molecular masses of the possible compounds are

(A) $CO_2 = 44$, (B) $CH_4 = 16$, (C) $C_5H_{12} = 72$, (D) $C_8H_{18} = 114$, (E) $C_6H_6 = 78$. C_8H_{18} is obviously the sample in this experiment.

13. **(E)** The Pauli exclusion principle states that no two electrons in an element can have the same set of four quantum numbers. This is another way of saying that no two electrons can occupy the same orbital with the same spin. Response (A) is false—electrons often have the same energy. Response (B) violates the rule that electrons in the same orbital must have paired spins. Response (D) is a statement of Hund's rule.

14. **(D)** Hydrogen bonds can form when a compound contains a F, an O, or a N atom with a hydrogen bonded to it. In this question, this criterion is fulfilled only by CH_3NH_2.

15. **(A)** The electron with no charge has tremendous penetrating ability. Many can pass through the earth without stopping.

16. **(A)** The Lewis structures of all molecules except PCl_4F are symmetric. The fact that two different atoms are attached to the central P atom immediately suggests that this molecule is not symmetric. We test it first by drawing its Lewis structure. We find that structure to be a trigonal bipyramid that is not symmetrical, and we predict it will be polar.

17. **(C)** All of the ions in this question are isoelectronic with the noble gas krypton except arsenic. Arsenic would have to be a 3– ion to be isoelectronic with krypton. It is written as a 5+ ion.

18. **(D)** Generally, with a pair of elements, the one closest to fluorine in the periodic table is negative (largest electronegativity), and the atom farthest from fluorine is positive (lowest electronegativity). The only one of the five pairs where the second element is closest to fluorine is the P—Cl pair.

19. **(A)** The equivalent weight of an acid is the mass of the acid that provides 1 mole of protons for a reaction. In this problem, the moles of KOH will equal the moles of protons neutralized.

moles of protons = $(0.0263 \text{ L}) (0.122 \text{ } M \text{ KOH}) = 3.21 \times 10^{-3}$ eq

equivalent weight = $(0.682 \text{ gram})/(3.21 \times 10^{-3} \text{ eq}) = 212 \text{ g eq}^{-1}$

20. **(E)** Hydrogen bonding is the strongest intermolecular attractive force and is listed first. Induced dipoles are the weakest attractive forces and last for a short period of time.

21. **(B)** An intermediate is defined as a substance that is neither a reactant nor a product. Intermediates are often difficult to detect. Catalysts fit the above description, but they are substances that are added to the reaction mixture and can be isolated afterward.

22. **(B)** Mathematically, we start with ? mol C_3H_8 = 6.2 g C_3H_8. We then use the conversion factor 1 mol C_3H_8 = 44 g C_3H_8 to convert g to mol.

$$? \text{ mol } C_3H_8 = 6.2 \text{ g } C_3H_8 \left[\frac{1 \text{ mol } C_3H_8}{44 \text{ g } C_3H_8} \right]$$

$$= 1.4 \text{ mol } C_3H_8$$

23. **(A)** Work can be determined from *PV* data. The equation is

$$\text{work } (w) = -P \, \Delta V.$$

We can use $P_1V_1 = P_2V_2$ to calculate the final volume in the problem above:

$$(2.4 \text{ atm})(2.0 \text{ L}) = (0.80 \text{ atm}) (x \text{ liters})$$
$$x \text{ liters} = 6.0 \text{ L}$$
$$\text{work} = -(0.80 \text{ atm})(6.0 \text{ L} - 2.0 \text{ L}) = -3.2 \text{ L atm}$$

24. **(C)** This is the best answer because all group 1A, 2A, and 3A metals, as well as some transition metals, have ions that are isoelectronic with a noble gas.

25. **(A)** From the integrated rate law we can derive that $\ln(2) = kt_{1/2}$, where $t_{1/2}$ is the half-life.

Converting 34 minutes to seconds yields $34 \times 60 = 2040$ seconds. The rate constant is calculated as

$$k = \ln(2)/2040 \text{ s}$$
$$= 0.693/2040 \text{ s}$$
$$= 3.4 \times 10^{-4} \text{ s}^{-1}$$

We round the final answer to two significant figures because 34 minutes has only two significant figures.

26. **(B)** In this molten salt the only possible products are magnesium and bromine. The question is at which electrode are these products found. Remembering that oxidation always occurs at the anode, we see that $2 \text{ Br}^- \rightarrow \text{Br}_2 + 2 \, e^-$ is the oxidation process. Therefore bromine is produced at the anode, and magnesium at the cathode.

27. **(D)** To calculate the oxidation number of an element in a polyatomic ion, the charges of the atoms must add up to the charge on the ion.

$$\text{Charge of ion} = (\text{ox. no. Cl}) + 4(\text{ox. no. O})$$
$$-1 = (\text{ox. no. Cl}) + 4(-2)$$
$$-1 + 8 = \text{ox. no. Cl}$$
$$+7 = \text{ox. no. Cl}$$

28. **(A)** The Gibbs free energy equation is $\Delta G° = \Delta H° - T \, \Delta S°$. To determine $\Delta G°$ at a temperature other than 298 K we need $\Delta H°$ and $\Delta S°$ so that the above equation can be solved at a temperature different from 298.

$$\Delta G° = -43.2 \text{ kJ mol}^{-1} - (1073 \text{ K})(22.0 \text{ J mol}^{-1} \text{ K}^{-1})$$
$$= -43.2 \text{ kJ mol}^{-1} - (23.6 \times 10^3 \text{ J mol}^{-1})$$
$$= -43.2 \text{ kJ mol}^{-1} - (23.6 \text{ kJ mol}^{-1})$$
$$= -66.8 \text{ kJ mol}^{-1}$$

29. **(D)** The symbol represents an isotope of chromium. Since these is no charge, it is not an ion that may have gained or lost electrons. Therefore the numbers of electrons and protons are equal (24 of each), which eliminates responses (C) and (E). Response (B) does not have 24 protons, and response (D) is the remaining one where the protons and neutrons add to 52.

30. **(C)** This response is true because of the large distance between gas particles. The other four responses are not necessarily true about any given gas sample.

31. **(D)** The definition of a Lewis acid is that it is an electron pair acceptor.

32. **(D)** This electron configuration has 20 electrons in the lowest possible energy levels according to the Aufbau ordering. Element 20 is calcium.

33. **(D)** The vapor pressure of a mixture is found to be

$$P_{total} = P_A X_A + P_B X_B$$
$$= 345 \text{ torr } (0.285) + 213 \text{ torr } (0.715)$$
$$= 98.3 \text{ torr } + 152.3 \text{ torr}$$
$$= 250.6 \text{ torr } = 251 \text{ torr (when rounded correctly)}$$

34. **(B)** The most important factor in entropy change is whether or not there is a change in volume as denoted by the change in the number of gas molecules in the chemical equation (Δn_g). For the reactions given, the Δn_g is: (A) 0, (B) +1, (C) 0, (D) –1, (E) –2. Reaction (B) has the greatest decrease in the moles of gas and should have the largest decrease in entropy.

35. **(C)** The conjugate acid of a base is obtained by adding one hydrogen atom and one positive charge to the base. Therefore, $H_2PO_4^- \rightarrow H_3PO_4$.

36. **(C)** Numbering the chain from left to right results in the lowest possible numbers for the chloro and bromo substituents.

37. **(E)** The nitrate ion has two oxygen atoms bonded to the nitrogen with single bonds, and one oxygen is bonded to the nitrogen with a double bond. This adds up to three sigma and one pi bond.

38. **(A)** The definition of molality is moles of solute per kilogram of solvent. In dilute solutions the mass of solute is negligible and the mass of one liter of water is one kilogram. Therefore, molarity and molality are essentially equal. We will calculate molarity as $(g/M)/(\text{L of solution})$

$$\text{molality} = \text{molarity} = (25 \times 10^{-3} \text{ g sucrose}/342 \text{ g mol}^{-1})/1.00 \text{ L}$$
$$= 7.31 \times 10^{-5} \text{ molar} = 7.3 \times 10^{-5} \text{ molal (when correctly rounded)}$$

39. **(C)** Silicon is a metalloid and is better known as a semiconductor material used in transistors.

40. **(E)** Mathematically, the fraction left is 1 microgram/10 kilograms, which is a ratio of $(10^{-6})/(10 \times 10^3)$ or $1/10^{-10}$. For half-lives, the fraction remaining is $(1/2)^n$, where n is the number of half-lives. By equating $(1/2)^n = (1/10^{-10})$, we can take the logarithm of both sides of the equation to get $n \log (1/2) = -10$. We can invert the fraction within the log term to get $n \log (2) = 10$. (Note that this operation changes the sign.) Finally, $n = 10/(\log 2) = 33.2$. This needs to be rounded to the next highest half-life, 34. This may be solved without a calculator by recalling that $\log (2) = 0.3$.

41. **(C)** Half-reactions 1 and 2 are reductions because the charge of the copper ions decreases in the process. The iron half-reaction is an oxidation because the oxidation number increases from +2 to +3.

42. **(B)** The difference in electronegativity 1.9. The others are (a) 1.1, (c) 0.5, (d) 0.6, (e) 0.4. This answer can be estimated since the P and F are more widely separated than the other pairs that are adjacent atoms in the periodic table.

43. **(D)** Because chromium has more than one isotope and this question does not specify which isotope, it is impossible to state the number of neutrons. The atomic mass in the periodic table is NOT an isotope mass and cannot be used for that purpose.

44. **(C)** The correct name is aluminum bromide. Aluminum is always 3+, and bromide is always 1–. The prefix *tri-* is used only for compounds composed of two nonmetals.

45. **(C)** $CH_3CCl_2CH_2CH_3$ We number the four carbons from the left side as shown.
    1 2 3 4

This results in the smallest number for the carbon that the chlorine is attached to. The chain length is four carbons, and this is a butane-type molecule. It has two chlorine atoms attached and is therefore a dichlorobutane. Finally, to fix the position of the chlorine atoms we write 2,2-dichlorobutane to show that both chlorine atoms are on the second carbon. This becomes more obvious if the structure is expanded to:

$$
\begin{array}{ccccc}
& H & Cl & H & H \\
& | & | & | & | \\
H- & C- & C- & C- & C- H \\
& | & | & | & | \\
& H & Cl & H & H
\end{array}
$$

Diagnostic Test 2

Directions: Answer the following multiple-choice questions. You may use a calculator and the periodic table on page 621, but no other information. Limit your time to 60 minutes. If you do not finish in 60 minutes, note the number of questions answered and then continue until all the remaining questions are answered. Record your total time. Score your test with the answer key at the end of the test. Also at the end of the test are tables to help diagnose your strengths and weaknesses. Review the topics that have the most errors and then continue with Diagnostic Test 3. For more of a challenge, do not use your calculator when answering these questions.

1. The functional group that represents an organic acid is written as (*Note:* R represents the rest of the molecule)

   (A) RCHO
   (B) ROH
   (C) RCOOH
   (D) RCl
   (E) (A), (B), and (C) because they all have an OH

2. Balance the following skeleton reaction in acid solution using the ion-electron method to obtain a reaction with the smallest possible whole-number coefficients. The sum of all the coefficients in the balanced equation is

   $$IO_3^- + Cl^- \rightarrow ClO^- + I_2$$

   (A) 8
   (B) 14
   (C) 16
   (D) 24
   (E) 46

3. The reaction of $Br_2(g)$ with $Cl_2(g)$ to form $BrCl(g)$ has an equilibrium constant of 15.0 at a certain temperature. If 10.0 grams of BrCl is initially present in a 15.0-liter reaction vessel, what will the concentration of BrCl be at equilibrium?

   (A) 3.8×10^{-3} mol/L
   (B) 5.77×10^{-3} mol/L
   (C) 1.97×10^{-3} M
   (D) 9.9×10^{-4} M
   (E) 1.16×10^{-5} mol/L

4. Which of the following is a balanced chemical reaction?

   (A) $Na_2SO_4 + Ba(NO_3)_2 \rightarrow$
       $$BaSO_4 + NaSO_4$$
   (B) $AgNO_3 + K_2CrO_4 \rightarrow$
       $$Ag_2CrO_4 + 2\ KNO_3$$
   (C) $5\ FeCl_2 + 8\ HCl + KMnO_4 \rightarrow$
       $$5\ FeCl_3 + MnCl_2 + 4\ H_2O + KCl$$
   (D) $Al(NO_3)_3 + 3\ KOH \rightarrow$
       $$Al(OH)_3 + KNO_3$$
   (E) $C_6H_{12}O_6 + 9\ O_2 \rightarrow$
       $$6\ H_2O + 6\ CO_2$$

5. A 50.0 mL sample of 0.0025 *M* HBr is mixed with 50.0 mL of 0.0023 *M* KOH. What is the pH of the resulting mixture?

(A) 1.00
(B) 4.00
(C) 5.00
(D) 11.00
(E) 13.00

6. How many milliliters of 0.250 *M* KOH does it take to neutralize completely 50.0 mL of 0.150 *M* H_3PO_4?

(A) 30.0 mL
(B) 27 mL
(C) 9.00 mL
(D) 270 mL
(E) 90.0 mL

7. The symbol for antimony is

(A) W
(B) Sb
(C) Fe
(D) An
(E) K

8. In acid solution the bromate ion, BrO_3^- can react with other substances, resulting in Br_2. Balance the half-reaction for bromate ions forming bromine. The balanced half-reaction has

(A) 6 electrons on the left
(B) 6 electrons on the right
(C) 3 electrons on the left
(D) 3 electrons on the right
(E) 10 electrons on the left

9. The quantum number ℓ signifies the

(A) relative distance of the electron from the nucleus
(B) orientation in space of a particular orbital
(C) shape of an orbital
(D) spin of the electron
(E) energy of the electron

10. When an ideal gas is allowed to expand isothermally, which one of the following is true?

(A) $q = 0$
(B) $w = 0$
(C) $E = 0$
(D) $q = w$
(E) $q = -w$

11. Which pair of reactants will have a net ionic equation in which all the ions cancel?

(A) $Na_2SO_3 + FeCl_2$
(B) $CaCl_2 + MnSO_4$
(C) $NH_4I + Pb(NO_3)_2$
(D) $KOH + HClO_4$
(E) $HCl + CaCO_3$

12. The rate law for the reaction of $2\,A + B \rightarrow 2\,P$ is

(A) impossible to determine without experimental data
(B) $[A]^2[B]$
(C) $k[A]^2[B]$
(D) second order with respect to A
(E) impossible without a catalyst

13. Hund's rule requires that

(A) no two electrons can have the same energy
(B) no two electrons with the same spin can occupy an orbital
(C) no two electrons can occupy separate orbitals
(D) no two electrons can pair up if there is an empty orbital at the same energy level available
(E) no two electrons can have the same four quantum numbers

14. Radon is a health hazard because

    (A) it is a gas that can be inhaled and then it may decay to a solid that resides in the lungs
    (B) it is a gas that is extremely soluble in the bloodstream, and it decays in vital organs
    (C) it is a gas that enters the body easily and targets the thyroid because it is chemically similar to iodine
    (D) it is a gas that enters the body easily and targets bones because it is chemically similar to calcium
    (E) it is a liquid that is easily absorbed through the skin if gloves are not worn

15. The collision theory of reaction rates includes

    (A) the number of collisions per second
    (B) the transition state
    (C) the energy of each collision
    (D) the orientation of each collision
    (E) only (A), (C), and (D) above

16. A 2.35-gram sample was dissolved in water, and the chloride ions were precipitated by adding silver nitrate ($Ag^+ + Cl^- \rightarrow AgCl$). If 0.435 g of precipitate was obtained, what is the percentage of chlorine in the sample?

    (A) 10.8%
    (B) 4.60%
    (C) 43.5%
    (D) 18%
    (E) 81.5%

17. What is the proper form of the Nernst equation for the following half-reaction?

    $$2\ ClO_2(aq)^- + 6\ e^- \rightarrow Cl_2(aq) + 2\ O_2(g)$$

    (A) $E = E° + 0.0591 \log \dfrac{[ClO^-][e^-]^6}{[Cl_2][O_2]^2}$

    (B) $E = E° - 0.0591 \log \dfrac{[Cl_2]P_{O_2}^2}{[ClO^-][e^-]^6}$

    (C) $E = E° + \dfrac{RT}{n\mathcal{F}} \log \dfrac{[Cl_2]P_{O_2}^2}{[ClO^-][e^-]^6}$

    (D) $E = E° - \dfrac{RT}{n\mathcal{F}} \ln \dfrac{[Cl_2][O_2]^2}{[ClO^-][e^-]^6}$

    (E) $E = E° - \dfrac{RT}{n\mathcal{F}} \ln \dfrac{[Cl_2][O_2]^2}{[ClO_2^-]^2}$

18. Which of the following has an octet of electrons around the central atom?

    (A) BF_3
    (B) NH_4^+
    (C) PF_5
    (D) SF_6
    (E) XeF_6

19. What is the pH of a solution made by dissolving 0.0300 mol of sodium ethanoate in enough water to make 50 mL of solution (K_a for ethanoic acid is 1.8×10^{-5})?

    (A) 7.00
    (B) 9.26
    (C) 4.74
    (D) 2.98
    (E) 11.02

20. The weight percent of sodium hydroxide dissolved in water is 50%. What is the mole fraction of sodium hydroxide?

    (A) 31.0%
    (B) 0.164
    (C) 0.311
    (D) 0.500
    (E) 0.690

21. Which of the following should be named using Roman numerals after the cation?

    (A) $CaCl_2$
    (B) $CuCl_2$
    (C) $AlBr_3$
    (D) $K_2Cr_2O_7$
    (E) NH_4Br

22. Carbon exists in various forms called allotropes. Which of the following is not an allotrope of carbon?

    (A) diamond
    (B) soot
    (C) buckminsterfullerene
    (D) graphite
    (E) all of the above are forms of carbon

23. What is the molecular mass of a gas that has a density of 2.05 g/L at 26.0 °C and 722 torr?

    (A) 53.0 g/mol
    (B) 46.7 g/mol
    (C) 4.67 g/mol
    (D) 2876 g/mol
    (E) 26.2 g/mol

24. *para*-dichlorobenzene is used as "mothballs." This compound can also be named

    (A) 1,2-dichlorobenzene
    (B) 1,3-dichlorobenzene
    (C) 2,4-dichlorobenzene
    (D) 1,4-dichlorobenzene
    (E) 6,4-dichlorobenzene

25. Under which conditions will a real gas most closely behave as an ideal gas?

    (A) high temperature and high pressure
    (B) high temperature and low pressure
    (C) high volume and high temperature
    (D) low temperature and low pressure
    (E) low temperature and high pressure

26. The equilibrium constant of a certain reaction is 2.6×10^8 at 25 °C. What is the value of $\Delta G°$?

    (A) −48.0 kJ/mol
    (B) 20.8 J mol^{-1}
    (C) 4.68×10^{-3} kJ/mol
    (D) −4.03 kJ mol^{-1}
    (E) −20.8 J/mol

27. Which of the following geometries and hybridizations corresponds to a substance that has five sigma bonds and one nonbonding pair of electrons?

    (A) tetrahedron, sp^3
    (B) square planar, d^2sp^3
    (C) octahedron, d^2sp^3
    (D) square pyramid, d^2sp^3
    (E) planar triangle, sp^2

28. The relationship between the vapor pressure of a liquid and the heat of vaporization is expressed in the

    (A) Rydberg equation
    (B) Nernst equation
    (C) Clausius-Clapeyron equation
    (D) Arrhenius equation
    (E) Gibbs free energy equation

29. What is the pH of a solution prepared by dissolving 0.150 mole of hypochlorous acid (HClO) in enough water to make 500 mL of solution? The K_a is 3.0×10^{-4}

    (A) 1.00
    (B) 2.00
    (C) 5.4×10^{-3}
    (D) 1.76
    (E) 2.56

30. Substances whose Lewis structures must be drawn with an unpaired electron are called

    (A) ionic compounds
    (B) free radicals
    (C) resonance structures
    (D) polar molecules
    (E) semicovalent

31. The following nuclear reaction is best completed with

$$^{20}_{9}F \longrightarrow \, ^{20}_{10}Ne + \underline{\hspace{1cm}}$$

    (A) a beta particle
    (B) an alpha particle
    (C) a neutron
    (D) a positron
    (E) a photon

32. An ideal solution is a

    (A) mixture where two solvents can be dissolved in all ratios
    (B) mixture that has the same physical properties as the individual solvents
    (C) mixture where the potential energy of the mixture is the same as that of the individual solvents
    (D) mixture that is colorless
    (E) mixture that is less of an environmental hazard than the individual solvents

33. Which of the following lists the electromagnetic spectral regions in order of decreasing wavelength?

    (A) ultraviolet, visible, infrared, X ray
    (B) X ray, visible, ultraviolet, infrared
    (C) X ray, ultraviolet, visible, infrared
    (D) visible, infrared, ultraviolet, X ray
    (E) infrared, visible, ultraviolet, X ray

34. Sodium fluoride dissolves in water and undergoes hydrolysis. What is the equilibrium law for the hydrolysis reaction?

    (A) $K = \dfrac{[HF][OH^-]}{[F^-]}$

    (B) $K = \dfrac{[HF][OH^-]}{[F^-][Na^+]}$

    (C) $K = \dfrac{[F^-][OH^-]}{[HF]}$

    (D) $K = \dfrac{[HF]}{[F^-][OH^-]}$

    (E) $K = \dfrac{[F^-]}{[OH^-][HF]}$

35. Of the following pairs of elements, which pair has the second element with the larger electronegativity based on its position in the periodic table?

    (A) oxygen, chromium
    (B) chlorine, iodine
    (C) calcium, cesium
    (D) sulfur, nitrogen
    (E) carbon, aluminum

36. The following nuclear reaction is best completed with

    $$^{87}_{36}Kr \longrightarrow \, ^{86}_{36}Kr + \underline{\quad}$$

    (A) a beta particle
    (B) an alpha particle
    (C) a neutron
    (D) a positron
    (E) a photon

37. Which of the following molecules is a strong electrolyte when dissolved in water?

    (A) CH_3COOH
    (B) $HC_2H_3O_2$
    (C) PCl_5
    (D) CO_2
    (E) HBr

38. When using the ideal gas law, standard conditions for temperature and pressure are

    (A) 0 K and 0 torr
    (B) 25 °C and 1 atmosphere pressure
    (C) 0 °C and 760 torr
    (D) 0 °F and 1 atmosphere pressure
    (E) 25 °C and 1 torr

39. The dimerization of $NO_2(g)$ to $N_2O_4(g)$ is an endothermic process. Which of the following will, according to LeChâtlier's principle, increase the amount of N_2O_4 in a reaction vessel?

    (A) decreasing the temperature
    (B) increasing the size of the reaction vessel
    (C) adding a selective catalyst
    (D) increasing the pressure by adding argon
    (E) making the reaction vessel smaller

40. Which of the following is expected to have the highest electronegativity?

    (A) Mg
    (B) Fe
    (C) W
    (D) Ag
    (E) S

41. When a solid melts, the entropy change and enthalpy changes expected are

    (A) positive enthalpy change and positive entropy change
    (B) negative entropy change and a negative enthalpy change
    (C) negative entropy change and positive enthalpy change
    (D) negative enthalpy change and positive entropy change
    (E) both enthalpy and entropy changes should be zero

42. What is the mass of one molecule of cholesterol ($C_{27}H_{46}O$, molecular mass = 386).

    (A) 6.41×10^{-22} g
    (B) 1.5×10^{21} g
    (C) 1.38×10^{-21} g
    (D) 3×10^{-23} g
    (E) 6.02×10^{-23} g

43. Which of the following pair of liquids is expected to be immiscible?

    (A) H_2O and CH_3OH
    (B) C_6H_6 and C_5H_{12}
    (C) $C_{10}H_{22}$ and $CH_2CH_2CH_2OH$
    (D) $CH_3CH_2NH_2$ and $CH_3CH_2CH_2OH$
    (E) $C_6H_5CH_3$ and $C_3H_6Cl_2$

44. The Arrhenius equation may be used to determine

    (A) the activation energy of a reaction
    (B) the rate constant at various temperatures
    (C) the shelf life of a consumer product or drug
    (D) all of the above
    (E) only (A) and (B)

45. Argon can be liquefied at low temperature because of

    (A) dipole-dipole attractive forces
    (B) hydrogen bonding
    (C) induced dipoles
    (D) ionic attractions
    (E) the very low temperature

Diagnostic Test 2
ANSWER KEY

1. C	10. E	19. B	28. C	37. E
2. C	11. B	20. C	29. B	38. C
3. A	12. A	21. B	30. B	39. E
4. C	13. D	22. B	31. A	40. E
5. B	14. A	23. A	32. C	41. A
6. E	15. E	24. D	33. E	42. A
7. B	16. B	25. B	34. A	43. C
8. E	17. E	26. A	35. D	44. D
9. C	18. B	27. D	36. C	45. C

EVALUATING YOUR RESULTS

Score your test using the answer key. Then complete the following tables. The first table is designed to find your general strengths and weaknesses based on four broad categories. The second table is a more specific diagnostic chart that will suggest which particular chapters you should concentrate your studies on. In combination, these two tables will help you focus your efforts on the material that needs the most study.

QUESTION CATEGORIES

QUESTION TYPE	QUESTIONS	NUMBER WRONG
Basic Facts	1, 7, 11, 21, 24, 30, 31, 33, 35, 36, 37, 38, 40	
Basic Concepts	4, 9, 10, 12, 13, 15, 18, 23, 25, 28, 32, 34, 39, 44, 45	
Calculations	2, 3, 5, 6, 8, 16, 19, 20, 23, 26, 29, 42	
Mixed Concepts	14, 17, 22, 27, 28, 41, 43, 44	

BREAKDOWN BY TOPICS

CHAPTER	QUESTIONS	NUMBER WRONG
1. Structure of the Atom	9, 13, 33	
2. Periodic Table	7, 35, 40	
3. Nuclear Chemistry	14, 31, 36	
4. Ionic Compounds and Reactions	4, 11, 21	
5. Covalent Compounds	18, 27, 30	
6. Stoichiometry	6, 16, 42	
7. Gases	23, 25, 38	
8. Liquids and Solids	28, 43, 45	
9. Solutions	20, 32, 37	
10. Equilibrium	3, 34, 39	
11. Kinetics	12, 15, 44	
12. Thermodynamics	10, 26, 41	
13. Redox and Electrochemistry	2, 8, 17	
14. Acids and Bases	5, 19, 29	
15. Organic Chemistry	1, 22, 24	

ANSWERS EXPLAINED

1. **(C)** The acid functional group is the —COOH grouping in organic molecules. Choice (A), RCHO, represents aldehydes. Choice (B), ROH, represents an alcohol. Choice (D), RCl, represents an organic chloride.

2. **(C)** The sum of the coefficients is 16. The reaction is balanced in steps. First write two half-reactions:

$$IO_3^- \rightarrow I_2$$
$$Cl^- \rightarrow ClO^-$$

Balance all atoms except H and O:

$$\mathbf{2}\ IO_3^- \rightarrow I_2$$
$$Cl^- \rightarrow ClO^-$$

Balance oxygens by adding water molecules:

$$2\ IO_3^- \rightarrow I_2 + 6\ H_2O$$
$$H_2O + Cl^- \rightarrow ClO^-$$

Balance hydrogens by adding H^+:

$$12\ H^+ + 2\ IO_3^- \rightarrow I_2 + 6\ H_2O$$
$$H_2O + Cl^- \rightarrow ClO^- + 2\ H^+$$

Balance charges with electrons:

$$10\ e^- + 12\ H^+ + 2\ IO_3^- \rightarrow I_2 + 6\ H_2O$$
$$H_2O + Cl^- \rightarrow ClO^- + 2\ H^+ + 2\ e^-$$

Multiply the second half-reaction by 5 so both equations have the same number of electrons (now the e^- will cancel):

$$10\ e^- + 12\ H^+ + 2\ IO_3^- \rightarrow I_2 + 6\ H_2O$$
$$5\ H_2O + 5\ Cl^- \rightarrow 5\ ClO^- + 10\ H^+ + 10\ e^-$$

Add the two equations:

$$10\ e^- + 12\ H^+ + 2\ IO_3^- + 5\ H_2O + 5\ Cl^- \rightarrow$$
$$I_2 + 6\ H_2O + 5\ ClO^- + 10\ H^+ + 10\ e^-$$

Cancel like items for the final equation:

$$2\ H^+ + 2\ IO_3^- + 5\ Cl^- \rightarrow I_2 + H_2O + 5\ ClO^-$$

Remember, I_2 and H_2O have coefficients of 1 although they are not explicitly written.

3. **(A)** The reaction is $Br_2(g) + Cl_2(g) \rightarrow 2\ BrCl(g)$

The equilibrium law is $K = \dfrac{[BrCl]^2}{[Br_2][Cl_2]}$.

The initial concentration of BrCl is = (10.0 g/115 g mol^{-1})/15 L = 5.8 × 10^{-3} mol/L.

If 2x moles of BrCl reacts, x moles each of Br_2 and Cl_2 will be formed. (You may want to set up an equilibrium table at this point.)

Substituting into the equilibrium law gives $K = \dfrac{(0.0058 - 2x)^2}{[x][x]} = 15.0$.

Take the square root of the entire equation:

$$3.87 = (0.0058 - 2x)/x$$
$$3.87x = 0.0058 - 2x$$
$$5.87x = 0.0058$$
$$x = (0.0058/5.87)$$
$$x = 9.9 \times 10^{-4} = [Br_2] = [Cl_2]$$
$$[Br\,Cl] = 0.0058 - 2\,(9.9 \times 10^{-4})$$
$$= 0.0038 \text{ mol/L}$$

4. **(C)** All of the compounds are written correctly, and all the elements have the same numbers of atoms on each side of the arrow. Na is not balanced in response (A), Ag is not balanced in response (B), the K is not balanced in response (D), and the oxygen atoms are not balanced in response (E).

5. **(B)** This is a strong acid reacting with a strong base. We need to find out how much of either the strong acid or the strong base is left after the reaction is complete. This is a type of limiting-reactant problem. The reaction is HBr + KOH → KBr + H_2O.

With solutions it is easiest to calculate the moles of each reactant as

$$? \text{ moles HBr} = 50.0 \text{ mL HBr} \left[\frac{0.0025 \text{ mol HBr}}{1000 \text{ mL HBr}} \right] = 1.25 \times 10^{-4} \text{ mol HBr}$$

$$? \text{ moles KOH} = 50.0 \text{ mL KOH} \left[\frac{0.0023 \text{ mol KOH}}{1000 \text{ mL KOH}} \right] = 1.15 \times 10^{-4} \text{ mol KOH}$$

Since 1 mole of HBr reacts with 1 mole of KOH, we can see that 1.15×10^{-4} mol of each will react. We can also see that ALL of the KOH is used up and must be the limiting reactant. The HBr is the excess reactant, and $(1.25 \times 10^{-4} - 1.15 \times 10^{-4}) = 1.0 \times 10^{-5}$ mol of HBr is left over. The molarity of HBr is the moles left over divided by the liters of solution, or

$$\text{molarity HBr} = \left[\frac{0.000010 \text{ mol HBr}}{0.100 \text{ L solution}} \right] = 0.00010 \text{ } M \text{ HBr}$$

Since HBr is a strong acid, it dissociates completely, and $[H^+] = 0.00010 \text{ } M$

The pH = $-\log (0.00010) = 4.00$.

6. **(E)** The question put into mathematical terms is

? mL KOH = 50.0 mL H_3PO_4

A balanced chemical equation is needed. The complete neutralization of H_3PO_4 is

H_3PO_4 + 3 KOH → K_3PO_4 + 3 H_2O

We change molarity units to mol/L or mol/1000 mL to use them as conversion factors:

0.150 M H_3PO_4 = $\left[\dfrac{0.150 \text{ mol } H_3PO_4}{1000 \text{ mL } H_3PO_4}\right]$ and

0.250 M KOH = $\left[\dfrac{0.250 \text{ mol KOH}}{1000 \text{ mL KOH}}\right]$

? mL KOH = 50.0 mL H_3PO_4 $\left[\dfrac{0.150 \text{ mol } H_3PO_4}{1000 \text{ mL } H_3PO_4}\right]\left[\dfrac{3 \text{ mol KOH}}{1 \text{ mol } H_3PO_4}\right]\left[\dfrac{1000 \text{ mL KOH}}{0.250 \text{ mol KOH}}\right]$

= 90.0 mL KOH

7. **(B)** The symbol for antimony is Sb. W is tungsten. Fe is iron. An does not exist. K is potassium.

8. **(E)** The bromate ion (by analogy with the chlorate ion ClO_3^-) is BrO_3^-, and we are told the product is Br_2. This is the basis of the half-reaction we need to balance.

$$BrO_3^- \rightarrow Br_2 \quad \text{Now balance Br atoms.}$$
$$2 \ BrO_3^- \rightarrow Br_2 \quad \text{Now balance O with } H_2O.$$
$$2 \ BrO_3^- \rightarrow Br_2 + 6 \ H_2O \quad \text{Now add } H^+ \text{ to balance H's.}$$
$$12 \ H^+ + 2 \ BrO_3^- \rightarrow Br_2 + 6 \ H_2O \quad \text{Now add electrons.}$$
$$10 \ e^- + 12 \ H^+ + 2 \ BrO_3^- \rightarrow Br_2 + 6 \ H_2O$$

This is the balanced half-reaction with ten electrons on the left.

9. **(C)** The ℓ quantum number designates the shape of an orbital, n designates the distance from the nucleus, and m_ℓ designates the arbitrary direction in space of the orbital.

10. **(E)** For an isothermal expansion the temperature does not change, and therefore the average kinetic energy stays the same. Because the gas is ideal, there is no change in potential energy. That means that $\Delta E = 0 = w + q$. Therefore $q = -w$, and the correct response is (E).

11. **(B)** Both $CaCl_2$ and $MnSO_4$ are soluble. The products of this reaction are $CaSO_4$ and $MnCl_2$, which are also soluble. This results in the cancellation of all ions. For the others: (A) will produce insoluble iron(II) sulfite, (C) produces PbI_2, (D) produces water as the product of a neutralization reaction, and (E) produces carbon dioxide.

12. **(A)** Rate laws cannot be determined from the reaction stoichiometry. Experimental data is always necessary to determine the rate law. One exception is a reaction that is an elementary step in a mechanism.

13. **(D)** This requires electrons to fill each available orbital in an energy level before pairing. Responses (B) and (E) are forms of the Pauli exclusion principle. Responses (A) and (C) have little meaning.

14. **(A)** The hazard of radon is that it is part of the radioactive decay scheme of uranium. If the radon disintegrates to solid polonium-214 while in the body, the solid may not be exhaled and will lodge close to tissues that can be damaged with subsequent disintegrations. It is a noble gas and is not similar to either iodine or calcium.

15. **(E)** The transition state is part of the transition state theory, not the collision theory.

16. **(B)** The percentage of chlorine in the sample is calculated as

$$\% \ Cl = \frac{grams\ chlorine}{grams\ of\ sample} \times 100$$

We know the grams of sample = 2.35 grams. We need to calculate the grams of chlorine from the 0.435 gram of AgCl precipitated. We start with ? g Cl = 0.435 g AgCl and add the conversion factors shown below.

$$? \ g\ Cl = 0.435\ g\ AgCl \left[\frac{1\ mol\ AgCl}{143\ g\ AgCl}\right]\left[\frac{1\ mol\ Cl^-}{1\ mol\ AgCl}\right]\left[\frac{35.5\ g\ Cl^-}{1\ mol\ Cl^-}\right]$$

$$= 0.108\ g\ Cl^-$$

Substitute the data into the above equation to get

$$\% \ Cl = \frac{0.108\ grams\ chlorine}{2.35\ grams\ of\ sample} \times 100 = 4.60\% \ Cl$$

17. **(E)** The concentration term must be the equilibrium law for the half-reaction. All the usual rules for writing equilibrium laws apply with an additional one that places the number of electrons in the prelog coefficient as the value of n. The only one that has the correct equilibrium law is (E). All the rest have electrons where they do not belong.

18. **(B)** The nitrogen on NH_4^+ has an octet of electrons. Boron forms compounds with only six electrons, and the remaining compounds are in period 3 or above and may utilize d orbitals to expand the octet on the central atom to more than eight electrons. The others all have expanded octets.

19. **(B)** $NaC_2H_3O_3 \rightarrow Na^+ + C_2H_3O_2^-$

    The ethanoate ion then hydrolyzes water in the reaction

    $$C_2H_3O_2^- + H_2O \rightarrow HC_2H_3O_2 + OH^-$$

    At this point we can see that only two answers can possibly be correct (B) and (E). (Some experience will teach that hydrolysis rarely makes solutions very alkaline, so the best guess is response (B).)

    $$K = K_w$$

    $$K_a = \frac{[OH^-][HC_2H_3O_2]}{[C_2H_3O_2^-]}$$

    The molarity of the salt is 0.300 mol/0.500 L $= 0.600$ mol/L.

    $$\frac{(1.0 \times 10^{-14})}{(1.8 \times 10^{-5})} = \frac{(x)(x)}{[0.600 - x]}$$

    Assume that $0.600 \gg x$, so that $0.600 - x = 0.600$

    $$x^2 = 3.33 \times 10^{-10}$$
    $$x = 1.82 \times 10^{-5} = [OH^-]$$
    $$pOH = 4.74$$
    $$pH = 9.26$$

20. **(C)** Taking 100 g of solution, 50 g will be NaOH and 50 g will be H_2O. The moles of each is

$$? \text{ mol NaOH} = 50 \text{ g NaOH} \left[\frac{1 \text{ mol NaOH}}{40 \text{ g NaOH}}\right] = 1.25 \text{ mol NaOH}$$

$$? \text{ mol } H_2O = 50 \text{ g } H_2O \left[\frac{1 \text{ mol } H_2O}{18 \text{ g } H_2O}\right] = 2.78 \text{ mol } H_2O$$

$$\text{mol fraction NaOH, } \chi = \frac{\text{moles NaOH}}{\text{moles NaOH} + \text{moles } H_2O}$$

$$= \frac{1.25 \text{ moles NaOH}}{1.25 \text{ moles NaOH} + 2.78 \text{ moles } H_2O}$$

$$= 0.311$$

21. **(B)** Roman numerals are used to indicate the oxidation state of a metal in a compound where the metal may have more than one possible oxidation state. Calcium, aluminum, potassium, and ammonium ions all have one possible charge. Only copper can have two oxidation states, and therefore it must be named using Roman numerals.

22. **(B)** Soot is not a pure form of carbon. It is residue from incomplete combustion. Soot contains many organic compounds and inorganic compounds in addition to fine particles of graphite. The composition of soot varies with the distance from the flame that produced it.

23. **(A)** We rearrange the ideal gas law $PV = nRT$ to $PV = (g/M)RT$

Dividing by PV and multiplying by the molar mass gives:

$$\boldsymbol{M} = (g/V)(RT/P)$$

We will use $R = 0.0821$ L atm mol^{-1}.

Therefore we need to convert 722 torr to atm:

$$? \text{ atm} = 722 \text{ torr} \left(\frac{1 \text{ atm}}{760 \text{ torr}}\right) = 0.950 \text{ atm}$$

$$K = {}^{\circ}C + 273 = 26 + 273 = 299 \text{ K}$$

Entering the data and solving, we get

$$\boldsymbol{M} = (2.05 \text{ g } L^{-1})(0.0821 \text{ L atm mol}^{-1})(299 \text{ K})/(0.950 \text{ atm})$$
$$= 53.0 \text{ g mol}^{-1}$$

24. **(D)** The prefix *para-* indicates bonding to opposite sides of the benzene ring. The smallest possible numbers are used in naming compounds. Therefore the 1, 4 terminology is correct. Responses (A) and (B) represent the *ortho-* and *meta-* orientations around the ring.

25. **(B)** At high temperature the molecules have the largest kinetic energy and pass each other rapidly to minimize interactions. The low pressure means few molecules in a given volume that increases the distance between molecules and decreases interactions. Response (C) is attractive, but high volume does not guarantee low pressure and minimal interactions.

26. **(A)** Conversion of the equilibrium constant into a free energy involves the equation

$$\Delta G° = -RT \ln K$$
$$= -(8.314 \text{ J mol}^{-1} \text{ K}^{-1}) (298 \text{ K}) \ln (2.6 \times 10^8)$$
$$= -48,000 \text{ J mol}^{-1} = -48.0 \text{ kJ mol}^{-1}$$

27. **(D)** The square pyramid is the shape of the molecule considering only the location of the atoms.

28. **(C)** The following describes the other equations: (A) the Rydberg equation concerns the energy emitted or absorbed by hydrogen atoms changing energy levels; (B) the Nernst equation concerns electrode potentials in a galvanic cell; (D) the Arrhenius equation concerns reaction kinetics; (E) the Gibbs free energy equation concerns the energy changes in reactions.

29. **(B)** Hypochlorous acid ionizes as follows: $HClO \rightarrow H^+ + ClO^-$. The equilibrium law is

$$K_a = \frac{[H^+][ClO^-]}{[HClO]}$$

The initial concentration of HClO is 0.150 mole/0.500 L = 0.333 *M* HClO.

The HClO will dissociate by some amount x, and therefore the hydrogen ion concentration and ClO^- will increase by x. Entering this information into the equilibrium law gives

$$K_a = \frac{xx}{0.333 - x} = 3.0 \times 10^{-4}$$

We will make the traditional assumption that $x \ll 0.333$, so that $0.333 - x = 0.333$.

Recasting the equilibrium law, we get

$$K_a = \frac{xx}{0.333} = 3.0 \times 10^{-4}$$
$$x^2 = (0.333)(3.0 \times 10^{-4}) = 1.0 \times 10^{-4}$$
$$x = 1.0 \times 10^{-2} \text{ (Note that our assumption is true.)}$$
$$pH = -\log(1.0 \times 10^{-2}) = -(-2.00) = 2.00$$

30. **(B)** Free radicals are substances with an unpaired electron that makes them highly reactive.

31. **(A)** The loss of a beta particle from the nucleus does not change the mass. However, it does increase the atomic number. The symbol for a beta particle is $_{-1}^{0}\beta$.

32. **(C)** The ideal solution has the same attractive forces, or potential energy, in the mixture as in the pure solvents. This also means that mixing is neither exothermic nor endothermic.

33. **(E)** We know that energy and wavelength are inversely proportional to each other. The lowest energy radiation listed is infrared, and the highest energy is the X ray. Therefore, their wavelengths go from high to low.

34. **(A)** Sodium fluoride dissolves in water to form $Na^+ + F^-$. The fluoride ion then reacts with water: $F^- + H_2O \leftrightarrow HF + OH^-$. The correct equilibrium constant is (A). Water is considered a constant that is part of K.

35. **(D)** Generally, elements closer to fluorine in the periodic table have larger electronegativities.

36. **(C)** The missing particle must be $_0^1 X$. (Note that the pre-superscripts must balance and the pre-subscripts must also balance to complete the reaction.) The subatomic particle that fits this profile having a mass of 1 and a charge of zero is the neutron.

37. **(E)** HBr is a strong acid when dissolved in water. This means that it is also a strong electrolyte.

38. **(C)** Standard temperature is defined as zero degrees Celsius when it comes to the gas laws. The standard pressure is one atmosphere of pressure that is also 760 torr (mm Hg).

39. **(E)** The reaction is: $HEAT + 2\ NO_2(g) \rightarrow N_2O_4(g)$

Decreasing the size of the vessel will increase the pressure of both substances. However, since there are 2 moles of gas on the left and 1 mole of gas on the right, the reaction will be forced to the right to make more product as desired.

Decreasing the temperature will favor reactants, as will increasing the vessel size. Adding argon has no effect on the process, and a catalyst will speed up both the forward and reverse reactions.

40. **(E)** In the periodic table sulfur is closest to fluorine and is expected to have the greatest electronegativity. Also, in general, metals tend to have lower electronegativities than nonmetals.

41. **(A)** We know that heat must be added to melt solids. Therefore the enthalpy must be positive. We can also readily visualize that the molecules in a solid are much more ordered and closer together than those in a liquid. Therefore the entropy change must be positive also.

42. **(A)** The question is ? g cholesterol = 1 molecule cholesterol

The important conversion factors in this problem are the number of grams per mole and the number of molecules per mole of cholesterol.

$$?g\ C_{27}H_{46}O = 1\ \text{molecule}\ C_{27}H_{46}O \left(\frac{1\ \text{mol}\ C_{27}H_{46}O}{6.02 \times 10^{23}\ \text{molecules}\ C_{27}H_{46}O} \right) \left(\frac{386\ g\ C_{27}H_{46}O}{1\ \text{mol}\ C_{27}H_{46}O} \right)$$

$$= 6.41 \times 10^{-22}\ g\ C_{27}H_{46}O$$

43. **(C)** Decane is nonpolar, and propanol is polar and forms hydrogen bonds. Their polarities are very different, and they are not expected to be miscible. All other pairs have similar polarities and therefore should mix readily.

44. **(D)** The Arrhenius equation is $k = A\ e^{-E_a/RT}$,

where k is the rate constant, E_a is the activation energy, T is the Kelvin temperature, and A is called the preexponential constant. All of these can be determined as long as the correct measurements are made. Therefore (A), (B), and (C) can be evaluated using kinetic methods.

45. **(C)** Argon has no polarity, and induced dipoles or instantaneous dipoles are the reason why it can be liquefied.

Diagnostic Test 3

Directions: Answer the following multiple-choice questions. You may use a calculator and the periodic table on page 621, but no other information. Limit your time to 60 minutes. If you do not finish in 60 minutes, note the number of questions answered and then continue until all the remaining questions are answered. Record your total time. Score your test with the answer key at the end of the test. Also at the end of the test are tables to help diagnose your strengths and weaknesses. Review those topics that have the most errors. For more of a challenge, do not use your calculator when answering these questions.

1. The most massive subatomic particle is the

(A) neutrino
(B) neutron
(C) alpha particle
(D) beta particle
(E) proton

2. The molar heat of vaporization of water is +43.9 kJ. What is the entropy change for the vaporization of water?

(A) 8.49 J mol^{-1} K^{-1}
(B) 4.184 J mol^{-1} K^{-1}
(C) 2.78 J mol^{-1} K^{-1}
(D) 118 J mol^{-1} K^{-1}
(E) 0.118 J mol^{-1} K^{-1}

3. The Pauli exclusion principle states that

(A) no two electrons can have the same energy
(B) no two electrons with the same spin can occupy an orbital
(C) no two electrons can occupy separate orbitals
(D) no two electrons can pair up if there is an empty orbital available
(E) no two electrons can have the same four quantum numbers

4. A liquid element that is a dark-colored, nonconducting substance at room temperature is

(A) mercury
(B) bromine
(C) iodine
(D) bismuth
(E) fluorine

5. A large positive value for the standard Gibbs free energy change ($\Delta G°$) for a reaction means

(A) the reaction is spontaneous with virtual complete conversion of reactants to products
(B) an extremely fast chemical reaction
(C) a reaction with a high exothermic heat of reaction
(D) a reaction with a very large increase in entropy
(E) none of the above

6. A metal is reacted with HCl to produce hydrogen gas. If 0.0623 gram of metal produces 28.3 mL of hydrogen at STP, the mass of the metal that reacts with one mole of hydrochloric acid is

   (A) 98.6 g
   (B) 493 g
   (C) 24.7 g
   (D) 49.3 g
   (E) 454 g

7. An element in its ground state

   (A) has all of its electrons in the lowest possible energy levels
   (B) is an element as found in nature
   (C) is an element that is unreactive and found free in nature
   (D) has all of its electrons paired
   (E) has all of its orbitals occupied

8. Which following pair of substances can be used to make a buffer solution?

   (A) NaCl and HCl
   (B) $HC_2H_3O_2$ and $KC_2H_3O_2$
   (C) NaBr and KBr
   (D) HIO_3 and $KClO_3$
   (E) NaH and NaOH

9. Which of the following indicates that a reaction is spontaneous?

   (A) at equilibrium there are more products that reactants
   (B) the value of $\Delta G°$ is greater than zero
   (C) the value of $\Delta S°$ is greater than zero
   (D) the value of K_{eq} is less than one
   (E) the value of $\Delta H°$ is greater than zero

10. Which of the following is expected to have two or more resonance structures?

   (A) CCl_2F_2
   (B) SO_3
   (C) PF_5
   (D) H_2O
   (E) $CaBr_2$

11. Which of the following is a radioactive element?

   (A) Na
   (B) Cr
   (C) Am
   (D) Al
   (E) Au

12. The units for the rate of a chemical reaction are

   (A) $L^2 \, mol^{-2} \, s^{-1}$
   (B) $mol \, L^{-1} \, s^{-1}$
   (C) s^{-1}
   (D) $L \, mol^{-1} \, s^{-1}$
   (E) it depends on the particular reaction

13. Which of the following is not a good measure of relative intermolecular attractive forces?

   (A) heat of fusion
   (B) boiling points
   (C) vapor pressures
   (D) heat of vaporization
   (E) surface tension

14. Which of the following is expected to be the least soluble in water?

   (A) NaBr
   (B) $NiSO_3$
   (C) $CrCl_3$
   (D) $Mn(NO_3)_2$
   (E) $Al_2(SO_4)_3$

15. The net ionic equation expected when solutions of NH_4Br and $AgNO_3$ are mixed together is

   (A) $Ag^+(aq) + Br^-(aq) \rightarrow AgBr(s)$
   (B) $NH_4^+ + Ag^+ \rightarrow Ag(NH_4)_2^+(aq)$
   (C) $Br^+ + NO_3^- \rightarrow NO_3Br(aq)$
   (D) $NH_4Br(aq) + NO_3^- \rightarrow$
   $$NH_4NO_3 + Br^-$$
   (E) all combinations of ions are soluble compounds and no net ionic equation can be written

16. Alpha particles are most often emitted by radioactive elements

    (A) with masses less than 18
    (B) with masses greater than 83
    (C) with odd numbers of protons
    (D) with odd numbers of neutrons
    (E) with atomic masses divisible by 4

17. Three half-lives after an isotope is prepared,

    (A) 25% of the isotope is left
    (B) 25% of the isotope decayed
    (C) 12.5% of the isotope is left
    (D) 12.5% of the isotope decayed
    (E) 1/3% of the isotope is left

18. In determining the order for reactant A in a certain reaction, the concentrations of all other reactants are held constant while the concentration of A is tripled from one experiment to another. The reaction rate is found to triple. The appropriate exponent for A in the rate law is

    (A) 1
    (B) 2
    (C) 3
    (D) 4
    (E) 0

19. The molecule with a tetrahedral shape is

    (A) PCl_4F
    (B) BF_3
    (C) CO_2
    (D) CBr_4
    (E) SCl_6

20. According to the kinetic-molecular theory of gases,

    (A) the average kinetic energy of a gas particle is directly related to the Kelvin temperature
    (B) ideal gas particles do not attract or repel each other
    (C) the atoms or molecules of an ideal gas have no volume
    (D) only (B) and (C) are part of the theory
    (E) (A), (B), and (C) are part of the theory

21. Of the following, the most important experimental information used to deduce the structure of the atom was

    (A) the density of each element
    (B) the specific heat capacity
    (C) the emission spectrum of the elements, particularly hydrogen
    (D) the x-rays emitted from each element
    (E) the radioactive decay of the elements

22. The units for R in the ideal gas law equation are

    (A) L atm mol^{-1} K^{-1}
    (B) J mol^{-1} K^{-1}
    (C) volt coulomb mol^{-1} K^{-1}
    (D) only (B) and (C)
    (E) (A), (B), and (C)

23. Which of the following is considered an acid anhydride?

    (A) HCl
    (B) H_2SO_3
    (C) SO_2
    (D) $Al(NO_3)_3$
    (E) CH_3CH_2OH

24. The standard state for redox reactions includes

    (A) the temperature is 25 °C
    (B) concentrations of soluble species are 1 molar
    (C) saturated solutions of slightly soluble species are the standard state
    (D) partial pressures of gases are 1 atmosphere
    (E) all of the above are true

25. What is the theoretical yield of ethanoate when 100 grams of ethanoic acid is reacted with 100 grams of ethyl alcohol?

$$CH_3COOH + CH_3CH_2OH \rightarrow$$
$$CH_3COOCH_2CH_3 + H_2O$$

    (A) 337 g
    (B) 147 g
    (C) 191 g
    (D) 45 g
    (E) 200 g

26. Iron(III) hydroxide has $K_{sp} = 1.6 \times 10^{-39}$. What is the molar solubility of this compound?

    (A) $5.9 \times 10^{-41}\ M$
    (B) $2.0 \times 10^{-10}\ M$
    (C) $7.4 \times 10^{-14}\ mol/L$
    (D) $9.4 \times 10^{-6}\ mol/L$
    (E) $8.8 \times 10^{-11}\ M$

27. When the dichromate ion is reacted, one of its most common products is Cr^{3+}. What is the oxidation state (oxidation number) of chromium in the dichromate ion? Does reduction or oxidation occur when dichromate forms Cr^{3+}?

    (A) 3+, reduction
    (B) 12+, reduction
    (C) 6+, reduction
    (D) 6+, oxidation
    (E) 7+, oxidation

28. Which of the following does not contain oxygen?

    (A) an aldehyde
    (B) an alkane
    (C) an alcohol
    (D) an ether
    (E) an ester

29. From the solubility rules, which of the following is true?

    (A) all chlorides, bromides, and iodides are soluble
    (B) all sulfates are soluble
    (C) all hydroxides are soluble
    (D) all ammonium-containing compounds are soluble
    (E) all silver compounds are soluble

30. Which of the following molecules is expected to have the highest normal boiling point?

    (A) $CH_3CH_2CH_2CH_3$
    (B) $CH_3CH_2CH_2CH_2OH$
    (C) $CH_3CH_2CH_2CH_2Cl$
    (D) $CH_3CH_2CH_2CH_2Br$
    (E) $CH_3CH_2CH_2CHO$

31. Which is correct about the calcium atom?

    (A) it contains 20 protons and neutrons
    (B) it contains 20 protons and 20 electrons
    (C) it contains 20 protons, neutrons, and electrons
    (D) all atoms of calcium have a mass of 40.078 u
    (E) all atoms of calcium give off red light

32. What is the theoretical yield of iron when 2.00 grams of carbon is reacted with 26.0 grams of Fe_2O_3?

    $$2\ Fe_2O_3 + 3\ C \rightarrow 4\ Fe + 3\ CO_2$$

    (A) 5.8 g
    (B) 12.4 g
    (C) 74.6 g
    (D) 30.6 g
    (E) 18.2 g

33. The freezing-point depression constant for camphor is $40.0\ °C\ molal^{-1}$. A 2.35-gram sample of an unknown molecular compound is mixed with 50.0 grams of camphor. The melting point of camphor is then observed to decrease from the normal 178.40 °C to 173.20 °C. What is the molecular mass of the unknown molecule?

    (A) 230 g/mol
    (B) 54.6 g/mol
    (C) 23.5 g/mol
    (D) 72.3 g/mol
    (E) 362 g/mol

34. An ester is made by reacting

    (A) an amine and an alcohol
    (B) two different alcohols
    (C) an alcohol and an acid
    (D) an acid and a base
    (E) an ether and an aldehyde

35. Which of the following salts is expected to produce an alkaline solution when one mole is dissolved in one liter of water?

    (A) $NaClO_4$
    (B) $CaCl_2$
    (C) NH_4Br
    (D) Na_2S
    (E) $Fe(NO_3)_3$

36. Which of the following is a temperature-dependent concentration unit?

    (A) molarity
    (B) mole fraction
    (C) weight percent
    (D) molality
    (E) weight fraction

37. When collecting a gas over water, it is important to

    (A) set the temperature at 0 °C
    (B) be sure the gas does not burn
    (C) wait until the barometer reads 760
    (D) correct for the vapor pressure of water
    (E) be sure that absolutely pure water is used.

38. Chlorine gas reacts most readily with

    (A) toluene
    (B) ethylene
    (C) ethanoic acid
    (D) ethane
    (E) cyclohexane

39. Sulfur dioxide reacts with oxygen to form sulfur trioxide in the presence of a catalyst. The equilibrium constant, K_p, at a certain temperature is 3.0×10^{22}. A 2.0-liter flask has enough SO_3 added to it to produce a pressure of 0.789 atm Hg. After the reaction comes to equilibrium, the expected partial pressure of O_2 will be

    (A) 2.88×10^{-6} torr
    (B) 3×10^{-18} mm Hg
    (C) 1100 mm Hg
    (D) 1.32×10^{-5} torr
    (E) 600 torr

40. The melting point of straight-chain hydrocarbons increases as the number of carbon atoms increase. The reason for this is the

    (A) increasing mass of the compounds
    (B) increasing polarity of the compounds
    (C) increasing number of induced dipoles per molecule
    (D) increased probability of hydrogen bonds
    (E) rigid chains of these hydrocarbons

41. What is the empirical formula of a compound that is 51.9% carbon, 4.86% hydrogen, and 43.2% bromine

    (A) C_7H_5Br
    (B) $C_6H_4Br_3$
    (C) C_8H_9Br
    (D) $C_{12}H_{22}Br$
    (E) $C_{10}H_{14}Br$

42. What is the mass of a biomolecule that produces an osmotic pressure of 0.890 torr when 20 g is dissolved in 400 mL of water at 4 °C?

    (A) 1.67×10^4 g/mol
    (B) 3.62×10^5 g mol^{-1}
    (C) 9.7×10^5 g/mol
    (D) 9.72 g mol^{-1}
    (E) 14 g/mol

43. The standard galvanic cell voltage, $E°_{cell}$

    (A) is equal to $E°_{reduction} - E°_{oxidation}$
    (B) can be used to calculate K_{eq}
    (C) can be used to calculate $\Delta G°$
    (D) can be used to determine if a reaction is spontaneous
    (E) all of the above

44. When will K_p and K_c have the same numerical value?

    (A) at absolute zero for all reactions
    (B) when the concentrations are at standard state
    (C) when the concentrations are all 1.00 molar
    (D) when the reaction exhibits no change in pressure at constant volume
    (E) when $\Delta G°$ is equal to zero

45. The rate of a chemical reaction is determined by

    (A) the equilibrium constant
    (B) the rate-determining or slow step of the mechanism
    (C) the reaction vessel pressure
    (D) the intermediates formed in the first step
    (E) all of the above

Diagnostic Test 3
ANSWER KEY

1. C	10. B	19. D	28. B	37. D
2. D	11. C	20. E	29. D	38. B
3. E	12. B	21. C	30. B	39. D
4. B	13. A	22. E	31. B	40. C
5. E	14. B	23. C	32. B	41. C
6. C	15. A	24. E	33. E	42. C
7. A	16. B	25. B	34. C	43. E
8. B	17. C	26. E	35. D	44. D
9. A	18. A	27. C	36. A	45. B

EVALUATING YOUR RESULTS

Score your test using the answer key. Then complete the following tables. The first table is designed to find your general strengths and weaknesses based on four broad categories. The second table is a more specific diagnostic chart that will suggest which particular chapters you should concentrate your studies on. In combination, these two tables will help you focus your efforts on the material that needs the most study.

QUESTION CATEGORIES

QUESTION TYPE	QUESTIONS	NUMBER WRONG
Basic Facts	1, 4, 11, 12, 16, 21, 23, 24, 28, 29, 31, 34, 36	
Basic Concepts	3, 5, 7, 8, 9, 13, 18, 19, 20, 22, 30, 31, 35, 38, 40, 44, 45	
Calculations	2, 6, 17, 25, 26, 32, 33, 39, 41, 42	
Mixed Concepts	10, 12, 14, 15, 19, 20, 27, 30, 35, 37, 43	

BREAKDOWN BY TOPICS

CHAPTER	QUESTIONS	NUMBER WRONG
1. Structure of the Atom	3, 21, 31	
2. Periodic Table	4, 7, 11	
3. Nuclear Chemistry	1, 16, 17	
4. Ionic Compounds and Reactions	14, 22, 29	
5. Covalent Compounds	10, 15, 19	
6. Stoichiometry	25, 32, 41	
7. Gases	6, 20, 37	
8. Liquids and Solids	13, 30, 40	
9. Solutions	33, 36, 42	
10. Equilibrium	26, 39, 44	
11. Kinetics	12, 18, 45	
12. Thermodynamics	2, 5, 9	
13. Redox and Electrochemistry	24, 27, 43	
14. Acids and Bases	8, 23, 35	
15. Organic Chemistry	28, 34, 38	

ANSWERS EXPLAINED

1. **(C)** The alpha particle is a helium nucleus with a mass of 4. The neutrino and the beta particle have very little mass. The proton and the neutron each have a mass of 1.

2. **(D)** The transformation of water from $H_2O(l) \leftrightarrow H_2O(g)$ at 100 °C has $\Delta G° = 0$. Therefore $\Delta H° = T\Delta S°$. Because we know the heat of vaporization and the boiling point of water, 100 °C (373 K), the entropy change can be calculated as

$$\Delta S° = (43,900 \text{ J mol}^{-1})/373 \text{ K}$$
$$= 118 \text{ J mol}^{-1} \text{ K}^{-1}$$

3. **(E)** The Pauli exclusion principle states that no two electrons in an element can have the same set of four quantum numbers. This is another way of saying that no two electrons may occupy the same orbital with the same spin. Response (A) is false—electrons often have the same energy. Response (B) violates the rule that electrons in the same orbital must have paired spins. Response (D) is a statement of Hund's rule.

4. **(B)** There are only two elements that are liquids at room temperature. They are bromine and mercury. Mercury is a silver-colored metal, and bromine is a brown nonmetal liquid. Nonmetals do not conduct electricity.

5. **(E)** None of the four answers fits. The positive free energy indicates a non-spontaneous system with more reactants than products. The large exothermic heat of reaction and the large increase in entropy are very unlikely with a large positive free energy.

6. **(C)** We set up the problem with the desired ratio to the left of the equal sign. To the right of the equal sign replace the given mass of metal in the numerator and the volume of H_2 in mL in the denominator. Since we already know the value of the final numerator, we just need to convert the denominator into moles of HCl.

$$? \frac{\text{grams of metal}}{\text{moles of HCl}} = \frac{0.0623 \text{ g metal}}{28.3 \text{ mL H}_2} \left[\frac{22,400 \text{ mL H}_2}{1 \text{ mol H}_2} \right] \left[\frac{1 \text{ mol H}_2}{2 \text{ mol HCl}} \right]$$
$$= 24.7 \text{ g metal/mol HCl}$$

Notice that we used the molar volume of an ideal gas at STP to convert the H_2 from mL to moles. Then we used the mole ratio that relates H_2 and HCl for the second conversion factor.

7. **(A)** The definition of the ground state is the lowest total energy. To have the lowest total energy an atom must have its electrons in their lowest possible energy levels.

8. **(B)** Buffers are prepared from a weak acid and its conjugate base or from a weak base and its conjugate acid. (A) NaCl and HCl; this pair has HCl which

is a strong acid. (B) $HC_2H_3O_2$ and $KC_2H_3O_2^-$; this pair has a weak acid and its conjugate base (the $C_2H_3O_2^-$ ion). (C) NaBr and KBr; this pair has no weak acid or base. (D) HIO_3 and $KClO_3$; this pair does not have a conjugate acid-base pair. (E) NaH and NaOH; this is not a conjugate acid-base pair.

9. **(A)** The responses for (B), (D) and (E) indicate a nonspontaneous process. The response in (C) may indicate a spontaneous process, but it is not sufficient for a spontaneous reaction. Only response (A) is universally true for a spontaneous reaction.

10. **(B)** The SO_3 molecule has three resonance structures. Each has one oxygen drawn with a double bond to the sulfur and the other two oxygens drawn with a single bond to the sulfur. The remaining compounds have only one possible Lewis structure.

11. **(C)** We expect elements with atomic numbers greater than 83 to be radioactive. Americium is the only one given that fulfills that criterion.

12. **(B)** All reaction rates have the same units of moles per liter per second ($mol\ L^{-1}\ s^{-1}$). This may refer to the rate of appearance of a product as the reaction progresses or the disappearance of a reactant.

13. **(A)** The heat of fusion is the energy needed to disrupt the crystal lattice but not completely separate the molecules. The remaining attractive forces may be significant. (B), (C), and (D) all involve vaporizing the liquid and indicate the total attractive force. Surface tension is similar to vaporization except that one side of the system is vapor and the other the liquid and the resulting forces are related to intermolecular attractive forces.

14. **(B)** The solubility rules specify that sulfites are one group of compounds that are generally insoluble, especially if the metal ion is a transition metal.

15. **(A)** Silver bromide is insoluble, whereas all other substances in the reaction are soluble. The ammonium ions and nitrate ions are spectator ions that cancel. Silver ions will form a complex with ammonia, NH_3, not with ammonium ions, NH_4^+.

16. **(B)** Elements with atomic masses greater than 83 are the most likely to emit alpha particles when they undergo radioactive decay.

17. **(C)** For three half-lives, $1/2 \times 1/2 \times 1/2 = 1/8 = 0.125$, or 12.5%. Therefore 12.5% of the original material will remain.

18. **(A)** Because the reaction rate increases by the same factor as the concentration of A, the reaction with respect to reactant A must be first order.

19. **(D)** (A) is trigonal bipyramidal; (B) is triangular planar; (C) is linear; (E) is octahedral.

20. **(E)** All three statements are considered to be parts of the kinetic-molecular theory of gases as stated by Clausius in 1857.

21. **(C)** The Bohr theory that preceded the quantum model of the atom relied on the atomic spectrum of hydrogen for important clues.

22. **(E)** All of the first three are units for R.

23. **(C)** An acid anhydride is an oxide of a nonmetal that dissolves in water to form an oxo acid. Sulfur dioxide dissolves in water to produce an acidic solution.

24. **(E)** All four of the statements define one aspect of standard state as applied to the study of redox reactions.

25. **(B)** We calculate the grams of ethanoic acid needed to react with the given amount of ethyl alcohol. Two important conversion factors are

$$1 \text{ mol } CH_3COOH = 60 \text{ g } CH_3COOH$$

and

$$1 \text{ mol } CH_3CH_2OH = 44 \text{ g } CH_3CH_2OH$$

Setup: $? \text{ g } CH_3COOH = 100 \text{ g } CH_3CH_2OH$

Applying the above conversion factors we get

$? \text{ g } CH_3COOH =$

$$= 100 \text{ g } CH_3CH_2OH \left[\frac{1 \text{ mol } C_2H_5OH}{44 \text{ g } C_2H_5OH}\right]\left[\frac{1 \text{ mol } HC_2H_5O_2}{1 \text{ mol } C_2H_5OH}\right]$$

$$\left[\frac{60 \text{ g } HC_2H_5O_2}{1 \text{ mol } HC_2H_5O_2}\right]$$

$$= 136 \text{ g } CH_3COOH$$

The problem only gave us 100 g CH_3COOH. Therefore CH_3COOH is the limiting reactant. We now calculate the mass of ethyl ethanoate formed from the GIVEN mass of 100 g CH_3COOH as follows.

Set up the question.

$$? \text{ g } CH_3CH_2OC{=}OCH_3 = 100 \text{ g } CH_3COOH$$

Apply conversion factors

$? \text{ g } CH_3CH_2OC{=}OCH_3 =$

$$= 100 \text{ g } CH_3COOH\left[\frac{1 \text{ mol } HC_2H_5O_2}{60 \text{ g } HC_2H_5O_2}\right]\left[\frac{1 \text{ mol } C_4H_8O_2}{1 \text{ mol } HC_2H_5O_2}\right]$$

$$\left[\frac{88 \text{ g } C_4H_8O_2}{1 \text{ mol } C_4H_8O_2}\right]$$

$$= 147 \text{ g } C_4H_8O_2 \text{ (or } CH_3CH_2OC{=}OCH_3)$$

NOTE: We get the same result if we convert each reactant to grams of ethyl ethanoate and then choose the smaller of the two results.

26. **(E)** The dissolution reaction is $Fe(OH)_3(s) \leftrightarrows Fe^{3+} + 3\ OH^-$, and the K_{sp} equation is $K_{sp} = [Fe^{3+}][OH^-]^3$.

If s mol/L of $Fe(OH)_3$ dissolves, the solution will contain s mol/L of Fe^{3+} and $3s$ mol/L of OH^-.

Substituting into the K_{sp} equation results in

$$K_{sp} = 1.6 \times 10^{-39} = (s)(3s)^3$$
$$= (s)(27s^3)$$
$$= 27s^4$$
$$s^4 = 1.6 \times 10^{-39}/27 = 5.93 \times 10^{-41}$$
$$s = 8.8 \times 10^{-11}$$

(the fourth root can be obtained by taking successive square roots or by raising the number to the 1/4 or 0.25 power using the y^x function found on most calculators.) Note that we kept one extra significant figure in step 4 and rounded to the correct number of significant figures only at the end.

27. **(C)** The dichromate ion is $Cr_2O_7^{2-}$. When we calculate the oxidation number for each chromium in the dichromate ion, we get 6+. Because the Cr^{3+} is only 3+, there is a decrease in oxidation number, and the process is called a reduction.

28. **(B)** Alkanes have just carbon and hydrogen atoms. All others have oxygen.

29. **(D)** It is risky to say that all compounds have any given property. However, given the solubility rules generally used in AP courses, ammonium ions for the most part indicate a soluble ionic substance. The other items listed all have well-known insoluble substances.

30. **(B)** Butan-1-ol (butanol), $CH_3CH_2CH_2CH_2OH$, can hydrogen-bond, whereas the other compounds cannot. London forces (instantaneous dipoles) are also present, but each compound is roughly the same length, and these forces will be similar for all five molecules.

31. **(B)** All calcium atoms contain 20 protons and 20 electrons. Depending on the isotope, a calcium atom may or may not have 20 neutrons. Finally, the mass in the periodic table is a weighted average of isotopes and is NOT the mass of any calcium atom. Response (E) implies that calcium in the ground state gives off light, which is not true.

32. **(B)** We are asked for the theoretical yield of iron. Usually, the theoretical yield is expressed in units of grams of product, as all the responses imply. This is a limiting-reactant problem because the mass of both reactants is given. We will solve it by calculating the mass of iron that can be made from 2.00 g C assuming that Fe_2O_3 is the excess reactant. Then we will calculate the mass of

iron that we can prepare from 26.0 g Fe_2O_3 assuming that carbon is the excess reactant. We set up the two equations as

$$? \text{ g Fe} = 2.00 \text{ g C} \left[\frac{1 \text{ mol C}}{12 \text{ g C}}\right]\left[\frac{4 \text{ mol Fe}}{3 \text{ mol C}}\right]\left[\frac{55.84 \text{ g Fe}}{1 \text{ mol Fe}}\right] = 12.4 \text{ g Fe}$$

$$? \text{ g Fe} = 26.0 \text{ g Fe}_2\text{O}_3 \left[\frac{1 \text{ mol Fe}_2\text{O}_3}{159.7 \text{ g Fe}_2\text{O}_3}\right]\left[\frac{4 \text{ mol Fe}}{2 \text{ mol Fe}_2\text{O}_3}\right]\left[\frac{55.84 \text{ g Fe}}{1 \text{ mol Fe}}\right] = 18.2 \text{ g Fe}$$

Carbon produces the smaller amount of iron. Therefore, carbon is the limiting reactant, and 12.4 g of iron is the theoretical yield.

33. **(E)** We use the freezing-point depression equation given in the table of equations:

$$\Delta T = ik_f m = ik_f \left[\frac{\text{g solute/molar mass solute}}{\text{kg solvent}}\right]$$

Since the unknown compound is described as molecular, we can assume it is nondissociating, and $i = 1$ and can be omitted from the rest of the calculations.

Rearranging this equation gives:

$$M \text{ solute} = \left[\frac{k_f \text{ (g solute)}}{\Delta T \text{ (kg solvent)}}\right]$$

$$= \left[\frac{40.0 \text{ °C/m (2.35 g)}}{5.20 \text{ (0.050 kg)}}\right]$$

$$= 362 \text{ g/mol}$$

34. **(C)** An alcohol and acid will react to form an ester.

35. **(D)** Salts made from a weak acid and a strong base will produce an anion that will hydrolyze water to form an alkaline solution. The only salt produced from a weak acid is Na_2S, and the sulfide ion hydrolyzes water according to the reaction

$$S^{2-} + H_2O \rightarrow HS^- + OH^-$$

36. **(A)** The definition of molarity is moles of solute dissolved in one liter of solution. The volume of almost all common solvents changes as the temperature changes. Therefore the molarity of a solution depends on the temperature. All of the rest are defined in terms of units that do not change as the temperature changes.

37. **(D)** The largest error will occur if the results are not corrected for the vapor pressure of water. The temperature of the water must be measured, but it does not have to be 0 °C. The barometric pressure is part of the ideal gas law and

does not need to be 760 torr, but it must be measured when the experiment is performed. The use of tap water instead of distilled water and the flammability properties of the gas will not have a significant effect on the gas.

38. **(B)** Chlorine and all halogens readily attack aliphatic double bonds, as in ethylene. Chlorine will react with the other compounds, however, the process is more difficult.

39. **(D)** We need to write the chemical reaction as

$$2 \, SO_2(g) + O_2(g) \rightarrow 2 \, SO_3(g)$$

The equilibrium law for this reaction is written in terms of partial pressures, measured in atmospheres, and the equilibrium constant is K_p.

$$K_p = \frac{P_{SO_3}^2}{P_{O_2} P_{SO_2}^2} = 3.0 \times 10^{22}$$

The initial partial pressure of SO_3 is converted from 600 torr to 0.789 atm. If $2x$ atm of the initial SO_3 decomposes, x atm of O_2 and $2x$ atm of SO_2 will form. This is summarized in the ICES table.

	2 SO$_2$(g) +	O$_2$(g)	→	2 SO$_3$(g)
Initial	0	0		0.789
Change	+2x	+x		−2x
Equilibrium (I + C)	2x	x		0.789 − 2x ≈ 0.789
Solution (sub. x into E)	3.46 × 10^{-8}	1.73 × 10^{-8}		0.789

We enter the information from the equilibrium line into the equilibrium law. We use the approximation that $0.789 - 2x = 0.789$ because the equilibrium constant is large.

$$K_p = \frac{(0.789 - 2x)^2}{x(2x)^2} = \frac{(0.789)^2}{x(2x)^2}$$

$$3.0 \times 10^{22} = \frac{(0.789)^2}{x(2x)^2}$$

$$12.0 \times 10^{22} x^3 = 0.6233$$

$$x = 1.73 \times 10^{-8} \, atm = 1.32 \times 10^{-5} \, torr$$

The value of x is very small, indicating that our assumption was correct. Notice how we needed to solve the problem using units of atmospheres.

40. **(C)** Straight-chain hydrocarbons are essentially nonpolar, and they interact only with London forces of attraction that include instantaneous dipoles and induced dipoles. As the chain becomes longer, each molecule has more of these forces attracting it to neighboring molecules. With long non-polar carbon chains, these instantaneous dipoles can be a strong attractive force.

41. **(C)** To determine the empirical formula we must calculate the simplest ratio of moles of each element in the formula. The percentages are converted to grams by assuming 100 g of sample and changing all percentages to grams. The moles are then calculated as

$$? \text{ mol C} = 51.9 \text{ g C} \left(\frac{1 \text{ mol C}}{12 \text{ g C}} \right) = 4.32 \text{ mol C}$$

$$? \text{ mol H} = 4.86 \text{ g H} \left(\frac{1 \text{ mol H}}{1 \text{ g H}} \right) = 4.86 \text{ mol H}$$

$$? \text{ mol Br} = 43.2 \text{ g Br} \left(\frac{1 \text{ mol Br}}{79.9 \text{ g Br}} \right) = 0.541 \text{ mol Br}$$

Divide each of the answers above by 0.541 to get 7.98 mol C, 8.98 mol H, and 1.00 mol Br. The empirical formula is C_8H_9Br.

42. **(C)** The osmotic pressure equation is the same as the ideal gas law equation and is written as

$$\Pi V = nRT$$

where Π represents the osmotic pressure. Because $R = 0.0821$ L atm mol^{-1} K^{-1}, we need to convert all terms to these units.

Therefore,

    400 mL = 0.400 L
    4 °C = 277 K
    0.890 torr/760 torr atm^{-1} = 1.17×10^{-3} atm
    $(1.17 \times 10^{-3}$ atm$)(0.400$ L$) = n(0.0821$ L atm mol^{-1} K$^{-1})(277$ K$)$
    $n = 2.06 \times 10^{-5}$ mol

Molecular mass = 20 g/2.06×10^{-5} mol = 9.72×10^5 g/mol

43. **(E)** All four responses are correct applications of $E°_{cell}$

44. **(D)** When the reaction shows no pressure changes, it indicates that $\Delta n_g = 0$. If a chemical reaction has the same number of moles of gas as reactants and products, then K_p will be equal to K_c.

45. **(B)** The rate-determining step governs the overall reaction rate. Proposed mechanisms must provide appropriate steps to include the slow step.

PART 1

STRUCTURE OF MATTER

Structure of the Atom

CHAPTER 1

- Atomic Theory
- Models of the Atom
- Structure of the Atom
- Protons, Electrons, and Neutrons
- Isotopes
- Atomic Spectra
- Wave Mechanical Atom
- Energy Levels
- Electronic Structure
- Electron Configurations
- Valence Electrons
- Hund's Rule
- Orbital Diagrams
- Pauli Exclusion Principle
- Quantum Numbers

A REVIEW OF IMPORTANT DISCOVERIES ABOUT THE ATOM

Major milestones in the development of chemistry were reached in 1774, when Antoine Lavoisier performed careful experiments and measurements that led to the **law of conservation of matter**, and in 1799, when Joseph Proust made measurements on chemical reactions and compounds and developed the **law of constant composition**. The first of these important laws states that in chemical reactions matter is neither created nor destroyed. The second law states that each pure chemical compound always has the same percentage composition of each element by mass. These two laws led John Dalton to develop his **atomic theory** over the years from 1803 to 1808. Dalton's atomic theory states that:

> Law of Conservation of Matter

> Law of Constant Composition

> Dalton's Atomic Theory

1. All matter is composed of tiny, indivisible particles, called atoms, that cannot be destroyed or created.
2. Each element has atoms that are identical to each other in all of their properties, and these properties are different from the properties of all other atoms.
3. Chemical reactions are simple rearrangements of atoms from one combination to another in small whole-number ratios.

Every scientific theory provides new predictions that may be tested by experiments to support or disprove the theory. [It is important to remember that scientists can never prove a theory to be true. Experiments may be used to support a theory, but not to prove it.]

The atomic theory led John Dalton to propose the **law of multiple proportions:**

> Law of Multiple Proportions

When two elements can be combined to make two different compounds, and if samples of these two compounds are taken so that the masses of one of the elements in the two compounds are the same in both samples, then the ratio of the masses of the other element in these compounds will be a ratio of small whole numbers.

This law was used to test Dalton's atomic theory.

In 1834 Michael Faraday showed that electric current could cause chemical reactions to occur, demonstrating the electric nature of the elements. Sir William Crookes in the 1870s developed what is known today as the **cathode ray tube**. He mistakenly thought that the cathode rays were negatively charged molecules instead of electrons. (Crookes was also the first to suggest the existence of isotopes.) In 1897 J. J. Thomson determined that cathode rays were a fundamental part of matter he called electrons. He also determined their **charge to mass ratio** ($e/m = -1.76 \times 10^8$ coulombs gram^{-1}) by measuring the deflection of the cathode rays in the presence of electric and magnetic fields. Twelve years later, in 1909, Robert Millikan performed his **oil drop experiment,** which allowed him to calculate the charge of the electron (-1.60×10^{-19} coulomb). Combined with Thomson's charge to mass ratio, the mass of the electron was calculated to be 9.11×10^{-28} gram. This information led to the **"plum pudding" model** of the atom, which had electrons bathed in a sea of positive charges similar to raisins in the famous English pudding.

At this time Ernest Rutherford was interested in radioactive materials and had identified alpha and beta particles in his research. Along with Hans Geiger and Ernest Marsden, he performed the **gold foil experiment,** in which heavy alpha particles were aimed at a thin gold foil. While most of the alpha particles went through the foil with no visible effect, a few of them were deflected from their path and some actually bounced back in the direction they came from. From these results Rutherford deduced the **nuclear model of the atom,** with an extremely small, dense, and positively charged nucleus surrounded by empty space sparsely occupied by electrons. Ten years later, in 1919, Rutherford discovered the basic unit of positive charge in the atom and named it the proton. The proton has a positive charge which is exactly equal in magnitude to the electron charge. It also has a mass of 1.67×10^{-24} gram, 1836 times heavier than the electron. In 1932 James Chadwick discovered a very penetrating form of radiation. He demonstrated that it was the third major particle that makes up the atom, and it was named the neutron since it is neutral (i.e., it has no charge). The neutron has a mass almost equal to that of the proton.

While the fundamental particles of the atom (Table 1.1) were being discovered, other physicists were performing experiments that laid the foundation for a fundamental revolution in the way all matter is viewed. In the mid-1800s physicists were interested in the interaction of light and matter. One of the interesting things they found was that each element, when heated or sparked with electricity, gives off characteristic colors. A spectroscope was used to show that these colors consist of discrete wavelengths of light (line spectra), not the uniform rainbow observed when white light is separated by a prism. The line spectra of most elements and compounds are very complex. Hydrogen, however, has a seemingly simple series of lines. In 1885 Johann Balmer found an empirical mathematical relationship between the wavelengths of the lines he observed in the visible region of the spectrum. When similar series of lines were found in the infrared (Paschen series) and ultraviolet (Lyman series) regions, Johannes Rydberg extended Balmer's equation so that all of the wavelengths could be predicted.

Michael Faraday

J. J. Thomson

Ernest Rutherford

ATOMIC MODELS

Solid particle model 400 BC

Plum pudding model 1909

Nuclear model Rutherford 1910

Solar system model Bohr 1913

Wave-mechanical model Schrödinger 1927

Johann Balmer

Johannes Rydberg

TABLE 1.1

Fundamental Parts of the Atom

Name	Symbol	Absolute Charge (coulombs)	Absolute Mass (grams)	Relative Charge	Relative Mass (U)
Electron	e or e⁻	-1.602×10^{-19}	9.109×10^{-28}	21	5.486×10^{-4}
Proton	p	$+1.602 \times 10^{-19}$	1.673×10^{-24}	11	1.0073
Neutron	n	0	1.675×10^{-24}	0	1.0087

In 1913 Niels Bohr completed his theory of how the hydrogen atom is constructed. He assumed that electrons move around the nucleus in circular orbits. Using a **solar system model,** he was able to duplicate the Rydberg equation from fundamental constants already known to physicists. Bohr's most important contribution was the concept that electrons exist in only certain "allowed orbits." This, along with Max Planck's description in 1900 of light as packets, or quanta, of energy, called photons, aided Bohr in developing the solar system model of the atom.

Louis de Broglie suggested in 1924 that, if light can be considered as particles, then the small particles such as electrons may have the characteristics of waves. In 1927 Erwin Schrödinger applied the equations for waves to the electrons in an atom and began the **wave-mechanical theory** of the atom. For hydrogen the results are very similar to Bohr's model except that the electron does not follow a precise orbit. The position of the electron in the wave-mechanical model is described by a probability of where it will be located. Also in the 1920s, Werner Heisenberg developed the uncertainty principle, which bears his name. It states that the position and the momentum of any particle cannot both be known exactly at the same time. As one is known more precisely, the other becomes less certain.

> Niels Bohr

> **TIP**
>
> Quantum mechanics has replaced classical physics for describing events at the atomic level

> Max Planck

> Louis De Broglie

> Erwin Schrödinger

> Werner Heisenberg

EXPERIMENTS DESCRIBING THE STRUCTURE OF THE ATOM

Charge to Mass Ratio of the Electron

Sir William Crookes designed an evacuated tube with two electrodes, as shown in Figure 1.1. When a high voltage was applied to the electrodes, a glow was noticed between them. When an object was placed in the path of the glow, it blocked part of the beam. This experiment showed that the beam must originate at the negative electrode (cathode) and flow toward the positive electrode (anode).

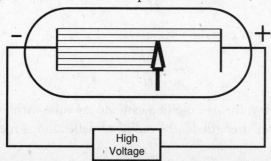

FIGURE 1.1. Cathode rays blocked by an object, showing direction of flow.

Cathode
Ray Tubes
(CRT)

Thus the glow was called cathode rays since the rays apparently came from the cathode, or negative terminal. The path of the cathode rays could be deflected by both electric fields (Figure 1.2) and magnetic fields. The fact that these rays were attracted toward the positive electric field and repelled by the negative electric field indicated that they had a negative charge.

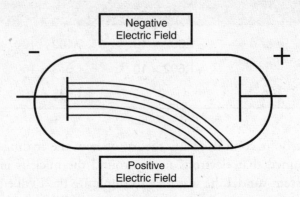

FIGURE 1.2. Deflection of cathode rays by an electric field.

In a magnetic field the magnetic force causes a negative particle to move in a circular motion with a radius of r according to the equation

$$Hev = \frac{mv^2}{r} \tag{1.1}$$

In this equation, H is the strength of the magnetic field, e is the charge of the electron, m is the mass of the electron, v is the velocity of the electron, and r is the radius of curvature caused by the magnetic field. Rearranging this equation yields

$$\frac{e}{m} = \frac{v}{rH} \tag{1.2}$$

The magnetic field H and the radius of curvature are easily measured. It is necessary to know the velocity of the electron to calculate the e/m ratio in Equation 1.2. If an electric field is applied to the beam so that it cancels the deflection of the magnetic field, the force of the electric field must be the same as the force of the magnetic field. Mathematically this is expressed as

$$Ee = Hev \tag{1.3}$$

where E is the electric field, e is the charge of the electron, and v is the velocity of the electron. Rearranging this equation gives the needed velocity of the electron.

$$v = \frac{E}{H} \tag{1.4}$$

Charge-to-
mass-ratio

Substituting this into Equation 1.2 gives

$$\frac{e}{m} = \frac{E}{H^2 r} \tag{1.5}$$

The experiment involves the use of a cathode ray tube with an applied magnetic field having a known strength H; the radius of deflection is measured. An electric field that may be varied is then applied to cancel the effect of the magnetic field. The value of the electric field needed to cancel the magnetic field is then used to calculate the *e/m* ratio. The value obtained was -1.76×10^8 coulombs per gram.

Millikan Oil Drop Experiment

Robert Millikan set up an apparatus as shown in Figure 1.3, where he could spray oil droplets (from his wife's perfume atomizer) so they would settle into a beam of X rays. The X rays caused the oil droplets to become charged with electrons. Using a small telescope, Millikan could measure the diameter of each droplet. In addition, he applied an electric voltage to the top and bottom of the chamber so that the droplet would stop falling and remain stationary. The positive plate attracted, and the negative plate repelled, the negatively charged droplet. Adjusting the voltage, Millikan was able to make the droplet stand still.

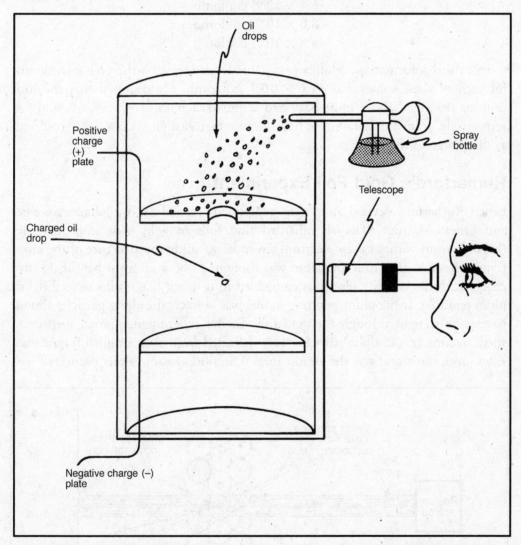

FIGURE 1.3. Diagram of apparatus used by Millikan to determine the charge of the electron.

Knowing the density of the oil, Millikan could calculate the mass, *m*, of each oil droplet (volume × density = mass). He knew that the equation for the force of gravitational attraction was *Force = mG*, where *G* is the acceleration due to gravity. The equation for electrical force is *Force = Ee*, where *E* is the applied voltage and *e* is the charge of the electron. These forces are equal when the voltage is adjusted, so that the oil droplet remains stationary and

$$m_{\text{droplet}}G = Ee \tag{1.6}$$

From this equation the value of the charge of the electron, *e*, can be easily calculated.

There was one problem: the X rays did not always give an oil droplet with only one negative charge. The result was that Millikan did not get one single value for the charge of the electron. Instead he obtained a series of results that may have looked like this:

$$-3.2 \times 10^{-19} \text{ coulomb}$$
$$-6.4 \times 10^{-19} \text{ coulomb}$$
$$-8.0 \times 10^{-19} \text{ coulomb}$$
$$-4.8 \times 10^{-19} \text{ coulomb}$$

From this information, Millikan was able to deduce that the common divisor for each of these values was -1.6×10^{-19} coulomb. He reasoned that the four droplets described above must have had acquired charges of −2, −4, −5, and −3, respectively, from the X rays. He had the insight to assign -1.6×10^{-19} coulomb as the charge of the electron.

Rutherford's Gold Foil Experiment

Ernest Rutherford devised an elegant experiment carried out by Johannes Geiger and Ernest Marsden. They bombarded thin foils of gold with alpha particles (helium atoms without their electrons) in order to study the structure of the atom (Figure 1.4). At that time the atom was thought to be a uniform ball of positive charge with the negative electrons embedded in it much like raisins in an English plum pudding. If the plum pudding model was correct, the alpha particles should have gone straight through the gold foil. To the experimenters' total surprise, a small fraction of the alpha particles were deflected from their original trajectories. Even more surprising was the almost total reflection of some alpha particles.

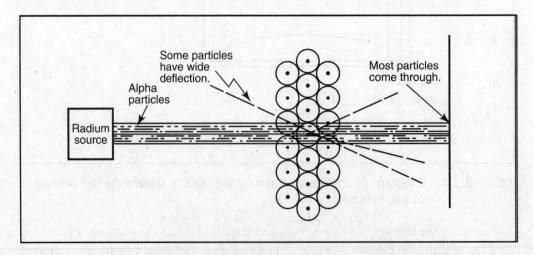

FIGURE 1.4. Diagram of Rutherford's gold foil apparatus.

These results suggested a very massive nucleus as compared to the alpha particle. Moreover, this nucleus must be positively charged to repel the positively charged alpha particle. The nuclear model of the atom was developed by Rutherford from this information. He concluded that almost all of the mass of the atom is concentrated in a positively charged nucleus and that the rest of the atom is empty space in which the negatively charged electrons move around the nucleus.

Discovery of the Proton

Using a cathode ray tube, experimenters drilled holes into the anode (Figure 1.5), and rays moving in opposite directions to the electron were discovered. Originally called "canal rays," they were recognized as atoms with one or more electrons removed. The measurements described for the *e/m* ratio of the electron were used to determine that the lightest of the canal rays, the proton, must have a mass 1800 times greater than that of the electron. It also has a charge equal to the electron but with the opposite sign (the proton has a positive charge).

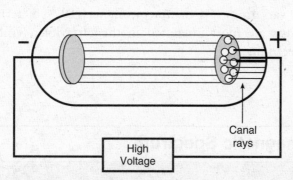

FIGURE 1.5. **Holes drilled in the anode of a Crookes tube reveal atomic ions, the lightest of which is the proton.**

Determination of Atomic Numbers

In 1913 Henry Moseley studied the X rays emitted by various elements in specially designed X-ray tubes. He found that the atomic number (number of protons in the nucleus) was proportional to the square root of the frequency of the X rays emitted by the element. Moseley's brilliant scientific career was cut short when he was killed in World War I.

Discovery of the Neutron

Another particle, called the neutron, was postulated by Rutherford and was finally discovered in 1932 by Sir James Chadwick. The neutron was very difficult to detect since it has no charge and therefore could not be manipulated with electric and magnetic fields, as was done with the electron and proton.

ATOMIC STRUCTURE

Light and the Atom

If an element is heated strongly with a flame or if it is subjected to an electrical discharge such as a high-voltage spark, the element will emit light. By studying the light emitted by the simplest of all elements, hydrogen, scientists formulated the first modern ideas of atomic structure. An understanding of visible light and other forms of electromagnetic radiation is important also in many other areas of chemistry.

THE ELECTROMAGNETIC SPECTRUM

Visible light is the most familiar form of electromagnetic radiation. All other electromagnetic radiation is invisible to the human eye but can be detected by a variety of instruments. Table 1.2 lists the various types of electromagnetic radiation, ranging from the cosmic rays with very high energy to the lowest energy radio waves. The wavelengths vary from less than a picometer (10^{-12} meter) to more than 1 kilometer. Chemists are most interested in the ultraviolet, visible, and infrared regions of the spectrum.

TABLE 1.2

Regions of the Electromagnetic Spectrum

Common Name	Wavelength Range, λ (meters)	Frequency Range, v (1/s)	Wavelength Range, λ (common units)
gamma rays	$<10^{-11}$	$>10^{19}$	<10 pm
X rays	2.5×10^{-11}–2.5×10^{-8}	1.2×10^{16}–1.2×10^{19}	25 pm–25 nm
ultraviolet	1.0×10^{-8}–3.5×10^{-7}	3.3×10^{17}–1.2×10^{-15}	10–350 nm
visible	3.5×10^{-7}–7×10^{-7}	1.2×10^{-15}–2.3×10^{-15}	350–700 nm
infrared	7×10^{-7}–5×10^{-5}	2.3×10^{-15}–1.7×10^{-10}	0.7–50 μm
microwaves	1×10^{-4}–5×10^{-2}	3.3×10^{-13}–1.7×10^{-10}	0.1–50 mm
radio waves	10^{-2}–10^{3}	3.3×10^{-11}–3.3×10^{-6}	1 cm–100 m

Figure 1.6 illustrates the electromagnetic spectrum with a logarithmic scale based on frequency and wavelength. In this representation the visible region is rather small and is surrounded by the ultraviolet and infrared regions. The names of the latter two spectral regions help to remind us that violet (or blue) is at the high-energy end of the visible spectrum and that red is at the low-energy end. The colors of the visible spectrum are violet (highest energy), blue, green, yellow, orange, and red (lowest energy).

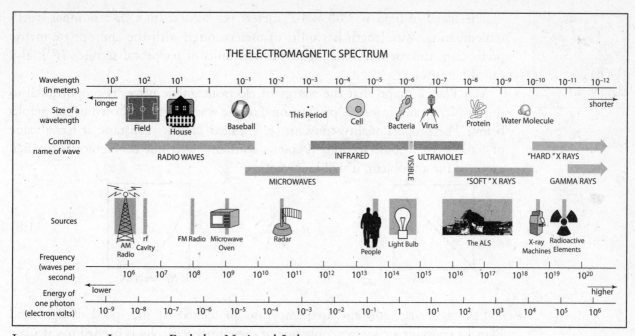

THE ELECTROMAGNETIC SPECTRUM

Image courtesy Lawrence Berkeley National Laboratory

FIGURE 1.6. **Pictorial representation of the electromagnetic spectrum emphasizing the narrow range of visible light.**

Wavelength, Frequency, and Energy of Light

All electromagnetic radiation may be considered as waves that are defined by the wavelength (λ) and frequency (ν). The wavelength (Figure 1.7) is the distance between two repeating points (either two minima or two maxima) on a sine wave. The frequency is defined as the number of waves that pass a point in space each second.

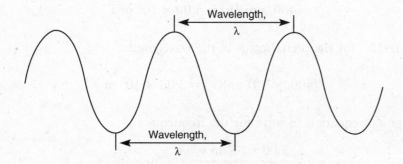

Wavelength, λ

Wavelength, λ

FIGURE 1.7. **Definition of wavelength.**

The wavelength and the frequency of light are inversely proportional to each other as shown in the graph and the equation below.

$$(\text{Wavelength})(\text{Frequency}) = \text{Speed of light} \qquad (1.7)$$
$$\lambda\nu = c$$

Speed of light in vacuum $3.00 \times 10^8\,\text{m s}^{-1}$

The speed of light is 3.00×10^8 meters per second (m s⁻¹), a number worth remembering. Wavelength has units of meters, often with the appropriate metric prefix (cm, μm, or nm), and frequency has units of reciprocal seconds (s⁻¹), also called hertz (Hz).

Max Planck found that the energy of electromagnetic waves is proportional to the frequency and inversely proportional to the wavelength as shown in the graphs below. The proportionality constant, h, is called Planck's constant; it has a value of 6.62×10^{-34} joule second. [Planck's constant need not be memorized; when required for a problem, it will be given.]

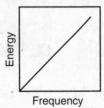

$$h\nu = E = h\frac{c}{\lambda}$$

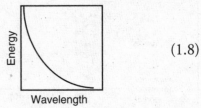

(1.8)

Energy, wavelength, and frequency are all related. If the speed of light and Planck's constant are known, only one of these three variables is needed to calculate the others.

EXERCISE 1.1

(In this exercise and all others in this book it is recommended that you cancel the units with a red pencil or pen to verify the calculations.)

What are the frequency and the energy of blue light that has a wavelength of 400. nm? (Planck's constant = 6.62×10^{-34} J s.)

Solution

Substitute the given values into the equation $\lambda\nu = c$

$$(400.\ \text{nm})(\nu) = 3.00 \times 10^8\ \text{m s}^{-1}$$

Substitute 10^{-9} for the prefix *nano-* in the wavelength

$$(400.\times 10^{-9}\,\text{m})(\nu) = 3.00 \times 10^8\ \text{m s}^{-1}$$

Rearrange this equation to solve for the frequency,

$$\nu = \frac{3.00 \times 10^8\ \text{m s}^{-1}}{400.\times 10^{-9}\ \text{m}} = 7.50 \times 10^{14}\ \text{s}^{-1}$$

The energy of this light may be calculated directly from the frequency or the wavelength by using Equation 1.8.

$$E = h\nu = (6.62 \times 10^{-34}\ \text{J s})(7.50 \times 10^{14}\ \text{s}^{-1})$$

$$= 49.65 \times 10^{-20}\ \text{J}$$

$$= 4.96 \times 10^{-19}\ \text{J}$$

or

$$E = h\frac{c}{\lambda} = (6.62 \times 10^{-34} \text{ J s})\left(\frac{3.00 \times 10^8 \text{ m s}^{-1}}{400. \times 10^{-9} \text{ m}}\right)$$

$$= 4.96 \times 10^{-19} \text{ J}$$

It is very important to be sure that the units cancel properly when solving these problems. The meter units cancel each other since there is one in the numerator and another in the denominator, and the seconds units cancel because s is present in the first term and s^{-1} in the numerator of the second.

EXERCISE 1.2

What are the wavelength and the energy of light that has a frequency of $1.50 \times 10^{15} \text{ s}^{-1}$?

Solution

The relationship between the wavelength and the frequency is $\lambda\nu = c$. Substitution of the given values yields

$$\lambda(1.50 \times 10^{15} \text{ s}^{-1}) = 3.00 \times 10^8 \text{ m s}^{-1}$$

Rearrange to solve for the wavelength, λ:

$$\lambda = \frac{3.00 \times 10^8 \text{ m s}^{-1}}{1.50 \times 10^{15} \text{ s}^{-1}}$$

$$= 2.00 \times 10^{-7} \text{ m} = 200. \times 10^{-9} \text{ m}$$

Use the metric prefix, 1 nm = 10^{-9} m, to simplify the answer:

$$\lambda = 200. \text{ nm}$$

The energy of this light is calculated from $E = h\nu$. Substituting given data yields

$$E = (6.62 \times 10^{-34} \text{ J s})(1.50 \times 10^{15} \text{ s}^{-1})$$

Cancelling units and solving gives the result (remember that $s \times s^{-1} = 1$)

$$E = 9.93 \times 10^{-19} \text{ J}$$

ATOMIC SPECTRA AND SPECTROSCOPY

Visible, ultraviolet, and infrared radiation can be separated into the various wavelengths by focusing the radiation through a triangular prism. The device used for this purpose is called a spectroscope if the light is observed by eye, and called a spectrograph if the light is detected by an electronic device and recorded on paper. A spectrometer (or spectrophotometer) is similar to a spectrograph except that the information is read from a meter. Modern instruments often have digital displays and store data in computer memories.

Using spectroscopes, experimenters found that when gaseous elements are heated they emit light. The light emitted consists of discrete wavelengths that can be measured with great accuracy. Each element emits a unique pattern of spectral wavelengths. These patterns, called atomic spectra, are used to identify elements. Figure 1.8 illustrates the spectrum of the hydrogen atom in the ultraviolet, visible, and infrared spectral regions.

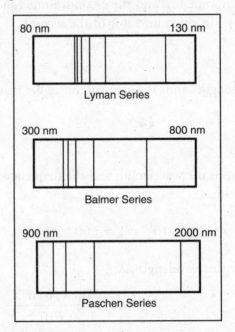

FIGURE 1.8.   Hydrogen spectra that are the basis of the Rydberg equation.

Balmer discovered an empirical mathematical equation for the spectrum that bears his name. When the Lyman and Paschen series were discovered later, Rydberg extended Balmer's equation to include them. One form of the Rydberg equation is as follows:

$$\nu = 3.2881 \times 10^{15} \text{ s}^{-1} \left(\frac{1}{n_1^2} - \frac{1}{n_2^2} \right) \tag{1.9}$$

Balmer's and later Rydberg's equations were intriguing. Using different integers for n_1 and n_2 enabled researchers to calculate all of the spectral lines of hydrogen very accurately. It was not known why this should be true until Niels Bohr came up with a solution showing that n_1 and n_2 refer to the energy levels of the electrons. We can use the energy levels given in Equation 1.10 to perform the same calculation as does the Rydberg equation. The AP exam does not use the Rydberg equation.

THE BOHR MODEL OF THE ATOM

Neils Bohr revolutionized concepts about the atom with his solar system model. This model of the atom required that the electrons be confined to specific allowed orbits. Using conventional physics the energy of an orbit with a number n is

$$E_n = \frac{-2\pi^2 me^4}{n^2 h^2} = \frac{-2.178 \times 10^{-18}}{n^2} \text{ joule} \qquad (1.10)$$

where m = mass of the electron, e = charge on the electron, h = Planck's constant, and n = orbit number. Later, the number n became known as the principal quantum number. Equation 1.10 is given on the AP Exam.

> **TIP**
>
> $$E_n = \frac{-2.178 \times 10^{-15}}{n^2} \text{ joule}$$
>
> is the format used on the AP Exam.

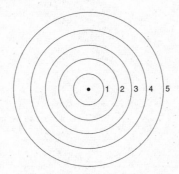

FIGURE 1.9. Diagram of Bohr orbits, showing numbering and relative sizes.

In Bohr's theory the symbol n represents the number of each orbit, starting with the one closest to the nucleus. This theory was readily accepted because it resulted in an equation identical in form to the empirical equation of Rydberg, and the value of the Rydberg constant calculated by Bohr was almost identical to the value determined by experiment. Bohr's theory also gave meaning to Rydberg's equation. Energy in the form of light is emitted from an atom when an electron moves from its initial orbit to an orbit with a lower value of n (Figure 1.10a). When an electron is promoted from a low orbit to a higher numbered orbit, energy must be added (Figure 1.10b). The energy difference between any two orbits is constant, and the same amount of energy needed to raise an electron from one orbit to another will be released when the electron drops back to the original orbit.

Line spectra result from the emission of light by atoms and therefore represent electrons in excited atoms dropping from high orbits to lower ones. To obtain a positive value for the frequency, the higher number orbit is assigned to n_2 and the lower numbered orbit to n_1 in the Rydberg equation. In the Lyman series n_2 is always equal to 1. The observed lines represent electrons dropping from the second, third, fourth, and fifth orbits down to the first orbit. In the Balmer series n_2 is always 2, and in the Paschen series $n_2 = 3$.

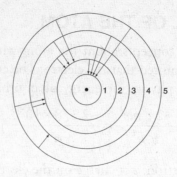

FIGURE 1.10a. **Electrons dropping from outer orbits to inner orbits. They emit light energy.**

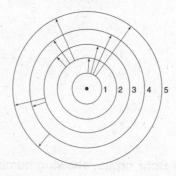

FIGURE 1.10b. **Electrons excited from inner to outer orbits. They require added energy.**

Another way of describing this process is the energy-level diagram, where the *y*-axis represents the energy of each orbit. A horizontal line, rather than the circles shown above, represents the energy level of an orbit. Arrows are drawn from one energy level to another to show where an electron starts and where it ends up. All energy-level diagrams for line spectra show electrons moving from high orbits (energy levels) to lower orbits.

The energy of each level in Figure 1.11 is calculated from Equation 1.9 as shown in Exercise 1.3. The energies have a negative sign since the electron loses energy as it drops toward the $n = 1$ level. The most stable position for the electron in the atom is the first level since the electron has lost the greatest possible amount of energy.

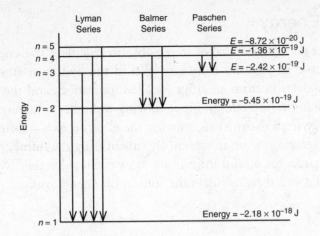

FIGURE 1.11. Energy level diagram for the Lyman, Balmer, and Paschen series of lines in the spectra of hydrogen.

EXERCISE 1.3

Determine the energy and wavelength of light associated with an electron moving from the second to the fourth energy level in a hydrogen atom.

Solution

There are two ways to solve this problem. One is by solving the Rydberg equation. (The Rydberg equation is not one of the equations given to you on the AP exam, and you would have to memorize it, including the constant of 3.2881×10^{15} s^{-1}, in order to use it.) The AP exam does give Equation 1.10 (see p. 87). To use Equation 1.10, we need to calculate E_n for the initial state where $n = 2$ and again for the final state where $n = 4$. The difference in energy is $\Delta E = E_{\text{final}}$ 2 E_{initial}.

$$E_2 = \frac{-2.178 \times 10^{-18}}{2^2} \text{ joule} = -5.445 \times 10^{-19} \text{ joule}$$

$$E_4 = \frac{-2.178 \times 10^{-18}}{4^2} \text{ joule} = -1.361 \times 10^{-19} \text{ joule}$$

$$\Delta E = E_4 - E_2$$
$$= (-1.361 \times 10^{-19} \text{ joule}) - (-5.445 \times 10^{-19} \text{ joule})$$
$$= +4.084 \times 10^{-19} \text{ joule}$$

(The positive sign indicates that energy must be put into the atom to achieve this transition.)

The wavelength of this energy is calculated from

$$\Delta E = \frac{hc}{\lambda}$$

rearranged to $\lambda = \dfrac{hc}{\Delta E} = \dfrac{(6.63 \times 10^{-34} \text{ J s})(3.00 \times 10^8 \text{ m s}^{-1})}{(4.084 \times 10^{-19} \text{ J})}$

$$= 4.87 \times 10^{-7} \text{ m} = 487 \text{ nm}$$

Ionization Energy

The Rydberg equation also makes it possible to calculate the energy needed to completely remove an electron from any level in the hydrogen atom. The process of removing an electron from an atom is called ionization, and the energy needed to do this is the ionization energy. Since moving to higher numbered orbits moves the electron away from the nucleus, moving the electron to n = infinity is the same as removing the electron from the atom. By substituting the number of the starting orbit of the electron for n_1 and infinity for n_2, we can solve the Rydberg equation for the frequency and then calculate the ionization energy using $E = h\nu$.

EXERCISE 1.4

What is the energy needed to ionize an electron from a hydrogen atom in its ground state? What is the ionization energy of a mole of these electrons? (1 mol = 6.02 × 10^{23} atoms.)

Solution

The ground state of any atom is when all of the electrons are in their lowest possible allowed orbitals. In this case, $n = 1$ for hydrogen. If the hydrogen atom is ionized, the electron is removed completely. This can be visualized as a transition to an energy level very far from the nucleus, or $n = \infty$ (infinity). Now that the initial and final states have been determined, the problem is solved as in Exercise 1.3.

$$E_1 = \frac{-2.178 \times 10^{-18}}{1^2} \text{ joule} = -2.178 \times 10^{-18} \text{ joule}$$

$$E_\infty = \frac{-2.178 \times 10^{-18}}{\infty^2} \text{ joule} = 0.00 \text{ joule}$$

$$\Delta E = E_\infty - E_1$$
$$= (0.00 \times 10^{-19} \text{ joule}) - (-2.178 \times 10^{-18} \text{ joule})$$
$$= +2.178 \times 10^{-18} \text{ joule per electron}$$

(The positive sign indicates that energy must be put into the atom to achieve this transition.)

For one mole of electrons we need to multiply this amount by 6.02 × 10^{23}.

$$? \text{ joule mol}^{-1} = +2.178 \times 10^{-18} \text{ J/e}^- (6.023 \times 10^{23} \text{ e}^- \text{ mol}^{-1})$$
$$= 1.312 \times 10^5 \text{ joule mol}^{-1}$$

(For reference, this is the amount of energy needed to completely vaporize 50 mL of water starting at 25 °C.)

The Size of the Atom

Bohr postulated that the momentum (mass × velocity) of the electron must be related to the size of the electron's orbit. The relationship he used was

$$mv = \frac{nh}{2\pi r} \qquad (1.11)$$

When Planck's constant, h, the electron mass, m, and the electron velocity, v, are entered into the equation for the first energy level, $n = 1$, a radius of r = 53 picometers (pm) is calculated. If $n = 2$, the orbit radius is 106 pm. The value 53 pm is often called the Bohr radius for the hydrogen atom. The radii of the other orbits are whole-number multiples of the Bohr radius. The Bohr radius gave chemists a theoretical value for the size of a hydrogen atom and confirmed that the atomic sizes determined by experiment were indeed reasonable.

The Wave-Mechanical Model of the Atom

Soon after Bohr's achievement (for which he received the Nobel Prize) Louis de Broglie suggested that the electron could behave as a wave as well as a particle. One way of looking at this duality involves the equation for the energy of a wave $\left(E = h\nu = \frac{hc}{\lambda} \right)$ and Einstein's well-known equation for the energy of a particle $(E = mc^2)$. Since the electron can have only one energy at any given moment, E in both equations must be the same, resulting in the equality

$$\frac{hc}{\lambda} = h\nu = mc^2 \qquad (1.12)$$

This shows the obvious relationship between the particle (mass) and wave (ν or frequency) nature of the electron. Since all particles do not move at the speed of light, c, we use the symbol v for the velocity of the particle to get $\frac{h}{\lambda} = mv$.

Describing the motion of an electron as a wave required the use of complex "wave equations." Actual use of these wave equations is left for higher level college chemistry courses. However, it is important to understand the results of using these wave equations, which can be summarized as follows:

1. The wave equations require three numbers, called quantum numbers, in order to reach a solution. They are the principal quantum number, n, the azimuthal quantum number, ℓ, and the magnetic quantum number, m_ℓ. In addition, to specify an electron completely and uniquely, a fourth quantum number, called the spin quantum number, m_s, is needed. There are specific rules for assigning the four quantum numbers to electrons.
2. The wave equations changed the picture of the atom drastically (Figure 1.12). In particular, the fixed orbits of the Bohr theory are replaced with a cloud of electrons around the nucleus. The modern orbit is the region of space in which the probability of finding the electron is highest.

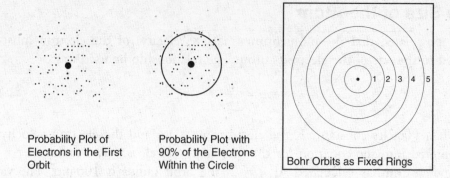

Probability Plot of Electrons in the First Orbit

Probability Plot with 90% of the Electrons Within the Circle

Bohr Orbits as Fixed Rings

FIGURE 1.12.   Comparison of the Bohr model and the wave model of the hydrogen atom.

3. The circular orbits of the Bohr theory have been replaced with spherical electron clouds. The wave equations have shown that the shapes of most electron clouds, although more complex than Bohr's orbits, are still simple geometric shapes.

4. The arrangement of electrons deduced from wave equations agrees well with the periodic table. Many physical and chemical properties of elements and compounds are more fully understood with the knowledge gained about the electronic structure and orbit shape.

5. The results of the wave equations agree completely with the Bohr model. Specifically, the energy change for an electron moving from one electron cloud to another agrees with Bohr's calculations. In addition, the identical 53-pm radius is found for the electron cloud in the wave model of the hydrogen atom.

6. The Heisenberg uncertainty principle is fundamental to the wave model of the atom. This principle states that both the position and the momentum of an electron cannot be exactly known at the same time. The more precisely that the position, x, of the electron is known, the more uncertainty exists as to its momentum, mv. Heisenberg's uncertainty equation is as follows:

| Heisenberg Uncertainty Principle

$$(\Delta x)(\Delta mv) \leq \frac{h}{4\pi} \tag{1.13}$$

This equation reads, "The uncertainty in the position times the uncertainty of the momentum is equal to Planck's constant divided by four pi." The concept of this equation is important, although *no questions involving a numerical solution will be asked on the AP Chemistry Exam.*

Structure of the Atom

The atom is put together in a regular fashion that is not extraordinarily complex. Difficulty often arises, however, from the fact that chemists use several different models of this structure. The fact that the symbols and terminology used in these various models are not always consistent causes the subject to appear more complex than it really is. The reason behind the different models is that chemists are interested in different features of the atom, each of which is best described in a unique manner.

Some chemists desire to describe all of the electrons in an atom. Others are interested only in the outermost electrons, which are called the valence electrons. A third group is interested only in the one electron that differentiates one atom from the immediately preceding atom in the periodic table. This one electron is called the differentiating electron. Other chemists wish to see how the atom is built, adding one electron after another. Still others wish to understand how atoms lose electrons to form ions.

The following sections will show the relationships between the ideas and terminologies for all of these points of view.

PRINCIPAL ENERGY LEVELS (SHELLS)

In the current model of the atom the positively charged nucleus is surrounded by one or more principal energy levels or electron clouds. Principal energy levels may also be called the principal shells with reference to Bohr's atomic model. In either case, the principal energy level, or principal shell, nearest the nucleus is assigned the number 1, and each succeeding energy level is numbered with consecutive integers. The largest element known needs only seven principal energy levels to hold all of its electrons. The number of the principal energy level is given the symbol n, and it is the same as the principal quantum number discussed later.

Since the principal energy levels become larger the further they are from the nucleus, they can hold correspondingly more electrons. Each can hold a maximum number of electrons equal to $2n^2$, where n is the principal energy level number. From this fact we can calculate that the first four principal energy levels can hold 2, 8, 18, and 32 electrons, respectively. The last three could hold 50, 72, and 98 electrons, but they are not completely filled.

SUBLEVELS (SUBSHELLS)

Each principal energy level within an atom contains one or more sublevels. These sublevels may be called subshells in the older terminology. The number of sublevels possible in each principal energy level is equal to the value of n for that energy level. For instance, the third principal energy level ($n = 3$) may contain a maximum of three sublevels. For the 109 known elements, only four sublevels are actually used. The fifth, sixth, and seventh sublevels (for $n = 5$, 6, and 7) are theoretically possible but are not currently needed.

Sublevels are numbered with consecutive whole numbers starting with zero. These numbers are the azimuthal quantum numbers, ℓ. The value of ℓ can never be greater than $n - 1$. In addition to the numbering system, the sublevels are also given corresponding letters of *s*, *p*, *d*, and *f*. Table 1.3 shows the sublevels possible for each energy level. It is important to remember that a sublevel will not exist unless the atom has enough electrons to occupy at least part of the sublevel.

Each principal energy level has an s sublevel, all except the first have *p* sublevels, and the *d* and *f* sublevels also are found in more than one principal energy level. To distinguish one sublevel from another, chemists usually combine the principal energy level number with the sublevel letter in order to indicate in which principal energy level the sublevel is located. With this method, the designation *4p* indicates a *p* sublevel in the fourth principal energy level.

TABLE 1.3

Sublevels in the Atom

Principal Level, n	Sublevel Number, ℓ	Sublevel Letter
1	0	*s*
2	0, 1	*s, p*
3	0, 1, 2	*s, p, d*
4	0, 1, 2, 3	*s, p, d, f*
5*	0, 1, 2, 3	*s, p, d, f*
6*	0, 1, 2	*s, p, d*
7*	0, 1	*s, p*

*Only sublevels used by the known elements are shown here.

TIP

The value of *n* also indicates the number of sublevels within each principal energy level.

TIP

Orbitals can hold only two electrons each.

ORBITALS

Each sublevel of the atom may contain one or more electron orbitals. An orbital is defined as a region of space that has a high electron density, and each orbital may contain a maximum of two electrons. To share an orbital, two electrons must have opposite spins. When two electrons share an orbital, they are said to be paired. Orbitals are designated as *s, p, d,* or *f* according to the sublevel they are in.

The number of orbitals that a sublevel may have depends on the azimuthal quantum number, ℓ, of the sublevel and is equal to $2\ell + 1$. From this fact we see that there is one *s* orbital in an *s* sublevel, three *p* orbitals in a *p* sublevel, five *d* orbitals in a *d* sublevel, and seven *f* orbitals in an *f* sublevel. Table 1.4 summarizes this information.

TABLE 1.4

Orbitals in the Atom

Sublevel Number, ℓ <	Sublevel Letter	Number of Orbitals, $2\ell + 1$	Number of Electrons per Sublevel
0	*s*	1	2
1	*p*	3	6
2	*d*	5	10
3	*f*	7	14

This table also shows why the principal energy levels contain 2, 8, 18, and 32 electrons, respectively. The first principal energy level has only one *s* orbital and therefore holds only 2 electrons. The second principal energy level has *s* and *p* orbitals that hold 2 and 6 electrons, respectively, or 8 for that energy level. The third principal energy level holds the sum of 2 + 6 + 10 or 18 electrons, and the fourth principal energy level holds 2 + 6 + 10 + 14 or 32 electrons. Orbitals take the same letter designation as the sublevel letter (*s*, *p*, *d*, or *f*). Each orbital is given a number, called the magnetic quantum number, m_ℓ. The possible values of m_ℓ range from $-\ell$ to $+\ell$, including zero, as Table 1.5 shows.

TABLE 1.5

Possible Values of m_ℓ	
Orbital	m_ℓ Values
s	0
p	−1, 0, +1
d	−2, −1, 0, +1, +2
f	−3, −2, −1, 0, +1, +2, +3

> **TIP**
>
> Orbital shapes are noted as a letter or a number for the quantum number, ℓ.

The designation of a sublevel also tells the chemist the shape of the orbitals within that sublevel. In *s* sublevels the shape of the electron cloud is spherical. For the *s* sublevel in the first principal energy level, the highest electron density is found within a sphere 53 pm from the nucleus, as Bohr predicted. In the second and higher numbered principal energy levels, the *s* sublevel has a high electron density at the expected distance from the nucleus ($n \times 53$ pm). These sublevels also have an appreciable electron density at all of the lower energy level radii. Figure 1.13 illustrates this feature of the *s* sublevels.

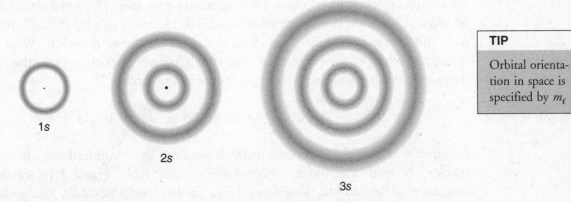

1s

2s

3s

> **TIP**
>
> Orbital orientation in space is specified by m_ℓ

FIGURE 1.13. Diagrams of the 1*s*, 2*s*, and 3*s* orbitals, showing that the highest radial density for the 2*s* and 3*s* orbitals occurs not only at the expected radius but also at intermediate levels.

The *p* orbitals have a dumbbell shape, with the electron density being greatest in two lobes on either side of the nucleus. There are three *p* orbitals in each *p* sublevel. Each is oriented along a different axis, as shown in Figure 1.14. These orbitals may be designated as p_x, p_y, and p_z.

The five d orbitals in a d sublevel have the shapes shown in Figure 1.15. They often have subscripts indicating the general locations of the orbitals on the x-, y-, and z-axes.

The seven f orbitals are slightly more complex in shape than the d orbitals. A knowledge of the exact shapes is not needed at this time.

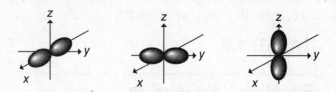

FIGURE 1.14. Diagrams of the three p orbitals aligned with the *x*-, *y*-, and *z*-axes.

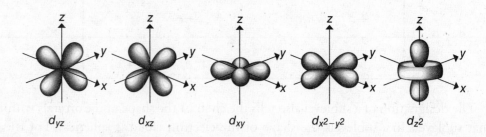

FIGURE 1.15. Shapes of the five *d* orbitals in the atom. Four of the shapes are identical; the fifth is very distinctive.

Electronic Structure of the Atom

At this point we need to see how the information given above is used to develop the complete picture of an atom. The underlying principle of how the electrons are arranged is based on the energy of each orbital. Electrons will fill the orbitals that have the lowest energy first, much as water always flows downhill. While the numerical values of the energies of the orbitals are not important here, the order, from lowest to highest energy is important. That sequence

$$1s\ 2s\ 2p\ 3s\ 3p\ 4s\ 3d\ 4p\ 5s\ 4d\ 5p\ 6s\ 4f\ 5d\ 6p\ 7s\ 5f\ 6d$$

is known as the aufbau, or energy order. It need not be memorized since it may be quickly obtained from the structure of the periodic table. Figure 1.16 shows the long form of the periodic table divided into shaded blocks labeled *s*, *p*, *d*, and *f* for the highest energy electron in each atom.

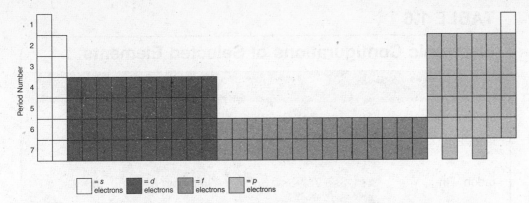

FIGURE 1.16. **Expanded periodic table divided to show the regions where *s*, *p*, *d*, and *f* electrons are the highest energy electrons in each atom.**

Reading across this periodic table, we can see that the energy ordering is an integral part of this table. The first period represents the 1*s* electrons. Filling the second period adds the 2*s* electrons on the left and then 2*p* electrons on the right. The third period fills with 3*s* and then 3*p* electrons, the fourth period with 4*s*, 3*d*, and 4*p* electrons, and the fifth with 5*s*, 4*d*, and 5*p* electrons. Notice that in the fourth and fifth periods a *d* electron has a number that is one less than the period it is in. The sixth period fills with 6*s*, 4*f*, 5*d*, and 6*p* electrons, and the seventh with 7*s*, 5*f*, and 6*d* electrons. Once again, notice that the *d* electrons have numbers that are one less than the period number. In addition, the *f* orbitals have numbers that are two less than the period number.

ELECTRONIC CONFIGURATIONS

Once the energy order of the orbitals is known, the electrons in any atom may be described as the electronic configuration. Chemists use two forms of this configuration to display information. The first is the complete electronic configuration, which lists all of the electrons in the atom. The second is an abbreviated version that lumps all of the inner electrons together and lists only the highest energy electrons.

COMPLETE ELECTRONIC CONFIGURATIONS, $n\ell^x$

In complete electronic configurations the electrons present are listed by designating the principal energy level (n) by number, the sublevel (ℓ) by letter, and the number of electrons (x) in each sublevel, x. Atoms are "built up" by adding electrons to the lowest energy sublevel possible.

To obtain an electronic configuration for an element, the number of electrons is determined from the element's atomic number. Then these electrons are placed into the sublevels, completely filling one sublevel before starting to fill the next one. Several examples are shown in Table 1.6.

TABLE 1.6

Electronic Configurations of Selected Elements

Element	Complete Electronic Configuration
sodium, Na	$1s^2\ 2s^2\ 2p^6\ 3s^1$
lead, Pb	$1s^2\ 2s^2\ 2p^6\ 3s^2\ 3p^6\ 4s^2\ 3d^{10}\ 4p^6\ 5s^2\ 4d^{10}\ 5p^6\ 6s^2\ 4f^{14}\ 5d^{10}\ 6p^2$
radon, Rn	$1s^2\ 2s^2\ 2p^6\ 3s^2\ 3p^6\ 4s^2\ 3d^{10}\ 4p^6\ 5s^2\ 4d^{10}\ 5p^6\ 6s^2\ 4f^{14}\ 5d^{10}\ 6p^6$
antimony, Sb	$1s^2\ 2s^2\ 2p^6\ 3s^2\ 3p^6\ 4s^2\ 3d^{10}\ 4p^6\ 5s^2\ 4d^{10}\ 5p^3$
cobalt, Co	$1s^2\ 2s^2\ 2p^6\ 3s^2\ 3p^6\ 4s^2\ 3d^7$
chlorine, Cl	$1s^2\ 2s^2\ 2p^6\ 3s^2\ 3p^5$

IMPORTANT EXCEPTIONS TO AUFBAU ORDERING

The above examples follow the aufbau order in the filling of the sublevels. Many elements, however, do not follow the aufbau ordering, as shown in the table of electron configurations in Appendix 2. The only exceptions to the aufbau ordering that are important at this time occur in the electron configurations for chromium, molybdenum, copper, silver, and gold. These are listed in Table 1.7. It is apparent that the completely filled *d* sublevel of Cu, Ag, and Au confers stability to the atom at the expense of "unfilling" the previous *s* sublevel. For Cr and Mo, it appears that a half-filled *d* sublevel also confers stability, once again at the expense of a previously filled *s* sublevel.

TABLE 1.7

Electronic Configurations of Selected Elements

Element	Electronic Configuration
copper, Cu	$1s^2\ 2s^2\ 2p^6\ 3s^2\ 3p^6\ 4s^1\ 3d^{10}$
silver, Ag	$1s^2\ 2s^2\ 2p^6\ 3s^2\ 3p^6\ 4s^2\ 3d^{10}\ 4p^6\ 5s^1\ 4d^{10}$
gold, Au	$1s^2\ 2s^2\ 2p^6\ 3s^2\ 3p^6\ 4s^2\ 3d^{10}\ 4p^6\ 5s^2\ 4d^{10}\ 5p^6\ 6s^1\ 4f^{14}\ 5d^{10}$
chromium, Cr	$1s^2\ 2s^2\ 2p^6\ 3s^2\ 3p^6\ 4s^1\ 3d^5$
molybdenum, Mo	$1s^2\ 2s^2\ 2p^6\ 3s^2\ 3p^6\ 4s^2\ 3d^{10}\ 4p^6\ 5s^1\ 4d^5$

ABBREVIATED ELECTRONIC CONFIGURATIONS

When atoms react with each other, their first point of contact is the outermost, highest energy electrons of each atom involved. In fact, the inner core of lower energy electrons and the nucleus play virtually no role in most chemical reactions. These inner electrons may be represented by the noble gas at the end of the period just before the period containing the element of interest. In a complete electronic configuration, all the electrons up to the last completely filled p^6 sublevel are the inner or core electrons. They may be replaced by the symbol, in brackets, for the appropriate noble gas. For iron, the complete and abbreviated electronic configurations are as follows:

$$Fe = 1s^2\ 2s^2\ 2p^6\ 3s^2\ 3p^6\ 4s^2\ 3d^6$$
$$Fe = [Ar]\ 4s^2\ 3d^6$$

> $[Ar] = 1s^2\ 2s^2\ 2p^6\ 3s^2\ 3p^6$

Notice that the last completely filled p orbital is the $3p$ and that argon has the electronic configuration of all of the electrons up to and including the $3p^6$ electrons.

VALENCE ELECTRONS

In many instances chemists are interested in the outermost electrons in an atom. These are the valence electrons. In a practical sense, these include only s and p electrons of the atom. For any given atom the principal energy level of the outer d electrons will always be one less than the principal energy level of the s and p electrons. Similarly, for the f electrons the principal energy level is always two less than the outer s and p electrons.

How to Count Valence Electrons

Determining the number of valence electrons for any atom involves counting the groups (columns) from the left of the periodic table to the element of interest. The d and f electrons shown in Figure 1.16 are not counted. Valence electrons are often shown as dots surrounding the symbol of an atom, as shown in Figure 1.17.

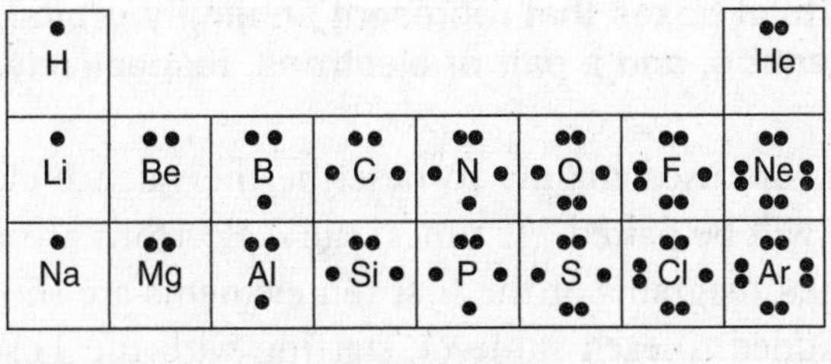

> **TIP**
>
> Only s or p electrons count as valence electrons.

FIGURE 1.17. Representation of valence electrons as dots around the atomic symbol, for the first three periods in the periodic table.

You may often see the electrons unpaired in groups IIA, IIIA, and IVA.

HUND'S RULE

Notice that the second valence electron (second column in Figure 1.17) is represented as a pair rather than placing them on opposite sides of the symbol. This is done to indicate that the electrons are paired in the completed s orbital. The third through eighth columns fill the p orbitals. Hund's rule requires that p, d, or f orbitals in a sublevel must all be filled with one electron each before a second electron is allowed to pair in any orbital. The three separate p orbitals fill with one electron each in boron, carbon, and nitrogen, as shown by the unpaired dots in Figure 1.17. The electrons for oxygen, fluorine, and neon then form pairs until all of the *p* orbitals are filled. Electrons will begin to pair up only if every orbital in the sublevel is first occupied with one electron.

Hund's rule makes sense since electrons repel each other strongly because of their negative charges. This repulsion forces the electrons into separate orbitals within a sublevel until every orbital is filled with one electron. Once this occurs, additional electrons must pair up to fill each orbital with two electrons. Pairing will occur, however, only if the electrons have opposite spins. A simple view of this is that a spinning electron produces a magnetic field. If the spins are aligned with the same spin, the magnetic fields add to the repulsion of the negative charges. If the spins are opposite each other, however, the magnetic fields will attract, thus reducing the repulsion slightly.

ORBITAL DIAGRAMS

Electronic configurations and valence electrons are useful for most purposes in describing the structure of the atom. However, since all of the different orbitals in a sublevel are lumped together, some detail is lost. To see that detail, orbital diagrams, which show each of the orbitals in the valence shell of the atom as a box or circle, are often used. Arrows, representing electrons, are placed in each orbital. The second electron in each orbital has the arrow facing in the opposite direction from the first, indicating that their spins are paired. Figure 1.18 shows the three possible situations for an orbital.

Figure 1.18.   Orbital boxes that represent an empty orbital, an unpaired electron, and a pair of electrons, respectively.

Orbital diagrams are used mainly to describe the valence electrons since all of the inner electrons will be paired. At times, the *d* electrons are also shown in these diagrams. The orbital diagrams of the first ten elements are shown in Figure 1.19. The electrons are added to each sublevel, starting with the 1*s* sublevel. The arrow pointing upward traditionally represents the first electron in each orbital until each orbital in a sublevel contains one electron. Then the downward arrows, representing electrons of opposite spin, are added to complete the sublevel before a new sublevel starts to fill.

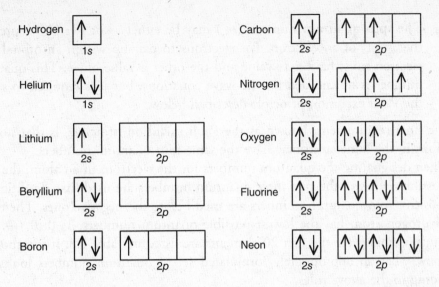

Figure 1.19. **Orbital diagrams for the first ten elements, showing only the valence electrons.**

In later chapters we will want to show energy differences between the orbitals along with the orbital diagrams. This can be done by drawing the orbitals as shown in Figure 1.20 to indicate that the 2p orbitals are higher in energy than the 2s orbitals.

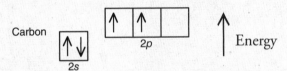

Figure 1.20. **Orbital diagram showing that the 2s electrons in carbon have a lower energy than the 2p electrons.**

Quantum Numbers

Erwin Schrödinger developed the wave equations describing the probabilities of where the electrons are located in the atom. As mentioned earlier, these equations require three integers, called quantum numbers, which describe each electron. The quantum numbers have definite rules for their possible values:

1. The **principal quantum number (n)** may have any integer value starting from 1. This represents the principal energy level of the atom in which the electron is located and is related to the average distance of the electron from the nucleus.
2. The **azimuthal quantum number (ℓ)** may have any number from 0 up to 1 less than the current value of n ($\ell = 0, \ldots, n - 1$). This designates the sublevel of the electron and also represents the shape of the orbitals in the sublevel.
3. The **magnetic quantum number (m_ℓ)** may be any integer, including 0 from $-\ell$, to $+\ell$. ($m_\ell = -\ell, \ldots, 0, \ldots, +\ell$.) This quantum number designates the orientation of an orbital in space. These orientations are shown in Figures 1.13–1.15.

4. The **spin quantum number** (m_s) may be either $+\frac{1}{2}$ or $-\frac{1}{2}$. This represents the "spin" of an electron. For electrons to pair up within an orbital, one electron must have a $+\frac{1}{2}$ value and the other a value of $-\frac{1}{2}$. This quantum number is not needed for the wave equations, but it is required to satisfy the Pauli exclusion principle described below.

The final requirement, known as the Pauli exclusion principle, is that no two electrons in the same atom may have the same four quantum numbers.

When designating the quantum numbers for the electrons in an atom, the lowest possible values of the first three quantum numbers are used first. Traditionally the positive spin quantum numbers are used before the negative ones. Therefore, the hydrogen atom has the lowest possible quantum numbers, 1, 0, 0, $+\frac{1}{2}$. The quantum numbers for the first 20 elements are listed in Table 1.8. It is important to know what an appropriately formulated set of quantum numbers looks like according to the above rules.

TABLE 1.8

Quantum Numbers for the Last Electron Added to Each of the First 20 Elements Using the Rules Above

Element	n	ℓ	m_ℓ	m_s
H	1	0	0	$+\frac{1}{2}$
He	1	0	0	$-\frac{1}{2}$
Li	2	0	0	$+\frac{1}{2}$
Be	2	0	0	$-\frac{1}{2}$
B	2	1	-1	$+\frac{1}{2}$
C	2	1	0	$+\frac{1}{2}$
N	2	1	$+1$	$+\frac{1}{2}$
O	2	1	-1	$-\frac{1}{2}$
F	2	1	0	$-\frac{1}{2}$
Ne	2	1	$+1$	$-\frac{1}{2}$
Na	3	0	0	$+\frac{1}{2}$
Mg	3	0	0	$-\frac{1}{2}$
Al	3	1	-1	$+\frac{1}{2}$
Si	3	1	0	$+\frac{1}{2}$
P	3	1	$+1$	$+\frac{1}{2}$
S	3	1	-1	$-\frac{1}{2}$
Cl	3	1	0	$-\frac{1}{2}$
Ar	3	1	$+1$	$-\frac{1}{2}$
K	4	0	0	$+\frac{1}{2}$
Ca	4	0	0	$-\frac{1}{2}$

EXERCISE 1.5

Designate each of the following sets of quantum numbers as possible or impossible according to the above rules.

(a) 1, 0, 0, $\frac{1}{2}$ (b) 1, 3, 0, $\frac{1}{2}$ (c) 3, 2, 0, $\frac{1}{2}$ (d) 2, 2, 2, $-\frac{1}{2}$

(e) 3, 2, 2, $-\frac{1}{2}$ (f) 3, 1, -1, $\frac{1}{2}$ (g) 4, 2, -2, $-\frac{1}{2}$ (h) 4, 4, 0, $\frac{1}{2}$

(i) 3, 2, 1, 0   (j) 1, 1, 1, $\frac{1}{2}$ (k) 6, 4, -4, $-\frac{1}{2}$ (l) 5, 3, -2, $\frac{1}{2}$

(m) 2, 0, 1, $-\frac{1}{2}$ (n) 5, 0, 0, $\frac{1}{2}$ (o) 3, 1, 2, $-\frac{1}{2}$

Solution

(a), (c), (e), (f), (g), (k), (l), and (n) are valid. The others disobey one or more of the rules on pages 101–102 as follows: (b), (d), (h), and (j) violate rule 2; (m) and (o) violate rule 3; (i) violates rule 4.

RELATIONSHIP OF QUANTUM NUMBERS TO THE PERIODIC TABLE

It must be remembered that each electron in an atom is described by a set of four quantum numbers. Calcium, for instance, will have 20 sets of quantum numbers, one set for each of its 20 electrons. In addition, the rules for the sequence in which the quantum numbers are assigned are somewhat arbitrary. Consider the electron in the hydrogen atom. There is no reason for its spin quantum number to be $+\frac{1}{2}$ In fact, since the hydrogen atoms are energetically equal, half of them will have a spin quantum number of $+\frac{1}{2}$ and half will be $-\frac{1}{2}$. In a similar fashion there is no requirement that the p orbitals fill with quantum number $m_\ell = -1$ first. In fact, all three values are equally probable. Energy levels that are the same are said to be degenerate, and all configurations are equally probable.

Referring to Figure 1.13 and the discussion about electron configurations, we can see that a knowledge of the first two quantum numbers, n and ℓ gives us the period and sublevel (s, p, d, or f) to which the electron belongs. Because the energies of the various possibilities of m_ℓ and m_s are degenerate, they do not help in identifying the atom to which the electron belongs. For example, consider an atom that contains an electron with the quantum numbers 3, 1, 0, $-\frac{1}{2}$. All that can be said about this electron is that it may exist in any element with an atomic number of 13 or greater. If this electron is specified as the last electron in a particular element, we may narrow the possibilities to the elements from aluminum to argon. Finally, strictly following the rules above, we would define this electron as the last one to fill the chlorine atom.

Any atom will have electrons that have all of the possible quantum numbers for the completely filled sublevels. If an atom has an incompletely filled sublevel, the electrons in that sublevel can have any of the quantum numbers associated with that sublevel as long as Hund's rule and the Pauli exclusion principle are followed. For instance, carbon has completely filled $1s$ and $2s$ sublevels, and each carbon atom has electrons with quantum numbers of 1, 0, 0, $+\frac{1}{2}$; 1, 0, 0, $-\frac{1}{2}$; 2, 0, 0, $+\frac{1}{2}$; and 2, 0, 0, $-\frac{1}{2}$. The two additional electrons in the incompletely filled p sublevel may be any two of the six possible sets of quantum numbers below as long as both spin quantum numbers are the same.

$$2, 1, -1, +\tfrac{1}{2} \qquad 2, 1, -1, -\tfrac{1}{2}$$
$$2, 1, 0, +\tfrac{1}{2} \qquad 2, 1, 0, -\tfrac{1}{2}$$
$$2, 1, 1, +\tfrac{1}{2} \qquad 2, 1, 1, -\tfrac{1}{2}$$

Hund's rule requires that the third quantum number in the selected pair cannot be the same. The Pauli exclusion principal requires that the two electrons cannot have all four quantum numbers the same and all unpaired electrons must have the same spin.

SIGNIFICANCE OF THE QUANTUM NUMBERS

The four quantum numbers are often looked upon as more data to be memorized along with the rules for obtaining valid sets. It is important to remember, however, that these numbers represent real physical properties of the atom that may be summarized in simplified form as follows:

1. The principal quantum number, n, represents the average distance of the electron from the nucleus, or the size of the principal energy level.
2. The azimuthal quantum number, ℓ, represents the shape(s) of the orbitals within the sublevel, as shown in Figures 1.13, 1.14, and 1.15.
3. The magnetic quantum number, m_ℓ, represents the orientation of each orbital in space.
4. The spin quantum number, m_s, represents the "spin" of the electron.

SUMMARY

This chapter has briefly reviewed the structure of the atom and the atomic theory. This includes some history about how the electron, proton, and neutron were discovered and characterized. The use of spectroscopy to discover the structure and energy levels of electrons within the atom is described along with the modern designation s of electron configuration. The use and meaning of the four quantum numbers is finally introduced. You should be able to discuss these discoveries and principles and work problems involving the electronic structure of the atom.

Important Concepts

Bohr model of the atom
Quantum mechanical model of the atom
Electronic configurations
Atomic theory

Important Equations

$$\lambda v = c$$
$$E = hv$$
$$E_n = \frac{-2.178 \times 10^{-18}}{n^2} \text{ joule}$$

Practice Exercises

Multiple-Choice

For the first three problems below, one or more of the following responses will apply; each response may be used more than once or not at all in these questions.

    I. Rydberg equation
    II. Heisenberg uncertainty principle
    III. Hund's rule
    IV. Pauli exclusion principle
    V. Bohr model

1. Our inability to know precisely the position and momentum of a subatomic particle is summarized by

    (A) I
    (B) II
    (C) III
    (D) IV
    (E) V

2. In filling the atom with electrons, the rule(s) that must be considered are

    (A) I and III
    (B) II and V
    (C) III and IV
    (D) IV
    (E) III

3. The best tool to use to calculate the ionization energy of a hydrogen atom is

    (A) I
    (B) II
    (C) III
    (D) IV
    (E) V

For the next three problems below, one or more of the following responses will apply; each response may be used more than once or not at all in these questions.

    I. n
    II. s
    III. m_l
    IV. m_s
    V. f

4. Which of these symbols represents an orbital?

    (A) I and II
    (B) II and V
    (C) I, II, and V
    (D) III and IV
    (E) V

5. This designation for an electron in calcium can occur only if calcium is in an excited state.

    (A) I and II
    (B) II and V
    (C) I, II, and III
    (D) III and IV
    (E) V

6. Which of these will tell you the shape of any orbital that an electron may occupy?

    (A) I
    (B) II
    (C) III
    (D) IV
    (E) V

7. Which of the following types of electromagnetic radiation has the highest energy?

   (A) visible
   (B) ultraviolet
   (C) microwave
   (D) infrared
   (E) X rays

8. What is the wavelength of light that has a frequency of 4.00×10^{14} s^{-1}? (The speed of light is 3.00×10^8 m s^{-1}.)

   (A) 7.5 nm
   (B) 1333 nm
   (C) 750. nm
   (D) 1.33 cm^{-1}
   (E) 1.2×10^{23} m

9. Which of the following elements has the greatest number of p electrons?

   (A) C
   (B) Si
   (C) Fe
   (D) Cl
   (E) As

10. An electron with the four quantum numbers 3, 2, -1, $-\frac{1}{2}$ may be an electron in an unfilled sublevel of

    (A) Ca
    (B) Fe
    (C) Al
    (D) Ar
    (E) Ag

11. For a d orbital:

    (A) the value of n must be 2
    (B) the value of m_s must be $+\frac{1}{2}$
    (C) the value of ℓ must be 3
    (D) the value of m_ℓ must be 3
    (E) the value of ℓ must be 2

12. The numbers of electrons, protons, and neutrons, respectively, in the ^{31}P isotope are

    (A) 15, 31, 15
    (B) 15, 15, 31
    (C) 31, 15, 16
    (D) 15, 15, 16
    (E) 31, 31, 16

13. Which element has an electronic configuration that does NOT follow the Aufbau principle?

    (A) Fe
    (B) Mg
    (C) Al
    (D) Ag
    (E) Ni

14. Which electronic configuration corresponds to that of a noble gas?

    (A) $1s^2\ 2s^2\ 2p^6\ 3s^2\ 3p^6\ 4s^1$
    (B) $1s^2\ 2s^2\ 2p^6\ 3s^2\ 3p^4$
    (C) $1s^2\ 2s^2\ 2p^6\ 3s^2\ 3p^6$
    (D) $1s^2\ 2s^2\ 2p^6\ 3s^1$
    (E) $1s\ 2s\ 2p\ 3s\ 3p$

15. Which electronic transition requires the addition of the most energy?

    (A) $n = 1$ to $n = 3$
    (B) $n = 5$ to $n = 2$
    (C) $n = 2$ to $n = 3$
    (D) $n = 4$ to $n = 1$
    (E) $n = 5$ to $n = 1$

16. Which of the following has only five valence electrons?

    (A) Rb
    (B) C
    (C) Si
    (D) P
    (E) B

17. Which was used to determine the charge of the electron?

    (A) the gold foil experiment
    (B) deflection of cathode rays by electric and magnetic fields
    (C) the oil drop experiment
    (D) electrolysis
    (E) the mass spectrometer

18. Which of the following principles is NOT part of Dalton's atomic theory?

    (A) Atoms are the smallest, indivisible particles in nature.
    (B) Chemical reactions are simple rearrangements of atoms.
    (C) Atoms follow the law of multiple proportions.
    (D) Each atom of an element is identical to every other atom of that element.
    (E) All matter is composed of atoms.

19. Which of the following atoms has the most unpaired electrons?

    (A) Mg
    (B) Ti
    (C) Al
    (D) Cu
    (E) Cr

20. Which quantum number describes the shape of an orbital?

    (A) n
    (B) ℓ
    (C) m_ℓ
    (D) m_s
    (E) s

21. You have just discovered a new, fundamental particle of nature. When measuring its mass, you obtain the following data for five samples:

    4.72×10^{-34} gram
    9.44×10^{-34} gram
    1.18×10^{-33} gram
    1.65×10^{-33} gram
    7.08×10^{-34} gram

    If you make the same assumptions that Millikan did, what is the maximum mass of the new particle?

    (A) 4.72×10^{-34} g
    (B) 1.18×10^{-34} g
    (C) 9.44×10^{-34} g
    (D) 2.36×10^{-34} g
    (E) 9.91×10^{-34} g

22. Which of the following is FALSE?

    (A) The $4d$ orbitals are in the fourth period of the periodic table.
    (B) The $7s$ orbitals are in the seventh period of the periodic table.
    (C) The $4f$ orbitals are in the sixth period of the periodic table.
    (D) The $6s$ orbitals are spherical in shape.
    (E) The $5p$ orbitals are dumbbell-shaped.

---

*This symbol indicates a challenge question, one that is more difficult than the typical exam questions.

23. The *f* sublevel may contain a maximum of

    (A) 2 electrons
    (B) 14 electrons
    (C) 6 electrons
    (D) 10 electrons
    (E) 8 electrons

24. The valence electrons are

    (A) all electrons in an atom beyond the preceding noble gas
    (B) all outermost electrons in a sublevel
    (C) *s* and any *p* electrons in the highest energy level or shell
    (D) electrons in the last unfilled sublevel
    (E) any electrons that can ionize

25. Which equation best expresses the energy of a photon?

    (A) $E = \frac{1}{2}mv^2$
    (B) $E = mc^2$
    (C) $E = IR$
    (D) $E = h\nu$
    (E) $E = E^0 - RT \ln K$

26. Bohr's equation for the energy of an electron in its orbit can be used

    (A) to calculate the energy released as an electron drops from a high orbit to a lower one
    (B) to calculate the ionization energy of hydrogen
    (C) to determine the energy needed to promote an electron to a higher energy level
    (D) to understand the spectra of the hydrogen atom
    (E) for all of the above purposes

ANSWER KEY

1. B	7. E	13. D	19. E	25. D
2. C	8. C	14. C	20. B	26. E
3. A	9. E	15. A	21. D	
4. B	10. B	16. D	22. A	
5. E	11. E	17. C	23. B	
6. C	12. D	18. C	24. C	

See Appendix 1 for explanations of answers.

Free-Response

Answer the following question concerning the structure of the atom, the use of light as a probe of atomic structure, and the development of the modern quantum mechanical view of the atom.

(a) Describe the significant experiments that led to the discovery of the properties of the electron, proton, and neutron. Describe the results and the interpretation of results that contributed to development of the atomic model.

(b) The bond energy of a carbon-carbon bond is approximately 350 kJ/mol. Explain why electromagnetic radiation with wavelengths shorter than those in the visible region of the spectrum is often called ionizing radiation.

(c) Sketch an outline of the periodic table and discuss how the four quantum numbers n, ℓ, m_ℓ, and m_s relate to the periodic table.

(d) What wavelength of light is expected if an electron drops from the fifth energy level to the second energy level in a hydrogen atom?

ANSWERS

(a) Mention the cathode ray tube that verified the charge, particle nature, and e/m ratio of the electron (using magnetic and electric fields). The Milliken oil-drop experiment was used to determine the electron charge. Mention Rutherford's alpha-particle experiment to discover the positively charged nucleus. Moseley used x-rays to determine atomic numbers, and Chadwick discovered the neutron.

(b) First we calculate the wavelength of light that has energy equivalent to the C–C bond. First calculate the energy per C–C bond by dividing 350 kJ/mol by 6.02×10^{23} to get 5.81×10^{-19} J. Use the relationship $E = hc/\lambda$ to determine the wavelength as

$$\lambda = hc/E = (6.63 \times 10^{-34} \text{ J s})(3.00 \times 10^8 \text{ m s}^{-1})/(5.81 \times 10^{-19} \text{ J})$$
$$= 3.42 \times 10^{-7} \text{ m} = 342 \text{ nm}$$

This calculation demonstrates that a photon with a wavelength of 342 nm has enough energy to break a C–C bond. The 342-nm wavelength is at the upper end of the ultraviolet region of the spectrum, and wavelengths below 342 nm have even more energy for breaking (ionizing) bonds.

(c) You want to point out that n represents the period of the periodic table while also noting that the d and f electrons have n quantum numbers that are 1 and 2 units less than the period in which they are found. Next, the ℓ quantum number represents various "blocks" within the periodic table similar to what is shown in Figure 1.16. Finally, the possible values of m_ℓ and m_s result in the number of elements that can occupy each of the blocks mentioned. These blocks can also be designated with the letters s, p, d, and f.

(d) Using the equation $E_n = -2.178 \times 10^{-18}$ J/n^2, substitute $n = 2$ and $n = 5$ and calculate two energies. Subtract the two energies to get 4.57×10^{-19} J. Use this value to calculate the wavelength as $\lambda = hc/E = (6.63 \times 10^{-34}$ J s$)$ $(3.00 \times 10^8$ m s$^{-1})/(4.57 \times 10^{-19}$ J$)$ and $\lambda = 4.35 \times 10^{-7}$ m $= 435$ nm. This is in the visible region of the spectrum.

The Periodic Table

CHAPTER 2

- Periodic Table
- Atomic Symbols
- Atomic Masses
- Atomic Numbers
- Isotope Masses, Mass Number
- Metals, Nonmetals, and Metalloids
- Ionization Energy
- Electron Affinity
- Diagonal Variation of Physical Properties

Chemists repeatedly refer to the **periodic table** to find specific information about the elements and to understand how the properties of the various elements are related to each other. In the table we find chemical and physical similarities between elements in the same group. We also find trends where these properties vary regularly.

This chapter reviews the periodic relationships that are necessary for a complete understanding of chemistry. The history and development of the periodic table are discussed at the end of the chapter.

THE MODERN PERIODIC TABLE

The periodic table summarizes a large amount of information useful to chemists and will be referred to many times throughout this book. A complete periodic table, updated in 2009 and similar to the one given with the AP Chemistry test, can be found on page 621. Basic information about its construction includes the following:

1. The **symbol** for each element is shown in a separate box, in order of increasing **atomic number,** from 1 to 116 with space for the undiscovered elements 113 and 115. Each box shows the atomic symbol of the element with the atomic number above the symbol and the **atomic mass** below the symbol as shown here:

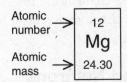

2. Each row of the periodic table is called a **period** and may contain from 2 to 32 elements. Periods with 32 elements are usually written with 14 of the elements placed below the main table.

3. Each column is called a **group,** and elements within groups have similar chemical and physical properties. Within a group, the closer elements are to each other, the more similar they are.

4. The groups of the periodic table are normally numbered; however, the groups are usually not numbered on the periodic table supplied with the AP exam.

Atomic Symbols

In the periodic table each element is designated by a one- or two-letter symbol. Most symbols for the elements are simple abbreviations of their English names. The symbols for 11 elements, however, are derived from the elements' names in other languages, mainly Latin. These elements are listed in Table 2.1, and both the symbol and the name of each element should be memorized for quick recall.

TABLE 2.1

Non-English Chemical Symbols					
sodium	Na	silver	Ag	gold	Au
potassium	K	tin	Sn	mercury	Hg
iron	Fe	antimony	Sb	lead	Pb
copper	Cu	tungsten	W		

TIP

These names and symbols must be memorized.

When chemists use atomic symbols for formulas, ions, and isotopes, they add subscripts and superscripts to the four corners of the symbol to represent various features of the atom. As shown below for the element calcium, the upper left-hand corner is reserved for the mass of an isotope (mass number) of the

$$^{A}_{Z}\text{Ca}^{\text{charge}}_{\text{subscript}}$$

TIP

The four corners of an atomic symbol have specific uses.

atom designated by the symbol A. The lower left-hand corner is reserved for the atomic number, Z. In the upper right-hand corner chemists place the charge of the ion that results when the element gains or loses electrons. Subscripts used in writing chemical formulas are placed in the lower right-hand corner. These subscripts indicate how many atoms of the element are present in a formula unit of a compound. For instance, N_2 means that a molecule of nitrogen contains two atoms of nitrogen.

Electrons, Protons, and Neutrons

The periodic table may be used to quickly determine the number of **electrons** and **protons** in a particular element. The number of **neutrons** may be calculated only if a specific isotopic mass of an element is known.

PROTONS

For any given element, the number of protons is *always* equal to the atomic number, Z, of the element:

$$\text{Number of protons} = Z \qquad (2.1)$$

ELECTRONS

For any given element, the number of electrons is equal to the atomic number:

$$\text{Number of electrons} = Z \qquad (2.2)$$

For an ion of an element, the number of electrons may be calculated as:

$$\text{Number of electrons} = Z - \text{charge of the ion} \qquad (2.3)$$

A positive ion (cation) has lost electrons, and a negative ion (anion) has gained electrons, compared to the neutral element.

NEUTRONS

The number of neutrons in an element depends on the specific **isotope** of the element in question. Since the atomic masses listed in the periodic table are the **weighted averages** of all the naturally occurring isotopes, the number of neutrons in an atom cannot normally be determined from the periodic table. If the specific isotope mass number is known, however, the number of neutrons can be calculated as the difference between the isotope mass and the atomic number:

$$\text{Number of neutrons} = A - Z \qquad (2.4)$$

EXERCISE 2.1

Using a periodic table and Equations 2.1–2.4, fill in the blanks in the following table:

Symbol	Atomic Number	Isotope Mass	Number of Protons	Number of Electrons	Number of Neutrons
Fe		56			
	60	144			
		102	45	45	
		59			31
Al		27			

Solution

Symbol	Atomic Number	Isotope Mass	Number of Protons	Number of Electrons	Number of Neutrons
Fe	26	56	26	26	30
Nd	60	144	60	60	84
Rh	45	102	45	45	57
Ni	28	59	28	28	31
Al	13	27	13	13	14

Isotopes

Dalton's atomic theory states that all atoms of a given element are identical in all of their properties. However, it is now known that atoms of an element may have two or more different masses. This is due to differing numbers of neutrons in the nucleus. For instance, a carbon atom may have a mass of 12, 13, or 14 atomic mass units (^{12}C, ^{13}C, and ^{14}C). Although all carbon atoms have six electrons and six protons, the different isotopes have 6, 7, and 8 neutrons, respectively. **The number of protons identifies an element. The electrons are responsible for an element's chemical properties.**

Atomic Masses

It is important to remember that the atomic masses listed in the periodic table are all relative to the mass of the ^{12}C isotope of carbon. For this reason, masses listed in the periodic table have no units. Laboratory chemists often assign units of grams to these **relative masses** for calculation purposes, as will be shown later. Other chemists may assign units of kilograms, pounds, or even tons to the masses for large industrial uses.

A second concept to remember is that chemists may refer to the exact mass of a particular isotope of an atom, or to the weighted average of the masses of all naturally occurring isotopes. This weighted average is the relative mass listed in the periodic table, and it usually does not represent the relative mass of any particular atom of that element since most naturally occurring elements have more than one isotope. Fortunately, the percentage of each isotope is relatively constant throughout the world. As a result, the measured atomic mass is the weighted average of the masses of the individual isotopes and their relative abundances.

A weighted average is calculated as the sum of all isotope masses, each multiplied by its **natural abundance.** The natural abundance of an isotope is the fraction of all atoms of an element that have the same number of neutrons.

$$\text{Weighted average} = \sum_{i=1}^{n} (\text{mass of isotope } i)(\text{abundance of isotope } i) \quad (2.5)$$

EXERCISE 2.2

Magnesium has three isotopes: ^{24}Mg, ^{25}Mg, and ^{26}Mg. They occur naturally with percentage abundances of 78.6%, 10.1%, and 11.3%, respectively. The exact masses of these isotopes are 23.9924, 24.9938, and 25.9898. What is the weighted average of the three isotopic masses?

Solution

The weighted average is calculated using Equation 2.5. In this equation the mass of each isotope is multiplied by the abundance of that isotope. These products are then added to obtain the weighted average:

$$
\begin{aligned}
\text{Weighted average} = & \ (\text{mass of } ^{24}Mg)(\text{abundance of } ^{24}Mg) \\
& + (\text{mass of } ^{25}Mg)(\text{abundance of } ^{25}Mg) \\
& + (\text{mass of } ^{26}Mg)(\text{abundance of } ^{26}Mg)
\end{aligned}
$$

Since the abundances are usually given as percentages, they must be converted to fractions by dividing by 100 before using them in the equation. When the appropriate numbers are substituted, the equation becomes

$$
\begin{aligned}
\text{Weighted average} &= (23.9924)(0.786) + (24.9938)(0.101) + (25.9898)(0.113) \\
&= \quad 18.86 \quad + \quad 2.52 \quad + \quad 2.94 \\
&= \quad 24.32
\end{aligned}
$$

You may have noticed that the atomic masses in the periodic table may have four, five, six, or seven significant figures. A greater number of significant figures suggests that an atomic mass is known with more certainty than one with fewer significant figures. Experimental difficulties are one source of uncertainty in determining atomic masses. Another source is the variation in natural isotopic abundance. If the natural isotopic abundance is fairly constant, more significant figures can be obtained for atomic masses.

EXERCISE 2.3

The atomic mass of bromine is listed as 79.9 in the periodic table. There is no isotope of bromine with a mass of 80. Suggest an explanation for this fact.

Solution

The masses listed in the periodic table rarely represent the masses of specific isotopes. The atomic mass of bromine (79.9) can be obtained from many possible combinations of isotope masses and their relative abundances. In this case bromine has only two natural isotopes (^{79}Br and ^{81}Br), which occur in almost equal amounts in nature.

PERIODIC PROPERTIES OF THE ELEMENTS

Two forms of the periodic table are shown in Figures 2.1 and 2.2. In Figure 2.1 the periodic table is arranged in the conventional manner with the **lanthanide** and

actinide series placed below the body of the table. Figure 2.2 places the lanthanide and actinide elements where they normally belong. However, the form in Figure 2.2 is rarely used because the boxes become too small to read.

Chemists often speak of groups of related elements such as the **alkali metals, alkaline earth metals, transition elements, halogens,** and **noble gases.** The locations of seven of these groupings are shown in Figure 2.1. A knowledge of the names of these groupings is important, and they will be referred to frequently in the following chapters.

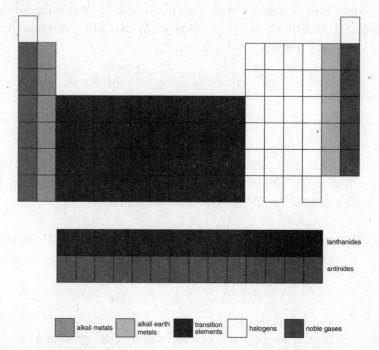

FIGURE 2.1. **Common form of the periodic table, showing groups of related elements.**

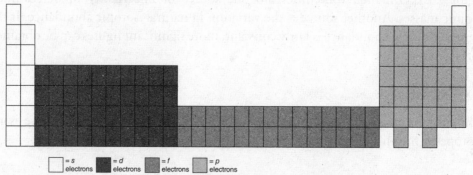

FIGURE 2.2. **Extended form of the periodic table. The shading shows the outermost electrons and their groupings.**

Chemical reactions occur when one atom collides with another. In these collisions the outermost electrons make the first contact between the atoms. This is the main reason why elements with similar electronic structures have similar chemical properties.

After the review of the electronic structure of the atom in Chapter 1, the reasons for the chemical similarities and differences of the elements become obvious. Figure 2.2 shows the blocks of the periodic table that have *s, p, d,* and *f* electrons as the **differentiating electrons.** Each column or group has the same number and type of outermost electrons, resulting in the chemical similarities of these elements. For example, each element in Group IA has a single ns^1 valence electron. For instance, all the noble gases have completely filled *s* and *p* sublevels, which give them extraordinary stability. The halogens are missing one *p* electron; otherwise, they would be electronically the same as the noble gases. The halogens tend to enter reactions that enable them to gain that one *p* electron. Similarly, the alkali metals and the alkaline earth metals have one and two *s* electrons, respectively. They readily lose these electrons to become electronically identical to (**isoelectronic** with) a noble gas.

> A *differentiating electron* is the electron in a neutral element that makes it different from the previous element.

> *Isoelectronic* refers to atoms and ions that have identical electron configurations.

Physical Properties of the Elements

Of the 114 elements in the periodic table, only two, mercury and bromine, are liquids under normal conditions. The noble gases, hydrogen, nitrogen, oxygen, fluorine, and chlorine are gases at room temperature. The remaining elements are solids.

Most of the elements in the periodic table may be considered as individual atoms. A few elements, however, exist naturally as diatomic molecules. These are H_2, O_2, N_2, and the halogens. Other elements, notably sulfur and phosphorus, exist in polyatomic units such as S_8 and P_4, but are commonly represented as single atoms in chemical reactions.

METALS AND METALLOIDS

Metals dominate the elements in the periodic table. Most periodic tables show a heavy line dividing the metals from the nonmetals, as in Figure 2.3.

Elements bordering this line are often termed **metalloids** since they exhibit some properties of metals and some properties of nonmetals. The metallic character of the elements increases from the top of the periodic table to the bottom, as the group headed by nitrogen shows clearly. Nitrogen and phosphorus are nonmetals, arsenic and antimony are metalloids, and bismuth is a metal.

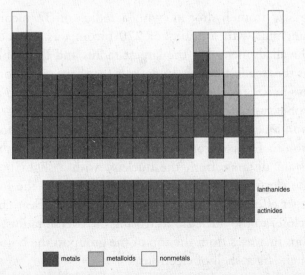

FIGURE 2.3. Location of metals and nonmetals in the periodic table. The heavy line divides the two. Elements along the line have metallic and nonmetallic properties and are called metalloids.

ALLOTROPES

Most elements in the periodic table exist in nature in only one form. Others may have multiple forms, called **allotropes,** which are distinctly different from each other. For instance, carbon exists as graphite, diamond, and the newly discovered buckminsterfullerene. These three allotropes of carbon have distinctly different properties. Graphite conducts electricity and is slippery enough to be used as a lubricant. The diamond allotrope of carbon is the hardest of all natural materials. Large, pure diamonds are gem stones, while small diamonds are produced commercially and are used in industrial grinding machines. Buckminsterfullerene is the allotrope of carbon that exists in clusters of 60 atoms, C_{60}.

Arsenic, oxygen, phosphorus, selenium, sulfur, and tin are other elements that have more than one allotropic form. Arsenic in its gray allotrope is metallic in character, while the yellow allotrope of As_4 molecules is nonmetallic. In the same group in the periodic table, phosphorus has several allotropes, one of which is white phosphorus (P_4), which burns spontaneously when exposed to air. The red and black allotropes of phosphorus are long chains of atoms that are more stable. Sulfur has many allotropes; one is the S_8 molecule, and others are long chains of sulfur atoms. Selenium, in the same group as sulfur, has an Se_8 allotrope that is nonmetallic and another, semimetallic allotrope. Tin has three allotropes, one that has a distinctly metallic look and two that are white crystals.

VARIATION OF PHYSICAL PROPERTIES

As mentioned earlier, the metallic character of the elements increases from the top to the bottom of a group. Many other properties also vary regularly. The melting and boiling points of metals tend to decrease from the top to the bottom of a group. Nonmetals, on the other hand, show an increase in their melting and boiling points. Similar trends in electrical properties, densities, and specific heats are also noted within each group.

ATOMIC RADII

Atoms range in size from hydrogen, with a radius of 37 picometers (1 pm $=$ 10^{-12} m), to francium, with a radius of 270 picometers. In each period of the periodic table the alkali metal has the largest radius and the noble gas at the end of the period has the smallest radius. In periods that have transition elements, there is a slight increase in size for the last transition elements and then a decrease in size to the smaller noble gas.

Of the atoms in a period, each succeeding atom has an additional electron and an additional proton. The electrons are added to the same shell and are located at a relatively constant distance from the nucleus, with a slight increase in atomic radius due to the electrons repelling each other. However, the increasing nuclear charge is a stronger effect, and this increase in effective nuclear charge attracts the electron clouds closer to the nucleus, decreasing the overall radius. In atoms within a group, the radius increases from the top of the group to the bottom because each period has another, larger shell of electrons.

IONIZATION ENERGY

Ionization energy is defined as the energy required to remove an electron from an atom. Energy is always required to remove electrons. Francium, in the lower left corner of the periodic table, has the lowest ionization energy, while helium (the upper right corner) has the highest ionization energy. In general, a line running from the lower left to the upper right corner of the periodic table defines this diagonal relationship. We may also conclude that the ionization energy decreases from the top to the bottom of any group in the periodic table, and it increases from left to right across a period in the table.

Table 2.2 shows the ionization energies for several metals. From this table, it is clear that very little energy is required to remove one electron from sodium and potassium, but much more energy is needed to remove a second electron. This confirms the observation that sodium and potassium easily form only Na^+ and K^+ ions. For calcium and magnesium, the table shows that ionization of the first two electrons to form Ca^{2+} and Mg^{2+} ions is relatively easy but removal of a third electron requires too much energy to be feasible. Ionization energies of other metals show similar trends.

TABLE 2.2

Ionization Energies (kJ mol^{-1}) of Selected Elements			
Metal	First Electron	Second Electron	Third Electron
Na	496	*4563*	*6913*
Mg	737	1450	*7731*
K	419	*3051*	*4411*
Ca	590	1145	*4912*

ELECTRON AFFINITY

Electron affinity is defined as the energy change that accompanies the addition of an electron to an atom. Some atoms readily attract electrons, and the electron affinity has a negative value, meaning that energy is released (see Chapter 12). Most atoms, however, do not accept additional electrons readily and the electron affinity is a positive value, indicating that energy must be used to add the electron.

Fluorine has the highest affinity for electrons and francium the lowest. Electron affinity varies diagonally across the periodic table. The atoms close to fluorine tend to accept electrons readily, and those close to francium do not.

ELECTRONEGATIVITY

The concept of **electronegativity** was developed by Linus Pauling to describe the attraction of electrons by individual atoms. Electronegativity is a combination of ionization energy, electron affinity, and other factors. Electronegativities show the same diagonal trend as do ionization energies and electron affinities. Fluorine has the highest electronegativity, and francium the lowest. The electronegativity concept is used in determining how electrons are distributed in molecules, as shown in later chapters. The periodic table in Figure 2.4 shows the electronegativities of the elements to illustrate the increasing trend from the lower left corner to the upper right corner of the table.

H 2.1																	
Li 1.0	Be 1.5											B 2.0	C 2.5	N 3.1	O 3.5	F 4.0	
Na 1.0	Mg 1.3											Al 1.5	Si 1.8	P 2.1	S 2.4	Cl 2.9	
K 0.8	Ca 1.1	Sc 1.2	Ti 1.3	V 1.5	Cr 1.6	Mn 1.6	Fe 1.7	Co 1.7	Ni 1.8	Cu 1.8	Zn 1.7	Ga 1.8	Ge 2.0	As 2.2	Se 2.5	Br 2.8	
Rb 0.8	Sr 1.0	Y 1.1	Zr 1.2	Nb 1.3	Mo 1.3	Tc 1.4	Ru 1.4	Rh 1.5	Pd 1.4	Ag 1.4	Cd 1.5	In 1.5	Sn 1.7	Sb 1.8	Te 2.0	I 2.5	
Cs 0.7	Ba 0.9	La 1.1	Hf 1.2	Ta 1.4	W 1.4	Re 1.5	Os 1.5	Ir 1.6	Pt 1.5	Au 1.4	Hg 1.5	Tl 1.5	Pb 1.6	Bi 1.7	Po 1.8	At 2.2	
Fr 0.7	Ra 0.9	Ac 1.0															

Figure 2.4. Periodic table showing the electronegativities of the elements.

When we consider the trends in the periodic table, they generally change from the lower left corner to the upper right corner. Ionization energy, electron affinity, atomic radii, and electronegativity are some of these properties. If we compare an element in Period 2 with an element in Period 3 (such as lithium and magnesium), we find that they have similar physical and chemical properties. The change expected by advancing from one group to the next is canceled to a large extent by dropping from Period 2 to Period 3. In the case of Li and Mg, as we move from group IA to Group IIA, we expect the atomic radius to decrease. However, as we move from Period 2 to Period 3, we expect the atomic radius to increase. The result is that the radium of Li is 152 pm and the radius of Mg is 160 pm. Other physical properties are similar, resulting in similar chemical properties.

IONIC RADII

There are two types of ions: cations and anions. Cations are atoms that have lost one or more electrons and carry a positive charge; anions are atoms that have gained one or more electrons and carry a negative charge.

Cations are always smaller than neutral atoms of the same element. For many cations an entire shell of electrons has been lost. In those instances, the cations are only about half the size of the neutral atoms. A further decrease in size is due to the fact that cations have more protons than electrons.

Anions are always larger than the neutral atoms; many are almost twice the size. Since the added electron(s) go into the same shell, the gain in electrons does not fully explain the increase in size observed. Chemists reason that the outer electrons in an atom effectively shield each other from the nuclear charge. This shielding decreases the attractive forces, and the size of the electron cloud expands, resulting in the large anion radius.

TIP

Sizes of ions:
Axions > Element
Cations < Element

SUMMARY

Chapter 2 focused on the periodic table and how to use it effectively. This table summarizes trends of varying physical and chemical properties from group to group and also from period to period. You should be able to describe these trends. In addition, there are diagonal relationships that are best summarized as trends that vary fairly uniformly from one corner of the periodic table to the other. Once again, the ability to recall and use these diagonal trends is very useful to the chemist. This chapter also delves into the structure of the atom in terms of electrons, protons, and neutrons. Finally, the concept of atomic mass based on the carbon-12 standard is presented. The concepts of relative atomic masses and the weighted average of isotopes to get an average atomic mass should be understood.

Important Concepts

Relationship of periodic table to electronic configuration
Chemical and physical similarities within groups
Variation of properties within groups
Diagonal relationships
Electronegativity relationships
Weighted average
Diagonal trends

Practice Exercises

Multiple-Choice

For the first five problems below, one or more of the following responses will apply; each response may be used more than once or not at all in these questions.

I. sodium
II. strontium
III. uranium
IV. bromine
V. bismuth

1. Trends in the periodic table show that elements become more metallic in character from the top of a group to the bottom. Which of these is an element whose properties are opposite those of the element at the top of its group?

(A) I
(B) II
(C) III
(D) IV
(E) V

2. The halogen in this group is

(A) I
(B) II
(C) III
(D) IV
(E) V

3. The element that contains 38 protons is

(A) I
(B) II
(C) III
(D) IV
(E) V

4. There are only two liquid elements at room temperature and atmospheric pressure. One of these is

    (A) I
    (B) II
    (C) III
    (D) IV
    (E) V

5. The atom with the largest radius is

    (A) I
    (B) II
    (C) III
    (D) IV
    (E) V

6. The differentiating electrons for transition elements are

    (A) *d* electrons
    (B) *s* electrons
    (C) *p* electrons
    (D) *f* electrons
    (E) valence electrons

7. The best way to estimate the boiling point of Pd is to

    (A) average the boiling points of Rh and Ag
    (B) average the boiling points of Ni and Pt
    (C) average the boiling points of Ir and Cu
    (D) average the boiling points of Co and Au
    (E) None of the above will work.

8. In which of the following pairs of elements is the element with the lower boiling point listed first?

    (A) Na, Cs
    (B) Te, Se
    (C) P, N
    (D) Ba, Sr
    (E) I, Br

9. In which of the following pairs is the first element expected to have a higher electronegativity than the second?

    (A) O, P
    (B) Cs, Rb
    (C) I, Br
    (D) Al, P
    (E) Sb, As

10. A solid element has two valence electrons. That element must be

    (A) a halogen
    (B) a noble gas
    (C) a radioactive element
    (D) an alkali metal
    (E) an alkaline earth metal

11. The number of electrons, protons, and neutrons in argon-40 is

    (A) 18 e, 18 p, 40 n
    (B) 40 e, 40 p, 18 n
    (C) 18 e, 22 p, 18 n
    (D) 18 e, 18 p, 22 n
    (E) 40 e, 18 p, 22 n

12. Which of the following is LEAST likely to have allotropes?

    (A) S
    (B) P
    (C) H
    (D) Se
    (E) Sn

13. Which of the following is expected to have the largest third ionization energy?

    (A) Be
    (B) B
    (C) C
    (D) N
    (E) Al

14. Which pair of elements is expected to have the most similar properties?

    (A) potassium and lithium
    (B) sulfur and phosphorus
    (C) silicon and carbon
    (D) lithium and magnesium
    (E) fluorine and iodine

15. Most elements in the periodic table are

    (A) nonmetals
    (B) liquids
    (C) gases
    (D) metals
    (E) metalloids

16. How many protons, neutrons, and electrons are in an atom of bromine?

    (A) 35 p, 45 n, 35 e
    (B) 35 p, 80 n, 35 e
    (C) 45 p, 35 n, 45 e
    (D) 80 p, 35 n, 80 e
    (E) Neutrons cannot be determined unless an isotope is specified.

17. Chemical properties of elements are defined by the

    (A) electrons
    (B) ionization energy
    (C) protons
    (D) neutrons
    (E) electronegativity

18. The chemical symbol for tin is

    (A) Sb
    (B) Ti
    (C) Sn
    (D) K
    (E) At

19. In which pair of elements is the larger atom listed first?

    (A) K, Ca
    (B) Na, K
    (C) Cl, S
    (D) Mg, Na
    (E) O, N

20. Which of the following is LEAST likely to be a metalloid?

    (A) As
    (B) Hg
    (C) Ge
    (D) Si
    (E) Sb

21. In general, atomic radii decrease from left to right across a period. The main reason for this behavior is

    (A) the number of neutrons in the nucleus increases
    (B) the number of electrons increases
    (C) the atomic mass increases
    (D) the effective nuclear charge increases
    (E) the polarizability increases

ANSWER KEY

1. E	6. A	11. D	16. E	21. D
2. D	7. B	12. C	17. A	
3. B	8. D	13. A	18. C	
4. D	9. A	14. D	19. A	
5. C	10. E	15. D	20. B	

See Appendix 1 for explanations of answers.

Free-Response

Correct interpretation of the periodic table is very helpful when considering chemical and physical properties of the elements. Answer the following questions in sufficient detail to justify any observations.

(a) What feature of the periodic table suggests stable configurations and what evidence supports this conclusion?

(b) Sketch a diagram of the periodic table and indicate where metals, metalloids, and nonmetals are found.

(c) Sketch another diagram of the periodic table and indicate where halogens, alkali metals, noble gases, transition elements, and alkaline earth metals are found.

(d) Starting at the top of any group, do the elements become more or less metallic in character? Justify your response with an appropriate example.

(e) Is the charge of a monatomic ion a diagonal relationship? Justify your response.

(f) What is a diagonal trend? List and briefly explain three diagonal trends in the periodic table.

ANSWERS

(a) Noble gases react in very rare instances; in fact, they were called inert gases until 1963. Electronic structure identical to noble gases suggests stability. This can be inferred from the structure of stable anions and cations of the elements that are isoelectronic with noble gases.

(b) Figure 2.3 illustrates the location of metals, nonmetals, and metalloids in the periodic table. Sketch a similar diagram.

(c) For this question you should to sketch and label a diagram of the periodic table similar to Figure 2.1.

(d) Elements become more metallic as we progress from the top of a group to the bottom. This is most apparent in the nitrogen group that starts as the nonmetal nitrogen, goes through the metalloids arsenic and antimony, and ends in the metal bismuth.

(e) The answer is NO. Ionic charges are dependent on the group and vary only from left to right, not from top to bottom of the periodic table.

(f) Diagonal trends are those that have one extreme of a physical or chemical property in one corner of the periodic table and a regular progression to the other extreme of the physical or chemical property in the opposing corner of the periodic table. Some examples are

ionization energy
electron affinity
electronegativity
metallic character
atomic radius

These diagonal trends all run along a diagonal from the lower left of the periodic table to the upper right corner, often excluding the noble gases.

Nuclear Chemistry

CHAPTER 3

- Alpha Particles
- Beta Particles
- Protons
- Neutrons
- Positrons
- Gamma Rays, X Rays
- Nuclear Reaction Balancing
- Natural Isotopes
- Radioactive Decay
- Half-life
- Archeological Dating
- Nuclear Fission, Fusion

RADIOACTIVITY

SUBATOMIC PARTICLES

Name	Symbol	Mass	Charge
Beta	$_{-1}^{0}\beta$ or $_{-1}^{0}e$	0	1−
Positron	$_{+1}^{0}\beta$	0	1+
Alpha	$_{2}^{4}He$ or $_{2}^{4}\alpha,$	4	2+
Proton	$_{1}^{1}H$ or $_{1}^{1}p$	1	1+
Neutron	$_{0}^{1}n$	1	0

Radioactivity is a property of matter whereby an unstable nucleus spontaneously emits small particles and/or energy in order to attain a more stable nuclear state. The process is called **radioactive decay**, and an isotope that contains an unstable nucleus is termed a **radioactive isotope** or **radioisotope.** One radioactive nucleus may decay to another radioactive nucleus, and then to another. Eventually all radioactive decay results in an isotope with a stable nucleus. Some radioactive isotopes are found to exist in nature and are called natural radioactive substances. Artificial radioactive isotopes, on the other hand, are created in the laboratory in nuclear experiments.

Radioactive isotopes, both natural and artificial, emit only a few types of **subatomic particles** as they disintegrate. These include the electron (beta particle), neutron, helium nucleus (alpha particle), and positron. When these particles are emitted in a radioactive decay process, the **nuclear mass** (nuclear mass = atomic mass = A) and/or **nuclear charge** (nuclear charge = atomic number = Z) of the nucleus changes. As a result, one isotope is converted into another with a different identity. Energy may also be released in the form of X rays or gamma rays. The energy released does not affect the identity of the isotope since these rays have neither mass nor nuclear charge. The characteristics of the particles emitted during radioactive decay are described below.

Alpha (α) particles are helium nuclei. They have a mass of 4 and a nuclear charge of $+2$. The symbol for the alpha particle may take many forms; the two most common are α and $_2^4\text{He}$. For example, radon-222 disintegration emits an alpha particle, and the reaction may be written as

$$_{86}^{222}\text{Rn} \rightarrow {}_{84}^{218}\text{Po} + {}_2^4\text{He}$$

or

$$_{86}^{222}\text{Rn} \rightarrow {}_{84}^{218}\text{Po} + \alpha$$

Alpha particles travel a few centimeters in air at the most before colliding with air molecules, losing kinetic energy and electrons to become ordinary helium. Alpha particles do not penetrate human skin.

Beta (β) particles are electrons. They have negligible mass and have a nuclear charge of -1. Like the alpha particle, the beta particle is indicated by a variety of symbols, including β and $_{-1}^0\text{e}$. A decay in beta particles involves the conversion of a neutron in the nucleus into a proton and a beta particle. The beta particle is emitted and the proton remains in the nucleus. A reaction in which this occurs is the decay of carbon-14 into nitrogen-14

$$_6^{14}\text{C} \rightarrow {}_7^{14}\text{N} + {}_{-1}^0\text{e}$$

or

$$_6^{14}\text{C} \rightarrow {}_7^{14}\text{N} + \beta$$

Beta particles, electrons, are very small and collide much less frequently with air molecules than alpha particles. The beta particle can travel up to 300 cm in air. Beta particles with very high energies can penetrate the skin but, like alpha particles, they normally do not.

Neutrons (n) are often emitted in nuclear reactions. The symbol for a neutron is usually n or $_0^1\text{n}$. A neutron has a mass of 1 and a nuclear charge of 0. When scandium-49 decays to calcium-50, a neutron and a beta particle are emitted:

$$_{49}^{21}\text{Sc} \rightarrow {}_{50}^{20}\text{Ca} + {}_0^1\text{n} + {}_{-1}^0\text{e}$$

Neutrons are very penetrating, easily passing through most materials because of their zero charge. The mass of the neutron means that there will be significant damage when collisions occur.

Positrons, are the positive equivalents of electrons or beta particles. The positron has essentially zero mass and a nuclear charge of $+1$. The symbol is either $_{+1}^0\beta$

or $_{+1}^{0}\text{e}$. Carbon-11 is radioactive and emits a positron when it decays to form boron-11:

$$_{6}^{11}\text{C} \quad \rightarrow \quad _{5}^{11}\text{B} \quad + \quad _{+1}^{0}\beta$$

The same overall reaction may occur by **electron capture.** In electron capture we visualize an electron combining with a proton to form a neutron. This reaction does not change the total mass because the electron is very light. However, converting a proton to a neutron results in decreasing the atomic number by 1. The following equation represents an electron capture process:

$$_{6}^{11}\text{C} \quad + \quad _{-1}^{0}\beta \quad \rightarrow \quad _{5}^{11}\text{B}$$

Positrons are not able to penetrate matter to a significant extent. In addition, they are considered antimatter and are annihilated when they encounter an electron.

Gamma rays (γ) are produced when some radioactive decay events occur that leave the nucleus with excess energy. When this excess energy is lost, it is sometimes in the form of gamma rays. Gamma rays have neither charge nor mass and therefore do not change the identity of the nucleus. The symbol for a gamma ray is γ. Gamma rays are highly penetrating electromagnetic radiation. Gamma rays pass easily through most objects.

X rays represent another form of energy that may be released in a radioactive decay event. Although X rays and gamma rays may be part of a radioactive decay event, they need not be written in the equations since they do not change the identity of the isotope. X rays are highly penetrating electromagnetic radiation. X rays penetrate through the body unless blocked with dense structures such as bones.

NUCLEAR REACTIONS AND EQUATIONS

Using the nuclear particles described above, we can write equations for nuclear reactions as for any other chemical reaction. In writing and balancing nuclear equations, careful attention must be paid to the total mass and the nuclear charge (atomic number) of each element. For this reason, in nuclear equations all reactants and products are usually written with *preceding* superscripts and subscripts to indicate the atomic masses, A, and atomic numbers, Z, respectively.

For example, radon-222 decays spontaneously into polonium-218 by emitting an alpha particle. This information allows us to write the nuclear equation as

$$_{86}^{222}\text{Rn} \quad \rightarrow \quad _{84}^{218}\text{Po} \quad + \quad _{2}^{4}\text{He}$$

> **TIP**
>
> Nuclear reactions require balancing atomic charges and atomic numbers.

If necessary, we can use the periodic table to determine the atomic number needed to write the preceding subscript (atomic number or nuclear charge) for each of the symbols. In this equation we see that the superscripts representing the atomic masses add up to 222 on either side of the arrow ($218 + 4 = 222$). The subscripts representing the atomic numbers add up to 86 on either side of the arrow ($84 + 2 = 86$).

Since the atomic masses and atomic numbers must be equal on the two sides of the arrow, it is possible to deduce the identity of a missing isotope or particle in a partial equation. For example, given the partial reaction below, we can identify the particle represented by the question mark:

$$^{238}_{92}U \rightarrow {}^{234}_{90}Th + ?$$

Since the total atomic mass on both sides must be 238, the question mark must represent a particle that has a mass of 4. Also, the atomic number must be 92 on both sides of the arrow, leading to the conclusion that the unknown particle has an atomic number of 2. The subatomic particle that has these characteristics is the helium nucleus, also known as the alpha particle (^4_2He or $^4_2\alpha$).

EXERCISE 3.1

Write the equations for the following nuclear reactions:

(a) Helium-4 and sodium-23 combine to form aluminum-27.
(b) An alpha particle combines with nitrogen-14 to produce oxygen-17 and a proton.
(c) Fluorine-20 decays to neon-20 by emitting a beta particle.
(d) Uranium-238 decays to thorium-234 by emitting an alpha particle.
(e) Krypton-87 decays to krypton-86 by emitting a neutron.

TIP

In nuclear equations, the pre-superscripts that represent mass must balance. The pre-subscripts that represent the number of protons must also balance.

Solution

(a) $^4_2He + {}^{23}_{11}Na \rightarrow {}^{27}_{13}Al$

(b) $^4_2He + {}^{14}_7N \rightarrow {}^{17}_8O + {}^1_1H$

(c) $^{20}_9F \rightarrow {}^{20}_{10}Ne + {}^{\ 0}_{-1}\beta$

(d) $^{238}_{92}U \rightarrow {}^{234}_{90}Th + {}^4_2\alpha$

(e) $^{87}_{36}Kr \rightarrow {}^{86}_{36}Kr + {}^1_0n$

EXERCISE 3.2

By balancing the masses and atomic numbers, predict the particle represented by the question mark in each of the following equations:

(a) $^{14}_7N \rightarrow ? + {}^1_1H$

(b) $^{54}_{27}Co \rightarrow {}^{54}_{26}Fe + ?$

(c) $^{220}_{86}Rn \rightarrow {}^4_2He + ?$

(d) $^{51}_{23}V + {}^2_1H \rightarrow ?$

(e) $^{55}_{25}Mn + {}^1_1H \rightarrow {}^1_0n + ?$

Solution

(a) $^{13}_6C$

(b) $^{\ 0}_0e$

(c) $^{216}_{84}Po$

(d) $^{53}_{24}Cr$

(e) $^{55}_{26}Fe$

PREDICTING RADIOACTIVITY AND RADIOACTIVE DECAY

It is possible to predict, with reasonable certainty, if an isotope will be radioactive by using a few general principles. First, bismuth (at. no. = 83) is the last element in the periodic table that has a stable, nonradioactive isotope. All elements with atomic numbers greater than 83 are radioactive, and their isotopes decay by emitting alpha particles and beta particles and by electron capture. Alpha particles are emitted by most of these isotopes. The heavy elements often decay to other radioactive elements, which then decay to still other radioactive elements. For instance, uranium-238 decays to lead-206 in a series of 14 steps called a **radioactive disintegration series.** For uranium, this sequence is called the **uranium series** since uranium is the first element in the series. Two other radioactive disintegration series are the thorium series and the actinium series.

While the heavy isotopes ($Z > 83$) often decay by emitting alpha particles, the lighter radioactive elements tend to decay by emitting beta particles or positrons. We may also generalize that, if the radioactive isotope of a light element has a mass greater than the average mass of that element, it will undergo a beta emission. If the mass of the isotope is less than the average mass of the element, a positron emission may be expected. For example, the average mass of carbon is 12. Carbon-14 decays by emitting a beta particle, while carbon-10 decays by emitting a positron:

$$^{14}C \rightarrow {}_{-1}^{0}e + {}^{14}N$$
$$^{10}C \rightarrow {}_{+1}^{0}e + {}^{10}B$$

Why some isotopes are radioactive and some are stable is beyond the scope of this book. However, the stability of isotopes with atomic numbers less than 83 seems to depend on the ratio of neutrons to protons (n/p). Up to atomic number 20, the stable isotopes have an n/p ratio very close to 1.00. From atomic number 21 up to atomic number 83, this ratio for stable isotopes increases slowly from 1.00 to 1.5.

This stability ratio helps us to understand why carbon-14 and carbon-10 decay as they do. We predicted above that carbon-14 (n/p = 1.33) would decay to nitrogen-14 (n/p = 1.00) by beta emission. The beta particle emission reduced the n/p ratio to the value expected for a stable isotope. For the decay of carbon-10 (n/p = 0.67) to boron-10 (n/p = 1.00) by positron emission, the n/p ratio increased to a more favorable value. The examples illustrate that isotopes decay by a process that brings their n/p ratios toward the stable value.

Instead of memorizing the stable n/p ratio for each element, we can use the relative atomic mass as an average mass of the stable nuclei of a given element. This allows us to estimate the stable n/p ratio for any element. For example, the atomic mass of mercury is 200 and its atomic number is 80; therefore the number of protons is 80 and the number of neutrons is 200 − 80 = 120. The estimated stable n/p ratio for mercury is 120/80 = 1.50. The characteristic values of the stable n/p ratios for isotopes from $Z = 21$ to $Z = 83$ may be estimated in the same manner.

In another measure of stability, we find that isotopes with even numbers of **nucleons** (protons plus neutrons) tend to be more stable than those with odd numbers of nucleons. In particular, approximately 60 percent of all stable isotopes have an even number of protons *and* an even number of neutrons. Only 1.5 percent of stable isotopes have an odd number of neutrons *and* an odd number of protons.

NATURAL RADIOACTIVE ISOTOPES

Naturally radioactive elements are those from polonium ($Z = 84$) to uranium ($Z = 92$). The 17 **transuranium** elements with atomic numbers 92–112 are artificially prepared and are also radioactive. In addition to the heavy radioactive elements, a very few of the lighter elements, with atomic numbers less than 83, have naturally occurring radioactive isotopes. These lighter isotopes include potassium-40, vanadium-50, and lanthanum-138. Several interesting naturally radioactive isotopes are described below.

Radon-222 is currently an important potential environmental hazard. It forms, as part of the uranium series, from the decomposition of uranium in many rocks, particularly granite. Radon migrates from the uranium-bearing granite rocks that underlie much of the United States. Often the easiest migration route is through cracks in the basements of many dwellings.

Radium-226 was one of the first radioactive elements associated with biological damage. Some radium salts are **phosphorescent** and glow in the dark. In the early 1900s, these salts were painted on watch dials by workers who licked the paint brushes to get the fine points needed for such work. Many of these workers developed cancers of the mouth.

Uranium-238 is implicated in the hazards of radon gas. However, it has been used for a constructive purpose, to estimate the age of the earth. Uranium eventually decays to lead-206.

Potassium-40 is one of the few radioactive light elements. One way it decays is by emitting a positron to form argon-40. Interestingly, most of the argon in the atmosphere is argon-40, and it is thought to have been produced from the radioactive decay of potassium.

ARTIFICIAL RADIOACTIVE ISOTOPES—TRANSMUTATION

Today, artificial radioactive isotopes are produced by bombarding a target element with nuclei of other elements that have been accelerated to high speeds. This process is carried out in a machine called a **cyclotron,** which accelerates the nuclei towards the target. Another method used to produce artificial radioactive isotopes is to bombard a target with beams of neutrons. Many new and useful radioisotopes have been generated by these methods. Some of these are listed in Table 3.1, where it will be noticed that many of the isotopes have biological applications.

Neutron activation analysis is the name given to a sophisticated analysis method using artificial isotopes. A sample is exposed to an intense beam of neutrons that causes many of the elements in the sample to become radioactive. These elements then decay back into the original elements at rates that are different for each element. When properly calibrated, a neutron activation analysis experiment can identify 20 or more elements in an hour or less. Often very low concentrations can be determined for trace analysis. One advantage of neutron activation analysis is that it is a nondestructive method; in other words, after the induced radioactivity decays to almost zero, the sample remains unchanged.

TABLE 3.1

Some Practical Uses of Radioisotopes

Isotope Symbol	Isotope Name	Typical Application
^{3}H	Tritium	Radiolabeled organic compounds and archaeological dating
^{14}C	Carbon-14	Radiolabeled organic compounds and archaeological dating
^{24}Na	Sodium-24	Circulatory system testing for obstruction
^{32}P	Phosphorus-32	Cancer detection, agricultural tracer
^{51}Cr	Chromium-51	Determination of blood volume
^{57}Co	Cobalt-57	*In vivo* vitamin B$_{12}$ assay
^{59}Fe	Iron-59	Measurements of red blood cell formation and lifetimes
^{60}Co	Cobalt-60	Cancer treatment
^{131}I	Iodine-131	Measurement of thyroid activity and treatment of thyroid disorders
^{153}Gd	Gadolinium-153	Measurement of bone density
^{226}Ra	Radium-226	Cancer treatment
^{235}U	Uranium-235	Nuclear reactors and weapons
^{238}U	Uranium-238	Archaeological dating
^{192}Ir	Iridium-192	Industrial tracer
^{241}Am	Americium-241	Smoke detectors
^{11}C	Carbon-11	Positron emission tomography

RATE OF RADIOACTIVE DECAY

Radioactive decay is a random event. We cannot predict when a single isotope will spontaneously decay. For a large group of radioactive atoms, however, we can apply the well-defined laws of kinetics (see Chapter 11). Radioactive decay is a first-order reaction process. The number of nuclei that disintegrate per second depends only on the number of radioactive nuclei in the sample. The equation for the number of radioactive disintegrations per second (dps) is as follows:

$$\text{Rate (dps)} = (\text{Constant})(\text{Number of radioactive nuclei})$$
$$= kN \tag{3.1}$$

This rate equation may be integrated by using simple calculus to obtain the expression

$$\ln\left(\frac{N_0}{N_t}\right) = kt \tag{3.2}$$

where N_0 is the original number of radioactive atoms, N_t is the number of radioactive atoms left after t seconds have elapsed, k is the **rate constant** with units of reciprocal seconds (s^{-1}), and t is the time in seconds from the start of the experiment. Since the number of atoms in a sample is directly proportional to the mass of the sample, we may interpret N_0 and N_t as the masses of radioactive atoms at the start and at time $= t$, respectively.

When half of the atoms have decayed, $N_0 = 2N_t$. Substituting $2N_t$ in place of N_0 in Equation 3.2 allows us to develop a relationship between the **half-life**, $t_{1/2}$, of an isotope and the specific rate constant, k:

$$\ln\left(\frac{2N_t}{N_t}\right) = kt_{1/2} \tag{3.3}$$

N_t cancels on the left side of this equation to give

$$\ln(2) = kt_{1/2} \tag{3.4}$$

Since the natural logarithm of 2 is 0.693,

$$t_{1/2} = \frac{0.693}{k} \tag{3.5}$$

> **TIP**
>
> The half-life is another way to express the rate constant.

In Equation 3.5 $t_{1/2}$ is called the half-life of the isotope.

The half-life is an important concept. It means that one-half of the radioactive isotope present at the start of an experiment will decompose in a length of time equal to the half-life. The bar graph in Figure 3.1 shows how the amount of radioactive material declines with each half-life. After one half-life half of the isotope is left; after two half-lives half of that half, or one-quarter of the original amount, is left. With the next half-life half of what remained after the second half-life decays, and only one-eighth is left. Note that each bar in Figure 3.1 is one-half the size of the one preceding it.

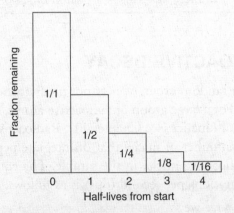

FIGURE 3.1. Bar graph illustrating how half of a radioactive material decomposes with each half-life.

We can calculate the fraction of the original sample left after a given number of half-lives by using Equation 3.6, where the fraction $^1/_2$ is raised to the power equal to the number of half-lives:

$$\text{Fraction left} = \left(\frac{1}{2}\right)^{\text{number of half-lives}} \qquad (3.6)$$

The decimal value of these fractions multiplied by 100 gives us the percentage of the radioactive isotope remaining. The percentage that has decayed is obtained by subtracting the percentage remaining from 100.

Radioactive Decay Calculations

Calculations involving the disintegration of radioactive isotopes require a thorough understanding of Equations 3.1, 3.2, 3.5, and 3.6. Usually such calculations require a combination of two or more of these equations.

EXERCISE 3.3

The half-life of ^{210}Pb is 25 years. In a sample starting with 50 μg of ^{210}Pb, how much is left after 100 years?

Solution

Since 25 years is one half-life, 100 years is equal to four half-lives for ^{210}Pb. There-fore $\frac{1}{16}$ of the original amount will remain (see Figure 3.1). To calculate the amount remaining, the $\frac{1}{16}$ or its decimal equivalent (0.0625) is multiplied by the original amount given. Mathematically we may calculate:

> **TIP**
>
> Half-lives can be used to estimate answers for first-order reactions.

$$? \text{ half-lives} \quad = \quad 100 \text{ yr} \left(\frac{1 \text{ half-life}}{25 \text{ yr}}\right) \quad = \quad 4 \text{ half-lives}$$

$$? \text{ fraction left} \quad = \quad \left(\frac{1}{2}\right)^4 \qquad\qquad = \quad 0.0625$$

$$? \text{ amount left} \quad = \quad 50 \text{ μg} (0.0625) \quad = \quad 3.1 \text{ μg}$$

Whenever a problem involves a whole number of half-lives, it is usually simpler to solve it as was done here, rather than using the more complex Equation 3.2.

EXERCISE 3.4

How many years will 50 μg of ^{210}Pb take to decrease to 5.0 μg ($t_{1/2}$ = 25 years)?

Solution

A quick calculation shows that 5.0 μg/50 μg is 0.1 or $\frac{1}{10}$ of the original amount. From the discussion above we know that after three half-lives (75 years) $\frac{1}{8}$ remains and that after four half-lives (100 years) $\frac{1}{16}$ remains. We may estimate that the answer should be between three and four half-lives (75 and 100 years). If a more precise answer is needed, Equation 3.2 must be used:

$$\ln\left(\frac{N_0}{N_t}\right) = kt$$

Here N_0 is the original number of atoms of the isotope, and N_t is the number remaining after time t has elapsed. We can use the mass of the isotope in place of the number of atoms since mass is directly proportional to number of atoms. Also, k is the rate constant for the radioactive disintegration, calculated from Equation 3.5. Then

$$t_{1/2} = \frac{0.693}{k}$$

Substitution yields:

$$k = \frac{0.693}{25 \text{ yr}} = 2.77 \times 10^{-2} \text{ yr}^{-1}$$

Entering our data into Equation 3.2 gives us

$$\ln\left(\frac{50 \text{ μg}}{5.0 \text{ μg}}\right) = (2.77 \times 10^{-2} \text{ yr}^{-1})t$$

$$\frac{2.30}{2.77 \times 10^{-2} \text{ yr}^{-1}} = t = 83 \text{ years}$$

As expected, the result fits the preliminary estimate of somewhere between 75 and 100 years. Effective use of good estimates cannot be overemphasized. At times an estimate allows us to select the correct answer to a multiple-choice problem. If not, the estimate serves as a check on our calculated answer.

EXERCISE 3.5

Ninety-nine percent of a radioactive element disintegrates in 36.0 hours. What is the half-life of this isotope?

Solution

First, it must be realized that the problem states how much of the element has disintegrated. Half-life calculations using our equations refer to the amount that is left (N_t). Subtracting 99 from 100%, we find only 1% (or $\frac{1}{100}$) left after 36.0 hours. Next we can estimate the half-life by continuing the sequence depicted in Figure 3.1. We find that $\frac{1}{32}$ is left after five half-lives, $\frac{1}{64}$ after six half-lives, and $\frac{1}{128}$ after seven half-lives. Therefore our sample has undergone at least six, but not quite seven, half-lives. Six half-lives in 36.0 hours means a half-life of 6 hours. Seven half-lives in 36.0 hours is approximately 5 hours for a half-life. Our estimate of the half-life is between 5 and 6 hours.

To calculate the exact half-life (if needed), we substitute the data into Equation 3.2:

$$\ln\left(\frac{N_0}{N_t}\right) = kt$$

$$\ln\left(\frac{100}{1}\right) = k(36.0 \text{ hr})$$

$$4.605 = k(36.0 \text{ hr})$$

$$k = \frac{4.605}{36.0} = 0.128 \text{ hr}^{-1}$$

Then we use Equation 3.5:

$$t_{1/2} = \frac{0.693}{k}$$

$$= \frac{0.693}{0.128 \text{ hr}^{-1}}$$

$$= 5.41 \text{ hr}$$

This answer agrees with our quick estimate. Often the estimate is sufficient to make an informed selection of the correct answer to a multiple-choice question. More important, it allows you to reject incorrect answers quickly so that you increase your chances if you must guess at an answer.

Dating Archaeological Samples

Equation 3.2, the integrated rate equation for radioactive decay, can be used to calculate the time elapsed as long as N_0, N_t, and k are known. The rate constant can be obtained from Equation 3.3 if the half-life, $t_{1/2}$, of an isotope is known. Therefore the problem of determining the dates of very old materials involves obtaining valid data for N_0 and N_t. Two examples are presented below to illustrate the different types of logic used.

Uranium-238 decays slowly through the uranium disintegration series to lead-206. The half-life of this conversion is 4.5×10^9 years. To determine the age of a rock, we need to determine the amount of uranium present when the rock solidified, N_0, and the amount of uranium in the rock today, N_t. The amount of ^{238}U present in the rock today can be measured. It is impossible, however, to go back billions of years to obtain a sample containing the original amount of ^{238}U, which is needed to determine the value for N_0. The solution to this dilemma lies in the fact that each atom of uranium ends up as an atom of lead. Therefore, if we take a rock sample and determine the number of atoms of ^{238}U and the number of atoms of ^{206}Pb, we can say that

$$N_0 = \text{atoms } ^{238}\text{U} + \text{atoms } ^{206}\text{Pb}$$

and

$$N_t = \text{atoms } ^{238}\text{U}$$

When we set up an experiment, we must consider all possibilities for error. Two obvious ones exist in this experiment. First, if the original rock may have contained some ^{206}Pb, this would make our value of N_0 too high. Second, if some of the original ^{238}U did not end up as ^{206}Pb, the value for N_0 would be too low. One way this could have occurred is if radon gas, part of the uranium series, escaped from the rock before it decayed to the next solid isotope. Scientists working on this type of project evaluate these possibilities and make corrections or do additional experiments as needed.

In determining the age of carbon-containing materials, the radioactive isotope carbon-14 is measured. ^{14}C is not a natural isotope; it is constantly formed in the upper atmosphere, where ^{14}N is bombarded with neutrons. This keeps the proportion of ^{14}C relatively constant in the biosphere. While alive, animals and plants maintain that same proportion of ^{14}C in their bodies because carbon is continuously recycled. When an organism dies, however, the ^{14}C is no longer replenished by the diet and the fraction of this isotope in the dead organic matter decreases with time.

Understanding the process whereby ^{14}C enters living matter allows us to obtain reasonable measures for N_0 and N_t for our calculations. In this case we assume that the fraction of ^{14}C in the biosphere today is the same as it was in prehistoric times. With that assumption, we can write these equations:

$$N_0 = \frac{\text{living g } ^{14}\text{C}}{\text{living g total C}}$$

and

$$N_t = \frac{\text{ancient g } ^{14}\text{C}}{\text{ancient g total C}}$$

With a value for the half-life of ^{14}C we can solve Equation 3.2 to obtain the age of an ancient sample.

In using radioactive dating techniques, the equation we solve involves the ratio N_0/N_t. This ratio is difficult to determine accurately if it is very close to 1.00 or close to 0.00. A rule of thumb for radioactive isotope dating of materials is that the age of the sample should be 0.3 to 3 half-lives of the isotope used for the dating. Uranium's half-life is close to the age of the earth and is appropriate to use to date the age of old rocks. The half-life of carbon-14 is 5730 years. As a result, we can most reliably determine the ages of biological materials that range from 1700 to 17,000 years old. Radiocarbon dating would certainly not be appropriate in verifying the age of a bottle of wine with a label date of 1865.

NUCLEAR FISSION

Most nuclear disintegrations involve the emission of small particles, none larger than a helium nucleus. In some nuclear reactions the nucleus of a large atom breaks nearly in half. Any process that yields two nuclei of almost equivalent mass is called **nuclear fission.** Nuclear fission does not occur spontaneously, but requires that the nucleus be bombarded with energetic neutrons. Incorporation of a neutron into a **fissile** nucleus such as $^{235}_{92}$U or $^{239}_{94}$Pu causes the nucleus to undergo fission while releasing several more neutrons.

A fissile nucleus is one that is capable of undergoing fission. The neutrons emitted in the fission process then can react with other nuclei to cause additional fission events. Since more neutrons are released in each fission than are needed to start it, a **chain reaction** occurs.

One of the fission reactions of uranium-235 is as follows:

$$^{235}_{92}U + ^{1}_{0}n \rightarrow ^{87}_{35}Br + ^{146}_{57}La + 3^{1}_{0}n$$

If uncontrolled, the first reaction produces 3 neutrons, which will cause three more disintegrations, and 9 neutrons. These 9 neutrons then cause nine more disintegrations and the release of 27 more neutrons. Eventually the reaction is out of control.

There are two ways to keep a fission process from becoming uncontrolled. First, if the sample is small enough, the released neutrons will not hit other uranium-235 nuclei to continue the chain reaction. A **critical mass** of several pounds is needed before a chain reaction will be sustained. Second, excess neutrons can be absorbed by certain materials such as graphite and paraffin. Control rods made of graphite are used in nuclear reactors to adjust the number of available neutrons and therefore the rate of the nuclear reactions. In this use, graphite effectively absorbs some of the neutrons and keeps them from continuing the chain reaction.

Plutonium-239 is another fissile isotope. It is important because it can be produced by bombarding nonfissionable uranium-238 with neutrons. Since only 1 percent of natural uranium is ^{235}U, and most of the rest is ^{238}U, the ability to produce ^{239}Pu means that raw material is more readily available. The reactor that produces plutonium is known as a breeder reactor.

Nuclear reactors are used mainly to generate electricity. The enormous heat generated in fission reactions is used to boil water. The steam is then used to run conventional turbines that produce the electricity. In practice, the nuclear power plant is very complex in order to keep the radioactive materials contained while extracting the desired energy. Uranium reactors require expensive enriched uranium ores. Breeder reactors do not require enriched fuel; since they produce weapons-grade plutonium, their use is highly regulated and discouraged.

NUCLEAR FUSION

The combination of two nuclei into a larger atom is called **nuclear fusion.** The reactions within the sun are fusion reactions that combine hydrogen nuclei to form a helium atom in a three-step reaction:

$$^{1}_{1}H + ^{1}_{1}H \rightarrow ^{2}_{1}H + ^{0}_{1}e$$
$$^{1}_{1}H + ^{2}_{1}H \rightarrow ^{3}_{2}He$$
$$^{3}_{2}He + ^{3}_{2}He \rightarrow ^{4}_{2}He + ^{1}_{1}H + ^{1}_{1}H$$

Since fusion takes simple, nonradioactive materials and produces helium, it should be a clean and inexpensive source of energy.

Instead of trying to duplicate the three-step reaction in the sun, current fusion research focuses on the reaction

$$^{3}_{1}H + ^{2}_{1}H \rightarrow ^{4}_{2}He + ^{1}_{0}n$$

Several designs for fusion reactors are being tested. Although some success has been achieved, it will be some time before this cheap, nonpolluting energy source will become a reality.

SUMMARY

This chapter on nuclear chemistry focuses on the important topic of how nuclear changes occur, the common particles found in nuclear reactions, and the idea that some nuclei are more stable than others. Methods for predicting what type of radioactive decay to expect from a radioactive nucleus are presented and should be understood. Information that will be useful for answering free-response questions concerning natural radioactive isotopes and artificial isotopes that are made for specific medical or technological purposes was discussed. The very important concept of radioactive decay is introduced, and calculations involving the half-life of an isotope are presented. The use of carbon-14 for dating organic matter and uranium/lead ratios for dating the age of rocks is described.

Important Concepts

Radioactivity
Nuclear fusion and fission
Balancing nuclear equations

Important Equations

decompositions per second (dps) = kN

$$\ln\left(\frac{N_0}{N_t}\right) = kt$$

$$t_{1/2} = \frac{0.693}{k}$$

Practice Exercises

Multiple-Choice

For the first four problems below, one or more of the following responses will apply; each response may be used more than once or not at all in these questions.

> I. alpha particle
> II. positron
> III. beta particle
> IV. gamma ray
> V. neutron

1. An electron is also called this in nuclear reactions.

   (A) I
   (B) II
   (C) III
   (D) IV
   (E) V

2. Radioactive elements with an atomic number greater than 83 often decompose with the emission of this particle.

   (A) I
   (B) II
   (C) III
   (D) IV
   (E) V

3. Emission of which of these in a nuclear reaction will result in no change in either the mass or the atomic number?

   (A) I
   (B) II
   (C) III
   (D) IV
   (E) V

4. The nuclear decay of uranium-235 into krypton-94 and barium-139 process is balanced with which of these?

   (A) I
   (B) II
   (C) III
   (D) IV
   (E) V

5. In which of the following are nuclear particles listed in order of increasing penetrating power?

   (A) alpha particles < beta particles < neutrons
   (B) beta particles < alpha particles < neutrons
   (C) neutrons < alpha particles < beta particles
   (D) beta particles < neutrons < alpha particles
   (E) neutrons < beta particles < alpha particles

6. What symbol should replace the question mark in the following nuclear reaction?

   $$^{234}U \rightarrow {}^{230}Th + ?$$

   (A) ^{4}He
   (B) $^{0}_{-1}e$
   (C) $^{1}_{1}H$
   (D) $^{0}_{1}P$
   (E) $^{0}_{+1}e$

7. Carbon-14 has a half-life of 5730 years. How old is a wooden object if 70 percent of the ^{14}C has decayed?

    (A) 2948 y
    (B) 1280 y
    (C) 9950 y
    (D) 4321 y
    (E) 7002 y

8. The half-life of ^{104}Ag is 16.3 min. What is the rate constant in units of reciprocal seconds (s^{-1})?

    (A) 0.0425
    (B) 7.1×10^{-4}
    (C) 2.55
    (D) 7.25
    (E) 1411

9. Uranium-235 undergoes a fission reaction, and one of the products is ^{139}Ba. One neutron is required to induce the fission, and three neutrons are emitted. What is the other fission product?

    (A) $^{96}_{34}Se$
    (B) $^{96}_{36}Kr$
    (C) $^{84}_{35}Br$
    (D) $^{94}_{36}Kr$
    (E) $^{90}_{38}Sr$

10. Copper-67 is a radioactive form of copper with a half-life of 58.5 hours. What particle does ^{67}Cu emit when it decays?

    (A) ^{4}He
    (B) $^{0}_{-1}e$
    (C) $^{1}_{1}H$
    (D) $^{0}_{1}p$
    (E) $^{0}_{+1}e$

11. Bromine-82 has a half-life of 35.7 hours. How many milligrams of ^{82}Br will remain if 2.30 g of ^{82}Br decays for exactly 1 week?

    (A) 4.40
    (B) 238
    (C) 24.6
    (D) 88.1
    (E) 0.0669

12. Which of the following is most probably a stable isotope?

    (A) $^{68}_{29}Cu$
    (B) $^{11}_{4}Be$
    (C) $^{24}_{14}Si$
    (D) $^{66}_{30}Zn$
    (E) $^{42}_{19}K$

13. A radioactive substance decays to 75 percent of its original activity in 3.24 hr. What is the rate constant in units of reciprocal hours?

    (A) 0.19
    (B) 0.43
    (C) 0.34
    (D) 0.039
    (E) 0.089

14. A radioactive sample of ^{141}Cs has a half-life of 32.5 days. In a sample that contains 8.00×10^{7} nuclei of ^{141}Cs, how many beta particles per second will be emitted?

    (A) 19.8
    (B) 1.7×10^{6}
    (C) 71,076
    (D) 1185
    (E) 2.47×10^{-7}

ANSWER KEY

1. C	4. E	7. C	10. B	13. E
2. A	5. A	8. B	11. D	14. A
3. D	6. A	9. D	12. D	

See Appendix 1 for explanations of answers.

Free-Response

Nuclear reactions and radioactive decay are important issues in today's society. They are also fundamental to many biochemical techniques used to follow events in metabolic pathways. Answer the following questions about nuclear processes that are also important in the study of chemistry.

(a) Describe how nuclear reactions are used to generate electricity.

(b) Nuclear weapons testing caused many elements in the soil to become radioactive, and dust particles were called radioactive "fallout." One of the elements that became radioactive was strontium-90. What chemical property of strontium makes it an important health concern?

(c) Holonium-168 has a half-life of 3.0 minutes. Starting with 1.00 g of pure Ho-168, how long will it take to decrease the amount of Ho-168 to 0.1 percent of its original amount? How long will it take until only one atom of Ho-168 is left?

(d) Why is radon-222 considered such a hazard in the home environment?

(e) Balance the following nuclear reaction:

$$^{235}U + {}^{1}n \rightarrow {}^{94}Sr + 2\,{}^{1}n + \underline{\hspace{2cm}}$$

ANSWERS

(a) Nuclear reactions produce tremendous quantities of heat with very small masses of nuclear reactants. This heat is used, eventually, to boil water to produce steam that turns the turbines that produce electricity.

(b) Strontium-90 is a problem because Sr is chemically similar to calcium. Strontium-90 deposits in the bones of exposed persons. The subsequent nuclear decays of Sr-90 are close to vital tissues and greatly increase the possibility of genetic damage leading to a variety of health problems.

(c) The decrease from 1.0 g to 0.1 percent of the original is a decrease to 0.001 g of Ho. This is 1/1000 of the original 1.00 g. Ten half-lives leaves 1/1024 of the original sample. Therefore it will take about 30 minutes for the original sample to decrease to 0.1 percent of the starting amount. To solve Equation 3.2, we need to calculate N_0, the number of atoms of ^{168}Ho in 1.00 g of ^{168}Ho. We also need to calculate k from $t_{1/2}$.

$$?\text{atoms } ^{168}\text{Ho} = 1.00 \text{ g } ^{168}\text{Ho} \left(\frac{1 \text{ mol } ^{168}\text{Ho}}{168 \text{ g } ^{168}\text{Ho}} \right) \left(\frac{6.02 \times 10^{23} \text{ atoms } ^{168}\text{Ho}}{1 \text{ mol } ^{168}\text{Ho}} \right)$$

$$= 3.65 \times 10^{21} \text{ atoms of } ^{168}\text{Ho} = N_0$$

$$k = \frac{0.693}{3.0 \text{ min}} = 0.231 \text{ min}^{-1}$$

Now the above data are entered into the integrated rate law as

$$\ln\left(\frac{N_0}{N_t}\right) = kt$$

$$\ln\left(\frac{3.65 \times 10^{21} \text{ atoms } ^{168}\text{Ho}}{1 \text{ atom } ^{168}\text{Ho}}\right) = \left(0.231 \text{ min}^{-1}\right)t$$

$$49.65 = \left(0.231 \text{ min}^{-1}\right)t$$

$$t = \left(49.65/0.231 \text{ min}^{-1}\right) = 215 \text{ minutes}$$

$$t = 3.6 \text{ hours}$$

(d) Radon-222 is a radioactive gas. When present, it is inhaled into and exhaled from the lungs. If the Ra-222 undergoes radioactive decay while inside the body, it becomes Po-94, a solid that is not exhaled. Po-94 then undergoes a series of radioactive decompositions that affect nearby cells, possibly causing genetic damage.

(e) $^{235}\text{U} + {}^{1}n \rightarrow {}^{94}\text{Sr} + 2 \, {}^{1}n + \underline{\hspace{2cm}}$

The unknown substance, X, must have a mass of 140 ((235 + 1) − (94 + 2)). The atomic number of a neutron is zero; therefore the atomic number of X must be the difference between uranium, 92, and strontium, 38. That is, 54, and element 54 is Xe. The unknown is Xe-140.

PART 2

CHEMICAL BONDING

Ionic Compounds, Formulas, and Reactions

CHAPTER 4

- Chemical Formulas
- Empirical Formulas
- Structural Formulas
- Molecular Formulas
- Chemical Reactions
- Equation Balancing
- Combustion Reactions
- Neutralization Reactions
- Single-replacement Reactions
- Double-replacement Reactions
- Formation Reactions
- Addition Reactions
- Decomposition Reactions

- Net Ionic Reactions
- Ionic Bonding
- Electron Configurations
- Formation of Cations and Anions
- Polyatomic Ions
- Constructing Ionic Formulas
- Naming Ionic Compounds
- Solutions of Ionic Compounds
- Solubility of Ionic Compounds
- Predicting Chemical Reactions
- Chemical Driving Forces
- Predicting REDOX Reactions

CHEMICAL FORMULAS

Chapters 1–3 show how the atom is constructed from protons, neutrons, and, most important, electrons. A good knowledge of the electronic makeup of the atom enables us to predict formulas and reactions of many chemical compounds rather than memorizing them. However, as in learning a new language, some basics must be memorized in order to use information properly and quickly.

We start with the chemical **formula,** which is a shorthand method of describing **compounds.** It uses the **atomic symbols** in the periodic table to identify the elements in a compound. If there is more than one atom of an element in the formula, a **subscript** is used to show how many atoms are present. For example, the compound potassium permanganate contains

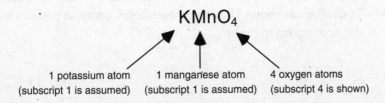

$KMnO_4$

1 potassium atom
(subscript 1 is assumed)

1 manganese atom
(subscript 1 is assumed)

4 oxygen atoms
(subscript 4 is shown)

Parentheses in chemical formulas are used to clarify and to provide additional information. A subscript placed after a closing parenthesis multiplies everything within the parentheses. For example, aluminum nitrate contains

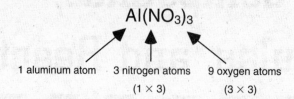

This formula could have been written as AlN_3O_9, which represents the same number of each atom as $Al(NO_3)_3$. However, the parentheses around the NO_3 gives the added information that the nitrogens and oxygens are in three groups of NO_3 units, called the nitrate group. (As will be seen later, the NO_3 should be properly written as the nitrate ion, NO_3^-, with a negative charge.)

The formula for ammonium phosphate is

$$(NH_4)_3PO_4$$

This compound contains 3 nitrogen, 12 hydrogen, 1 phosphorus, and 4 oxygen atoms. Here 3 ammonium groups (actually NH_4^+ ions) are shown in the formula by use of the parentheses.

Another type of formula is used for compounds called **hydrates.** These compounds have fixed numbers of water molecules, called the **water of hydration,** in their crystal lattices. To show the water of hydration clearly in the chemical formula, it is written after a dot that is placed in the middle of the line. The dot links two separate compounds into one unit. Two examples are shown below.

The hexahydrate of cobalt(II) chloride is written as

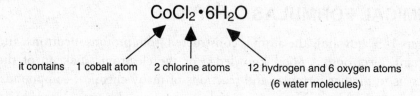

The compound

$$Na_2SO_4 \cdot 10H_2O$$

contains 2 sodium, 1 sulfur, 14 oxygen, and 20 hydrogen atoms.

The names of these compounds are cobalt(II) chloride hexahydrate and sodium sulfate decahydrate. The term *hexahydrate* indicates 6 water molecules in the formula, while *decahydrate* shows 10 water molecules. Common prefixes for hydrates are listed in the accompanying table.

TABLE 4.1

Prefixes Used with Hydrates

1	mono-
2	di-
3	tri-
4	tetra-
5	penta-
6	hexa-
7	hepta-
8	octa-
9	nona-
10	deca-

The compounds discussed above are ionic, and their formulas represent the simplest ratio of the atoms in a crystal of the substance. This simplest formula is called an **empirical formula.**

For compounds that have covalent bonds **molecular formulas** are used. Benzene has a molecular formula of C_6H_6, and ethanoic acid has a formula of $HC_2H_3O_2$. These are not empirical formulas; they represent the actual number of each atom present in a single molecule of each of these compounds. A formula written this way is called a condensed formula.

Another form of the molecular formula is the **structural formula.** A structural formula shows a chemist the way the atoms are connected with each other and the covalent bonds between the atoms. The structural formula for ethanoic (acetic) acid is shown in Figure 4.1.

$C_2H_4O_2$

or

$HC_2H_3O_2$

(a)

CH_3COOH

or

(b)

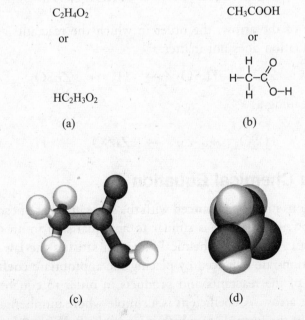

(c)   (d)

FIGURE 4.1. Representations of ethanoic (acetic) acid in several formats: (a) condensed formulas, (b) structural formulas, (c) ball-and-stick model, and (d) space-filling model.

Modern organic chemistry uses structures that attempt to show the three-dimensional shape of moleculae. These may be a ball-and-stick model or a space-filling model as shown in Figure 4.1.

EXERCISE 4.1

Determine the number of each different atom represented in each of the following chemical formulas:

(a) $NaClO_4$
(b) $NH_4C_2H_3O_2$
(c) LiH_2AsO_3
(d) $Ca(C_3H_5O_3)_2 \cdot 5H_2O$
(e) $Cu(NH_3)_4SO_4 \cdot H_2O$

Solution

(a) 1 Na, 1 Cl, 4 O
(b) 1 N, 7 H, 2 C, 2 O
(c) 1 Li, 2 H, 1 As, 3 O
(d) 1 Ca, 6 C, 20 H, 11 O
(e) 1 Cu, 4 N, 14 H, 1 S, 5 O

CHEMICAL REACTIONS AND EQUATIONS

All chemical reactions are essentially the same, with **reactants** being converted into **products.** A chemical equation is written to describe the reaction process. The formulas of the reactants are placed on the left side, and the products on the right side, of an arrow that indicates that the reactants are converted into products:

$$REACTANTS \rightarrow PRODUCTS$$

On each side of the arrow, the order in which the reactants and products are written in an equation does not matter.

$$Zn + H_2SO_4 \rightarrow H_2 + ZnSO_4$$

has the same meaning as

$$H_2SO_4 + Zn \rightarrow ZnSO_4 + H_2$$

Balancing a Chemical Equation

Chemical equations must be **balanced** with the same number of each atom on both sides of the arrow. (The arrow is similar to an equal sign in an ordinary mathematical equation.) A balanced chemical equation satisfies the law of conservation of matter. Equations are balanced by placing the appropriate **coefficients** in front of the formulas of the reactants and products in order to equalize the atoms on both sides of the arrow. A coefficient is a simple whole number, and it multiplies all of the atoms in the formula to which it is attached. *Neither the formulas of compounds nor their subscripts are altered to balance an equation.*

One common reaction type is a **combustion** reaction. An example is the burning of propane fuel ($CH_3CH_2CH_3$) in a barbecue grill:

$$CH_3CH_2CH_3 \; + \; O_2 \; \rightarrow \; CO_2 \; + \; H_2O \quad \text{(unbalanced)}$$

As written, this expression simply gives the reactants ($CH_3CH_2CH_3 + O_2$) and the products ($CO_2 + H_2O$). It may be balanced by using the appropriate coefficients for the two reactants and two products. There are two methods for determining these coefficients. One is the **inspection method** and the other is the **ion-electron method,** which is used for complex oxidation-reduction equations. The ion-electron method will be discussed in Chapter 13.

The first step in the inspection method of balancing equations involves counting the number of each atom in the equation on the reactant side and then on the product side. This requires care and attention to detail since the smallest mistake ruins the entire effort.

The next step is to balance one atom at a time by adding a coefficient where needed and recounting the atoms. Adding coefficients and recounting continue until the same number of atoms is present on each side of the arrow.

Chemists find the process simpler if they balance the most complex molecule first, leaving the simple compounds and elements until last. Also, elements that appear in more than one compound on either the reactant or product side are left to the end. Finally, it is faster to balance groups of atoms, such as the sulfate or nitrate ions discussed later, as if they were individual atoms.

Returning to the combustion of propane, we count 3 carbon, 8 hydrogen, and 2 oxygen atoms on the reactant side. On the product side we have 1 carbon, 2 hydrogen, and 3 oxygen atoms. Since propane is the most complex molecule in the reaction, it is used as the starting point. Its 3 carbon atoms can be balanced by adding a coefficient of 3 to CO_2:

$$CH_3CH_2CH_3 \; + \; O_2 \; \rightarrow \; \mathbf{3}CO_2 \; + \; H_2O \quad \text{(still unbalanced)}$$

There are now 3 carbon atoms on both sides, but the numbers of hydrogen and oxygen atoms are still not equal.

Next, the 8 hydrogen atoms in propane can be balanced by adding a coefficient of 4 to the water molecules:

$$CH_3CH_2CH_3 \; + \; O_2 \; \rightarrow \; \mathbf{3}CO_2 \; + \; \mathbf{4}H_2O \quad \text{(still unbalanced)}$$

Recounting the atoms, we find that there are now 3 carbon, 8 hydrogen, and 2 oxygen atoms on the reactant side and 3 carbon, 8 hydrogen, and 10 oxygen atoms on the product side. The carbon and hydrogen atoms are balanced, and only the oxygen atoms remain unequal. The equation can be balanced by using a coefficient of 5 for the O_2 molecule:

$$CH_3CH_2CH_3 \; + \; \mathbf{5}O_2 \; \rightarrow \; \mathbf{3}CO_2 \; + \; \mathbf{4}H_2O \quad \text{(balanced)}$$

The equation now has 3 carbon, 8 hydrogen, and 10 oxygen atoms on both the reactant and the product side. The coefficient for $CH_3CH_2CH_3$ is 1, but it is not written.

TIP

You may wish to make a table of the atoms to keep track of your progress.

Element	Reactant	Product
Cr	2	1
S	3	1
O	13	7
K	1	2
H	1	3

Another type of reaction is the **double-replacement** reaction. The reaction of chromium(III) sulfate with potassium hydroxide illustrates this class of reactions.

$$Cr_2(SO_4)_3 + KOH \rightarrow Cr(OH)_3 + K_2SO_4 \text{ (unbalanced)}$$

In this unbalanced form there are 2 chromium, 3 sulfur, 13 oxygen, 1 hydrogen, and 1 potassium atoms on the reactant side and 1 chromium, 1 sulfur, 7 oxygen, 3 hydrogen, and 2 potassium atoms on the product side. The chemist would see 2 chromium atoms, 3 sulfate ions (the SO_4^{2-} unit), 1 potassium atom, and 1 hydroxide ion (the OH^- unit) on the reactant side and 1 chromium atom, 1 sulfate ion, 2 potassium atoms, and 3 hydroxide ions on the product side.

Focusing on $Cr_2(SO_4)_3$, we can balance the chromium atoms with a 2 in front of $Cr(OH)_3$:

$$Cr_2(SO_4)_3 + KOH \rightarrow \mathbf{2}Cr(OH)_3 + K_2SO_4 \text{ (unbalanced)}$$

TIP

In almost all cases, balanced equations with the smallest whole-number coefficients are preferred.

Next, the 6 hydroxide ions in $2Cr(OH)_3$ can be balanced by placing a 6 in front of KOH:

$$Cr_2(SO_4)_3 + \mathbf{6}KOH \rightarrow 2Cr(OH)_3 + K_2SO_4 \text{ (unbalanced)}$$

Then the 6 potassium atoms in 6 KOH can be balanced with a 3 in front of K_2SO_4. This also balances the sulfate ions and the equation is balanced:

$$Cr_2(SO_4)_3 + 6KOH \rightarrow 2Cr(OH)_3 + \mathbf{3}K_2SO_4 \text{ (balanced)}$$

As this point, you should have noticed that balancing combustion and double-replacement reactions use slightly different approaches. It will be helpful to recall the appropriate method depending upon the type of reaction you are balancing. Later we will use a third method for balancing oxidation-reduction reactions.

SIMPLEST COEFFICIENTS

The reaction of hydrogen with oxygen to form water can be balanced as

$$2H_2 + O_2 \rightarrow 2H_2O$$

It will also be balanced if written as

$$4H_2 + 2O_2 \rightarrow 4H_2O$$
$$16H_2 + 8O_2 \rightarrow 16H_2O$$
$$H_2 + \tfrac{1}{2}O_2 \rightarrow H_2O$$

The last three reactions are technically balanced since they have the same numbers of hydrogen atoms and oxygen atoms on both sides of the arrow, but they are not in the best form. **Properly balanced equations have the smallest whole-number coefficients possible.** For the three reactions,

the first one should be divided by 2.
the second should be divided by 8.
the third should be multiplied by 2.

An equation may be multiplied or divided as necessary, but it should be remembered that all coefficients in the equation must be multiplied or divided by the same factor. Balancing reactions requires practice to develop skill and speed.

EXERCISE 4.2

Balance the following reactions by inspection:

C_6H_6 + O_2 → CO_2 + H_2O

$MgCl_2$ + $AgNO_3$ → $AgCl$ + $Mg(NO_3)_3$

Al + O_2 → Al_2O_3

CaO + H_2SO_4 → H_2O + $CaSO_4$

Al + Fe_3O_4 → Fe + Al_2O_3

NO_2 + O_2 → N_2O_5

HCl + $CaCO_3$ → $CaCl_2$ + H_2O + CO_2

O_2 + $C_4H_9NH_2$ → CO_2 + H_2O + N_2

Mg + HCl → H_2 + $MgCl_2$

Zn + $Cu(NO_3)_2$ → Cu + $Zn(NO_3)_2$

$CoCl_3$ + $Ba(OH)_2$ → $BaCl_2$ + $Co(OH)_3$

$Ba(OH)_2$ + H_3PO_4 → H_2O + $Ba_3(PO_4)_2$

C_6H_8 + H_2 → C_6H_{12}

SO_2 + O_2 → SO_3

C_4H_{10} + O_2 → CO_2 + H_2O

H_2S + $AuCl_3$ → Au_2S_3 + HCl

Solution

The balanced equations are given in the solution to Exercise 4.3, page 154.

Reaction Types

Many chemical reactions fall into distinct groups with definite similarities. By classifying chemical reactions, it is possible to compare the properties of the reactants and products. In addition, such classification often serves as a shorthand method in place of writing a complete chemical reaction. The combustion of propane illustrates one such classification. By calling the reaction a combustion process, it is immediately known that one of the reactants is oxygen and that the products are carbon dioxide and water. Some other reaction types are described below.

COMBUSTION REACTIONS

In these reactions, an organic (carbon-containing) compound reacts with oxygen to form carbon dioxide and water. If the organic compound contains elements other than carbon, hydrogen, and oxygen, it is often assumed that those elements end up in the elemental state as products. (Sulfur is an exception. It will form SO_2.) Two typical combustion reactions are

$$C_5H_{12} + 8O_2 \rightarrow 5CO_2 + 6H_2O$$
$$2C_4H_9SH + 15O_2 \rightarrow 8CO_2 + 10H_2O + 2SO_2$$

SINGLE-REPLACEMENT REACTIONS

In some reactions, an element may react with a compound to produce a different element and a new compound. A typical reaction of this sort is

$$2AgNO_3 + Cu \rightarrow 2Ag + Cu(NO_3)_2$$

In this reaction the copper replaces the silver in the silver nitrate. This type of reaction is also known as a single-displacement reaction.

DOUBLE-REPLACEMENT REACTIONS

In these reactions, two compounds react and the cation in one compound replaces the cation in the second compound, and vice versa. A double replacement reaction is

$$MgSO_4 + Ba(NO_3)_2 \rightarrow Mg(NO_3)_2 + BaSO_4$$

In this type of reaction, the magnesium replaces the barium and the barium replaces the magnesium—thus the term *double replacement*.

NEUTRALIZATION REACTIONS

These reactions are a special type of double-replacement reaction in which one reactant is an acid and the other is a base. The products are a salt and water. A typical neutralization reaction is

$$HCl + LiOH \rightarrow LiCl + H_2O$$

SYNTHESIS REACTIONS

Reactions of two or more elements to form a compound are often called synthesis reactions. One such reaction is the formation of rust, Fe_2O_3:

$$4Fe + 3O_2 \rightarrow 2Fe_2O_3$$

FORMATION REACTIONS

A formation reaction is the same as a synthesis reaction except that the product must have a coefficient of 1. The reactants are the elements in their normal state at room temperature and atmospheric pressure. The formation reaction for $Fe(NH_4)_2(SO_4)_2$ is

$$Fe + N_2 + 4H_2 + 2S + 4O_2 \rightarrow Fe(NH_4)_2(SO_4)_2$$

If necessary, the use of fractional coefficients for the reactants is permitted in a formation reaction.

ADDITION REACTIONS

In these reactions a simple molecule or an element is added to another molecule, as in the addition of HCl to pentene:

$$HCl \ + \ C_5H_{10} \ \rightarrow \ C_5H_{11}Cl$$

DECOMPOSITION REACTIONS

These reactions result when a large molecule decomposes into its elements or into smaller molecules. When sucrose is heated strongly, in the absence of O_2, this reaction occurs:

$$C_{12}H_{22}O_{11} \ \rightarrow \ 12C \ + \ 11H_2O$$

NET IONIC REACTIONS

When ionic compounds react in aqueous solution, (symbolized by (*aq*) in a reaction) usually only one ion from each compound reacts. The other ions are "spectator ions" and do not react. Writing a reaction in ionic form focuses attention on the actual reaction and allows the chemist to find substitute reactants to achieve the same result. For instance, the reaction of silver nitrate with sodium chloride produces a precipitate of silver chloride in the molecular equation

$$NaCl(aq) \ + \ AgNO_3(aq) \ \rightarrow \ AgCl(s) \ + \ NaNO_3(aq)$$

Written as a net ionic equation, this becomes

$$Cl^-(aq) \ + \ Ag^+(aq) \ \rightarrow \ AgCl(s)$$

In this form the chemist now knows that any soluble chloride salt (KCl, $MgCl_2$, etc.) and any soluble silver salt ($AgClO_4$, Ag_2SO_4, etc.) will also give AgCl as the product. In ionic reactions the charges must balance as well as the atoms.

HALF-REACTIONS

These reactions are used extensively with oxidation-reduction reactions and in describing electrochemical processes in Chapter 13. The half-reaction is a reduction reaction if electrons are on the reactant side and an oxidation reaction if the electrons are products.

$$I_2 \ + \ 2e^- \ \rightarrow \ 2I^- \quad \text{(reduction half reaction)}$$

$$Fe^{2+} \ \rightarrow \ Fe^{3+} \ + \ e^- \quad \text{(oxidation half-reaction)}$$

Half-reactions may be combined to make a complete oxidation-reduction reaction as long as the electrons all cancel.

OXIDATION-REDUCTION REACTIONS

These reactions involve the loss of electrons by one compound or ion and the subsequent gain of the same electrons by another compound or ion. The two half-reactions above may be added (after multiplying the second reaction by 2 and cancelling the electrons) to obtain the oxidation-reduction reaction

$$I_2 + 2Fe^{2+} \rightarrow 2Fe^{3+} + 2I^-$$

The combustion and single-replacement reactions discussed above are also oxidation-reduction reactions.

NUCLEAR REACTIONS

We have already seen this type of reaction in Chapter 3. To review these, see Exercises 3.1 and 3.2.

EXERCISE 4.3

Classify each of the reactions balanced in Exercise 4.2 as one of the reaction types described in this section.

Solution

$2C_6H_6$ + $15O_2$	$\rightarrow$	$12CO_2$ + $6H_2O$			(combustion)
$MgCl_2$ + $2AgNO_3$	$\rightarrow$	$2AgCl$ + $Mg(NO_3)_2$			(double replacement)
$4Al$ + $3O_2$	$\rightarrow$	$2Al_2O_3$			(synthesis)
CaO + H_2SO_4	$\rightarrow$	H_2O + $CaSO_4$			(double replacement)
$8Al$ + $3Fe_3O_4$	$\rightarrow$	$9Fe$ + $4Al_2O_3$			(single replacement)
$4NO_2$ + O_2	$\rightarrow$	$2N_2O_5$			(combustion or addition)
$2HCl$ + $CaCO_3$	$\rightarrow$	$CaCl_2$ + H_2O + CO_2			(double replacement)
$27O_2$ + $4C_4H_9NH_2$	$\rightarrow$	$16CO_2$ + $22H_2O$ + $2N_2$			(decomposition)
Mg + $2HCl$	$\rightarrow$	H_2 + $MgCl_2$			(single replacement)
Zn + $Cu(NO_3)_2$	$\rightarrow$	Cu + $Zn(NO_3)_2$			(single replacement)
$2CoCl_3$ + $3Ba(OH)_2$	$\rightarrow$	$3BaCl_2$ + $2Co(OH)_3$			(double replacement)
$3Ba(OH)_2$ + $2H_3PO_4$	$\rightarrow$	$6H_2O$ + $Ba_3(PO_4)_2$			(neutralization)
C_6H_8 + $2H_2$	$\rightarrow$	C_6H_{12}			(addition)
$2SO_2$ + O_2	$\rightarrow$	$2SO_3$			(addition or combustion)
$2C_4H_{10}$ + $13O_2$	$\rightarrow$	$8CO_2$ + $10H_2O$			(combustion)
$3H_2S$ + $2AuCl_3$	$\rightarrow$	Au_2S_3 + $6HCl$			(double replacement)

BONDING

When elements combine with each other to form compounds, a chemical bond is formed. An understanding of how and why bonds are formed helps the chemist to predict many physical and chemical properties of molecules and compounds, including chemical reactivity, shape, solubility, physical state, and polarity. The key to bond formation is the behavior of the outermost, or valence, electrons. When two atoms share valence electrons to form a bond, the bond is known as a **covalent bond.** When one atom loses electrons and another gains electrons, ions are formed. The attraction between ions to form a compound is called an **ionic bond.**

The underlying principle of chemical bonding can be explained using the electronic configurations of the noble gases. Noble gases are very unreactive elements. Until 1962, when Neil Bartlett produced the first noble gas compound, they were considered inert and most periodic tables called them "inert gases." Except for helium, all noble gases have valence shells in which the outermost s and p sublevels are completely filled (ns^2, np^6), as shown in bold type in Table 4.2. This configuration gives unusual stability to the noble gases and also to atoms that can lose, gain, or share electrons to attain the same configuration.

Ionic Substances

The basis of the ionic bond is the attraction of a positively charged ion (cation) toward a negatively charged ion (anion). It is necessary to understand which elements tend to form ions and which do not. For elements that form ions we also want to develop methods to determine what kind of ion should be expected. Finally, once the ions are known, we can use that information to predict chemical formulas and chemical reactions.

TABLE 4.2

Electronic Configurations of the Noble Gases

Noble Gas	Electronic Configuration
He	$\mathbf{1s^2}$
Ne	$1s^2\ \mathbf{2s^2}\ \mathbf{2p^6}$
Ar	$1s^2\ 2s^2\ 2p^6\ \mathbf{3s^2}\ \mathbf{3p^6}$
Kr	$1s^2\ 2s^2\ 2p^6\ 3s^2\ 3p^6\ \mathbf{4s^2}\ 3d^{10}\ \mathbf{4p^6}$
Xe	$1s^2\ 2s^2\ 2p^6\ 3s^2\ 3p^6\ 4s^2\ 3d^{10}\ 4p^6\ \mathbf{5s^2}\ 4d^{10}\ \mathbf{5p^6}$
Rn	$1s^2\ 2s^2\ 2p^6\ 3s^2\ 3p^6\ 4s^2\ 3d^{10}\ 4p^6\ 5s^2\ 4d^{10}\ 5p^6\ \mathbf{6s^2}\ 4f^{14}\ 5d^{10}\ \mathbf{6p^6}$

MONATOMIC IONS OF THE REPRESENTATIVE ELEMENTS

The representative elements are those found within the s and p blocks of the periodic table as shown in Figure 1.16. These elements have regular properties that follow basic chemical principles with very few exceptions. In the case of ion formation, the principle is that the ions of the representative elements will have electronic configurations identical to those of the noble gases.

Representative metals will lose electrons to form positively charged ions called cations. The equation

$$M \rightarrow M^{n+} + ne^-$$

where n is the number of electrons lost by the metal, M, is used to represent the formation of all cations.

The electronic configuration allows the chemist to determine the number of electrons that a representative metal will lose. For instance, sodium has the electronic configuration $1s^2\ 2s^2\ 2p^6\ 3s^1$. When the element forms the Na$^+$ ion, the $3s^1$ electron is lost. The electronic configuration for the Na$^+$ ion is then $1s^2\ 2s^2\ 2p^6$ which is the same as the electronic configuration of neon.

TIP

Cations of the representative metals are formed by removing outer s and p electrons to achieve an electronic structure the same as that of a noble gas.

Barium has the electronic configuration

$$\text{Ba} = 1s^2 \ 2s^2 \ 2p^6 \ 3s^2 \ 3p^6 \ 4s^2 \ 3d^{10} \ 4p^6 \ 5s^2 \ 4d^{10} \ 5p^6 \ \mathbf{6s^2}$$

When the Ba^{2+} ion is formed, the two $6s$ electrons are lost, giving the barium ion the configuration

$$\text{Ba}^{2+} = 1s^2 \ 2s^2 \ 2p^6 \ 3s^2 \ 3p^6 \ 4s^2 \ 3d^{10} \ 4p^6 \ \mathbf{5s^2} \ 4d^{10} \ \mathbf{5p^6}$$

which is identical to the electronic configuration of xenon:

$$\text{Xe} = 1s^2 \ 2s^2 \ 2p^6 \ 3s^2 \ 3p^6 \ 4s^2 \ 3d^{10} \ 4p^6 \ \mathbf{5s^2} \ 4d^{10} \ \mathbf{5p^6}$$

The representative metals will lose all of their valence s and p electrons. The electronic configuration of a metal will then be identical to that of the preceding noble gas in the periodic table.

In periods 4, 5, and 6 the metals that contain outer s and p electrons may form a second ion by losing only their p electrons. We find that gallium, indium, and thallium form 1+ and 3+ ions. Tin and lead form 2+ and 4+ ions, and bismuth forms 3+ and 5+ ions.

The representative nonmetals gain electrons to form negatively charged ions called anions. The general reaction for this process is

$$\text{X} + ne^- \rightarrow \text{X}^{n-}$$

where n represents the number of electrons gained by the nonmetal represented by X.

As an example, bromine has the electronic configuration

$$\text{Br} = 1s^2 \ 2s^2 \ 2p^6 \ 3s^2 \ 3p^6 \ 4s^2 \ 3d^{10} \ \mathbf{4p^5}$$

When an electron is added to the $4p$ subshell, the bromide ion, Br^-, is formed. Its electronic configuration becomes

$$\text{Br}^- = 1s^2 \ 2s^2 \ 2p^6 \ 3s^2 \ 3p^6 \ 4s^2 \ 3d^{10} \ \mathbf{4p^6}$$

which is the same as that of the noble gas krypton:

$$\text{Kr} = 1s^2 \ 2s^2 \ 2p^6 \ 3s^2 \ 3p^6 \ \mathbf{4s^2} \ 3d^{10} \ \mathbf{4p^6}$$

All halogens will gain one electron to form anions with one negative charge. Oxygen, sulfur, and selenium gain two electrons each to form 2− ions. We will see that these elements often participate in covalent bonding as well. The remaining nonmetals, nitrogen, phosphorus, and carbon, usually bond using covalent bonds, but they can form N^{3-}, P^{3-}, and C^{4-} ions. These are called the nitride, phosphide, and carbide ions, respectively.

MONATOMIC IONS OF THE NONREPRESENTATIVE ELEMENTS

The nonrepresentative elements are the remaining d block and f block metals in the periodic table. These elements are characterized by the fact that many of them may have more than one possible cation and they often form polyatomic anions. In general, it is not possible to predict with certainty the charge of the cations for these elements. However, none of these elements forms monatomic anions.

TIP

Anions are formed by adding electrons to complete an electronic shell.

TIP

Ions and atoms that have identical electron configurations are said to be isoelectronic.

TIP

Cations of nonrepresentative elements can be understood by arranging electrons by their principal quantum number.

Appreciating what happens to the transition elements is not always simple, but some hints may be obtained from their electronic configurations. For these elements the electronic configurations are arranged by their principal quantum number, rather than in the aufbau order. For instance, the complete electronic configuration for lead is as follows:

$$Pb = 1s^2\ 2s^2\ 2p^6\ 3s^2\ 3p^6\ 4s^2\ 3d^{10}\ 4p^6\ 5s^2\ 4d^{10}\ 5p^6\ 6s^2\ 4f^{14}\ 5d^{10}\ 6p^2$$

Grouping the electrons by shell or principal quantum number gives

$$Pb = 1s^2\ 2s^2\ 2p^6\ 3s^2\ 3p^6\ 3d^{10}\ 4s^2\ 4p^6\ 4d^{10}\ 4f^{14}\ 5s^2\ 5p^6\ 5d^{10}\ \mathbf{6s^2\ 6p^2}$$

and allows us to see more clearly which electrons are in the outermost shell of the atom. We can now see that the Pb^{2+} ion is formed when the two $6p$ electrons are removed and that the Pb^{4+} ion forms when all of the $6s$ and $6p$ electrons are removed.

Similarly, the 2+ and 4+ ions of titanium may be deduced from the electronic structure

$$Ti\ =\ 1s^2\ 2s^2\ 2p^6\ 3s^2\ 3p^6\ 4s^2\ 3d^2$$

Regrouping by shell gives

$$Ti\ =\ 1s^2\ 2s^2\ 2p^6\ 3s^2\ 3p^6\ \mathbf{3d^2\ 4s^2}$$

Removal of just the two $4s$ electrons yields the Ti^{2+} ion, and removal of the two $3d$ and two $4s$ electrons gives the Ti^{4+} ion.

Iron is another example worth considering. It forms Fe^{2+} and Fe^{3+} ions. From the electronic configuration arranged by shells we get

$$Fe\ =\ 1s^2\ 2s^2\ 2p^6\ 3s^2\ 3p^6\ 3d^6\ \mathbf{4s^2}$$

Removal of the two $4s$ electrons gives the Fe^{2+} ion. To obtain the Fe^{3+} ion one more electron must be removed. Obviously it is one of the six $3d$ electrons that are now the outermost electrons. As we saw in Chapter 1, a d subshell that contains one electron in each orbital is a stable state. Removal of one of the $3d$ electrons results in the electronic structure

$$Fe^{3+}\ =\ 1s^2\ 2s^2\ 2p^6\ 3s^2\ 3p^6\ \mathbf{3d^5}$$

which has the stable half-filled $3d$ subshell.

Not all ions in the nonrepresentative group can be rationalized without much more sophisticated reasoning. However, in the absence of additional information the logic used above provides for a reasonable first approximation of why certain ions form and others do not.

POLYATOMIC IONS

Many elements combine with oxygen (and sometimes hydrogen and nitrogen) to form a charged group of atoms called a **polyatomic ion**. (In older texts polyatomic ions are called radicals.) Polyatomic ions are unusually stable groups of atoms that tend to act as single units in many chemical reactions. The formulas, names, and charges of the common polyatomic ions are listed in Table 4.3 and **should be**

memorized. Note that all of these are anions (negatively charged ions) except for the ammonium ion, NH_4^+.

The atoms in the polyatomic ions (Table 4.3) are bound to each other with covalent bonds, which will be described later. Polyatomic ions form ionic compounds by combining with other ions of opposite charge.

TABLE 4.3

Common Polyatomic Ions	
Ion Formula	Ion Name
NH_4^+	ammonium ion
CO_3^{2-}	carbonate ion
HCO_3^-	bicarbonate ion (hydrogen carbonate)
PO_4^{3-}	phosphate ion
ClO^-	hypochlorite ion
ClO_2^-	chlorite ion
ClO_3^-	chlorate ion
ClO_4^-	perchlorate ion
NO_2^-	nitrite ion
NO_3^-	nitrate ion
SO_3^{2-}	sulfite ion
HSO_3^-	bisulfite ion (hydrogen sulfite)
SO_4^{2-}	sulfate ion
HSO_4^-	bisulfate ion (hydrogen sulfate)
MnO_4^-	permanganate
$Cr_2O_7^{2-}$	dichromate ion
CrO_4^{2-}	chromate ion
$S_2O_3^{2-}$	thiosulfate ion
PO_4^{3-}	phosphate ion
HPO_4^{2-}	hydrogen phosphate ion
$H_2PO_4^-$	dihydrogen phosphate ion

TIP

Polyatomic ion formulas and names must be memorized.

IONIC FORMULAS

Ionic compounds are formed when cations are attracted to anions because of their opposing charges. The formulas for ionic compounds can be deduced because no compound can have a net charge. In other words, the total positive charge of the cations must be exactly canceled by the negative charge of the anions in the chemical formula. Some chemists refer to this as the **law of electroneutrality.** In addition, every ionic compound has a formula that represents the simplest ratio of the elements needed to obey the law of electroneutrality. As mentioned earlier in this chapter, this simplest ratio is called the **empirical formula.**

When the anion and cation have the same, but opposite, charges, the compound is written with only one atom of each element. This is another way of saying that the formulas for all ionic compounds are empirical formulas.

$$Na^+ \quad + \quad F^- \quad \rightarrow \quad NaF \qquad \text{sodium fluoride}$$
$$Mg^{2+} \quad + \quad O^{2-} \quad \rightarrow \quad MgO \qquad \text{magnesium oxide}$$
$$Fe^{2+} \quad + \quad S^{2-} \quad \rightarrow \quad FeS \qquad \text{iron(II) sulfide}$$
$$Al^{3+} \quad + \quad N^{3-} \quad \rightarrow \quad AlN \qquad \text{aluminum nitride}$$
$$La^{3+} \quad + \quad PO_4^{3-} \quad \rightarrow \quad LaPO_4 \qquad \text{lanthanum(III) phosphate}$$
$$NH_4^+ \quad + \quad NO_3^- \quad \rightarrow \quad NH_4NO_3 \qquad \text{ammonium nitrate}$$

When the charges of the anion and cation are not equal and opposite, it is necessary to adjust the numbers of the ions so that the total charge adds up to zero. The most convenient way to do this is to use the charge of the cation for the subscript of the anion and the charge of the anion (without the minus sign) as the subscript for the cation, as shown below. Remember that, when a subscript is 1, it is not written.

$$Ca^{2+} \quad + \quad 2Cl^- \quad \rightarrow \quad CaCl_2 \qquad \text{calcium chloride}$$
$$2Al^{3+} \quad + \quad 3S^{2-} \quad \rightarrow \quad Al_2S_3 \qquad \text{aluminum sulfide}$$
$$3Na^+ \quad + \quad PO_4^{3-} \quad \rightarrow \quad Na_3PO_4 \qquad \text{sodium phosphate}$$
$$Pb^{4+} \quad + \quad 4Cl^- \quad \rightarrow \quad PbCl_4 \qquad \text{lead(IV) chloride}$$

When a subscript must be used with a polyatomic ion, it is necessary to place parentheses around the polyatomic ion before adding the subscript, as shown in the following equations:

$$Ca^{2+} \quad + \quad 2NO_3^- \quad \rightarrow \quad Ca(NO_3)_2 \qquad \text{calcium nitrate}$$
$$2NH_4^+ \quad + \quad SO_4^{2-} \quad \rightarrow \quad (NH_4)_2SO_4 \qquad \text{ammonium sulfate}$$
$$2Al^{3+} \quad + \quad 3SO_4^{2-} \quad \rightarrow \quad Al_2(SO_4)_3 \qquad \text{aluminum sulfate}$$

NAMING IONIC COMPOUNDS

Ionic compounds contain a metal and a nonmetal. [The ammonium ion (NH_4^+) is considered a metal, and polyatomic anions are considered nonmetals for this purpose.] These compounds are also known as salts. Names are created by giving the name of the cation first and then the name of the anion.

For cations that have only one possible charge, the name is the same as that of the element. For cations that may have more than one charge, such as the lead and titanium discussed previously, the element name is followed by parentheses enclosing the charge written in roman numerals. Two examples are lead(II) and lead(IV). This method, known as the Stock system, is the preferred method for naming ionic compounds.

In the old naming system, the higher charged cation was given the suffix *-ic* and the lower charged cation had the *-ous* ending. Table 4.4 lists some of these old names for reference. They are most useful for understanding the older chemical literature.

TABLE 4.4

Old Nomenclature for Some Common Ions

Ion	Stock Name	Old Name
Fe^{2+}	iron(II)	ferrous
Fe^{3+}	iron(III)	ferric
Sn^{2+}	tin(II)	stannous
Sn^{4+}	tin(IV)	stannic
Pb^{2+}	lead(II)	plumbous
Pb^{4+}	lead(IV)	plumbic
Cu^+	copper(I)	cuprous
Cu^{2+}	copper(II)	cupric
Hg^+ or Hg_2^{2+}	mercury(I)	mercurous
Hg^{2+}	mercury(II)	mercuric

TIP

Use the Stock system when naming ionic compounds on the AP exam.

Naming anions depends on whether the anion is a monatomic ion or a polyatomic anion. Monatomic anions are named by taking the root or first portion of the element name and then changing the ending to *-ide*, as shown in the accompanying table of examples. Polyatomic anions have unique names, given in Table 4.3, which must be memorized.

COMMON ANION NAMES	
sul**fur**	sul**fide**
oxy**gen**	ox**ide**
chlor**ine**	chlor**ide**
nitro**gen**	nit**ride**
brom**ine**	brom**ide**
fluor**ine**	fluor**ide**

The entire compound is named by writing the name of the cation followed by the name of the anion as a separate word; for example, MgF_2 is magnesium fluoride and FeI_3 is iron(III) iodide. Names of chemical compounds are not capitalized except at the beginning of a sentence. Examples of the names of representative compounds have been given in the preceding discussion of formula writing.

To name a compound which contains a metal that may have more than one possible charge, we must know the charge on the ion. Ionic charge is determined by "taking apart" the formula unit to find out what the charges of the ions were before the ions combined. We will know the charge of the anion, which will be either a representative monatomic anion or one of the polyatomic anions in Table 4.3. If the charge of one anion and the number of anions are known, the charge on the cation can be deduced since the formula must always have a net charge of zero. Review how formulas are determined and see how the process can be reversed.

EXERCISE 4.4

Name each of the following compounds:

$MgCl_2$ Na_2CrO_4 $TiBr_4$ $HgSO_4$

MnO_2 $Fe(ClO_2)_2$ Na_3PO_4 $SnCl_4$

Cr_2O_3 $Hg(NO_3)_2$ $(NH_4)_2SO_3$ BiF_3

Ca_3N_2 $Al(NO_3)_3$

Solution

$MgCl_2$	magnesium chloride	$Al(NO_3)_3$	aluminum nitrate
MnO_2	manganese(IV) oxide	$TiBr_4$	titanium(IV) bromide
Cr_2O_3	chromium(III) oxide	Na_3PO_4	sodium phosphate
Ca_3N_2	calcium nitride	$(NH_4)_2SO_3$	ammonium sulfite
Na_2CrO_4	sodium chromate	$HgSO_4$	mercury(II) sulfate
$Fe(ClO_2)_2$	iron(II) chlorite	$SnCl_4$	tin(IV) chloride
$Hg(NO_3)_2$	mercury(II) nitrate	BiF_3	bismuth(III) fluoride

EXERCISE 4.5

Write the formula for each of the following compounds:

aluminum sulfate	gold(III) nitrate
magnesium oxide	lithium sulfite
vanadium(III) bromide	ammonium phosphate
barium nitrite	strontium fluoride
cobalt(II) chloride	lead(IV) carbonate

Solution

Writing formulas from names is often easier than writing names from formulas since the charges of the nonrepresentative elements are given in parentheses in the names. Remember the requirement for electric neutrality or no net charge on any chemical compound.

aluminum sulfate	$Al_2(SO_4)_3$	gold(III) nitrate	$Au(NO_3)_3$
magnesium oxide	MgO	lithium sulfite	Li_2SO_3
vanadium(III) bromide	VBr_3	ammonium phosphate	$(NH_4)_3PO_4$
barium nitrite	$Ba(NO_2)_2$	strontium fluoride	SrF_2
cobalt(II) chloride	$CoCl_2$	lead(IV) carbonate	$Pb(CO_3)_2$

Ionic Reactions

IONS IN SOLUTION

Most ionic compounds dissolve in water, and in the process the compound separates into the cations and anions. This solution process may be written as

$$NaBr(s) \rightarrow Na^+(aq) + Br^-(aq)$$

TIP
(s) = solid
(g) = gas
(l) = pure liquid
(aq) = aqueous

The symbol in parentheses designates the state of each substance in the reaction: (*s*) means that the substance is a solid, and (*aq*) that the substance is in an aqueous solution. Other symbols used are (*l*) for liquid and (*g*) for gas.

Chromium(III) nitrate dissolves according to the equation

$$Cr(NO_3)_3(s) \rightarrow Cr^{3+}(aq) + 3NO_3^-(aq)$$

One chromium(III) ion and three nitrate ions are obtained from one formula unit of chromium(III) nitrate.

The following general principles apply to the dissolution of ionic compounds.

1. Only one cation and one anion are formed. Compounds containing three or more different atoms will break apart into the appropriate polyatomic ion(s). (Exceptions are discussed in higher level chemistry courses.)
2. The charges of the ions obey the same rules as discussed above. In particular, the charges of all of the ions must add up to zero, which is the charge of any compound.
3. The subscripts of monatomic ions become coefficients for the ions. For polyatomic ions, only the subscripts after parentheses become coefficients.

EXERCISE 4.6

Write the ions expected when the following compounds are dissolved in water:

K_2S	$MgBr_2$	$Na_2Cr_2O_7$	K_3PO_4
$FeCl_3$	$AlCl_3$	$(NH_4)_2S$	$Ti(NO_3)_4$

Solution

$K_2S(s)$	$\rightarrow$	$2K^+(aq)$	$+$	$S^{2-}(aq)$
$FeCl_3(s)$	$\rightarrow$	$Fe^{3+}(aq)$	$+$	$3Cl^-(aq)$
$MgBr_2(s)$	$\rightarrow$	$Mg^{2+}(aq)$	$+$	$2Br^-(aq)$
$AlCl_3(s)$	$\rightarrow$	$Al^{3+}(aq)$	$+$	$3Cl^-(aq)$
$Na_2Cr_2O_7(s)$	$\rightarrow$	$2Na^+(aq)$	$+$	$Cr_2O_7^{2-}(aq)$
$(NH_4)_2S(s)$	$\rightarrow$	$2NH_4^+(aq)$	$+$	$S^{2-}(aq)$
$K_3PO_4(s)$	$\rightarrow$	$3K^+(aq)$	$+$	$PO_4^{3-}(aq)$
$Ti(NO_3)_4(s)$	$\rightarrow$	$Ti^{4+}(aq)$	$+$	$4NO_3^-(aq)$

Notice that the charges on Fe^{3+} and Ti^{4+} must be calculated from the known negative charge of the anion and the fact that the total charge must add up to zero. Also observe which subscripts have become coefficients and which remained as part of a polyatomic ion. Finally, this entire process is just the reverse of the method used to determine the formulas of ionic compounds.

Any ionic compound can be broken apart into its cations and anions in this manner. Whether or not an ionic compound will dissolve to an appreciable extent in water depends on which cations and anions make up the compound. Some general guidelines for predicting solubility should be remembered.

> ## Solubility Rules
>
> 1. Compounds containing alkali metal cations or the ammonium ion are **soluble**.
> 2. Compounds containing NO_3^-, ClO_4^-, ClO_3^-, and $C_2H_3O_2^-$ anions are **soluble**.
> 3. Chlorides, bromides, and iodides are **soluble except** those containing Ag^+, Pb^{2+}, or Hg_2^{2+}.
> 4. Sulfates are **soluble except** those containing Hg_2^{2+}, Pb^{2+}, Sr^{2+}, Ca^{2+}, or Ba^{2+}.
> 5. Hydroxides are **insoluble except** compounds of the alkali metals, Ca^{2+}, Sr^{2+}, and Ba^{2+}.
> 6. Compounds containing PO_4^{3-}, S^{2-}, CO_3^{2-}, and SO_3^{2-} ions are **insoluble except** those that also contain alkali metals or NH_4^+.

> **TIP**
>
> Solubility rules must be memorized for the AP exam.

DOUBLE-REPLACEMENT REACTIONS

Predicting Products

If we know how to determine which ions make up an ionic compound, we can then take two ionic compounds, mix them together, and predict the possible products. As the name suggests, two replacements occur in these reactions.

In one replacement the cation of the first salt replaces the cation of the second salt.

In the second replacement the cation of the second salt replaces the cation of the first salt.

For example, if we mix together solutions containing $AgNO_3$ and Na_2CrO_4, what products should we predict? The first step is to determine the ions that make up the two reacting compounds. They are Ag^+, NO_3^-, $2Na^+$, and CrO_4^{2-}. The next step is to pair up these four ions in various ways to make two new ionic compounds that will be the predicted products. It is worthwhile to look at all of the possible pairs to see how we arrive at our conclusions.

Ag^+ $+ NO_3^- \rightarrow AgNO_3$ We already have this as a reactant.
Ag^+ $+ Na^+ \rightarrow$ cannot form an ionic compound from two positive ions
$2Ag^+$ $+ CrO_4^{2-} \rightarrow Ag_2CrO_4$ **This is a possibility.**
Ag^+ $+ Ag^+ \rightarrow$ cannot form an ionic compound from two identical ions
NO_3^- $+ Na^+ \rightarrow NaNO_3$ **This is a possibility.**
NO_3^- $+ CrO_4^{2-} \rightarrow$ cannot form an ionic compound from two negative ions
NO_3^- $+ NO_3^- \rightarrow$ cannot form an ionic compound from two identical ions
$2Na^+$ $+ CrO_4^{2-} \rightarrow Na_2CrO_4$ We already have this as a reactant.
Na^+ $+ Na^+ \rightarrow$ cannot form an ionic compound from two identical ions
CrO_4^{2-} $+ CrO_4^{2-} \rightarrow$ cannot form an ionic compound from two identical ions

All of the possible combinations of the four ions are given above, with the reasons why they are good or bad choices. Only two of the combinations give reasonable new compounds. Every other combination leads to either an impossible situation or back to the original compounds. Using the only reasonable results as the products, we may begin to construct a chemical equation:

$$AgNO_3 \ + \ Na_2CrO_4 \ \rightarrow \ Ag_2CrO_4 \ + \ NaNO_3$$

The final step is to balance the equation so it has the same number of each atom on both sides of the arrow:

$$2AgNO_3 \ + \ Na_2CrO_4 \ \rightarrow \ Ag_2CrO_4 \ + \ 2NaNO_3$$

Reviewing what was done to predict this equation, we see that only the positions of the silver and sodium ions have been switched on the reactant and product sides. In switching the positions of the metal atoms, we were careful to write the new formulas properly, based on the charges of the ions.

EXERCISE 4.7

Predict the products obtained from the following pairs of ionic compounds. Then write a balanced chemical equation for each pair.

(a) KCl and $Pb(NO_3)_2$
(b) $CaCl_2$ and $MgSO_4$
(c) $NaOH$ and $Fe_2(SO_4)_3$

Solution

(a) The ions involved are K^+, Cl^-, Pb^{2+}, and $2NO_3^-$. The possible new combinations are KNO_3 and $PbCl_2$. Placing these into a reaction and balancing it gives

$$2KCl \ + \ Pb(NO_3)_2 \ \rightarrow \ 2KNO_3 \ + \ PbCl_2$$

(b) The ions involved are Ca^{2+}, $2Cl^-$, Mg^{2+}, and SO_4^{2-}. The two new compounds are $CaSO_4$ and $MgCl_2$. The balanced chemical reaction is

$$CaCl_2 \ + \ MgSO_4 \ \rightarrow \ CaSO_4 \ + \ MgCl_2$$

(c) The ions are Na^+, OH^-, $2Fe^{3+}$, and $3SO_4^{2-}$. Note that the charge of the iron is determined by calculation. The new compounds are Na_2SO_4 and $Fe(OH)_3$. The balanced chemical reaction is

$$6NaOH \ + \ Fe_2(SO_4)_3 \ \rightarrow \ 3Na_2SO_4 \ + \ 2Fe(OH)_3$$

We can write these chemical reactions, but the major question is whether or not a chemical reaction will actually occur if the compounds are mixed together in the laboratory. In the next sections some ways in which the chemist can predict if an actual reaction will occur are presented.

Chemical Driving Forces

So far, we can predict the products of any mixture of two ionic compounds. However, not all such mixtures react. Chemists rely on three fundamental principles to make an educated guess about the possibility for a reaction to occur in a double-replacement reaction. These principles are sometimes described as **driving forces**.

> **TIP**
>
> Formations of:
>
>   water
>   weak electrolyte
>   precipitate
>   gas
>
> are the main driving forces.

1. The formation of water is perhaps the strongest driving force. In an ionic reaction where water is a product, it is almost a certainty that a double-replacement reaction is occurring.
2. Formation of a precipitate (insoluble compound) is another indicator of a strong driving force.
3. The formation of a nonionic (covalent) compound from ionic reactants is another driving force. Many of these nonionic compounds are organic acids (ethanoic, formic, benzoic acids) or gases such as NH_3, SO_2, and CO_2.

Some common examples of driving forces are as follows:

$HCl(aq)$ $+ NaOH(aq)$ $\rightarrow H_2O(aq)$ $+ NaCl(aq)$ (water formed)
$Na_2SO_4(aq)$ $+ Ba(NO_3)_2(aq)$ $\rightarrow BaSO_4(s)$. $+ 2NaNO_3(aq)$ (precipitate formed)
$KC_2H_3O_2(aq)$ $+ HCl(aq)$ $\rightarrow HC_2H_3O_2(aq)$ $+ KCl(aq)$ (covalent compound formed)
$K_2SO_3(aq)$ $+ 2HNO_3(aq)$ $\rightarrow 2KNO_3(aq)$ $+ H_2O + SO_2(g)$ (gas formed)

In some reactions two of these driving forces may be present. As mentioned above, the driving force to form water is especially strong and will overcome another force that may be driving the reaction in the opposite direction. In the equation below, the formation of water overcomes the fact that CaO is a solid. Since CaO is on the reactant side of the equation, it is driving the equation toward the reactants. The production of water, however, is a stronger driving force, and the net result is that this reaction actually occurs:

$$CaO(s) \;+\; 2HCl(aq) \;\rightarrow\; CaCl_2(aq) \;+\; H_2O(l)$$

Net Ionic Equations

In the process of determining the products of a reaction between ionic compounds, the ions for each substance were determined. In fact, in aqueous solution only the soluble compounds appear as ions, while insoluble compounds (precipitates), gases, and covalent compounds are written as molecules in the equation. It is possible to take a balanced double-replacement reaction and convert it into a net ionic equation that shows the actual reactants, if any, for a given reaction.

For example, a simple neutralization reaction is

$$HCl(aq) \;+\; NaOH(aq) \;\rightarrow\; NaCl(aq) + H_2O(l)$$

The ionic reaction is obtained by writing all of the soluble ionic compounds as ions.

$$H^+(aq) + Cl^-(aq) + Na^+(aq) + OH^-(aq) \rightarrow Na^+(aq) + Cl^-(aq) + H_2O(l)$$

Since the Na^+ and the Cl^- are identical on both sides of the equation, they can be canceled to give the net ionic equation:

$$H^+(aq) \quad + \quad OH^-(aq) \quad \rightarrow \quad H_2O(l)$$

This allows the chemist to show that it is the H^+ and OH^- ions that are the active components of the reaction.

A reaction does not occur if potassium chloride and sodium nitrate solutions are mixed. We can demonstrate that no reaction occurs by deducing the reaction products as sodium chloride and potassium nitrate and then writing the net ionic equation. First we write the molecular equation

$$KCl(aq) + NaNO_3(aq) \rightarrow NaCl(aq) + KNO_3(aq) \quad \text{(molecular equation)}$$

Next, the ionic equation is written by separating each of the compounds into its ions:

$$K^+(aq) + Cl^-(aq) + Na^+(aq) + NO_3^-(aq) \rightarrow Na^+(aq) + Cl^-(aq) + K^+(aq) + NO_3^-(aq)$$
$$\text{(ionic equation)}$$

Finally, after identical ions are canceled from both sides of this equation, nothing remains. This means that there is no net ionic equation and no reaction occurs:

NO NET IONIC EQUATION POSSIBLE

Taking a close look at the molecular equation, we see also that no driving force is present. No water, no precipitate, no covalent molecule, and no gas is formed.

A common laboratory experiment is the determination of sulfate ions by precipitation with barium ions. The precipitate is carefully collected, dried, and weighed in this experiment. The reaction between potassium sulfate and barium nitrate may be predicted to produce barium sulfate and sodium nitrate. The balanced equation is

$$K_2SO_4(aq) \quad + \quad Ba(NO_3)_3(aq) \quad \rightarrow \quad BaSO_4(s) \quad + \quad 2KNO_3(aq)$$

The ionic equation is

$$2K^+(aq) + SO_4^{2-}(aq) + Ba^{2+}(aq) + 2NO_3^-(aq) \rightarrow BaSO_4(s) + 2K^+(aq) + 2NO_3^-(aq)$$

In the ionic equation the two potassium and two nitrate ions may be canceled, resulting in

$$SO_4^{2-}(aq) \ + \ Ba^{2+}(aq) \ \rightarrow \ BaSO_4(s)$$

This balanced net ionic equation represents the reaction implied above by the words "determination of sulfate ions by precipitation with barium ions." In chemical analysis, the chemist is usually interested in a specific ion, such as the sulfate ion in this example. The net ionic equation shows us how to isolate the sulfate ion from all other ions by precipitation with barium ions.

Reactions that evolve gases are a bit more complex. Archaeologists typically carry a small bottle of hydrochloric acid on field trips. Carbonate rocks can be quickly identified since they will give off carbon dioxide (evidenced by bubbling and fizzing) when a few drops of HCl are placed on them. Most of these rocks are made of calcium carbonate. Using the techniques for a double-replacement reaction, we may predict the products to be calcium chloride and carbonic acid:

$$2HCl(aq) \ + \ CaCO_3(s) \ \rightarrow \ CaCl_2(aq) \ + \ H_2CO_3(aq)$$

The ionic equation is

$$2H^+(aq) + 2Cl^-(aq) + CaCO_3(s) \rightarrow Ca^{2+}(aq) + 2Cl^-(aq) + H_2CO_3(aq)$$

In this reaction only the Cl^- ions will cancel:

$$2H^+(aq) \ + \ CaCO_3(s) \ \rightarrow \ Ca^{2+}(aq) \ + \ H_2CO_3(aq)$$

This equation does not show any carbon dioxide gas. The key is that H_2CO_3 may also be written as $CO_2 + H_2O$. When this is substituted, the final reaction is

$$2H^+(aq) \ + \ CaCO_3(s) \ \rightarrow \ Ca^{2+}(aq) \ + \ H_2O(l) \ + \ CO_2(g)$$

Table 4.5 lists the common gases and their equivalents when dissolved in water. The gas and aqueous forms are interchangeable in reactions as needed.

TABLE 4.5

Common Gases and Their Equivalents in Aqueous Solution

Gas Name	Aqueous Form	Gas Form
carbon dioxide	$H_2CO_3(aq)$	$CO_2(g) + H_2O(l)$
sulfur dioxide	$H_2SO_3(aq)$	$SO_2(g) + H_2O(l)$
hydrogen sulfide	$H_2S(aq)$	$H_2S(g)$
ammonia	$NH_4OH^*(aq)$	$NH_3(g) + H_2O(l)$

*NH_4OH does not actually exist, and should always be written as $NH_3(aq) + H_2O$.

EXERCISE 4.8

Write the balanced equation for each of the following pairs of ionic substances. Then write the ionic and net ionic equations for these reactions. Use (*aq*) to show soluble substances, (*s*) for insoluble compounds, (*l*) for liquids, and (*g*) for gases.

(a) $AgNO_3$ and $CaCl_2$
(b) Na_2CO_3 and $Fe(NO_3)_3$
(c) CaF_2 and HCl
(d) NH_4Cl and KOH

Solutions

(a) $2AgNO_3(aq) + CaCl_2(aq)$ $\rightarrow 2AgCl(s) + Ca(NO_3)_2(aq)$
$2Ag^+(aq) + 2NO_3^-(aq) + Ca^{2+}(aq) + 2Cl^-(aq)$ $\rightarrow 2AgCl(s) + Ca^{2+}(aq) + 2NO_3^-(aq)$
$Ag^+(aq) + Cl^-(aq)$ $\rightarrow AgCl(s)$

(b) $3Na_2CO_3(aq) + 2Fe(NO_3)_3(aq)$ $\rightarrow Fe_2(CO_3)_3(s) + 6NaNO_3(aq)$
$6Na^+(aq) + 3CO_3^{2-}(aq) + 2Fe^{3+}(aq) + 6NO_3^-(aq)$ $\rightarrow Fe_2(CO_3)_3(s) + 6Na^+(aq) + 6NO_3^-(aq)$
$3CO_3^{2-}(aq) + 2Fe^{3+}(aq)$ $\rightarrow Fe_2(CO_3)_3(s)$

(c) $CaF_2(aq) + 2HCl(aq)$ $\rightarrow 2HF(aq) + CaCl_2(aq)$
$Ca^{2+}(aq) + 2F^-(aq) + 2H^+(aq) + 2Cl^-(aq)$ $\rightarrow 2HF(aq) + Ca^{2+}(aq) + 2Cl^-(aq)$
$F^-(aq) + H^+(aq)$ $\rightarrow HF(aq)$

(d) $NH_4Cl(aq) + KOH(aq)$ $\rightarrow NH_4OH(aq) + KCl(aq)$
$NH_4^+(aq) + Cl^-(aq) + K^+(aq) + OH^-(aq)$ $\rightarrow NH_3(g) + H_2O(l) + K^+(aq) + Cl^-(aq)$
$NH_4^+(aq) + OH^-(aq)$ $\rightarrow NH_3(g) + H_2O(l)$

Note that the NH_4OH obtained from the double-replacement technique was replaced by $NH_3(g)$ and $H_2O(l)$ in the ionic equations since NH_4OH does not exist. It should not appear in the first equation either.

SINGLE-REPLACEMENT REACTIONS

A single-replacement reaction may be described as the reaction between an element and an ionic compound (or ions in solution) to form a different element and a new ionic compound. The products of these reactions, like those of double-replacement reactions, may be predicted. In one type of single-replacement reaction the element used as a reactant may be a metal that becomes a cation as a product. The second type of single-replacement reaction involves a nonmetal as the elemental reactant that then forms an anion. Several reactions in which the reacting metal forms a cation are shown below.

One of these reactions is

$$Cu(s) + 2AgNO_3(aq) \rightarrow 2Ag(s) + Cu(NO_3)_2(aq)$$

We can write the ionic equation by breaking the $AgNO_3$ and $Cu(NO_3)_2$ into their ions:

$$Cu(s) + 2Ag^+(aq) + 2NO_3^-(aq) \rightarrow 2Ag(s) + Cu^{2+}(aq) + 2NO_3^-(aq)$$

Cancelling the two nitrate ions from both sides gives the net ionic equation:

$$Cu(s) + 2Ag^+(aq) \rightarrow 2Ag(s) + Cu^{2+}(aq)$$

Active metals (i.e., the alkali metals) react with water. This is observed in the explosive reaction of potassium when it is placed in water.

$$2K(s) + H_2O(l) \rightarrow H_2(g) + 2KOH(aq)$$

We can write the ionic equation, which, since no ions cancel, is also the net ionic equation:

$$2K(s) + H_2O(l) \rightarrow H_2(g) + 2K^+(aq) + 2OH^-(aq)$$

Less active metals react with acids, as we observe with zinc and hydrochloric acid:

$$Zn(s) + 2HCl(aq) \rightarrow H_2(g) + ZnCl_2(aq)$$

the ionic equation is

$$Zn(s) + 2H^+(aq) + 2Cl^-(aq) \rightarrow H_2(g) + Zn^{2+}(aq) + 2Cl^-(aq)$$

and the net ionic equation is

$$Zn(s) + 2H^+(aq) \rightarrow H_2(g) + Zn^{2+}(aq)$$

In general, the metal reactant will form its ion, and the cation of the reactant will become the element. These reactions may be predicted when the metal reactant has only one possible cation. If, however, the metal can form several differently charged cations, as is true of lead, tin, or iron, the cation formed must be specified before the reaction can be completed.

Nonmetals that react to form anions are usually limited to the halogens. One of these reactions is

$$Cl_2(g) + 2KBr(aq) \rightarrow Br_2(l) + 2KCl(aq)$$

The net ionic reaction is

$$Cl_2(g) + 2Br^-(aq) \rightarrow Br_2(l) + 2Cl^-(aq)$$

In this reaction, the reactants Cl_2 and KBr are virtually colorless (Cl_2 is slightly yellow) and the product Br_2 produces a dark yellow or brown solution that allows us to directly observe that a reaction has occurred.

As we saw with double-replacement reactions, we can write equations for any mixture of an element and an ionic compound. To determine which reactions actually occur, we need to know whether a given element will displace an ion in a single-replacement reaction. Commonly this information is first given as the **activity series** of the elements. Later, when discussing oxidation-reduction reactions, we will find the same information in the table of standard reduction potentials. In this book we will use only an abbreviated table of standard reduction potentials (Table 4.6). We will describe the use of this table in more detail when discussing oxidation-reduction reactions. For the time being, it is important only to remember how to use this table effectively to predict whether or not single-replacement reactions will occur.

TABLE 4.6

Standard Reduction Potentials, 25 °C

Half-Reaction					$E°$ (volts)
$F_2(g)$ +		$2e^-$	→	$2F^-$	2.87
Co^{3+} +		e^-	→	Co^{2+}	1.82
Au^{3+} +		$3e^-$	→	Au	1.50
$Cl_2(g)$ +		$2e^-$	→	$2Cl^-$	1.36
$O_2(g)$ +	$4H^+$ +	$4e^-$	→	$2H_2O$	1.23
$Br_2(g)$ +		$2e^-$	→	$2Br^-$	1.07
$2Hg^{2+}$ +		$2e^-$	→	Hg_2^{2+}	0.92
Ag^+ +		e^-	→	Ag	0.80
Hg_2^{2+} +		$2e^-$	→	Hg	0.79
Fe^{3+} +		e^-	→	Fe^{2+}	0.77
I_2 +		$2e^-$	→	$2I^-$	0.53
Cu^+ +		e^-	→	Cu	0.52
Cu^{2+} +		$2e^-$	→	Cu	0.34
Cu^{2+} +		e^-	→	Cu^+	0.15
Sn^{4+} +		$2e^-$	→	Sn^{2+}	0.15
S +	$2H^+$ +	$2e^-$	→	H_2S	0.14
$2H^+$ +		$2e^-$	→	H_2	0.00
Pb^{2+} +		$2e^-$	→	Pb	−0.13
Sn^{2+} +		$2e^-$	→	Sn	−0.14
Ni^{2+} +		$2e^-$	→	Ni	−0.25
Co^{2+} +		$2e^-$	→	Co	−0.28
Tl^+ +		e^-	→	Tl	−0.34
Cd^{2+} +		$2e^-$	→	Cd	−0.40
Cr^{3+} +		e^-	→	Cr^{2+}	−0.41
Fe^{2+} +		$2e^-$	→	Fe	−0.44
Cr^{3+} +		$3e^-$	→	Cr	−0.74
Zn^{2+} +		$2e^-$	→	Zn	−0.76
Mn^{2+} +		$2e^-$	→	Mn	−1.18
Al^{3+} +		$3e^-$	→	Al	−1.66
Be^{2+} +		$2e^-$	→	Be	−1.70
Mg^{2+} +		$2e^-$	→	Mg	−2.37
Na^+ +		e^-	→	Na	−2.71
Ca^{2+} +		$2e^-$	→	Ca	−2.87
Sr^{2+} +		$2e^-$	→	Sr	−2.89
Ba^{2+} +		$2e^-$	→	Ba	−2.90
Rb^+ +		e^-	→	Rb	−2.92
K^+ +		e^-	→	K	−2.92
Cs^+ +		e^-	→	Cs	−2.92
Li^+ +		e^-	→	Li	−3.05

TIP

Reading and using the standard reduction potential table is second only to reading and using the periodic table.

TIP

The standard reduction potential table on the AP exam usually starts with the most negative $E°$ (i.e., it is upside down).

The table of standard reduction potentials supplied with the AP Chemistry Examination is usually arranged with the largest negative reduction potential at the top and the largest positive reduction potential at the bottom (the reverse of Table 4.6). On the exam, it is necessary only to locate the two reactants (one an element and the other an ion) in the net ionic equations listed in the table and then draw a line between the two. If the drawn line has a positive (upward) slope, the reaction will occur. If the line has a negative (downward) slope, the reaction will not occur. (In Table 4.6 a negative slope indicates a reaction will occur.) The reason why this procedure works is explained in Chapter 13 on oxidation-reduction reactions. It is instructive to take all of the single-replacement reactions above and verify, using Table 4.6, that each has a downward slope.

EXERCISE 4.9

A major portion of the free-response section of the AP test involves writing chemical equations from a description of the reactions. This exercise is typical of the AP questions and may be answered using the information given in this chapter.

Write the chemical formulas (ions or molecules as appropriate) for each of the following. Each reaction does occur. Where appropriate, write net ionic equations. Equations must be balanced, and an aqueous solution is assumed unless otherwise stated.

(a) A saturated solution of Br_2 in water is added to a solution containing potassium iodide.
(b) A piece of magnesium metal is placed in a solution of sulfuric acid.
(c) Sodium bromide solution is added to a silver perchlorate solution.
(d) Iron(III) nitrate solution is mixed with a potassium sulfite solution.
(e) Solid magnesium carbonate is reacted with hydrochloric acid.
(f) Ammonium chloride solution is added to a barium hydroxide solution.
(g) Chromium(III) chloride solution reacts with solid magnesium.
(h) Sodium carbonate solution is mixed with aluminum sulfate.
(i) Calcium hydroxide and sodium phosphate solutions are mixed.
(j) Carbon dioxide is bubbled into a solution of iron(III) nitrate.
(k) Solid lead(IV) oxide reacts with hydrochloric acid.
(l) Ethyl alcohol is burned in excess oxygen.
(m) Hydrogen sulfide gas is bubbled into a solution containing cobalt(II) chloride.
(n) Solid calcium oxide reacts with sulfuric acid.
(o) Chlorine gas is bubbled into a solution of potassium iodide.

Solutions

The following reactions are balanced. Each solution gives the formulas for the reactants, the molecular equation, and the correct ionic equation if needed. The symbol (*aq*) is omitted for soluble ions and molecules. The gas, liquid, and solid states are noted with (*g*), (*l*), and (*s*), respectively.

(a) Reactants: Br_2 and KI

$$Br_2 + 2KI \rightarrow I_2 + 2KBr$$
$$Br_2 + 2I^+ \rightarrow I_2 + 2Br^-$$

(b) Reactants: Mg and H_2SO_4

$$Mg(s) + H_2SO_4 \rightarrow MgSO_4 + H_2(g)$$
$$Mg(s) + 2H^+ \rightarrow Mg^{2+} + H_2(g)$$

(c) Reactants: NaBr and $AgClO_4$

$$NaBr + AgClO_4 \rightarrow AgBr(s) + NaClO_4$$
$$Br^- + Ag^+ \rightarrow AgBr(s)$$

(d) Reactants: $Fe(NO_3)_3$ and K_2SO_3

$$2Fe(NO_3)_3 + 3K_2SO_3 \rightarrow Fe_2(SO_3)_3(s) + 6KNO_3$$
$$2Fe^{3+} + 3SO_3^{2-} \rightarrow Fe_2(SO_3)_3(s)$$

(e) Reactants: $MgCO_3$ and HCl

$$MgCO_3(s) + 2HCl \rightarrow MgCl_2 + H_2CO_3$$

(Remember that H_2CO_3 can be written as $H_2O + CO_2$.)

$$MgCO_3(s) + 2H^+ \rightarrow Mg^{2+} + H_2O + CO_2(g)$$

(f) Reactants: NH_4Cl and $Ba(OH)_2$

$$2NH_4Cl + Ba(OH)_2 \rightarrow BaCl_2 + 2NH_4OH$$

(Remember NH_4OH should always be written as $NH_3(g) + H_2O$.)

$$NH_4^+ + OH^- \rightarrow NH_3(g) + H_2O$$

(g) Reactants: $CrCl_3$ and Mg

$$2CrCl_3 + 3Mg(s) \rightarrow 3MgCl_2 + 2Cr(s)$$
$$2Cr^{3+} + 3Mg(s) \rightarrow 3Mg^{2+} + 2Cr(s)$$

(h) Reactants: Na_2CO_3 and $Al_2(SO_4)_3$

$$3Na_2CO_3 + Al_2(SO_4)_3 \rightarrow 3Na_2SO_4 + Al_2(CO_3)_3(s)$$
$$3CO_3^{2-} + 2Al^{3+} \rightarrow Al_2(CO_3)_3(s)$$

(i) Reactants: $Ca(OH)_2$ and Na_3PO_4

$$3Ca(OH)_2 + 2Na_3PO_4 \rightarrow Ca_3(PO_4)_2(s) + 6NaOH$$
$$3Ca^{2+} + 2PO_4^{3-} \rightarrow Ca_3(PO_4)_2(s)$$

(j) Reactants: CO_2, H_2O, and $Fe(NO_3)_3$

$$CO_2(g) + 3H_2O + 2Fe(NO_3)_3 \rightarrow Fe_2(CO_3)_3(s) + 6HNO_3$$
$$CO_2(g) + 3H_2O + 2Fe^{3+} \rightarrow Fe_2(CO_3)_3(s) + 6H^+$$

(k) Reactants: PbO_2 and HCl

$$PbO_2(s) + 4HCl \rightarrow PbCl_4 + 2H_2O$$
$$PbO_2(s) + 4H^+ \rightarrow Pb^{4+} + 2H_2O$$

(l) Reactants: CH_3CH_2OH and O_2

$$CH_3CH_2OH + 3O_2(g) \rightarrow 2CO_2(g) + 3H_2O$$

No ionic reaction is possible.

(m) Reactants: H_2S and $CoCl_2$

$$H_2S(g) + CoCl_2 \rightarrow CoS(s) + 2HCl$$
$$H_2S(g) + Co^{2+} \rightarrow CoS(s) + 2H^+$$

(n) Reactants: CaO and H_2SO_4

$$CaO(s) + H_2SO_4 \rightarrow CaSO_4(s) + H_2O$$
$$CaO(s) + 2H^+ + SO_4^{2-} \rightarrow CaSO_4(s) + H_2O$$

(o) Reactants: Cl_2 and KI

$$Cl_2(g) + 2KI \rightarrow I_2 + 2KCl$$
$$Cl_2(g) + 2I^- \rightarrow I_2 + 2Cl^-$$

SUMMARY

This chapter focuses on ionic compounds and their reactions. Ionic compounds have simple, often predictable formulas based on information in the periodic table. The structure of ions that make up ionic compounds is presented, and we note that ions of the representative elements are often isoelectric with noble gases. The nature of other ions can also be understood based on electronic structures that are stable. Ionic compounds also react in a process called double replacement as long as a precipitate, a gaseous product, or a weak electrolyte is part of the chemical equation. Ionic reactions are often best represented as balanced net ionic equations. This chapter introduces a set of solubility rules and also a summary of how reactions may be classified. A logical method for naming ionic compounds is also presented.

Important Concepts

Predicting ionic charges
Balancing reactions
Writing ionic formulas and electroneutrality of compounds
Writing ionic and net ionic equations
Naming ionic compounds
Solubility rules
Polyatomic ions

Practice Exercises

Multiple-Choice

For the first four problems below, one or more of the following responses will apply; each response may be used more than once or not at all in these questions.

   I. ammonium ion
  II. calcium ion
 III. bromide ion
  IV. iron(III)
   V. phosphate ion

1. Which of these carries a single negative charge?

    (A) I
    (B) II
    (C) III
    (D) IV
    (E) V

2. Which pair of these will form a compound?

    (A) I and II
    (B) II and IV
    (C) III and V
    (D) I and IV
    (E) II and V

3. How many different compounds can be made from these ions?

    (A) 1
    (B) 2
    (C) 4
    (D) 6
    (E) 9

4. Which compound will be insoluble?

    (A) I and V
    (B) III and I
    (C) II and III
    (D) III and IV
    (E) IV and V

5. Which of the following compounds is soluble?

    (A) $MgCO_3$
    (B) $Al(OH)_3$
    (C) Cr_2S_3
    (D) K_2CrO_4
    (E) $NiSO_3$

6. Which of the following compounds is insoluble?

    (A) $Ca(OH)_2$
    (B) Fe_2S_3
    (C) Na_2CO_3
    (D) H_2SO_3
    (E) $AuCl_3$

7. Which of the following is the permanganate ion?

    (A) ClO_4^-
    (B) PO_4^{3-}
    (C) MnO_2
    (D) SO_4^{2-}
    (E) MnO_4^-

8. When ammonium oxalate, $(NH_4)_2C_2O_4$, is dissolved in water, the ions formed are

   (A) $2N^{3-}(aq) + 8H^+(aq) +$ $2C^{4+}(aq) + 4O^{2-}(aq)$
   (B) $(NH_4)^{2+}(aq) + C_2O_4^{2-}(aq)$
   (C) $2NH_4^+(aq) + C_2O_4^{2-}(aq)$
   (D) $NH_4^{2+}(aq) + C_2O_4^{2-}(aq)$
   (E) $2NH_4^+(aq) + 2CO_2^-(aq)$

9. The correct name for $Fe(NO_3)_3$ is

   (A) iron nitrite
   (B) iron(II) nitrate
   (C) ferrous nitroxide
   (D) iron(III) sulfate
   (E) iron(III) nitrate

10. When the combustion reaction for benzene, C_6H_6, is properly balanced with the smallest whole-number coefficients, the sum of the coefficients is

    (A) 15
    (B) 12
    (C) 35
    (D) 17.5
    (E) 12

11. The potassium ion is isoelectronic to which noble gas?

    (A) He
    (B) Ne
    (C) Ar
    (D) Kr
    (E) Xe

12. Which two atoms will form isoelectronic ions?

    (A) Cl and Na
    (B) Cl and F
    (C) Na and F
    (D) S and Br
    (E) Fe and Ca

13. What is the formula for an ionic compound formed from aluminum and chlorine?

    (A) AlCl
    (B) Al_3Cl
    (C) Al_3Cl_3
    (D) $AlCl_3$
    (E) Al_2Cl_3

14. The electronic configuration $1s^2\ 2s^2\ 2p^6\ 3s^2\ 3p^6$ corresponds to the electronic configuration of

    (A) S^{2-}
    (B) Ca^{2+}
    (C) Cl^-
    (D) K^+
    (E) all of these

15. Which of the following is NOT correctly named?

    (A) Cl^- chloride ion
    (B) ClO^- hypochlorite ion
    (C) ClO_4^- perchlorate ion
    (D) ClO_2^- chlorous ion
    (E) ClO_3^- chlorate ion

16. Which of the following is NOT a correct chemical formula?

    (A) $SrBr_2$
    (B) Ca_2O_3
    (C) Mg_3N_2
    (D) Na_2S
    (E) AlI_3

17. Which of the following is a correct formula?

    (A) NH_4SO_3
    (B) $CaC_2H_3O_2$
    (C) Na_2ClO_4
    (D) $Ba(CO_3)_2$
    (E) KH_2PO_4

18. Which of the following is a single-replacement reaction?

    (A) sodium chloride with potassium nitrate
    (B) chlorine gas with sodium metal
    (C) aluminum metal with hydrobromic acid
    (D) ethanol with oxygen
    (E) magnesium oxide with sulfur trioxide

19. Which of the following is NOT true of a net ionic equation?

    (A) All of the nonreacting (spectator) ions have been canceled.
    (B) It shows the actual reactants in an equation.
    (C) It allows the chemist to substitute reactants in a logical manner.
    (D) It is used to determine which compounds are insoluble.
    (E) It must have the charges as well as the atoms balanced.

20. What is the balanced molecular equation when $FeCl_3(aq)$ is mixed with $Ba(OH)_2(aq)$?

    (A) $Ba(OH)_2(aq) + FeCl_3(aq) \rightarrow Fe(OH)_2(aq) + BaCl_3(aq)$
    (B) $2Ba(OH)_2(aq) + 3FeCl_3(aq) \rightarrow 2BaCl_2(aq) + 3Fe(OH)_3(s)$
    (C) $3Ba(OH)_2(aq) + 2FeCl_3(aq) \rightarrow 3BaCl_2(s) + 2Fe(OH)_3(s)$
    (D) $3Ba(OH)_2(aq) + 2FeCl_3(aq) \rightarrow 3BaCl_2(aq) + 2Fe(OH)_3(s)$
    (E) $Ba(OH)_2(aq) + 2FeCl_3(aq) \rightarrow BaCl_2(aq) + 2Fe(OH)_3(s)$

21. What is the net ionic equation for the following reaction that takes place in aqueous solution?

    $CrCl_3 + Pb(NO_3)_2 \rightarrow PbCl_2 + Cr(NO_3)_2$

    (A) $Cr^{2+}(aq) + Pb^{2+}(aq) \rightarrow CrPb(s)$
    (B) $Cr^3(aq) + 3NO_3(aq) \rightarrow Cr(NO_3)_3(s)$
    (C) $Pb^{2+}(aq) + 2Cl^-(aq) \rightarrow PbCl_2(s)$
    (D) $Pb(NO_3)(s) + 2Cl^-(aq) \rightarrow PbCl_2(s) + 2 NO_3(aq)$
    (E) $2CrCl_3(s) + 3Pb^{2+}(aq) \rightarrow 2Cr^{3+}(aq) + 3PbCl_2(s)$

ANSWER KEY

1. C	4. E	7. E	10. C	13. D	16. B	19. D
2. E	5. D	8. C	11. C	14. E	17. E	20. D
3. D	6. B	9. E	12. C	15. D	18. C	21. C

See Appendix 1 for explanations of answers.

Free-Response

Answer the following questions considering the chemical and physical properties of ionic substances.

(a) A large majority of the monatomic ions of the representative elements all have one feature in common. What is that feature? Give a specific example to illustrate it.

(b) Write the molecular equation, the ionic equation, and the net ionic equation for the aqueous reaction of potassium chloride with lead nitrate. Indicate the phase of each of the reactants and products.

(c) What is the name for $Cr(NO_3)_3$ and what is the formula for copper(II) sulfate pentahydrate?

(d) Although it has not been mentioned, by analogy, what is the formula for the iodate ion?

(e) What is potassium permanganate used for? Does it have distinguishing physical properties?

ANSWERS

(a) Monatomic ions of the representative elements contain a single atom of an element in the s-block or p-block of the periodic table. Most of these ions are formed be gaining or losing electrons to become isoelectronic with its nearest noble gas.

(b) The molecular equation writes the complete formulas of the reactants and products as

$$2\ KCl(aq) + Pb(NO_3)_2(aq) \rightarrow PbCl_2(s) + 2\ KNO_3(aq)$$

The ionic equation writes all soluble species as their aqueous ions:

$$2\ K^+(aq) + 2\ Cl^-(aq) + Pb^{2+}(aq) + 2\ NO_3^-(aq) \rightarrow$$
$$PbCl_2(s) + 2\ K^+(aq) + 2\ NO_3^-(aq)$$

Finally, the net ionic equation is obtained by canceling the potassium and nitrate ions to get

$$2\ Cl^-(aq) + Pb^{2+}(aq) \rightarrow PbCl_2(s)$$

(c) $Cr(NO_3)_3$ contains Cr^{3+} and three NO_3^- ions. Chromium is a transition element (d-block) and has multiple oxidation states, we also identify the NO_3^- ion as the nitrate ion. Therefore we must use the Stock system of naming. We start by converting the 3+ charge to (III) and adding it to the name of the metal to get chromium(III). Because we have a polyatomic ion, all we have to do is add its name to the metal to get chromium(III) nitrate.

The second compound is copper(II) sulfate pentahydrate. This contains the copper(II) ion or Cu^{2+}. It also contains the sulfate ion, SO_4^{2-}. The charges, 2+ and 2– add to zero so the main part of the formula is simply $CuSO_4$. The name includes pentahydrate. Penta translates as 5, and hydrate indicates water molecules and we have $5H_2O$ in the formula also. Combined it is written as $CuSO_4 \cdot 5H_2O$.

(d) We have named the chlorine polyatomic ions as hypochlorite, ClO^-, chlorite, ClO_2^-, chlorate, ClO_3^-, and perchlorate, ClO_4^-. Bt substituting iodine for each chlorine we can obtain the formulas for the hypoiodite, iodite, iodate, and periodate ions. Therefore the iodate ion is IO_3^-.

(e) Potassium permanganate is $KMnO_4$. It is a common oxidizing agent used in many redox reactions. In fact, it is one of the strongest oxidizing agents. By virtue of the presence of a potassium ion we can deduce that this compound must be soluble in aqueous solution. In addition, $KMnO_4$ is highly colored with a characteristic purple color.

CHAPTER 5

Covalent Compounds, Formulas, and Structure

- Covalent Molecules
- Covalent Bonds
- Lewis Structures
- Octet Rule
- Multiple Bonds
- Resonance Structures
- Formal Charges
- Benzene Structure
- Electronegativity
- Bond Polarity
- Ionic Character
- Bond Strength
- Bond Energy
- Bond Length
- Bond Order
- Naming Molecules

- VSEPR Theory
- Molecular Geometry
- Planar Triangle
- Tetrahedron
- Triangular Bipyramid
- Octahedron
- Bond Angles
- Molecular Polarity
- Valence Bond Theory
- Hybrid Orbitals
- sp Hybrid Orbitals
- sp^2 Hybrid Orbitals
- sp^3 Hybrid Orbitals
- sp^3d Hybrid Orbitals
- sp^3d^2 Hybrid Orbitals
- Sigma and Pi Bonds

COVALENT MOLECULES

Formation of **covalent bonds** represents another way in which elements may combine to form compounds and at the same time attain a noble-gas electronic configuration. In contrast to ionic bonding, where electrons are transferred from atom to atom, covalent bonding occurs when electrons are shared between two or more atoms. In addition, whereas ionic bonding is simply electrical attraction of oppositely charged ions, in covalent compounds the atoms are physically attached to each other. Compounds composed of covalently bonded groups of atoms are called **molecules.**

To conveniently show this sharing of electrons, chemists draw structures of covalent molecules using **Lewis electron-dot structures.** In the Lewis representation the outermost s and p electrons (valence electrons) are shown as dots arranged around the atomic symbol. **Lewis structures** for 18 elements are shown in Figure 5.1.

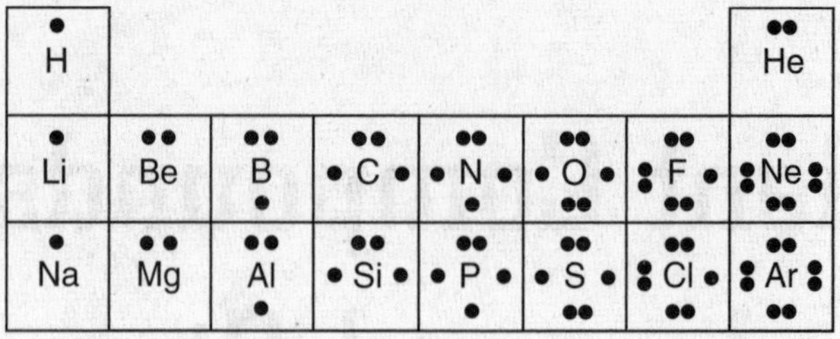

FIGURE 5.1. Elements normally represented with Lewis electron-dot symbols. (Sometimes the paired electrons in Be, B, C, Mg, Al, and Si are separated.)

When electrons are shared between two atoms, one atom donates one electron and the other atom donates the second electron. The shared pair of electrons represents a covalent bond. If two pairs of electrons are shared between two atoms, a double bond exists. When two atoms share three pairs of electrons, a triple bond exists.

The sharing of electrons follows the same basic principle as prevails in the formation of ions. This fact means that the atoms are trying to attain noble-gas electronic configurations. Referring to the noble-gas configurations, we see that there are two *s* electrons and six *p* electrons in complete sublevels of each noble gas. These eight electrons represent the octet of the octet rule, which governs covalent compounds. The **octet rule** states that the noble-gas configuration will be achieved if the Lewis structure shows eight electrons around each atom. Hydrogen is an exception; its "octet" is two electrons, corresponding to the two outermost electrons in the noble gas helium.

Figure 5.2 shows the Lewis electron dot structures for the **diatomic** gases. In these structures the electrons of one atom are represented as dots and those of the other atom as circles in order to illustrate that each atom contributes the same number of electrons to the bond.

FIGURE 5.2. Lewis structures of the diatomic elements.

In these molecules, each atom considers the shared electrons (those electrons between the two atoms) as its own. Each hydrogen in H_2 thinks it has two electrons and an electronic configuration the same as that of helium. Nitrogen molecules have triple bonds with three pairs of shared electrons. Oxygen molecules may be represented with a double bond. Fluorine and chlorine look very similar, as expected, since both are halogens.

Molecular oxygen has a simple Lewis structure, shown in Figure 5.2, which turns out to be incorrect. Experimental evidence shows that oxygen is paramagnetic and must contain unpaired electrons. More sophisticated molecular orbital methods must be used to obtain the correct bonding showing these unpaired electrons.

Magnetic properties of atoms, ions, and molecules arise from unpaired electrons. The electron spin generates a small magnetic field. When electrons are paired, these magnetic fields cancel, and when unpaired, they impart magnetic properties to the substance. The more unpaired electrons, the stronger the magnetic effect. Substances that have unpaired electrons are said to be **paramagnetic**. They are attracted toward an external magnet, and the force of the attraction is proportional to the number of unpaired electrons. **Diamagnetic** substances have all of their electrons paired, and they exhibit no magnetic properties.

LEWIS STRUCTURES OF MOLECULES

Lewis structures of other molecules and polyatomic ions may be drawn using the basic octet rule. For larger molecules it is first necessary to determine the general arrangement of the atoms, often called the "skeleton." In most molecules this skeleton consists of a central atom with surrounding atoms bonded to it. A few general rules apply to determining the skeleton.

1. Carbon is usually a central atom in the structure. In compounds with more than one carbon atom, the carbon atoms are joined in a chain to start the skeleton.
2. Hydrogen is never a central atom because it can form only one covalent bond.
3. Halogens form only a single covalent bond when oxygen is not present, and therefore a halogen will generally not be a central atom.
4. Oxygen forms only two covalent bonds and is rarely a central atom. However, it may link two carbon atoms in a carbon chain.
5. In the simpler molecules, the atom that appears only once in the formula will often be the central atom.

Once the molecular skeleton is determined, the available valence electrons must be arranged in octets around each atom. This is done in steps as follows:

1. The valence electrons of all atoms are added together.
2. If the substance is a polyatomic ion, we must take into account the electrons used to form the ion. For anions, the charge represents additional electrons that must be added to the total of the valence electrons. For cations, the charge represents missing electrons that must be subtracted from the valence electrons.
3. A pair of electrons is placed between each two atoms in the skeletal structure to represent a covalent bond. These electrons are called the **bonding pairs.**
4. The remaining electrons are used to complete the octets of all outer atoms in the skeleton. These electrons are **nonbonding pairs** of electrons (also called **lone pairs**).
5. If any electrons are left over, they are added in pairs to the central atom. These electrons are also nonbonding pairs (lone pairs).
6. When all electrons have been placed, the outer atoms will all have octets. The central atom may have an octet, or it may have more or fewer than eight electrons.
   a. If the central atom has an octet, the structure is complete (see the section on formal charges, page 187).
   b. It is all right if the central atom has fewer than eight electrons, provided that the atom is boron. For other central atoms, double bonds must be constructed to obtain an octet. This is done by taking a nonbonding pair of electrons from an outer atom and placing them as a bonding pair to

TIP

Drawing and interpreting Lewis structures is the key to understanding many chemical and physical properties of covalent compounds.

make a double bond. Enough double bonds are constructed to give the central atom an octet.

c. The central atom may have more than eight electrons (as in PCl_5 shown in Figure 5.7) only if it is in period 3–7 of the periodic table. If the central atom is in period 2, it cannot have more than an octet of electrons.

For example, we may construct the Lewis diagram for methane, CH_4, in the following manner. First we draw the skeleton with carbon in the center and the hydrogen atoms arranged symmetrically around it. Next we count the valence electrons. There are four valence electrons on the carbon atom and one on each of the four hydrogen atoms, for a total of eight electrons. In the next step we add the bonding pairs of electrons to the structure. This uses up all of the electrons, and we check to see whether each atom has an octet. (Remember that for hydrogen an "octet" is simply one pair of electrons.) Finally, to simplify the structure, we can replace the bonding pairs of electrons by a line representing a bond. These steps are shown in Figure 5.3.

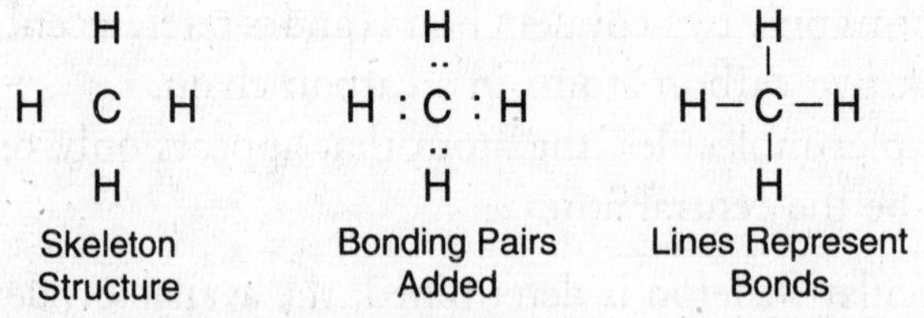

FIGURE 5.3. **Steps used to construct the Lewis structure for methane. Because of its simplicity, the structure is complete once the bonding pairs are added.**

The most common **electron-deficient** molecules involve boron. The structure of boron trifluoride, BF_3, is constructed as illustrated in Figure 5.4.

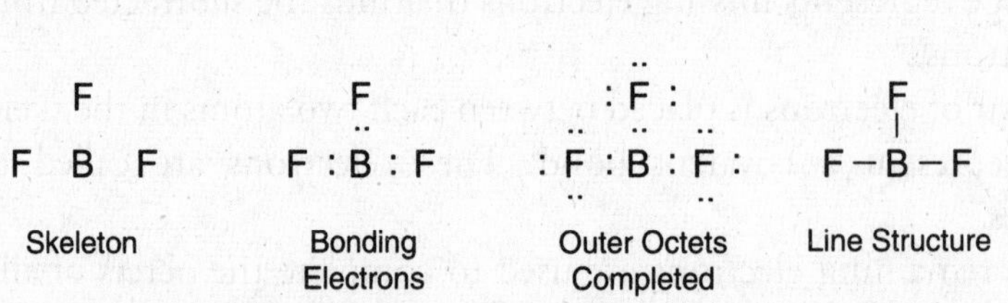

FIGURE 5.4. **Steps used to construct the BF_3 molecule. Compared to methane, this construction requires the additional step of constructing octets around the fluorine atoms.**

The skeleton is arranged with boron as the central atom. Adding up the valence electrons, we get $3 + 7 + 7 + 7 = 24$ electrons for the boron and three fluorine atoms. Next, bonding pairs are added between each fluorine and the boron atom, using six electrons and leaving 18 to be placed. Six more electrons are placed around each fluorine atom to complete its octet. All of the remaining electrons are now utilized. Since boron is commonly found with an electron-deficient structure (i.e., with less than an octet of electrons), the structure is complete. The line structure may be drawn for simplicity. The nonbonding pairs of the outer atoms are generally not shown in a line structure.

Phosphorus forms two compounds with chlorine, PCl_3 and PCl_5. Examining their Lewis structures, we find first that PCl_3 is constructed as shown in Figure 5.5.

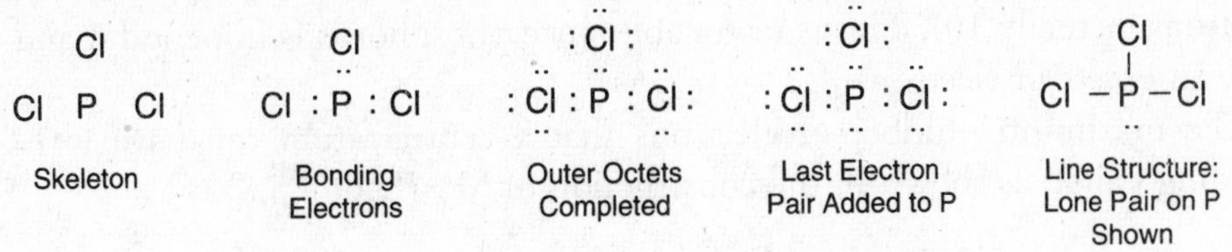

FIGURE 5.5. **Steps used to construct the Lewis structure for PCl₃. Compared to BCl₃, this construction requires the addition of a pair of nonbonding electrons to the central atom in step 4.**

As before, the skeleton is drawn and the valence electrons counted. There are $5 + 7 + 7 + 7 = 26$ electrons. The bonding electron pairs and the outer octets are completed in the next two steps, leaving two electrons unused. These are placed on the central phosphorus atom as shown in Figure 5.5, completing its octet. All atoms are checked to see that each has an octet of electrons. One way to do this is to draw a circle around all of the electrons adjacent to each atom, as shown in Figure 5.6.

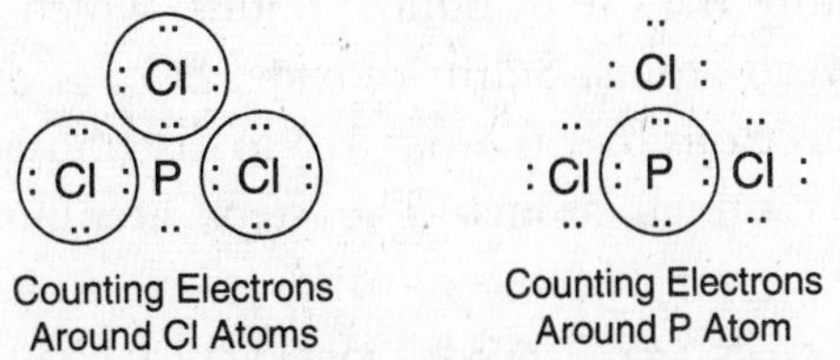

FIGURE 5.6. **Diagram illustrating how to count electrons around each atom by drawing a circle that includes all nonbonding electrons and all bonding electrons for that atom.**

> **TIP**
>
> Circling electrons helps identify octets.

Finally, the line structure may be used. Since nonbonding pairs on the central atom are important in determining the molecular geometry, they are shown in the line structure (see the last structure in Figure 5.5).

In constructing the PCl_5 molecule, we observe some differences, as shown in Figure 5.7.

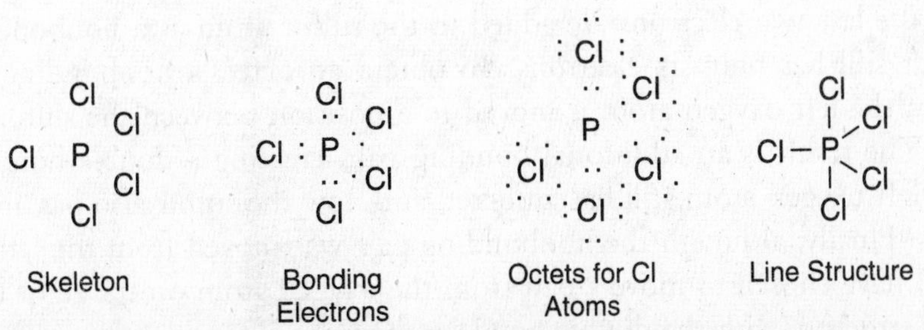

FIGURE 5.7. **Steps used to construct the PCl₅ molecule. The phosphorus atom has 10 bonding electrons (five pairs), which are allowed since phosphorus is in period 3 of the periodic table.**

There are a total of 40 valence electrons in PCl$_5$, which need to be placed. The skeleton is drawn, bonding pairs are added, and then the octets for chlorine are completed. This step uses up all 40 electrons, and the line structure may be drawn for clarity. Notice that the phosphorus atom has more than an octet of electrons (actually 10). This is reasonable since phosphorus is in period 3 and may have an excess of electrons.

The maximum number of electrons that a central atom can have is 12 (six bonding pairs), as shown in the construction of SF$_6$ (Figure 5.8).

| Skeleton | Bonding Pairs Added | Octets Around F Atoms | Line Structure |

FIGURE 5.8. **Steps used to construct the SF$_6$ molecule. This structure shows 12 bonding electrons (six pairs) around the sulfur atom.**

Multiple Covalent Bonds

Some compounds require the use of double bonds, which are represented by two pairs of electrons between atoms. Sulfur dioxide, SO$_2$, is one of those substances. There are 18 valence electrons ($6 + 6 + 6 = 18$) to distribute on the O S O skeleton so as to obtain octets on all atoms. The steps are illustrated in Figure 5.9.

| Skeleton | Bonding e$^-$ Added | Octets Around O Completed |

| Nonbonding Pair Added to S | Nonbonding Pair from Left O Are Moved to Make a Bonding Pair | Line Structure |

FIGURE 5.9. **Steps used to construct the Lewis structure for the SO$_2$ molecule. This involves the formation of a double bond as shown in the fifth diagram.**

TIP

SO$_2$ and NO$_2$ are not linear molecules. The section on molecular geometry describes how to determine actual shapes of molecules.

After the last two electrons are added to the sulfur atom as a nonbonding pair, the sulfur still has only six electrons. To obtain an octet, a nonbonding electron pair from the left oxygen atom is moved to a position between the sulfur and the oxygen. The result is an additional bonding pair, creating a double bond. Notice that the left oxygen atom still has an octet, and now the sulfur also has an octet of electrons. Finally, although the nonbonding pair was moved from the left oxygen, we could have chosen to move a pair from the oxygen atom on the right to obtain a similar structure with the double bond on the right. These two, equally probable structures (Figure 5.10) are called **resonance structures** (see page 190).

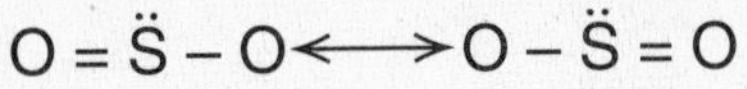

$$O = \ddot{S} - O \longleftrightarrow O - \ddot{S} = O$$

FIGURE 5.10. The two resonance structures of SO_2. These molecules differ only in the position of the double bond.

In addition, SO_2 is shown here as a linear molecule. See the Molecular Geometry section on page 198 to see how the correct shape is deduced.

Some molecules contain triple bonds. Common examples are nitrogen, N_2, acetylene, C_2H_2, and the cyanide ion, CN^-. A triple bond is formed by first constructing a double bond as described above. If the atoms still do not have octets, a second pair of nonbonding electrons may be moved to a bonding position.

Lewis Structures of Ions

In addition to the Lewis structures for molecules, we may also draw Lewis structures for covalently bonded polyatomic ions. The methods are the same as those for molecules except that we must account for the electrons that give an ion its charge. One electron is added to the valence electrons for each negative charge on an ion and one electron is subtracted from the valence electrons for each positive charge. The structure of the nitrite ion, NO_2^-, is an example. As with molecules, we count the valence electrons, $6 + 6 + 5 = 17$. The negative charge adds one electron for a total of 18. As shown in Figure 5.11, we construct the skeleton, add the bonding electrons, complete the octets around the oxygen atoms, add one nonbonding pair to nitrogen to use the remaining electrons, and finally construct a double bond so that all atoms have octets. When writing Lewis structures for ions, it is necessary to enclose the ion in brackets and indicate the charge of the ion as shown.

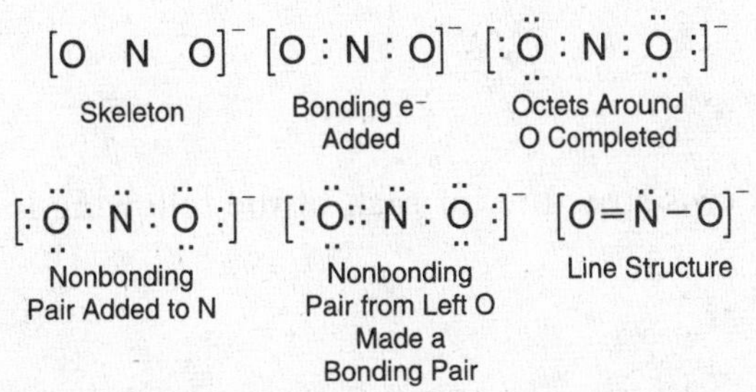

$$[O \ \ N \ \ O]^- \quad [O : N : O]^- \quad [:\ddot{O}:N:\ddot{O}:]^-$$

Skeleton	Bonding e⁻ Added	Octets Around O Completed

$$[:\ddot{O}:\ddot{N}:\ddot{O}:]^- \quad [:\ddot{O}::\ddot{N}:\ddot{O}:]^- \quad [O=\ddot{N}-O]^-$$

Nonbonding Pair Added to N	Nonbonding Pair from Left O Made a Bonding Pair	Line Structure

FIGURE 5.11. Steps used to construct the Lewis structure of the nitrite ion.

We notice a distinct similarity in the structures of the SO_2 molecule and the NO_2^- ion. The electrons are arranged in exactly the same manner. Since the nitrogen atom has one less electron than sulfur does, the nitrite ion requires an extra electron, which results in its -1 charge. As with the SO_2, the linear shape of the NO_2^- shown here is incorrect. The next step is to determine the proper shape (see pages 198–205).

Summary: By now, we see a pattern developing for the construction of Lewis structures. The table below summarizes the logical steps involved in obtaining a Lewis structure.

TABLE 5.1

Steps Used to Construct Lewis Structures

1. Arrange atoms in a skeleton structure. Remember that H and halogens cannot be central atoms and the oxo-anions and oxo-acids have H bonded to the oxygen atoms.

2. Count all valence electrons of all atoms. For anions, add one electron for each negative charge. For cations, subtract one electron for each positive charge. Remember to bracket ions with the overall charge of the ion.

3. Add pairs of electrons between atoms in the skeleton structure to form (sigma) bonds.

4. Then add electrons to all atoms to complete their octets (recall that an octet for H = 2 and for B = 6).

5. (a) If all atoms have octets and electrons are still left, add the extra electrons to the central atom as nonbonding (lone) pairs.

   (b) If all electrons have been added and the central atom does not have an octet, move a pair of nonbonding electrons from an outer atom to make a double bond. Repeat this step as needed.

EXERCISE 5.1

1. Construct the Lewis structure for each of the following compounds.

   (a) CH_3Cl
   (b) CS_2
   (c) PH_3
   (d) SiF_4
   (e) H_2S

2. Construct the Lewis structure for each of the following ions.

   (a) NO_3^-
   (b) CO_3^{2-}
   (c) PO_4^{3-}
   (d) SO_3^{2-}
   (e) ClO_4^-

Solution

1. (a)

$$H : \overset{\overset{\displaystyle H}{|}}{\underset{\underset{\displaystyle H}{|}}{C}} : \overset{..}{\underset{..}{Cl}} :$$

(c)

$$H : \overset{\overset{\displaystyle H}{}}{\underset{..}{P}} : H$$

(e)

$$H : \overset{..}{\underset{..}{S}} : H$$

(b) $: \overset{..}{S} :: C :: \overset{..}{S} :$

(d)

$$: \overset{..}{\underset{..}{F}} : \overset{\overset{\displaystyle : \overset{..}{F} :}{}}{\underset{\underset{\displaystyle : \overset{..}{F} :}{}}{Si}} : \overset{..}{\underset{..}{F}} :$$

2. (a) $[\,\mathrm{NO_3}\,]$ (c) $[\,\mathrm{PO_4}\,]^{3-}$ (e) $[\,\mathrm{ClO_3}\,]^{-}$

   (b) $[\,\mathrm{CO_3}\,]^{2-}$ (d) $[\,\mathrm{SO_3}\,]^{2-}$

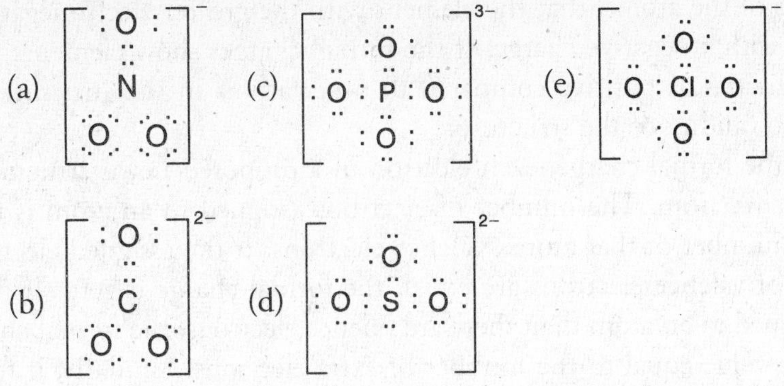

Lewis Structures of Odd Electron Compounds

Some compounds have formulas in which the total number of valence electrons is an odd number. In such cases it is impossible to construct a Lewis structure with an octet around each atom. Nitrogen dioxide, NO_2, is one such compound. One possible Lewis structure is shown in Figure 5.12.

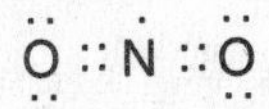

$$\ddot{\mathrm{O}} :: \mathrm{N} :: \ddot{\mathrm{O}}$$

FIGURE 5.12.   Lewis structure of the NO$_2$ molecule, showing the unpaired electron on the nitrogen atom.

Molecules that have Lewis structures with an unpaired electron are often called **free radicals.** The unpaired electron makes the molecule unusually reactive. Free radicals have been implicated in such biological processes as aging and cancer. In an effort to pair up the single electron, free radicals may also form **dimers** or pairs of molecules. For example, the molecule NO_2 dimerizes to produce the N_2O_4 molecule in the reaction

$$2NO_2 \rightarrow N_2O_4$$

Formal Charges

How do we know whether or not a Lewis structure is reasonable? Experimental evidence, such as bond length, is the best verification of structures. Without experimental data, calculating of the **formal charge** on each atom is one technique that may be used to make this judgment. The formal charge is the difference between the number of electrons an atom has in a Lewis structure and its number of valence electrons. This calculation is described in detail below.

First it is important to understand the concept of formal charges. In a covalent compound an atom shares some of its valence electrons to form bonds, while the rest of the valence electrons remain as nonbonding electron pairs. If we count these nonbonding electrons and the electrons that an atom shares to form bonds, we should end up with a number equal to the valence of the atom. A different number means that the atom has lost or gained one or more electrons, an unlikely event since the loss or gain of electrons implies ionic behavior. Technically the formal charge is a separation of charge. In fact, some Lewis structures do require charge separation, but it is the minimum possible. In addition, we know, based on the electronegativities of the atoms, that the element with the greater electronegativity will be the atom with a negative charge. If the formal charges show elements with large electronegativities as positive compared to other atoms in the structure, we must question the validity of the structure.

In calculating the formal charge, each electron in a proposed Lewis structure is assigned to a specific atom. The number of electrons assigned to an atom is then compared to the number of that atom's valence electrons. If the assigned electrons and the number of valence electrons are equal, the formal charge is zero. If more electrons are assigned to an atom than there are valence electrons, the formal charge will be a negative value equal to the number of extra electrons. Similarly, if fewer electrons are assigned to an atom than there are valence electrons, the atom has a positive formal charge equal to the number of missing electrons. To calculate the formal charge on each atom in a Lewis structure the following steps are taken:

1. For each atom count all electrons not used for bonding by the atom.
2. Count half of the atom's bonding electrons.
3. Add steps 1 and 2 to obtain the electrons assigned to that atom.
4. Subtract the assigned electrons from the valence electrons to obtain the formal charge.

In equation form the formal charge is calculated as follows:

Formal charge =
Valence e^- – [Number of nonbonding e^- + $\frac{1}{2}$ (Number of bonding e^-)]   (5.1)

The formal charge calculations may be quickly checked since the formal charges on the atoms in a molecule must add up to zero, obeying the law of electroneutrality. For a polyatomic ion the formal charges must add up to the charge on the ion.

A molecule or polyatomic ion with the lowest possible formal charge on each atom is usually judged to be a more probable structure than one where the formal charge is larger. Consider the sulfate ion, SO_4^{2-}. Its Lewis structure may be drawn as shown in Figure 5.13.

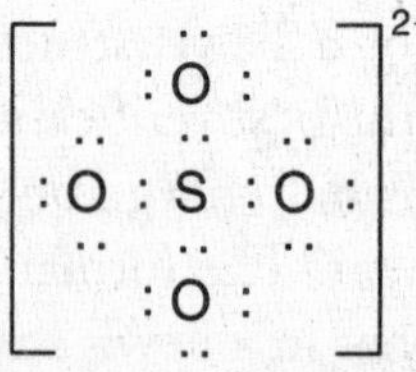

FIGURE 5.13. Lewis structure of the sulfate ion, constructed by the procedures described above.

From this structure we calculate the formal charges on sulfur and oxygen:

Formal charge on sulfur   = $6 - 0 - \frac{1}{2}(8) = +2$
Formal charge on **each** oxygen   = $6 - 6 - \frac{1}{2}(2) = -1$

We find that the formal charge on each oxygen is -1 and the formal charge on sulfur is $+2$. These charges add up to the charge of the ion, as they should ($+2 -1 -1 -1 -1 = -2$). These formal charges are large, and an alternative structure should be sought. The structure shown in Figure 5.14 is one possible variation.

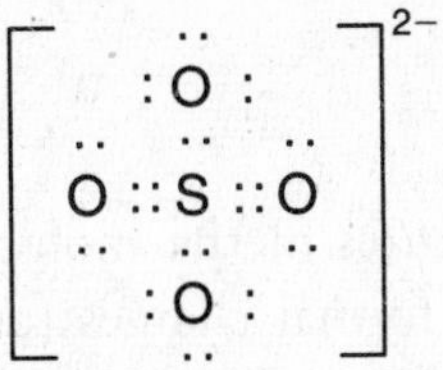

FIGURE 5.14.   Alternative Lewis structures for the sulfate ion.

There are now two types of oxygen-sulfur bonds (two with a single bond and two with a double bond). When the formal charges are calculated, we have these results:

Formal charge on sulfur = $6 - 0 - \frac{1}{2}(12) = 0$
Formal charge on each oxygen with double bond = $6 - 4 - \frac{1}{2}(4) = 0$
Formal charge on each oxygen with single bond = $6 - 6 - \frac{1}{2}(2) = -1$

This is the preferred structure since it has the minimum formal charges. The formal charges add up to the charge of the ion ($-1 - 1 = -2$) and cannot be any lower.

Formal charges may also be used to deduce the appropriate structure for a compound that has many possible Lewis structures. For instance, the NOCl molecule can be drawn with the four structures shown in Figure 5.15.

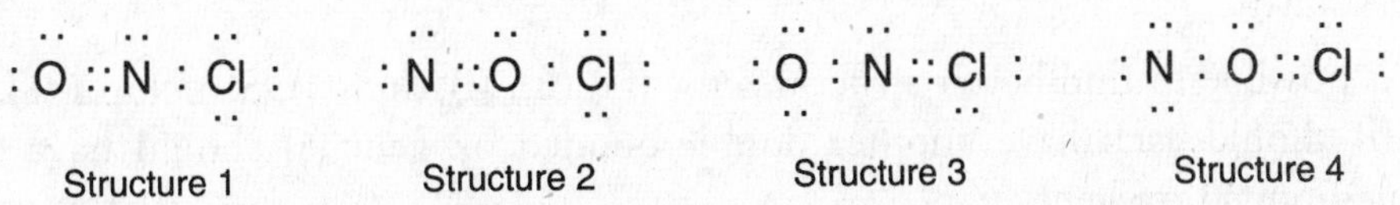

FIGURE 5.15.   Possible Lewis structures for the NOCl molecule.

To determine which of these structures is the most reasonable, we calculate the formal charges. Using the rules for determining these charges, we obtain the values in Table 5.1.

TABLE 5.2

Formal Charges on NOCl Structures				
Element	Structure 1	Structure 2	Structure 3	Structure 4
Oxygen	0	+1	−1	+1
Nitrogen	0	−1	0	−2
Chlorine	0	0	+1	+1

From Table 5.2 we see that the first structure has the lowest formal charges on all atoms, and it is preferred. Structure 2 has larger formal charges, but also has a negative charge on nitrogen even though nitrogen is less electronegative than oxygen. This is not a reasonable situation since the more electronegative atom is expected to have the more negative formal charge. Structures 3 and 4 also have negative charges on the less electronegative elements as well as formal charges greater than zero. We therefore conclude that structure 1 is the most reasonable structure.

EXERCISE 5.2

In Exercise 5.1 the Lewis structures of the molecules and ions listed below were constructed. Now calculate the formal charges for each of those structures. Also, suggest which ones may have better structures, and draw them.

(a) CH_3Cl (c) PH_3 (e) H_2S (g) CO_3^{2-} (i) SO_3^{2-}

(b) CS_2 (d) SiF_4 (f) NO_3^- (h) PO_4^{2-} (j) ClO_4^-

Solution

(a) all zero

(b) all zero

(c) all zero

(d) all zero

(e) all zero

(f) N = +1   O(single bonded) = −1   O(double bonded) = 0

(g) C = 0   O(single bonded) = −1   O(double bonded) = 0

(h) P = +1   O = −1

(i) S = +1   O = −1

(j) Cl = +3   O = −1

It is possible to find better structures for (f), (h), (i), and (j). Structures (f), (h), and (i) should each have another double-bonded oxygen; (j) should have three double-bonded oxygens.

Resonance Structures

In the discussion above we found that, when there are several possible structures, the most reasonable one can be selected by using the concept of formal charges. At times we can construct several Lewis structures for a substance that are totally equivalent, even down to the formal charges on the atoms. Chemists call these structures **resonance structures.**

Most resonance structures are very similar for a given substance, usually differing only in the geometry of the molecule or ion. It is found experimentally that none of the resonance Lewis structures properly describes the molecule. The true properties of the substance are found by blending all of the resonance structures together. For example, the resonance of the SO_3 molecule is shown in Figure 5.16 with the three possible Lewis structures.

FIGURE 5.16. Three resonance structures of SO_3. They differ only in the position of the double bond.

It is important to understand the nature of resonance. In Figure 5.16, each SO_3 has two single bonds and one double bond. In a variety of experiments it is found that all of the sulfur-oxygen bonds are identical. The measured properties of these bonds indicate that they are neither purely single bonds nor purely double bonds. Sulfur-oxygen bonds have characteristics in between those of the single bond and those of the double bond. To properly visualize the SO_3 molecule we must think of three identical sulfur-oxygen bonds as a blend of bonds that is approximately two-thirds single bond and one-third double bond in character.

When discussing resonance, the benzene ring must be mentioned. The formula for benzene is C_6H_6. The molecule is a ring of six carbon atoms with one hydrogen atom attached to each carbon. Assigning electrons to this skeleton results in the resonance structures shown in Figure 5.17.

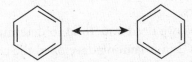

FIGURE 5.17.   Structural formulas for the two resonance forms of benzene.

These structures are often summarized as shown in Figure 5.18. Each corner of the hexagon is assumed to represent a carbon atom, and a hydrogen atom is assumed to be attached to each carbon.

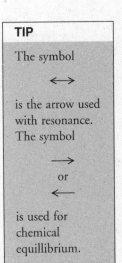

FIGURE 5.18.   Abbreviated resonance structures of benzene.

Finally, since we know that the actual structure of the benzene molecule is not properly represented by either resonance structure, but rather involves a blending of the two, organic chemists often represent the benzene ring as shown in Figure 5.19.

FIGURE 5.19.   Structure of benzene with a central circle representing the resonant nature of the molecule.

The circle within the ring reminds us that the double bonds are distributed (delocalized) over the entire molecule.

Benzene is one of many organic compounds classified as **aromatic molecules.** They definitely have a smell, but current chemical terminology recognizes the term *aromatic* as meaning that the structure contains one or more benzene rings. The benzene ring is found in such diverse compounds as aspirin, morphine, nicotine, proteins, saccharin, and many plastics. Benzene and many other aromatic compounds are considered carcinogens; however, a large number of beneficial compounds with aromatic character are not carcinogenic. Prudence dictates, however, that all aromatic compounds be treated with care in the laboratory.

EXERCISE 5.3

In Exercises 5.1 and 5.2 simple Lewis structures were constructed and then modified, based on the formal charges. Use those structures to draw the resonance structures of the following ions:

(a) NO_3^-
(b) CO_3^{2-}
(c) PO_4^{3-}
(d) SO_3^{2-}
(e) ClO_4^-

Solution

You should find three resonance structures each for (a), (b), and (d) and four resonance structures each for (c) and (e).

Covalent Bond Polarity and Electronegativity

Electrons are shared equally only in a covalent bond between two identical atoms (e.g., H_2, F_2, and N_2). If the electrons are not shared equally by two atoms, they will spend more time localized near one atom or the other. The result is that the atom that attracts the electrons will be relatively more negative than the other atom. When this occurs, we say that the bond between the atoms is **polar** with a positive end and a negative end. Understanding bond polarities allows the chemist to explain many physical properties of chemical compounds. We will use this concept frequently in later chapters.

To understand polarities, we need a method to determine how effectively different atoms attract electrons. Linus Pauling developed the concept of **electronegativity** to numerically represent the ability of an atom to attract electrons. Figure 5.20 shows the periodic table with the electronegativity value for each element. We see that electronegativity increases from left to right in each period of the periodic table. In addition, the electronegativity within any group increases from the bottom of the group to the top. In general, the electronegativity increases from the lower left corner of the periodic table up to the upper right corner. This is one of the important **diagonal trends** in the periodic table.

PERIODIC TREND

EN increases from lower left to upper right corner of periodic table.

H 2.1																
Li 1.0	Be 1.5											B 2.0	C 2.5	N 3.1	O 3.5	F 4.0
Na 1.0	Mg 1.3											Al 1.5	Si 1.8	P 2.1	S 2.4	Cl 2.9
K 0.8	Ca 1.1	Sc 1.2	Ti 1.3	V 1.5	Cr 1.6	Mn 1.6	Fe 1.7	Co 1.7	Ni 1.8	Cu 1.8	Zn 1.7	Ga 1.8	Ge 2.0	As 2.2	Se 2.5	Br 2.8
Rb 0.8	Sr 1.0	Y 1.1	Zr 1.2	Nb 1.3	Mo 1.3	Tc 1.4	Ru 1.4	Rh 1.5	Pd 1.4	Ag 1.4	Cd 1.5	In 1.5	Sn 1.7	Sb 1.8	Te 2.0	I 2.5
Cs 0.7	Ba 0.9	La 1.1	Hf 1.2	Ta 1.4	W 1.4	Re 1.5	Os 1.5	Ir 1.6	Pt 1.5	Au 1.4	Hg 1.5	Tl 1.5	Pb 1.6	Bi 1.7	Po 1.8	At 2.2
Fr 0.7	Ra 0.9	Ac 1.0														

FIGURE 5.20. Periodic table showing the electronegativities of the elements.

The values of electronegativities are generally not given on the AP Exam. However, the diagonal trend of the electronegativities allows you to estimate quickly which end of a bond is negative and which end is positive. In a bond, the element closest to fluorine will be relatively negative and the element furthest from fluorine will be relatively positive. Polarities are indicated by the symbols $\delta+$ and $\delta-$ for partially positive and partially negative atoms respectively. Full positive and negative charges are used only for ions. For example, the boron-oxygen bond would be written as $^{\delta+}B—O^{\delta-}$ to show that the boron atom is more positive than the oxygen atom.

EXERCISE 5.4

Without using Figure 5.20, indicate the positive and negative ends of each of the following bonds by the symbols $\delta+$ and $\delta-$:

(a) S—O
(b) C—N
(c) S—P
(d) C—F
(e) Si—O
(f) H—Br
(g) H—O

Solution

Using the diagonal relationships in the periodic table, we identify the positive and negative ends as follows:

(a) $^{\delta+}S—O^{\delta-}$
(b) $^{\delta+}C—N^{\delta-}$
(c) $^{\delta-}S—P^{\delta+}$
(d) $^{\delta+}C—F^{\delta-}$
(e) $^{\delta+}Si—O^{\delta-}$
(f) $^{\delta+}H—Br^{\delta-}$
(g) $^{\delta+}H—O^{\delta-}$

The electronegativity table in Figure 5.20 gives numerical data that may be used to evaluate the magnitude of bond polarity. This is done by taking the absolute value of the difference in the electronegativities of the two atoms participating in a bond. We may call this value delta EN or, in written form, ΔEN:

$$\Delta EN = \left(\begin{array}{c} \text{Atom with largest} \\ \text{electronegativity} \end{array} \right) - \left(\begin{array}{c} \text{Atom with smallest} \\ \text{electronegativity} \end{array} \right) \quad (5.2)$$

The larger the value of ΔEN, the greater the polarity of the bond. If ΔEN is zero, the bond is considered to be nonpolar.

EXERCISE 5.5

For the bonds in Exercise 5.4, determine the ΔEN values, and predict which bond is the least polar and which is the most polar.

Solution

(a) 1.7
(b) 0.6
(c) 0.3
(d) 1.5
(e) 1.7
(f) 0.7
(g) 1.4

The S—O and Si—O bonds are the most polar, and the S—P bond is the least polar.

Without an electronegativity table, it is possible to determine which of two bonds is the more polar if the bonds have one atom in common. The more polar bond will be the one where the second atom is furthest in the periodic table from the common atom. For instance, the nitrogen-fluorine bond is more polar than the oxygen-fluorine bond since nitrogen is further from fluorine than is oxygen.

EXERCISE 5.6

Using a periodic table, but not a table of electronegativities, estimate, for each of the following pairs, which bond is more polar:

(a) C—N or C—O
(b) H—Cl or H—Br
(c) S—O or S—Br
(d) H—S or H—O
(e) P—Br or S—Br

Solution

The more polar bond in each pair belongs to (a) C—O, (b) H—Cl, (c) S—O, (d) H—O, and (e) P—Br.

Dipole Moments

As shown in the previous section, we can very successfully estimate the polarity of a bond between two atoms. That is, we can tell which atom will attract electrons more easily than another. As a result, the atom that attracts electrons will have a partial negative charge, $d-$. The other atom will have a partial positive change, $d+$. From a table of electronegativities or by even just estimating from the periodic table, we can make good estimates of how polar a bond is. A better measure is the **dipole moment**. The dipole moment is a measure of the difference in charge, q, on two covalently bonded atoms and the distance, r, between the two nuclei.

$$\text{Dipole moment} = q \times r$$

The units for the dipole moment are coulomb-meters, and a common unit for dipole moments is the debye.

$$1 \text{ debye} = 3.34 \times 10^{-30} \text{ Cm}$$

You will often see dipole moments expressed as debyes. Another feature of dipole moments is that they are treated mathematically as vectors. As we will see shortly, molecules with polar bonds can be nonpolar if opposing dipole moments cancel each other.

Electronegativity and Ionic Character

At one end of the polarity scale are the completely **nonpolar** bonds between diatomic elements. We may visualize ionic compounds as being at the other end of the polarity scale since the electrons are actually transferred from one atom to another. From the table of electronegativities in Figure 5.2, we see that the largest ΔEN is 3.3 for the ionic compound FrF. The well-known ionic compounds NaCl and $CaBr_2$ have ΔEN values of 3.0 and 1.7, respectively. The ΔEN of a bond has been used to estimate the percentage ionic character of a bond. Chemists say that a ΔEN of 1.7 represents a bond that is 50 percent ionic and 50 percent covalent in character. A bond with a ΔEN of 1.7 or greater is considered ionic. Bonds with ΔENs of less than 1.7 are polar covalent, and those for which ΔEN is zero are nonpolar covalent bonds.

Bond Order

Bond order is a term that refers to the average number of bonds that an atom makes in all of its bonds to other atoms. From the Lewis structures of the diatomic elements in Figure 5.2, we see that fluorine (F_2) and chlorine (Cl_2) have one bond each and a bond order of 1. Oxygen (O_2) has a double bond and a bond order of 2, and nitrogen (N_2) has a triple bond and thus a bond order of 3. In the SO_3 resonance structures there is a total of four bonds: two single bonds and one double bond. Since the sulfur is bonded to three oxygen atoms, the average number of bonds that sulfur has with its oxygen atoms is $\dfrac{\text{four bonds}}{\text{three O atoms}}$, and the bond order of sulfur is $^4/_3$. In benzene, the bond order for each carbon atom is $^3/_2$.

EXERCISE 5.7

Determine the bond order of the central atom of each of the following compounds. You may use the structures determined in Exercises 5.1 and 5.2.

(a) CH_3Cl	(c) PH_3	(e) H_2S	(g) CO_3^{2-}	(i) SO_3^{2-}
(b) CS_2	(d) SiF_4	(f) NO_3^-	(h) PO_4^{3-}	(j) ClO_4^-

Solution

If we use the simple Lewis structures, we obtain

(a) 1	(c) 1	(e) 1	(g) $^4/_3$	(i) 1
(b) 2	(d) 1	(f) $^4/_3$	(h) 1	(j) 1

If we use the best formal charge structures, we obtain

(a) 1	(c) 1	(e) 1	(g) $^4/_3$	(i) $^4/_3$
(b) 2	(d) 1	(f) $^5/_3$	(h) $^5/_4$	(j) $^7/_4$

Bond Strength, Bond Energy, and Bond Length

Just as two ropes are twice as strong as one rope, a double bond is almost twice as strong as a single bond. The strength of a covalent bond is expressed as its bond energy. When two atoms are covalently bonded together, they vibrate in the same way that a spring vibrates. The frequency of this vibration is related to the two masses attached to the ends of the spring and to the strength of the spring itself. Since we know the masses of the two atoms, the frequency of vibration will be related to the strength of the bond. Frequency is related to bond energy (bond strength) by Equation 5.3, where the energy is equal to Planck's constant times the frequency of vibration.

$$E = h\nu \tag{5.3}$$

Bond vibrations may be observed in the infrared spectral region by using an infrared spectrometer. Infrared spectra confirm that single bonds have the lowest energy, double bonds have a higher energy, and triple bonds have the highest energy.

Another method used to measure bond energies involves measurement of the energy released when organic compounds are burned. Ethane, C_2H_6, contains only single bonds; ethylene, C_2H_4, contains a double bond; and acetylene, C_2H_2, contains a triple bond. When burned, ethane yields 1540 kilojoules, ethylene 1387 kilojoules, and acetylene 1305 kilojoules of energy. The fact that the lowest energy is released by acetylene is taken to indicate that this compound's bonds are already in the highest energy configuration. The high energy of combustion for ethane, on the other hand, indicates that its bonds have lower energy to start with and can release more energy upon combustion.

The length of a covalent bond may be measured in several ways. One way is by X-ray crystallography, as described in Chapter 8. The positions of the atoms in a crystal can be determined based on the diffraction of X rays by a crystal. Another method involves the fact that the frequency of vibration and the length of the vibrating medium are related. As described above, these vibrations are measured using infrared spectroscopy. Whatever method is used to measure bond lengths, consistent results are obtained. In particular, a single bond has the longest length and a triple bond has the shortest.

We have discussed resonance structures and the fact that the actual structure of a molecule is a blend of all equivalent resonance structures. In these molecules we do not have pure single, double, or triple bonds. We can, however, calculate bond order, which represents the average number of bonds per atom. Bond order is related to bond strength and length also. The general rule is that the greater the bond order, the shorter the bond length. Table 5.3 lists some typical bond lengths and energies.

TABLE 5.3

Typical Bond Lengths and Energies

Bond	Bond Order	Bond Length (pm)	Bond Energy (kJ mol^{-1})
C—C	1	154	347
C=C	2	134	612
C≡C	3	120	820
N—N	1	145	159
N=N	2	123	418
N≡N	3	110	914
C—O	1	143	351
C=O	2	120	715

Nomenclature

The naming of covalently bonded binary molecules (these are predominately molecules with two nonmetals) is quite different from the naming of ionic compounds, and the two methods should not be confused. In addition, many of these covalent substances were discovered long before the modern method of naming compounds was developed. Common covalent compounds often have trivial (older) and systematic (modern) names. Some important trivial names are shown in Table 5.4, along with their modern equivalents.

TABLE 5.4

Trivial and Systematic Names of Some Common Covalent Molecules

Molecule	Trivial Name	Systematic Name
H_2O	water	dihydrogen monoxide*
CH_4	methane	carbon tetrahydride*
N_2O	nitrous oxide	dinitrogen oxide
NO	nitric oxide	nitrogen monoxide
N_2O_3	nitrous anhydride	dinitrogen trioxide
N_2O_5	nitric anhydride	dinitrogen pentoxide
NH_3	ammonia	nitrogen trihydride*
AsH_3	arsine	arsenic trihydride
H_2O_2	hydrogen peroxide	dihydrogen dioxide*
N_2H_4	hydrazine	dinitrogen tetrahydride

TIP

Names of binary molecular compounds use prefixes.

*These names are almost never used.

In the systematic naming of covalent compounds, the prefixes listed in Table 4.1 indicate the number of each atom present in a molecular formula. Some of these prefixes are indicated in the systematic names in Table 5.4. Other examples of the use of these prefixes can also be given. The formula for carbon dioxide is CO_2, and the name indicates one carbon atom and, because of the prefix *di–*, two oxygen atoms. Sulfur forms two compounds, SO_2 and SO_3, named sulfur dioxide and sulfur trioxide, respectively. Again the prefixes *di–* and *tri–* indicate the number of oxygen atoms bound to the sulfur in the two compounds. The compound N_2O_4 has two nitrogen and four oxygen atoms and is named dinitrogen tetroxide.

EXERCISE 5.8

Name each of the following covalent molecules:

(a) SO_2 (c) N_2O_3 (e) PBr_5 (g) $BrCl_3$

(b) P_4O_{10} (d) SF_4 (f) XeF_4 (h) S_4N_4

Solution

(a) sulfur dioxide   (e) phosphorus pentabromide

(b) tetraphosphorus decaoxide   (f) xenon tetrafluoride

(c) dinitrogen trioxide   (g) bromine trichloride

(d) sulfur tetrafluoride   (h) tetrasulfur tetranitride

EXERCISE 5.9

Give the formula for each of the following compounds:

(a) diboron tetrabromide   (e) diphosphorus pentoxide

(b) boron trifluoride   (f) carbon disulfide

(c) carbon tetrafluoride   (g) sulfur trioxide

(d) carbon monoxide   (h) nitrogen triiodide

Solution

(a) B_2Br_4 (c) CF_4 (e) P_2O_5 (g) SO_3

(b) BF_3 (d) CO (f) CS_2 (h) NI_3

Additional rules and names for compounds are given in Chapter 14 on acids and bases and in Chapter 15 on common organic compounds.

MOLECULAR GEOMETRY

Once a valid Lewis structure has been determined, the overall geometry of a simple molecule with one central atom can be established. The process can be extended to very large macromolecules, such as proteins and DNA, by determining the geometries around individual atoms and then combining them to obtain the entire structure. This overall geometry is extremely important in understanding the properties of chemical compounds. The key to the discovery of DNA's double helix was the geometric structures of the four bases, which must hydrogen-bond to each other in order to hold the total structure together.

The **valence-shell electron-pair repulsion theory** (VSEPR theory) allows us to determine the three-dimensional shapes of covalently bonded molecules with a minimum of information. This theory states that the geometry around each atom will depend on the repulsion of the valence-shell electrons (bonding electrons and nonbonding pairs) away from each other. This repulsion, due to the negative charges, results in electron pairs being aligned as far away from each other as possible.

Basic Structures

To determine the three-dimensional geometry around a central atom, A, all we need to know is how many atoms, X, are covalently bonded to it. The one restriction is that central atom A must have no nonbonding pairs of electrons. Table 5.5 lists the **basic structures** for the six possible geometries. In this table we use the AX_n notation to represent the number of atoms, X, attached to the central atom, A, with no nonbonding electrons on A.

TABLE 5.5

Basic Structures for Six Geometries			
Notation	Shape	Example	Angle(s)
AX^*	Linear	HBr	—
AX_2	Linear	CS_2	180°
AX_3	Planar triangle	BCl_3	120°
AX_4	Tetrahedron	CCl_4	109.5°
AX_5	Trigonal bipyramid	PCl_5	120°, 90°
AX_6	Octahedron	XeF_6	90°

*This structure is trivial but is included for completeness.

The angles listed in this table are the angles between the bonds, assuming that the central atom is the vertex of the angle. For structures AX to AX_4 every bond is equidistant from every other one. For the AX_5 and AX_6 structures bond angles are measured between the nearest neighbors. For the AX_5 structure, the 120° angle is for the three equatorial atoms and the 90° angle is the angle between the axial atoms and the equatorial atoms. Figure 5.21 illustrates these shapes in diagram format. The AX structure is omitted from many texts as being trivial since any molecule that contains only two atoms must be linear.

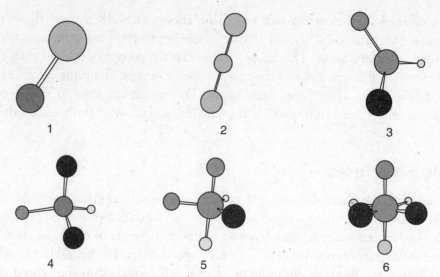

FIGURE 5.21. **Perspective diagrams of the six basic geometric structures. Darkest atoms are closest to the viewer, structures are tilted to show all atoms. 1—linear diatomic; 2—linear triatomic; 3— planar triangle; 4—tetrahedron; 5—trigonal bipyramid; 6— octahedron. For diagrams 5 and 6 the axial atoms are at the top and bottom of the figures, while the equatorial atoms are in the center.**

The geometry around a central atom that does not have any nonbonding electron pairs is determined by counting the atoms bonded to it. For instance, three atoms bound to a central atom with no nonbonding pairs must be a planar triangle. Five atoms bound to a central atom must have the shape of a trigonal bipyramid.

Derived Structures

Nonbonding electron pairs on the central atom take up space, just as an atom does. In fact, a nonbonding electron pair takes up slightly more space than an atom. As a result, we count the number of nonbonding pairs, as well as the atoms attached to the central atom, to determine the *basic structure* as we did above. Since the nonbonding pairs of electrons are not "seen" when a structure is drawn, the actual geometries of the atoms we do see will be only part of this basic structure. The part that we see is called the **derived structure**. The possible derived structures are listed in Table 5.6. The symbol E in the notation represents a nonbonding pair of electrons on the central atom.

TIP

Simply stated, atom and electron pairs arrange themselves so that they are as far from each other as possible.

TABLE 5.6

Derived Structures with Nonbonding Electron Pairs on the Central Atom

Basic Structure Notation	Derived Structure Notation	Derived Structure Shape	Example	Derived Structure Angle(s)
A	A*	Single atom	None	None
AX_2	AXE*	Linear diatomic	CN^-	None
AX_3	AX_2E	Bent	$SnCl_2$	120°
AX_4	AX_3E	Triangular pyramid	NH_3	109.5°
AX_4	AX_2E_2	Bent	H_2O	109.5°
AX_5	AX_4E	Distorted tetrahedron	SF_4	120°, 90°
AX_5	AX_3E_2	T-shape	ICl_3	90°
AX_5	AX_2E_3	Linear	I_3^-	180°
AX_6	AX_5E	Square pyramid	IF_5	90°
AX_6	AX_4E_2	Square planar	XeF_4	90°

*These structures are trivial but are included for completeness.

Since the first two entries in Table 5.6 represent single atoms and diatomic substances, their structures need not be drawn. The geometries that correspond to the other derived structures are shown in Figures 5.22–5.25.

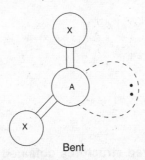

Bent

FIGURE 5.22. Bent AX_2E derived structure, showing an electron pair occupying the space formerly occupied by an atom in the basic AX_3 (trigonal planar) structure.

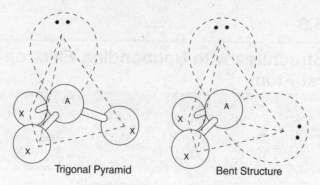

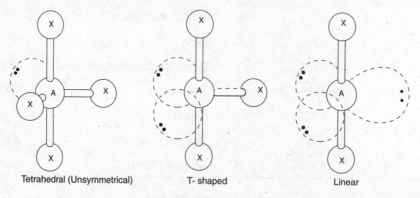

FIGURE 5.23. **Pictorial representation of the AX$_3$E (triangular pyramid) and AX$_2$E$_2$ (bent) derived structures that are derived from the AX$_4$ (tetrahedral) basic structure.**

FIGURE 5.24. **The three possible derived structures obtained from the AX$_5$ (trigonal bipyramid) basic structure. Notice that the equatorial atoms are replaced by electron pairs.**

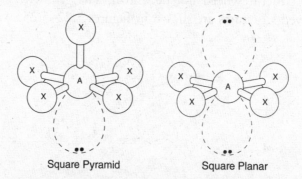

FIGURE 5.25. **The two derived structures obtained from the AX$_6$ (octahedral) basic structure. Note that the second atom replaced by an electron pair is on the opposite side of the molecule, so that the electron clouds have the extra space they need.**

In Table 5.6 and Figures 5.22 and 5.23 we find two bent structures, AX$_2$E and AX$_2$E$_2$. The bond angles for these two structures will be very different, and we can distinguish the structures on the basis of their angles. Since the AX$_2$E bent structure is derived from the trigonal planar AX$_3$ structure, we expect its angle to be approximately 120°. The AX$_2$E$_2$ structure is derived from the tetrahedral AX$_4$ basic structure, and bond angles of approximately 109.5° are expected for structures related to the tetrahedron.

In forming the derived structures from the basic structures, we are, in effect, replacing one or more atoms with pairs of nonbonding electrons. Up to the AX_4 structure, it does not matter which atom is replaced by an electron pair; we get the same derived structure. However, AX_5 and AX_6 will have different shapes, when we make the derived structures, depending on which atom is replaced by a nonbonding electron pair. The actual shapes of the molecules can be explained by the fact that a nonbonding pair of electrons takes up relatively more space than a bonded atom. In the AX_5 structure, the nonbonding electron pairs will replace the equatorial atoms since more room is available (120° between the atoms compared to 90° for the axial atoms) for the electron cloud. In the AX_6 structure, it does not matter which atom is replaced by the first nonbonding electron pair; the result is always a square pyramid. The second nonbonding electron pair always replaces the atom opposite the first nonbonding electron pair. This positioning allows the nonbonding electron pairs the most room possible on the molecule.

When determining the geometry around a central atom, it is necessary to count the number of atoms bound to the central atom and the number of nonbonding pairs of electrons, if any. We now understand why it is so important to show nonbonding pairs of electrons in Lewis structures. Of particular importance are the nonbonding pairs on the central atom.

EXERCISE 5.10

Construct the Lewis structure and predict the shape of each of the following molecules and ions:

(a) CH_3Cl (d) SiF_4 (g) CO_3^{2-} (i) SO_3^{2-} (k) SO_2

(b) CS_2 (e) H_2S (h) PO_4^{3-} (j) ClO_4^- (l) NO_2^-

(c) PH_3 (f) NO_3^-

Solution

The shapes are as follows:

(a) tetrahedron   (e) bent   (i) tetrahedron
(b) linear   (f) triangular planar   (j) tetrahedron
(c) triangular pyramid   (g) triangular planar   (k) bent
(d) tetrahedron   (h) tetrahedron   (l) bent

These shapes are the same whether we use the simple Lewis structure or the structures optimized for the best formal charges.

Complex Structures

Geometries of more complex molecules are constructed by determining the geometry around each atom in sequence and then stringing the geometries together. Organic (carbon-based) compounds often have complex structures where these geometries are very important. The three-dimensional structures of these compounds often help define chemical, physical, and biological properties.

Carbon, with its four valence electrons, can form a maximum of four covalent bonds with four other atoms. It can also bond to three atoms as long as one of the bonds is a double bond. In bonding to two atoms, carbon will form either two double bonds, as in carbon dioxide, CO_2, or one single bond and one triple bond, as in hydrogen cyanide, HCN. In all instances, carbon never has a nonbonding pair of electrons. As a result, a carbon bonded to four atoms is tetrahedral; if bonded to three atoms, trigonal planar; and if bonded to two atoms, linear.

For instance, the molecule of ethylene, CH_2CH_2, has the structure shown in Figure 5.26.

$$
\begin{array}{ccc}
H & & H \\
\diagdown & & \diagup \\
& C = C & \\
\diagup & & \diagdown \\
H & & H
\end{array}
$$

FIGURE 5.26. Ethylene molecule.

Since each carbon atom is bonded to only three atoms (two hydrogen and one carbon), the carbon atoms must each have a trigonal planar geometry. We will see the reason later, but both planes are lined up so that this molecule is perfectly flat. We can also predict that the benzene ring shown in Figure 5.19 must be flat also since each of its six carbon atoms is trigonal planar.

When we have a molecule such as butane, $CH_3CH_2CH_2CH_3$, we may draw the structures shown in Figure 5.27.

$$
\begin{array}{ccccccccc}
& H & & H & & H & & H & \\
& | & & | & & | & & | & \\
H - & C & - & C & - & C & - & C & - H \\
& | & & | & & | & & | & \\
& H & & H & & H & & H &
\end{array}
$$

(a)   (b)

FIGURE 5.27. Structure of butane, C_4H_{10} (a), with all H shown and (b) with lines only. Each end and vertex represents a carbon atom.

Each carbon atom is bonded to four other atoms; therefore, the geometries of the carbon atoms are all tetrahedral. Placing the four tetrahedral structures together, we obtain the three-dimensional structure illustrated in Figure 5.28.

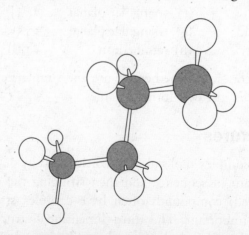

FIGURE 5.28. Three-dimensional computer-generated structure for butane, showing the tetrahedral arrangement of atoms around each carbon (shaded circles).

Since three-dimensional structures are difficult to draw on paper, organic chemists often find it convenient to build models of these structures so that they can inspect their features more easily.

Oxygen atoms in organic compounds always have two nonbonding pairs of electrons. An oxygen bonded to two atoms will have an AX_2E_2 derived structure (bent), and a double-bonded oxygen with an AXE_2 derived structure (linear) will be bonded to only a single atom. Nitrogen atoms in organic compounds will have one nonbonding pair of electrons. The nitrogen atom will have an AX_3E structure (triangular pyramid) if bonded to three other atoms. When bonded to only two atoms, one with a double bond, it will have an AX_2E structure (bent).

EXERCISE 5.11

Predict the geometry around each of the carbon atoms in this molecule:

$$CH_3CH_2CH = \underset{\underset{OH}{|}}{\overset{\overset{CH_3}{|}}{C}}CH_2C = O$$

Solution

From left to right along the main chain, the geometries are tetrahedral, tetrahedral, triangular planar, triangular planar, tetrahedral, triangular planar, respectively. The CH_3 above the molecule is tetrahedral. Using this information, it is possible to make a more realistic drawing, or molecular model, of the compound.

MOLECULAR POLARITY

Bond polarities depend on the electronegativities of the two elements bonded together. Very few molecules are diatomic, meaning that for most molecules more than one bond must be considered in determining the polarity of the molecule as a whole. These bonds are arranged geometrically in space as described in the preceding section. The result is that, even if a bond is polar, the molecule as a whole may or may not be polar. There are four general rules for determining whether a molecule is polar.

> **TIP**
>
> Molecular polarity is determined by combining bond polarity and molecular geometry.

1. A molecule that is symmetrical is nonpolar. It does not matter how polar the individual bonds are.
2. A nonsymmetrical molecule is polar if the bonds are polar.
3. A molecule with more than one type of atom attached to the central atom is often nonsymmetrical and therefore polar.
4. A central atom with nonbonding electron pairs is often nonsymmetrical and polar.

When determining the polarity of a molecule, we must remember that there is only one positive end and one negative end, directly opposite each other. We recognize the CH_3Cl and CBr_4 molecules as tetrahedral structures. As Figure 5.29 indicates, CH_3Cl is polar because it is not symmetrical, and CBr_4 is symmetrical and nonpolar. We can also calculate that the four equal dipole moments will add up to zero because of the tetrahedral geometry.

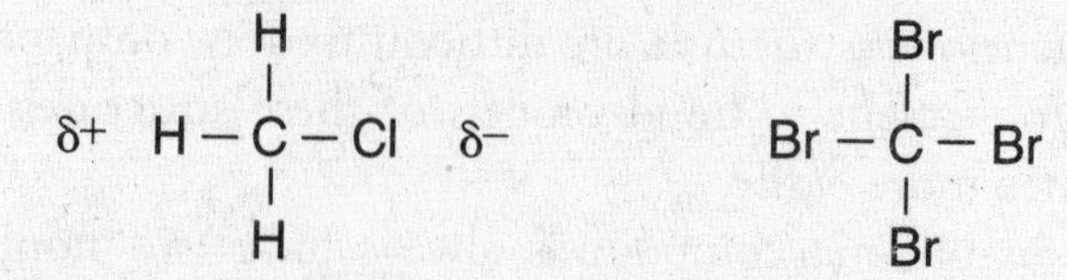

$$\delta+ \quad H - \overset{\displaystyle H}{\underset{\displaystyle H}{\overset{|}{\underset{|}{C}}}} - Cl \quad \delta- \qquad\qquad Br - \overset{\displaystyle Br}{\underset{\displaystyle Br}{\overset{|}{\underset{|}{C}}}} - Br$$

FIGURE 5.29.   Structures of CH₃Cl, showing the polarity of the molecule, and of CBr₄, showing the symmetry and nonpolarity.

From our knowledge of electronegativities we may predict that the negative end of the CH_3Cl molecule is located near the chlorine atom and that the positive end is opposite the chlorine atom in a region of space between the three hydrogen atoms.

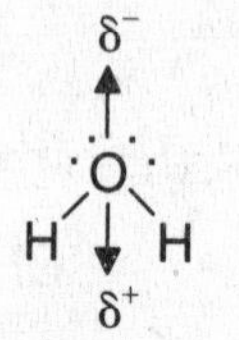

FIGURE 5.30.   Structure of water and its polarity.

For the water molecule we may draw the structure shown in Figure 5.30. The electronegativity of the oxygen indicates that it is the negative end of the molecule and that the region of space opposite the oxygen, between the two hydrogen atoms, is the positive end.

EXERCISE 5.12

Construct the Lewis structure and predict the polarity of each of the following:

(a) CH_3Cl (c) PH_3 (e) H_2S (g) CO_3^{2-} (i) SO_3^{2-}

(b) CS_2 (d) SiF_4 (f) NO_3^- (h) PO_4^{3-} (j) ClO_4^-

Solution

(a) polar with Cl the negative end. (e) polar with S the negative end
(b) nonpolar (f–j) nonpolar because of resonance;
(c) polar with P the negative end   charge on ion is distributed evenly
(d) nonpolar   over the ion. These ions are charged
    but not polar.

COVALENT BOND FORMATION

Wave Mechanics and Covalent Bond Formation

TIP

The AP exam does not ask questions on molecular orbitals.

Up to now, we have constructed molecules only according to the octet rule. We have determined the shapes and structures based on the very successful VSEPR theory. We now turn our attention to the wave mechanical nature of the covalent bond. Two successful theories approach the formation of the covalent bond in different ways. The **valence bond theory** (VB theory) considers a covalent bond to be the overlapping of two atomic orbitals when the electron spins are paired. The **molecular orbital theory** (MO theory) considers that a molecule is similar to an atom in that they both have distinct energy levels that can be populated by electrons. Figure 5.31 attempts to illustrate the idea that the electrons and nuclei of two hydrogen atoms are rearranged into molecular orbitals.

FIGURE 5.31. Combination of two hydrogen atoms to form the hydrogen molecule. The electrons in the hydrogen atoms have opposing spins so that they can pair in the molecular orbital.

In atoms these levels are called atomic orbitals, and in molecules they are molecular orbitals. Both the VB theory and the MO theory have been refined to produce similar results. The AP Exam tends to focus on the valence bond approach, and it is the subject of the next section.

Valence Bond Theory

In the VB theory, illustrated in Figure 5.32, two hydrogen atoms approach and interact with an overlap of atomic orbitals with electrons of opposing spin (the opposing spins are indicated by the different shading of the hydrogen atoms).

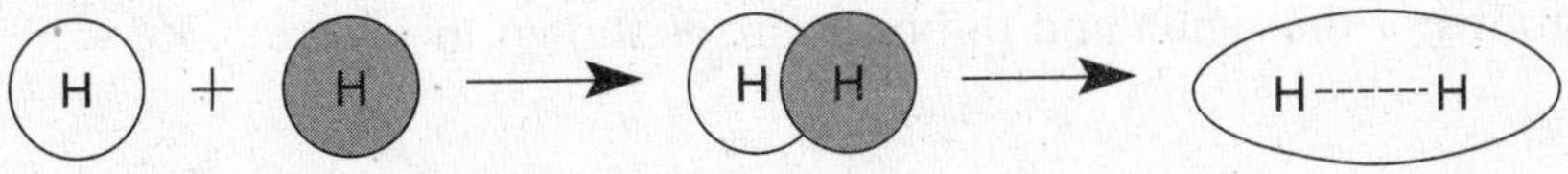

FIGURE 5.32. Overlap of two s atomic orbitals to form a molecular orbital in hydrogen.

As the bond is formed, the paired electrons spread out over the molecule to form the final electron cloud surrounding the nuclei. The results of the MO theory and the VB theory are the same (Figures 5.31 and 5.32), with a high electron density along the internuclear axis. The formation of H_2 and many other compounds and bonds is described as the overlap of two *s* orbitals to form a sigma bond.

Sigma bonds may also be formed by the overlap of an *s* orbital and a *p* orbital (Figure 5.33), as in the formation of hydrogen fluoride, HF, or by the overlap of two *p* orbitals (Figure 5.34), as in F_2.

FIGURE 5.33. Overlap of an s orbital and a p orbital to form a sigma bond in a substance such as HF.

FIGURE 5.34. Overlap of two p orbitals to form a sigma bond in a molecule such as F_2.

Orbital Overlap Model (Pi Bonds)

We have described the three important ways in which sigma bonds are formed. Every covalent bond has one and only one sigma bond. If a compound has a double or triple covalent bond, additional overlap of orbitals is needed. Such a bond, called a **pi bond** (π bond), is formed by the sideways overlap of two *p* orbitals as shown in Figure 5.35.

FIGURE 5.35.   Sideways overlap of p orbitals to form a pi bond.

The **pi bond** has its electron density arranged in two electron clouds, one above and one below the internuclear axis (dashed line). When arranged in this manner, the electrons in the pi bond do not interfere with the electrons in the sigma bond.

A double bond involves one sigma bond and one pi bond. A triple bond between two atoms may be formed by adding a second pi bond, which has its two electron clouds centered behind and in front of the two nuclei.

If we place one atom in front of the other and look down the internuclear axis, the positions of the sigma and pi bonds are as shown in Figure 5.36.

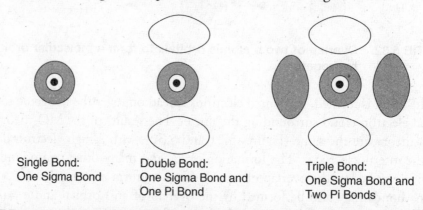

Single Bond:
One Sigma Bond

Double Bond:
One Sigma Bond and
One Pi Bond

Triple Bond:
One Sigma Bond and
Two Pi Bonds

FIGURE 5.36.   End view of the single, double, and triple bonds, looking down the internuclear axis. The dot and white circle represent the two nuclei. The shaded circle represents the sigma bond present in all three bonds. The two white ovals represent one of the pi bonds, and the two shaded ovals represent the second pi bond.

From this discussion we see that a single covalent bond is always a sigma bond. A double covalent bond has one sigma and one pi bond, while a triple covalent bond has one sigma and two pi bonds. All of these bonds are arranged so that their electron clouds do not interfere with each other.

Hybrid Orbital Model

The overlap of *s* and *p* orbitals to form sigma and pi bonds works well to describe some features of the covalent bond and for molecules with two and sometimes three atoms. Larger molecules require another model of bond formation.

To understand why a new model is needed, we need to review the implications of the overlap model. First, *p* orbitals are oriented at 90° from each other and *s* orbitals are spherical, having no directionality. If all covalent compounds were formed from the overlap of these orbitals, we would expect all covalent molecules to have 90° bond angles. As we have seen, however, few molecular geometries have angles of 90°. Second, even a simple molecule such as methane, CH_4, cannot be adequately explained by the overlap model. We know that methane is a tetrahedral molecule with four totally equivalent C—H bonds. Using the overlap method, we see that the carbon in methane has only two unpaired *p* electrons (the *s* electrons are paired), and we would expect the formation of the CH_2 molecule. The bonds would be oriented at a 90° angle since *p* orbitals are 90° apart. If we allowed the two *s* electrons to unpair so that they could also form bonds, we could obtain the CH_4 molecule. However, the bond angles would still not be correct, and we would expect two distinctly different C—H bond types in methane, one from the overlap of *s* orbitals and the other from the overlap of *p* orbitals. Our new model must be able to explain correctly the molecular geometries and bonding in larger molecules.

sp^3 Hybrid Orbitals

The problem posed by the CH_4 molecule requires that we develop a better model of sigma bond formation. To construct this model, we postulate the formation of hybrid orbitals. A **hybrid orbital** may be defined as a set of orbitals with identical properties formed from the combination of two or more different orbitals with different energies. The orbital diagram of carbon is presented in Figure 5.37, along with the conversion of the *s* and *p* electrons into hybrid orbitals called sp^3 orbitals.

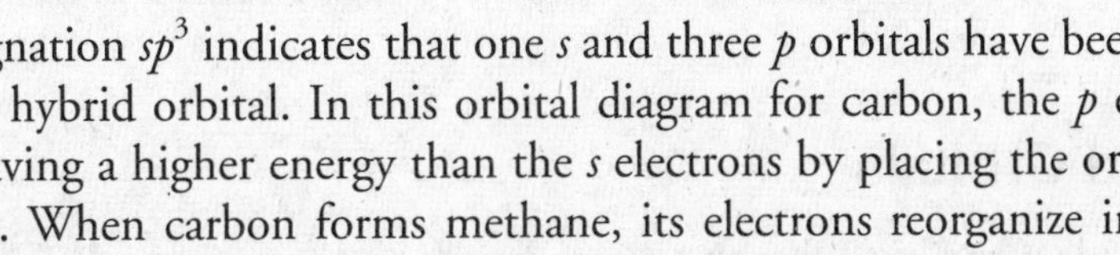

FIGURE 5.37. **The *s* and *p* electrons of the carbon atom and their conversion into the sp^3 hybrid orbitals used in bonding.**

The designation sp^3 indicates that one *s* and three *p* orbitals have been combined to form the hybrid orbital. In this orbital diagram for carbon, the *p* electrons are shown as having a higher energy than the *s* electrons by placing the orbitals at different levels. When carbon forms methane, its electrons reorganize into the four identical sp^3 hybrid orbitals shown on the right. The energy of the electrons in the hybrid orbitals lies between the original *s* and *p* energies, as shown. When the sp^3 hybrid orbitals form, their orientation is tetrahedral. Overlap of the four identical sp^3 electrons with electrons from hydrogen atoms forms the tetrahedral methane molecule as we know it. Any molecule whose basic structure is the tetrahedron will have sp^3 hybrid orbitals. This includes the CH_4, NH_3, and H_2O molecules described previously.

sp^2 Hybrid Orbitals

Formaldehyde, CH_2O, is a carbon compound having single bonds to the hydrogenatoms and a double bond with the oxygen atom. Carbon has three sigma bonds and one pi bond in this compound. The structure is triangular planar because there are no nonbonding electron pairs on carbon. The orbitals in this compound are designated as sp^2 hybrids. The formation of these orbitals can be diagrammed as shown in Figure 5.38. Here we see that three electrons are in three identical orbitals, called sp^2 hybrids. The remaining electron stays in an unhybridized p orbital and overlaps with a p orbital on the oxygen atom to form the pi bond in the C=O double bond.

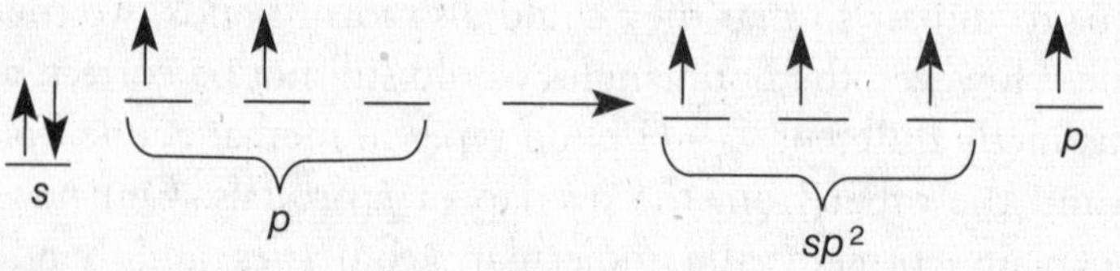

FIGURE 5.38. Formation of the sp^2 hybrid orbitals for carbon.

In the earlier discussion of the ethylene and benzene molecules, it was stated that these molecules are totally flat. Figure 5.39 shows why. In order for the p orbitals to overlap, they must be aligned as shown. This requirement fixes the remaining sp^2 bonds in one plane, resulting in the planar ethylene molecule. Proper alignment of the p orbitals in benzene forces this molecule also to be planar.

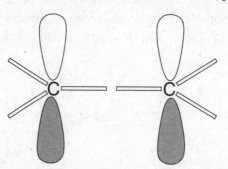

FIGURE 5.39. Two carbon atoms with sp^2 hybridization. The thin lines are the triangular planar sp^2 bonding orbitals. The large orbitals are the unhybridized p orbitals that overlap to form a pi bond.

sp Hybrid Orbitals

Carbon dioxide, O=C=O, has two sigma bonds and two pi bonds. The hybridization for this molecule is shown in Figure 5.40.

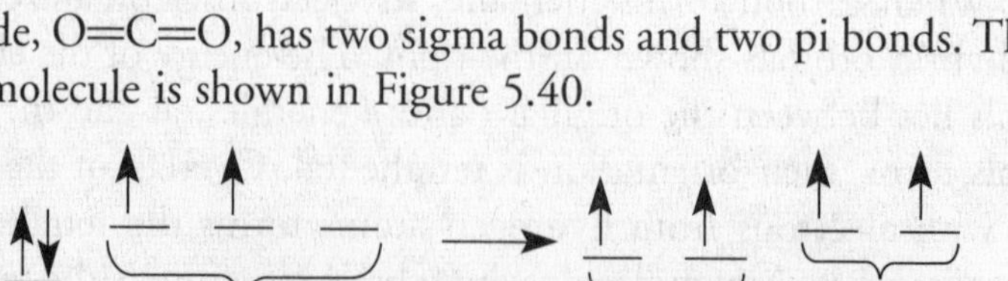

FIGURE 5.40. Hybridization of carbon to produce the sp hybrid orbitals. The two unhybridized p electrons are available to form pi bonds.

In forming *sp* hybrid orbitals, we obtain two equivalent electrons that can form sigma bonds. The two remaining, unhybridized *p* electrons can overlap with *p* electrons from the oxygen atoms to form the required double bonds. The hybridized and unhybridized orbitals in carbon's *sp*² hybrid may be pictured as shown in Figure 5.41. This diagram shows the *p* orbitals available for pi bonding. Since there are two *p* orbitals, two additional pi bonds can form. These two pi bonds can be directed toward different atoms to form compounds, such as O=C=O, with two double bonds. They can also be directed toward the same atom to form a triple bond, as in the cyanide ion, C≡N⁻.

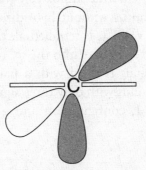

FIGURE 5.41. **The *sp* hybrid orbitals, shown as thin lines. The remaining two *p* orbitals are shown as larger lobes. These p orbitals overlap with other *p* orbitals to form two pi bonds to the carbon.**

*sp*³*d* Hybrid Orbitals

The *sp*³*d* notation indicates a hybrid that has five equivalent orbitals with at least one electron in each orbital. Since carbon has only four valence electrons, it cannot form a *sp*³*d* hybrid. In addition, only atoms that have available *d* orbitals can form these hybrids. This requirement means that an element must be in period 3 or higher. Phosphorus, which has five valence electrons and is in period 3, is a typical element that can form the *sp*³*d* hybrid. The orbital diagram for phosphorus in Figure 5.42 shows the 3*d* orbitals, even though they are empty before hybridization.

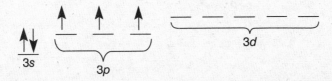

FIGURE 5.42. **Orbital diagram for a phosphorus atom. Energy levels are shown by the positions of the orbitals.**

When the hybrid orbitals are formed, they may be represented as shown in Figure 5.43. The five electrons in the *sp*³*d* hybrid will form five sigma bonds in a covalent compound. The basic structure for the *sp*³*d* hybrid is the trigonal bipyramid.

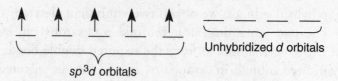

FIGURE 5.43. Five identical sp^3d hybrid orbitals for phosphorus. Four unoccupied d orbitals are not used and remain unhybridized.

Not all compounds of phosphorus will be sp^3d hybrids. When phosphorus forms PCl_3, for example, the three chlorine atoms can readily combine with the three unpaired electrons in phosphorus without hybridization involving the d orbitals. However, since the structure of PCl_3 is tetrahedral, there must be some hybridization. Here the s and p orbitals hybridize into the sp^3 form, as shown in Figure 5.44. Since one of the hybrid orbitals is filled with a pair of electrons, this hybrid of phosphorus will form compounds with only three sigma bonds, such as phosphorus trichloride. PCl_3 has an AX_3E configuration and a triangular pyramid shape (a derived structure).

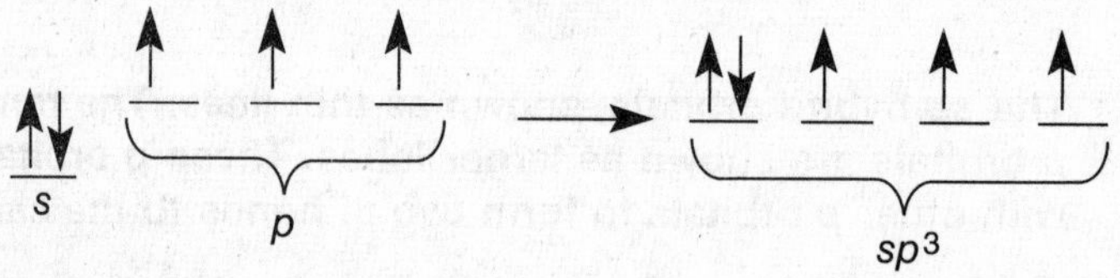

FIGURE 5.44. The sp^3 hybridization for phosphorus.

sp^3d^2 Hybrid Orbitals

As is true for sp^3d hybrids, an element must be in period 3 or higher to be able to participate in sp^3d^2 hybrids. At least six valence electrons are required. Sulfur is one element that forms sp^3d^2 hybrids. Sulfur's orbital configuration and the sp^3d^2 hybrid are shown in Figure 5.45.

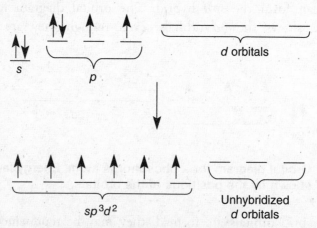

FIGURE 5.45. Orbital diagram of the valence electrons in sulfur and the conversion to the sp^3d^2 hybrid orbitals.

The sp^3d^2 hybrid allows sulfur to form six covalent bonds with an octahedral structure. One of those compounds is sulfur hexafluoride. In addition, we have seen a variety of other sulfur compounds that result from other types of hybridization.

Sulfur dichloride, SCl_2, uses sp^3 hybridization, as shown in Figure 5.46. In SCl_2, two unpaired electrons form sigma bonds with the chlorine atoms, and there are two pairs of nonbonding electrons. This sp^3 hybrid gives an AX_2E_2 structure, which also corresponds to a basic tetrahedral configuration. Since the two nonbonding pairs of electrons are not seen, SCl_2 has a bent shape with an angle close to 109.5°. Whereas sulfur can form both sp^3d^2 and sp^3 hybrids, oxygen in period 2 does not have available d orbitals to form the sp^3d^2 hybrid and can form only single bonded compounds with an sp^3 hybrid.

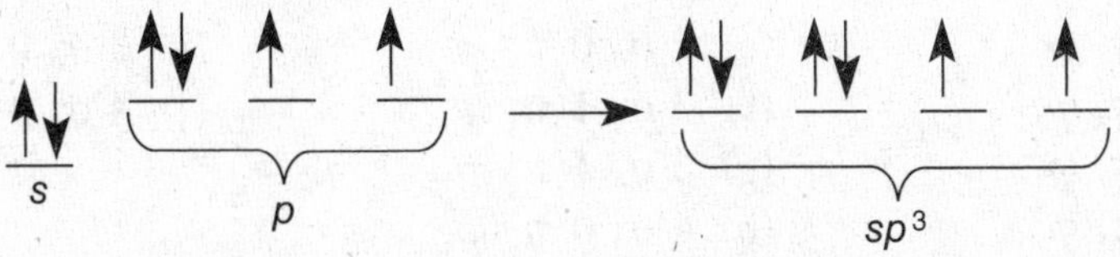

FIGURE 5.46. **Hybridization used to explain the structure of molecules such as SCl_2.**

There is a direct correspondence between hybridization and structure, which is shown in Table 5.7. If the basic structure is known, the hybridization can be determined; similarly, if the hybridization is known, the structure is likewise known.

TABLE 5.7

Correspondence Between Hybridization and Structure

Basic Structure	Derived Structure	Hybrid	Bonding e⁻ Pairs	Nonbonding e⁻ Pairs
Linear		sp	2	0
Planar triangle		sp^2	3	0
Planar triangle	Bent	sp^2	2	1
Tetrahedron		sp^3	4	0
Tetrahedron	Triangular pyramid	sp^3	3	1
Tetrahedron	Bent	sp^3	2	2
Trigonal bipyramid		sp^3d	5	0
Trigonal bipyramid	Distorted tetrahedron	sp^3d	4	1
Trigonal bipyramid	T-shape	sp^3d	3	2
Trigonal bipyramid	Linear	sp^3d	2	3
Octahedron		sp^3d^2	6	0
Octahedron	Square pyramid	sp^3d^2	5	1
Octahedron	Square planar	sp^3d^2	4	

EXERCISE 5.13

Determine the total number of sigma and pi bonds in each of the following. Using the simple Lewis structure, also determine the hybridization for each.

(a) CH_3Cl (c) PH_3 (e) H_2S (g) CO_3^{2-} (i) SO_3^{2-}

(b) CS_2 (d) SiF_4 (f) NO_3^- (h) PO_4^{3-} (j) ClO_4^-

Solution

(a) $4\,\sigma,\ 0\,\pi,\ sp^3$ (e) $2\,\sigma,\ 0\,\pi,\ sp^3$ (i) $3\,\sigma,\ 0\,\pi,\ sp^2$

(b) $2\,\sigma,\ 2\,\pi,\ sp$ (f) $3\,\sigma,\ 1\,\pi,\ sp^2$ (j) $4\,\sigma,\ 0\,\pi,\ sp^3$

(c) $3\,\sigma,\ 0\,\pi,\ sp^3$ (g) $3\,\sigma,\ 1\,\pi,\ sp^2$

(d) $4\,\sigma,\ 0\,\pi,\ sp^3$ (h) $4\,\sigma,\ 0\,\pi,\ sp^3$

Before leaving the topic of hybrid orbitals, we must recognize that this model is used to explain experimental results. For a molecule that has a particular shape, the concept of hybrid orbitals may be used to explain that shape. **The reverse is not true.** We say that H_2O has sp^3 hybridization because it is a bent structure with a bond angle of 104°, which is close to the 109.5° bond angle expected for a tetrahedral structure. Experiments show, however, that the similar molecule H_2S has a bond angle of 90°. In this case the simple overlap of the p orbitals of sulfur with the s orbitals of hydrogen is sufficient to explain the structure. Hybridization is not needed, and is apparently unwarranted, in this example.

SUMMARY

After discussing ionic compounds in Chapter 4, this chapter covers molecular compounds that are characterized by covalent bonds. In order to understand molecular compounds and their structures, the chapter starts by reviewing the logical methods for drawing Lewis structures. Equivalent Lewis structures are described as resonance structures that don't resemble any one structure but are a blend of each contributing structure. Formal charges help us decide which Lewis structures may be better than others. The chapter then discusses bond order and bond strength along with electronegativity and bond polarity.

From here the chapter looks at the structures of the molecules using the VSEPR theory. There are five basic structures of matter that need to be remembered. Several additional structures are derived from the five basic structures. From the three-dimensional structures we find that molecules will be nonpolar if they are totally symmetrical (even if their bonds are polar). Nonsymmetric molecules are often polar. Because the geometry of the orbitals is different from the geometry of the molecules, hybrid orbitals are introduced to explain this observation. The relationships between hybrid orbitals and structure are also developed in this chapter. Finally, valence bond theory is used to describe the formation of sigma and pi bonds in molecular compounds.

Important Concepts

Octet rule and when it can be disobeyed
Lewis structures and formal charges
Bond polarity
Electronegativity
Dipole moment
Molecular geometry and molecular polarity
Hybrid orbitals

Practice Exercises

Multiple-Choice

For the first five problems below, one or more of the following responses will apply; each response may be used more than once or not at all in these questions.

I. CBr_4
II. PF_5
III. NH_3
IV. SO_3
V. HCN

1. Which of these molecules has a shape related to a tetrahedron (or has sp^3 bonding)?

   (A) I and III
   (B) II
   (C) III and V
   (D) IV
   (E) V

2. Which of these molecules has the largest bond angle?

   (A) I
   (B) II
   (C) III
   (D) IV
   (E) V

3. Which of these molecules has the most pi bonds?

   (A) I
   (B) II
   (C) III
   (D) IV
   (E) V

4. Which of these molecules have all atoms lying in the same plane?

   (A) I and III
   (B) II
   (C) III and V
   (D) IV
   (E) IV and V

5. Which of these molecules uses more than an octet of electrons in its Lewis structure?

   (A) I and III
   (B) II
   (C) III and V
   (D) IV
   (E) V

6. In which of the following are the elements listed in order of increasing electronegativity?

   (A) Ba, Zn, C, Cl
   (B) N, O, S, Cl
   (C) N, P, As, Sb
   (D) K, Ba, Si, Ga
   (E) Li, K, Na, Ca

7. Which of the following bonds is expected to have the largest dipole moment?

   (A) C—Si
   (B) C—N
   (C) O—C
   (D) S—C
   (E) H—C

8. For which of the following may we draw both polar and nonpolar Lewis structures?

   (A) $CHCl_3$
   (B) NH_3
   (C) $.BF_3$
   (D) SF_2Cl_4
   (E) PCl_5

9. Which of the following has the fewest pi bonds and is nonpolar?

   (A) HCCH
   (B) CO_2
   (C) CO_3^{2-}
   (D) N_2
   (E) SO_2

10. The SF_5^- ion has a square pyramid structure. The hybridization of the *s* orbitals in sulfur is

    (A) sp^3d
    (B) sp
    (C) sp^3d^2
    (D) sp^3
    (E) sp^2

11. Which of the following is NOT a linear structure?

    (A) I_2
    (B) I_3^-
    (C) CO_2
    (D) H_2S
    (E) $H—C{\equiv}C—H$

12. The Lewis structure of the cyanide ion most closely resembles

    (A) N_2
    (B) O_2
    (C) CO_2
    (D) NO
    (E) C_2H_2

13. In which of the following pairs are the two items NOT properly related?

    (A) sp^3 and 109.5°
    (B) trigonal planar and 120°
    (C) octahedral and sp^3d
    (D) sp and 180°
    (E) square planar and sp^3d^2

14. How many resonance structures are possible for the SO_3 molecule?

    (A) none
    (B) 2
    (C) 3
    (D) 4
    (E) $^4/_3$

15. Which of the following has a nonbonding pair of electrons on the central atom?

    (A) BCl_3
    (B) NH_3
    (C) CCl_2Br_2
    (D) PF_5
    (E) SO_4^{2-}

16. Which of the following is true when the C=C and C≡C bonds are compared?

    (A) The triple bond is shorter than the double bond.
    (B) The double bond vibrates at a lower frequency than the triple bond.
    (C) The double-bond energy is lower than the triple-bond energy.
    (D) Both are composed of sigma and pi bonds.
    (E) All of the above are true.

17. How many electrons are available to construct the Lewis structure of the sulfite ion?

    (A) 24
    (B) 18
    (C) 26
    (D) 22
    (E) 20

18. The correct name for N_2O_3 is

    (A) dinitrogen tetroxide
    (B) dinitrogen trioxide
    (C) dinitrogen oxide
    (D) trinitrogen dioxide
    (E) nitric anhydride

19. Which angle is NOT expected in any simple molecule?

    (A) 60°
    (B) 90°
    (C) 109.5°
    (D) 120°
    (E) All of these are reasonable angles.

20. Sulfur forms the following compounds: SO_2, SF_6, SCl_4, SCl_2. Which form of hybridization is NOT represented by these molecules?

    (A) sp
    (B) sp^2
    (C) sp^3
    (D) sp^3
    (E) sp^3d^2

21. Which of the following is least related to the strength of a covalent bond?

    (A) vibrational frequency
    (B) bond order
    (C) bond length
    (D) bond direction
    (E) bond energy

ANSWER KEY

1. A	6. A	11. D	16. E	21. D
2. E	7. C	12. A	17. C	
3. E	8. D	13. C	18. B	
4. E	9. C	14. C	19. A	
5. B	10. C	15. B	20. A	

See Appendix 1 for explanations of answers.

Free-Response

Answer the following questions regarding the concepts and properties concerning the structure and geometry of molecular compounds.

(a) What name should be given to a molecule with the formula N_2O_5? What formula should be written for the compound named sulfur hexafluoride? Explain your answers briefly.

(b) Draw the Lewis structure for NCl_3. Explain the geometric shape of this molecule based on the VSEPR theory. Explain what hybrid orbitals are needed to produce this structure. Why is the molecule NCl_5 impossible?

(c) Use the results for part (b) to answer this question. Based on this structure, are the bonds in this molecule polar? If so, indicate the positive and negative poles in an appropriate manner. Explain your reasoning.

(d) Use the results for part (b) to answer this question. Finally, based on this structure, is this molecule polar? If so, indicate the positive and negative poles in an appropriate manner. Explain your reasoning.

(e) Explain the difference between sigma and pi bonds using the appropriate theories and examples.

ANSWERS

(a) The molecule N_2O_5 should be named dinitrogen pentoxide. The prefix *di-* is used to indicate two nitrogen atoms, and the prefix *penta-* indicates five oxygen atoms in the formula. Sulfur hexafluoride is SF_6 because the prefix *hexa-* stands for the number 6.

(b) The structure of NCl_3 is very similar to the PCl_3 structure illustrated in Figure 5.6. This molecule is an AX_3E shape, where three atoms and a nonbonding electron pair are on the central atom, N. The VSEPR theory states that these four will repel each other and form an arrangement where the atoms and electron pairs are as far apart as possible. This results in a tetrahedral shape. If the electrons are not "seen," then the shape of the atoms alone is a triangular pyramid. To obtain the tetrahedral shape, a set of four sp^3 hybrid orbitals must form. This results when the 2s (one orbital) and 2p (three orbitals) energy levels combine to form one energy level with four equivalent orbitals.

The formation of NCl_5 would require the formation of an sp^3d^2 hybrid. However, there are no available *d*-orbitals to use to make these hybrids, and therefore NCl_5 is impossible.

(c) The N–Cl bonds in this molecule are polar because a significant difference in electronegativity is expected between Cl and N. We should indicate that the polarity of the bond is

$$\delta+ \; N\text{---}Cl \; \delta-$$

The partial sign, δ, is used to symbolize a partial charge because a complete positive or complete negative charge is not formed (if so, these would be called ions).

(d) The structure and direction of the polar bonds in a molecule are arranged to see if the bond polarity cancels. Generally, bond polarity will cancel and a molecule will be nonpolar if the molecule is perfectly symmetrical. In this case, the molecule is not symmetrical, and the polarities of the individual N–Cl bonds do not cancel. The molecule has only two poles, one negative, in space between the three Cl atoms. The positive end resides near the nitrogen atom. Again, only partial charges are formed, and we may write

$$\delta+ \; N\text{---}Cl_3 \; \delta-$$

(e) Sigma bonds form when there is a high electron density between two nuclei. The line between the two nuclei is called the internuclear axis. The greater the density of electrons along the internuclear axis, the stronger the bond. Because the region in space is taken up with sigma bonding electrons, the pi bond forms with increased electron density on either side of the internuclear axis. Each pi bond has two electron clouds, one on either side of the internuclear axis. Reproduce Figure 5.36 to illustrate the positioning of the sigma bond and the pi bond.

The nature of sigma and pi bonds can be explained on the basis of valence bond theory. Valence bond theory considers bonds forming from the overlap of orbitals. Direct overlap of two *s*-orbitals, an *s*- and a *p*-orbital, or two *p*-orbitals along the internuclear axis results in a sigma bond (Figures 5.31 to 5.34). Sidewise overlap of two *p*-orbitals results in a pi bond as shown in Figure 5.35.

Stoichiometry

CHAPTER 6

- Conversion Calculations Using the Dimensional Analysis Method
- Metric Units and Prefixes
- Converting Complex Units
- Mole Concept
- Avogadro's Number
- Molar Mass
- Mole Relationships in Formulas
- Mole Relationships in Equations
- Stoichiometry Sequence
- Mole Conversions
- Gram-to-Gram Conversions
- Theoretical Yield Calculations
- Limiting-Reactant Calculations
- Titration Methods
- Titration Stoichiometry
- Percent Composition
- Empirical Formulas
- Molecular Formulas

Stoichiometry (measurement of the elements) is the name given to the quantitative relationships between the compounds in a chemical reaction. These quantitative relationships allow chemists to calculate the amounts of reactants needed for a reaction and to predict the quantity of product. Using stoichiometric methods, chemists can determine the formulas of compounds and can simplify procedures in chemical analysis. This chapter discusses and illustrates stoichiometric calculations and the fundamental concepts that make stoichiometry the most important topic in chemistry.

Quantitative calculations in chemistry fall into two groups. The first group involves taking a memorized equation, entering data for all but one variable, and then solving for the remaining variable. The second group involves converting information with one set of units into an answer with another set of units. All stoichiometric calculations are the conversion type.

The **conversion factor method** (also called dimensional analysis) has become the predominant method taught for solving stoichiometry problems. Over the years many methods have been used to perform stoichiometric conversions. Many of these methods were limited in their utility, were complex, and involved a large amount of memorization. Today, the conversion factor method is used in most courses. It has the advantage of minimizing memorization while applying basic chemical concepts to define the conversion process.

CONVERSION CALCULATIONS USING THE DIMENSIONAL ANALYSIS METHOD

Defining Conversion Factors

In this section, the methods used for dimensional analysis are developed in detail. If you are adept at using this method, you can skip to page 229.

A conversion problem changes the units of a measurement but not its magnitude. For instance, a distance of 2.0 yards may be converted into 72 inches by multiplying 2.0 by the factor 36. This does not indicate how the units of yards became units of inches. If, instead of using just the factor 36, we multiply by the factor with its units, it is clear how the yards became inches:

$$2.0 \text{ yards} \left(\frac{36 \text{ inches}}{1 \text{ yard}} \right) = 72 \text{ inches}$$

In this equation the yard units cancel, and it is clear that the remaining units are inches. The ratio $\left(\dfrac{36 \text{ inches}}{1 \text{ yard}} \right)$ is known as the **conversion factor.** Using labels along with the numerical factors enables us to keep track of the units and to produce an answer with the correct units. Proper use of conversion factors also tells us when to multiply and when to divide.

A conversion factor is derived from a defined relationship between two sets of units. The conversion factor above was obtained from the definition of 1 yard:

$$1 \text{ yard} = 36 \text{ inches}$$

Two conversion factors can be obtained from every defined equality. For example, dividing both sides of the above equivalence by 1 yard gives

$$\frac{1 \text{ yard}}{1 \text{ yard}} = \frac{36 \text{ inches}}{1 \text{ yard}} = 1$$

Conversion Factor

Dividing both sides by 36 inches gives

$$\frac{1 \text{ yard}}{36 \text{ inches}} = \frac{36 \text{ inches}}{36 \text{ inches}} = 1$$

Conversion Factor

The two possible conversion factors are the inverse of each other. In addition, they are both equal to 1. When a measurement is multiplied by a conversion factor, it is multiplied, in effect, by 1, and its true value does not change, although the units do change.

Using Conversion Factors

TIP

Conversion calculations, just like driving directions, need a starting point and an end destination.

Every conversion problem must start with two pieces of information: (1) the number and the units that need to be converted and (2) the units of the answer. This is set up as

$$? \text{ answer units} = xxx \text{ given units}$$

On the right side is *xxx*, which represents the number given in the problem, and its units, which will be converted. The left side of the equal sign reminds us of the units needed for the solution to the problem.

After this step, it is necessary to find the equalities that can be used to produce the conversion factors needed to solve the problem. For example, we might be required to convert 35 yards into feet in part (a) and to convert 624 feet into yards in part (b). The solution to part (a) starts with the initial setup of the question with the desired units as the unknown in an equation and the given data as the starting point:

$$? \text{ feet} = 35 \text{ yards} \quad \text{(initial setup)}$$

Being familiar with the English system of measurement, we know that 1 yard is defined as being 3 feet in length:

$$1 \text{ yard} = 3 \text{ feet}$$

From this equality two conversion factors, $\left(\dfrac{3 \text{ feet}}{1 \text{ yard}} \right)$ and $\left(\dfrac{1 \text{ yard}}{3 \text{ feet}} \right)$, can be written.

The correct conversion factor allows us to cancel the yard units, leaving units of feet. This conversion factor is used to multiply the given 35 yards:

$$? \text{ feet} = 35 \text{ yards} \left(\frac{3 \text{ feet}}{1 \text{ yard}} \right)$$

After being sure that the units properly cancel, we calculate the answer as 105 feet. The following equation illustrates what happens if the wrong form of the conversion factor is chosen:

$$? \text{ feet} = 35 \text{ yards} \left(\frac{1 \text{ yard}}{3 \text{ feet}} \right) = 11.7 \text{ yards}^2 \text{ feet}^{-1} \quad \text{(wrong conversion factor chosen)}$$

This conversion will not work since the units do not cancel; in fact the final units are the meaningless yard^2 per foot. The correct answer for part (a) is

$$? \text{ feet} = 35 \text{ yards} \left(\frac{3 \text{ feet}}{1 \text{ yard}} \right) = 105 \text{ feet}$$

IN ALL OF THE FOLLOWING EXAMPLES, THE CANCELLATION OF THE UNITS IS NOT SHOWN. IT IS SUGGESTED THAT YOU TAKE A COLORED PENCIL AND PERFORM ALL OF THE CANCELLATIONS TO ASSURE YOURSELF THAT EACH CONVERSION FACTOR DOES INDEED CANCEL PROPERLY.

In part (b) of the question, the reverse calculation, from feet to yards, is requested. Starting as before with the question and the data supplied, we write

$$? \text{ yards} = 624 \text{ feet} \quad \text{(initial setup)}$$

Selecting the correct conversion factor gives

$$? \text{ yards} = 624 \text{ feet} \left(\frac{1 \text{ yard}}{3 \text{ feet}} \right) = 208 \text{ yards}$$

For this conversion the same equality, but a different conversion factor, was used. One of these conversion factors converts from yards to feet, and the other converts feet to yards.

Many conversions require more than one step to reach the desired units. These problems may be solved stepwise, one conversion factor at a time, or the conversion factors may be combined in one large equation. Both methods are illustrated in Exercise 6.1.

At this point go back to the start of this chapter and cancel units with a red pencil to reinforce the cancellation process.

EXERCISE 6.1

A typical school year includes 180 days of classes. How many minutes are there in those days?

Solution

We start with the question and the given information to obtain

$$? \text{ minutes} = 180 \text{ days}$$

Next, we find the conversion equalities that may be useful. These are as follows:

$$1 \text{ day} = 24 \text{ hours}$$
$$1 \text{ hour} = 60 \text{ minutes}$$

The given data have units of days, and the only equality that also has units of days is the first one. The ratio needed for the conversion must have the day units in the denominator so that these units will cancel. Multiplying by this ratio yields

$$? \text{ minutes} = 180 \text{ days} \left(\frac{24 \text{ hr}}{1 \text{ day}} \right) = 4320 \text{ hr}$$

TIP

Paying attention to the units of conversion factors helps to minimize errors.

This result still does not have the desired units, so another step is needed. Starting with the result obtained above, we write

$$? \text{ minutes} = 4320 \text{ hr}$$

In our second conversion equality there is a relationship between hours and minutes, 1 hour = 60 minutes. When the next conversion factor is inserted, the hour units cancel:

$$? \text{ minutes} = 4320 \text{ hr} \left(\frac{60 \text{ min}}{1 \text{ hr}} \right) = 2.59 \text{ n } 10^5 \text{ min}$$

Since the units of this answer match the units that the question requires, the problem is solved. This answer is rounded off to 2.59×10^5 min.

Solving this same problem with one large equation involves writing all of the conversion factors needed until the units match:

$$? \text{ minutes} = 180 \text{ days} \left(\frac{24 \text{ hr}}{1 \text{ day}} \right) \left(\frac{60 \text{ min}}{1 \text{ hr}} \right) = 2.59 \times 10^5 \text{ min}$$

The step-by-step and the combined methods result in exactly the same answer, and either method is correct.

Conversion of Metric Units

Chemistry students in the United States are hampered by the fact that the metric system is used in chemistry, whereas the English system of measurement governs everyday life. The examples above were done in **English units** because these units are often more familiar. From this point onward, however, only **metric units** will be used. The original metric system was developed in 1790 in France. Our modern version of the metric system is called the Système International, or S.I. The seven base units of the S.I. are defined in Table 6.1.

TABLE 6.1

Seven Base S.I. Units

Property Defined	Unit Name	Abbreviation
Mass	kilogram	kg
Length	meter	m
Time	second	s
Temperature	kelvin	K
Quantity	mole	mol
Electric current	ampere	A
Light intensity	candela	cd

These seven base units may be combined in a variety of ways to obtain other common units. For example, area can be expressed as square meters (m^2), and volume as cubic meters (m^3). These base units are often too large or too small for practical use in a laboratory. The base units may be modified by the use of a metric prefix. Each **metric prefix** represents a number that multiplies the base unit. The most commonly used prefixes for metric units are listed in Table 6.2.

TABLE 6.2

Metric Prefixes

Prefix Name	Prefix Symbol	Exponential Value
mega-	M	10^6
kilo-	k	10^3
deci-	d	10^{-1}
centi-	c	10^{-2}
milli-	m	10^{-3}
micro-	μ	10^{-6}
nano-	n	10^{-9}
pico-	p	10^{-12}

TIP

Knowing these prefixes saves valuable time on the exam.

It is essential to remember the first five of the metric base units in Table 6.1 and all of the metric prefixes in Table 6.2. Conversion factors between a unit with a prefix and the corresponding metric base unit may be quickly obtained by first writing an equality:

$$1 \text{ cm} = 1 \text{ cm}$$

and then replacing one of the prefixes with the corresponding exponent:

$$1 \text{ cm} = 1 \times 10^{-2} \text{ m}$$

This equality can be used to write the two possible conversion factors as $\left(\dfrac{1 \text{ cm}}{10^{-2} \text{ m}} \right)$ and $\left(\dfrac{10^{-2} \text{ m}}{1 \text{ cm}} \right)$. Conversion between a prefix and a base unit is a one-step calculation, but conversion between two different prefixes requires two steps.

EXERCISE 6.2

Convert 2.38 cm to meters and millimeters.

Solution

For the first conversion one of the conversion factors above may be used:

$$? \text{ m} = 2.38 \text{ cm (setup)}$$

$$= 2.38 \text{ cm} \left(\frac{10^{-2} \text{ m}}{1 \text{ cm}} \right)$$

$$= 0.0238 \text{ m}$$

For the second conversion, centimeters to millimeters, two conversion factors are needed. The first converts the prefix to the base unit, and the second converts the base unit to the new prefix.

$$? \text{ mm} = 2.38 \text{ cm (setup)}$$

$$= 2.38 \text{ cm} \left(\frac{10^{-2} \text{ m}}{\text{cm}} \right) \left(\frac{\text{mm}}{10^{-3} \text{ m}} \right)$$

$$= 23.8 \text{ mm}$$

The answers to both parts of this exercise were written in exponential notation in order to show that the number 2.38 does not change in these conversions, but the exponent does.

Conversion of Complex Units

Many times the data used in chemistry involve complex units. There are area measurements that have squared units such as square kilometers (km^2), square meters (m^2), or square centimeters (cm^2). There are volume measurements with cubic units such as cubic centimeters (cm^3), cubic millimeters (mm^3), and cubic meters (m^3). For velocity the units may be meters per second, which is abbreviated as m/s or $m \, s^{-1}$. Acceleration has units of meters per second squared ($m \, s^{-2}$), and energy has units of kilogram meters squared per second squared ($kg \, m^2 \, s^{-2}$).

The following exercises illustrate the conversion method used when the units have squared or cubed terms and when they involve a ratio.

EXERCISE 6.3

How many square centimeters are there in 180 m^2?

Solution

Setting up the problem as before, we have

$$? \text{ cm}^2 = 180 \text{ m}^2$$

The units are square centimeters and square meters. For clarity it is best to write the setup of this problem as

$$? \text{ cm} \times \text{cm} = 180 \text{ m} \times \text{m}$$

We can use the equality that says that 10^{-2} m is equal to 1 cm to write the needed conversion factor. Using the ratio that will cancel out the meter units, we obtain

$$? \text{ cm} \times \text{cm} = 180 \text{ m} \times \text{m} \left(\frac{1 \text{ cm}}{10^{-2} \text{ m}} \right)$$

By using the conversion ratio only once, we have canceled only one of the meter units, leaving the mixed units of centimeter meter. Applying the conversion ratio a second time cancels all of the meter units and leaves us with the desired square centimeters:

$$? \, cm \times cm = 180 \, m \times m \left(\frac{1 \, cm}{10^{-2} \, m} \right) \left(\frac{1 \, cm}{10^{-2} \, m} \right)$$

Once the units cancel properly, the answer can be calculated as $1.80 \times 10^6 \, cm^2$. For volumes, which have cubic units, each conversion factor is used three times.

When a ratio of units need to be converted, as illustrated in Exercise 6.4, the units in the numerator are converted into the units required by the problem. Then the units in the denominator are converted.

EXERCISE 6.4

A train is moving at a speed of 25 km hr^{-1}. How fast is it moving in units of meters per minute?

Solution

Set up the problem as before, with the desired units as the question and the given speed as the starting point in the conversion.

$$? \, \frac{m}{min} = \frac{25 \, km}{h} \, (setup)$$

The two equalities that must be used to construct conversion factors for this problem are

$$1 \, kilometer = 10^3 \, meters$$

$$1 \, hour = 60 \, minutes$$

Applying the factor obtained from the first equality converts the kilometers to meters:

$$? \, \frac{m}{min} = \frac{25 \, km}{h} \left(\frac{10^3 \, m}{1 \, km} \right)$$

Next, the second equality is used to convert the hour units in the denominator to the minutes required:

$$? \, \frac{m}{min} = \frac{25 \, km}{h} \left(\frac{10^3 \, m}{1 \, km} \right) \left(\frac{1 \, h}{60 \, min} \right)$$

The units cancel properly, leaving the desired meters per minute units. The answer is then calculated to be $4.2 \times 10^2 \, m \, min^{-1}$.

Exercises 6.1–6.4 illustrate three important principles about the conversion factor method. First, it is necessary to know the equalities required to obtain the proper conversion factors. Second, there is often a proper sequence in which to use these conversion factors. Third, if the units for the given value are in the form of a ratio, they must be converted into units that also represent a ratio. Similarly, if the problem requests an answer where the units must be a ratio, the starting data must have units in the form of a ratio.

EXERCISE 6.5

Convert each of the following:

(a) 8.89 nm to mm
(b) 3.89×10^5 cm^2 to μm^2
(c) 2.43×10^2 kg m^2 s^{-2} to g cm^2 s^{-2}

Solution

(a) 8.89×10^{-6} mm
(b) 3.89×10^{13} μm^2
(c) 2.43×10^9 g cm^2 s^{-2}

CHEMICAL EQUALITIES AND RELATIONSHIPS

The Mole and Avogadro's Number

The **mole** (mol) is the central unit of measurement in chemistry. Numerically one mole represents 6.02×10^{23} units of a chemical substance. For the elements 1 mole represents 6.02×10^{23} atoms of the element. In compounds such as CH_4 or CO_2, 1 mole represents 6.02×10^{23} molecules. For ionic compounds it represents 6.02×10^{23} empirical formula units of a substance such as NaCl, or $MgBr_2$. The value 6.02×10^{23} is called **Avogadro's number** in honor of that chemist-physicist's pioneering work in stoichiometry.

In mathematical equations the number of moles is given the symbol n. This symbol is also used for many other purposes, however, and only a thorough understanding of any equation will tell whether n represents moles or some other quantity.

In general,

$$1 \text{ mole of X} = 6.02 \times 10^{23} \text{ units of X} \qquad (6.1)$$

Some specific chemistry examples are as follows:

$$1 \text{ mole of argon atoms} = 6.02 \times 10^{23} \text{ Ar atoms}$$

$$1 \text{ mole of } CH_4 \text{ molecules} = 6.02 \times 10^{23} \text{ } CH_4 \text{ molecules}$$

$$1 \text{ mole of } Mg^{2+} \text{ ions} = 6.02 \times 10^{23} \text{ } Mg^{2+} \text{ ions}$$

$$1 \text{ mole of NaCl formula units} = 6.02 \times 10^{23} \text{ NaCl formula units}$$

Molar Mass

The periodic table used for the AP Chemistry Exam lists the **relative atomic mass** of each element directly underneath the chemical symbol. Relative atomic mass has no units and simply indicates the mass of one element as compared to that of another. In chemistry it is customary to add grams units to the atomic masses listed in the periodic table, calling them **gram-atomic masses.** One mole of an element is equal to the gram-atomic mass of that element:

$$1 \text{ mole of an element} = \text{gram-atomic mass of that element} \quad (6.2)$$

For a chemical compound the gram-molar mass is equal to the sum of the gram-atomic masses of all atoms in the chemical formula:

$$\text{gram-molar mass of a compound} = \Sigma(\text{gram-atomic masses in formula}) \quad (6.3)$$

Similar to the cases for elements, the **gram-molar mass** of a compound is equal to 1 mole of that compound. Gram-molar masses of ions are determined by the atom(s) present in the ion since the gain or loss of electrons has virtually no effect on the total mass.

$$1 \text{ mole of a compound} = \text{gram-molar mass of that compound} \quad (6.4)$$

Some specific examples of the molar-mass relationships are as follows:

$$1 \text{ mole of argon} = 39.948 \text{ grams of argon}$$
$$1 \text{ mole of uranium} = 238.029 \text{ grams of uranium}$$
$$1 \text{ mole of } CH_4 = 16.043 \text{ grams of } CH_4$$
$$1 \text{ mole of NaCl} = 58.4424 \text{ grams of NaCl}$$
$$1 \text{ mole of } Mg^{2+} = 24\ 3050 \text{ grams of } Mg^{2+}$$

For most calculations, masses may be rounded off to whole numbers. In addition, standard abbreviations are used for the equalities. The five examples above are more commonly written as

$$1 \text{ mol Ar} = 40 \text{ g Ar}$$
$$1 \text{ mol U} = 238 \text{ g U}$$
$$1 \text{ mol } CH_4 = 16 \text{ g } CH_4$$
$$1 \text{ mol NaCl} = 58 \text{ g NaCl}$$
$$1 \text{ mol } Mg^{2+} = 24 \text{ g } Mg^{2+}$$

Although it is correct to refer to the gram-atomic mass of an element and the gram-molar mass of a compound, it is common usage to refer to these as simply the atomic mass, A, and molar mass, respectively. Also note that the formula mass is the same as the molar mass, while the term molecular mass should be reserved for molecular substances.

The equalities defined in this section can be used to produce the appropriate conversion factors for conversion calculations. Currently more than ten million compounds are known, and these definitions will give twice as many conversion factors.

EXERCISE 6.6

Using the periodic table, determine the molar mass of each of the following compounds and round to two decimal places:

(a) $Cd(NO_3)_2$
(b) $CH_3(CH_2)_4Br$
(c) $(NH_4)_2SO_4$
(d) $(CH_3CH_2CH_2)_2O$
(e) $CuSO_4 \cdot 5H_2O$

Solution

(a) 236.42
(b) 151.04
(c) 132.14
(d) 102.18
(e) 249.69

Conversion Factors from Chemical Formulas

Because of its structure, the formula for an ionic compound is an empirical formula representing the simplest ratio of atoms. Formulas for molecular compounds give the numbers and types of all the atoms that make up one molecule. All chemical formulas represent the ratio of atoms within the formula. This fact allows the chemist to write relationships between a formula as a whole and the individual atoms in that formula. For example, common table sugar is sucrose with the formula $C_{12}H_{22}O_{11}$. On the atomic scale this gives the following relationships, where the equal sign is read as "is chemically equivalent to":

$$1 \text{ molecule of } C_{12}H_{22}O_{11} = 12 \text{ atoms of carbon}$$
$$1 \text{ molecule of } C_{12}H_{22}O_{11} = 22 \text{ atoms of hydrogen}$$
$$1 \text{ molecule of } C_{12}H_{22}O_{11} = 11 \text{ atoms of oxygen}$$
$$12 \text{ atoms of carbon} = 22 \text{ atoms of hydrogen}$$
$$12 \text{ atoms of carbon} = 11 \text{ atoms of oxygen}$$
$$22 \text{ atoms of hydrogen} = 11 \text{ atoms of oxygen}$$

In short, one sucrose molecule gives the chemist six different relationships with which to construct conversion factors. (It should be emphasized that these are not true equalities. The equal signs in these equations should be read as "is chemically equivalent to"; then the first relationship actually says that 1 molecule of $C_{12}H_{22}O_{11}$ "is chemically equivalent to" 12 atoms of carbon.)

It is rare that the chemist thinks of these relationships in terms of molecules and atoms. Rather they are viewed as moles of molecules or moles of atoms. In essence, each side of the relationships given above is multiplied by Avogadro's number to obtain the chemical equivalences:

$$1 \text{ mol } C_{12}H_{22}O_{11} = 12 \text{ mol C}$$
$$1 \text{ mol } C_{12}H_{22}O_{11} = 22 \text{ mol H}$$
$$1 \text{ mol } C_{12}H_{22}O_{11} = 11 \text{ mol O}$$
$$12 \text{ mol C} = 22 \text{ mol H}$$
$$12 \text{ mol C} = 11 \text{ mol O}$$
$$22 \text{ mol H} = 11 \text{ mol O}$$

Finally, it is important to remember that these relationships are true only for the specified compound. They may be different for other compounds.

EXERCISE 6.7

How many different chemical equivalences may be written for the following formulas?

(a) NaCl
(b) $FeCl_3$
(c) $NiSO_4$
(d) $(NH_4)_3PO_4$
(e) O_3

Solution

(a) 3
(b) 3
(c) 6
(d) 10
(e) 1

Conversion Factors from Balanced Chemical Equations

While the chemical formula of a compound provides many relationships for conversion factors, these relationships are limited to that single compound. A balanced chemical equation, however, gives a group of relationships that can be used to generate additional conversion factors. Consider the balanced equation for the combustion of benzene, C_6H_6:

$$2C_6H_6 + 15O_2 \rightarrow 12CO_2 + 6H_2O$$

On a mole basis the following equivalences can be obtained:

$$2 \text{ mol } C_6H_6 = 15 \text{ mol } O_2$$
$$2 \text{ mol } C_6H_6 = 12 \text{ mol } CO_2$$
$$2 \text{ mol } C_6H_6 = 6 \text{ mol } H_2O$$
$$15 \text{ mol } O_2 = 12 \text{ mol } CO_2$$
$$15 \text{ mol } O_2 = 6 \text{ mol } H_2O$$
$$12 \text{ mol } CO_2 = 6 \text{ mol } H_2O$$

Once again, the equal sign does not represent mathematical equality but it should be read as "is equivalent to." These equalities apply only to the balanced equation from which they are derived.

Concentration and Density as Conversion Factors

The concentration of a solution can also be used as a conversion factor for conversion calculations. The most common concentration unit used in chemistry is molarity. **Molarity** (M) is the number of moles of a solute that are dissolved in 1 liter of solution. When a problem gives the concentration of a solution, such as 0.250 molar NaOH (also written as 0.250 M NaOH), this value can be made into a conversion factor by specifying its units:

$$0.250\ M\ \text{NaOH} = \frac{0.250\ \text{mol NaOH}}{1\ \text{L NaOH}}$$

This ratio is a conversion factor for conversions between moles of NaOH and liters of NaOH solution. As with all conversion factors, its inverse is also a conversion factor:

$$\frac{1\ \text{L NaOH}}{0.250\ \text{mol NaOH}}$$

In many instances the volume is expressed in milliliters (mL). Then the two conversion factors can be written as

$$\frac{0.250\ \text{mol NaOH}}{1000\ \text{mL NaOH}} \quad \text{and} \quad \frac{1000\ \text{mL NaOH}}{0.250\ \text{mol NaOH}}$$

since there are 1000 mL in each liter.

The density of a substance may be used as a conversion factor for conversions between volume and mass. Most densities in chemistry are given in units of grams per cubic centimeter (g cm^{-3}). If an organic liquid has a density of $0.741\ \text{g cm}^{-3}$, this fact may be written as the conversion factor:

$$\text{Density} = \frac{0.741\ \text{g}}{\text{cm}^3}$$

As with molarity, the inverse of this ratio is the other conversion factor obtained from the density:

$$\frac{\text{cm}^3}{0.741\ \text{g}}$$

Since 1 cm^3 is the same as 1 mL, we can interchange the two terms as needed to obtain

$$\frac{\text{mL}}{0.741\ \text{g}} \quad \text{and} \quad \frac{0.741\ \text{g}}{\text{mL}}$$

TIP
Remember:
$1.0\ \text{mL} = 1.0\ \text{cm}^3$

Other Conversion Factors

Another useful equality for constructing a conversion factor is the relationship between the moles of a gas and the volume of a gas at standard temperature and pressure (see Chapter 4). Standard temperature is 0 °C, and standard pressure is 1 atmosphere of pressure. Under these conditions 1 mole of a gas occupies 22.4 liters. The equality and its two conversion factors are as follows:

$$1\ \text{mol gas} = 22.4\ \text{L gas}$$

$$1 = \frac{22.4\ \text{L}}{1\ \text{mol gas}}$$

and

$$1 = \frac{1\ \text{mol gas}}{22.4\ \text{L}}$$

While this expression specifically refers to an ideal gas, this conversion factor can be used for calculations involving most real gases with little error.

Finally, there are several equalities that are very useful to remember:

$$1 \text{ cm}^3 \quad = \quad 1 \text{ mL}$$

$$1 \text{ L} \quad = \quad 1000 \text{ mL} \quad = \quad 1000 \text{ cm}^3$$

$$1 \text{ cm}^3 \text{ (H}_2\text{O)} \quad = \quad 1 \text{ g H}_2\text{O}$$

THE CONVERSION SEQUENCE

Once the various equalities and relationships are known, they must be used in the correct way to perform stoichiometric calculations. It is important to understand the sequence of operations required to perform any conversion successfully and efficiently.

Figure 6.1 illustrates how conversions in chemistry are related to each other. This diagram shows the sequence of conversions from any given item of chemical information to any other that may be desired. The notations along the arrows indicate the type of conversion factor needed to perform each conversion. At most, a conversion will require three steps, not including any conversions of metric prefixes.

To start analyzing this diagram, note that the left-side boxes all refer to "SUBSTANCE A," and those on the right side to "SUBSTANCE B." The examples that follow in the text will be divided into two groups. The first group involves only the conversion of units of one substance and therefore uses only the SUBSTANCE A side of the diagram. A typical question might be "How many atoms of iron are in a 2.00-gram sample of Fe?" The second group involves problems in which a given amount of substance A is converted into an equivalent amount of substance B. These conversions start on the SUBSTANCE A side of the diagram and end on the SUBSTANCE B side. A typical question might be "How many grams of carbon are there in 10.0 grams of $Fe_2(CO_3)_3$?" In most problems the given information will be found as one of the boxes on the SUBSTANCE A side of the diagram. The box representing the desired units of the answer is then located, and conversions are made step by step, using the indicated conversion factors. In some problems, the given information may not be exactly in the form shown in Figure 6.1. For instance, you may be given milligrams, mg, of a substance instead of grams, g. In those cases, you will need to convert the given units to the needed starting units.

Calculations Involving One Substance

Some stoichiometric questions involve converting from one set of units to another set for the same chemical substance. In this case we focus entirely on the left side of Figure 6.1. Some sample questions are given below. To help clarify the method, use a colored pen or pencil to perform the unit cancellations in the following problems.

EXERCISE 6.8

How many grams of $FeCl_3$ (molar mass $= 162.3$) need to be weighed to have 0.456 mol of $FeCl_3$?

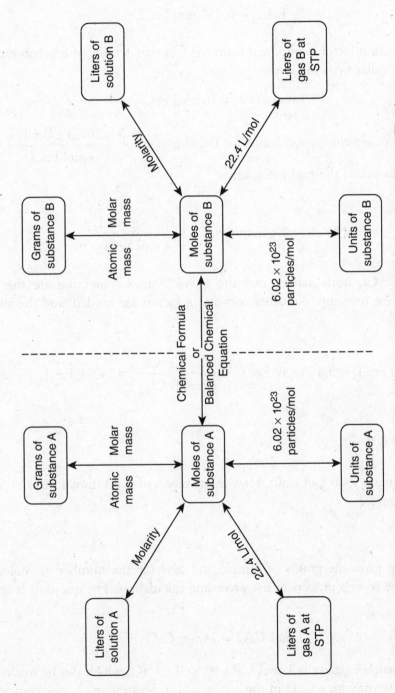

FIGURE 6.1. Diagram of the sequence of steps used in stoichiometry calculations. The box corresponding to the given data is found on the SUBSTANCE A side of the diagram. The box corresponding to the desired data is then located. The notations alongside the arrows tell what information is used to construct the required conversion factor.

Solution

This problem gives the number of moles of $FeCl_3$ and asks for grams. This is a one-step conversion that uses the equality between the molar mass of $FeCl_3$ and the moles of $FeCl_3$ as the conversion factor. The question to be answered is set up as

$$? \text{ g } FeCl_3 = 0.456 \text{ mol } FeCl_3$$

The conversion factor is obtained from the fact that 1 mole of any substance is equal to the molar mass in grams:

$$1 \text{ mol } FeCl_3 = 162.3 \text{ g } FeCl_3$$

The appropriate conversion factor for the conversion is $\left(\dfrac{162.3 \text{ g } FeCl_3}{1 \text{ mol } FeCl_3} \right)$ since it allows us to cancel the mol $FeCl_3$ units:

$$? \text{ g } FeCl_3 = 0.456 \text{ mol } FeCl_3 \left(\frac{162.3 \text{ g } FeCl_3}{1 \text{ mol } FeCl_3} \right)$$

The mol $FeCl_3$ units cancel, and the g $FeCl_3$ units remaining are the ones requested in the question. No other conversion factors are needed, and the answer is calculated as

$$? \text{ g } FeCl_3 = 0.456 \text{ mol } FeCl_3 \left(\frac{162.3 \text{ g } FeCl_3}{1 \text{ mol } FeCl_3} \right) = 74.0 \text{ g } FeCl_3$$

EXERCISE 6.9

A sample contains 24.6 g of CaO. How many moles of CaO (molar mass = 56.0) are in this sample?

Solution

This problem gives the grams of sample and asks for the number of moles. In effect, it is the reverse process of the preceding calculation. The question is set up as

$$? \text{ mol } CaO = 24.6 \text{ g } CaO$$

The conversion equality is 1 mol CaO = 56.0 g CaO, which can be made into a conversion factor with g CaO in the denominator. Multiplying by the conversion factor gives

$$? \text{ mol } CaO = 24.6 \text{ g } CaO \left(\frac{1 \text{ mol } CaO}{56.0 \text{ g } CaO} \right)$$

When the g CaO units are canceled, the mol CaO units remain. These are the desired units, and the result is then calculated as

$$? \text{ mol CaO} = 24.6 \text{ g CaO} \left(\frac{1 \text{ mol CaO}}{56.0 \text{ g CaO}} \right) = 0.439 \text{ mol CaO}$$

EXERCISE 6.10

A solution has a molarity of 0.658 mol $MgBr_2 L^{-1}$. How many moles of $MgBr_2$ are in 0.400 L of this solution?

Solution

This problem requires that the moles of $MgBr_2$ be determined, but two numerical items of information are given. From Figure 6.1 it is seen that the molarity is used as a conversion factor, and therefore the starting point is the liters of solution given. The question is set up as

$$? \text{ mol } MgBr_2 = 0.400 \text{ L } MgBr_2$$

The molarity is already a ratio:

$$0.658 \, M \, MgBr_2 = \frac{0.658 \text{ mol } MgBr_2}{1 \text{ L } MgBr_2}$$

and may be used as the conversion factor:

$$? \text{ mol } MgBr_2 = 0.400 \text{ L } MgBr_2 \left(\frac{0.658 \text{ mol } MgBr_2}{1 \text{ L } MgBr_2} \right)$$

Canceling the units and solving give the answer:

$$? \text{ mol } MgBr_2 = 0.263 \text{ mol } MgBr_2$$

> **TIP**
>
> It is often assumed that the 1 L in this type of problem is an exact number and does not affect the number of significant figures in the answer.

EXERCISE 6.11

The same $MgBr_2$ solution as in Exercise 6.10 must be used to obtain 0.500 mol of $MgBr_2$. How many milliliters of this solution are needed?

Solution

Now we must calculate the volume from the number of moles given, and the setup starts with

$$? \text{ mL } MgBr_2 = 0.500 \text{ mol } MgBr_2$$

Once again, the molarity is the conversion factor. However, it cannot be used directly since the units will not cancel. The molarity ratio is inverted and then used in the equation as

$$? \text{ mL MgBr}_2 = 0.500 \text{ mol MgBr}_2 \left(\frac{1 \text{ L MgBr}_2}{0.658 \text{ mol MgBr}_2} \right)$$

Although the mol units cancel properly, the answer will be calculated in liters, not the milliliters requested. We must use an additional conversion factor to change the prefix of the liter units. The appropriate conversion factor is $\left(\dfrac{1 \text{ mL}}{10^{-3} \text{ L}} \right)$. Multiplying by this conversion factor and canceling the L units, we obtain the desired mL units:

$$? \text{ mL MgBr}_2 = 0.500 \text{ mol MgBr}_2 \left(\frac{1 \text{ L MgBr}_2}{0.658 \text{ mol MgBr}_2} \right) \left(\frac{1 \text{ mL}}{10^{-3} \text{ L}} \right)$$

The answer is 760 mL of $MgBr_2$ solution.

Exercises 6.8–6.11 demonstrated the one-step conversion of data to and from mole units. Many common calculations, however, involve two steps, as shown below.

EXERCISE 6.12

How many grams of KCl (molar mass = 74.6) are there in 0.250 L of a 0.300 molar solution of KCl?

Solution

From Figure 6.1 we see that to get from the given volume of the solution to the grams required for the answer involves two steps. The first step uses the molarity as a conversion factor to convert to moles, and then the second step uses the molar mass conversion factor to convert to grams. The problem starts with the volume of the solution:

$$? \text{ g KCl} = 0.250 \text{ L KCl}$$

Next the molarity is used to convert to moles:

$$? \text{ g KCl} = 0.250 \text{ L KCl} \left(\frac{0.300 \text{ mol KCl}}{1 \text{ L KCl}} \right)$$

Then the conversion factor for the molar mass is used:

$$? \text{ g KCl} = 0.250 \text{ L KCl} \left(\frac{0.300 \text{ mol KCl}}{1 \text{ L KCl}} \right) \left(\frac{74.6 \text{ g KCl}}{1 \text{ mol KCl}} \right)$$

After canceling the L KCl and mol KCl units, we have the desired g KCl units, and the calculation, can be made:

$$? \text{ g KCl} = 0.250 \text{ L KCl} \left(\frac{0.300 \text{ mol KCl}}{1 \text{ L KCl}} \right) \left(\frac{74.6 \text{ g KCl}}{1 \text{ mol KCl}} \right)$$

$$= 5.60 \text{ g KCl}$$

EXERCISE 6.13

What is the mass of one molecule of CH_4 (molar mass $= 16.0$ g CH_4/mol CH_4)?

Solution

This problem involves another two-step calculation, starting with one molecule of CH_4 and ending with the number of grams of CH_4. The conversion involves the use of Avogadro's number and the molar mass of CH_4 as the conversion factors. The question is set up as

$$? \text{ g } CH_4 = 1 \text{ molecule } CH_4$$

Figure 6.1 shows that the first conversion uses Avogadro's number, 6.02×10^{23}, as a conversion factor. The units for Avogadro's number in this problem are molecules of CH_4, and the equality used is:

$$1 \text{ mol } CH_4 = 6.02 \times 10^{23} \text{ molecules } CH_4$$

Setting up the conversion factor properly, so that the units of molecules CH_4 cancel, gives

$$? \text{ g } CH_4 = 1 \text{ molecule } CH_4 \left(\frac{1 \text{ mol } CH_4}{6.02 \times 10^{23} \text{ molecules } CH_4} \right)$$

The next step is to use the molar mass to convert from moles to grams:

$$? \text{ g } CH_4 = 1 \text{ molecule } CH_4 \left(\frac{1 \text{ mol } CH_4}{6.02 \times 10^{23} \text{ molecules } CH_4} \right) \left(\frac{16.0 \text{ g } CH_4}{1 \text{ mol } CH_4} \right)$$

Since all units cancel properly, the calculation can now be performed to obtain

$$? \text{ g } CH_4 = 2.66 \times 10^{-23} \text{ g } CH_4$$

as the mass of one molecule of CH_4.

EXERCISE 6.14

How many molecules of CO_2 are contained in a 3.00-L flask at standard temperature and pressure? (Assume that CO_2 behaves as an ideal gas.)

Solution

This problem starts with the volume of a gas and ends with the number of molecules of CO_2. In the first step the volume of gas is converted to moles, using the fact that

1 mol of a gas occupies 22.4 L; then the moles are converted to molecules using Avogadro's number. The setup of the question is

$$? \text{ molecules } CO_2 = 3.00 \text{ L } CO_2$$

This is multiplied by the ratio $\left(\dfrac{1 \text{ mol } CO_2}{22.4 \text{ L } CO_2} \right)$:

$$? \text{ molecules } CO_2 = 3.00 \text{ L } CO_2 \left(\frac{1 \text{ mol } CO_2}{22.4 \text{ L } CO_2} \right)$$

The next step is to convert to molecules, being sure that the units of the ratio cancel properly:

$$? \text{ molecules } CO_2 = 3.00 \text{ L } CO_2 \left(\frac{1 \text{ mol } CO_2}{22.4 \text{ L } CO_2} \right) \left(\frac{6.02 \times 10^{23} \text{ molecules } CO_2}{1 \text{ mol } CO_2} \right)$$

The units cancel properly, and the answer is calculated as

$$? \text{ molecules } CO_2 = 8.06 \times 10^{22} \text{ molecules } CO_2$$

EXERCISE 6.15

Perform each of the following conversions:

(a) 26.5 g of $MgCl_2$ to moles of $MgCl_2$
(b) 3.456 mol of CH_4 to grams of CH_4
(c) 6.57×10^{18} atoms of Fe to moles of Fe
(d) 2.22 mol of O_2 to molecules of O_2
(e) 1.45 mol of KCl to liters of KCl with a molarity of 0.135
(f) 23.5 mL of 0.766 M HF to moles of HF
(g) 1.46 L of CO_2 at STP to moles of CO_2
(h) 0.025 mol of N_2 to liters of N_2
(i) 26.5 g of $MgCl_2$ to liters of 0.200 M $MgCl_2$ solution
(j) 3.456 g of CH_4 to liters of CH_4 at STP
(k) 6.57×10^{18} atoms of Fe to grams of Fe
(l) 2.22 g of O_2 to molecules of O_2
(m) 1.45 L of HCl at STP to liters of HCl with a molarity of 0.135
(n) 23.5 mL of 0.766 M HF to liters of HF gas at STP
(o) 0.025 g of N_2 to liters of N_2
(p) 1.46 L of CO_2 at STP to molecules of CO_2

Solution

(a) 0.278 mol $MgCl_2$
(b) 55.44 g CH_4
(c) 1.09×10^{-5} mol Fe
(d) 1.34×10^{24} molecules O_2
(e) 10.7 L KCl
(f) 0.0180 mol HF

(g) 0.0652 mol CO_2

(h) 0.56 L N_2

(i) 1.39 L $MgCl_2$

(j) 4.83 L CH_4

(k) 6.10×10^{-4} g Fe

(l) 4.18×10^{22} molecules O_2

(m) 0.479 L HCl(aq)

(n) 0.403 L HF(g)

(o) 0.0200 L N_2

(p) 3.92×10^{22} molecules CO_2

Calculations Involving Two Substances

To this point all the sample calculations have started and ended with the same substance. In Figure 6.1, all of the conversions took place between the boxes labeled "SUBSTANCE A." When we start with one substance and end up with a different one, however, the conversion **must always** include the central conversion from moles of substance A to moles of substance B. There is simply no other possible way to perform the conversions. These conversions must use information obtained from a given chemical formula or from a balanced chemical reaction. The conversion factors used here are often called mole ratios.

The simplest two-substance conversions are mole-to-mole conversions, as shown in Exercises 6.16 and 6.17.

EXERCISE 6.16

How many moles of nitrogen atoms are there in 6.50 mol of ammonium phosphate, $(NH_4)_3PO_4$?

Solution

The chemical formula gives the information for the conversion factor; this compound has three nitrogen atoms for each unit of ammonium phosphate. The setup states the question as

$$? \text{ mol N} = 6.50 \text{ mol } (NH_4)_3PO_4$$

This is then multiplied by the conversion factor $\left(\dfrac{3 \text{ mol N}}{1 \text{ mol } (NH_4)_3PO_4} \right)$:

$$? \text{ mol N} = 6.50 \text{ mol } (NH_4)_3PO_4 \left(\dfrac{3 \text{ mol N}}{1 \text{ mol } (NH_4)_3PO_4} \right)$$

The mol $(NH_4)_3PO_4$ units cancel, leaving the mol N units that the problem requests. The answer is calculated as

$$? \text{ mol N} = 19.5 \text{ mol N}$$

EXERCISE 6.17

How many moles of water will be formed in the complete combustion of 2.50 mol of methane, CH_4?

Solution

This problem starts with one substance, methane, and asks a question about a second very different substance, water. A chemical reaction will be needed to solve the problem. Every combustion reaction has oxygen as a reactant and carbon dioxide and water as the products:

$$CH_4 + 2O_2 \rightarrow CO_2 + 2H_2O$$

This equation tells us that 1 mol of methane will form 2 mol of water, and this information is used to construct the conversion factor, $\left(\dfrac{2 \text{ mol } H_2O}{1 \text{ mol } CH_4} \right)$. The question is written as

$$? \text{ mol } H_2O = 2.50 \text{ mol } CH_4$$

Multiplying this by the conversion factor yields

$$? \text{ mol } H_2O = 2.50 \text{ mol } CH_4 \left(\dfrac{2 \text{ mol } H_2O}{1 \text{ mol } CH_4} \right)$$

After cancelling the mol CH_4 units and verifying that the proper units, mol H_2O, have been obtained to satisfy the question, the answer is calculated:

$$? \text{ mol } H_2O = 2.50 \text{ mol } CH_4 \left(\dfrac{2 \text{ mol } H_2O}{1 \text{ mol } CH_4} \right) = 5.00 \text{ mol } H_2O$$

More complex calculations involve adding a step before the mole-to-mole conversion and a step afterwards. In all of these calculations the units for the given information in the problem are found in one of the five boxes on the left or SUBSTANCE A side of Figure 6.1. Then the units requested by the problem are found on the SUBSTANCE B side of the diagram. Proceeding from the given units to the requested units defines the sequence of conversions, and the conversion factors, that are needed to obtain the correct answer.

One of the more common calculations involves calculating the mass of a compound given the mass of another compound and the balanced chemical reaction. Figure 6.1 shows that this procedure involves three steps:

1. Convert the starting mass to moles, using the molar mass of the given compound.
2. Convert the moles of compound A to moles of compound B, using the equivalencies derived from the balanced chemical reaction.
3. Convert from moles back to grams, using the molar mass of the requested compound.

EXERCISE 6.18

Propane, C_3H_8, is a common heating and cooking fuel in rural areas of the country. Propane is also the fuel used in outdoor grills and in small handheld torches. If 100. g of propane is burned in excess oxygen, how many grams of oxygen will be needed? In addition, how many grams of carbon dioxide and water will be formed in the reaction?

Solution

This problem asks for the calculation of three quantities, O_2, CO_2, and H_2O. These will be calculated as three separate problems. First, a balanced chemical equation is needed to determine the relationships between the reactants and products. The reactants and products are listed in Equation (a):

$$C_3H_8 + O_2 \rightarrow CO_2 + H_2O \qquad (a)$$

This is then balanced to obtain

$$C_3H_8 + 5O_2 \rightarrow 3CO_2 + 4H_2O \qquad (b)$$

To calculate the oxygen needed, the question is set up as

$$? \, g \, O_2 = 100. \, g \, C_3H_8$$

The first conversion uses the molar mass of the C_3H_8 molecule, which is $3(12.0) + 8(1.0) = 44.0$:

$$? \, g \, O_2 = 100. \, g \, C_3H_8 \left(\frac{1 \, mol \, C_3H_8}{44.0 \, g \, C_3H_8} \right)$$

The next conversion factor is obtained from Equation (b), which says that 1 mol of C_3H_8 is equivalent to 5 mol of O_2:

$$? \, g \, O_2 = 100. \, g \, C_3H_8 \left(\frac{1 \, mol \, C_3H_8}{44.0 \, g \, C_3H_8} \right) \left(\frac{5 \, mol \, O_2}{1 \, mol \, C_3H_8} \right)$$

The final conversion factor uses the molar mass of the O_2 molecule, $16.0 + 16.0 = 32.0$:

$$? \, g \, O_2 = 100. \, g \, C_3H_8 \left(\frac{1 \, mol \, C_3H_8}{44.0 \, g \, C_3H_8} \right) \left(\frac{5 \, mol \, O_2}{1 \, mol \, C_3H_8} \right) \left(\frac{32.0 \, g \, O_2}{1 \, mol \, O_2} \right)$$

The units cancel to leave only units of g O_2. The answer is calculated as

$$? \, g \, O_2 = 364 \, g \, O_2$$

The remaining question in the statement of the problem is answered by using the following setups:

$$? \text{ g CO}_2 = 100. \text{ g C}_3\text{H}_8 \left(\frac{1 \text{ mol C}_3\text{H}_8}{44.0 \text{ g C}_3\text{H}_8} \right) \left(\frac{3 \text{ mol CO}_2}{1 \text{ mol C}_3\text{H}_8} \right) \left(\frac{44.0 \text{ g CO}_2}{1 \text{ mol CO}_2} \right)$$

$$= 300. \text{ g CO}_2$$

$$? \text{ g H}_2\text{O} = 100. \text{ g C}_3\text{H}_8 \left(\frac{1 \text{ mol C}_3\text{H}_8}{44.0 \text{ g C}_3\text{H}_8} \right) \left(\frac{4 \text{ mol H}_2\text{O}}{1 \text{ mol C}_3\text{H}_8} \right) \left(\frac{18.0 \text{ g H}_2\text{O}}{1 \text{ mol H}_2\text{O}} \right)$$

$$= 164 \text{ g H}_2\text{O}$$

> **TIP**
>
> All of our conversions must obey the law of conservation of matter.

Reviewing this problem, we see that it started with 100. grams of C_3H_8, which was found to require 364 grams of O_2 for complete combustion. The masses of the products were calculated as 300. grams of CO_2 and 164 grams of H_2O.

The total mass of the reactants, 100. g C_3H_8 and 364 g O_2, is 464 grams. At the end of the reaction, the total mass of the products, 300. g CO_2 and 164 g H_2O, is also 464 grams. The law of conservation of mass states that matter cannot be created or destroyed in a chemical reaction. Since this law cannot be violated, we expect that the results of our calculations will obey it. Obtaining the same total mass for the reactants and the products indicates that the law of conservation of mass was not violated.

It is always an advantage to be able to estimate an answer to a problem to assure that no errors were made in calculations. In stoichiometry problems, only mass to mass conversions allow us to do this easily. In 95 percent of these problems, the calculated mass is between one-fifth and five times the given mass. For the problem above, this means that the answers should be between 20. and 500. grams. All of our results fell in that range, giving added confidence that the conversions were correctly done. If the results did not fit the estimates, we would be well advised to carefully recheck our calculations. The 5 percent of reactions that do not follow this general rule are those in which the molar masses of the compounds are very different, as, for example, those of H_2 and Zn.

In some instances the information given is in the form of the volume and molarity of a reactant that produces a precipitate. These calculations also involve a three-step conversion:

1. Start with the given volume and convert to moles, using the molarity of the given solution.
2. Use the balanced chemical reaction to calculate the moles of product.
3. Convert the moles of product to grams, using the molar mass.

EXERCISE 6.19

A 45.0 mL sample of 0.300 molar $FeCl_3$ is reacted with enough NaOH solution to precipitate all the iron as $Fe(OH)_3$. How many grams of $Fe(OH)_3$ will be precipitated?

Solution

First a balanced chemical reaction must be written:

$$FeCl_3 + 3NaOH \rightarrow Fe(OH)_3 + 3NaCl \qquad \text{(c)}$$

The question is set up as

$$? \text{ g Fe(OH)}_3 = 45.0 \text{ mL FeCl}_3$$

The first conversion factor is the molarity, written as

$$0.300 \; M \text{ FeCl}_3 = \frac{0.300 \text{ mol FeCl}_3}{1000 \text{ mL FeCl}_3}$$

The denominator of this conversion factor includes the conversion from liters to milliliters without using another conversion factor:

$$? \text{ g Fe(OH)}_3 = 45.0 \text{ mL FeCl}_3 \left(\frac{0.300 \text{ mol FeCl}_3}{1000 \text{ mL FeCl}_3} \right)$$

The next conversion factor is obtained from the relationships in the chemical reaction shown in Equation (c), where 1 mol of $FeCl_3$ is equivalent to 1 mol of $Fe(OH)_3$:

$$? \text{ g Fe(OH)}_3 = 45.0 \text{ mL FeCl}_3 \left(\frac{0.300 \text{ mol FeCl}_3}{1000 \text{ mL FeCl}_3} \right) \left(\frac{1 \text{ mol Fe(OH)}_3}{1 \text{ mol FeCl}_3} \right)$$

Finally, the moles of $Fe(OH)_3$ are converted to grams by using the molar mass, 107, for $Fe(OH)_3$:

$$? \text{ g Fe(OH)}_3 =$$

$$45.0 \text{ mL FeCl}_3 \left(\frac{0.300 \text{ mol FeCl}_3}{1000 \text{ mL FeCl}_3} \right) \cdot \left(\frac{1 \text{ mol Fe(OH)}_3}{1 \text{ mol FeCl}_3} \right) \left(\frac{107 \text{ g Fe(OH)}_3}{1 \text{ mol Fe(OH)}_3} \right)$$

Since the units cancel, the answer is calculated as

$$? \text{ g Fe(OH)}_3 = 1.44 \text{ g Fe(OH)}_3$$

In many reactions it is important to know the volume of one reactant that will react with a given volume of a second reactant. The molarities of both reactants must be given for this type of problem to be solved.

EXERCISE 6.20

How many milliliters of a 0.250 M NaOH solution are needed to completely neutralize 65.0 mL of a 0.400 M solution of sulfuric acid?

Solution

A balanced equation is required. Since the problem states that the sulfuric acid, H_2SO_4, is completely neutralized, both protons on the sulfuric acid react with the NaOH:

$$H_2SO_4 + 2NaOH \rightarrow Na_2SO_4 + 2H_2O \tag{d}$$

Figure 6.1 indicates another three-step calculation:

1. Convert milliliters of H_2SO_4 to moles of H_2SO_4, using the molarity of H_2SO_4.
2. Convert moles of H_2SO_4 to moles of NaOH using the balanced chemical equation.
3. Convert moles of NaOH to milliliters, using the molarity of NaOH.

The initial setup of the question is

$$? \text{ mL NaOH} = 65.0 \text{ mL } H_2SO_4$$

Using the molarity of the H_2SO_4 as a conversion factor gives

$$? \text{ mL NaOH} = 65.0 \text{ mL } H_2SO_4 \left(\frac{0.400 \text{ mol } H_2SO_4}{1000 \text{ mL } H_2SO_4} \right)$$

Then Equation (d) is used to convert to moles NaOH:

$$? \text{ mL NaOH} = 65.0 \text{ mL } H_2SO_4 \left(\frac{0.400 \text{ mol } H_2SO_4}{1000 \text{ mL } H_2SO_4} \right)\left(\frac{2 \text{ mol NaOH}}{1 \text{ mol } H_2SO_4} \right)$$

Finally, the molarity of the NaOH is used to convert moles NaOH to milliliters NaOH:

$$? \text{ mL NaOH} =$$

$$65.0 \text{ mL } H_2SO_4 \left(\frac{0.400 \text{ mol } H_2SO_4}{1000 \text{ mL } H_2SO_4} \right)\left(\frac{2 \text{ mol NaOH}}{1 \text{ mol } H_2SO_4} \right)\left(\frac{1000 \text{ mL NaOH}}{0.250 \text{ mol NaOH}} \right)$$

Since the units cancel properly, the answer may be calculated as 208 mL.

In some chemical reactions a gas is evolved as one of the products. The most common cases are the reactions of active metals with mineral acids and the reactions of carbonates with acids. It is possible to calculate the volume of gas from a chemical reaction at standard temperature and pressure (STP = 0 °C and 1 atm). (If the final conditions are not at STP, see Chapter 4 on gases for further calculations using the ideal gas law.) The volume of gas evolved from a given mass or volume of reactant is calculated in Exercises 6.21–6.23.

EXERCISE 6.21

Metallic copper reacts with concentrated nitric acid to produce nitrogen dioxide. Calculate the volume of NO_2 that will form at STP when 1.25 g of copper is completely reacted according to the equation

$$Cu(s) + 4HNO_3(aq) \rightarrow Cu(NO_3)_2(aq) + 2NO_2(g) + 2H_2O(l) \qquad (e)$$

Solution

We have the chemical reaction, and need to follow three steps to convert the grams of copper to the volume of $NO_2(g)$ formed:

TIP

Solving all problems of this type requires a start and end set of units.

1. Convert grams Cu to moles Cu, using the atomic mass.
2. Convert moles Cu to moles NO_2, using the balanced equation.
3. Convert moles NO_2 to volume, using the molar volume of an ideal gas.

We start with the setup of the question:

$$? \, L \, NO_2 = 1.25 \, g \, Cu$$

Using the atomic mass of copper as a conversion factor, we get

$$? \, L \, NO_2 = 1.25 \, g \, Cu \left(\frac{1 \, mol \, Cu}{63.55 \, g \, Cu} \right)$$

The next conversion factor involves the balanced chemical equation given in Equation (e), which states that 1 mol Cu will form 2 mol NO_2. The equation becomes

$$? \, L \, NO_2 = 1.25 \, g \, Cu \left(\frac{1 \, mol \, Cu}{63.55 \, g \, Cu} \right) \left(\frac{2 \, mol \, NO_2}{1 \, mol \, Cu} \right)$$

Finally, the fact that 1 mol of an ideal gas at STP occupies 22.4 L is used as a conversion factor to obtain

$$? \, L \, NO_2 = 1.25 \, g \, Cu \left(\frac{1 \, mol \, Cu}{63.55 \, g \, Cu} \right) \left(\frac{2 \, mol \, NO_2}{1 \, mol \, Cu} \right) \left(\frac{22.4 \, L \, NO_2}{1 \, mol \, NO_2} \right)$$

Since all of the units cancel properly, the answer may be calculated as

$$? \, L \, NO_2 = 0.881 \, L \, NO_2 \text{ at STP}$$

EXERCISE 6.22

Blackboard chalk is almost 100% calcium carbonate, $CaCO_3$. What volume of carbon dioxide, CO_2, will be evolved at STP if an excess of chalk is reacted with 35.0 mL of 0.888 molar hydrochloric acid?

Solution

The chemical reaction may be obtained from the facts given in the problem. Knowing that the reactants are HCl and $CaCO_3$ and that one of the products is CO_2, we can readily deduce the other products, H_2O and $CaCl_2$:

$$2HCl(aq) + CaCO_3(s) \rightarrow CO_2(g) + H_2O(l) + 2CaCl_2(aq) \qquad (f)$$

The starting point for the calculation is

$$? \, L \, CO_2 = 35.0 \, mL \, HCl$$

The molarity of the HCl is used to convert to moles HCl:

$$? \, L \, CO_2 = 35.0 \, mL \, HCl \left(\frac{0.888 \, mol \, HCl}{1000 \, mL \, HCl} \right)$$

Next the chemical reaction shown in Equation (f) is used to obtain a conversion factor for the conversion from moles HCl to moles CO_2:

$$? \, L \, CO_2 = 35.0 \, mL \, HCl \left(\frac{0.888 \, mol \, HCl}{1000 \, mL \, HCl} \right) \left(\frac{1 \, mol \, CO_2}{2 \, mol \, HCl} \right)$$

Finally, the molar volume of a gas is used to convert to the volume of CO_2 formed:

$$? \, L \, CO_2 = 35.0 \, mL \, HCl \left(\frac{0.888 \, mol \, HCl}{1000 \, mL \, HCl} \right) \left(\frac{1 \, mol \, CO_2}{2 \, mol \, HCl} \right) \left(\frac{22.4 \, L \, CO_2}{1 \, mol \, CO_2} \right)$$

The units cancel properly, and the answer is calculated as

$$? \, L \, CO_2 = 0.348 \, L \, CO_2 \, at \, STP$$

EXERCISE 6.23

(a) How many grams of water are obtained when 35.6 g of benzene, C_6H_6, are burned in excess oxygen? How many liters of CO_2 at STP will be produced in the same reaction?

(b) How many milliliters of 0.248 *M* HCl are needed to react with 1.36 g of zinc to produce hydrogen gas? How many milliliters of hydrogen gas are expected at STP?

Solution

(a) The balanced reaction is

$$2C_6H_6 + 15O_2 \rightarrow 12CO_2 + 6H_2O$$

$$? \text{ g } H_2O = 35.6 \text{ g } C_6H_6 \left(\frac{1 \text{ mol } C_6H_6}{78.0 \text{ g } C_6H_6} \right) \left(\frac{6 \text{ mol } H_2O}{2 \text{ mol } C_6H_6} \right) \left(\frac{18.0 \text{ g } H_2O}{1 \text{ mol } H_2O} \right)$$

$$= 16.4 \text{ g } H_2O$$

$$? \text{ L } CO_2 = 35.6 \text{ g } C_6H_6 \left(\frac{1 \text{ mol } C_6H_6}{78.0 \text{ g } C_6H_6} \right) \left(\frac{12 \text{ mol } CO_2}{2 \text{ mol } C_6H_6} \right) \left(\frac{22.4 \text{ L } CO_2}{1 \text{ mol } CO_2} \right)$$

$$= 61.3 \text{ L } CO_2$$

(b) The balanced reaction is

$$2HCl + Zn \rightarrow ZnCl_2 + H_2$$

$$? \text{ mL } HCl = 1.36 \text{ g } Zn \left(\frac{1 \text{ mol } Zn}{65.38 \text{ g } Zn} \right) \left(\frac{2 \text{ mol } HCl}{1 \text{ mol } Zn} \right) \left(\frac{1000 \text{ mL } HCl}{0.248 \text{ mol } HCl} \right)$$

$$= 168 \text{ mL } HCl$$

$$? \text{ mL } H_2 = 1.36 \text{ g } Zn \left(\frac{1 \text{ mol } Zn}{65.38 \text{ g } Zn} \right) \left(\frac{1 \text{ mol } H_2}{1 \text{ mol } Zn} \right) \left(\frac{22400 \text{ mL } H_2}{1 \text{ mol } H_2} \right)$$

$$= 466 \text{ mL } H_2$$

Limiting-Reactant Calculations

When chemicals are mixed together under the appropriate conditions, a chemical reaction is started. The reaction will stop when one of the reactants is completely used up. The reactant that is totally consumed, stopping the reaction, is called the **limiting reactant** or **limiting reagent.** The other reactant(s) are called the excess reactant(s). The amount of the limiting reactant determines how much of the other reactant(s) react and how much of each product is formed. Up to this point, only the amount of one reactant has been given in a problem, and this reactant was assumed to be the limiting reactant. When the amounts of two or more reactants are given, special procedures for limiting-reactant calculations must be used.

To understand the concept of a limiting reactant more fully, consider a vending machine that accepts quarters and dimes only and does not give any change. If an item in that machine costs 45 cents, the only way it can be purchased is with two dimes and one quarter. If you have ten dimes and ten quarters, only five 45-cent items can be obtained from this machine since the dimes will run out before the quarters do. Figure 6.2 illustrates this example.

TIP

Prepare a different analogy that will help you recall the concept of a limiting reactant.

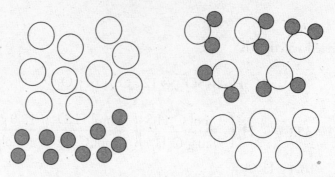

FIGURE 6.2. The vending machine example of a limiting reactant. On the left are the ten quarters (large circles) and ten dimes (small circles). On the right are five groups of one quarter and two dimes each used for 45-cent purchases. Also on the right are five leftover quarters. The dimes are the limiting reactant.

A variety of problems can be solved in the context of the limiting-reactant concept. These include the determination of which reactant is the limiting reactant, the amount of product formed, and the amount of excess reactant that does not react. Examples of these calculations are shown in Exercises 6.24–6.29.

> PROCEDURES FOR SOLVING LIMITING-REACTANT PROBLEMS MUST ALWAYS BE USED WHEN THE AMOUNTS OF TWO OR MORE REACTANTS ARE GIVEN IN THE STATEMENT OF THE PROBLEM.

FINDING THE LIMITING REACTANT:

EXERCISE 6.24

For the reaction below, determine the limiting reactant if 100. g of $FeCl_3$ is reacted with 50.0 g of H_2S:

$$2FeCl_3(aq) + 3H_2S(g) \rightarrow Fe_2S_3(s) + 6HCl(aq)$$

Solution

The first thing to remember is that the compound present in the smallest amount is not necessarily the limiting reactant. To determine the limiting reactant, the amount given for one reactant is converted into the amount of the other reactant that is needed to react with it. Using the procedures above, we can convert the 50 g of H_2S into the number of grams of $FeCl_3$ needed to react with it as follows:

$$? \text{ g } FeCl_3 = 50.0 \text{ g } H_2S \left(\frac{1 \text{ mol } H_2S}{34.0 \text{ g } H_2S} \right) \left(\frac{2 \text{ mol } FeCl_3}{3 \text{ mol } H_2S} \right) \left(\frac{162 \text{ g } FeCl_3}{1 \text{ mol } FeCl_3} \right)$$

$$= 159 \text{ g } FeCl_3$$

From this result we see that 159 g of $FeCl_3$ is needed to react with all 50.0 g of H_2S. However, the amount of $FeCl_3$ given in the problem is only 100. g. The conclusion must be that we will run out of $FeCl_3$ before all of the H_2S can be reacted. Since the $FeCl_3$ is used up first, it is the limiting reactant.

(If the problem had stated that 200. g of $FeCl_3$, instead of 100. g, was available, it would be apparent that we had more than enough $FeCl_3$ to react all of the H_2S. Then we would have concluded that H_2S was the limiting reactant.)

> ONCE A LIMITING REACTANT IS IDENTIFIED, ALL FURTHER CALCULATIONS ARE BASED ON THE AMOUNT OF THE LIMITING REACTANT GIVEN IN THE ORIGINAL STATEMENT OF THE PROBLEM.

DETERMINING THE THEORETICAL YIELD

The term **theoretical yield** refers to the maximum amount of product formed in a reaction based on the amounts of reactants used. This amount is a theoretical yield since no laboratory work is done. Many events can occur in the laboratory that result in less than the theoretical amount of product. One is that the reactants do not combine completely. Another is the possibility of side reactions that produce different products. Products can also be lost by poor lab techniques or in the purification process. In most cases, the theoretical yield refers to the maximum mass of product that can be produced.

EXERCISE 6.25

What is the theoretical yield of Fe_2S_3 that can be obtained from 100. g of $FeCl_3$ and 50.0 g of H_2S?

Solution

These are the same data as in Exercise 6.24, but the question has changed. To solve the problem, we need to know the limiting reactant. In this case it has already been determined as the $FeCl_3$. We now use the 100. g of $FeCl_3$ given in the problem to calculate the mass of the Fe_2S_3.

$$? \text{ g } Fe_2S_3 = 100. \text{ g } Fe_2Cl_3 \left(\frac{1 \text{ mol } FeCl_3}{162.5 \text{ g } FeCl_3} \right) \left(\frac{1 \text{ mol } Fe_2S_3}{2 \text{ mol } FeCl_3} \right) \left(\frac{208 \text{ g } Fe_2S_3}{1 \text{ mol } Fe_2S_3} \right)$$

$$= 64.0 \text{ g } Fe_2S_3$$

(It is essential that the 100. g of $FeCl_3$ given in the problem be used for this calculation. If we had used the 159 grams calculated in Exercise 6.24, a totally incorrect answer would have been obtained.)

Another question that can be asked in a limiting-reactant problem is how much of the excess reactant is left over when the reaction stops. In the vending machine example in Figure 6.2 this can be done by counting the five quarters that combine with two dimes and then subtracting them from the ten quarters we started with. In the chemical calculation the amount of the excess reactant that reacts with the limiting reactant is calculated. This is then subtracted from the starting amount of the excess reactant as given in the problem. Exercise 6.26 illustrates the procedure.

EXERCISE 6.26

When 100. g of $FeCl_3$ is reacted with 50.0 g of H_2S, how many grams of which reactant will be left over when the reaction is complete?

Solution

To determine which reactant is left over, the limiting reactant is identified and then all other reactants become the excess reactants. Since the data are the same for this problem as for Exercises 6.24 and 6.25, we know that $FeCl_3$ is the limiting reactant and, therefore, H_2S is the excess reactant. To determine how much H_2S is left over, we first calculate the number of the grams of H_2S that react:

$$? \text{ g } H_2S = 100. \text{ g } Fe_2Cl_3 \left(\frac{1 \text{ mol } FeCl_3}{162.5 \text{ g } FeCl_3} \right) \left(\frac{3 \text{ mol } H_2S}{2 \text{ mol } FeCl_3} \right) \left(\frac{34.0 \text{ g } H_2S}{1 \text{ mol } H_2S} \right)$$

$$= 31.4 \text{ g } H_2S$$

Since the problem started with 50.0 g of H_2S and 31.5 g reacted, the remaining amount of H_2S must be

$$? \text{ g } H_2S \text{ (left)} = 50.0 \text{ g } H_2S \text{ (start)} - 31.4 \text{ g } H_2S \text{ (reacted)}$$

$$= 18.6 \text{ g } H_2S$$

To complete all of the information about this reaction, we can calculate the theoretical yield of HCl in the same way that the theoretical yield of Fe_2S_3 was calculated.

EXERCISE 6.27

When 100. g of $FeCl_3$ is reacted with 50.0 g of H_2S, how many grams of HCl are formed?

Solution

As before, the limiting reactant must be determined; we already know that it is $FeCl_3$. The 100. grams of $FeCl_3$ given in the problem is used as the starting point for the calculation:

$$? \text{ g } HCl = 100. \text{ g } FeCl_3$$

$$= 100. \text{ g } FeCl_3 \left(\frac{1 \text{ mol } FeCl_3}{162.5 \text{ g } FeCl_3} \right) \left(\frac{6 \text{ mol } HCl}{2 \text{ mol } FeCl_3} \right) \left(\frac{36.5 \text{ g } HCl}{1 \text{ mol } HCl} \right)$$

$$= 67.4 \text{ g } HCl$$

This calculation follows the same principles as the calculation of the amount of Fe_2S_3 in Exercise 6.25.

With this calculation we have determined the masses of all reactants and products in the chemical equation. Listing the masses of the substances before and after reaction demonstrates once again that the law of conservation of matter is obeyed.

REACTION:	$2FeCl_3$	+	$3H_2S$	$\rightarrow$	Fe_2S_3	+	$6HCl$
Start	100 g		50.0 g		0 g		0 g
End	0 g		18.6 g		64.0 g		67.4 g

In this table we can add up the masses of all of the substances at the start of the reaction to get a total of 150 grams. At the end of the reaction the masses again add up to 150 grams. These results show that the law of conservation of mass has been obeyed since total mass at the start and at the end of the reaction is the same.

EXERCISE 6.28

Silver tarnishes in air because of a complex reaction with oxygen and hydrogen sulfide, H_2S, in the air, which may be written as

$$2Ag + O_2 + H_2S \rightarrow Ag_2S + 2H_2O$$

What is the theoretical yield of silver sulfide, Ag_2S, that can be produced from a mixture of 0.200 g of silver, 1.50 L of oxygen at STP, and 65.0 mL of 0.350 molar H_2S solution?

Solution

This is a limiting-reactant problem since the amounts of three different reactants are given. An added complexity is that each of these amounts has a different type of unit. To begin with, the identity of the limiting reactant must be determined. Because there are three reactants, Ag, O_2, and H_2S, a process of elimination is used. First, one pair of reactants is selected to determine which *might be* the limiting reactant, while eliminating the excess reactant. Next, the third reactant and the possible limiting reactant from the first step are used to determine the actual limiting reactant. Choosing silver and oxygen as the first pair, we have

$$? \text{ L O}_2 = 0.200 \text{ g Ag}\left(\frac{1 \text{ mol Ag}}{108 \text{ g Ag}}\right)\left(\frac{1 \text{ mol O}_2}{2 \text{ mol Ag}}\right)\left(\frac{22.4 \text{ L O}_2}{1 \text{ mol O}_2}\right)$$

$$= 0.0207 \text{ L O}_2$$

Since the problem gives the amount of oxygen as 1.5 L, there is plenty of this reactant and it cannot be the limiting reactant, but silver may be. The next step determines the amount of hydrogen sulfide needed to react with the silver given in the problem.

$$? \text{ mL H}_2S = 0.200 \text{ g Ag}\left(\frac{1 \text{ mol Ag}}{108 \text{ g Ag}}\right)\left(\frac{1 \text{ mol H}_2S}{2 \text{ mol Ag}}\right)\left(\frac{1000 \text{ mL H}_2S}{0.350 \text{ mol H}_2S}\right)$$

$$= 2.65 \text{ mL H}_2S$$

Since the question states that there is 65.0 mL of H_2S solution and all that is needed is 2.65 mL, H_2S cannot be the limiting reactant. Because neither H_2S nor O_2 can be the limiting reactant, it must be the silver. Now the theoretical yield of Ag_2S can be calculated, based on the 0.200 g of silver given in the original problem:

$$? \text{ g } Ag_2S = 0.200 \text{ g } Ag \left(\frac{1 \text{ mol } Ag}{108 \text{ g } Ag} \right) \left(\frac{1 \text{ mol } Ag_2S}{2 \text{ mol } Ag} \right) \left(\frac{248 \text{ g } Ag_2S}{1 \text{ mol } Ag_2S} \right)$$

$$= 0.230 \text{ g } Ag_2S$$

EXERCISE 6.29

Silver nitrate, $AgNO_3$ (molar mass = 170.), reacts with sodium chromate, Na_2CrO_4 (molar mass = 162), to form silver chromate (molar mass = 332) and sodium nitrate. If 45.5 mL of 0.200 M $AgNO_3$ is mixed with 35.8 mL of 0.436 M Na_2CrO_4, what is the theoretical yield of the precipitate silver chromate? How many grams of which reactant are left over?

Solution

The reaction is

$$2AgNO_3(aq) + Na_2CrO_4(aq) \times Ag_2CrO_4(s) + 2NaNO_3(aq)$$

To determine the limiting reactant, the calculation is

$? \text{ mL } Na_2CrO_4 =$

$$45.5 \text{ mL } AgNO_3 \left(\frac{0.200 \text{ mol } AgNO_3}{1000 \text{ mL } AgNO_3} \right) \left(\frac{1 \text{ mol } Na_2CrO_4}{2 \text{ mol } AgNO_3} \right) \left(\frac{1000 \text{ mL } Na_2CrO_4}{0.436 \text{ mol } Na_2CrO_4} \right)$$

$$= 10.4 \text{ mL } Na_2CrO_4$$

> **TIP**
>
> The limiting reactant always produces fewer moles or grams of product than the excess reactant.

Since we were given 35.8 mL of Na_2CrO_4, the conclusion is that $AgNO_3$ is the limiting reactant, and all further calculations are based on the given amount of $AgNO_3$:

$? \text{ g } Ag_2CrO_4 =$

$$45.5 \text{ mL } AgNO_3 \left(\frac{0.200 \text{ mol } AgNO_3}{1000 \text{ mL } AgNO_3} \right) \left(\frac{1 \text{ mol } Ag_2CrO_4}{2 \text{ mol } AgNO_3} \right) \left(\frac{332 \text{ g } Ag_2CrO_4}{1 \text{ mol } Ag_2CrO_4} \right)$$

$$= 1.51 \text{ g } Ag_2CrO_4$$

Since $AgNO_3$ is the limiting reactant, Na_2CrO_4 is the excess reactant. The amount of Na_2CrO_4 that reacts is calculated and subtracted from the given amount:

$? \text{ g } Na_2CrO_4 =$

$$45.5 \text{ mL } AgNO_3 \left(\frac{0.200 \text{ mol } AgNO_3}{1000 \text{ mL } AgNO_3} \right) \left(\frac{1 \text{ mol } Na_2CrO_4}{2 \text{ mol } AgNO_3} \right) \left(\frac{162 \text{ g } Na_2CrO_4}{1 \text{ mol } Na_2CrO_4} \right)$$

$$= 0.737 \text{ g } Na_2CrO_4 \text{ reacts}$$

The original number of grams of Na_2CrO_4 is calculated as

$$? \, g \, Na_2CrO_4 \;=\; 35.8 \, mL \, Na_2CrO_4 \left(\frac{0.436 \, mol \, Na_2CrO_4}{1000 \, mL \, Na_2CrO_4} \right) \left(\frac{162 \, g \, Na_2CrO_4}{1 \, mol \, Na_2CrO_4} \right)$$

$$=\; 2.529 \, g \, Na_2CrO_4 \text{ initially present}$$

The amount of sodium chromate left over is calculated by subtraction:

$$? \, g \, Na_2CrO_4 \text{ left over} = 2.529 \, g \, Na_2CrO_4 - 0.737 \, g \, Na_2CrO_4$$

$$= 1.792 \, g \, Na_2CrO_4$$

Titrations

The titration technique used for chemical analysis utilizes the reactions of two solutions. One reactant solution is placed in a beaker, and the other in a buret, which is a long, graduated tube with a stopcock. The stopcock is a valve that allows the chemist to add controlled amounts of solution from the buret to the beaker. An indicator, that is, a compound that changes color when the reaction is complete, is added to the solution in the beaker. The chemist reads the volume of solution in the buret at the start of the experiment and again at the point where the indicator changes color. The difference in these volumes represents the volume of reactant delivered from the buret. Figure 6.3 illustrates the experimental setup.

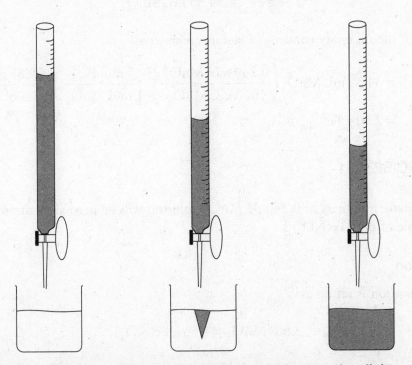

FIGURE 6.3. **Titration experiment showing the initial setup, the slight color change of the indicator before the endpoint is reached, and the colored solution at the endpoint.**

The crucial point about the titration experiment is that the indicator is designed to change color when the amount of reactant delivered from the buret is exactly the amount needed to react with the solution in the beaker. From this experiment, a variety of calculations may be made as shown below.

A classic chemical reaction is the one between Fe^{2+} and the permanganate ion MnO_4^-:

$$5Fe^{2+} + MnO_4^- + 8H^+ \rightarrow Mn^{2+} + 5Fe^{3+} + 4H_2O \qquad (6.5)$$

In titrations, the purple permanganate ion is the indicator of the point where the correct amount has been added to completely react all of the Fe^{2+} ions in the sample. All of the calculations in Exercises 6.30–6.32 refer to Equation 6.5.

EXERCISE 6.30

It takes 34.35 mL of a 0.240 *M* solution of $KMnO_4$ (0.240 *M* MnO_4^-) to titrate an unknown sample of Fe^{2+} to its endpoint. How many grams of Fe^{2+} are in the sample?

Solution

This problem gives the amount of MnO_4^-, and the grams of Fe^{2+} are to be calculated. Figure 6.1 shows the sequence of steps, and the required conversion factors can be determined. The question is set up as

$$? \text{ g Fe}^{2+} = 34.35 \text{ mL MnO}_4^-$$

Then the necessary conversion factors are entered:

$$? \text{ g Fe}^{2+} = 34.35 \text{ mL MnO}_4^- \left(\frac{0.240 \text{ mol MnO}_4^-}{1000 \text{ mL MnO}_4^-} \right) \left(\frac{5 \text{ mol Fe}^{2+}}{1 \text{ mol MnO}_4^-} \right) \left(\frac{55.85 \text{ g Fe}^{2+}}{1 \text{ mol Fe}^{2+}} \right)$$

$$= 2.30 \text{ g Fe}^{2+}$$

EXERCISE 6.31

How many milliliters of 0.240 *M* MnO_4^- solution will be needed to titrate a 1.56 g sample of pure $Fe(NO_3)_2$?

Solution

The question is set up as

$$? \text{ mL MnO}_4^- = 1.56 \text{ g Fe(NO}_3)_2$$

The grams are converted to moles by using the molar mass of $Fe(NO_3)_2$:

$$? \text{ mL MnO}_4^- = 1.56 \text{ g Fe(NO}_3)_2 \left(\frac{1 \text{ mol Fe(NO}_3)_2}{180. \text{ g Fe(NO}_3)_2} \right)$$

Using the ionization reaction

$$\text{Fe(NO}_3)_2 \rightarrow \text{Fe}^{2+} + 2\text{NO}_3^-$$

we can apply the conversion factor $\left(\dfrac{1 \text{ mol Fe}^{2+}}{1 \text{ mol Fe(NO}_3)_2} \right)$ to convert to the Fe^{2+} ion:

$$? \text{ mL MnO}_4^- = 1.56 \text{ g Fe(NO}_3)_2 \left(\frac{1 \text{ mol Fe(NO}_3)_2}{180. \text{ g Fe(NO}_3)_2} \right) \left(\frac{1 \text{ mol Fe}^{2+}}{1 \text{ mol Fe(NO}_3)_2} \right)$$

Equation 6.5 is used to convert to moles of MnO_4^-, and then the molarity is used to convert to milliliters of MnO_4^-.

$$? \text{ mL MnO}_4^- =$$

$$1.56 \text{ g Fe(NO}_3)_2 \left(\frac{1 \text{ mol Fe(NO}_3)_2}{180. \text{ g Fe(NO}_3)_2} \right) \left(\frac{1 \text{ mol Fe}^{2+}}{1 \text{ mol Fe(NO}_3)_2} \right) \left(\frac{1 \text{ mol MnO}_4^-}{5 \text{ mol Fe}^{2+}} \right) \left(\frac{1000 \text{ mL MnO}_4^-}{0.240 \text{ mol MnO}_4^-} \right)$$

$$= 7.22 \text{ mL MnO}_4^-$$

This exercise demonstrates that the stoichiometry calculations described above can be used also for calculations in titration experiments. There is one calculation, however, that the stoichiometry calculations do not address—conversion of the molarity of one solution to the molarity of another. Since the answer desired is molarity, and molarity is not one of the possible starting points in Figure 6.1, a new method for using conversion factors must be developed.

EXERCISE 6.32

What is the molarity of an Fe^{2+} solution if 4.53 mL of a 0.687 M MnO_4^- solution is required to titrate 30.00 mL of the Fe^{2+}-containing solution to the endpoint?

Solution

The desired information is the molarity of the Fe^{2+} solution. The units of molarity are a ratio, and we look for a ratio of units with which to start the problem. The appropriate term is the molarity of the MnO_4^- solution. The setup of the question is as follows:

$$? \frac{\text{mol Fe}^{2+}}{\text{L Fe}^{2+}} = \frac{0.687 \text{ mol MnO}_4^-}{1 \text{ L MnO}_4^-}$$

We need conversion factors to convert the numerator from mol MnO_4^- to mol Fe^{2+} and the denominator from L MnO_4^- to L Fe^{2+}. The conversion factor for the numerator comes from Equation 6.31. We obtain the conversion factor for the denominator from the ratio of the volumes of the two solutions at the endpoint of the titration:

$$? \frac{\text{mol Fe}^{2+}}{\text{L Fe}^{2+}} = \left(\frac{0.687 \text{ mol MnO}_4^-}{1000 \text{ mL MnO}_4^-} \right) \left(\frac{5 \text{ mol Fe}^{2+}}{1 \text{ mol MnO}_4^-} \right) \left(\frac{4.53 \text{ mL MnO}_4^-}{30.00 \text{ mL Fe}^{2+}} \right)$$

$$= \frac{0.519 \text{ mol Fe}^{2+}}{1000 \text{ mL Fe}^{2+}} = 0.519 \text{ } M \text{ Fe}^{2+}$$

EXERCISE 6.33

A solution of HCl is titrated to the endpoint with 23.4 mL of 0.216 M NaOH. (a) How many grams of HCl were in the titrated sample? (b) If the volume of the HCl sample is 50.0 mL, what is the molarity of the HCl solution?

Solution

The reaction is

$$HCl + NaOH \rightarrow NaCl + H_2O$$

(a) To calculate the grams of HCl, we use the equation

$$? \text{ g HCl} = 23.4 \text{ mL NaOH} \left(\frac{0.216 \text{ mol NaOH}}{1000 \text{ mL NaOH}} \right) \left(\frac{1 \text{ mol HCl}}{1 \text{ mol NaOH}} \right) \left(\frac{36.461 \text{ g HCl}}{1 \text{ mol HCl}} \right)$$

$$= 0.184 \text{ g HCl}$$

(b) Calculating the molarity of the HCl solution involves units that are a ratio. The starting point is the molarity of the NaOH, which is also a ratio of units:

$$? \frac{\text{mol HCl}}{\text{L HCl}} = \left(\frac{0.216 \text{ mol NaOH}}{1 \text{ L NaOH}} \right) \left(\frac{1 \text{ mol HCl}}{1 \text{ mol NaOH}} \right) \left(\frac{23.4 \text{ mL NaOH}}{50.0 \text{ mL HCl}} \right)$$

$$= \frac{0.101 \text{ mol HCl}}{\text{L HCl}} = 0.101 \text{ } M \text{ HCl}$$

PERCENT COMPOSITION

The percent composition of a chemical substance tells the chemist how much of each element, or polyatomic ion, is present in a compound on a percent basis. In other words, the percent composition is the mass of an element or polyatomic ion that is present in 100 grams of a chemical compound.

EXERCISE 6.34

What is the percentage of Fe, O, H, and the OH$^-$ polyatomic ion in Fe(OH)$_3$?

Solution

Each of these calculations will start with 100 g of $Fe(OH)_3$, and we calculate the grams of each element in turn to obtain their percentages:

$$? \text{ g Fe} = 100 \text{ g Fe(OH)}_3 \left(\frac{1 \text{ mol Fe(OH)}_3}{106.8 \text{ g Fe(OH)}_3} \right) \left(\frac{1 \text{ mol Fe}}{1 \text{ mol Fe(OH)}_3} \right) \left(\frac{55.8 \text{ g Fe}}{1 \text{ mol Fe}} \right)$$

$$= 52.2 \text{ g Fe}$$

$$\% \text{ Fe} = \frac{52.2 \text{ g Fe}}{100 \text{ g Fe(OH)}_3} \times 100 = 52.2\% \text{ Fe}$$

$$? \text{ g O} = 100 \text{ g Fe(OH)}_3 \left(\frac{1 \text{ mol Fe(OH)}_3}{106.8 \text{ g Fe(OH)}_3} \right) \left(\frac{3 \text{ mol O}}{1 \text{ mol Fe(OH)}_3} \right) \left(\frac{16.0 \text{ g O}}{1 \text{ mol O}} \right)$$

$$= 44.9 \text{ g O}$$

$$\% \text{ O} = \frac{44.9 \text{ g O}}{100 \text{ g Fe(OH)}_3} \times 100 = 44.9\% \text{ O}$$

$$? \text{ g H} = 100 \text{ g Fe(OH)}_3 \left(\frac{1 \text{ mol Fe(OH)}_3}{106.8 \text{ g Fe(OH)}_3} \right) \left(\frac{3 \text{ mol H}}{1 \text{ mol Fe(OH)}_3} \right) \left(\frac{1.0 \text{ g H}}{1 \text{ mol H}} \right)$$

$$= 2.8 \text{ g H}$$

$$\% \text{ H} = \frac{2.8 \text{ g H}}{100 \text{ g Fe(OH)}_3} \times 100 = 2.8\% \text{ H}$$

$$? \text{ g OH}^- = 100 \text{ g Fe(OH)}_3 \left(\frac{1 \text{ mol Fe(OH)}_3}{106.8 \text{ g Fe(OH)}_3} \right) \left(\frac{3 \text{ mol OH}^-}{1 \text{ mol Fe(OH)}_3} \right) \left(\frac{17.0 \text{ g OH}^-}{1 \text{ mol OH}^-} \right)$$

$$= 47.8 \text{ g OH}^-$$

$$\% \text{ OH}^- = \frac{47.8 \text{ g OH}^-}{100 \text{ g Fe(OH)}_3} \times 100 = 47.8\% \text{ OH}^-$$

The percentages of the elements add up to 99.9%. The sum should equal 100%, but does not because the numbers were rounded to one decimal place.

Another method for calculating the percentage composition is to use the formula

$$\text{percent of element} = \frac{\left(\begin{array}{c} \text{atomic mass} \\ \text{of element} \end{array} \right) \left(\begin{array}{c} \text{number of atoms} \\ \text{in formula} \end{array} \right)}{\text{molar mass of compound}} \times 100 \qquad (6.6)$$

A close look at the stoichiometry equations in Exercise 6.34 shows that Equation 6.6 is just a summary of these stoichiometric calculations.

EXERCISE 6.35

What is the percentage of each element in (a) $Ca(NO_3)_2$ and (b) $CH_3CH_2NH_2$? Express your answer to four significant figures.

Solution

(a) Ca = 34.52%, N = 24.13%, O = 41.35%
(b) C = 53.28%, H = 15.65%, N = 31.07%

EMPIRICAL FORMULAS

Given a chemical formula and the atomic masses of the elements, it is possible to calculate the percent composition of a compound. More important is the fact that, with a knowledge of the percent composition and atomic masses, we can deduce the empirical formula of a compound, that is, the simplest ratio of atoms in the compound. Butyric acid has the molecular formula $HC_4H_7O_2$. Its empirical formula is C_2H_4O. Benzene, C_6H_6, has the empirical formula CH. The empirical formula of the sugar glucose, $C_6H_{12}O_6$, is CH_2O. All sugars have the same empirical formula, which is why they are also called carbohydrates (carbo- for the carbon and -hydrate for the water molecule).

To determine the empirical formula of a compound, the simplest ratio of the moles of the atoms in the compound is found by using the following steps:

1. From the given data, calculate the number of moles of each element in the compound.
2. Divide the number of moles of each element obtained in step 1 by the smallest value found to obtain whole-number subscripts for the empirical formula.
3. Note that, if step 2 does not give whole numbers (within ±0.1), the decimal portion of each number will be close to a rational fraction; for example, $0.5 = \frac{1}{2}$, $0.33 = \frac{1}{3}$, $0.67 = \frac{2}{3}$. In this situation, multiply each item of the data by the denominator of the rational fraction to remove the fraction and end up with a whole-number subscript for the empirical formula.

EXERCISE 6.36

What is the empirical formula of a compound that contains 4.0 g of calcium and 7.1 g of chlorine?

Solution

The first step involves converting the grams of Ca and Cl to moles:

$$? \text{ mol Ca} = 4.0 \text{ g Ca} \left(\frac{1 \text{ mol Ca}}{40.0 \text{ g Ca}} \right) = 0.10 \text{ mol Ca}$$

$$? \text{ mol Cl} = 7.1 \text{ g Cl} \left(\frac{1 \text{ mol Cl}}{35.5 \text{ g Cl}} \right) = 0.20 \text{ mol Cl}$$

The second step is to divide both of these answers by the smaller value, 0.10:

$$\frac{0.10 \text{ mol Ca}}{0.10} = 1 \text{ mol Ca}$$

$$\frac{0.20 \text{ mol Cl}}{0.10} = 2 \text{ mol Cl}$$

This tells us that the empirical formula contains 1 mol Ca and 2 mol Cl and is written as $CaCl_2$.

EXERCISE 6.37

A compound containing carbon, hydrogen, and oxygen is found to contain 9.1% H and 54.5% C. What is its empirical formula?

Solution

Since the percentage of an element is the number of grams per 100 g of compound, we may assume a 100 g sample of compound and convert the percent sign directly to gram units. Also, since the percent hydrogen and percent carbon do not add up to 100%, we may conclude that the missing 36.4% is oxygen. Step 1 of the procedure on page 260 is used to calculate the number of moles of each element:

$$? \, mol \, C = 54.5 \, g \, C \left(\frac{1 \, mol \, C}{12.0 \, g \, C} \right) = 4.54 \, mol \, C$$

$$? \, mol \, H = 9.1 \, g \, H \left(\frac{1 \, mol \, H}{1.0 \, g \, H} \right) = 9.1 \, mol \, H$$

$$? \, mol \, O = 36.4 \, g \, O \left(\frac{1 \, mol \, O}{16.0 \, g \, O} \right) = 2.28 \, mol \, O$$

Using step 2 and dividing by the smallest number, 2.28, we obtain

$$\frac{4.54 \, mol \, C}{2.28} = 1.99 \, mol \, C$$

$$\frac{9.1 \, mol \, H}{2.28} = 3.99 \, mol \, H$$

$$\frac{2.28 \, mol \, O}{2.28} = 1.00 \, mol \, O$$

Since these numbers are within ±0.1 of a whole number, they are rounded to 2 mol C, 4 mol H, and 1 mol O, and the empirical formula is written as C_2H_4O.

EXERCISE 6.38

A compound is analyzed and found to contain 74.1% oxygen and 25.9% nitrogen. What is its empirical formula?

Solution

Assuming a 100-g sample, we convert the percentages directly to gram units, and the moles of N and O are calculated as

$$? \text{ mol N} = 25.9 \text{ g N} \left(\frac{1 \text{ mol N}}{14 \text{ g N}} \right) = 1.85 \text{ mol N}$$

$$? \text{ mol O} = 74.1 \text{ g N} \left(\frac{1 \text{ mol O}}{16 \text{ g O}} \right) = 4.63 \text{ mol O}$$

Dividing both answers by 1.85 (step 2) gives

$$\frac{1.85 \text{ mol N}}{1.85} = 1.00 \text{ mol N}$$

$$\frac{4.63 \text{ mol O}}{1.85} = 2.50 \text{ mol O}$$

The 2.50 mol O cannot be rounded to a whole number, but the decimal 0.50 represents the rational fraction $\frac{1}{2}$. We must use step 3, in which all of the data are multiplied by the denominator of the rational fraction—in this case, 2.

$$1.00 \text{ mol N} \times 2 = 2.00 \text{ mol N}$$

$$2.50 \text{ mol O} \times 2 = 5.00 \text{ mol O}$$

These whole numbers are used to write the empirical formula N_2O_5.

EXERCISE 6.39

Determine the empirical formula for each of the following compounds, given its composition:

(a) A compound composed of 17.72 g Cl and 3.10 g P.
(b) A compound that is 24.74% K, 40.50% O, and 34.76% Mn.
(c) A compound containing carbon, hydrogen, and oxygen that is 40.0% C and 6.66% H.

Solution

(a) PCl_5
(b) $KMnO_4$
(c) CH_2O

DETERMINING AN EMPIRICAL FORMULA

1. Determine the mass of each element in the compound. You may have to use one or more of the following techniques:

   a. Convert percentages to grams.
   b. Convert the mass of compound, obtained experimentally, to the mass of an element (e.g., g CO_2 to g C).
   c. Calculate the mass of a missing element (usually oxygen).

2. Convert the mass of each element to moles.

3. Divide moles in step 2 by the smallest number of moles.

4. If the results in step 2 are integers, use them as subscripts.

5. If results in step 2 are not integers, multiply by the appropriate number to obtain integers (usually a small number such as 2, 3, 4, or 5) Use trial and error or the decimal portion of the numbers in step 2 to determine the multiplier.

MOLECULAR FORMULAS

An empirical formula, which gives the simplest ratio of atoms in a molecule, is used to represent an ionic compound. For a molecular, covalent compound, however, the actual molecular formula may be the empirical formula or some whole-number multiple of the empirical formula. Once the empirical formula has been determined as shown in the preceding section, the molar mass of the compound can be used to determine the molecular formula.

The number of empirical formula units in the molecular formula of a compound is determined by dividing the molar mass of the compound by the empirical formula mass. This will result in a small whole number:

$$\frac{\text{molar mass}}{\text{empirical formula mass}} = \text{small whole number} \qquad (6.7)$$

Each and every subscript in the empirical formula is then multiplied by this small whole number to obtain the molecular formula.

EXERCISE 6.40

A compound has an empirical formula of CH_2O, and its molar mass is determined in a separate experiment to be 180 g mol^{-1}. What is the molecular formula of this compound?

Solution

The number of CH_2O units in the molecule is determined from Equation 6.7:

$$\frac{180 \text{ g mol}^{-1}}{30 \text{ g emp. form.}^{-1}} = 6 \text{ empirical formula units per mole}$$

Each subscript in the empirical formula is multiplied by 6 to obtain $C_6H_{12}O_6$. For the AP Exam, you should be familiar with several methods to determine the molar mass.

EXERCISE 6.41

The following empirical formulas were determined, and their molar masses are given in parentheses after the formulas. Determine the molecular formulas.

(a) C_3H_7 (86)
(b) CH_2 (70)
(c) $C_4H_3O_2$ (166)
(d) BH_3 (27.7)
(e) CH_2ON (176)

Solution

(a) C_6H_{14}
(b) C_5H_{10}
(c) $C_8H_6O_4$
(d) B_2H_6
(e) $C_4H_8O_4N_4$

OTHER STOICHIOMETRIC EQUATIONS

Chemists become very familiar with the dimensional analysis method and see shortcuts in the calculation of the moles of a substance from a variety of units, as shown below:

$$\text{moles} = \frac{\text{grams}}{\text{molar mass}} \tag{6.8}$$

$$\text{moles} = \text{molarity} \times \text{liters of solution} \tag{6.9}$$

$$\text{moles} = \frac{\text{liters of gas}}{22.4} \tag{6.10}$$

$$\text{moles} = \frac{\text{molecules or atoms}}{6.02 \times 10^{23}} \tag{6.11}$$

$$\text{moles} = (\text{molarity})(\text{liters}) \tag{6.12}$$

These relationships can speed the calculations, but must be used with care to ensure that the proper units are chosen in all instances.

SUMMARY

This chapter summarizes approximately one half of all the calculations in chemistry. These are stoichiometric calculations, and they involve converting from one set of units to another using a logical system. The favored system is the conversion factor method, which is also called dimensional analysis. This chapter reminds you that three things are needed for a successful stoichiometric calculation. First is an understanding of both the starting point and where you need to end up. Second is a logical sequence of conversions. And third, you need the appropriate conversion factors to apply at each step. Many calculations can be performed using the three-step approach. This includes calculating the mass of an atom or molecule, mass-to-mass calculations, and limiting-reactant calculations. The chapter also shows that the conversion factor method is applicable to titration calculations.

In addition to the above calculations this chapter shows how simple mass data allowed early chemists to deduce the formulas of compounds. Methods for calculating percent composition, empirical formulas, and molecular formulas are presented here.

Important Concepts

Dimensional analysis (conversion factor) method
Equalities from: Avogadro's number, molar masses, chemical formulas, and chemical equations
Stoichiometric conversion sequence
Limiting reactant calculations
Percent composition and empirical formulas
Chemical analysis by titration

Important Equations

moles = grams/molar mass
molarity = moles/liter
moles = molecules or atoms/6.02×10^{23}
moles = liters of gas at STP/22.4

Practice Exercises

Multiple-Choice

For the first four problems below, one or more of the following responses will apply; each response may be used more than once or not at all in these questions.

    I. molarity
    II. mass percentage
    III. molar mass
    IV. empirical formula
    V. molecular formula

1. For the compound $FeCl_3$ we can, with the aid only of a periodic table determine its

    (A) I and III
    (B) II
    (C) III and V
    (D) II, III, and IV
    (E) V

2. The only item on this list that changes as the temperature changes is

    (A) I
    (B) II
    (C) III
    (D) IV
    (E) V

3. We can determine the _____ if we know the _____.

    (A) I, III
    (B) II, III
    (C) III, V
    (D) IV, II
    (E) V, I

4. We can determine the _____ if we know the _____ and _____.

    (A) I, III, II
    (B) II, I, V
    (C) V, III, IV
    (D) IV, I, II
    (E) V, III, I

5. What weight of $KClO_3$ (molar mass = 122.5) is needed to make 200 mL of a 0.150 M solution of this salt?

    (A) 2.73 g
    (B) 3.68 g
    (C) 27.3 g
    (D) 164 g
    (E) 3.69 kg

6. In an experiment 35.0 mL of 0.345 M HNO_3 is titrated with 0.130 M NaOH. What volume of NaOH will have been used when the indicator changes color?

    (A) 35.0 mL
    (B) 13.2 mL
    (C) 26.4 mL
    (D) 50.0 mL
    (E) 92.9 mL

7. In the reaction

$$CaCO_3 + 2HCl \rightarrow H_2O + CO_2 + CaCl_2$$

how many grams of $CaCO_3$ (molar mass = 100.) are needed to produce 3.00 L of CO_2 at STP?

(A) 13.4
(B) 9.11
(C) 5.89
(D) 300
(E) 7.47

8. What is the simplest formula for a compound composed of only carbon and hydrogen and containing 14.3% H?

(A) CH
(B) CH_4
(C) C_4H
(D) CH_2
(E) CH_3

9. To two decimal places, what is the molar mass of $Al(NO_3)_3$?

(A) 165.00
(B) 56.99
(C) 213.00
(D) 88.99
(E) 184.99

10. How many milligrams of Na_2SO_4 (molar mass = 142) are needed to prepare 100. mL of a solution that is 0.00100 M in Na^+ ions?

(A) 28.4
(B) 14,200
(C) 1.00
(D) 7.1
(E) 14.2

11. The wavelength of blue light is 400 nm. What is the wavelength in centimeters?

(A) 4.00×10^{-5} cm
(B) 400×10^{-9} cm
(C) 400×10^{-2} cm
(D) 2.5×10^{6} cm
(E) 2.5×10^{8} cm

12. In the reaction below, how many moles of aluminum will produce 1.0 mol of iron?

$$8Al + 3Fe_3O_4 \rightarrow 9Fe + 4Al_2O_3$$

(A) 1
(B) $3/4$
(C) $9/8$
(D) $8/9$
(E) $4/3$

13. In the following reaction:

$$2KOH + H_2SO_4 \rightarrow K_2SO_4 + 2H_2O$$

35.4 mL of 0.125 M KOH is required to neutralize completely 50.0 mL of H_2SO_4. What is the molarity of the H_2SO_4 solution?

(A) 0.0883 M
(B) 0.100 M
(C) 0.0443 M
(D) 0.125 M
(E) 0.177 M

14. A substance has an empirical formula of CH_2. Its molar mass is determined in a separate experiment as 83.5. What is the most probable molecular formula for this compound?

(A) C_2H_4
(B) C_6H_2
(C) C_4H_2
(D) CH_{12}
(E) C_6H_{12}

15. The mass of one atom of iron is

    (A) 1.66×10^{-24} g
    (B) 2.11×10^{-22} g
    (C) 3.15×10^{-22} g
    (D) 9.28×10^{-23} g
    (E) 3.36×10^{25} g

16. How many grams of the gas SO_2 are there in a 4.00-L sample of SO_2 at STP?

    (A) 256.2
    (B) 11.4
    (C) 358.7
    (D) 2.78×10^{-3}
    (E) 2.86

17. What is the percentage of potassium in K_3PO_4?

    (A) 14.6%
    (B) 29.2%
    (C) 18.4%
    (D) 55.2%
    (E) 39.1%

18. A 0.200-g sample of a compound containing only carbon, hydrogen, and oxygen is burned, and 0.357 g of CO_2 and 0.146 g of H_2O are collected. What is the percentage of carbon in this compound?

    (A) 56.0%
    (B) 73.0%
    (C) 48.7%
    (D) 24.3%
    (E) 43.2%

19. Which of the following is NOT a base SI unit?

    (A) meter
    (B) gram
    (C) mole
    (D) second
    (E) ampere

20. In the reaction

    $$2AgNO_3 + CaCl_2 \rightarrow 2AgCl + Ca(NO_3)_2$$

    how many grams of AgCl (molar mass = 143.5) will precipitate when a solution containing 20.0 g $AgNO_3$ (molar mass = 170) is reacted with a solution containing 15.0 g $CaCl_2$ (molar mass = 111)?

    (A) 16.9
    (B) 38.8
    (C) 33.8
    (D) 8.45
    (E) 67.6

21. In the reaction

    $$2AgNO_3 + CaCl_2 \rightarrow 2AgCl + Ca(NO_3)_2$$

    how many grams of which reactant will remain when a solution containing 20.0 g $AgNO_3$ (molar mass = 170) is reacted with a solution containing 15.0 g $CaCl_2$ (molar mass = 111)?

    (A) 6.53 g $CaCl_2$
    (B) 6.53 g $AgNO_3$
    (C) 45.9 g $CaCl_2$
    (D) 8.47 g $CaCl_2$
    (E) 25.9 g $AgNO_3$

22. A 50.0-g sample of impure $CaCl_2$ is reacted with excess $AgNO_3$ according to the reaction

    $$2AgNO_3 + CaCl_2 \rightarrow 2AgCl + Ca(NO_3)_2$$

    If 5.86 g of AgCl (molar mass = 143.5) precipitates, what is the percentage of chlorine (molar mass = 35.5) in the sample?

    (A) 1.45%
    (B) 2.90%
    (C) 3.80%
    (D) 11.7%
    (E) 8.53%

23. How many liters of air are needed to completely burn 1 mol of methane in air (20% oxygen) at STP according to the reaction

$$CH_4 + 2O_2 \rightarrow CO_2 + 2H_2O$$

(A) 22.4
(B) 44.8
(C) 11.2
(D) 224
(E) 64.0

24. Determine the empirical formula for a compound that is 25% hydrogen and 75% carbon.

(A) CH
(B) CH_2
(C) CH_4
(D) C_2H_8
(E) C_4H

25. One granule of sucrose ($C_{12}H_{22}O_{11}$, molar mass = 342) weighs 2.5 micrograms. How many sucrose molecules are in that granule?

(A) 2.5×10^{17}
(B) 4.4×10^{15}
(C) 6.02×10^{17}
(D) 4.4×10^{21}
(E) 3.42×10^{18}

ANSWER KEY

1. D	6. E	11. A	16. B	21. D
2. A	7. A	12. D	17. D	22. B
3. D	8. D	13. C	18. C	23. D
4. C	9. C	14. E	19. B	24. C
5. B	10. D	15. D	20. A	25. B

See Appendix 1 for explanations of answers.

Free-Response

The following questions involve stoichiometry problems frequently encountered by the chemist in theoretical and laboratory situations. Use the appropriate stoichiometric methods to answer the following questions.

(a) One beaker holds a solution that contains 4.65 grams of sodium sulfide. A second beaker holds a solution that contains 8.95 grams of lead(II) nitrate. When the two solutions are mixed, what mass of precipitate forms?

(b) Perform the calculations to determine the empirical formula of a CHNO compound that is analyzed and found to contain 52.63 percent carbon, 7.02 percent hydrogen, and 12.28 percent nitrogen.

(c) If the compound in part (b) has a molar mass of 228, what is the molecular formula?

(d) A neutralization reaction uses a 0.125 molar solution of sodium hydroxide to titrate 50.0 mL of an unknown sulfuric acid solution. If the reaction takes 23.5 mL of the sodium hydroxide to completely neutralize the sulfuric acid, what is the molarity of the sulfuric acid solution?

ANSWERS

(a) Virtually every stoichiometry problem requires a balanced chemical equation, so we use previous knowledge of double-replacement reactions and the solubility rules to write

$$Na_2S(aq) + Pb(NO_3)_2(aq) \rightarrow PbS(s) + 2NaNO_3(aq)$$

This tells that PbS is the precipitate and is the mass we want to determine. We start to write ? g PbS =, and we find that we have two starting points, the mass of lead(II) nitrate and the mass of sodium sulfide. That means that this is a limiting-reactant problem. We can solve it by calculating the mass of PbS obtained from each reactant and choosing the smaller mass as the limited amount that is actually produced. We write

? g PbS = 4.65 g Na_2S and ? g PbS = 8.95 g $Pb(NO_3)_2$

We then use the dimensional analysis method to perform the calculations, calculating molar masses as we go.

$$? \text{ g PbS} = 4.65 \text{ g } Na_2S \left(\frac{1 \text{mol } Na_2S}{78.1 \text{ g } Na_2S} \right) \left(\frac{1 \text{mol PbS}}{1 \text{mol } Na_2S} \right) \left(\frac{239.3 \text{ g PbS}}{1 \text{mol PbS}} \right)$$

? g PbS = 14.2 g PbS (from Na_2S)

? g PbS =

$$8.95 \text{ g } Pb(NO_3)_2 \left(\frac{1 \text{mol } Pb(NO_3)_2}{331.2 \text{ g } Pb(NO_3)_2} \right) \left(\frac{1 \text{mol PbS}}{1 \text{mol } Pb(NO_3)_2} \right) \left(\frac{239.3 \text{ g PbS}}{1 \text{mol PbS}} \right)$$

? g PbS = 6.47 g PbS (from $Pb(NO_3)_2$)

The limiting reactant limits the product to the smaller of these two results. Therefore the mass of PbS is 6.47 g.

(b) First we see that the percentages do not add up to 100% and that a percentage for oxygen is not mentioned. We obtain the percentage from oxygen as (this is an application of the law of conservation of mass)

% oxygen = $100 - 52.63 - 7.02 - 12.28 = 28.07\%$ O

Now we assume the sample size is 100 g and convert the percentages to grams. Next the number of moles of each element is calculated.

$$? \text{ mol O} = 28.07 \text{ g O} \left(\frac{1 \text{mol O}}{16.0 \text{ g O}} \right) = 1.754 \text{ mol O}$$

$$? \text{ mol C} = 52.62 \text{ g C} \left(\frac{1 \text{mol C}}{12.0 \text{ g C}} \right) = 4.385 \text{ mol C}$$

$$? \text{ mol H} = 7.02 \text{ g H} \left(\frac{1 \text{mol H}}{1.00 \text{ g H}} \right) = 7.02 \text{ mol H}$$

$$? \text{ mol N} = 12.28 \text{ g N} \left(\frac{1 \text{mol N}}{14.0 \text{ g N}} \right) = 0.8771 \text{ mol N}$$

Divide the entire list by the smallest number 0.8771 to get

$? \text{ mol O} = 1.754 \text{ mol O}/0.8771 \quad = 1.999$

$? \text{ mol C} = 4.385 \text{ mol C}/0.8771 \quad = 4.999$

$? \text{ mol H} = 7.02 \text{ mol H}/0.8771 \quad = 8.004$

$? \text{ mol N} = 0.8771 \text{ mol N}/0.8771 = 1.00$

All these numbers are very close to integers, and we can use them as subscripts for the empirical formula $C_5H_8O_2N$.

(c) The mass of the empirical formula is 114. Because the molar mass is 228, we can see that $228/114 = 2$, which means that two empirical formula units comprise the entire molecule. We double the subscripts to write the molecular formula as $C_{10}H_{16}O_4N_2$.

(d) We write and balance the chemical equation for the complete neutralization as

$$H_2SO_4 + 2NaOH \rightarrow Na_2SO_4 + 2H_2O$$

We can start with the molarity of the sodium hydroxide solution and convert it to the molarity of the sulfuric acid.

$$? \ \frac{\text{mol } H_2SO_4}{\text{L } H_2SO_4} = \frac{0.125 \text{ mol NaOH}}{1 \text{ L NaOH}} \left(\frac{1 \text{ mol } H_2SO_4}{2 \text{ mol NaOH}} \right) \left(\frac{0.0235 \text{ L NaOH}}{0.0500 \text{ L } H_2SO_4} \right)$$

$$? \ \frac{\text{mol } H_2SO_4}{\text{L } H_2SO_4} = 0.0234 \ M \ H_2SO_4$$

Notice that the last conversion factor could have been written using mL units $\left(\text{e.g., } \dfrac{23.5 \text{ mol NaOH}}{50.0 \text{ mL } H_2SO_4} \right)$ and produced the same result.

PART 3

STATES OF MATTER

Gases

CHAPTER 7

- Ideal Gas Law
- Boyle's Law
- Charles's Law
- Gay-Lussac's Law
- Avogadro's Principle
- Universal Gas Constant, *R*
- Standard Temperature and Pressure
- Molar Mass Calculations
- Gas Density
- Molar Volume
- Kinetic Molecular Theory

- Graham's Law of Effusion
- Root Mean Square Velocity
- Average Kinetic Energy
- Kinetic Energy Relates to Temperature
- Real Gases
- Dalton's Law of Partial Pressures
- Gas Measurements
- Manometers
- Gas Collection

While the chemical properties of gases vary widely, the physical behavior of all gases is extraordinarily similar, to the extent that one equation, the **ideal gas law,** defines the relationships among the volume, pressure, temperature, and moles of gas in any sample. Similarly, only one theory, the **kinetic molecular theory,** is commonly used to describe the behavior of all gases. A thorough understanding of the ideal gas law is necessary for numerical calculations based on gases, and a similar understanding of the kinetic molecular theory provides the basis for explaining why gases behave as they do.

DEVELOPMENT OF THE IDEAL GAS LAW (*PV* = *nRT*)

Historically, the ideal gas law was formulated from a combination of Boyle's law, Charles's law, Gay-Lussac's law, and Avogadro's principle.

In 1660 Robert Boyle discovered the **inverse relationship** between the pressure and volume of a gas, which is given below in equation and in graphical form:

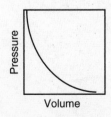

$$P \; \alpha \; \frac{1}{V} \quad \text{or} \quad PV = \text{constant} \qquad (7.1)$$

275

If a sample of gas starts with initial conditions of pressure and volume, and an experiment is done to change those conditions (without changing T or the amount of gas), then the relationship

$$P_i V_i = P_f V_f \qquad (7.2)$$

is obtained. The subscripts i and f represent the initial and the final conditions, respectively. This form of **Boyle's law** shows that, if the pressure on a gas is increased, the volume must correspondingly decrease. If the pressure decreases, the volume will increase.

In 1787, over 100 years later, Jacques Charles discovered the **direct relationship** between the volume of a gas and its temperature, as illustrated in the following equations and graph:

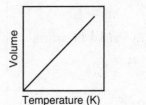

$$V \, \alpha \, T \quad \text{or} \quad \frac{V}{T} = \text{constant} \qquad (7.3)$$

If a gas sample starts with initial conditions of volume and temperature that are changed to some final conditions (while the pressure and the amount of gas do not change), **Charles's law** may be reformulated as

$$\frac{V_i}{T_i} = \frac{V_f}{T_f} \qquad (7.4)$$

Another way of stating this law is to say that, as the temperature of a gas decreases, its volume will also decrease.

Absolute zero is the lowest possible temperature. It is zero on the Kelvin and -273 degrees on the Celsius temperature scales. One method of determining absolute zero is to construct a graph of the volume of a gas as its temperature is changed and to extrapolate the data to the temperature that corresponds to zero volume of the gas, as shown in Figure 7.1.

The exact value for absolute zero is -273.15 °C.

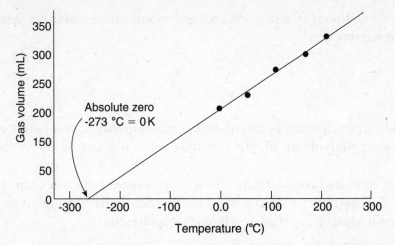

FIGURE 7.1. Using Charles's law to determine absolute zero. Each dot represents an experimental measurement. The line is the best straight line through the data, and the intercept is at –273 °C.

At about the same time, Gay-Lussac, along with Charles, discovered the direct relationship between pressure and temperature:

$$P \; \alpha \; T \quad \text{or} \quad \frac{P}{T} = \text{constant} \qquad (7.5)$$

If initial conditions of P and T are changed to some final conditions, **Gay-Lussac's law** requires that

$$\frac{P_i}{T_i} \; = \; \frac{P_f}{T_f} \qquad (7.6)$$

Finally, in 1811, Avogadro suggested the principle that equal volumes of gases contain equal numbers of molecules or atoms (i.e., moles of a gas) under identical conditions of temperature and pressure. This direct relationship between the number of moles and volume is written in equation form and shown graphically as:

$$V \; \alpha \; n \quad \text{or} \quad \frac{n}{V} = \text{constant} \qquad (7.7)$$

If initial conditions of n or V are changed to some final conditions, **Avogadro's principle** requires that

$$\frac{n_i}{V_i} = \frac{n_f}{V_f} \tag{7.8}$$

Avogadro's principle can be interpreted as meaning that, at constant temperature and pressure, the volume of the container must increase as the moles of gas increase.

Each of these laws considers the relationship between only two of the four variables, P, V, T, and n, that affect gases. This fact means that the other two variables must remain constant, or the law will not be applicable.

GAS LAWS

Boyle's Law	PV = constant
Charles's Law	V/T = constant
Gay-Lussac's Law	P/T = constant
Avogadro's Principle	n/V = constant

IDEAL GAS LAW

The four laws of Boyle, Charles, Gay-Lussac, and Avogadro are combined into the ideal gas law:

$$PV = nRT \tag{7.9}$$

where P is the pressure, V is the volume, n is the number of moles of gas, and T is the temperature in kelvins. The constant, R, called the **universal gas constant,** is needed to make all of the relationships fit together. Table 7.1 gives some of the possible values and units for R. This constant appears in many different equations, and selection of the proper value and corresponding units for it is often critical. The selection of R is based on the units given in the problem. In many problems not all of the data are given in units compatible with the units of R. For instance, most problems give the temperature in degrees Celsius, which must *always* be converted to kelvin. In other instances the pressure or volume may need to be converted to the appropriate units. Units must be watched very carefully in gas law calculations.

TABLE 7.1

Different Formulations of the Universal Gas Constant, R (These are given with problems and need not be memorized.)

Value	Units
0.0821*	$L\ atm\ mol^{-1}\ K^{-1}$
8.314*	$J\ mol^{-1}\ K^{-1}$
6.24×10^4	$L\ mm\ Hg\ mol^{-1}\ K^{-1}$
1.99	$cal\ mol^{-1}\ K^{-1}$
8.314*	$volt\ coulomb\ mol^{-1}\ K^{-1}$

*These are given on the AP Exam.

Instead of memorizing the four laws developed by Boyle, Charles, Gay-Lussac, and Avogadro, it is more convenient to use only the ideal gas law equation for all problems. There are two ways to use the ideal gas law. First, a problem may give three of the four variables and ask that the fourth be calculated. This involves direct substitution of data into the ideal gas law equation. The second way to use the law involves taking a gas under certain initial conditions of P, V, T, and n and changing to some different, final conditions of these four variables. To solve this type of problem the ratio of the ideal gas law equations for the initial and final conditions is written as

$$\frac{P_i V_i}{P_f V_f} = \frac{n_i R T_i}{n_f R T_f} \qquad (7.10)$$

Variables that the problem states (or implies) as remaining constant are canceled, along with R, and then the appropriate substitutions are made from the given data to perform the calculation. Using this approach, it can be seen that the ideal gas law becomes Boyle's law if n and T are constant (i.e., they cancel), and becomes Charles's law if n and P are held constant. If V and n are held constant, Equation 7.10 cancels to become Gay-Lussac's law, and it becomes Avogadro's principle if P and T are held constant. Only one equation need be remembered for all ideal gas law calculations!

DERIVATION OF THE IDEAL GAS LAW

One method for deriving the ideal gas law uses Avogadro's principle, $\dfrac{n}{V} = C$, and Gay-Lussac's law, $\dfrac{P}{T} = C'$. Since C and C' are not the same constants, we multiply Avogadro's principle by a constant, R, so that $CR = C'$. Then

$$\frac{Rn}{V} = RC = C' \quad \text{and} \quad \frac{P}{T} = C'$$

then

$$\frac{nR}{V} = \frac{P}{T}$$

Rearranging the last two terms in this expression yields

$$PV = nRT$$

the ideal gas law, and R is the universal gas constant.

EXERCISE 7.1

A gas occupies 250 mL, and its pressure is 550 mm Hg at 25 °C.

(a) If the gas is expanded to 450 mL, what is the pressure of the gas now?

(b) What temperature is needed to increase the pressure of the gas to exactly 1 atmosphere and 250 mL?

(c) How many moles of gas are in this sample?

(d) The sample is an element and has a mass of 0.525 g. What is it?

Solution

(a) Use the ratio of two ideal gas law equations as shown in Equation 7.10:

$$\frac{P_i V_i}{P_f V_f} = \frac{n_i R T_i}{n_f R T_f}$$

Cancel all but the P and V terms since no change is specified for the other terms and they are assumed to be constant:

$$\frac{P_i V_i}{P_f V_f} = 1$$

Assign the data to the variables: $P_i = 550$ mm Hg, $V_i = 250$ mL, and $V_f = 450$ mL, enter the data in the equation, and solve:

$$\frac{(550 \text{ mm Hg})(250 \text{ mL})}{P_f(450 \text{ mL})} = 1$$

$$\frac{(550 \text{ mm Hg})(250 \text{ mL})}{450 \text{ mL}} = P_f$$

$$306 \text{ mm Hg} = P_f$$

(b) Use the same procedure as in part (a) but cancel n, R, and V since they remain constant:

$$\frac{P_i V_i}{P_f V_f} = \frac{n_i R T_i}{n_f R T_f}$$

$$\frac{P_i}{P_f} = \frac{T_i}{T_f}$$

Assign the data: $P_i = 550$ mm Hg, $T_i = 25 + 273 = 298$ K, and $P_f = 1$ atm $= 760$ mm Hg:

$$\frac{550 \text{ mm Hg}}{760 \text{ mm Hg}} = \frac{298 \text{ K}}{T_f}$$

$$T_f = 298 \text{ K} \left(\frac{760 \text{ mm Hg}}{550 \text{ mm Hg}}\right)$$

$$= 412 \text{ K} = 139 \text{ }^\circ\text{C}$$

(c) To solve this question, we use the ideal gas law equation and substitute the given values:

$$PV = nRT$$

To use the constant $R = 0.0821$ L atm mol^{-1} K^{-1}, the data must be converted to the proper units so that

$$P = 550 \text{ mm Hg} \left(\frac{1 \text{ atm}}{760 \text{ mm Hg}}\right) = 0.724 \text{ atm},$$

$$V = 250 \text{ mL} \left(\frac{1 \text{ L}}{1000 \text{ mL}}\right) = 0.250 \text{ L, and}$$

$$T = 25 \text{ }^\circ\text{C} \ 298 \text{ K}.$$

Enter these data and solve:

$$(0.724 \text{ atm})(0.250 \text{ L}) = n(0.0821 \text{ L atm mol}^{-1} \text{ K}^{-1})(298 \text{ K})$$

$$n = \frac{(0.724 \text{ atm})(0.250 \text{ L})}{(0.0821 \text{ L atm mol}^{-1} \text{ K}^{-1})(298 \text{ K})}$$

$$= 7.40 \times 10^{-3} \text{ mol}$$

(d) We can identify a gas by its molar mass. The molar mass is

$$\text{molar mass} = \frac{\text{mass of sample}}{\text{moles of sample}}$$

$$= \frac{0.525 \text{ g}}{7.40 \times 10^{-3} \text{ mol}}$$

$$= 70.9 \text{ g mol}^{-1}$$

The element with a molar mass closest to 70.9 is chlorine, Cl_2.

STANDARD TEMPERATURE AND PRESSURE (STP)

The ideal gas law has four variables, P, V, n, and T, along with the constant R. By defining the standard pressure as exactly 1 atmosphere* and the standard temperature as exactly 0 degrees Celsius, two of the variables can be stated quickly and easily. Therefore, any gas at **STP** is understood to have $P = 1.00$ atm and $T = 273$ K, and only n and V need be stated in a problem.

Besides making the statement of problems simpler, gases at STP can be compared to each other. The next section shows how the molar mass and density of a gas and the molar volume of a gas at STP are calculated. Differences between the densities and molar volumes calculated from the ideal gas law and the densities and molar volumes determined by experimental measurements indicate to scientists the difference between ideal gases and real gases.

MOLAR MASS, DENSITY, AND MOLAR VOLUME

In the ideal gas law, n represents the number of moles of gas. The number of moles is calculated as $n = \dfrac{\text{grams of gas}}{\text{molar mass}}$. Substituting this equivalency into the ideal gas law yields

$$PV = \frac{g}{\text{molar mass}} RT \qquad (7.11)$$

This equation can be used to determine the **molar mass** of a gas if P, V, g, and T are known for a given sample.

Rearranging Equation 7.11 algebraically yields another useful relationship:

$$P(\text{molar mass}) = \frac{g}{V} RT \qquad (7.12)$$

In this equation, $\dfrac{g}{V}$ is the density of the gas in grams per liter. The **density** of a gas can be determined if P, T, and the molar mass are known.

---

*In 1999, the International Union of Pure and Applied Chemistry, IUPAC, decided that the standard pressure would be 100 kilopascals (kPa). The difference between the current definition of 1 atmosphere or 760 torr and the IUPAC definition is approximately 1%.

Finally, at standard temperature, 0 degrees Celsius, and pressure, 1.0 atmosphere, the ideal gas law can be solved to calculate the **molar volume,** $\frac{V}{n}$, as

$$\frac{V}{n} = \frac{RT}{P} = \frac{(0.0821 \text{ L atm mol}^{-1} \text{ K}^{-1})(273 \text{ K})}{1.00 \text{ atm}}$$

$$= 22.4 \text{ L mol}^{-1} \tag{7.13}$$

This indicates that 1 mole of an **ideal gas** at STP has a volume of 22.4 liters, a fact that is useful in stoichiometric calculations. Table 7.2 lists the molar volumes of some real gases at STP. They are all close to 22.4 liters indicating that they behave as ideal gases under these conditions.

TABLE 7.2

Molar Volumes of Some Gases at STP

Gas	Symbol	Molar (liters) Volume
argon	Ar	22.401
carbon dioxide	CO_2	22.414
helium	He	22.398
hydrogen	H_2	22.410
nitrogen	N_2	22.413
oxygen	O_2	22.414

EXERCISE 7.2

What are the expected densities of argon, neon, and air at STP?

Solution

Each density is calculated, using Equation 7.12, as

$$\text{density} = \frac{g}{V} = \frac{(\text{molar mass})P}{RT}$$

At STP, $P = 1.00$ atm and $T = 273$ K. Substituting these data gives

$$\text{density Ar} = \frac{(39.95 \text{ g mol}^{-1})(1.00 \text{ atm})}{(0.0821 \text{ L atm mol}^{-1} \text{ K}^{-1})(273 \text{ K})} = 1.78 \text{ g L}^{-1}$$

$$\text{density Ne} = \frac{(20.18 \text{ g mol}^{-1})(1.00 \text{ atm})}{(0.0821 \text{ L atm mol}^{-1} \text{ K}^{-1})(273 \text{ K})} = 0.900 \text{ g L}^{-1}$$

$$\text{density air} = \frac{(28.8 \text{ g mol}^{-1})(1.00 \text{ atm})}{(0.0821 \text{ L atm mol}^{-1} \text{ K}^{-1})(273 \text{ K})} = 1.28 \text{ g L}^{-1}$$

TIP

This is similar to the calculation of the average atomic mass of an element. See page 114.

The effective molar mass of air is approximated from the fact that air is 80 percent nitrogen and 20 percent oxygen:

$$\text{effective molar mass} = (0.80)\left(\frac{28 \text{ g N}_2}{\text{mol}}\right) + (0.20)\left(\frac{32 \text{ g O}_2}{\text{mol}}\right)$$
$$= 28.8 \text{ g/mol air}$$

We may also conclude that a balloon full of neon will rise in air, whereas an argon-filled balloon will sink to the floor.

KINETIC MOLECULAR THEORY

The ideal gas law describes the relationships among *P, V, T,* and *n* for ideal gases. The **kinetic molecular theory** describes gases at the level of individual particles. This theory, developed largely by Boltzmann, Clausius, and Maxwell between 1850 and 1880, is often stated as five postulates:

1. Gases consist of molecules or atoms in continuous random motion.
2. Collisions between these molecules and/or atoms in a gas are elastic.
3. The volume occupied by the atoms and/or molecules in a gas is negligibly small.
4. The attractive or repulsive forces between the atoms and/or molecules in a gas are negligible.
5. The average kinetic energy of a molecule or atom in a gas is directly proportional to the Kelvin temperature of the gas.

The concept of gas **pressure** is important to understand since it is central to the kinetic molecular theory. Pressure is defined in physics as the force exerted per unit area. The English units for pressure, pounds per square inch, are familiar. For gases, the force is generated by collisions of the gas particles with the container walls. Each collision has a certain force, which is related to the velocity of the gas particle. The total force is the sum of the forces of all the collisions occurring each second per unit area. Thus the pressure is dependent on the velocity of the gas particles and the collision frequency. In turn, the collision frequency depends on the velocity of the gas particle and the distance to the container walls. Changing the temperature changes the force of the collisions, as well as the frequency of collision. The frequency of collision can be also changed by altering the size of the container, but the force of the collisions will not change.

Based on this understanding of pressure, the molecular meanings of the gas laws can be appreciated as follows.

Boyle's law states the inverse relationship between pressure and volume $\left(P \ \alpha \ \frac{1}{V}\right)$. The kinetic molecular theory agrees with this observation. If the volume of a gas is decreased, the gas particles will strike the walls of the container more frequently. Increasing the frequency of these collisions increases the observed pressure of the gas.

Gay-Lussac's law finds a direct relationship between temperature and pressure $(P \ \alpha \ T)$. In this case, an increase in temperature increases the kinetic energy of the gas particles and their average velocities. This increase has two effects. First, the force of each collision is greater; second, since the average velocity of the gas particles increases with temperature, the frequency of collisions with the container walls also increases. Thus the kinetic molecular theory predicts an increase in gas pressure as temperature rises.

Charles's law predicts a direct relationship between temperature and volume ($V \propto T$). An increase in temperature increases both the force of each collision and the frequency of collisions as in Gay-Lussac's law. Both of these effects increase the pressure. If the pressure of the gas is to remain constant, the volume must increase to correspondingly decrease the frequency of collisions with the walls of the container because the gas molecules will have to travel farther to collide with a wall.

Finally, the kinetic molecular theory also explains **Graham's law of effusion**. The last of the five postulates given above may be stated as the following equation:

$$\overline{KE} = cT = \frac{1}{2}m\overline{v^2} \tag{7.14}$$

where $\overline{KE}$ is the average **kinetic energy**, c is a constant that is the same for all gases, T is the temperature in Kelvin, m is the mass of the gas atom or molecule, and $\overline{v^2}$ is the average of the squares of the velocities of the gas molecules. Since cT will be the same for all gases at the same temperature, the average kinetic energy of any two gases at the same temperature will also be the same:

$$\overline{KE}_1 = \overline{KE}_2$$

and

$$\frac{1}{2}m_1\overline{v_1^2} = \frac{1}{2}m_2\overline{v_2^2}$$

Then the root mean square velocity is defined as $v_{rms} = \sqrt{\overline{v^2}}$:

$$\sqrt{\frac{m_1}{m_2}} = \frac{v_{rms2}}{v_{rms1}} \tag{7.15}$$

Equation 7.15 is Graham's law of effusion. When the rates at which two gases will effuse through a pinhole in a container are compared, they are found to be inversely related to the square roots of the masses of the gas particles. **Effusion** through a pinhole requires that a gas particle hit the pinhole just right to pass through it. The more collisions a gas has with the walls of a container, the higher the probability is that it will hit the pinhole and go through it.

Equation 7.15 is also the mathematical statement of the relative rates of **diffusion** of gases. All gases will expand to fill a container, but filling it may take some time, depending on the conditions present. This equation illustrates that the root mean square velocity of the gas will be inversely proportional to the square root of the masses of the individual gas particles. Since this is a ratio and the units cancel, m_1 and m_2 may be either the actual mass or the molar mass of the gas particles. Heavier gases would be expected to diffuse more slowly than lighter gases, and they do.

EXERCISE 7.3

Helium leaks through a very small hole into a vacuum at a rate of 3.22×10^{-5} mol s^{-1}. How fast will oxygen effuse through the same hole under the same conditions?

Solution

Graham's law of effusion is

$$\sqrt{\frac{m_1}{m_2}} = \frac{v_{rms2}}{v_{rms1}}$$

and we assign the masses as $m_1 = 4$ and $m_2 = 32$. Since the rate at which molecules effuse through a pinhole in moles per second is directly proportional to the v_{rms}, we write $v_{rms1} = 3.22 \times 10^{-5}$ mol s^{-1} while v_{rms2} is the unknown:

$$\sqrt{\frac{4}{32}} = \frac{v_{rms2}}{3.22 \times 10^{-5} \text{ mol s}^{-1}}$$

$$v_{rms2} = 1.14 \times 10^{-5} \text{ mol s}^{-1}$$

AVERAGE KINETIC ENERGIES AND VELOCITIES

Much experimentation has shown that the average kinetic energy is directly proportional to the temperature of a gas, some gas molecules will have kinetic energies above the average, and some will have kinetic energies below the average. Figure 7.2 shows the distribution of the kinetic energies of gas particles at two different temperatures. It should also be noted from Figure 7.2 that the average kinetic energy is not the same as the most probable kinetic energy. The reason is that the curves in Figure 7.2 are not symmetrical and the point at which half of the molecules have higher kinetic energies and half of the molecules have lower kinetic energies lies to the right of the peak.

Since

$$KE = \frac{1}{2}mv^2$$

Figure 7.2 can also be drawn with the x-axis representing the square of the molecular velocity. Some molecules would have velocities above the average, and others would have lower than average velocities.

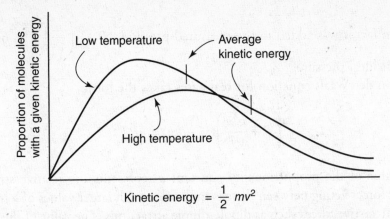

FIGURE 7.2. **Kinetic energy distribution diagrams for gas particles at two different temperatures. Note that average kinetic energy is not at the curve maximum. The maximum is the most probable kinetic energy.**

REAL GASES

The ideal gas law works very well for most gases; however, the law does not work well for gases under high pressures or gases at very low temperatures. These conditions are the same conditions used to condense gases, and therefore it may be generalized that any gas close to its boiling point (condensation point) will deviate significantly from the ideal gas law. The kinetic molecular theory makes two fundamental assumptions about the properties of gas particles themselves: (1) gases have no volume, and (2) they exhibit no attractive or repulsive forces. These two assumptions define an **ideal gas.** They work quite well, since gas particles are often widely separated and what little volume they have is relatively unimportant. Similarly, because of the large average distance between gas particles the attractive forces and repulsive forces are very weak. As a **real gas** is cooled and/or compressed, the distance between the particles decreases dramatically, and these real volumes and forces can no longer be ignored.

Johannes van der Waals developed a modification of the ideal gas law to deal with the nonideal behavior of real gases. He reasoned that, if the gas particles each occupied some volume, there would be a net decrease in the useful volume of the container. (Imagine a fishbowl filled with marbles. Little useful room would be left for the fish.) This decrease in volume must be proportional to the moles of gas particles present. By using the letter b for the proportionality constant, the volume term in the ideal gas law could be replaced by $(V - nb)$.

In evaluating the effect of intermolecular attractions, it was reasoned that, if gas particles attracted each other, they would have curved paths and therefore would take longer to collide with the container walls. The result would be a decreased collision frequency and a lower pressure for the real gas than the pressure of an ideal gas under the same conditions. The frequency of attractions between gas particles would be expected to rise with greater concentrations of the gas. The best correction factor was found to be based on the square of the concentration of the gas, $(n/V)^2$. The proportionality constant was given the symbol a, and the entire correction factor was added to the measured pressure $\left[P + a\left(\dfrac{n}{V}\right)^2 \right]$ to obtain the equivalent ideal pressure.

The van der Waals equation for real gases takes the form

$$\left(P + \frac{n^2 a}{V^2} \right)(V - bn) = nRT \tag{7.16}$$

The value of the proportionality constant a represents the relative strength of attractive forces acting between the gas molecules, with larger values of a indicating stronger attractive forces such as dipole-dipole attractions. The value of the constant b represents the relative size of the gas molecule. The larger the value of b, the larger the size of the molecule.

The size of gas molecules does not vary greatly from one molecule to another, and this contribution to deviations from the ideal gas law is similar for many molecules. Attractive forces, however, vary greatly, depending on molecular polarity, and contribute the most to deviations from the ideal gas law.

DALTON'S LAW OF PARTIAL PRESSURES

Dalton's law of partial pressures is based on the fact that, when two gases are mixed together, the gas particles tend to act independently of each other. The result is that, for a mixture of gases, the total pressure is equal to the sum of the pressures of all of the components of the mixture:

$$P_{\text{total}} = P_1 + P_2 + \cdots \tag{7.17}$$

In this equation, the lowercase p stands for the **partial pressure** of each individual gas. The ellipsis (three dots) at the end of the equation indicates that, if more than two gases are mixed, the equation should be expanded to include the additional components.

EXERCISE 7.4

A mixture of gases contains 2.00 mol of O_2, 3.00 mol of N_2, and 5.00 mol of He. The total pressure of the mixture is 850 torr. What is the partial pressure of each gas?

Solution

The ideal gas law can be interpreted to mean that the pressure is proportional to the number of moles of gas present when T and V are constant. In this problem we have a total of 10.00 mol of gases. Since there are 2.00 mol of O_2, 2.00/10.00 of all moles of gas are O_2; therefore, the same ratio applies to the partial pressure of O_2:

$$P_{O_2} = 850 \text{ torr} \left(\frac{2.00 \text{ mol } O_2}{10.00 \text{ mol total}} \right) = 170 \text{ torr}$$

Using similar calculations for N_2 and He gives

$$P_{N_2} = 850 \text{ torr} \left(\frac{3.00 \text{ mol } N_2}{10.00 \text{ mol total}} \right) = 255 \text{ torr}$$

$$P_{He} = 850 \text{ torr} \left(\frac{5.00 \text{ mol He}}{10.00 \text{ mol total}} \right) = 425 \text{ torr}$$

To check the calculations, we determine the total pressure from the three partial pressures:

$$P_{total} = 170 \text{ torr} + 255 \text{ torr} + 425 \text{ torr}$$
$$= 850 \text{ torr}$$

This result agrees with the total pressure given in the problem.

GAS MEASUREMENTS

The physical properties of gases depend on the four variables of volume, temperature, pressure, and number of moles. Each of these quantities may be measured independently. Often, however, it is easier to measure three of these four variables and to calculate the last one from the ideal gas law equation. The measurement methods for the variables are reviewed briefly below.

Volume is expressed in liters (sometimes milliliters) and may be determined in several ways, including:

1. Careful measurement of the dimensions of the container followed by the appropriate geometric calculations.
2. Measurement of the mass of a liquid of known density that fills the container to capacity.

Temperature is measured with standard thermometers. Special instruments may be used if high accuracy and precision are needed or if the temperature of the gas is extremely high or low. Since gases have very low densities, sufficient time must be allowed for a thermometer to reach the correct temperature. Often, the temperature of the gas is determined by measuring the temperature of its surroundings after sufficient time has elapsed for the container and the surroundings to reach the same temperature (thermal equilibrium). The container may be submersed in a liquid such as water to thermostat the system (i.e., keep its temperature constant) and also to speed the attainment of thermal equilibrium.

Pressure is the force exerted per unit area. Pressures may be recorded in SI units of kilopascals (kPa) or in experimental units, such as torr (also known as millimeters of mercury) or atmospheres (1 atm = 760 torr). Pressure may be determined either by direct reading of an instrument called a pressure gauge or by using a **manometer**. Two types of manometer, the closed-end and open-end, are available. Both are U-shaped pieces of glass with one end open so that it can be attached to a vessel containing a gas. The other end may be open or closed, as the names suggest, and as shown in Figure 7.3. Both manometers are read by finding the difference in height of the Hg level. These pressures are recorded as mm Hg, and 1 mm Hg = 1 torr.

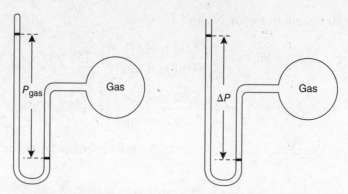

FIGURE 7.3. **On the left, a closed-end manometer measures the gas pressure directly. On the right, an open-end manometer measures the difference between the atmospheric pressure and the gas pressure.**

The closed-end manometer (also called a **eudiometer**) is completely filled with mercury. When the instrument is set up, the difference in the mercury levels indicates the pressure of whatever system is attached to the apparatus. If this device is not attached to an experimental setup, but is left open to the atmosphere, it measures atmospheric pressure and is then called a **barometer**.

The open-end manometer is usually filled with mercury, and the two ends are connected to gases at different pressures. The difference in pressure will be equal to the difference in the level of the mercury in the two sides of the manometer. This method gives only the difference in pressure between the two sides of the manometer. If one side is left open to the atmosphere, as shown in Figure 7.3, the difference represents the difference between the gas pressure and the atmospheric pressure.

If a fluid other than mercury is used in a manometer, the difference in the heights of this fluid represents the pressure. However, for comparison with a normal mercury manometer, the readings must be corrected for the relative densities of the fluid used and of mercury by using the equation

$$\text{mm Hg} = \text{mm fluid}\left(\frac{\text{density fluid}}{\text{density Hg}}\right) \qquad (7.18)$$

EXERCISE 7.5

Medical devices for anesthesiology must not restrict the air flow to the patient by more than 2.50 mm Hg. Suggest a method for making these measurements accurately.

Solution

It is difficult to measure 2.50 mm Hg with accuracy in a mercury-filled manometer. If a water manometer is used, the difference in liquid levels will be much greater and will be easier to measure. For example, the difference in water levels equal to 2.50 mm Hg is calculated from Equation 7.18 as

$$2.50 \text{ mm Hg} = \text{mm H}_2\text{O}\left(\frac{1.00 \text{ g H}_2\text{O/mL H}_2\text{O}}{13.6 \text{ g Hg/mL Hg}}\right)$$

$$\text{mm H}_2\text{O} = 34 \text{ mm H}_2\text{O}$$

It is much easier to measure 34 mm of water accurately than to measure 2.5 mm of mercury accurately.

EXPERIMENTS INVOLVING GASES

As a candidate for advanced placement in chemistry, you should be familiar with a variety of experiments for producing, collecting, and manipulating gases in the laboratory. A few of the more familiar reactions for producing gases are these:

$$2\text{KClO}_3(s) \rightarrow 2\text{KCl}(s) + 3\text{O}_2(g) \quad \text{(with heat and MnO}_2 \text{ as a catalyst)}$$

$$\text{NH}_4^+(aq) + \text{OH}^-(aq) \rightarrow \text{NH}_3(g) + \text{H}_2\text{O}(l)$$

$$\text{CaCO}_3(s) + \text{HCl}(aq) \rightarrow \text{CO}_2(g) + \text{CaCl}_2(aq) + \text{H}_2\text{O}(l)$$

$$2\text{HCl}(aq) + \text{Zn}(s) \rightarrow \text{ZnCl}_2(aq) + \text{H}_2(g)$$

In addition to knowledge of the chemical reactions, the use of a **pneumatic trough** for collecting gas samples over water should be familiar. Figure 7.4 illustrates a pneumatic trough with a bottle full of water submersed. The gas from the chemical reaction bubbles into the bottle, displacing the water as it is collected.

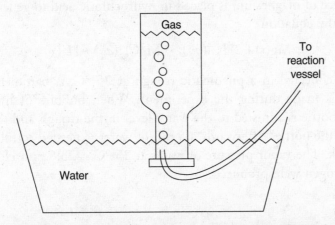

FIGURE 7.4. Collection of gases in a pneumatic trough. When the reaction is complete, the position of the bottle is adjusted so that the water level in the bottle is equal to the water level in the trough. At that point $P_{gas} = P_{atm}$.

The pressure of the gas inside the collecting bottle must be determined in order to solve the ideal gas equation. When the reaction is complete, the gas pressure is determined by moving the bottle in the water until the two liquid levels coincide. At this point the pressure inside the bottle is the same as the barometric pressure outside the bottle.

When gases are collected over water, there will always be some water vapor in the collecting bottle. The pressure of the water vapor depends on only the temperature and is obtained from the appropriate reference table. (The AP chemistry test does not include this table; the information will be given in the problem if needed.) The pressure of the gas that was generated is calculated from Dalton's law of partial pressures as

$$P_{gas} = P_{atm} - P_{water} \qquad (7.19)$$

The volume of gas may be determined by marking the jar when the liquid levels are equal and then measuring the jar's volume to the mark at a later time. This is done by filling the jar to the mark with water and then carefully pouring the water into a graduated cylinder to determine the volume.

Temperature is measured with a laboratory thermometer, and atmospheric pressure is determined from a barometer for use in Equation 7.19. These measurements give P, V, and T for the ideal gas law equation, and the amount of gas produced, n, can be calculated.

The *number of moles of gas* is given the symbol "n." This is determined by measuring the mass of the gas and by using the molecular mass to convert it to moles (n = mol = g/molar mass). Measuring the mass of any gas involves evacuating a vessel to a very low pressure using a vacuum pump. The evacuated vessel is weighed and then the gas sample is introduced to the vessel and it is weighed again to determine the increase in mass due to the gas.

EXERCISE 7.6

A 0.060-g piece of magnesium is placed in hydrochloric acid to generate hydrogen according to the equation

$$Mg(s) + 2HCl(aq) \rightarrow MgCl_2(aq) + H_2(g)$$

The gas is collected in a pneumatic trough at 25 °C. A barometer reading of 755 mm Hg is made during the experiment. When bubbles of hydrogen cease forming, the bottle is adjusted to the water level in the trough and the water level is marked on the bottle. Afterwards, 65 mL of water is needed to fill the bottle to the same mark. The vapor pressure of water at 25 °C is 23.8 mm Hg. How many moles of hydrogen were produced?

Solution

The volume of the gas is 0.065 L, its temperature is 298 K, and its pressure is calculated from Equation 7.19 as

$$P_{H_2} = 755 \text{ mm Hg} - 23.8 \text{ mm Hg} = 731 \text{ mm Hg}$$

This is converted to 0.962 atm. The ideal gas law is then used to calculate n, the number of moles of hydrogen:

$$(0.962 \text{ atm})(0.065 \text{ L}) = n(0.0821 \text{ L atm mol}^{-1} \text{ K}^{-1})(298 \text{ K})$$

$$n = 2.6 \times 10^{-3} \text{ mol H}_2(g)$$

SUMMARY

The chapter on gases introduces the start of the discussion on the states of matter and their properties. Gases all obey the ideal gas law, and various solutions to this equation are demonstrated in the chapter. An ideal gas is defined as one that has no molecular volume and no attractive forces between gas particles. The ideal gas law also allows one method for determining the molar mass of a compound. Gases are an important part of many experiments. This chapter describes in detail how to make measurements on gases and how to collect a gas in the laboratory. The kinetic molecular theory is introduced along with a discussion of how it can be used to explain all the gas laws. The chapter ends with a discussion of how a real gas differs from an ideal gas.

Important Concepts

Ideal gas law and ideal gas
Standard temperature and pressure, STP
Universal gas law constant, R
Dalton's law of partial pressures
Kinetic molecular theory
Average kinetic energy
Graham's law of effusion

Important Equations

$$PV = nRT$$

$$\sqrt{\frac{m_1}{m_2}} = \frac{\bar{v}_2}{\bar{v}_1}$$

$$P_{\text{total}} = P_1 + P_2 + \cdots$$

Practice Exercises

Multiple-Choice

For the first five problems below, one or more of the following responses will apply; each response may be used more than once or not at all in these questions.

I. increased volume
II. increased temperature
III. increased average kinetic energy
IV. increased effusion rate
V. increased pressure

1. Considering Boyle's law, a gas will have an _____ if the pressure is reduced.

   (A) I
   (B) II
   (C) III
   (D) IV
   (E) II and III

2. There will be an _____ if methane, CH_4, rather than ethane, CH_3CH_3 is the gas used under comparable conditions.

   (A) I and III
   (B) II
   (C) III and V
   (D) IV
   (E) V

3. The kinetic molecular theory postulates a direct relationship between

   (A) I and III
   (B) II and III
   (C) III and V
   (D) IV
   (E) V

4. Which of the following pairs are inversely proportional to each other, assuming all other conditions are kept constant?

   (A) I, V
   (B) II, III
   (C) III, V
   (D) IV, II
   (E) V, II

5. A gas will behave more like an ideal gas if we have an _____ and an _____.

   (A) I, II
   (B) II, III
   (C) III, V
   (D) IV, V
   (E) V, II

6. What volume will 2.50 mol of N_2 occupy at 45 °C and 1.50 atm of pressure?

   (A) 43.5 L
   (B) 6.08 L
   (C) 0.0233 L
   (D) 56.00 L
   (E) 14.9 L

7. How many moles of helium are needed to fill a balloon that has a volume of 6.45 L and a pressure of 800 mm Hg at a room temperature of 24 °C? Assume ideal gas behavior.

   (A) 0.288
   (B) 214
   (C) 0.278
   (D) 2.65×10^3
   (E) 0.255

8. If ideal gas behavior is assumed, what is the density of neon at STP?

(A) 1.11 g L^{-1}
(B) 448 g L^{-1}
(C) 0.009 g L^{-1}
(D) 0.901 g L^{-1}
(E) 1.25 g L^{-1}

9. A sample of CO has a pressure of 58 mm Hg and a volume of 155 mL. When the CO is quantitatively transferred to a 1.00-L flask, the pressure of the gas will be

(A) 374 mm Hg
(B) 8990 mm Hg
(C) 111 mm Hg
(D) 8.99 mm Hg
(E) 2.67 mm Hg

10. At 30 °C a sample of hydrogen is collected over water $[P_{H_2O} = 31.82$ mm Hg at 30 °C$]$ in a 500-mL flask. The total pressure in the collection flask is 745 mm Hg. What will be the percent of error in the amount of hydrogen reported if the correction for the vapor pressure of water is not made?

(A) 0.0%
(B) +4.5%
(C) −4.5%
(D) +4.3%
(E) −4.3%

11. What will the total pressure be in a 2.50-L flask at 25 °C if it contains 0.016 mol of CO_2 and 0.035 mol of CH_4?

(A) 31.4 mm Hg
(B) 380 mm Hg
(C) 0.041 mm Hg
(D) 935 mm Hg
(E) 1.23 atm

12. The carbon dioxide from the combustion of 1.50 g of C_2H_6 is collected over water at 25 °C. The pressure of CO_2 in the collection flask is 746 mm Hg, and the volume is 2.00 L. How much of the CO_2 formed apparently dissolved in the water of the pneumatic trough?

(A) 0.0814 mol
(B) 0.87 g
(C) 1.79 g
(D) 0.100 mol
(E) 2.55 g

13. In which of the following is it impossible to predict whether the pressure of a gas will increase, decrease, or stay the same?

(A) A gas sample is heated.
(B) A gas sample is heated, and the volume is increased.
(C) A gas sample is cooled, and some gas is withdrawn.
(D) Additional gas is added to a sample of gas.
(E) A gas sample is cooled, and the volume is increased.

14. The kinetic molecular theory predicts that at a given temperature

(A) all gas molecules have the same kinetic energy
(B) all gas molecules have the same average velocity
(C) only real gas molecules collide with each other
(D) on the average, heavier molecules move more slowly
(E) elastic collisions result in the loss of energy

15. The effect that increasing the temperature has on the pressure may be explained by the kinetic molecular theory as due to

    (A) the increase in force with which the gas molecules collide with the container walls
    (B) the increase in rotational energy of the gas molecules
    (C) the increase in average velocity of the gas molecules, which causes a corresponding increase in the frequency of collisions with the container walls
    (D) a decrease in the attractive forces between gas molecules
    (E) a combination of (A) and (C)

16. Ideal gases

    (A) have no volume
    (B) have no mass
    (C) have no attractive forces between them
    (D) have a combination of (A) and (C)
    (E) have a combination of (A) and (B)

17. Under which conditions will a real gas behave most like an ideal gas?

    (A) high pressure and high temperature
    (B) low pressure and low temperature
    (C) low volume and high temperature
    (D) low pressure and high temperature
    (E) high pressure and low temperature

18. Compared to ideal gases, real gases tend to have

    (A) larger volumes
    (B) greater kinetic energies
    (C) lower average kinetic energies
    (D) lower pressures
    (E) both (A) and (D)

19. Under identical conditions gaseous CO_2 and CCl_4 are allowed to effuse through a pinhole. If the rate of effusion of the CO_2 is 6.3×10^{-2} mol s^{-1}, what is the rate of effusion of the CCl_4?

    (A) 6.3×10^{-2} mol s^{-1}
    (B) 2.2×10^{-1} mol s^{-1}
    (C) 1.8×10^{-2} mol s^{-1}
    (D) 3.4×10^{-2} mol s^{-1}
    (E) 1.2×10^{-1} mol s^{-1}

20. A gas has a density, at STP, of 3.48 g L^{-1}. The most reasonable formula for this compound is

    (A) C_2H_6
    (B) HF
    (C) CCl_4
    (D) C_6H_6
    (E) CaF_2

21. The number of moles of an ideal gas in a 2.50-L container at 300 K and a pressure of 0.450 atm is ($R = 0.0821$ L atm mol^{-1} K^{-1})

    (A) 0.0457
    (B) 21.9
    (C) 4.93×10^{-5}
    (D) 2.03×10^{4}
    (E) 6.02

22. A gas mixture contains twice as many moles of O_2 as N_2. Addition of 0.200 mol of argon to this mixture increases the pressure from 0.800 atm to 1.10 atm. How many moles of O_2 are in the mixture?

    (A) 0.355
    (B) 0.178
    (C) 0.533
    (D) 0.200
    (E) 0.0750

23. A gas in a 1.50-L container has a pressure of 245 mm Hg. When the gas is transferred completely to a 350. mL container at the same temperature, the pressure will be

    (A) 1.05 mm Hg
    (B) 1.05 atm
    (C) 2.14 mm Hg
    (D) 1050 mm Hg
    (E) 9.5×10^{-4} atm

24. At STP a 5.00-L flask filled with air has a mass of 543.251 g. The air in the flask is replaced with another gas, and the mass of the flask is then determined to be 566.107 g. The density of air is 1.290 g L^{-1}. What is the gas that replaced the air?

    (A) Ne
    (B) O_2
    (C) Ar
    (D) Xe
    (E) He

25. What volume of hydrogen gas, at STP, will a 0.100-g sample of magnesium (molar mass = 24.31) produce when reacted with an excess of HCl?
    ($Mg + 2HCl \rightarrow MgCl_2 + H_2$)

    (A) 92.1 mL
    (B) 46.1 mL
    (C) 184 mL
    (D) 9.2 mL
    (E) 4.6 L

ANSWER KEY

1. A	6. A	11. B	16. D	21. A
2. D	7. C	12. B	17. D	22. A
3. B	8. D	13. B	18. E	23. D
4. A	9. D	14. D	19. D	24. D
5. A	10. B	15. E	20. D	25. A

See Appendix 1 for explanations of answers.

Free-Response

Answer the following questions concerning the properties of gases and the theories used to explain their behavior.

(a) Describe an ideal gas and compare it to a real gas. Why do many real gases behave as ideal gases?

(b) The kinetic molecular theory is used to explain the behavior of gases. Briefly describe this theory and two of its applications to gases.

(c) There are at least two methods for determining molecular masses of gases. What are they? Give the applicable mathematical equations and state what measurements are needed to obtain molar masses.

(d) A sample of carbon dioxide in a 2.50-liter flask at 785 torr and 35 °C is transferred to a 1.00-liter flask at 405 K. What is the final pressure of the system in atmospheres of pressure?

ANSWERS

(a) An ideal gas is composed of particles in continuous, rapid motion that neither attract nor repel each other. An ideal gas cannot be condensed to a liquid (or a solid). In addition, ideal gas particles occupy no volume. Real gas molecules occupy volume and may attract or repel each other. Cooling a real gas can slow their motions so that attractive forces can cause them to condense. In many cases, especially those that approach low pressure and high temperature, the attractive forces and the relative volumes of real gases are so low that they act very much like ideal gases.

(b) The kinetic molecular theory states that gases are molecules or atoms in constant random motion and that all collisions are elastic. The volume of gases is negligible compared to the total volume, and attractive forces are also negligible. The average kinetic energy of a gas is directly proportional to the temperature.

We can see how this explains Boyle's law, which states that the product of pressure and volume is a constant: PV = constant. If the volume of a container is increased, the frequency at which the gas molecules strike the walls decreases because the molecules must travel further to reach each wall. Pressure is proportional to the frequency with which the gas molecules strike the container walls and the average force of each collision. The forces of the collisions do not change because the kinetic energy does not change unless the temperature changes. Therefore the pressure must decrease as the volume increases, all other factors being kept constant.

The kinetic molecular theory also explains Charles's law, which states that the volume of a gas and the Kelvin temperature are directly proportional to each other: V/T = constant. Consider increasing the volume of a gas container. To maintain the pressure, the number and force of collisions with the walls must be kept constant. When the temperature increases, the number of collisions with the walls increases along with the force because of an increase in the average kinetic energy. In total, the pressure remains constant. The other gas laws can also be explained using the kinetic molecular theory.

(c) Two methods that give the molar mass use the density of a gas and the ideal gas law equation. The second uses Graham's law of effusion.

The number of moles, n, in the ideal gas law is calculated as the mass of the sample divided by the molecular mass, **M**. $n = m$/molar mass. Substitution into the ideal gas law yields molar mass = mRT/PV. To obtain the molar mass of a gas we need to know its mass and volume (or density = m/V), the Kelvin temperature, and the pressure in atmospheres.

Graham's law of effusion is Equation 7.15. It requires measurement of the rate of effusion for a known gas and then measuring the rate of effusion for the gas with an unknown molar mass.

(d) To solve this problem we start by writing the ratio of two gas law equations: $\dfrac{P_1V_1}{P_2V_2} = \dfrac{nRT_1}{nRT_1}$. The constant terms n and R are canceled to give $\dfrac{P_1V_1}{P_2V_2} = \dfrac{T_1}{T_2}$.

Values are then converted to the required units and then substituted into the equation. Convert 35 °C to 308 K, and 785 torr to 1.033 atm. These are substituted to obtain $\dfrac{(1.033 \text{ atm})(2.50 \text{ L})_1}{P_2(1.00 \text{ L})} = \dfrac{308 \text{ K}}{405 \text{ K}}$. Solving this equation for the final pressure, P_2, we obtain.

$$P_2 = 1.033 \text{ atm} \left(\frac{2.50 \text{ L}}{1.00 \text{ L}} \right) \left(\frac{405 \text{ K}}{308 \text{ K}} \right)$$

By estimating the answer, we get

$$P_2 \approx (1 \text{ atm})(2.5)\left(\frac{4}{3} \right)$$

$$\approx (1 \text{ atm})\left(\frac{10}{3} \right)$$

$$\approx 3.3 \text{ atm}$$

The actual calculated answer is 3.40 atm.

Liquids and Solids

8

- Intermolecular Forces
- Dipole-Dipole Attractions
- Hydrogen Bonding
- London Forces
- Dispersion Forces
- Instantaneous and Induced Dipoles
- Surface Tension
- Viscosity
- Evaporation
- Vapor Pressure
- Boiling Points
- Heats of Vaporization
- Clausius-Clapeyron Equation
- Crystal Types
- Metallic Crystals
- Ionic Crystals
- Molecular Crystals
- Network Crystals
- Amorphous Substances
- Phase Changes
- Heating and Cooling Curves
- Phase Diagrams
- Triple Point
- Critical Point
- Supercritical Fluid

COMPARISON OF LIQUIDS AND SOLIDS TO GASES

Liquids and solids are distinctly different from the gases discussed in Chapter 7. First, liquids and solids are much more dense than gases. Inorganic liquids and solids have densities that range from 1 to 8 $g \, cm^{-3}$; a few have densities up to 20 $g \, cm^{-3}$. Most organic liquids and solids have densities from 0.7 to 2.0 $g \, cm^{-3}$. In contrast, gas densities at STP are generally between 10^{-2} and $10^{-4} \, g \, cm^{-3}$. Second, gases expand to fill all available space and must be kept in an enclosed container, while a liquid fills any container from the bottom up to a level dictated only by the mass of liquid present. Liquids also conform to the shape of the container. Solids maintain their shape without any container. Third, and most important, is the lack of significant attractive forces in gases and the presence of significant attractive forces in liquids and solids. The obvious physical differences between the three states of matter are explained on the basis of these forces.

INTERMOLECULAR FORCES

In discussing gases, we found that the ideal gas law can be used to describe gases for two reasons. First, the volume of a gas molecule is so small that 99 percent of a gas is empty space. Second, the gas molecules are so far apart that there is no significant intermolecular attraction. If gases were truly ideal (zero volume and zero attractive forces), it would be impossible to condense them to liquids.

For condensation to occur, intermolecular attractive forces must overcome the kinetic energy of the gas molecules. A real gas can be condensed to a liquid by increasing the pressure and/or decreasing the temperature. Pressure is increased to force the gas molecules closer together, thereby increasing the attractive forces. Temperature is decreased to lower the average kinetic energy.

By understanding the attractive forces between molecules, it is possible to appreciate many of the physical properties of both liquids and solids, including the condensation process. We can identify several of these intermolecular forces of attraction as dipole-dipole attractive forces, London forces, and hydrogen bonding. These forces are described below.

Dipole-Dipole Attractive Forces

Molecular compounds share electrons in a covalent bond. This electron sharing is rarely equal, particularly between dissimilar elements. Consequently, the electrons may congregate at one end of the molecule, giving it polarity. Polar molecules are also called **dipoles** to remind us that there is only one positive and only one negative end to each molecule. The positive end has a partial positive charge, indicated as $\delta+$. Similarly, the negative end of a molecule is only partially negative and is designated as $\delta-$. Polar molecules are attracted toward each other, with the negative end of one molecule attracted to the positive end of another molecule. It may seem logical that since there are always the same number of $\delta+$ and $\delta-$ ends to any molecule that the attractive and repulsive forces would cancel. However, dipoles move away from repulsion orientations and tend to maintain attractive orientations. This results in an overall attraction between dipoles.

One of the simplest dipoles is hydrogen chloride. In Figure 8.1 the electron clouds around the nonpolar H_2 and the polar HCl molecules are compared.

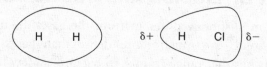

FIGURE 8.1. Representations of the electron clouds around the nonpolar H_2 and the polar HCl molecules.

In the gaseous state, polar molecules show little attraction for each other because they are so far apart (about 3000 pm). The molecules in solids and liquids, however, are approximately ten times closer (about 300 pm). Attractive forces between dipoles may be represented by Equation 8.1:

$$\text{Force} = \frac{(\delta+)(\delta-)}{r^2} \tag{8.1}$$

This equation shows that the attractive force is inversely proportional to the square of the distance, r, between two polar molecules. In gases r is so large that the attractive force is negligibly small. In liquids, where the distance between molecules is much smaller, these forces are significant.

For a gas to become a liquid, the attractive forces must overcome the kinetic energy of the moving gas molecule. Equation 8.1 indicates that decreasing the distance between molecules will increase the attractive force. Increasing the pressure on a gas forces the molecules closer together, and cooling a gas reduces its average kinetic energy. Therefore, decreasing the temperature of a gas and/or increasing the pressure on it will help condense the gas to the liquid phase. The boiling point, which is also the same as the condensation point, is an indication of the attractive forces between molecules since it is a measure of how much the kinetic energy must be increased so that it can overcome the attractive forces in a liquid. Low boiling points indicate low attractive forces, and high boiling points indicate higher attractive forces.

In the condensed state of a liquid the dipole-dipole forces define many of the observed properties. For instance, highly polar molecules have higher boiling points than molecules with lower polarities. The vapor pressure, surface tension, viscosity, and solubilities of liquids are also based on considerations involving attractive forces, as described in the following sections.

London Forces of Attraction

Dipole-dipole interactions are used to explain how and why polar molecules may be condensed to the liquid state. It remained for Fritz London, in 1928, to give a logical explanation of how nonpolar gases develop the forces necessary for condensation. He postulated that nonpolar atoms and molecules may become momentarily polar when an unsymmetrical distribution of their electrons results in the formation of instantaneous dipoles. These instantaneous dipoles provide weak attractive forces in nonpolar substances. **London forces** may also be called **dispersion forces, instantaneous dipole forces,** or **induced dipole forces.**

To describe how London forces develop, consider a noble gas such as argon. Previously, argon was described as an atom with 14 electrons arranged in symmetrical orbitals around the 14 protons in its nucleus. The electrons around the argon nucleus are in constant motion. This motion results in a high probability that, at any moment in time, the electrons will not be arranged symmetrically. To illustrate this, several "instantaneous snapshots" of the argon atom are shown in Figure 8.2.

FIGURE 8.2. **Random distribution of electrons around an argon nucleus. The first two obviously have more electrons on one side (the upper left quadrant). The third looks symmetrical but a close examination shows more electrons in the lower left quadrant.**

When the electrons are not evenly distributed, argon will be a dipole for an instant before the electrons move to new positions. This instantaneous dipole may be attracted to another nearby instantaneous dipole, or it may induce another dipole in a neighboring atom by distorting the neighboring atom's electron cloud. The result is a very weak, attractive force, allowing argon to condense. Since such forces are very weak, argon and the other noble gases have very low boiling points.

The halogens are like the noble gases in having no permanent dipoles. Yet iodine is a solid, and bromine is a liquid, at room temperature and all of the halogens have much higher boiling points than the neighboring noble gases. The explanation for this seeming paradox lies in the **polarizability** of the electron clouds of the halogen molecules. Polarizability refers to the ease with which the electron cloud around an atom or molecule can be deformed into a dipole. Small atoms and molecules, with their electrons tightly held near the nucleus, have a low polarizability. Large atoms or molecules, with many loosely held electrons, have electron clouds with high polarizability. The difference may be visualized by comparing a small, hard golf ball and a large, soft sponge basketball. Since the large electron clouds of bromine and iodine are easily polarized, they have much higher boiling points than the neighboring noble gases.

We can explain the behavior of many molecules on the basis of London forces. For instance, methane, CH_4, is a nonpolar tetrahedral molecule. Instantaneous dipoles are used to explain why methane condenses to a liquid. Ethane, C_2H_6, has a higher boiling point than methane because the six hydrogen atoms may become instantaneous dipoles, resulting in a stronger attractive force. The related propane, C_3H_8, and butane, C_4H_{10}, molecules, have increasingly more hydrogens to form more instantaneous dipoles; therefore, they have higher boiling points. These four compounds are the first four in a series of compounds called the **normal alkanes.** All *n*-alkanes have the general formula C_nH_{2n+2}. Figure 8.3, with the number of carbon atoms on one axis and the boiling point on the other axis, shows that boiling points rise because of increased instantaneous dipole forces.

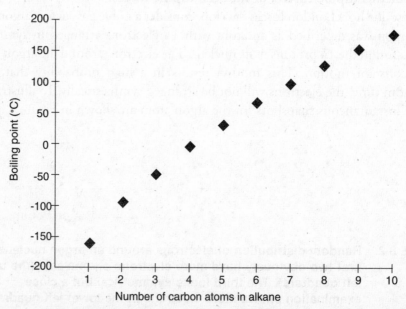

FIGURE 8.3. **Plot of boiling points of the normal alkanes versus the number of carbon atoms in each *n*-alkane. The number of hydrogen atoms is proportional to the number of carbon atoms.**

In general, we may conclude that the more atoms in a molecule, the more opportunity there is to form instantaneous dipoles. The result is to increase the attractive forces and to raise the boiling point.

Hydrogen Bonding

Figure 8.3 shows a plot of the boiling points of the *n*-alkanes. These compounds are called a **homologous series** because their formulas vary in a regular fashion. For the *n*-alkanes in the graph, we added one more carbon and two more hydrogen atoms for each compound. Chemists make similar plots of other homologous series of compounds to help visualize trends in physical properties.

Figure 8.4 illustrates such a plot for the hydrogen compounds of the elements in the four groups of the periodic table headed by fluorine, oxygen, nitrogen, and carbon. We see that the compounds headed by carbon all fall on a reasonably straight line, and we conclude that they all act in a similar manner. We see also that the first compound, H_2O, NH_3, and HF, in each of the other three groups in the periodic table has a much greater boiling point than expected based on the boiling points of the other compounds in these groups.

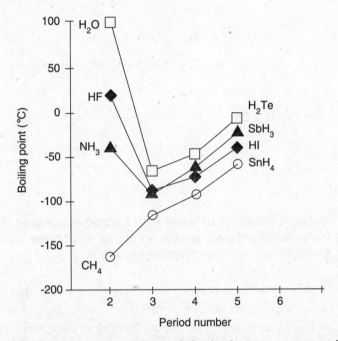

FIGURE 8.4. **Plot of the boiling points of the hydrogen compounds in the groups headed by fluorine (HF, HCl, HBr, and HI), oxygen (H_2O, H_2S, H_2Se, H_2Te), nitrogen (NH_3, PH_3, AsH_3, SbH_3), and carbon (CH_4, SiH_4, GeH_4, SnH_4). Only the first and last compounds of each group are shown on the graph.**

This behavior may be attributed to the large electronegativity difference (ΔEN) between hydrogen and fluorine, oxygen, and nitrogen. This large ΔEN means that H_2O, NH_3, and HF are very polar molecules with very strong dipole-dipole forces. These extraordinarily large dipole-dipole forces are given a special name, **hydrogen bonds.**

The hydrogen-bonded liquid states of HF, NH_3, and H_2O are somewhat structured. In hydrogen fluoride, the hydrogen of one HF molecule is attracted to the fluorine on another HF molecule. This attraction can extend for many HF units, creating a chainlike structure (Figure 8.5).

$$\cdots H-F\cdots H-F\cdots H-F\cdots H-F\cdots H-F\cdots H-F\cdots H-F\cdots$$

FIGURE 8.5. A structure illustrating the chain structure of the HF hydrogen bonding. The dotted lines indicate hydrogen bonds.

For ammonia also, a chainlike structure can form. Although NH_3 has three hydrogen atoms, it has only one nonbonding pair of electrons on its nitrogen, limiting it to a chainlike structure similar to that of HF. The increase in boiling point over the expected boiling point for both HF and NH_3 is similar, although it is slightly greater for HF since HF has a larger electronegativity difference.

Water is different. It has two hydrogen atoms, which can participate in two hydrogen bonds with neighboring oxygen atoms. In addition, the oxygen atoms have two lone pairs of electrons, which can hydrogen-bond with two hydrogen atoms. Water can form a large network structure, as diagrammed in Figure 8.6.

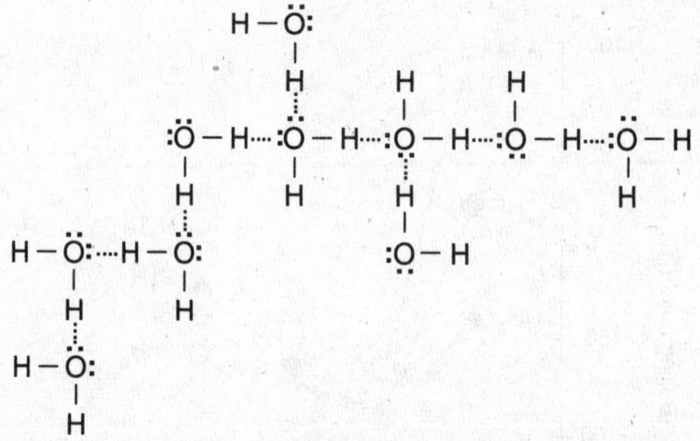

FIGURE 8.6. Network structure of water with hydrogen bonding. Tetrahedral water molecules are shown as planar structures for clarity. Dotted lines represent hydrogen bonds.

As a result of this network structure, water has the greatest increase in boiling point compared to its expected boiling point. The boiling point increase is higher in water than in HF, even though HF has a larger electronegativity difference.

EXERCISE 8.1

Determine the boiling points that would be expected for HF, H_2O, and NH_3 if hydrogen bonding did not exist.

Solution

Using Figure 8.4, we extrapolate each group to period 2 and we read the temperature at that point. The approximate answers are HF = -105 °C, H_2O = -95 °C, $NH_3 = -120$ °C.

Hydrogen bonding is not limited to HF, H_2O, and NH_3. This effect is observed whenever hydrogen is covalently bonded to fluorine, oxygen, or nitrogen. In the case of fluorine, there is only one compound that hydrogen-bonds: HF itself. Many compounds, however, contain the O–H bond. These include alcohols, sugars, organic acids, and phenol-type compounds. In addition to ammonia, primary and secondary amines are nitrogen containing compounds that form hydrogen bonds. These compounds are described in more detail in Chapter 15.

Hydrogen bonding is a phenomenon that has wide-reaching effects. It causes water to be a liquid at the temperatures normally encountered on Earth. It causes solid water, ice, to be less dense than liquid water, so that ice floats. If ice did not float on water, the entire planet would be ice-covered all year, winter, spring, summer, and fall. In addition, it is hydrogen bonding that holds the two strands of the double helix together in DNA, and gives structure to proteins such as hemoglobin and antibodies. Life as we know it is dependent on hydrogen bonding.

EXERCISE 8.2

Predict the type of intermolecular forces expected for each of the following compounds:

C_6H_6 (benzene), CH_3OH, CH_3NH_2, CF_4, SO_3, XeF_6, $C_{12}H_{26}$, CH_3OCH_3.

Solution

All compounds will have London attractive forces, and they should always be mentioned. In small polar molecules dipole-dipole or hydrogen-bonding attractive forces may predominate. In large organic molecules, London forces can be substantial and be the dominant force of attraction. For the compounds listed, the following are the dominant forces: C_6H_6, London forces; CH_3OH, hydrogen bonding; CH_3NH_2, hydrogen bonding; CF_4, London forces; SO_3, London forces; XeF_6, London forces; $C_{12}H_{26}$, London forces; and CH_3OCH_3, dipole-dipole forces.

SUMMARY OF ATTRACTIVE FORCES

London forces. These forces (also called van der Waals dispersion forces, instantaneous dipoles, induced dipoles, etc.) are very weak attractive forces because of the unequal distribution of electrons around an atom.

Dipole-dipole forces. The attraction between the partial positive end of one dipole and the partial negative end of another dipolar molecule. The molecules can be the same or different substances.

Hydrogen bonding. Very strong dipole-dipole attractive forces observed exclusively in compounds that have an F, N, or O bonded directly to a hydrogen atom.

Note that some chemists combine all of the above forces under the general term van der Waals forces.

PHYSICAL PROPERTIES OF LIQUIDS

We may use the forces discussed above to describe the liquid state. Some of the physical properties of liquids are surface tension, viscosity, evaporation, vapor pressure, boiling point, and heat of vaporization.

Surface Tension

Surface tension is due to an increase in the attractive forces between molecules at the surface of a liquid compared to the forces between molecules in the center, or bulk, of the liquid. This property causes fluids to minimize their surface areas. As a result, small droplets of liquids tend to form spheres. Surface tension produces a "skin" on a liquid surface that allows small insects to literally stand on water. Also, carefully placed small iron objects such as needles and pins can float on the surface of water even though iron is almost eight times as dense as water.

When considering surface tension, we first look at a molecule of the liquid in the interior of the sample (the bulk solvent). In the interior, the solvent molecule is surrounded by other solvent molecules on all sides as shown in Figure 8.7a. At the interface between the liquid and the gas phase, some of the molecules surrounding the solvent have been removed as in Figure 8.7b. Since potential energy must be added to remove those surrounding molecules, the surface molecules will compensate by attracting neighboring molecules more strongly to reduce that added potential energy. This stronger attraction results in the molecules at the surface being closer to each other and the effect we note as surface tension.

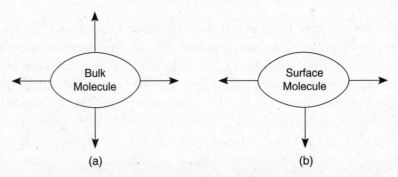

(a)   (b)

FIGURE 8.7. **Diagram illustrating attractive forces for molecules (a) in the bulk of a liquid compared to those at the surface (b). Arrows indicate attractive forces; front and rear arrows are not shown.**

Surface tension determines whether or not droplets of a liquid will bead up, as on a freshly waxed car, or spread out when placed on a flat surface. To describe this phenomenon, we define **cohesive forces** as attractions between identical molecules in the liquid, and **adhesive forces** as attractions between different molecules, such as those in the liquid and a flat surface. If the cohesive forces of the liquid are strong compared to the adhesive forces between that liquid and the flat surface, the liquid will retain its shape and form beads. If, however, the adhesive forces between the flat surface and the liquid are strong enough, the liquid will spread uniformly.

Very clean glass has low adhesive forces, and water beads readily on it. In dishwashers these beads of water evaporate, leaving undesired spots on glasses and dishes. For this reason dishwasher detergents contain chemicals called **surfactants.** A surfactant decreases the cohesive forces, and therefore the surface tension, of liquids. The surfactant in dishwasher detergent lowers the cohesive forces of water so that the adhesive forces between water and glass are small. Then water does not bead up on glassware, and the spotting problem is reduced.

Viscosity

Viscosity refers to a liquid's resistance to flow. A liquid such as water has a fairly low viscosity and flows easily. Pancake syrup, on the other hand, has a high viscosity, particularly when cold, and flows slowly. The attractive forces within the liquid are responsible for viscosity. In order for a liquid to flow, the molecules must move past each other. Molecules are able to move more freely in solutions that have relatively low attractive forces. The liquid alkanes have lower viscosities than water because alkanes are attracted to each other only by London forces. Water is more viscous because of hydrogen bonding. Syrup is very viscous since its bulky sugar molecules contain many —OH groups, which hydrogen-bond to the water in the mixture.

Viscosity usually decreases as the temperature of a liquid is increased. Pancake syrup flows much more easily at room temperature than it does when first taken from the refrigerator. The reason is that at the higher temperature the molecules have a higher kinetic energy since $KE = kT$. This increase in kinetic energy weakens the intermolecular forces, thus decreasing the viscosity.

Evaporation

Evaporation is a familiar process in which a liquid in an open container is slowly converted into a gas. Some liquids evaporate more rapidly than others. For example, a beaker of gasoline evaporates in a few hours, whereas a beaker of water may take a day or two. In addition, the rate at which a liquid evaporates increases as the temperature increases.

Evaporation may be explained by considering the attractive forces involved and the kinetic energy needed to overcome these forces. This process is the reverse of condensation. In order for a molecule to be converted from a liquid to a gas, it must have sufficient kinetic energy to overcome its attractive forces. In any group of molecules the average kinetic energy is proportional to the Kelvin temperature. The actual kinetic energies are distributed as shown in Figure 8.8. Some molecules have low, and others have high, kinetic energies. The escape energy is defined as the minimum kinetic energy needed for a molecule to escape from the liquid into the gas phase. All molecules with kinetic energies greater than the escape energy are capable of evaporating.

The area under the curve in Figure 8.8 represents the total number of molecules. The shaded area represents the number of molecules that have kinetic energies equal to or exceeding the escape energy. The ratio of these two areas is a constant as long as the temperature is constant. Molecules that are close enough to the surface and traveling in the correct direction will escape as gas molecules. At constant temperature, the proportion of molecules with enough kinetic energy to escape will remain constant and the liquid will evaporate at a uniform rate until all molecules have entered the gas phase.

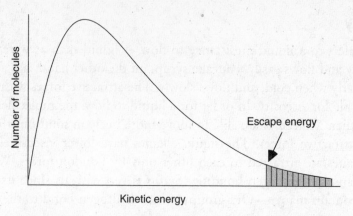

FIGURE 8.8. Distribution of kinetic energies of a group of molecules at a given temperature. Total area under the curve represents all molecules, and shaded area represents molecules with kinetic energies greater than the escape energy.

Since only molecules near the surface may escape into the gas phase, the surface area of the liquid is a factor in evaporation (Figure 8.9). Ten milliliters of a liquid in a narrow test tube will evaporate more slowly than the same 10 milliliters in a beaker. If the liquid is poured into an evaporating dish, it will evaporate even more quickly. The reason is found in the increased proportion of molecules that are close enough to the surface to escape readily.

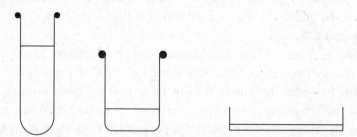

FIGURE 8.9. Surface areas of a test tube, beaker, and evaporating dish, each holding 10 mL of liquid. The increasing surface area indicates that evaporation will be slowest from the test tube and will be fastest from the evaporating dish.

Temperature is another important factor in the evaporation process. An increase in temperature increases the proportion of the molecules that have kinetic energies above the escape energy. Lowering the temperature decreases the proportion with enough kinetic energy to escape.

The temperature of a beaker of liquid evaporating on a lab bench is usually constant because it can absorb heat from its surroundings easily and quickly. If the beaker is insulated from the surroundings, however, the liquid cools and the rate of evaporation decreases. These phenomena are explained using curves similar to the one in Figure 8.8. First, when a molecule escapes from the liquid into the gas phase, the average kinetic energy of the remaining molecules decreases. A decrease in the average kinetic energy means that the temperature also decreases, explaining the cooling observed. Second, as the average kinetic energy decreases, the proportion of molecules that have kinetic energies above the escape energy decreases. Since fewer molecules have enough energy to escape, the rate of evaporation decreases.

Conversely, increasing the temperature of a liquid increases the rate of evaporation. This rise occurs because a greater proportion of all molecules now have a kinetic energy greater than the escape energy. When the temperature is increased sufficiently, boiling occurs. Boiling is recognized as the formation, throughout the solution, of gas bubbles which then rise to the surface. At the boiling point, the molecules do not have to reach the surface to enter the gas phase. Enough molecules, with the appropriate escape energy, can come together within the solution to form the bubbles we observe.

Vapor Pressure

Vapor pressure is the pressure that develops in the gas phase above a liquid when the liquid is placed in a closed container. Evaporation of molecules from the liquid still occurs in the closed container, but the gas molecules cannot escape to the surroundings. As more molecules enter the gas phase, the pressure increases, finally stopping at a level that is dependent only on the temperature. This final pressure is called the vapor pressure.

When a liquid is placed in a closed container, it starts evaporating just as it would in an open beaker. In the closed container, however, gas molecules cannot escape. As the gas molecules move, they collide with the walls of the container, the liquid in the lower part of the container being one of these "walls." When the gas molecules collide with the liquid, very few bounce off; almost all condense to the liquid state again. Initially the rate at which the molecules evaporate is much greater than the rate at which they condense. As the gas molecules increase in number, they collide with the liquid surface more frequently. Eventually the rate at which the liquid molecules evaporate is equal to the rate at which the gas molecules condense. Under these conditions the liquid and gas are said to be in **dynamic equilibrium.** Figure 8.10 illustrates this process.

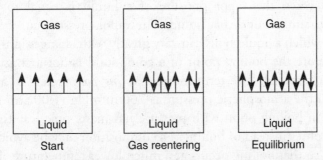

FIGURE 8.10. **Diagrams illustrating the establishment of the equilibrium vapor pressure. At the start, molecules leave the liquid into empty space. The middle frame shows some gas reentering the liquid but not as rapidly as molecules leave. At equilibrium, molecules leave and enter the liquid at the same rate.**

At equilibrium, the rate at which molecules leave the liquid must equal the rate at which they reenter the liquid. The rate at which molecules escape the liquid (evaporate) depends on the temperature. The rate at which they enter the liquid (condense) depends on the frequency at which the gas molecules collide with the liquid "wall" of the container. In Chapter 7 on gases, it was shown that the frequency of collision is part of the definition of gas pressure. As a result, the vapor pressure depends only on the nature of the liquid (attractive forces) and the temperature (kinetic energy). As the temperature increases, the vapor pressure increases as shown in Figure 8.11.

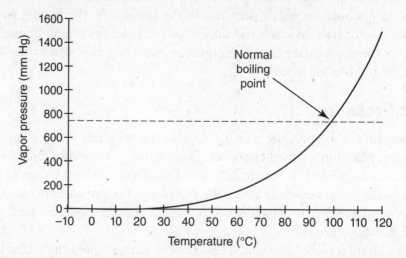

FIGURE 8.11. **Vapor pressure curve for water. Dashed line at 760 mm Hg intersects the curve at 100 °C, the normal boiling point of water.**

A vapor pressure curve such as the one shown in Figure 8.11, or the tabular form of the data, may be used to determine the vapor pressure of a liquid at any temperature. Therefore chemists do not have to repeatedly determine the vapor pressure; it may be looked up in a convenient source.

Vapor pressure data point to some interesting facts about the boiling process.

Boiling Point

Boiling occurs when the vapor pressure of a liquid is equal to the prevailing atmospheric pressure around that liquid. The vapor pressure curve shows that the temperature at which a liquid boils can vary greatly with changes in the atmospheric pressure. Therefore the boiling point of a compound is not a constant unless the pressure is also specified. The term *normal boiling point* refers to a boiling point measured when the atmospheric pressure is 760 mm Hg (1.00 atm.).

The change in boiling point with pressure has many practical aspects. Suppose a liquid decomposes instead of boiling. Decomposition may be avoided if the pressure is reduced so that boiling occurs at a much lower temperature. This technique is used in **vacuum distillation** to purify heat-sensitive materials.

The normal atmospheric pressure in Denver, Colorado, is much lower than 760 millimeters of mercury because of the mile-high altitude of the city. The result is that water boils at a lower temperature. At high elevations, longer heating times are needed to cook food properly. For this reason, many people in Denver (and elsewhere) use pressure cookers. These covered pots increase the pressure and therefore the boiling point of water. At the increased temperature, foods cook faster.

Heat of Vaporization

The **heat of vaporization** is the energy needed to convert 1 gram of liquid into 1 gram of gas at a temperature equal to the normal boiling point of the liquid. The units for the heat of vaporization are joules per gram ($J\ g^{-1}$). If 1 mole of liquid is vaporized, we call the energy the molar heat of vaporization and we use the units joules per mole ($J\ mol^{-1}$). In either case the symbol is ΔH_{vap}. Since energy must always be added to a liquid to cause it to vaporize, ΔH_{vap} is always positive. In Chapter 12 on thermodynamics, a positive ΔH_{vap} is defined as indicating an endothermic process. The reverse process, condensation, requires the gas to give off heat in an exothermic process. Since vaporization and condensation describe the same process from different directions, their heats are related by the equation

$$\Delta H_{vap} = -\Delta H_{cond} \tag{8.2}$$

Table 8.1 lists some heats of vaporization for compounds with different types of intermolecular forces.

TABLE 8.1

Heats of Vaporization of Representation Compounds

Compound	Formula	Heat of Vaporization (kJ-mol^{-1})	Attractive Force
water	H_2O	+43.9	hydrogen bonding
ammonia	NH_3	+21.7	hydrogen bonding
hydrogen fluoride	HF	+30.2	hydrogen bonding
hydrogen chloride	HCl	+15.6	dipole-dipole
hydrogen sulfide	H_2S	+18.8	dipole-dipole
fluorine	F_2	+5.9	london
chlorine	Cl_2	+10.0	london
bromine	Br_2	+15.0	london
methane	CH_4	+8.2	london
ethane	C_2H_6	+15.1	london
propane	C_3H_8	+16.9	london

There are differences in the heats of vaporization that can be related to the intermolecular attractive forces. For similar-size molecules, hydrogen-bonded substances have the largest ΔH_{vap} values. Polar substances have higher heats of vaporization than similar-size nonpolar substances. Molecules that have the same intermolecular attractive forces also show trends in their heats of vaporization. Water has the highest ΔH_{vap}, and ammonia has the lowest, among hydrogen-bonded molecules. Fluorine, chlorine, and bromine show a regular increase in ΔH_{vap}, and methane, ethane, and propane also have increasing ΔH_{vap} values because of increasing London forces.

The amount of heat required to vaporize a liquid is very large. For example, the heat energy needed to vaporize 1 gram of water could be used to raise the temperature of six times as much water from zero to 100 °C. This fact explains why water can be quickly raised to its boiling point, but a long time is needed to boil away all of the water.

EXERCISE 8.3

For each of the following pairs, predict which compound will have (1) the lower boiling point, (2) the higher heat of vaporization, (3) the higher evaporation rate, and (4) the lower vapor pressure.

(a) C_6H_{14} or C_8H_{18} (c) HF or HCl   (e) $C_6H_{13}OH$ or $C_3H_7OC_3H_7$

(b) C_6H_{14} or $C_6H_{13}OH$ (d) HBr or HCl   (f) PBr_3 or PBr_5

Solution

For these pairs we have to assess the relative attractive forces. The one with the greater attractive forces will have the higher boiling point, higher heat of vaporization, lower evaporation rate, and lower vapor pressure. For the six pairs, the substances with the greater attractive forces are as follows: (a) C_8H_{18} since it has more hydrogen atoms for greater London forces; (b) $C_6H_{13}OH$ since it has the —OH group, which forms hydrogen bonds; (c) HF since it forms hydrogen bonds; (d) HBr since its electron cloud is more polarizable; (e) $C_6H_{13}OH$ since it forms hydrogen bonds; (f) PBr_5 since it has more polarizable bromine atoms (the polarity of PBr_3 is a minor factor). Once the attractive forces are determined, the answers can be obtained:

(a) 1. C_6H_{14} (c) 1. HCl   (e) 1. $C_3H_7OC_3H_7$
      2. C_8H_{18} 2. HF   2. $C_6H_{13}OH$
      3. C_6H_{14} 3. HCl   3. $C_3H_7OC_3H_7$
      4. C_8H_{18} 4. HF   4. $C_6H_{13}OH$

(b) 1. C_6H_{14} (d) 1. HCl   (f) 1. PBr_3
      2. $C_6H_{13}OH$ 2. HBr   2. PBr_5
      3. C_6H_{14} 3. HCl   3. PBr_3
      4. $C_6H_{13}OH$ 4. HBr   4. PBr_5

Clausius-Clapeyron Equation

The heat of vaporization and the vapor pressure are both measures of the intermolecular forces that attract molecules together in the liquid state. Because they describe the same forces, there is a mathematical relationship, called the **Clausius-Clapeyron equation,** between the two:

$$\ln P = \frac{-\Delta H_{vap}}{RT} + C \tag{8.3}$$

In this equation P represents the vapor pressure, ΔH_{vap} is the heat of vaporization, R is the universal gas law constant, T is the Kelvin temperature, and C is a constant. A plot of the natural logarithm of the vapor pressure versus $1/T$ will produce a straight-line graph in which the slope of the line will be $-\Delta H_{vap}/R$.

Another form of the Clausius-Clapeyron equation relates the vapor pressures at two temperatures to ΔH_{vap} as shown in Equation 8.4.

$$\ln\left(\frac{P_1}{P_2}\right) = \frac{\Delta H_{vap}}{R}\left(\frac{1}{T_2} - \frac{1}{T_1}\right) \tag{8.4}$$

In this form the constant C has been eliminated. Equation 8.4 has five variables: two temperatures, two pressures, and ΔH_{vap}. It can be solved for any one of these if the other four variables are known. For instance, ΔH_{vap} can be determined by measuring the vapor pressures at two different temperatures. Or, if the heat of vaporization is known, the vapor pressure at any temperature can be determined if the vapor pressure is known for any other temperature. Finally, the normal boiling point of a liquid can be determined from the ΔH_{vap} value and a known vapor pressure at any temperature. This calculation is possible since the normal boiling point is always at 760 millimeters of mercury, or 1.00 atmospheres, of pressure.

EXERCISE 8.4

What are the heat of vaporization (in kJ mol^{-1}) and the normal boiling point of a liquid that has a vapor pressure of 254 mm Hg at 25 °C and a vapor pressure of 648 mm Hg at 45 °C? ($R = 8.314$ J mol^{-1} K^{-1})

Solution

The Clausius-Clapeyron equation, Equation 8.3, is used first to determine ΔH_{vap}, and then to determine the normal boiling point at 760 mm Hg. At this point, the necessary units are considered. The temperature must be converted to the Kelvin temperature units. The pressures may have any units as long as they are identical since the units for pressure will cancel in the ratio P_1/P_2.

$$\ln\left(\frac{P_1}{P_2}\right) = \frac{\Delta H_{vap}}{R}\left(\frac{1}{T_2} - \frac{1}{T_1}\right)$$

In solving this equation, we assign 298 K and 254 mm Hg to T_1 and P_1, respectively. T_2 and P_2 are 318 K and 648 mm Hg. Substituting these values into the equation, we have

$$\ln\left(\frac{254 \text{ mm Hg}}{648 \text{ mm Hg}}\right) = \frac{\Delta H_{vap}}{8.314 \text{ J mol}^{-1} \text{ K}^{-1}}\left(\frac{1}{318} - \frac{1}{298}\right)$$

Solving yields

$$-0.937 = \Delta H_{vap}(-2.53 \times 10^{-5} \text{ J}^{-1} \text{ mol})$$

Then

$$\Delta H_{vap} = \frac{-0.937}{-2.53 \times 10^{-5} \text{ J}^{-1} \text{ mol}}$$

$$= +36{,}911 \text{ J mol}^{-1}$$

$$= +36.9 \text{ kJ mol}^{-1}$$

Once the heat of vaporization has been determined, one of the two vapor pressure measurements can be combined with it to calculate the normal boiling point. Assigning $P_1 = 648$ mm Hg, $P_2 = 760$ mm Hg, and $T_1 = 318$ K, we enter the data into the equation and we calculate T_2, the normal boiling point.

Before actually solving the equation, we can estimate the answer. First, it must be above 318 K since 648 torr is less than 760 torr. Second, the normal boiling point cannot be very much greater than 318 K since 648 torr is not far from 760 torr.

Entering the data into the Clausius-Clapeyron equation we have

$$\ln\left(\frac{648 \text{ mm Hg}}{760 \text{ mm Hg}}\right) = \frac{36,900 \text{ J mol}^{-1}}{8.314 \text{ J mol}^{-1} \text{ K}^{-1}}\left(\frac{1}{T_2} - \frac{1}{318}\right)$$

We can solve this for T_2, which will be the boiling temperature:

$$-0.519 = 4438 \text{ K}\left(\frac{1}{T_2} - 3.14 \times 10^{-3} \text{ K}^{-1}\right)$$

$$= \frac{4438 \text{ K}}{T_2} - 13.96$$

$$13.44 = \frac{4438 \text{ K}}{T_2}$$

$$T_2 = \frac{4438 \text{ K}}{13.78}$$

$$= 330 \text{ K } (49 \text{ °C})$$

The value of 330 K for this part of the problem fits well with the estimate (slightly above 318 K) made beforehand.

SOLIDS

At room temperature and atmospheric pressure, many substances exist as solids. In the periodic table, there are two liquids and 11 gases; the remaining 96 elements are solids. Solids have the property of retaining their shapes with or without a container. This occurs because solids have rigid crystal structures. These solid structures may be defined based on the attractive forces that hold them together or on the arrangement of the atoms in the crystals themselves.

Crystal Types Based on Attractive Forces

METALLIC CRYSTALS

All metals in the periodic table are solids at 25 °C, except mercury. The **metallic crystal** is visualized as a rigid structure of metal nuclei and inner electrons. The valence electrons are thought to be very mobile in the structure, moving freely from atom to atom. These mobile electrons act to bond metal atoms together with widely varying degrees of force. Metals such as iron, chromium, cobalt, gold, platinum, and copper have melting points above 1000 °C. Others, for example, mercury and gallium, have melting points near or below room temperature. Melting points are one measure of the attractive forces since melting disrupts the crystal bonding, producing a liquid. The energy needed to disrupt a crystal is often called the lattice energy.

The mobile valence electrons provide an explanation for the ability of metals to conduct electricity and heat. In both cases, the electrons can quickly carry charge (electricity) and thermal energy (heat) throughout the metal. Also, the interaction of light with these electrons is responsible for the characteristic metallic luster. Most metals have a color similar to that of silver or aluminum. A few, notably copper and gold, are yellow.

Metals such as lead, gold, sodium, and potassium are soft and can be cut with a knife. Other metals, for example, tin and zinc, are somewhat brittle. Most metals are malleable and can be formed into various shapes with a hammer or extruded into thin wires. These properties are due to the metallic crystal structure, which allows the atoms to move from one position to another without a major disruption of the crystal. The softness, hardness, and brittleness of metals can be altered by preparing solutions of one metal dissolved in another. These solutions are known as alloys.

The atoms in metallic crystals are arranged in the most compact form possible. As a result, the atoms usually crystallize in either the face-centered cubic structure or the body-centered cubic structure. Both of these are described below.

IONIC CRYSTALS

The attraction of a cation (positive ion) toward an anion (negative ion) is the strongest attractive force known in chemistry. The result is that almost all ionic compounds are solids with rigid crystalline structures (lattices). Because of these strong attractions, a large amount of energy, called the lattice energy, is required to separate the ions. The high lattice energy of ionic compounds gives them very high melting and boiling points compared to molecular compounds of similar size and molar mass. For example, sodium chloride melts at 801 °C and boils at 1413 °C, while butane (C_4H_{10}, molar mass = 58) melts at −135 °C and boils at approximately 0 °C. The melting points of ionic crystals are consistently high in contrast to the variability evident in the metals.

An **ionic crystal** has a regular structure, or lattice, of alternating positive and negative ions. Many of these crystals are cubic structures, which will be described later. Other structures may provide shapes seen in many natural minerals. It is relatively simple to describe the crystal structures of the metals since all of the atoms are the same size. Ionic crystals, however, usually have ions of different sizes, which affect the manner in which they pack. These sizes also may limit the closeness with which the ions approach each other.

Ionic bonding causes these crystals to be rigid and brittle. To understand this property, we visualize a simple ionic substance such as NaCl with alternating sodium cations and chloride anions in a crystal lattice. The strong attraction of the positive and negative charges holds the crystal rigidly together. Hitting an ionic crystal with a hammer has a very different result compared to hitting a metallic crystal with the same force. In both, the atoms can be forced to move. In a metallic crystal, the atoms shift their positions but the metallic bond is not disrupted. In an ionic crystal, however, movement of the atoms by as little as one ionic diameter will cause positive ions to be aligned with positive ions and negative ions to be aligned with negative ions. The repulsion between like-charged ions is so great that the crystal shatters, as diagrammed in Figure 8.12.

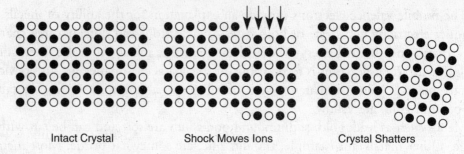

Intact Crystal　　　Shock Moves Ions　　　Crystal Shatters

FIGURE 8.12.   Illustration of why ionic crystals shatter. Light circles are cations and dark circles are anions.

MOLECULAR CRYSTALS

Molecular crystals may be composed of either atoms of the nonmetals or of covalent molecules. These crystals are held together by London forces, dipole-dipole attractions, hydrogen bonding, or a mixture of these. All of these forces are much weaker than the attractive forces between ions in ionic crystals. As a result, molecular crystals tend to be soft, with low melting points. Some substances that form molecular crystals with London forces holding the crystal together are neon, xenon, CO_2, sulfur, fluorine, methane (CH_4), and decane ($C_{10}H_{22}$). Dipole-dipole attractive forces hold crystals of SO_2, $CHCl_3$, and other polar molecules together. Hydrogen bonding is responsible for the attractive forces in crystals of H_2O and NH_3. In many molecules there may be a combination of attractive forces. For example, *n*-decanol has the structure

$$CH_3CH_2CH_2CH_2CH_2CH_2CH_2CH_2CH_2CH_2OH$$

The long carbon chain is responsible for London forces, while the —OH at the end of the molecule provides hydrogen bonding.

NETWORK (COVALENT) CRYSTALS

A **network crystal** has a lattice structure in which the atoms are covalently bonded to each other. The result is that the crystal is one large molecule with a continuous network of covalent bonds. A diamond is pure carbon with each carbon atom covalently bonded to four other carbon atoms in a tetrahedral (sp^3) geometry. The totality of this network of covalent bonds makes the diamond the hardest natural substance known. SiO_2 is the empirical formula for sand and quartz. Silicon dioxide forms a covalent crystal with each silicon forming bonds to four oxygen atoms and each oxygen bonding to two silicon atoms with a tetrahedral geometry. Silicon carbide is another network crystal similar to diamond with alternating tetrahedral silicon and carbon atoms. It is very hard and is used as an industrial substitute for diamonds. Network crystals, like ionic substances, are represented by their empirical formulas.

Graphite is another form (allotrope) of carbon in a covalent crystal. In graphite each carbon atom is covalently bonded to three other carbon atoms in a trigonal planar (sp^2) geometry that gives graphite its structure of flat sheets. The extra p electron that is not used in the sp^2 bonding holds these sheets together in a manner similar to that seen in a metallic crystal. The weak bonding of the p electrons allows the flat sheets to slide over each other easily and is responsible for the slippery feel of graphite. In addition, these *p* electrons are responsible for the ability of graphite to conduct electricity.

AMORPHOUS (NONCRYSTALLINE) SUBSTANCES

Some materials are **amorphous** and do not form crystals. One characteristic of a noncrystalline substance is that it does not have a distinct, sharp melting point. Rather, these materials soften gradually over a large temperature range. Ordinary glass is an example. Although glass is composed mainly of SiO_2, the atoms are not arranged in a network crystal as discussed above. Glass has often been described as a supercooled liquid. Many plastics (polymers) have combined characteristics; they are partially crystalline and partially amorphous.

PHASE CHANGES AND PHASE DIAGRAMS

Solids, liquids, and gases can be converted from one phase to another by temperature and pressure changes, and we can make qualitative and quantitative observations about these conversions. There are two ways to represent these changes: the heating or cooling curve and the phase diagram. If the pressure is held constant, the effect of heat may be explained by a heating or cooling curve. If we are interested in the effects of both temperature and pressure, a phase diagram is used.

Heating and Cooling Curves

Starting with a solid material well below its melting point and adding heat at a constant rate will produce the following effects:

1. The temperature of the solid will increase at a constant rate until the solid starts to melt.
2. When melting begins, the temperature stops rising and remains constant until all of the solid is converted into a liquid.
3. The temperature of the liquid starts increasing at a constant rate until boiling starts.
4. When boiling begins, the temperature stops rising and remains constant until all of the liquid has been converted into gas.
5. The temperature of the gas increases at a constant rate.

This process is often summarized in a **heating curve** as shown in Figure 8.13.

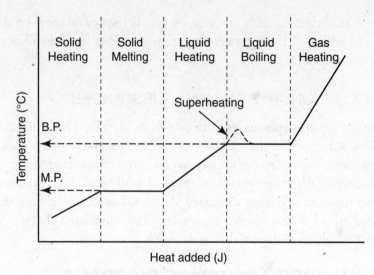

FIGURE 8.13. Typical heating curve, bringing a sample from the solid state on the left to the gaseous state on the right.

In a heating curve there are several points of interest. Adding heat to a pure solid, liquid, or gas phase increases its temperature. The **heat capacity** (J °C^{-1}) of a solid, liquid, or gas is the reciprocal of the slope of the curve (change in joules divided by the change in temperature) in those regions where the temperature increases as heat is added. The **specific heat** (J g^{-1} °C^{-1}) is the heat capacity divided by the number of grams of sample used. The plateaus represent the melting and boiling processes. The temperature does not change during melting and boiling, and two phases are present. Superheating may occur where the temperature of the liquid exceeds the boiling point. See dashed line in Figure 8.13. During melting both the solid and the liquid are in equilibrium; during boiling both the liquid and the gas phase are in equilibrium. The **molar heat of fusion** (melting) may be determined as the length of the first (solid-melting) plateau, which represents the heat added, divided by the number of moles of sample. The **molar heat of vaporization** is the length of the second (liquid-boiling) plateau divided by the number of moles of sample. Both the heat of fusion and the heat of vaporization have units of joules per mole (J mol^{-1}). These are also called the enthalpy of fusion and enthalpy of vaporization, respectively.

The **cooling curve** is the reverse of the heating curve, as shown in Figure 8.14. All of the features are the same except that we start with a gas and we end with a solid as the heat is removed from the sample. The gas condenses to a liquid at the same temperature at which the liquid boils, and the liquid crystallizes at the same temperature at which the solid melts. The terms **condensation point** and **crystallization point** are sometimes used in place of boiling point and melting point.

Some liquids exhibit an ability to be supercooled. **Supercooling** is observed when a liquid that is cooled to a temperature below its melting point remains a liquid. A supercooled liquid is described as being in a metastable state. A metastable liquid will crystallize rapidly if sufficiently disturbed by shaking or by adding a seed crystal to initiate crystallization. In Figure 8.14 the dashed line on the curve shows the alternative path taken by a substance that supercools. Gases do not supercool as some liquids can.

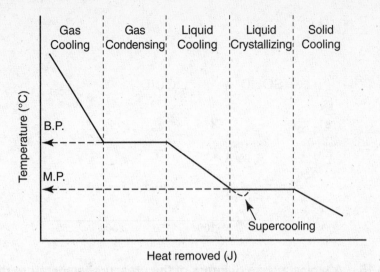

FIGURE 8.14. A cooling curve, showing the effect of removing heat from a gas to form first a liquid and then a solid. The dashed line shows an alternative path for a liquid that supercools.

EXERCISE 8.6

List all of the information that can be obtained from a heating curve.

Solution

We can obtain the following data from a heating curve: (1) melting point; (2) boiling point; (3) specific heats and heat capacities of the solid, liquid, and gas; (4) heat of fusion of the solid; (5) heat of vaporization of the liquid.

Phase Diagrams

We can draw a diagram, called a **phase diagram,** that shows the relationship between pressure and temperature and the three states of matter. A phase diagram of a substance allows us to make many predictions about the physical behaviors of a wide variety of materials. Most phase diagrams are similar to the one shown in Figure 8.15. Pressure is plotted on the *y*-axis, and temperature is plotted on the *x*-axis. The diagram is divided into the three physical states by three lines that meet at the **triple point.** Solids exist at pressures and temperatures in the upper left of the diagram. Liquids exist at pressures and temperatures in the upper right, and gases exist in the lower part, of the diagram. Along each line in the diagram there is an **equilibrium** mixture of the two phases on the two sides of that line. At the triple point all three phases exist together in equilibrium. The liquid-solid line is usually a straight line, while the gas-solid and liquid-gas lines are curved upward, as shown.

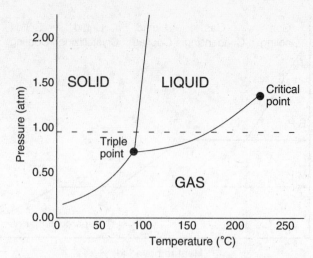

FIGURE 8.15. **A typical phase diagram of the pressure-temperature regions where gas, liquid, and solid phases exist. Each line represents an equilibrium between two phases, and the triple point indicates the pressure and temperature at which all three phases are in equilibrium.**

Temperature and pressure are the two variables that define a point in the phase diagram. We may draw a horizontal line representing a selected pressure and a vertical line representing a selected temperature. The intersection of the two lines on the phase diagram will indicate the phase(s) of the substance at that particular combination of temperature and pressure.

The **triple point** is the only point where all three phases are in equilibrium with each other. At this point each phase has the same temperature and vapor pressure. For each substance there is only one temperature and one pressure at which the triple point occurs. (Helium is the only substance that does not have a triple point since it has no solid phase.) The triple point of water is at 0.01 °C and 4.58 mm Hg.

Three equilibrium lines radiate from the triple point in the phase diagram. The pressures and temperatures along these lines indicate the pressures and temperatures where two adjoining phases will be in equilibrium.

The solid-liquid equilibrium line:

$$SOLID \rightleftharpoons LIQUID$$

points upward from the triple point. This line is nearly vertical because changes in pressure have very little effect on solids and liquids (these two phases are only slightly compressible in comparison to gases). For almost all compounds, the line slopes slightly toward higher temperatures. For water, however, it slopes in the opposite direction. The slight tilt of this equilibrium line may be explained based on the relative densities of the solid and liquid phases. When the pressure is increased, the equilibrium will be shifted toward the denser phase. To counter this shift, the temperature must change to reestablish the equilibrium. Most solids are denser than their liquids. Increasing the pressure forces the reaction toward the formation of solid. To maintain equilibrium, the temperature must be increased; therefore, the line slopes slightly toward the right. Liquid water, however, is denser than solid ice. As a result, an increase in pressure favors the formation of liquid water. To maintain equilibrium, the temperature must decrease, and therefore the solid-liquid equilibrium line tilts slightly toward the left. The normal melting point is the point on this line at 760 millimeters of mercury (1 atmosphere) of pressure. We can change the melting point only slightly by changing the pressure.

The liquid-gas equilibrium line:

$$\text{LIQUID} \rightleftharpoons \text{GAS}$$

extends in a curved manner upward and to the higher temperature side from the triple point. This line shows a pronounced curvature because of the compressibility of gases. We find that this line ends abruptly at the **critical point,** that is, the maximum temperature at which any liquid can exist. Above the critical point, the differences between gases and liquids disappear and the substance is often called a **supercritical fluid.** The normal boiling point of a liquid is the point where the liquid-gas equilibrium line crosses the dashed line, indicating a pressure of 760 millimeters of mercury. Unlike the melting point, the boiling point may vary greatly with pressure.

The solid-gas equilibrium line:

$$\text{SOLID} \rightleftharpoons \text{GAS}$$

extends downward and to lower temperatures from the triple point. This line is also highly curved because of the compressibility of gases. The process of converting a solid directly into a gas is called **sublimation.** In this portion of the diagram, pressures are usually at very low levels not normally encountered in the laboratory. Some substances, such as iodine, have solid-gas equilibrium lines at atmospheric pressure. Solid iodine readily sublimes from the solid to a vapor without any liquid phase.

EXERCISE 8.7

Determine the temperatures and pressures for the critical point, triple point, normal boiling point, and normal melting point for the substance described in Figure 8.15.

Solution

We may estimate the critical point at 220 °C and 1.4 of atmospheres pressure. The triple point is at 95 °C and 0.75 of atmospheres pressure. The normal boiling and melting points must be at 1.00 atm of pressure, and they are 180 °C and 100 °C, respectively.

SUMMARY

When molecules occupy most of the available space and there are relatively strong attractive forces, we get the condensed phases, liquids and solids, described in this chapter. Intermolecular forces including hydrogen bonding, dipole-dipole forces, and the very weak London forces that arise from instantaneous dipoles are described and used to explain many observed properties of matter. Melting points, boiling points, viscosity, vapor pressure, and the behavior of liquids in contact with solids are all understandable based on knowledge of these few intermolecular forces. In turn, the forces that influence the properties above are easily understood when we know the molecular structure and polarity that were discussed in Chapter 5. Methods to describe the phases of matter include heating and cooling curves, vapor pressure diagrams, and phase diagrams.

Important Concepts

Intermolecular forces, dipole-dipole, London and hydrogen bonds
Physical properties and intermolecular forces
Solid crystals and crystal structures
Heating and cooling curves
Phase diagrams

Practice Exercises

Multiple-Choice

1. In which of the following are the intermolecular forces listed from the weakest to the strongest?

    (A) dipole-dipole > London > hydrogen bonds
    (B) London < dipole-dipole < hydrogen bonds
    (C) hydrogen bonds < dipole-dipole < London
    (D) London > hydrogen bonds > dipole-dipole
    (E) London < hydrogen bonds < dipole-dipole

2. Which of the following consistently have the highest melting points?

    (A) metals
    (B) salts
    (C) molecular crystals
    (D) alkanes
    (E) hydrogen-bonded compounds

3. Which physical transformation occurs in the process called sublimation?

    (A) gas to liquid
    (B) gas to solid
    (C) solid to liquid to gas
    (D) solid to gas
    (E) solid to liquid

4. Which element is expected to have the greatest polarizability?

   (A) Fe
   (B) Ca
   (C) Ne
   (D) S
   (E) Al

5. Which of the following compounds will NOT form hydrogen-bonds?

   (A) CH_2F_2
   (B) CH_3OH
   (C) $H_2NCH_2CH_2CH_3$
   (D) $HOCH_2CH_2OH$
   (E) HClO

6. When the following compounds are kept at the same temperature, the compound expected to evaporate most quickly is

   (A) C_8H_{18}
   (B) $C_8H_{17}OH$
   (C) $C_8H_{17}NH_2$
   (D) C_6H_{14}
   (E) $C_7H_{15}COOH$

7. Which of the following will have the greatest *change* in its boiling point if the pressure at which it is boiled is changed from 1.00 atm to 0.900 atm? (The numbers in parentheses are the heats of vaporization in kilojoules per mole.)

   (A) water (43.9)
   (B) ammonia (21.7)
   (C) methane (8.2)
   (D) bromine (15.0)
   (E) fluorine (5.9)

8. Which of the following statements is NOT consistent with the crystal properties of the substance?

   (A) SiC is used to grind metal parts to shape.
   (B) Tungsten is drawn into thin wires.
   (C) Aluminum is used to cut glass.
   (D) Graphite is used to lubricate locks.
   (E) MgF_2 shatters when dropped.

9. What causes $C_{30}H_{62}$ to be a nonpolar compound that is a solid at room temperature?

   (A) ionic attractive forces
   (B) dipole-dipole attractions
   (C) London forces
   (D) the fact that the $C_{30}H_{62}$ is a very heavy molecule
   (E) hydrogen bonding involving the 62 hydrogen atoms

10. Which physical property can be determined if the unit cell and its dimensions are determined by X ray diffraction?

    (A) heat capacity
    (B) heat of fusion
    (C) boiling and melting points
    (D) vapor pressure
    (E) density

Questions 11 and 12 refer to the heating curve below.

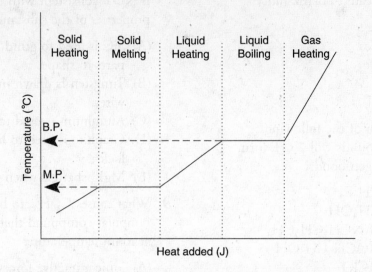

11. On the basis of this heating curve, which of the following statements is true?

(A) The heat of fusion and heat of vaporization are about equal.
(B) The heat capacities of the solid, liquid, and gas are approximately equal.
(C) The heat capacity of the gas is greater than that of the liquid.
(D) The heat capacity of the gas is greater than the heat of fusion.
(E) The heat of vaporization is much less than the heat of fusion.

12. On the basis of this heating curve, which of the following statements is correct about the substance?

(A) The substance supercools easily.
(B) The gas is a metastable state.
(C) The substance must be a salt that dissociates on heating.
(D) The density of the liquid is greater than that of the solid.
(E) The specific heat can be determined if the mass is known.

Questions 13–16 refer to the phase diagram below.

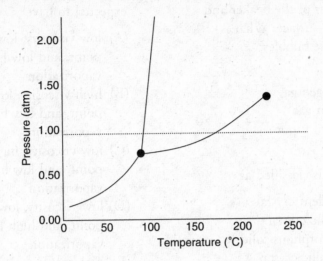

13. Determine, from the phase diagram, the maximum temperature at which this compound will sublime.

    (A) 100 °C
    (B) 230 °C
    (C) 115 °C
    (D) 95 °C
    (E) It cannot sublime.

14. What is the normal boiling point of this compound?

    (A) 95 °C
    (B) 100 °C
    (C) 50 °C
    (D) 180 °C
    (E) 150 °C

15. At 125 °C and 1.50 atm of pressure this substance will be

    (A) a liquid
    (B) a solid
    (C) a gas
    (D) a solid and liquid in equilibrium
    (E) a gas and liquid in equilibrium

16. Which pair of temperatures and pressures will produce a supercritical fluid?

    (A) 150 °C and 2.00 atm
    (B) 95 °C and 1.00 atm
    (C) 145 °C and 230 atm
    (D) 10 °C and 0.15 atm
    (E) 250 °C and 2.00 atm

17. A student observed that a small amount of acetone sprayed on the back of the hand felt very cool compared to a similar amount of water. Your explanation of this phenomena should be that

    (A) all organic compounds do this
    (B) acetone has a lower viscosity and transfers heat quanta better
    (C) water has a higher heat capacity than acetone, therefore retaining more heat
    (D) the higher vapor pressure of acetone results in more rapid evaporation and heat loss
    (E) the observed effect is not real and is only imagined

18. When a beaker of ethanol boils on the stove, small bubbles form at the bottom of the beaker and rise to the surface. What is inside these bubbles?

    (A) steam
    (B) hydrogen gas
    (C) oxygen gas
    (D) ethanol vapor
    (E) air

19. Diamond is classified as

    (A) a covalent crystal
    (B) an ionic crystal
    (C) an amorphous solid
    (D) a metallic crystal
    (E) a molecular crystal

20. A liquid substance that exhibits low intermolecular attractions is expected to have

    (A) low viscosity, low boiling point, and low heat of vaporization
    (B) high viscosity, low boiling point, and low heat of vaporization
    (C) low viscosity, high boiling point, and low heat of vaporization
    (D) low viscosity, low boiling point, and high heat of vaporization
    (E) high viscosity, high boiling point, and high heat of vaporization

ANSWER KEY

1. B	5. A	9. C	13. D	17. D
2. B	6. D	10. E	14. D	18. A
3. D	7. A	11. A	15. A	19. A
4. A	8. C	12. E	16. E	20. A

See Appendix 1 for explanations of answers.

Free-Response

Use the concepts and information in this chapter and in preceding chapters to answer the following questions.

(a) Why must helium be cooled to 4 K before it will condense? Explain your answer using fundamental concepts.

(b) Use the concepts in this chapter to explain why trans fats are considered harmful whereas cis fats are not.

(c) Use a phase diagram to illustrate why carbon dioxide sublimes.

(d) Sketch a plot of boiling points versus period number for the hydrides of the oxygen group, the nitrogen group, and the halogen group. Explain what this demonstrates.

(e) List and describe the intermolecular forces we use to explain physical properties.

ANSWERS

(a) Helium, the second lightest element, has the second highest average root mean square velocity at any temperature for the gases. For attractive forces to be effective in causing condensation, the molecules must move slowly enough for these attractions to occur. Helium must be cooled to extremely low temperatures so that the very weak London forces can take effect.

(b) Trans fats are essentially linear long-chain hydrocarbons and solidify readily because of multiple sites where London forces take effect. Cis acids are "V"-shaped. These molecules rarely line up effectively and do not solidify easily. Ease of solidification is thought to be related to the development of plaque and clots in arteries.

(c) Sketch a phase diagram as in Figure 8.15. However, adjust the pressure scale so that 1.0 atm pressure is below the triple point. Using this, we see that increasing the temperature of dry ice, at normal atmospheric pressures, will result in a direct transition from solid to gas. This is called sublimation.

(d) For this we need to reproduce the graph in Figure 8.4. This does not have to be exact. However, it will show that the boiling points of HF, H_2O, and NH_3 are all significantly greater than expected because of hydrogen bonding.

(e) The intermolecular forces are dipole-dipole attractive forces. They arise when a molecule has a fixed dipole. The information in Chapter 5 helps us decide if a molecule will be polar or nonpolar. Molecules containing a H–O–, H–F, or H–N– bond will form hydrogen bonds, a particularly strong dipole-dipole attraction. For nonpolar molecules the presence of very weak forces of attraction due to "instantaneous dipoles" was proposed by London and these forces are often called London forces (other names are dispersion forces or van der Waals forces). These very weak attractions are present in virtually all substances. They predominate in nonpolar molecules.

Solutions

CHAPTER 9

- Solution Terminology
- Solvent/Solute
- Dilute/Concentrated Solutions
- Unsaturated/Saturated and Super-saturated Solutions
- The Solution Process
- Strong Electrolytes
- Weak Electrolytes
- Nonelectrolytes
- Concentration Units
- Molarity
- Molality
- Mole Fraction
- Mass Fraction
- Parts per Million
- Parts per Billion

- Concentration Calculations
- Temperature-Independent Concentration Units
- Temperature-Dependent Concentration Units
- Effect of Temperature on Solubility
- Effect of Pressure on Solubility
- Colligative Properties
- Raoult's Law
- Ideal Solutions
- Boiling-Point Elevation
- Freezing-Point Depression
- Osmotic Pressure
- Colligative Properties of Electrolytes
- van't Hoff Factor

INTRODUCTION AND DEFINITIONS

Chemical reactions occur only when atoms, molecules, or ions directly interact, or collide, with each other. Reactants separated by as much as a few molecular diameters may as well be separated by miles. Consequently, most reactions are carried out either in liquid solutions or in the gas phase where freely moving molecules or ions may effectively collide millions of times per second.

Every **solution** is a uniform mixture of one or more **solutes** dissolved in a **solvent.** Of all the possible mixtures of substances, only a few form solutions. These include all gas mixtures, gases, liquids, or solids dissolved in a liquid, and many **alloys,** which are solid solutions of two or more metals.

In a gas the attractive forces are very small; and, as stated above, gaseous molecules or atoms interact only when they collide. The absence of major attractive forces means that gases can be mixed in all proportions, and there are no gases that are "insoluble" in other gases. Liquid solutions are more complex and more common. The presence of attractive forces is the basis of our understanding of liquid-phase solutions, which are described in detail in this chapter. Alloys are generally described in higher level chemistry courses.

To describe a solution in **quantitative,** or numerical, terms, chemists specify the amount of solute dissolved in a specified amount of solvent. This is generally called the **concentration** of a solution. The various concentration units used by chemists are defined later in this chapter. **Solubility** is the quantitative term for the maximum amount, in units of grams per liter, of solute that can dissolve in a solvent. **Molar solubility** is the maximum amount, in moles per liter, of solute that can dissolve. Solubility depends on the temperature of the solution, as described later.

Chemists also use several **qualitative,** or nonnumerical, terms to describe solutions. In a **concentrated** solution a large amount of solute is dissolved in the solvent. A **dilute** solution has a small amount of solute dissolved in the solvent. Solutions that have the maximum amount of solute dissolved in the solvent are called **saturated solutions.** A saturated solution may be identified as one in which the solution is in equilibrium with undissolved solute. An **unsaturated solution** has less than the maximum amount of solute dissolved in the solvent. An unsaturated solution never has any undissolved solute present. In a **supersaturated solution** more than the maximum amount of solute is dissolved. With many solutes a supersaturated solution can be obtained by preparing a saturated solution at a high temperature and then carefully cooling the liquid to avoid crystallizing the excess solute. Supersaturated solutions are **metastable,** meaning that the excess solute will crystallize if the solution is shaken or if a seed crystal is added to start the crystallization process.

The terms *concentrated* and *dilute* have no relationship to the terms *saturated*, *supersaturated*, and *unsaturated*. Some saturated solutions may be very dilute. An example is silver chloride, which is saturated at a concentration of 1.0×10^{-5} mol AgCl per liter. On the other hand, 10 moles of sucrose dissolved in a liter of solution is a concentrated solution, but it is not saturated at 100 °C.

SOLUBILITY AND THE SOLUTION PROCESS

TIP

To review how to determine if a molecule is polar or nonpolar, read pages 205–206.

Whether or not a substance will dissolve in a solvent is of major importance in chemistry. Separating and purifying compounds often depends on the difference in solubility between two compounds. The effectiveness of drugs depends on their solubilities in water and body fats. The ability of plants to use nutrients in the soil depends on the solubilities of the nutrients in water.

Since all gases are soluble in each other in all proportions and the consideration of alloys is outside the scope of this book, we shall focus on liquid-phase solutions. With liquids, the general rule for solubility is often stated as **"like dissolves like."** This means that solutions can be made from solutes and solvents with similar polarities, but not from solutes and solvents that have very different polarities.

Focusing on the solution process more closely, we note that the dissolution of one substance in another involves three distinct steps, as shown in Figure 9.1. First, the solute molecules are separated so that the solvent can fit in between them (step A energy is used to break attractions between solute molecules). Second, the solvent molecules are separated so that the solute can fit between them (step B energy is used to break attractions between solvent molecules). Third, the separated solute and solvent molecules are combined to form the solution (step C energy is released when new attractions are formed between solute and solvent molecules). When more energy is released in step C than is used in steps A and B, the solvent will dissolve the solute. This excess energy is released as heat when the solute and solvent are mixed, and an increase in temperature is observed.

In addition to the breaking and making of attractions among molecules, Figure 9.1 illustrates that the dispersal of molecules is another factor in the solution process. In diagram form, the solute is shown as an ordered crystal. In the solution, the solute is shown with its molecules arranged in many ways. This increase in states or arrangements that a solute can have is called **entropy** and helps the solution process. In some solutions, the energy released in step C is slightly less than the energy used in steps A and B. When this occurs, the solution cools as the solvent and solute are mixed. For the solute to dissolve, however, the shortfall in energy must be compensated by the increased entropy of the solution. Solutes that are solids have the greatest increase in entropy when dissolved. Liquids have a moderate increase, and gases have virtually no increase, in entropy when dissolved.

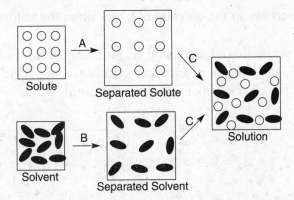

FIGURE 9.1. **The three steps involved in forming a solution. Step A separates the solute, step B separates the solvent, and step C combines the separated solvent and solute into a solution.**

Examples of the Solution Process

DISSOLVING IONIC COMPOUNDS

This example describes why ionic compounds dissolve to form aqueous solutions. Ionic compounds have the strongest attractive forces holding the ions in the crystal lattice. The energy needed to disrupt the crystal (step A in the solution process) is known as the lattice energy. Lattice energies are so high that dissolution of an ionic compound seems impossible. However, water, with its high polarity, is strongly attracted to the charged ions in the solution. As a result, many water molecules are attracted to each ion and enough energy is released (step C of the solution process) so that the dissolution is favored. The increase in entropy due to the breakup of the crystal structure also aids the solution process. Still, many ionic compounds are insoluble in water because the hydration of the ions cannot offset the lattice energy. Chapter 4 lists the solubility rules for ionic compounds.

GAS MIXTURES

In a second example we see that gases mix freely with other gases solely on the basis of the entropy increase since there are virtually no attractive forces between gas molecules. When a gas is allowed to expand into a vacuum, it does so because the increased volume means more ways for the molecules to arrange themselves, as shown in Figure 9.2.

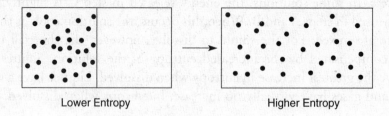

Lower Entropy   Higher Entropy

FIGURE 9.2. Increase in the entropy of a gas when the volume is increased.

When two gases are allowed to expand (Figure 9.3), each into the volume occupied by the other, the entropy of each gas increases. Since there are no attractive forces, each gas "thinks" it is expanding into a vacuum.

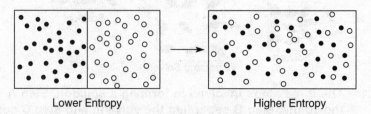

Lower Entropy   Higher Entropy

FIGURE 9.3. Mixing of two gases, illustrating the increase in entropy for each.

DETERGENTS

As a final example, the solubility effects of compounds called detergents are described. Detergents are compounds that have two types of attractive forces. As shown in Figure 9.4, one end of a long-chain carbon molecule contains an ionic group and the other end is a long-chain hydrocarbon. The ionic end causes a detergent molecule to be soluble in water. The hydrocarbon end can dissolve in nonpolar substances. An aqueous solution of a detergent can remove nonpolar stains from clothes because fats and greases dissolve in the hydrocarbon end of the detergent molecule.

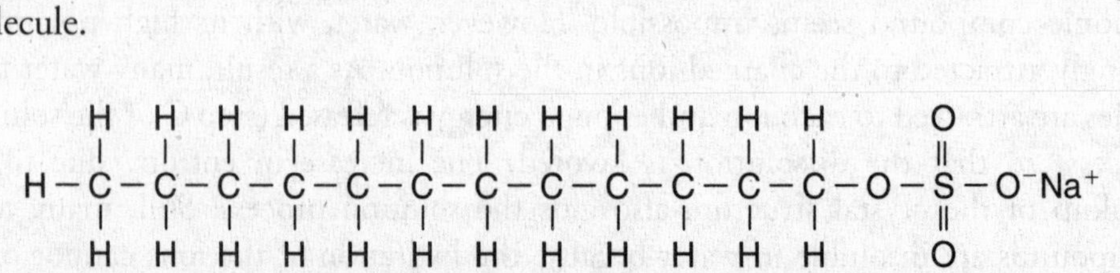

$$\begin{array}{c} \text{H} \quad \text{H} \quad \text{H} \quad \text{H} \quad \text{H} \quad \text{H} \quad \text{H} \quad \text{H} \quad \text{H} \quad \text{H} \quad \text{H} \quad \text{H} \qquad \quad \text{O} \\ | \quad | \quad | \quad | \quad | \quad | \quad | \quad | \quad | \quad | \quad | \quad | \qquad \quad || \\ \text{H}-\text{C}-\text{C}-\text{C}-\text{C}-\text{C}-\text{C}-\text{C}-\text{C}-\text{C}-\text{C}-\text{C}-\text{C}-\text{O}-\text{S}-\text{O}^-\text{Na}^+ \\ | \quad | \quad | \quad | \quad | \quad | \quad | \quad | \quad | \quad | \quad | \quad | \qquad \quad || \\ \text{H} \quad \text{H} \quad \text{H} \quad \text{H} \quad \text{H} \quad \text{H} \quad \text{H} \quad \text{H} \quad \text{H} \quad \text{H} \quad \text{H} \quad \text{H} \qquad \quad \text{O} \end{array}$$

FIGURE 9.4. Structure of the detergent sodium dodecyl sulfate. The ionic end dissolves in water, and the hydrocarbon end dissolves in nonpolar substances such as fats and grease.

Rates of Dissolution

In addition to the attractive forces that determine whether or not solutes will dissolve, we are interested in the rate at which solutes dissolve. Almost intuitively, we heat solutions to increase the speed of dissolution. We also grind chemicals into small particles and stir solutions vigorously to accelerate the solution process. All three of these operations speed the rate at which solvent molecules can reach the surface of a solid and start the dissolution process. Heating accelerates the motion of the solvent molecules so that they collide with the solid more rapidly. Grinding a solute into fine particles allows more of the solute to be exposed to the solvent. Stirring a solution moves dissolved solute away from the surface of the solute and brings less concentrated solvent in contact with the solute. While heating, grinding, and stirring increase the rate of dissolution, it must be remembered that the maximum concentration achieved depends only on the temperature of the solution and the chemical identity of the solute.

AQUEOUS SOLUTIONS

Classification of Solutes

Solutions in which water is the solvent are known as **aqueous** solutions. Ionic substances **dissociate** (break apart) completely into ions when dissolved in water. Some molecular compounds, notably HCl, also break apart into ions when dissolved in water in a process called **ionization.** Their solutions conduct electricity very well, and these compounds are called **electrolytes.** Other compounds dissociate or ionize only slightly when dissolved in water; their solutions conduct electricity, but not well. These compounds are called **weak electrolytes.** The rest of the soluble compounds form absolutely no ions when dissolved in water. Their solutions do not conduct electricity, and these compounds are called **nonelectrolytes.**

Dissociation is a term usually reserved for the separation of ions in an ionic compound when it is dissolved. *Ionization* is a term that should be reserved for molecular (nonionic) compounds that form ions upon dissolution. Often you may see that the terms dissociation and ionization are used interchangeably.

Strong Electrolytes

Strong electrolytes, or simply electrolytes, are ionic compounds, such as NaCl, KBr, and $Mg(NO_3)_2$, that are soluble in water. In addition, there are three electrolytes that are molecular (covalent) compounds in the gas phase but ionize completely when dissolved in water. These compounds are HCl, HBr, and HI. A typical reaction is

$$HCl(g) + H_2O(\ell) \rightarrow H_3O^+(aq) + Cl^-(aq)$$

Weak Electrolytes

Weak electrolytes are soluble molecular compounds that only partially ionize into ions when dissolved in water. Most weak electrolytes ionize less than 10 percent, meaning that more than 90 percent of the molecules remain intact. The most common examples of weak electrolytes are the weak acids and weak bases. Some weak acids are listed in Table 9.1.

TABLE 9.1

Some Common Weak Acids		
Name	Formula	Alternative Formula
ethanoic acid	$HC_2H_3O_2$	CH_3COOH
methanoic acid	$HCHO_2$	$HCOOH$
propionic acid	$HC_3H_5O_2$	CH_3CH_2COOH
benzoic acid	$HC_7H_5O_2$	C_6H_5COOH
hypochlorous acid	$HClO$	$HOCl$
chlorous acid	$HClO_2$	$HOClO$
chloric acid	$HClO_3$	$HOClO_2$
hydrosulfuric acid	H_2S	None
hydrofluoric acid	HF	None
phosphoric acid	H_3PO_4	None
water	H_2O	HOH

All weak bases are related to the weak base ammonia. Examples of some weak bases are methylamine, CH_3NH_2; diethylamine, $(CH_3CH_2)_2NH$; and ethylenediamine, $H_2NCH_2CH_2NH_2$. The reaction of a weak electrolyte with water results in an **equilibrium,** which is shown by a double arrow in a reaction:

$$NH_3(g) + H_2O(l) \rightleftharpoons NH_4^+(aq) + OH^-(aq)$$

Nonelectrolytes

Nonelectrolytes are molecular compounds that dissolve but show no tendency to form ions. Sugars such as glucose, $C_6H_{12}O_6$, and sucrose, $C_{12}H_{22}O_{11}$, and alcohols such as methanol, CH_3OH; ethanol, CH_3CH_2OH; and propanol, $CH_3CH_2CH_2OH$ are some examples of soluble nonelectrolytes.

CONCENTRATIONS OF SOLUTIONS

Definitions and Units

MOLARITY (*M*)

The most common unit of concentration used by chemists is **molarity.** It is defined as the number of moles of solute dissolved in 1 liter of solution and is abbreviated as the uppercase letter *M*:

$$\text{molarity } (M) = \frac{\text{number of moles of solute}}{1 \text{ L of solution}} \tag{9.1}$$

A molar solution is prepared by measuring the solute in an appropriate manner and quantitatively transferring it to a volumetric flask of the desired size. Enough solvent is then added to fill the flask about half full, and the mixture is shaken to dissolve all of the solute. Once the solute is dissolved, solvent is added exactly to the mark on the volumetric flask and mixing is repeated.

Concentration Terms

Name	Symbol	Formula
molarity	M	mol solute/L solution
molality	m	mol solute/kg solvent
mole fraction	χ	mol solute/total moles
mole percent	$\chi\%$	$\chi \times 100$
mass fraction	wt/wt	mass solute/mass solution
mass percent	%wt/wt	wt/wt $\times$ 100
parts per million	ppm	wt/wt $\times 10^6$
parts per billion	ppb	wt/wt $\times 10^9$

MOLALITY (*m*)

Molality is defined as the number of moles of solute dissolved in 1 kilogram of solvent and is abbreviated as the lowercase letter *m*:

$$\text{molality } (m) = \frac{\text{number of moles of solute}}{1 \text{ kg of solvent}} \qquad (9.2)$$

Preparation of solutions with molality units involves measuring the moles of solute and mixing the solute with the required mass of solvent.

Care must be used in distinguishing the terms *molarity* and *molality*, which have very similar spellings and abbreviations. Chemists use the uppercase M only for molarity and the lowercase m only for molality. In addition, the denominator for molarity is the total volume of the solvent and solute after mixing, while the denominator for molality is the mass of the solvent only, as shown in Equation 9.2.

MOLE FRACTION (X)

The **mole fraction,** X, of one component, A, of a mixture is defined as the number of moles of that component divided by the sum of all of the moles in the solution:

$$X_A = \frac{mol_A}{mol_A + mol_B + \cdots} = \frac{mol_A}{total\ moles} \tag{9.3}$$

The three dots (an ellipsis) in the denominator means that we must also add in the moles for any additional components of the solution. The mole fraction does not distinguish the solute from the solvent. For any solution, we can calculate the mole fraction of each compound in the mixture. The mole fraction always has a value from 0.00 to 1.00, and the sum of the mole fractions of all the compounds in a solution must add up to 1.00. At times, chemists may use a variation of this and report the **mole percent,** which is the mole fraction multiplied by 100.

A solution with mole fraction units is prepared by carefully measuring the desired number of moles of each of the components and then mixing.

MASS (WEIGHT) FRACTION (WT/WT)

The **mass fraction** is also known as the weight fraction. It may be used in situations in which it is inconvenient or impossible to use mole units. One such situation occurs when the molar mass of a compound has not been accurately determined.

The mass fraction, usually symbolized as wt/wt, is the ratio of the mass of the solute, A, to the mass of the entire solution:

$$\frac{wt}{wt} = \frac{mass_A}{mass_A + mass_B + \cdots} = \frac{mass_A}{total\ mass} \tag{9.4}$$

Like the mole fraction, the mass fraction must be a value between 0.00 and 1.00, and the sum of all mass fractions must add up to 1.00. The mass percent is obtained by multiplying the mass fraction by 100.

Mass-fraction solutions are prepared by weighing each component and then mixing the components together.

PARTS PER MILLION AND PARTS PER BILLION

A value expressed in any of the units above that has the word *fraction* in its name (mole fraction, mass fraction, mass-volume fraction, and volume fraction) may be converted to a percent (literally parts per hundred) by multiplying it by 100. In addition, the mass fraction and the volume fraction are often multiplied by 1 million (10^6) to obtain parts per million units or by 1 billion (10^9) to obtain parts per billion units. This is particularly true for solutions that contain only trace quantities of solutes.

To perform any calculations, however, the percent, parts per million, and parts per billion units must be converted back to their fractional forms first. This is done by dividing these units by 100, 10^6, and 10^9, respectively. The fractional form must always have a value between 0.00 and 1.00.

EXERCISE 9.1

Determine the concentration of each of the following solutions:

(a) A solution prepared by dissolving 25.0 g of $MgCl_2$ in enough water to make 450 mL of solution. Calculate the molarity of the solute.

(b) A solution is prepared by mixing 85.0 g of hexane, C_6H_{14}, and 45.0 g of decane, $C_{10}H_{22}$. Calculate the molality, the mole fraction of decane, and the mass fraction of hexane.

Solution

(a) The number of moles of $MgCl_2$ is calculated:

$$? \text{ mol } MgCl_2 = 25.0 \text{ g } MgCl_2 \left(\frac{1 \text{ mol } MgCl_2}{95.21 \text{ g } MgCl_2} \right) = 0.263 \text{ mol } MgCl_2$$

From the definitions

$$\text{molarity } (M) = \frac{0.263 \text{ mol } MgCl_2}{0.450 \text{ L solution}} = 0.584 \text{ } M \text{ } MgCl_2$$

(b) The number of moles of each compound is calculated:

$$? \text{ mol } C_6H_{14} = 85.0 \text{ g } C_6H_{14} \left(\frac{1 \text{ mol } C_6H_{14}}{86.0 \text{ g } C_6H_{14}} \right) = 0.988 \text{ mol } C_6H_{14}$$

$$? \text{ mol } C_{10}H_{22} = 45 \text{ g } C_{10}H_{22} \left(\frac{1 \text{ mol } C_{10}H_{22}}{142 \text{ g } C_{10}H_{22}} \right) = 0.317 \text{ mol } C_{10}H_{22}$$

The molality is calculated by assuming that the hexane is the solvent since it is the major component in the mixture

$$\text{molality } (m) = \frac{0.317 \text{ mol } C_{10}H_{22}}{0.085 \text{ kg } C_6H_{14}} = 3.73 \text{ molal } C_{10}H_{22}$$

$$X_{\text{decane}} = \frac{0.317 \text{ mol } C_{10}H_{22}}{0.317 \text{ mol } C_{10}H_{22} + 0.988 \text{ mol } C_6H_{14}} = 0.243$$

$$\text{mass fraction}_{\text{hexane}} = \frac{85.0 \text{ g } C_6H_{14}}{45.0 \text{ g } C_6H_{14} + 85.0 \text{ g } C_{10}H_{22}} = 0.654$$

CONCENTRATION-CONVERSION CALCULATIONS

Concentration units may be classified as temperature dependent or temperature independent. The temperature-dependent units are molarity, volume fraction, and mass-volume fraction. Since the volume of liquid changes slightly with temperature, temperature-dependent concentrations are identified as those that have volume units in their definitions. Molality, mole-fraction, and mass-fraction units are not affected by temperature changes. These units are defined on the basis of mass or moles, which do not change with temperature.

The two types of conversions between units may be summarized as follows.

1. Conversions from one temperature-independent unit to another temperature-independent unit.
2. Conversions between temperature-dependent units and temperature-independent units.

Each of these has its own techniques and logic, which are described below.

Conversions Between Temperature-Independent Concentration Units

When converting one temperature-independent concentration into another, the given concentration is used to construct a table showing the masses and moles of each component of the solution. This table of information is then used to calculate the desired concentration from its definition.

EXERCISE 9.2

Given an aqueous solution that is 0.500 molal in NaCl, what are the corresponding mass and mole fractions of NaCl?

Solution

To solve all problems involving conversions of concentration units from one to another, it is worthwhile to set up a table similar to the following one. This helps us keep track of all the information. It will have one column for each solute, another for the solvent, and a last column for the volume of the solution. The rows are labeled mass and moles. The last row is for molar masses of the solute and solvent and for the density of the solution.

> **TIP**
>
> The concentration table helps to keep track of the components in a mixture.

	Solute	Solvent	Solution	Volume of Solution
Mass, g				
Moles				
Constants	*MM =	*MM =	Density =	

*MM = molar mass

For our problem, we set up and label the table as shown. Then we enter the data that the molality immediately tells us. Since we can assume 1 kg of solvent, we must therefore have 0.500 mol of solute. We enter the numbers in the correct boxes.

	Solute, NaCl	Solvent, H₂O	Solution	Volume of Solution
Mass, g		**1 kg**		
Moles	**0.500**			
Constants	*MM = 58.44	*MM = 18.02	Density =	

*MM = molar mass

To complete the table, we calculate the mass of NaCl and the moles of water. The table now has all of the data needed for our calculation.

	Solute, NaCl	Solvent, H₂O	Solution	Volume of Solution
Mass, g	**28.22**	1 kg		
Moles	0.500	**55.5**		
Constants	*MM = 58.44	*MM = 18.02		

*MM = molar mass

From the information now available in the table, the mass-fraction and mole-fraction definitions can be used to calculate these values:

$$\text{mass fraction} = \frac{\text{g NaCl}}{\text{g NaCl} + \text{g H}_2\text{O}} = \frac{28.22 \text{ g}}{28.22 \text{ g} + 1000 \text{ g}} = 0.0274$$

$$\chi = \frac{\text{mol NaCl}}{\text{mol NaCl} + \text{mol H}_2\text{O}} = \frac{0.500 \text{ mol}}{0.500 \text{ mol} + 55.5 \text{ mol}}$$

$$= 0.00893$$

EXERCISE 9.3

In another example, an aqueous solution of glucose, $C_6H_{12}O_6$ (molar mass = 180), has a mole fraction of 0.100 for glucose. What are the molality and the mass fraction of this solution?

Solution

A table is constructed for the solution components:

	Solute, Glucose	Solvent, Water	Solution	Volume of Solution
Mass, g				
Moles				
Constants	*MM = 180	*MM = 18.02		

*MM = molar mass

The definition of the mole fraction

$$\chi = \frac{\text{mol glucose}}{\text{mol glucose} + \text{mol H}_2\text{O}} = 0.100$$

shows that, if the denominator is assumed to equal 1.00, then the numerator, which is the number of moles of glucose, must equal 0.100. At the same time, since the denominator is assumed to equal 1.00, 1.00 = mol glucose + mol H_2O. Therefore, there must be 0.900 mol of water. Entering these values in the table gives

	Solute, Glucose	Solvent, Water	Solution	Volume of Solution
Mass, g				
Moles	0.100	0.900	1.00	
Constants	*MM = 180	*MM = 18.02		

*MM = molar mass

From the molar masses of water and glucose, the grams of each can be calculated and entered in the table.

	Solute, Glucose	Solvent, Water	Solution	Volume of Solution
Mass, g	18.0	16.2		
Moles	0.100	0.900	1.00	
Constants	*MM = 180	*MM = 18.02		

*MM = molar mass

The mass fraction and molality are calculated from the tabulated data as

$$\text{mass fraction} = \frac{\text{g glucose}}{\text{g glucose} + \text{g H}_2\text{O}} = \frac{18.0 \text{ g}}{18.0 \text{ g} + 16.2 \text{ g}}$$

$$= 0.526$$

$$\text{molality } (m) = \frac{\text{mol glucose}}{\text{kg water}} = \frac{0.100 \text{ mol}}{0.0160 \text{ kg}} = 6.25 \text{ molal}$$

Conversions Between Temperature-Dependent and Temperature-Independent Concentration Units

Temperature-dependent units have volume units in their definitions, and temperature-independent concentrations have masses or moles. At some time during the conversion of units, there will be a conversion between mass and volume. *Density is the factor that must be used for such a conversion.* Aside from the need for density to convert between mass and volume, the calculations for this type of problem are a combination of those for the two exercises discussed above.

EXERCISE 9.4

The molarities of concentrated reagents such as ammonia, hydrochloric acid, sulfuric acid, and nitric acid are not specified when these substances are purchased from a supplier. Instead, these reagents have labels that list the weight percent (% wt/wt) and density. For instance, commercial hydrochloric acid is 36% (wt/wt) hydrogen chloride and has a density of 1.18 g mL^{-1}. Determine the molarity of commercial HCl.

Solution

The 36% (wt/wt) indicates that every 100 g of commercial acid contains 36 g of HCl. Setting up a stoichiometry-type conversion for this problem yields

$$? \, M \text{ HCl} = \frac{36 \text{ g HCl}}{100 \text{ g solution}}$$

Converting the symbol M to its actual units gives

$$? \, \frac{\text{mol HCl}}{\text{L sol'n}} = \frac{36 \text{ g HCl}}{100 \text{ g sol'n}}$$

The molar mass of HCl is used to convert g HCl to mol HCl, and the density is the conversion factor needed to convert grams of solution into liters:

$$? \, \frac{\text{mol HCl}}{\text{L sol'n}} = \left(\frac{36 \text{ g HCl}}{100 \text{ g sol'n}} \right) \left(\frac{1 \text{ mol HCl}}{36.46 \text{ g HCl}} \right) \left(\frac{1.18 \text{ g sol'n}}{1 \text{ mL sol'n}} \right) \left(\frac{1000 \text{ mL sol'n}}{1 \text{ L sol'n}} \right)$$

$$= 12 \, M \text{ HCl}$$

EXERCISE 9.5

The molarity of a solution of potassium fluoride, KF, is 0.748.

(a) What is the molality of this solution if the density is 1.022 g mL^{-1}?
(b) What is the mole fraction of KF?

Solution

Since we are going to calculate many different concentration units from the data, it is worthwhile to construct a table to keep track of all of the important data. To start, we enter the data given in the problem and constants such as molar masses. In this case, we assume exactly 1 liter of solution. That will contain 0.748 moles of KF as entered in the table.

	Solute, KF	Solvent, H_2O	Solution	Volume of Solution
Mass, g				1.00 L
Moles	0.748			
Constants	*MM = 58.09	*MM = 18.02	d = 1.022	

*MM = molar mass

Next we complete the information in the table. We can calculate the mass of KF from the given moles and the molar mass. We calculate the mass of the solution by multiplying 1000 mL by the density 1.022 g/mL. We can then subtract to obtain the mass of water, 977.9 g, and then convert that to the moles of water shown. We can also add the moles of KF and the moles of water to obtain the total moles in the mixture. The table below shows the results.

	Solute, KF	Solvent, H_2O	Solution	Volume of Solution
Mass, g	44.1	977.9	1022	1.00 L
Moles	0.748	54.3	55.0	
Constants	*MM = 58.09	*MM = 18.02	†d = 1.022	

*MM = molar mass
†d = density

(a) To calculate the molality, we need moles of solute and kg of solvent.

$$m = (0.748 \text{ mol KF})/(0.9779 \text{ kg } H_2O) = 0.765$$

(b) To calculate the mole fraction of KF, we need the moles of KF and the total moles in the mixture.

$$X = (0.748 \text{ mol KF})/(55.0 \text{ total mol}) = 0.0136$$

In Exercise 9.5 an aqueous solution that had a molarity of 0.748 was calculated to have a molality of 0.765. This is less than a 3.0 percent difference. We may generalize that for dilute aqueous solutions (less than 0.5 *M* or 0.5 *m*) the molality and molarity are virtually identical.

> IN DILUTE AQUEOUS SOLUTIONS
> MOLARITY ≈ MOLALITY

The same concepts that yielded the molarity-molality simplification are used to conclude that in dilute solutions the mass fraction is very close to the mass-volume fraction. This is particularly true for solutions in which the solute is present in trace quantities and the concentrations are expressed as parts per million or parts per billion.

EFFECT OF TEMPERATURE ON SOLUBILITY

This topic is rather complex, and the details are usually left for more advanced courses. For us, the important principle is that a system will follow Le Châtlier's principle. In particular, endothermic processes will favor the products when heat is added. In addition, processes that have an increase in entropy with increasing temperature will favor the products.

Entropy of States of Matter

SOLIDS << LIQUIDS << GASES

As a general rule, increasing temperatures drive molecules toward the state with the larger entropy. Thus an increase in temperature usually increases the solubility of solids in a liquid and often decreases the solubility of gases in a liquid. Decreasing temperatures have the opposite effect.

Most solids are more soluble in hot solvents than in cold solvents. Figure 9.5 illustrates the changes in solubility of a few solid compounds as temperature is increased. A simple explanation is that increasing the temperature increases the disorder of all molecules. If two states are present, molecules will tend to move from the lower entropy state to a higher entropy state. A more precise explanation is left for more advanced chemistry courses.

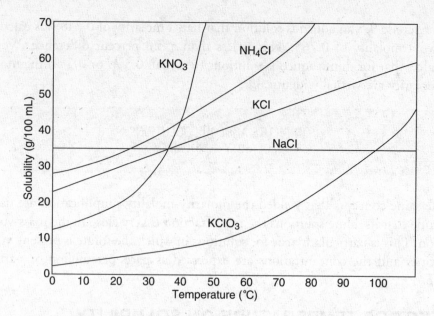

FIGURE 9.5. **Temperature dependence of solubility for several representative ionic compounds.**

In agreement with the general principle above, the solubility of gases in liquids, particularly water, usually decreases as the temperature is increased. You may have noticed that a warm bottle of soda fizzes more than a cold bottle of soda. In this case, the increased entropy caused by an increase in temperature favors molecules leaving the liquid phase and entering the gas phase. This occurs because molecules have more possible states in the gas phase than in the liquid phase.

EFFECT OF PRESSURE ON SOLUBILITY

External pressure has no significant effect on the solubility of solids or liquids because solids and liquids are not appreciably compressed when pressure is increased. Gases, on the other hand, are easily compressed. Compression increases the frequency with which gas molecules hit the liquid phase and enter it, thereby increasing the solubility.

The effect of pressure on the solubility of gases is expressed in Henry's law:

HENRY'S LAW:

Solubility of gases

$$\text{solubility}_{\text{gas}} = kP_{\text{gas}} \qquad (9.5)$$

Solubility$_{\text{gas}}$ is usually expressed as molarity or as wt/vol concentration units, k is the Henry's law proportionality constant, and P_{gas} is the partial pressure of the gas.

If the solubility of a gas is known at one pressure, Henry's law can be used to determine the solubility at another pressure as long as the temperature is the same in both cases.

EXERCISE 9.6

The solubility of oxygen in water is 1.25×10^{-3} M at 25 °C at sea level (1.00 atm). What will the solubility of oxygen be in Denver, where the atmospheric pressure is 0.800 atm?

Solution

The solution requires two steps. First, the Henry's law constant is determined by inserting the sea level information into the equation and solving for k. Next, the value of k along with the atmospheric pressure in Denver is used to calculate the solubility.

The value of k is calculated as

$$k(1.00 \text{ atm}) = 1.25 \times 10^{23} \ M \ O_2$$

$$k = 1.25 \times 10^{23} \ M \ O_2 \ \text{atm}^{-1}$$

Using this value of k, we can calculate the solubility of oxygen at Denver's atmospheric pressure:

$$\text{solubility}_{O_2} = (1.25 \times 10^{-3} \ M \ O_2 \ \text{atm}^{-1})(0.800 \text{ atm})$$

$$= 1.00 \times 10^{-3} \ M \ O_2$$

Alternative Solution

Another method for solving this type of problem is to write the Henry's law equation for two states, indicated by the subscripts 1 and 2:

$$\text{solubility}_1 = kP_1 \quad \text{and} \quad \text{solubility}_2 = kP_2$$

The k does not need a subscript since it is constant as long as the temperature does not change. Next, the ratio of these two equations is taken:

$$\frac{\text{solubility}_1}{\text{solubility}_2} = \frac{kP_1}{kP_2}$$

The constants k cancel, leaving

$$\frac{\text{solubility}_1}{\text{solubility}_2} = \frac{P_1}{P_2}$$

For ease of calculation we define solubility_1 as the solubility in Denver and $P_1 = 0.800$ atm. Then $\text{solubility}_2 = 1.25 \times 10^{-3} \ M$ and $P_2 = 1.00$ atm. Entering this gives

$$\frac{\text{solubility}_1}{1.25 \times 10^{-3} \ M} = \frac{0.800 \text{ atm}}{1.00 \text{ atm}}$$

$$\text{solubility}_1 = \frac{(0.800 \text{ atm})(1.25 \times 10^{-3} \ M)}{1.00 \text{ atm}} = 1.00 \times 10^{-3} \ M \ O_2$$

In this method the algebra was done with the symbolic variables before any data were entered or any calculations performed.

COLLIGATIVE PROPERTIES OF SOLUTIONS

Methods for Determining Molar Mass

Description	Equation (molar mass =)
Density of a gas (m/V) uses ideal gas law	$= (g_{gas}RT)/(PV)$
Osmotic pressure	$= (g_{solute}RT)/(PV)$
Freezing-point depression	$= (\Delta T \times kg_{solvent} \times g_{solute})/K_f$
Boiling-point elevation	$= (\Delta T \times kg_{solvent} \times g_{solute})/K_b$
Graham's law of effusion	$= *m_2 v_2^2/v_1^2$

*m_2 is the molar mass of the comparison gas, v = effusion rate

The physical properties of pure liquids are altered when solutes are dissolved in these liquids. The word *colligative*, meaning collective, is used to describe physical properties that are changed by the presence of any solute particles. The chemical identity of the solute is not important. Only the concentration of the solute particles is significant in altering physical properties. The physical properties that change when a solute is present include vapor pressure, boiling point, melting point, and osmotic pressure.

Vapor-Pressure Lowering by Nonelectrolytes

If a nonvolatile nonelectrolyte (often called a nondissociating solute) is dissolved in a liquid, it causes the vapor pressure to decrease in proportion to its concentration. This effect is explained on the basis that some of the nonvolatile solute molecules take up space at the surface of the liquid and decrease the effective surface area of the solute available for evaporation. The result is to decrease the evaporation rate and the vapor pressure.

In the 1880s F.M. Raoult was the first to show that the vapor pressure is proportional to the mole fraction of the solvent in a solution. This is entirely reasonable since the mole fraction of the solvent represents the proportion of solvent molecules at the surface of the liquid. **Raoult's law** states that the actual vapor pressure, P_{actual}, is equal to the vapor pressure of the pure solvent, P^0, multiplied by the mole fraction of the solvent:

$$P_{actual} = P^0 \chi_{solvent} \qquad (9.6)$$

EXERCISE 9.7

Many people use 2 teaspoons (18 g each) of sugar in each cup (250 mL) of coffee. The vapor pressure of water at 80 °C is 355 mm Hg. What is the vapor pressure of sugared coffee at the same temperature? Assume that the coffee itself has a very low concentration.

Solution

Sucrose, $C_{12}H_{22}O_{11}$, has a molar mass of 342 g mol^{-1}, and 2 teaspoons represents 0.105 mol. The cup of coffee has a volume of 250 mL; and since the density of water is 1.00 g mL^{-1}, the mass of the solvent is 250 g, or 13.89 mol. The mole fraction of solvent is

$$\chi = \frac{\text{mol } H_2O}{\text{mol sucrose} + \text{mol } H_2O}$$

$$= \frac{13.89 \text{ mol } H_2O}{0.108 \text{ mol sucrose} + 13.89 \text{ mol } H_2O}$$

$$= 0.992$$

The vapor pressure of the sugared coffee is then

$$P = (355 \text{ mm Hg})(0.992)$$

$$= 352 \text{ mm Hg}$$

We see from this result that the small mole fraction of sucrose, $X = 0.008$, has only a small effect on the vapor pressure. Larger amounts will of course have a larger effect.

Vapor Pressures of Volatile Liquid Mixtures

When one liquid is dissolved in another, the vapor pressure of the entire mixture depends on the vapor pressures of the individual liquids and the proportions in which these liquids are mixed. We may use the same logic as for the nonvolatile solute to understand what is happening. In this case the solute liquid occupies space at the surface of the solution, inhibiting the vaporization of the solvent and decreasing its vapor pressure. Now, however, the solute itself is volatile, and the space it occupies at the surface allows it to vaporize. By using Raoult's law for both volatile components of the solution, we obtain

$$P_{\text{total}} = \chi_1 P_1^0 + \chi_2 P_2^0 + \cdots \tag{9.7}$$

The subscripts 1 and 2 represent the two different volatile liquids in the mixture, and the ellipsis indicates that for mixtures of more than two volatile liquids similar terms are added. At this point we notice that, if one solute is not volatile ($P_2^0 = 0$), Equation 9.7 becomes the same as Equation 9.6.

EXERCISE 9.8

Calculate the vapor pressure of a mixture of equal weights of water and ethyl alcohol, C_2H_5OH, at 35 °C, when the vapor pressures of the two substances are 42.0 and 100 mm Hg, respectively.

Solution

If there are equal masses of the two liquids, let us assume that 100 g of one liquid is mixed with 100 g of the other. This gives us 5.56 mol of water and 2.17 mol of ethyl alcohol.

The mole fraction of water is

$$\chi_1 = \frac{5.56 \text{ mol H}_2\text{O}}{5.56 \text{ mol H}_2\text{O} + 2.17 \text{ mol C}_2\text{H}_5\text{OH}} = 0.719$$

The mole fraction of ethyl alcohol is

$$\chi_2 = \frac{2.17 \text{ mol C}_2\text{H}_5\text{OH}}{5.56 \text{ mol HO} + 2.17 \text{ mol C}_2\text{H}_5\text{OH}} = 0.281$$

Entering this information, along with the corresponding vapor pressures of the pure liquids, into the Raoult's law equation yields

$$P_{\text{total}} = (42 \text{ mm Hg})(0.719) + (100 \text{ mm Hg})(0.281)$$
$$= 58.3 \text{ mm Hg}$$

From Raoult's law for mixtures of liquids, we can make the generalization that the total vapor pressure of a liquid-liquid solution will always lie somewhere between the lowest and highest vapor pressures of the two liquids.

Figure 9.6 is a pictorial representation of Raoult's law for a binary mixture of water and ethyl alcohol.

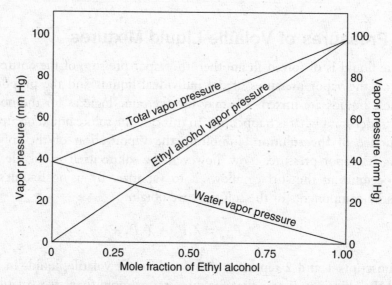

FIGURE 9.6. Graphical representation of Raoult's law for a binary liquid mixture of water and ethyl alcohol at 35 °C, assuming that the mixture behaves as an ideal solution.

Ideal Solutions and Raoult's Law

The preceding discussion of Raoult's law assumes that the solutions are ideal solutions. An ideal solution is one in which the energy used to disrupt the attractive forces in the solvent and solute is exactly balanced by the energy released when the solution forms. In other words, the attractive forces in the pure solute and solvent are the same as the attractive forces in the solution. If the solution is not ideal, deviations from Raoult's law are found. These deviations may be summarized as follows:

1. ***Positive deviation:*** The vapor pressure of the solution is greater than expected from Raoult's law because the solution has weaker attractive forces than those in the pure solute and solvent. A solution with positive deviation cools as the mixture is prepared since the extra energy needed to break the attractions is absorbed from the surroundings.

2. ***Negative deviation:*** The vapor pressure is lower than expected from Raoult's law because the attractive forces in the solution are greater than those in the pure solute and solvent. These solutions produce extra energy that is given off as heat. A mixture of this type becomes warmer when the solution is prepared.

Figure 9.7 illustrates the effects of positive and negative deviations from Raoult's law.

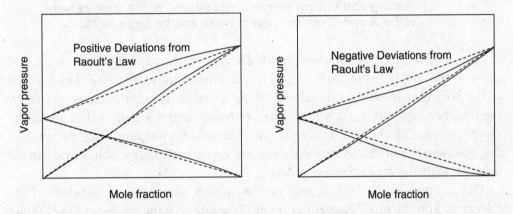

FIGURE 9.7.   Positive and negative deviations from Raoult's law. Dashed lines indicate Raoult's law for ideal solutions.

Boiling-Point Elevation and Freezing-Point Depression of Nonvolatile Nonelectrolyte Solutions

A solute decreases the vapor pressure of a solvent according to Raoult's law. The phase diagram (see Chapter 8 for a complete description of phase diagrams) of a solution is similar to the phase diagram for the pure solvent except that the decrease in vapor pressure lowers the liquid-gas equilibrium line. Figure 9.8 illustrates the change in the position of the liquid-gas equilibrium line, which establishes a new triple point and also shifts the solid-liquid equilibrium line for the solution as compared to the pure solvent.

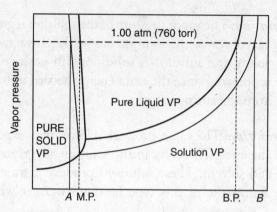

FIGURE 9.8. Change in the phase diagram of a solvent when a nonvolatile solute is dissolved in it. Heavy lines represent the phase diagram for the pure solvent. Lighter lines represent changes to the solid-liquid and liquid-gas equilibrium lines. M.P. and B.P. indicate the melting and boiling points, respectively, of the pure solvent, while *A* and *B* are the new melting and boiling points.

From Figure 9.8 it is apparent that a decrease in the vapor pressure must also cause the boiling point to increase and the melting point to decrease. The increase in the boiling point is directly related to the decrease in vapor pressure caused by the presence of a solute. As a result, higher temperatures are needed to make the vapor pressure of the solution equal to atmospheric pressure. The decrease in the melting point arises because the attractive forces between the solute and solvent interfere with the crystallization process.

The changes in the boiling and melting points of a solution depend on the solvent and on the molal concentration of the solute. For the increase in the boiling point

$$\Delta T = K_b m \qquad (9.8)$$

where m is the molality of the solution and K_b is the boiling-point-elevation constant for the solvent. For the decrease in the melting point, the equation is similar to Equation 9.8 except that it has a negative sign to indicate a decrease in temperature and a different constant, K_f, called the freezing-point-depression constant:

$$\Delta T = -K_f m \qquad (9.9)$$

In either boiling or freezing, the calculated ΔT must be added to the original boiling point or freezing point of the pure solvent to determine the actual boiling and freezing points. Table 9.2 lists some common solvents with their boiling-point-elevation and freezing-point-depression constants.

TABLE 9.2

Freezing-Point-Depression and Boiling-Point-Elevation Constants for Some Common Solvents

Solvent	Freezing Point (°C)	K_f (°C m^{-1})	Boiling Point (°C)	K_b (°C m^{-1})
ethanoic acid	16.66	3.90	117.90	2.53
benzene	5.50	5.10	80.10	2.53
cyclohexane	6.50	20.2	80.72	2.75
camphor	178.40	40.0	207.42	5.61
p-dichlorobenzene	53.10	7.1	174.1	6.2
naphthalene	80.29	6.94	217.96	6.2
water	0.00	1.86	100.0	0.52

EXERCISE 9.9

What are the freezing and boiling points of naphthalene when 10.0 g of nicotine, $C_{10}H_{14}N_2$, is dissolved in 50.0 g of naphthalene?

Solution

The constants K_f and K_b are given in Table 9.2. The molality of nicotine in this solution must be calculated to obtain ΔT:

$$? \ \frac{\text{mol } C_{10}H_{14}N_2}{\text{kg naphthalene}} = \frac{10.0 \text{ g } C_{10}H_{14}N_2/162 \text{ g mol}^{-1}}{0.050 \text{ kg naphthalene}} = 1.23 \ m \ C_{10}H_{14}N_2$$

Using the proper K_f and K_b, along with this molality, gives the temperature changes:

$$\Delta T_f = -(6.94 \text{ °C } m^{-1})(1.23 \ m) \quad \text{and} \quad \Delta T_b = (6.2 \text{ °C } m^{-1})(1.23 \ m)$$

$$= -8.54 \text{ °C} \qquad \text{and} \qquad = 7.63 \text{ °C}$$

These changes are then added to the freezing and boiling temperatures of the pure solvent to obtain the freezing point and boiling point of the solution:

$$\text{freezing point} = 80.29 \text{ °C} - 8.54 \text{ °C} = 71.75 \text{ °C}$$
$$\text{boiling point} = 217.96 \text{ °C} + 7.63 \text{ °C} = 225.59 \text{ °C}$$

Uses of Boiling-Point Elevation and Freezing-Point Depression

Chemists take advantage of the freezing-point depression to make cooling baths below 0 °C. This is done by dissolving large quantities of salt in water and then adding ice. Daniel Fahrenheit used this method to make the coldest solution he could and defined it as zero on his temperature scale. Everyone takes advantage of antifreeze in his or her car engine. Antifreeze lowers the freezing point of water by adding a high concentration of ethylene glycol. Antifreeze solutions simultaneously increase the boiling point of water to allow engines to run above 100 °C without boiling over.

The most important use of the freezing-point-depression and boiling-point-elevation phenomenon is in the determination of molar masses. The molality used to calculate ΔT_f and ΔT_b is defined as the number of moles of solute per kilogram of solvent. The number of moles of solute is equal to the mass of the solute divided by its molar mass (g solute/molar mass solute). The equations for ΔT_f and ΔT_b may be expanded to read as

$$\Delta T_f = -K_f \left(\frac{\text{g solute/molar mass solute}}{\text{kg solvent}} \right) \tag{9.10}$$

$$\Delta T_b = K_b \left(\frac{\text{g solute/molar mass solute}}{\text{kg solvent}} \right) \tag{9.11}$$

To obtain the molar mass of a substance, we need to know the number of grams of solute dissolved in each kilogram of solvent, the change in either the boiling point or the freezing point, and the appropriate constant, K_f or K_b, for the solvent.

EXERCISE 9.10

When 10.0 g of elemental sulfur is dissolved in 100.0 g of cyclohexane, the freezing point of this solution is found to be −1.40 °C. What is the molar mass of sulfur?

Solution

Expanded Equation 9.10 will be used:

$$\Delta T_f = -K_f \left(\frac{\text{g solute/molar mass solute}}{\text{kg solvent}} \right)$$

Since the freezing point of the solution is −1.40 °C and the normal freezing point of cyclohexane is 6.50 °C, $\Delta T_f = -7.90$ °C. From Table 9.2, the freezing-point-depression constant is 20.2 °C m^{-1}. Entering the data in the equation yields

$$-7.90\,°C = -20.2\,°C\ m^{-1}\left(\frac{10.0\ \text{g S/molar mass S}}{0.100\ \text{kg cyclohexane}}\right)$$

$$= \frac{-2020\,°C\ \text{g S/mol S}}{\text{molar mass S}}$$

$$\text{molar mass of sulfur} = \frac{-2020\,°C\ \text{g S/mol S}}{-7.90\,°C}$$

$$= 256\ \text{g S/mol S}$$

This result tells us that one molecule of sulfur has a mass of 256 g. Since the atomic mass of sulfur is 32 in the periodic table, there must be eight sulfur atoms in one molecule of sulfur ($256/32 = 8$). This means that this allotrope of elemental sulfur exists as S_8 molecules.

Osmotic Pressure

Freezing-point-depression and boiling-point-elevation measurements have constants in the range of a few degrees Celsius per molal. Therefore, to attain any reasonable measurements, solutions must be fairly concentrated, since small temperature changes are often difficult to measure.

Modern chemistry is often interested in extremely large molecules such as polymers, proteins, peptides, and DNA. These molecules, which have molar masses exceeding $100,000\ \text{g mol}^{-1}$, are impossible to dissolve in sufficient quantities to obtain temperature changes that can be accurately measured. However, dilute solutions produce an osmotic pressure that is easily measured even in very dilute solutions.

The principle of **osmosis** explains the flow of solvent through semipermeable membranes. It is the basic mechanism that transports water from the roots to the upper branches of trees that may be over 100 feet tall.

In explaining osmosis, we start by observing that two solutions of different concentrations will diffuse into each other until the concentrations are uniform. There is a natural tendency for solutes and solvents to diffuse from regions of high concentration to regions of lower concentration until the concentration is uniform throughout the system.

A semipermeable membrane placed between two solutions with different concentrations will allow only the solvent to diffuse. Solute molecules, which are often larger than solvent molecules, cannot pass through the semipermeable membrane. Diffusion of the solvent through the semipermeable membrane will stop when the concentrations have been equalized or when an opposing force is applied to stop the diffusion. This opposing force is known as **osmotic pressure.**

The osmotic pressure experiment is diagrammed in Figure 9.9. At the start of the experiment, two compartments are separated by a semipermeable membrane; one compartment holds a solution containing the solute, and the other holds a pure solvent. The heights of the two liquids in the measuring tubes are equal at the start, indicating that the same atmospheric pressure is acting on both compartments.

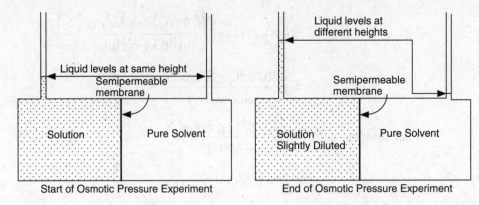

FIGURE 9.9. Start of an osmotic pressure experiment with equal pressures acting on both compartments. At the end of the experiment, the pressure developed is the osmotic pressure.

Since the solution is more concentrated than the solvent, solvent immediately starts diffusing from the solvent side to the solution side in an attempt to equalize the concentrations. This dilutes the solution by increasing its volume, and the liquid level in the measuring tube for the solution increases. Conversely, the loss of solvent from the solvent compartment makes the liquid level in its measuring tube decrease. The difference in liquid levels between the two compartments represents the opposing force, called the osmotic pressure, that will stop the diffusion at some point. When the liquid levels stop changing, the osmotic pressure is the difference between the heights of the two liquid levels.

The equation that governs this process is the ideal gas law equation from Chapter 7:

$$PV = nRT \tag{9.12}$$

The symbol Π replaces P for an osmotic pressure experiment:

$$\Pi V = nRT \tag{9.13}$$

EXERCISE 9.11

Calculate the osmotic pressure that would result from dissolving 2 teaspoons of sucrose in a cup of coffee at 80 °C. ($R = 0.0821$ L atm mol^{-1} K^{-1}.)

Solution

In Exercise 9.7 we determined that one cup of coffee is 0.250 L and 2 teaspoons of sucrose is equivalent to 0.105 mol of sucrose. Since the temperature is given in degrees Celsius, it must be converted to kelvins: 353 K. Substituting these data into the osmotic pressure equation yields

$$\Pi(0.250 \text{ L}) = (0.105 \text{ mol})(0.0821 \text{ L atm mol}^{-1} \text{ K}^{-1})(353 \text{ K})$$

Solving, we obtain an osmotic pressure of 12.17 atm or 9251 mm Hg (9.2 m of mercury or 125 m of water!).

This exercise shows that the osmotic pressure is very large for a 0.1 molar solution. A 10 micromolar solution will result in more than a centimeter difference in pressure for aqueous solutions. This fact makes osmotic pressure measurements ideal for very dilute solutions. Of particular interest is the use of osmotic pressure to determine molar mass of large molecules, as shown in the next exercise.

EXERCISE 9.12

A large polymer is synthesized, and 100 mL of a solution containing 1.00 g of this polymer is tested in an osmotic pressure experiment. The resulting pressure is measured as 55 mm of water at 25 °C. From this information calculate the molar mass of the polymer.

Solution

First, the pressure must be converted to millimeters of mercury and then to atmospheres since we intend to use a value of R that is 0.0821 L atm mol^{-1} K^{-1}. This is done by using the ratio of the densities of water and mercury: (NOTE: Converting between mm H$_2$O and mm Hg pressure units is not used on the AP Exam. It is presented here to illustrate the actual osmotic pressure experiment.)

$$? \text{ atm} = 55 \text{ mm H}_2\text{O}\left(\frac{1.00 \text{ g H}_2\text{O mL}^{-1}}{13.5 \text{ g Hg mL}^{-1}}\right) = 4.07 \text{ mm Hg}$$

$$= 4.07 \text{ mm Hg}\left(\frac{1 \text{ atm}}{760 \text{ mm Hg}}\right) = 5.36 \times 10^{-3} \text{ atm}$$

Entering the data into the osmotic pressure equation yields the number of moles of polymer:

$$(5.36 \times 10^{-3} \text{ atm})(0.100 \text{ L}) = n(0.0821 \text{ L atm mol}^{-1} \text{ K}^{-1})(298 \text{ K})$$

$$n = 2.19 \times 10^{-5} \text{ mol}$$

Since the number of moles of any substance is the mass divided by the molar mass, we can write

$$n = \frac{\text{g compound}}{\text{molar mass}}$$

or

$$\text{molar mass compound} = \frac{\text{g compound}}{n}$$

Entering the data, we obtain

$$\text{molar mass compound} = \frac{1.00 \text{ g compound}}{2.19 \times 10^{-5} \text{ mol compound}}$$

$$= 4.57 \times 10^4 \text{ g mol}^{-1} \text{ or } 45.7 \text{ kg mol}^{-1}$$

Since this solution is very dilute, the molarity and molality can be considered to be the same. In aqueous solution, the freezing-point depression would have been 4×10^{-4} °C. This is a temperature change that is difficult to measure precisely, and it indicates the utility of the osmotic pressure measurement.

COLLIGATIVE PROPERTIES OF ELECTROLYTES

As defined earlier, strong electrolytes are compounds that dissociate completely into two or more ions when dissolved in water. Each ion acts as a discrete particle as far as the colligative properties are concerned. The result is that NaCl solutions should have twice the effect on vapor pressure, boiling and freezing points, and osmotic pressure as the same concentration of a nondissociating solute such as sucrose. Compounds such as $MgCl_2$ and $(NH_4)_2SO_4$ should have three times the effect since they dissociate into three ions each. $AlCl_3$ and Na_3PO_4 should each dissociate into four ions.

EXERCISE 9.13

(a) Calculate the expected osmotic pressure of a 3.5×10^{-3} M aqueous solution of the electrolyte $Al(NO_3)_3$ at 30 °C,
(b) What are the expected normal boiling and melting points of this solution?

Solution

(a) The dissociation equation for $Al(NO_3)_3$ is

$$Al(NO_3)_3 \rightleftharpoons Al^{3+} + 3NO_3^-$$

Four ions are produced for each $Al(NO_3)_3$ dissolved. The concentration of all ions is $4 \times 3.5 \times 10^{-3}$ $M = 1.4 \times 10^{-2}$ M ions. Converting 30 °C to 303 K and substituting into the osmotic pressure equation gives

$$\Pi V = nRT$$

Rearranging and noting that $\dfrac{n}{V} = M$ yields

$$\Pi = \left(\frac{n}{V}\right)RT = MRT$$

$$= (14 \times 10^{-3})(0.082 \text{ L atm mol}^{-1} \text{ K}^{-1})(303 \text{ K})$$

$$= 0.348 \text{ atm} = 264 \text{ mm Hg}$$

(b) To calculate the freezing-point depression and the boiling-point elevation, we first assume that $M = m$ since the concentration is so small. From this assumption we obtain

$$\Delta T_f = -(1.86 \text{ °C } m^{-1})(1.4 \times 10^{-2}) = -0.026 \text{ °C}$$

$$\Delta T_b = (0.52 \text{ °C } m^{-1})(1.4 \times 10^{-2}) = 0.0073 \text{ °C}$$

This gives us a melting point of $0.00 - 0.026 = -0.026$ °C and a boiling point of $100 + 0.0073 = 100.0073$ °C. The osmotic pressure is obviously easier to measure.

Real solutions of both weak and strong electrolytes do not often give the results expected when measuring colligative properties, especially when the concentration exceeds 0.001 molal or 0.001 molar. Strong electrolytes show a behavior called ion pairing that reduces the effective concentration of particles in a solution. Weak electrolytes ionize to a small extent and increase the number of particles in a solution. Both these effects are summarized by the van't Hoff factor, i, defined as

$$i = \frac{\text{measured value of electrolyte solution}}{\text{expected value from same conc. of nonelectrolyte}}$$

For example, a 0.100 molal solution of NaCl has a freezing-point depression of –0.348 °C, whereas the expected decrease in the freezing point is –0.186 °C. The van't Hoff factor in this case is 1.87. If there were no ion pairing, we would expect the van't Hoff factor for NaCl to be 2.00. Similarly, ethanoic acid in a 0.100 molal solution has a van't Hoff factor of 1.05.

EXERCISE 9.14

Calculate the concentration of NaCl ion pairs using the information above. Also, calculate the percent ionization of ethanoic acid from the above information.

Solution

(a) When there is no ion pairing, we can write

$$i = (0.100 \ m \ Na^+ + 0.100 \ m \ Cl^-)/ \ 0.100 \ m \ NaCl = 2$$

The ion pairing reaction is $Na^+ + Cl^- \rightarrow NaCl$, and if x moles of sodium and chloride ions combine, they produce x moles of sodium chloride ion pairs. Because colligative properties depend just on the number of particles, not on their chemical identity, we give the Na^+, Cl^-, and NaCl the symbol P for particles. We rewrite the equation above as

$$i = (0.100 \ m \ P + 0.100 \ m \ P)/ \ 0.100 \ m \ NaCl = 2$$

When x molal of ion pairs are formed, we can then write

$$i = ((0.100 \ m - x)P + (0.100 \ m - x)P + x \ P)/ \ 0.100 \ m \ NaCl = 1.87$$

where the $(0.100 \ m - x)P$ represent the decrease in sodium and chloride ion concentrations and the last x P represents the particles of ion pairs present. Subtracting the two equations yields

$$(2x \ P - x \ P)/ \ 0.100 \ m \ NaCl = x \ P/ \ 0.100 \ m \ NaCl = 0.13$$

Multiplying both sides of the equation by the denominator yields

$$0.013 \text{ molal NaCl ion pairs.}$$

(b) For ethanoic acid, if it did not ionize at all, we would have molecular particles of $HC_2H_3O_2$, once again called P, so that

$$i = 1.00 = (0.100 \ m \ P)/(0.100 \ m \ HC_2H_3O_2)$$

If *x* molal of the ethanoic acid ionizes, we will produce an additional *x* molal of particles, so that

$$i = 1.05 = (0.100 \ m \ P + x \ m \ P)/(0.100 \ m \ HC_2H_3O_2)$$

Subtracting the first equation from the second yields

$$0.05 = x \ m \ P/(0.100 \ m \ HC_2H_3O_2)$$

Multiplying through by the denominator yields 0.005 *m* of ethanoic acid that has broken apart into ions.

SUMMARY

After discussing pure matter in Chapters 7 and 8, solutions are covered in this chapter. While solutions can be prepared from almost any two substances (gas and solid solutions are possible), this chapter focuses on solutions that have at least one liquid phase. Factors that allow one substance to dissolve in another are described. Again, molecular structure and polarity give us information to understand this process. In many instances, polarity estimates help decide if two substances will form a solution.

Properties of solutions depend upon the concentration of solute particles and are called colligative properties. Colligative property effects include increases in boiling point and decreases in freezing point, changes in vapor pressure, and changes in osmotic pressures of solutions. Many of these effects also allow us to determine the molecular mass of solutes. It is important to understand that osmotic pressure measurements are good for molecules with very large molecular masses and that the other methods are good for small molecules. There are also limitations to the methods that are mentioned in the text.

Important Concepts

Molecular view of the solution process
Attractive forces and randomness in the solution process
Solubility related to attractive forces
Concentration units, particularly molarity and molality
Colligative properties
Osmotic pressure

Important Equations

M = moles/liter solution
m = moles/kg solvent
$\Delta T = -K_f m$
$\Delta T = K_b m$
$\Pi V = nRT$
d = grams/cm^3

Practice Exercises

Multiple-Choice

For the first four problems below, one or more of the following responses will apply; each response may be used more than once or not at all in these questions.

    I. osmotic pressure
   II. freezing-point depression
  III. vapor pressure
  IV. Raoult's law
   V. Henry's law

1. Which of these best explains why a soda bottle fizzes when opened?

    (A) I
    (B) II
    (C) III
    (D) IV
    (E) V

2. Which of these is the method of choice to determine the molecular mass of large biomolecules?

    (A) I
    (B) II
    (C) III
    (D) IV
    (E) V

3. Which two items are most closely related to each other?

    (A) I and III
    (B) II and V
    (C) III and IV
    (D) IV and V
    (E) V and I

4. The extent of ion pairing in a solution of an electrolyte can be best estimated by using which of these?

    (A) I and III
    (B) II
    (C) III and V
    (D) IV
    (E) V

5. The solubility of cadmium chloride, $CdCl_2$, is 140 g per 100 mL of solution. What is the molar solubility (molarity) of a saturated solution of $CdCl_2$?

    (A) 0.765 M
    (B) 1.31 M
    (C) 7.65 M
    (D) 12.61 M
    (E) 0.131 M

6. The vapor pressure of an ideal solution is 456 mm Hg. If the vapor pressure of the pure solvent is 832 mm Hg, what is the mole fraction of the nonvolatile solute?

    (A) 0.548
    (B) 0.354
    (C) 0.645
    (D) 1.82
    (E) 0.452

7. All of the following physical properties change as solute is added to the solution. Which is least likely to be used to determine the molar mass of a compound?

    (A) boiling point
    (B) surface tension
    (C) vapor pressure
    (D) melting point
    (E) osmotic pressure

8. Which of the following is expected to be the most soluble in hexane, C_6H_{14}?

   (A) KCl
   (B) C_2H_5OH
   (C) C_6H_6
   (D) H_2O
   (E) $HC_2H_3O_2$

9. Molarity units are most appropriate in which of the following experiments?

   (A) freezing-point depression
   (B) vapor pressure
   (C) boiling-point elevation
   (D) surface tension
   (E) osmotic pressure

10. All of the following may be used to determine molar masses. Which one requires an ideal solution for accurate results?

    (A) freezing-point depression
    (B) boiling-point elevation
    (C) osmotic pressure
    (D) vapor pressure
    (E) gas density

11. To make a solution, 3.45 mol of $C_6H_{13}Cl$ and 1.26 mol of C_5H_{12} are mixed. Which of the following is needed, but not readily available, to calculate the molarity of this solution?

    (A) the density of the solution
    (B) the densities of $C_6H_{13}Cl$ and C_5H_{12}
    (C) the temperature
    (D) the molar masses of $C_6H_{13}Cl$ and C_5H_{12}
    (E) the volumes of $C_6H_{13}Cl$ and C_5H_{12}

12. Which of the following, when added to 1.00 kg H_2O, is expected to give the greatest increase in the boiling point of water? ($k_b = 0.052 °C\ m^{-1}$)

    (A) 1.25 mol sucrose
    (B) 0.25 mol iron(III) nitrate
    (C) 0.50 mol ammonium chloride
    (D) 0.60 mol calcium sulfate
    (E) 1.00 mol ethanoic acid

13. Ethyl alcohol, C_2H_5OH, and water become noticeably warmer when mixed. This is due to

    (A) the decrease in volume when they are mixed
    (B) smaller attractive forces in the mixture than in the pure liquids
    (C) the hydrogen bonding of the two liquids
    (D) the change in vapor pressure observed
    (E) stronger attractive forces in the mixture than in the pure liquids

14. Which is the most appropriate method for determining the molar mass of a newly discovered enzyme?

    (A) freezing-point depression
    (B) osmotic pressure measurements
    (C) boiling-point depression
    (D) gas density measurements
    (E) vapor pressure measurements

15. A polluted pond contains 25 ppb of lead ions. What is the concentration of lead ions in molarity units?

    (A) $1.2 \times 10^8\ M$
    (B) $1.2 \times 10^{-7}\ M$
    (C) $2.5 \times 10^{-8}\ M$
    (D) $0.121\ M$
    (E) $1.2 \times 10^{-10}\ M$

16. When algae decay in a pond, the process uses up the available oxygen. Which of the following factors will also contribute to a decrease in oxygen in a pond?

    (A) decreasing salinity (salt concentration)
    (B) increasing acidity due to acid rain
    (C) increasing temperature
    (D) increasing surface tension of the water
    (E) increasing atmospheric pressure

17. Liquid A has a vapor pressure of 437 mm Hg, and liquid B has a vapor pressure of 0.880 atm at 85 °C. Which of the following represents a possible solution of the two liquids?

    (A) a mixture with a vapor pressure of 345 mm Hg
    (B) a mixture with a vapor pressure of 0.750 atm
    (C) a mixture with a boiling point of 165 °C
    (D) a mixture with a vapor pressure of 1106 mm Hg
    (E) a mixture with a boiling point of 85 °C

18. The freezing-point-depression constant for water is 1.86 °C m^{-1}. When 100 g of a compound is dissolved in 500 g H$_2$O, the freezing point is −10.0 °C. Of the five possibilities below, which is the most reasonable identity of the compound?

    (A) Mg(NO$_3$)$_2$
    (B) KCl
    (C) Na$_2$SO$_4$
    (D) HCOOH
    (E) HF

19. Which of the following compounds is incorrectly classified?

    (A) NaF is an electrolyte.
    (B) CH$_3$OH is a weak electrolyte.
    (C) Mg(C$_2$H$_3$O$_2$) is an electrolyte.
    (D) CH$_3$CH$_2$COOH is a weak electrolyte.
    (E) glucose is a nonelectrolyte.

20. The k_f and k_b values for water are 1.86 and 0.52 °C m^{-1}, respectively. A solution boils at 107.5 °C. At what temperature does this solution freeze?

    (A)   7.5 °C
    (B)   −7.5 °C
    (C)   0.0 °C
    (D) −26.8 °C
    (E) −284.5 °C

21. If equal numbers of moles of each of the following are dissolved in 1 kg of distilled water, the one with the lowest boiling point will be

    (A) NaF
    (B) AlCl$_3$
    (C) Mg(C$_2$H$_3$O$_2$)$_2$
    (D) CH$_3$CH$_2$COOH
    (E) glucose

22. The solubility of acetylene, CHCH, in water at 30 °C is 0.975 g L^{-1} when the pressure of acetylene is 1.00 atm. What is the solubility, at the same temperature, when the pressure of acetylene above the water is reduced to 0.212 atm?

    (A) 4.60 g L^{-1}
    (B) 0.207 g L^{-1}
    (C) 0.975 g L^{-1}
    (D) 0.212 g L^{-1}
    (E) The answer cannot be determined from the data given.

23. When KCl dissolves in water, the solution cools noticeably to the touch. It may be concluded that

    (A) the solvation energy is greater than the lattice energy
    (B) KCl is relatively insoluble in water
    (C) the entropy decreases when KCl dissolves
    (D) the boiling point of the solution will be less than 100 °C
    (E) the entropy increase overcomes the unfavorable heat of dissolution

24. If 20.0 g of ethanol (molar mass = 46) and 30.0 g of water (molar mass = 18) are mixed together, the mole fraction of ethanol in this mixture is

    (A) 0.207
    (B) 0.261
    (C) 0.739
    (D) 0.793
    (E) 4.83

25. A 0.12% solution of a small protein in water results in an osmotic pressure of 2.25 torr at 4 °C. What is the molar mass of the protein?

    (A) 922 g mol^{-1}
    (B) 12 g mol^{-1}
    (C) $9.2 \times 10^3 \text{ g mol}^{-1}$
    (D) 135 g mol^{-1}
    (E) $9.3 \times 10^5 \text{ g mol}^{-1}$

ANSWER KEY

1. E	6. E	11. A	16. C	21. E
2. A	7. B	12. A	17. B	22. B
3. C	8. C	13. E	18. B	23. E
4. B	9. E	14. B	19. B	24. A
5. C	10. D	15. B	20. D	25. C

See Appendix 1 for explanations of answers.

Free-Response

Use the fundamental concepts of this chapter to consider solutions and their behaviors.

(a) Use the concepts in this chapter to explain the rule "like dissolves like."

(b) Explain how a nonvolatile, nondissociating solute affects the boiling point of a solvent.

(c) The freezing-point depression constant of cyclohexane is 20.2 °C per molal. The melting point of cyclohexane is 6.50 °C. What is the freezing point of a hexane solution prepared by dissolving 20.0 g of $C_{18}H_{38}$ in 100 g of cyclohexane?

(d) Solutes in aqueous systems are usually classified based on how electricity is conducted. Name these classifications and give examples of each.

(e) A solution is prepared by dissolving 15.2 milligrams of $CrCl_3$ in 2.50 liters of water. What is the concentration of chromium in parts per million?

ANSWERS

(a) Like dissolves like refers to the fact that substances with similar polarities tend to dissolve in each other. We see the reason for this in the dissolution diagrams on page 333. It requires energy to separate solvent and solute molecules, and that energy must be recovered when the solution is formed. If the attractive forces are both due to dipole-dipole interactions, the large amount of energy needed to separate dipoles is regained when new pairs of dipoles are formed. For nonpolar compounds little energy is needed to separate molecules, and little is recovered in forming new interactions. However, the solution is still formed because there is no need for a large amount of energy. If one solute is polar and the other is not, the polar compound requires a large amount of energy that is not recovered when new interactions are formed. Because the required energy is not present, the solution will not be possible.

(b) A nonvolatile solute reduces the vapor pressure of the solution because it decreases the fraction of the solute with sufficient kinetic energy to vaporize. Because the vapor pressure is decreased, it therefore requires a higher temperature to achieve a vapor pressure of 1.0 atmosphere than is required for boiling.

(c) The molality of the solution is = 0.0787 mol/0.100 kg = 0.787. The freezing-point depression is $\Delta T = -(0.787\ m)(20.2\ °C\ m^{-1}) = -15.91\ °C$. We now combine this with the given melting point to get $T = 6.50 - 15.91 = -9.41\ °C$.

(d) The classifications are strong electrolytes, weak electrolytes, and nonelectrolytes. Strong electrolytes are virtually 100 percent ionized when dissolved in water, and their solutions conduct electricity well. Weak electrolytes are generally ionized to a maximum of 10 percent and conduct electricity, but poorly. Nonelectrolytes do not ionize and do not conduct electricity.

(e) We need to calculate the mass of chromium in the 15.2 mg and determine the mass fraction. The mass fraction is multiplied by 10^6 to get the ppm.

$$? \text{g Cr} = 15.2 \times 10^{-3}\ \text{g } CrCl_3 \left(\frac{1 \text{ mol } CrCl_2}{158.4 \text{ g } CrCl_2} \right) \left(\frac{1 \text{ mol Cr}}{1 \text{ mol } CrCl_2} \right) \left(\frac{52.0 \text{ g Cr}}{1 \text{ mol Cr}} \right)$$

$$= 4.99 \times 10^{-3}\ \text{g Cr}$$

$$\text{mass fraction (wt/wt) Cr} = \left(\frac{4.99 \times 10^{-3} \text{ g Cr}}{2.5 \text{ kg } H_2O} \right) = 2.00 \times 10^{-3}$$

$$\text{ppm} = 2.00 \times 10^{-3} \times 10^6 = 2.00 \times 10^3$$

PART 4

PHYSICAL CHEMISTRY

Chemical Equilibrium

The concept of a **dynamic equilibrium** is central to many aspects of chemistry. In a dynamic equilibrium, chemicals are reacting rapidly at the molecular scale, while their concentrations remain constant on the macroscopic scale. Many physical and chemical processes participate in dynamic equilibria. For example, (1) in a closed container gas molecules condense just as fast as liquid molecules evaporate; (2) a saturated solution involves a solid dissolving at the same rate as solute crystallizes; and (3) in a weak electrolyte solution the ions recombine into molecules just as fast as other molecules ionize. In all of these chemical and physical processes, the rate in one direction is exactly equal to the rate in the other direction. It is important that, although reactions never stop in a dynamic equilibrium, the overall concentrations remain constant.

Figure 10.1 shows that a chemical reaction has two well-defined regions in time, and these regions are studied and measured in very different ways. When compounds are first mixed in a chemical reaction, they interact to form other compounds. During the reaction process, the concentrations of the reactants decrease and the concentrations of the products increase. While the concentrations are changing, the reaction is studied using the principles of **chemical kinetics,** which are reviewed in Chapter 11. At some point in time, the concentrations of the reactants and products stop changing. Although reactions do not stop at the molecular level, at the macroscopic level the concentrations of compounds in a dynamic equilibrium remain constant. At this point, the compounds are in a dynamic chemical equilibrium with each other, and they are studied and described using the concepts of **chemical equilibrium.**

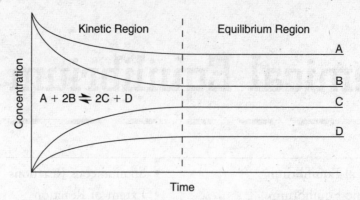

FIGURE 10.1. **The two regions of chemical reactions. On the left, in the kinetic region, concentrations are changing with time. On the right, in the equilibrium region, the concentrations, on a macroscopic or laboratory scale, no longer appear to change.**

<table>
<tr><td>

TIP

The equilibrium law is also called the law of mass action.

</td></tr>
</table>

THE EQUILIBRIUM LAW

In a chemical reaction the actual concentrations of the reactants and products, at equilibrium, are dependent on the initial concentrations of the reacting mixture. When several experiments with different initial concentrations of reactants are performed, they result in equilibrium mixtures with different concentrations. Although these mixtures may be different, they all obey the **equilibrium law.** This law states that the concentrations of all of the products multiplied together, divided by the concentrations of all the reactants multiplied together, will be equal to a number called the **equilibrium constant,** K. The value of the equilibrium constant depends only on the specific reaction and the temperature of the reaction mixture when equilibrium is reached.

The uppercase letter K is reserved as the symbol for the equilibrium constant. To describe the type of equilibrium constant a subscript is often used after the K. The symbol K_c represents the equilibrium constant when concentration is expressed in molarity units (mol L^{-1}). The symbol K_p is used when the partial pressures of gases represent the amounts of reactants and products. Special symbols for the equilibrium constant are K_{sp} for the solubility product, K_a for the acid ionization constant, K_b for the base ionization constant, K_f for the formation constant of complexes, and K_d for the dissociation constant in complexation reactions. These special forms of K are described in later sections of this chapter.

The equilibrium law depends on the chemical equation for the reaction under study. A general equilibrium reaction may be written as

$$aA + bB \rightleftharpoons pP + nN \tag{10.1}$$

and its equilibrium law will be written as

$$K_c = \frac{[P]^p[N]^n}{[A]^a[B]^b} \tag{10.2}$$

In all cases, the chemical reaction is balanced with the smallest possible whole-number coefficients. All tabulated values for equilibrium constants refer to equations with the simplest coefficients.

A more specific example of the formulation of the equilibrium law can be shown using the reaction between hydrogen and chlorine to form hydrogen chloride:

$$H_2(g) + Cl_2(g) \rightleftharpoons HCl(g) + HCl(g) \qquad (10.3)$$

or

$$H_2(g) + Cl_2(g) \rightleftharpoons 2HCl(g) \qquad (10.4)$$

The equilibrium law for this reaction is written as

$$K_c = \frac{[HCl] \times [HCl]}{[H_2] \times [Cl_2]} = \frac{[HCl]^2}{[H_2] \times [Cl_2]} \qquad (10.5)$$

Writing the chemical reaction with two separate HCl molecules as in Equation 10.3 illustrates that the concentration of HCl, written with square brackets as [HCl], should be multiplied by itself in the equilibrium law. Writing the chemical equation in the form of Equation 10.4 illustrates that the coefficient in a chemical equation will be used as an exponent in the equilibrium law.

For the combustion of propane, C_3H_8, the reaction is

$$C_3H_8(g) + 5O_2(g) \rightleftharpoons 3CO_2(g) + 4H_2O(g) \qquad (10.6)$$

and the equilibrium law for this equation is written as

$$K_c = \frac{[CO_2]^3 \times [H_2O]^4}{[C_3H_8] \times [O_2]^5} \qquad (10.7)$$

The coefficients of the reactants and products are written as exponents of the concentrations in the equilibrium law.

Any substance that has a constant concentration during a reaction, including solids and all pure liquids, is not written as part of the equilibrium law. In dilute solutions, the solvent concentration is also constant and is not written in the equilibrium law.

Three examples are a reaction where water is a pure liquid:

$$CH_4(g) + 2O_2(g) \rightleftharpoons CO_2(g) + 2H_2O(l) \qquad (10.8)$$

$$K_c = \frac{[CO_2]}{[CH_4] \times [O_2]^2} \qquad (10.9)$$

a reaction where AgCl is a solid:

$$AgCl(s) \rightleftharpoons Ag^+(aq) + Cl^-(aq) \qquad (10.10)$$

$$K_c = [Ag^+] \times [Cl^-] \qquad (10.11)$$

and a reaction where water is the solvent for dilute solutions:

$$NH_3(g) + H_2O(l) \rightleftharpoons NH_4^+(aq) + OH^-(aq) \qquad (10.12)$$

$$K_c = \frac{[NH_4^+] \times [OH^-]}{[NH_3]} \qquad (10.13)$$

In most cases a solution is considered to be dilute when the concentration of the solute is less than 1 mole per liter.

EXERCISE 10.1

Write the equilibrium law for each of the following chemical equations:

(a) $HF(aq)$ + $H_2O(l)$ $\rightleftharpoons$ $F^-(aq)$ + $H_3O^+(aq)$
(b) $2NH_3(aq)$ + $3I_2(s)$ $\rightleftharpoons$ $2NI_3(s)$ + $3H_2(g)$
(c) $CO(g)$ + $H_2O(g)$ $\rightleftharpoons$ $CO_2(g)$ + $H_2(g)$
(d) $Ba^{2+}(aq)$ + $SO_4^{2-}(aq)$ $\rightleftharpoons$ $BaSO_4(s)$
(e) $C_2H_4(g)$ + $3O_2(g)$ $\rightleftharpoons$ $2CO_2(g)$ + $2H_2O(g)$

Solution

(a) $K_c = \dfrac{[F^+][H_3O^-]}{[HF]}$

(b) $K_c = \dfrac{[H_2]^3}{[NH_3]^2}$

(c) $K_c = \dfrac{[CO_2][H_2]}{[CO][H_2O]}$

(d) $K_c = \dfrac{1}{[Ba^{2+}][SO_4^{2-}]}$

(e) $K_c = \dfrac{[CO_2]^2[H_2O]^2}{[C_2H_4][O_2]^3}$

MANIPULATING THE EQUILIBRIUM LAW

The equilibrium law is written directly from the balanced chemical equation. On paper, chemical equations are easily manipulated. We can reverse the direction of the reaction by writing the reactants as products and the products as reactants. The coefficients of an equation can all be multiplied or divided by a constant factor. Equations can be added and subtracted. Each of these operations results in a different equilibrium law and a different value for the equilibrium constant.

Reversing a Chemical Equation

The chemical equation with the corresponding K_c

$$O_2(g) + 2SO_2(g) \rightleftharpoons 2SO_3(g) \qquad K_c = \frac{[SO_3]^2}{[O_2] \times [SO_2]^2} \qquad (10.14)$$

can be written in the reverse direction as

$$2SO_3(g) \rightleftharpoons O_2(g) + 2SO_2(g) \qquad K'_c = \frac{[O_3] \times [SO_2]^2}{[SO_3]^2} \qquad (10.15)$$

The two equilibrium constants, K_c and K'_c, are inversely related to each other:

$$K_c = \frac{1}{K'_c} \qquad (10.16)$$

If a chemical reaction is reversed, the value of the new equilibrium constant will be the inverse of the original equilibrium constant.

Multiplying or Dividing Coefficients by a Constant

Taking the same reaction, Equation 10.14, of sulfur dioxide with oxygen:

$$O_2(g) + 2SO_2(g) \rightleftharpoons 2SO_3(g) \qquad K_c = \frac{[SO_3]^2}{[O_2] \times [SO_2]^2}$$

we can multiply each of the coefficients by 3 to obtain another balanced equation:

$$3O_2(g) + 6SO_2(g) \rightleftharpoons 6SO_3(g) \qquad K''_c = \frac{[SO_3]^6}{[O_2]^3 \times [SO_2]^6} \qquad (10.17)$$

The relationship between K_c and K''_c can be shown since

$$\frac{[SO_3]^6}{[O_2]^3[SO_2]^6} = \left(\frac{[SO_3]^2}{[O_2][SO_2]^2} \right) \left(\frac{[SO_3]^2}{[O_2][SO_2]^2} \right) \left(\frac{[SO_3]^2}{[O_2][SO_2]^2} \right)$$

$$= \left(\frac{[SO_3]^2}{[O_2][SO_2]^2} \right)^3 \qquad (10.18)$$

$$K_c'' = K_c K_c K_c = K^3 \tag{10.19}$$

We see that the original equilibrium constant is raised to the power equal to the factor used in the multiplication.

Dividing an equation by 2 is the same as multiplying the equation by $^1/_2$. Therefore, when an equation is divided by 2, the new equilibrium constant, K_c''', is the square root of the original K_c:

$$K_c''' = K_c^{1/2} = \sqrt{K_c} \tag{10.20}$$

Finally, reversing a reaction may also be considered mathematically the same as multiplying it by -1. The result is that

$$K_c' = K_c^{-1} = \frac{1}{K_c} \tag{10.21}$$

which agrees with Equation 10.16.

Adding Chemical Reactions

Chemical reactions are added by adding all reactants and all products in two equations and writing them as one equation. For example:

$$2S(g) + 2O_2(g) \rightleftharpoons 2SO_2(g) \quad K_1 = \frac{[SO_2]^2}{[S]^2 \times [O_2]^2} \tag{10.22}$$

$$O_2(g) + 2SO_2(g) \rightleftharpoons 2SO_3(g) \quad K_2 = \frac{[SO_3]^2}{[O_2]3[SO_2]^2} \tag{10.23}$$

Adding these two equations and canceling the $2SO_2(g)$, which are identical on both sides, yields

$$2S(g) + 3O_2(g) \rightleftharpoons 2SO_3(g) \quad K_{overall} = \frac{[SO_3]^2}{[S]^2 \times [O_2]^3} \tag{10.24}$$

Mathematically we find that

$$K_{overall} = K_1 \times K_2 \tag{10.25}$$

When equations are added, the overall equilibrium constant will be the product of the equilibrium constants of the reactions that were added.

EXERCISE 10.2

Given the following two reactions and their equilibrium constants:

$$Ag^+(aq) + Cl^-(aq) \rightleftharpoons AgCl(s) \qquad K_c = 1.0 \times 10^{10} \qquad \text{(a)}$$

$$Ag^+(aq) + 2NH_3(aq) \rightleftharpoons Ag(NH_3)_2^+(aq) \qquad K_c = 1.6 \times 10^7 \qquad \text{(b)}$$

what is the equilibrium constant of the reaction

$$AgCl(s) + 2NH_3(aq) \rightleftharpoons Ag(NH_3)_2^+(aq) + Cl^-(aq) \quad K_{overall} = ? \qquad \text{(c)}$$

Show how Equations (a) and (b) are added to obtain Equation (c).

Solution

To add the given equations, it is necessary to reverse Equation (a) to make $AgCl(s)$ a reactant and Cl^- a product as required in the overall reaction:

$$AgCl(s) \rightleftharpoons Ag^+(aq) + Cl^-(aq) \qquad K_c = 1/1.0 \times 10^{10} = 1.0 \times 10^{-10} \qquad \text{(d)}$$

The equilibrium constant is inverted, as shown, when a reaction is reversed. Equation (d) is added to Equation (b), and the $Ag^+(aq)$ ions cancel. When reactions are added, the equilibrium constants are multiplied:

$$K_{overall} = (1.0 \times 10^{-10})(1.6 \times 10^7) = 1.6 \times 10^{-3}$$

DETERMINING THE VALUE OF THE EQUILIBRIUM CONSTANT

The equilibrium constant, K, has a numerical value that may be determined by a variety of methods. The most direct method is to measure the concentration of each reactant and product in the mixture. For example, when sulfur dioxide reacts with oxygen to produce sulfur trioxide, the reaction and equilibrium law may be written as in Equation 10.14:

$$O_2(g) + 2SO_2(g) \rightleftharpoons 2SO_3(g) \quad K_c = \frac{[SO_3]^2}{[O_2] \times [SO_2]^2}$$

If the concentrations at equilibrium are determined as $[O_2] = 2.0 \times 10^{-8}\ M$, $[SO_2] = 3.4 \times 10^{-9}\ M$, and $[SO_3] = 0.971\ M$, they can be substituted into the equilibrium law to calculate the value of the equilibrium constant:

$$K_c = \frac{(0.971)^2}{(2.0 \times 10^{-8}) \times (3.4 \times 10^{-9})^2}$$

$$= 4.2 \times 10^{24}$$

This method of determining the value of an equilibrium constant requires three measurements, and, if K_c is very large or very small, it can lead to considerable experimental error. In addition, it may be impossible to analyze the mixture for all possible components. Other ways to determine the equilibrium constant, based on chemical stoichiometry, are described later in this chapter.

USING THE EQUILIBRIUM LAW

Extent of Reaction and Spontaneous Reactions

The value of the equilibrium constant indicates the extent to which reactants are converted into products in a chemical reaction. If the equilibrium constant is large, it indicates that the amount of products present at equilibrium is much greater than the amount of reactants. A very large equilibrium constant, greater than 10^{10}, for example, means that for all intents and purposes the reaction goes to completion. Conversely, when K is very small, less than 10^{-10}, very little product is formed and virtually no visible reaction occurs. If $K = 1$, the equilibrium mixture contains approximately equal amounts of reactants and products. These general ideas allow us to quickly estimate the composition of an equilibrium mixture.

In a **spontaneous reaction** products are formed when reactants are mixed, without any additional assistance. Chemists generally define a spontaneous reaction as one with an equilibrium constant greater than 1.00. A reaction with an equilibrium constant less than 1.00 is called a **nonspontaneous reaction.** The fact that $K' = 1/K_c$ when a reaction is reversed tells the chemist that a reaction that is nonspontaneous in one direction is spontaneous if written in the opposite direction.

EXERCISE 10.3

Which of the following reactions is spontaneous? List the reactions in order from the greatest extent of reaction to the lowest.

(a) $H_2(g) + I_2(g) \rightleftharpoons 2HI$ $K_c = 49$
(b) $Br_2 + Cl_2 \rightleftharpoons 2BrCl$ $K_c = 6.9$
(c) $HF(aq) + H_2O(\ell) \rightleftharpoons F^-(aq) + H_3O^+(aq)$ $K_c = 6.8 \times 10^{-4}$
(d) $2H_2(g) + O_2(g) \rightleftharpoons 2H_2O(g)$ $K_c = 9.1 \times 10^{80}$
(e) $2N_2O(g) \rightleftharpoons 2N_2(g) + O_2(g)$ $K_c = 7.0 \times 10^{34}$

Solution

All the reactions except (c) have equilibrium constants greater than 1 and are spontaneous.

The extent of reaction follows the order (d) > (e) > (a) > (b) > (c), based on the magnitude of the equilibrium constants.

The Reaction Quotient and Predicting the Direction of a Reaction

The **reaction quotient, Q,** is defined as the number obtained by entering all of the required concentrations into the equilibrium law and calculating the result. For the sulfur dioxide reaction discussed previously (Equation 10.14):

$$O_2(g) + 2SO_2(g) \rightleftharpoons 2SO_3(g)$$

$$K_c = \frac{[SO_3]^2}{[O_2] \times [SO_2]^2} \text{ (equilibrium law)} \qquad (10.26)$$

$$Q = \frac{[SO_3]^2}{[O_2] \times [SO_2]^2} \text{ (reaction quotient)} \qquad (10.27)$$

> **TIP**
>
> The equation for the reaction quotient, Q, is given on the AP Exam.

The equilibrium constant is the numerical value of K_c when the reaction is at equilibrium. If the chemicals in the reaction are not in equilibrium, the numerical value of the equilibrium law is called the reaction quotient, Q. Q has exactly the same form as the equilibrium law except that K_c has been replaced by Q.

Four principles may be ascribed to the value of Q:

1. If Q does not change with time, the reaction is in a state of equilibrium and $Q = K_c$.

2. If $Q = K_c$, the reaction is in a state of equilibrium.

3. If $Q < K_c$, the reaction will move in the forward direction (to the right) in order to reach equilibrium.

4. If $Q > K_c$, the reaction will move in the reverse direction (to the left) in order to reach equilibrium.

The first principle tells us how to determine whether a chemical reaction has reached equilibrium. It is necessary to measure the concentrations at different times (e.g., 1 hour, 2 hours, and 5 hours after the reaction has started). If the value of Q does not change, the system is in equilibrium and $Q = K_c$.

The second principle tells us that, if K_c is known from a prior experiment, the determination of Q will tell us whether the reactants and products have reached equilibrium. In particular, if $Q = K_c$, the system is in equilibrium.

The third and fourth principles tell us what will happen if Q is not equal to K_c. A value of Q that is less than the value of K_c means that the numerator of the equilibrium law must increase, while the denominator decreases, to raise the value of Q up to that of K_c. The fact that the numerator represents the concentrations of the products indicates that the reaction must proceed toward the product side of the reaction, or in the forward direction.

A value of Q that is larger than the value of K_c indicates that the numerator must decrease, while the denominator increases, to reach equilibrium. Again, the numerator represents the concentrations of products, and a decrease in products indicates that the reaction must proceed in the reverse direction.

If the value of K_c is known, we can tell whether or not a reaction is at equilibrium by determining Q and comparing it to K_c. If the reaction is not at equilibrium, we can also predict in which direction it will go in order to reach equilibrium.

EXERCISE 10.4

Using the equilibrium constants and reactions below, determine whether or not each of the following systems is in equilibrium. If the system is not in equilibrium, predict whether it will proceed in the forward or reverse direction.

(a) $H_2(g) + I_2(g) \rightleftharpoons 2HI(g)$ $K_c = 49$
$[H_2] = 0.10\ M$; $[I_2] = 0.10\ M$; $[HI] = 0.70\ M$

(b) $Br_2 + Cl_2 \rightleftharpoons BrCl$ $K_c = 6.9$
$[Br_2] = 0.10\ M$; $[Cl_2] = 0.20\ M$; $[BrCl] = 0.45\ M$

(c) $HF(aq) + H_2O(l) \rightleftharpoons F^-(aq) + H_3O^+(aq)$ $K_c = 6.8 \times 10^{-4}$
$[HF] = 0.20\ M$; $[F^-] = 2.0 \times 10^{-4}\ M$; $[H_3O^+] = 2.0 \times 10^{-4}\ M$

(d) $2H_2(g) + O_2(g) \rightleftharpoons 2H_2O(g)$ $K_c = 9.1 \times 10^{80}$
$[H_2] = 3.0 \times 10^{-30}\ M$; $[O_2] = 4.0 \times 10^{-26}\ M$; $[H_2O] = 0.0180\ M$

(e) $2N_2O(g) \rightleftharpoons 2N_2(g) + O_2(g)$ $K_c = 7.0 \times 10^{34}$
$[N_2O] = 2.4 \times 10^{-18}\ M$; $[N_2] = 0.0360\ M$; $[O_2] = 0.0090\ M$

Solution

The values of Q are calculated as follows: (a) 49; (b) 10.1; (c) 2×10^{-7}; (d) 9.1×10^{80}; (e) 2.0×10^{30}.

Reactions (a) and (d) are in equilibrium since $Q = K_c$. Reaction (b) has $Q > K_c$, indicating that the reaction must go in the reverse direction to reach equilibrium. Reactions (c) and (e) have $Q < K_c$, so the reaction must go forward to attain equilibrium.

EQUILIBRIUM CALCULATIONS

The Equilibrium Table

In solving equilibrium problems, it is convenient to organize the data in a logical format so that the proper conclusions may be drawn. To do this we construct a table of information called an **equilibrium table.** The basic table has five lines: the first line is used for the balanced chemical reaction; the next line is for the initial conditions stated in the problem; the third line represents the stoichiometric relationships, showing how the conditions on line 2 will change; the fourth line is the sum of the second and third lines and represents the concentrations at equilibrium; and the fifth will show the actual equilibrium amounts, calculated by solving the expression on line 4.

To illustrate such a table, we start with a chemical reaction, for example:

$$NO_2(g) + SO_2(g) \rightleftharpoons NO(g) + SO_3(g) \tag{10.28}$$

The table to be constructed will look like this:

REACTION	$NO_2(g)$	+	$SO_2(g)$	$\rightleftharpoons$	$NO(g)$	+	$SO_3(g)$
INITIAL							
CONDITIONS							
CHANGE							
EQUILIBRIUM							
ANSWER							

In this table, the first line is always the balanced chemical REACTION, as shown. The second and last lines will contain numerical data with units of molarity or partial pressure. All entries on these lines must have identical units. The CHANGE line represents the stoichiometric relationships inherent in the reaction. In this line the change is represented as the unknown, x. The coefficient for the x in each column is the same as the coefficient of the substance in the chemical reaction at the top of the same column. The algebraic signs of all reactants are the opposite of the signs of the products of the reaction on the third line. (Mathematically it does not matter which x's are positive and which are negative as long as all reactants have the same sign and all products have the opposite sign.) The EQUILIBRIUM line is the sum of the INITIAL CONDITIONS and CHANGE lines. Finally, the ANSWER and EQUILIBRIUM lines are mathematically equal to each other.

To illustrate a table filled out with the data given in a problem, let us assume that a 4.00-liter flask is filled with 1 mole of each of the four compounds in the given reaction. The molarity of each compound is 1.00 mole/4.00 liters = 0.250 M. The table will look like this:

REACTION	$NO_2(g)$	+	$SO_2(g)$	$\rightleftharpoons$	$NO(g)$	+	$SO_3(g)$
INITIAL							
CONDITIONS	0.250 M		0.250 M		0.250 M		0.250 M
CHANGE	$+x$		$+x$		$-x$		$-x$
EQUILIBRIUM	$0.250 + x$		$0.250 + x$		$0.250 - x$		$0.250 - x$
ANSWER							

Using this table, we can then perform the appropriate algebraic calculations.

Calculation of Equilibrium Constants

In the example on page 373, the equilibrium constant was determined by measuring all of the concentrations of a mixture when equilibrium was established. Using stoichiometric relationships provides an easier way to do the same thing. To determine the equilibrium constant for the reaction in Equation 10.28:

$$NO_2(g) + SO_2(g) \rightleftharpoons NO(g) + SO_3(g)$$

an experiment can be set up where the initial amount of every compound is 0.250 M. The equilibrium table is the same as before:

REACTION	$NO_2(g)$ +	$SO_2(g)$ ⇌	$NO(g)$ +	$SO_3(g)$
INITIAL CONDITIONS	0.250 M	0.250 M	0.250 M	0.250 M
CHANGE	$+x$	$+x$	$-x$	$-x$
EQUILIBRIUM	$0.250 + x$	$0.250 + x$	$0.250 - x$	$0.250 - x$
ANSWER				

When equilibrium is reached, the concentration of NO_2 is measured as 0.261 M. This value may be entered in the ANSWER row of the table under the NO_2 column as shown below:

REACTION	$NO_2(g)$ +	$SO_2(g)$ ⇌	$NO(g)$ +	$SO_3(g)$
INITIAL CONDITIONS	0.250 M	0.250 M	0.250 M	0.250 M
CHANGE	$+x$	$+x$	$-x$	$-x$
EQUILIBRIUM	$0.250 + x$	$0.250 + x$	$0.250 - x$	$0.250 - x$
ANSWER	**0.261 M**			

Since the values in the last two lines of the $NO_2(g)$ column are mathematically equal, we write

$$0.261 \ M = 0.250 \ M + x$$

and

$$x = 0.011 \ M$$

Since the value of x is now known, the ANSWER line can be completed for all of the other compounds in the reaction by evaluating the expression on the EQUILIBRIUM line:

REACTION	$NO_2(g)$	+	$SO_2(g)$	$\rightleftharpoons$	$NO(g)$	+	$SO_3(g)$
INITIAL CONDITIONS	0.250 *M*		0.250 *M*		0.250 *M*		0.250 *M*
CHANGE	+*X*		+*X*		−*X*		−*X*
EQUILIBRIUM	0.250 + *X*		0.250 + *X*		0.250 − *X*		0.250 − *X*
ANSWER	**0.261 *M***		**0.261 *M***		**0.239 *M***		**0.239 *M***

These values are then entered into the equilibrium law to determine the value of K_c:

$$K_c = \frac{[NO][SO_3]}{[NO_2][SO_2]}$$
$$= \frac{(0.239)(0.239)}{(0.261)(0.261)}$$
$$= 0.8389$$

(10.29)

Only one measurement, the concentration of NO_2, was needed to determine the value of K_c, instead of four measurements.

EXERCISE 10.5

In the example above, rewrite the table so that all reactants are −x and all products +x on the CHANGE line. Verify that the same answer is obtained.

Solution

The signs of x on the EQUILIBRIUM line must also be changed. In this form $x = -0.011$, and the calculated value of K_c is the same as before.

DETERMINATION OF EQUILIBRIUM CONCENTRATIONS BY DIRECT ANALYSIS

When a reaction is at equilibrium, it obeys the equilibrium law. The concentration of any compound in a reaction can be determined by measuring the concentrations of all the other compounds involved in the reaction. For instance, the reaction between H_2 and I_2 to form HI has an equilibrium constant equal to 49. At equilibrium, if $[I_2] = 0.200$ *M* and $[HI] = 0.050$ *M*, we can calculate the concentration of H_2:

$$H_2 + I_2 \rightleftharpoons 2HI$$

(10.30)

$$K = \frac{[HI]^2}{[H_2] \times [I_2]}$$

$$49 = \frac{(0.050)^2}{[H_2] \times (0.200)} \qquad (10.31)$$

$$[H_2] = \frac{(0.050)^2}{(49) \times (0.200)}$$

$$= 2.6 \times 10^{-4}$$

DETERMINATION OF EQUILIBRIUM CONCENTRATIONS FROM INITIAL CONCENTRATIONS AND STOICHIOMETRIC RELATIONSHIPS

Given initial concentrations and a known value for the equilibrium constant, we can determine the equilibrium concentrations of all compounds in a reaction. For example, the reaction

$$Br_2 + Cl_2 \rightleftharpoons 2BrCl \qquad (10.32)$$

has an equilibrium constant of 6.90. If 0.100 mole of BrCl is introduced into a 500-milliliter flask, the equilibrium concentrations of Br_2, Cl_2, and BrCl can be calculated.

The solution starts with setting up the equilibrium table. The initial concentration of BrCl is 0.100 mole/0.500 liter = 0.200 M; and since no Br_2 or Cl_2 is added to the flask, their concentrations are entered as zero. On the CHANGE line, a positive x for Br_2 and Cl_2 is chosen because their concentrations cannot be less than zero. The change in BrCl must then be entered as $-2x$. If the signs of all the x's were reversed, the same answer would be obtained. Writing the table as suggested, however, indicates a better understanding of the equilibrium process.

REACTION	$Br_2(g)$	+	$Cl_2(g)$	$\rightleftharpoons$	$2BrCl(g)$
INITIAL CONDITIONS	0 M		0 M		0.200 M
CHANGE	$+x$		$+x$		$-2x$
EQUILIBRIUM	$+x$		$+x$		$0.200 - 2x$
ANSWER					

The algebraic expressions on the EQUILIBRIUM line are entered into the equilibrium law:

$$K_c = \frac{[BrCl]^2}{[Br_2] \times [Cl_2]}$$

$$(10.33)$$

$$6.90 = \frac{(0.200 - 2x)^2}{(x)(x)}$$

Taking the square root of both sides of this equation yields

$$2.63 = \frac{0.200 - 2x}{x}$$

$$2.63x = 0.200 - 2x$$

$$4.63x = 0.200$$

$$x = 0.0432$$

Once a value for x is determined, we return to the table and we calculate the values for the ANSWER line:

REACTION	$Br_2(g)$	+	$Cl_2(g)$	$\rightleftharpoons$	$2BrCl(g)$
INITIAL CONDITIONS	0 M		0 M		0.200 M
CHANGE	+x		+x		−2x
EQUILIBRIUM	+x		+x		0.200 − 2x
ANSWER	**0.432 M**		**0.432 M**		**0.411 M**

In another example using the same reaction, let us mix together 0.200 M Br$_2$ and 0.300 M Cl$_2$ and determine the concentrations at equilibrium. We set up the equilibrium table and enter the initial concentrations:

REACTION	$Br_2(g)$	+	$Cl_2(g)$	$\rightleftharpoons$	$2BrCl(g)$
INITIAL CONDITIONS	**0.200 M**		**0.300 M**		**0 M**
CHANGE	−x		−x		+2x
EQUILIBRIUM	0.200 − x		0.300 − x		+2x
ANSWER					

Since BrCl cannot have a concentration less than zero, the change for it must be +2x and Br$_2$ and Cl$_2$ must have −x for their changes since the reactants must decrease if the product increases. We make the appropriate entries in the CHANGE line. Adding the INITIAL CONDITIONS to the CHANGE, we obtain the expressions for the EQUILIBRIUM line, using Equation 10.33, and we enter them into the equilibrium law:

$$K_c = \frac{[\text{BrCl}]^2}{[\text{Br}_2][\text{Cl}_2]}$$

$$6.90 = \frac{(2x)^2}{(0.200 - x)(0.300 - x)}$$

The square root cannot be taken to simplify this equation since the denominator is not an exact square. Multiplying out the denominator of the ratio yields

$$6.90 = \frac{(2x)^2}{x^2 - 0.5x + 0.06}$$

$$6.90x^2 - 3.45x + 0.414 = 4x^2$$

$$2.90x^2 - 3.45x + 0.414 = 0$$

This is a quadratic equation of the form $ax^2 + bx + c = 0$, which is solved for x by using the quadratic formula:

$$x = \frac{-b \pm \sqrt{b^2 - 4ac}}{2a} \tag{10.34}$$

Making the substitutions into the quadratic equation yields

$$x = \frac{-(-3.45) \pm \sqrt{(-3.45)^2 - 4(2.90)(0.414)}}{2(2.90)}$$

$$= \frac{3.45 \pm 2.66}{5.8}$$

There are two roots to this equation:

$$x = 1.05 \text{ or } 0.136$$

If the root 1.05 is chosen, we calculate a negative value for the Br_2 and Cl_2 concentrations, which is clearly impossible. The root 0.136 gives the reasonable results shown in the complete table:

REACTION	Br$_2$(g)	+	Cl$_2$(g)	⇌	2BrCl(g)
INITIAL CONDITIONS	0.200 *M*		0.300 *M*		0 *M*
CHANGE	−x		−x		+2x
EQUILIBRIUM	0.200 − x		0.300 − x		+2x
ANSWER	0.064 *M*		0.164 *M*		0.272 *M*

In the two examples above, the answers were obtained by standard mathematical methods. Such solutions are called **analytical solutions.** In other problems, the mathematics becomes very time-consuming, if not impossible, to solve analytically. Sometimes, however, as shown below, the chemist can make simplifying assumptions that result in answers that are accurate to better than ±10 percent. Other problems, which will not be described further, can be solved by making logical estimates of the answer. One way of making logical estimates is called the **method of successive approximations.** Other methods use sophisticated computer programs to make repeated estimates.

When the equilibrium constant is very small and the initial concentrations of the reactants are given, quadratic equations, or even higher order equations, may be avoided by making some assumptions based on our knowledge of the meaning of K_c. For example, the reaction in Equation 10.15:

$$2SO_3(g) \rightleftharpoons 2SO_2(g) + O_2(g)$$

has an equilibrium constant, $K_c = 2.4 \times 10^{-25}$. An equilibrium constant this small indicates that very little SO_3 will react to form the products. If 2.00 moles of SO_3 is placed in a 1.00-liter flask, we can calculate the concentrations of all three molecules when the system comes to equilibrium. First the equilibrium table is constructed:

REACTION	$2SO_3$	$\rightleftharpoons$	$2SO_2$	$+$	O_2
INITIAL CONDITIONS	**0.200 *M***		**0 *M***		**0 *M***
CHANGE	$-2x$		$+2x$		$+x$
EQUILIBRIUM	$0.200 - 2x$		$+2x$		$+x$
ANSWER					

Entering the expressions from the EQUILIBRIUM line into the equilibrium law gives

$$K_c = \frac{[SO_2]^2[O_2]}{[SO_3]^2}$$

$$2.4 \times 10^{-25} = \frac{(2x)^2(x)}{(2.00 - 2x)^2}$$

(10.35)

Equation 10.35 is a cubic equation and may be solved using advanced methods. However, because K_c is very small, it is possible to simplify this equation with an assumption based on our knowledge that this reaction produces very little product. The assumption is that $2x$ will be very small and that, when $2x$ is subtracted from 2.00 *M*, we will still have 2.00 *M* SO_3 left.

ASSUME: $2x \ll 2.00$ *M* so that 2.00 *M* $- 2x = 2.00$ *M*

The assumption allows us to simplify the denominator in the equation to

$$2.4 \times 10^{-25} = \frac{(2x)^2(x)}{(2.00)^2}$$

A solution can now be obtained with much simpler mathematical operations. Multiplying both sides by $(2.00)^2$ and placing the unknown x on the left, we obtain

$$4x^3 = 9.6 \times 10^{-25}$$
$$x^3 = 2.4 \times 10^{-25}$$
$$x = 6.2 \times 10^{-9} \; M$$

Before using the calculated x to fill in the ANSWER line in the table, we check the assumption to be sure it was valid. The assumption is true since $2.00 - 6.2 \times 10^{-9}$ is equal to 2.00. (The actual value is 1.9999999938, which rounds off to 2.00.) Once x is shown to be reasonable, we can complete the table:

REACTION	$2SO_3$	$\rightleftharpoons$	$2SO_2$	$+$	O_2
INITIAL CONDITIONS	0.200 *M*		0 *M*		0 *M*
CHANGE	−2x		+2x		+x
EQUILIBRIUM	0.200 − 2x		+2x		+x
ANSWER	**200 M**		**1.24 × 10⁻⁹ M**		**6.2 × 10⁻⁹ M**

Simplifying assumptions are used only for items on the EQUILIBRIUM line of the table. In addition, these assumptions can only be used with terms that are themselves sums or differences. In the table above it is impossible to make any simplifying assumptions regarding the $+2x$ for SO_2 or the $+x$ for O_2.

In another example, the reverse of the reaction in Equation 10.15 is

$$O_2 + 2SO_2 \rightleftharpoons 2SO_3 \qquad (10.36)$$

and its equilibrium constant will be

$$\frac{1}{2.4 \times 10^{-25}} = 4.2 \times 10^{24}$$

If the initial concentration of SO_3 is 2.00 *M*, the concentrations of O_2 and SO_2 can be determined by the same method as in the preceding example, which gives exactly the same results.

A general principle about equilibrium calculations and the assumptions used can be deduced from the two preceding examples. If the initial concentrations of reactants are given for a reaction with a very small equilibrium constant, we may assume that these concentrations will not change significantly when equilibrium is reached. Similarly, if the initial concentrations are given for the products of a reaction with a large equilibrium constant, we may assume that the concentrations of the products will not change significantly when equilibrium is reached.

TIP

Use simplifying assumptions only on terms with a + or − sign in them.

The above principle indicates also, that, if the equilibrium constant is large and the initial concentrations of the reactants are given, there will be a significant change in concentration that cannot be ignored. For example, we will use the reaction in Equation 10.36 with its equilibrium constant $K_c = 4.2 \times 10^{24}$, and determine the equilibrium concentrations if 2.00 moles of O_2 and 2.00 moles of SO_2 are placed in a 1.00-liter flask.

The equilibrium table can be set up as follows:

REACTION	$O_2(g)$	+	$2SO_2(g)$	$\rightleftharpoons$	$2SO_3(g)$
INITIAL CONDITIONS	0.200 *M*		2.00 *M*		0 *M*
CHANGE	$-x$		$-2x$		$+2x$
EQUILIBRIUM	$2.00 - x$		$2.00 - 2x$		$+2x$
ANSWER					

The equilibrium law for this case is written as

$$4.2 \times 10^{24} = \frac{(2x)^2}{(2.00 - x)(2.00 - 2x)^2}$$

Because the equilibrium constant is so large, we cannot ignore the x in the $2.00 - x$ and $2.00 - 2x$ terms on the EQUILIBRIUM line. This equation is too complex to solve analytically, and estimation methods are too time consuming. The method used to solve this problem requires us to assume that the reaction goes to completion. Calculating the concentrations that would be found if the reaction went to completion provides a new set of initial concentrations that allow us to solve the problem easily. The following example illustrates how this is done.

This reaction is a limiting reactant problem since the concentrations of two reactants are given. Using the equilibrium table, we can simplify the calculation by remembering that the limiting reactant is completely consumed in the reaction. If a reactant is completely used up, its concentration will be zero on the ANSWER line of the table. We can test whether O_2 is the limiting reactant by entering 0 for it on the ANSWER line of the table:

> **TIP**
>
> Limiting reactant problems can be solved using the equilibrium tables.

REACTION	$O_2(g)$	+	$2SO_2(g)$	$\rightleftharpoons$	$2SO_3(g)$
INITIAL CONDITIONS	0.200 *M*		2.00 *M*		0 *M*
CHANGE	$-x$		$-2x$		$+2x$
EQUILIBRIUM	$2.00 - x$		$2.00 - 2x$		$+2x$
ANSWER	**0 *M***				

With 0 on the ANSWER line for O_2 the other concentrations can be calculated since $0\ M = 2.00 - x$ and therefore $x = 2.00$. Using that value of x, we get the results in the next table.

REACTION	$O_2(g)$	+	$2SO_2(g)$	$\rightleftharpoons$	$2SO_3(g)$
INITIAL CONDITIONS	0.200 *M*		2.00 *M*		0 *M*
CHANGE	$-x$		$-2x$		$+2x$
EQUILIBRIUM	$2.00 - x$		$2.00 - 2x$		$+2x$
ANSWER	**0 *M***		**−2.00 *M***		**4.00 *M***

It is impossible for O_2 to be the limiting reactant since the concentration of SO_2 would then have to be negative as shown in the table above. Next we try SO_2 as the limiting reactant:

REACTION	$O_2(g)$	+	$2SO_2(g)$	$\rightleftharpoons$	$2SO_3(g)$
INITIAL CONDITIONS	0.200 *M*		2.00 *M*		0 *M*
CHANGE	$-x$		$-2x$		$+2x$
EQUILIBRIUM	$2.00 - x$		$2.00 - 2x$		$+2x$
ANSWER			**0 *M***		

Since $0\ M = 2.00 - 2x$, then $x = 1.00$ and the remaining items on the ANSWER line can be calculated:

REACTION	$O_2(g)$	+	$2SO_2(g)$	$\rightleftharpoons$	$2SO_3(g)$
INITIAL CONDITIONS	0.200 *M*		2.00 *M*		0 *M*
CHANGE	$-x$		$-2x$		$+2x$
EQUILIBRIUM	$2.00 - x$		$2.00 - 2x$		$+2x$
ANSWER	**1.00 *M***		**0 *M***		**2.00 *M***

These results are reasonable. To continue with the solution of the problem, the concentrations on the ANSWER line of the table above are entered in a new table as the initial concentrations, and new CHANGE and EQUILIBRIUM lines are written:

REACTION	$O_2(g)$	+	$2SO_2(g)$	$\rightleftharpoons$	$2SO_3(g)$
INITIAL CONDITIONS	1.00 *M*		0 *M*		2.00 *M*
CHANGE	+x		+$2x$		−$2x$
EQUILIBRIUM	1.00 − x		+$2x$		2.00 − $2x$
ANSWER					

The terms on the EQUILIBRIUM line are then entered into the equilibrium law:

$$4.2 \times 10^{24} = \frac{(2.00 - 2x)^2}{(1.00 + x)(2x)^2}$$

This equation is simplified by using two assumptions based on the fact that SO_3 is not expected to react to a large extent:

ASSUME $1.00 \gg x$ so that $1.00 + x = 1.00$

ASSUME $2.00 \gg 2x$ so that $2.00 - 2x = 2.00$

Then the equilibrium law becomes

$$4.2 \times 10^{24} = \frac{(2.00)^2}{(1.00)(2x)^2}$$

Solving this equation yields

$$4x^2 = \frac{(2.00)^2}{(1.00)(4.2 \times 10^{24})}$$

$$= 9.5 \times 10^{-25}$$

$$x^2 = 2.4 \times 10^{-25}$$

$$x = 4.9 \times 10^{-13}$$

The assumption is valid, and the table with answers entered becomes as follows:

REACTION	$O_2(g)$	+	$2SO_2(g)$	$\rightleftharpoons$	$2SO_3(g)$
INITIAL CONDITIONS	1.00 *M*		0 *M*		2.00 *M*
CHANGE	+*x*		+2*x*		−2*x*
EQUILIBRIUM	1.00 − *x*		+2*x*		2.00 − 2*x*
ANSWER	**1.00 *M***		**4.9×10^{-13} *M***		**2.00 *M***

This problem illustrates another principle of chemical equilibrium. As long as the same number of moles of each *element* is present initially in different reacting mixtures, and these mixtures have identical volumes, the final composition of these equilibrium mixtures will always be the same.

There are a variety of methods for solving problems involving initial concentrations. If the equilibrium constant is very large or very small, approximations and stoichiometric calculations can be used to simplify the mathematics. In general, a very large equilibrium constant is one where the value of K_c is at least 100 times greater than the initial concentrations. The value of K_c is considered to be very small when the equilibrium constant is less than one one-hundredth (<0.01) of the initial concentrations. If K_c is neither very large nor very small, assumptions cannot be used and analytical solutions are then sought.

K_p, an Equilibrium Constant for Gas-Phase Reactions

The ideal gas law

$$PV = nRT \tag{10.37}$$

can be rearranged to read

$$\frac{n}{V} = \frac{P}{RT} \tag{10.38}$$

Since the n/V term is the concentration in moles per liter, the pressure of a gas, at a constant temperature, is directly proportional to its concentration. Therefore, the equilibrium law may be written using the partial pressures of the gaseous reactants and products. Under these conditions the equilibrium constant is given the symbol K_p. Usually only reactions that are entirely in the gas phase are written in this manner.

One gas-phase reaction is

$$H_2(g) + I_2(g) \rightleftharpoons 2HI(g)$$

and its equilibrium law is written as

$$K_p = \frac{P_{HI}^2}{P_{H_2} P_{Br_2}} \tag{10.39}$$

where P represents the partial pressure of each gas. All calculations and procedures illustrated with the examples using K_c are done in exactly the same manner when the equilibrium constant is K_p.

EXERCISE 10.6

For the reaction of gaseous sulfur with oxygen at high temperatures:

$$2S(g) + 3O_2(g) \rightleftharpoons 2SO_3(g)$$

when the system reaches equilibrium the partial pressures are measured as $P_S = 0.0035$ atm, $P_{SO_3} = 0.0050$ atm, and $P_{O_2} = 0.0021$ atm. What is the value of K_p under these conditions?

Solution

This problem is solved by writing the correct equilibrium law. Since all of the chemicals in the reaction are gases and are measured in partial pressures, the equilibrium law should be written with the constant K_p:

$$K_p = \frac{P_{SO_2}^2}{P_S^2 P_{O_2}^3}$$

With this equilibrium law, the values for the partial pressures are entered and the solution is calculated:

$$K_p = \frac{(0.0050)^2}{(0.0021)^3 (0.0035)^2}$$
$$= 2.2 \times 10^8$$

EXERCISE 10.7

With the known value of $K_p = 2.2 \times 10^8$ from Exercise 10.6, determine whether each of the following systems is in equilibrium. If a system is not, determine in which direction the reaction will proceed.

System 1: $P_{SO_3} = 1.25$ atm $P_{O_2} = 0.256$ atm $P_S = 0.0112$ atm
System 2: $P_{SO_3} = 0.00677$ atm $P_{O_2} = 0.122$ atm $P_S = 0.212$ atm
System 3: $P_{SO_3} = 0.123$ atm $P_{O_2} = 0.00145$ atm $P_S = 0.0332$ atm

Solution

The values for the pressures are entered into the equilibrium-law equation to calculate the reaction quotient, Q, which is then compared to the known value of K_p:

System 1:

$$Q = \frac{P_{SO_3}^2}{P_S^2 P_{O_2}^3} = \frac{(1.25)^2}{(0.256)^3 (0.0112)^2} = 7.4 \times 10^5$$

System 2:

$$Q = \frac{P^2_{SO_3}}{P^2_S P^3_{O_2}} = \frac{(0.00677)^2}{(1.22)^3(0.212)^2} = 5.6 \times 10^{-1}$$

System 3:

$$Q = \frac{P^2_{SO_3}}{P^2_S P^3_{O_2}} = \frac{(0.123)^2}{(0.00145)^3(0.0332)^2} = 4.5 \times 10^9$$

None of the three systems has a value of Q equal to the known K_p of 2.2×10^8, indicating that none of the systems is in equilibrium.

In System 1 and System 2 Q is smaller than the known K_p. Q must increase as these systems approach equilibrium, and the numerator of the ratio will increase while the denominator decreases. Since the numerator contains the products and the denominator contains the reactants, the products must increase and the reactants decrease. Therefore the reaction in these systems must proceed in the forward direction.

System 3 has a value of Q that is greater than the known K_p. The ratio must decrease for the reaction to reach equilibrium. Products must be used to form more reactants, and the reaction in this system must proceed in the reverse direction.

Relationship Between K_p and K_c

TIP

This equation is given on the AP Exam.

The equilibrium constant for a gas-phase reaction can be written as K_p or K_c, and the two can be converted from one to another. Equation 10.38 shows that $\left(\dfrac{P}{RT}\right)$ can be substituted for the molar concentrations in the equilibrium law, resulting in the relationship

$$K_p = K_c(RT)^{\Delta n_g} \tag{10.40}$$

In this equation R is the universal gas law constant (0.0821 L atm mol^{-1} K^{-1}), T is the Kelvin temperature, and Δn_g is the change in the number of moles of gas in the balanced reaction:

$$\Delta n_g = \text{moles of gas products} - \text{moles of gas reactants} \tag{10.41}$$

In Exercise 10.4 (page 378) the reaction

$$I_2(g) + H_2(g) \rightleftharpoons 2HI(g)$$

had a $K_c = 49$. The equivalent K_p at 100 °C is calculated by determining Δn_g:

$$\Delta n_g = 2 \text{ mol HI} - (1 \text{ mol I}_2 + 1 \text{ mol H}_2)$$
$$= 0$$

If $\Delta n_g = 0$, then $K_p = K_c$ for this reaction.

In the reaction of sulfur with oxygen the value of Δn_g is not zero:

$$\Delta n_g = (2 \text{ mol } SO_3) - (2 \text{ mol } S + 3 \text{ mol } O_2)$$

$$= -3$$

With $K_p = 2.2 \times 10^8$ at 573 K, we can calculate K_c as

$$2.2 \times 10^8 = K_c[(0.0821)(573)]^{-3}$$

$$K_c = (2.2 \times 10^8)[(0.0821)(573)]^3$$

$$= 2.3 \times 10^{13}$$

> **TIP**
>
> If a value of an equilibrium constant is given without specifying K_p or K_c, you can assume it is K_p unless the chemical equation requires one or more substances in solution, such as (*aq*).
>
> Similarly, when K is calculated from $\Delta G°$ or $E°_{cell}$, it is K_p unless the chemical equation requires one or more substances in solution.

EXERCISE 10.8

The value of K_c for the following reaction:

$$2NO(g) + O_2(g) \rightleftharpoons 2NO_2(g)$$

is 5.6×10^{12} at 290 K. What is the value of K_p?

Solution

In this problem we are required to solve Equation 10.40. Since K_p is the unknown, we need values for K_c, R, T, and Δn_g. All of these are given in the problem except Δn_g. To determine the value for Δn_g, we note that this reaction has 3 mol of gases as reactants and 2 mol of gases as products. This is a decrease of 1 mol of gas in going from reactants to products, and therefore Δn_g is -1. Entering the numbers into the equation, we get

$$K_p = 5.6 \times 10^{12}[(0.081)(290)]^{-1}$$

Rearranging the equation, we calculate the value of K_p as

$$K_p = \frac{5.6 \times 10^{12}}{(0.081)(290)} = 2.4 \times 10^{11}$$

Units of Equilibrium Constants

In the exact derivation of equilibrium constants there are no units, and we say that the equilibrium constants are dimensionless quantities. At times, however, it is convenient to assign dimensions to an equilibrium constant. To assign units to K_p or K_c, we determine the value of Δn as

$$\Delta n = \text{moles of products} - \text{moles of reactants} \tag{10.42}$$

which represents the sum of the coefficients of the reactants subtracted from the sum of the coefficients of the products in a balanced chemical reaction.

With Δn determined, the units are assigned as follows:

$$\text{units for } K_p = (\text{atm})^{\Delta n}$$

and

$$\text{units for } K_c = (M)^{\Delta n} = \left(\frac{\text{mol}}{\text{L}}\right)^{\Delta n}$$

SPECIAL EQUILIBRIUM CONSTANTS

Solubility Product

In Chapter 4 rules for determining solubility were given. These rules are used to determine whether an ionic compound will dissolve to an appreciable extent in water. In quantitative terms, it is loosely agreed that a salt is soluble if at least 0.1 mol of it will dissolve in 1 liter of water (0.1 M solution). Saturated solutions of insoluble salts have concentrations that are less than 0.1 molar. However, most insoluble salts do dissolve to a small extent.

The solution process may be written in a form similar to a chemical reaction. For solid $Fe_2(OH)_3$ we have

$$Fe(OH)_3(s) \rightleftharpoons Fe^{3+}(aq) + 3OH^-(aq) \tag{10.43}$$

The equilibrium law for this is written as

$$K_c = [Fe^{3+}][OH^-]^3 \tag{10.44}$$

$Fe(OH)_3(s)$ does not appear in the equilibrium law because it is a solid. In the special case of the solubility of slightly soluble compounds, the equilibrium law always represents the product of the ions produced when the compound dissolves. The equilibrium constant is called the **solubility product** constant and is given the symbol K_{sp}:

$$K_{sp} = [Fe^{3+}][OH^-]^3 \tag{10.45}$$

Questions involving K_{sp} are solved in the same way as other equilibrium problems. For example, K_{sp} has a value of 1.6×10^{-39} for $Fe(OH)_3$. The molar solubility is calculated by setting up an equilibrium table:

REACTION	$Fe(OH)_3$	$\rightleftharpoons$	Fe^{3+}	+	$3OH^-$
INITIAL CONDITIONS	Solid		0 M		0 M
CHANGE	$-x$		$+x$		$+3x$
EQUILIBRIUM	Not used		$+x$		$+3x$
ANSWER					

Entering the information from the EQUILIBRIUM line into the K_{sp} equilibrium law gives

$$1.6 \times 10^{-39} = (x)(3x)^3$$
$$= (x)(27x^3)$$
$$= 27x^4$$
$$x^4 = 5.9 \times 10^{-41}$$
$$x = 8.8 \times 10^{-11}$$

The value of x is used to calculate the molar concentrations of Fe^{3+} and OH^- in the solution. We enter these data in the table:

REACTION	$Fe(OH)_3$	$\rightleftharpoons$	Fe^{3+}	+	$3OH^-$
INITIAL CONDITIONS	Solid		0 M		0 M
CHANGE	$-x$		$+x$		$+3x$
EQUILIBRIUM	Not used		$+x$		$+3x$
ANSWER			8.8×10^{-11} M		2.6×10^{-10} M

The solubility of $Fe(OH)_3$ may be deduced from this table also. The CHANGE line indicates that $-x$ of the compound dissolves, and therefore x represents the solubility of $Fe(OH)_3$, or 8.8×10^{-11} mol L^{-1}.

If $Fe(OH)_3$ is dissolved in a solution that already contains the Fe^{3+} ion, the **common ion effect** is observed. The common ion effect is a decrease in the solubility of a compound when it is dissolved in a solution that already contains an ion in common with the salt being dissolved. As an example, we can calculate the solubility of $Fe(OH)_3$ in a solution that already has a 6.5×10^{-5} M concentration of Fe^{3+}. The equilibrium table is constructed as follows:

REACTION	$Fe(OH)_3$	$\rightleftharpoons$	Fe^{3+}	+	$3OH^-$
INITIAL	**Solid**		6.5×10^{-5} M		0 M
CONDITIONS	Solid		0 M		0 M
CHANGE	$-x$		$+x$		$+3x$
EQUILIBRIUM	Not used		$6.5 \times 10^{-5} + x$		$+3x$
ANSWER					

Entering the information into the K_{sp} equation yields

$$1.6 \times 10^{-39} = (6.5 \times 10^{-5} + x)(3x)^3$$

To avoid a very complex fifth-order equation, we can use the results of the preceding problem to assume that x will be very small compared to 6.5×10^{-5}:

$$\text{ASSUME: } x \ll 6.5 \times 10^{-5} \text{ so that } 6.5 \times 10^{-5} = x = 6.5 \times 10^{-5}$$

Entering this assumption into the equation gives

$$1.6 \times 10^{-39} = (6.5 \times 10^{-5})(3x)^3$$

$$27x^3 = \frac{1.6 \times 10^{-39}}{6.5 \times 10^{-5}}$$

$$x^3 = 9.1 \times 10^{-37}$$

$$= 9.7 \times 10^{-13}$$

This result satisfies the simplifying assumption and may be entered in the table:

REACTION	$Fe(OH)_3$	$\rightleftharpoons$	Fe^{3+}	+	$3OH^-$
INITIAL					
CONDITIONS	Solid		6.5×10^{-5} M		0 M
CHANGE	$-x$		$+x$		$+3x$
EQUILIBRIUM	Not used		$6.5 \times 10^{-5} + x$		$+3x$
ANSWER			6.5×10^{-5} M		2.9×10^{-12} M

In addition we may conclude that the solubility of $Fe(OH)_3$ is 9.7×10^{-13} M. To determine the value of the solubility product, a variety of experiments may be performed. In one of the simplest, the molar solubility of a compound is determined by measuring the amount that dissolves in 1 liter of water. For example, if the molar solubility of MgF_2 is determined to be 0.00118 mol per liter, what is its K_{sp}? When solid MgF_2 dissolves, the reaction is

$$MgF_2(s) \rightleftharpoons Mg^{2+}(aq) + 2\,F^-(aq) \tag{10.46}$$

and the K_{sp} expression is

$$K_{sp} = [Mg^{2+}][F^-]^2 \tag{10.47}$$

The equilibrium table is set up as follows:

REACTION	MgF_3	$\rightleftharpoons$	Mg^{2+}	+	$2F^-$
INITIAL					
CONDITIONS	Solid		0 M		0 M
CHANGE	$-x$		$+x$		$+2x$
EQUILIBRIUM	Not used		$+x$		$+2x$
ANSWER					

The solubility given in the problem is the amount dissolved, which is also x. Knowing x, we can immediately complete the ANSWER line:

REACTION	MgF$_3$	$\rightleftharpoons$	Mg^{2+}	+	2F$^-$
INITIAL CONDITIONS	Solid		0 M		0 M
CHANGE	$-x$		$+x$		$+2x$
EQUILIBRIUM	Not used		$+x$		$+2x$
ANSWER			**0.00118 M**		**0.00236 M**

Substituting the values from the ANSWER line into the K_{sp} expression yields

$$K_{sp} = (0.00118)(0.00236)^2$$
$$= 6.6 \times 10^{-9}$$

Weak-Acid and Weak-Base Equilibria

Weak acids and weak bases are compounds that ionize only slightly when dissolved in water, as shown for autic acid and ammonia in the following equations:

$$CH_3COOH(aq) + H_2O(\ell) \rightleftharpoons CH_3COO^-(aq) + H_3O^+(aq) \qquad (10.48)$$

$$NH_3(aq) + H_2O(\ell) \rightleftharpoons NH_4^+(aq) + OH^-(aq) \qquad (10.49)$$

The equilibrium laws for these two reactions are as follows:

$$K_c = K_a = \frac{[CH_3COO^-][H_3O^+]}{[CH_3COOH]} \qquad (10.50)$$

and

$$K_c = K_b = \frac{[NH_4^+][OH^-]}{[NH_3]} \qquad (10.51)$$

> **TIP**
>
> Equations for K_a, K_b, and K_w are given on the AP Exam.

For weak acids the equilibrium constant is called the **acid ionization constant** and is given the symbol K_a. Weak bases have a corresponding **base ionization constant,** K_b. Since all weak acids ionize to form an anion and the hydronium ion, H_3O^+, all of their ionization reactions have the same form as the one for ethanoic acid and their K_a expressions are similar. Weak bases all ionize the same way as ammonia, and their base ionization expressions are similar to that for ammonia. Acid and base equilibrium problems are approached in exactly the same manner as other equilibrium problems.

For example, when 0.100 mole of NH_3 is dissolved in 1 liter of solution, what are the equilibrium concentrations of all solutes? First, the equation for the ionization of ammonia is written, along with its equilibrium law, as shown in Equations 10.49 and 10.51. Next, K_b will be needed and can be found in a table of ionization constants of the bases; for ammonia $K_b = 1.8 \times 10^{-5}$. Then an equilibrium table is constructed to summarize the given information and the stoichiometric relationships:

REACTION	NH_3	+	H_2O	$\rightleftharpoons$	NH_4^+	+	OH^-
INITIAL			**Pure**				
CONDITIONS	**0.100 M**		**Solvent**		**0 M**		**0 M**
CHANGE	−x		−x		+x		+x
EQUILIBRIUM	0.100 − x		Not used		+x		+x
ANSWER							

For ethanoic acid K_a is 1.8×10^{-5}, and this value can be used in the equilibrium law, along with the expressions on the EQUILIBRIUM line of the equilibrium table, to obtain

$$1.8 \times 10^{-5} = \frac{(x)(x)}{0.100 - x}$$

Since the equilibrium constant is small, it is expected that x will be much smaller than 0.100:

ASSUME: $x \ll 0.100$ so that $0.100 - x = 0.100$

Using this simplifying assumption, we have

$$1.8 \times 10^{-5} = \frac{(x)(x)}{0.100}$$
$$x^2 = 1.8 \times 10^{-6}$$
$$x = 1.3 \times 10^{-3}$$

Since x is much smaller than 0.100, the equilibrium table can be completed as follows:

REACTION	NH_3	+	H_2O	$\rightleftharpoons$	NH_4^+	+	OH^-
INITIAL			Pure				
CONDITIONS	0.100 M		Solvent		0 M		0 M
CHANGE	−x		−x		+x		+x
EQUILIBRIUM	0.100 − x		Not used		+x		+x
ANSWER	.099				0.0013 M		0.0013 M

There are many different calculations involving acids and bases that interest chemists. Chapter 14 provides the details for weak-acid and weak-base calculations, buffer calculations, and hydrolysis calculations. All of these calculations are based on the equilibrium concept developed here.

Formation Constants

Metal ions can react with anions and molecules to form chemical species called **complexes.** Ammonia complexes with copper ions in solution, turning the solution from light blue to a much darker blue. The reaction is

$$Cu^{2+}(aq) + 4NH_3(aq) \rightleftharpoons Cu(NH_3)_4^{2+} \tag{10.52}$$

$Cu(NH_3)_4^{2+}$ is a complex ion of copper and ammonia. Since this reaction is written as an equilibrium, its equilibrium law is written as

$$K_c = K_f \quad \frac{[Cu(NH_3)_4^{2+}]}{[Cu^{2+}][NH_3]^4} \tag{10.53}$$

The reaction represents the formation of the complex, and the equilibrium constant is known as the **formation constant,** K_f.

When the complexation reaction is reversed, it represents the dissociation of the complex into its parts. The equilibrium constant for the dissociation is often called the **dissociation constant,** K_d. K_f and K_d are inversely proportional to each other:

$$K_f = \frac{1}{K_d} \tag{10.54}$$

Problems involving complexation equilibria are solved in exactly the same fashion as other equilibrium problems.

LE CHÂTELIER'S PRINCIPLE

In 1888 Henry Le Châtelier proposed his fundamental principle of chemical equilibrium. He observed that chemical systems react until they reach a state of equilibrium. He also observed that, if chemicals in a state of equilibrium are disturbed in some manner, they will react again until equilibrium is reestablished, if possible.

> ### LE CHÂTELIER'S PRINCIPLE
>
> Whenever a system in dynamic equilibrium is disrupted by changes in chemical concentrations or physical conditions, the system will respond with internal physical and chemical changes to reestablish a new equilibrium state, if possible.

Chemical changes to a system involve the addition or removal of one or more of the products or reactants. Physical changes to a system include changes in temperature, pressure, and volume. An understanding of how these factors affect chemical equilibria allows chemists to adjust experimental conditions to maximize the desired products and minimize waste.

Effect of Concentration

Changing the concentration of any reactant or product in a chemical reaction will alter the concentrations of the other chemicals present as the system reacts to reestablish equilibrium. Figure 10.2 illustrates how this process works. A two-compartment container is set up. Dividing the compartments is a very porous barrier such as a window screen. When a liquid is added to the container, the fluid flows easily to equal heights in both compartments to establish Equilibrium 1. If some more liquid is then added to the reactant (R) compartment, the equilibrium is momentarily disturbed, as shown in the middle diagram. However, the liquid flows through the barrier and reestablishes a new equilibrium condition, Equilibrium 2, in the last diagram.

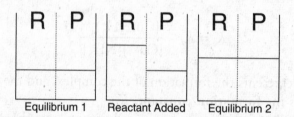

Equilibrium 1   Reactant Added   Equilibrium 2

FIGURE 10.2. Diagrams illustrating an initial equilibrium of two liquids with a porous barrier, a disturbance to the equilibrium by adding liquid to one side, and finally the reestablishment of equilibrium.

Figure 10.2 clearly illustrates the action of a chemical system. Adding a reactant to an equilibrium system disturbs it by raising the reactant concentration. This disturbance causes the reaction to produce a greater amount of product and to reduce the amount of reactant in order to reestablish equilibrium. Also important is the fact that the final equilibrium has different amounts of reactants and products than were present in the initial equilibrium state.

Illustrations similar to Figure 10.2 can be used to visualize other possible concentration changes and their effects. Increasing the concentration of a product will cause an increase in reactant formation (reverse reaction). Decreasing the concentration of a reactant will cause more reactant to form (reverse reaction), and decreasing the concentration of a product will cause more product to form (forward reaction).

Increasing the concentration of a reactant or product is a simple experimental process of adding more chemicals to the reaction mixture. Decreasing the concentration of a product or reactant, however, is experimentally more difficult. Some techniques that are used to remove products or reactants from a reaction mixture are described below.

In the reaction to form ammonia:

$$N_2(g) + 3H_2(g) \rightleftharpoons 2NH_3(g)$$

the ammonia gas produced is very soluble in water:

$$NH_3(g) \xrightarrow{H_2O} NH_3(aq)$$

so that a little water in the reaction system effectively removes the product NH_3. Another method used to remove a gas from a reaction is to condense it into a pure liquid.

Substances in solution can be removed by causing them to precipitate as solids. Additional chemicals may be added to a reaction mixture in order to cause precipitation.

Formation of a complex is an effective method for removing metal ions from solution. Complexing a metal ion changes it into a distinctly different substance, the complex. For example, the dissolution of AgCl with ammonia is a complexation reaction. The $Ag(NH_3)_2^+$ complex is very soluble, and it removes Ag^+ from the solution so that more AgCl may dissolve. The two separate steps of the reaction are as follows:

$$AgCl(s) \rightleftharpoons Ag^+(aq) + Cl^-(aq)$$

and

$$Ag^+(aq) + NH_3(aq) \rightleftharpoons Ag(NH_3)_2^+(aq)$$

which add up to

$$AgCl(s) + 2NH_3(aq) \rightleftharpoons Ag(NH_3)_2^+(aq) + Cl^-(aq)$$

The effects of changing reactant and product concentrations are summarized in Table 10.1.

TABLE 10.1

Effect of Changing Concentrations	
Concentration Change	Observed Effect
Increase reactant	Favors products
Decrease reactant	Favors reactants
Increase product	Favors reactants
Decrease product	Favors products

Effect of Pressure

An increase in pressure easily compresses gases but has little effect on solids and liquids. Increasing the pressure of a gas increases its molar concentration. Changing the pressures of individual gaseous reactants by adding or removing a gas follows the same principles as changing the concentrations discussed in the preceding section. Changing the pressure of a system by adding an inert gas does nothing, however, since the gases originally present still have the same partial pressures and concentrations.

Increasing the pressure of a gaseous reaction system by decreasing its volume will have an effect on the equilibrium only if Δn_g is not zero (Δn_g is the difference between the moles of gaseous products and of gaseous reactants, used previously to make conversions between K_c and K_p). If $\Delta n_g = 0$, there will be no shift in the equilibrium since there is the same number of moles of gaseous products and of gaseous reactants. When $\Delta n_g > 0$, however, the reaction will be forced toward the reactant side because the larger number of moles of gaseous product will be compressed to a higher concentration than the reactants. Similarly, if $\Delta n_g < 0$, there will be more moles of gaseous reactant and the reaction will be forced toward producing more product. Decreasing the pressure will have the opposite effects.

These effects are summarized in Table 10.2.

TABLE 10.2

Effect of Pressure on Gaseous Reactions

Value of Δn_g	Increasing Pressure	Decreasing Pressure
Positive	Favors reactants	Favors products
Zero	No effect	No effect
Negative	Favors products	Favors reactants

TIP

Only temperature changes can cause changes in *K*.

Effect of Temperature

The only experimental variable that has any effect on the value of the equilibrium constant is temperature. For some reactions the equilibrium constant increases as the temperature increases; for others the equilibrium constant decreases. Which direction the equilibrium constant changes depends on whether the reaction is exothermic (ΔH is negative) or endothermic (ΔH is positive). An exothermic reaction gives off heat to the surroundings, and an endothermic reaction absorbs heat from the surroundings.

An exothermic reaction may be represented as a reaction in which one of the products is heat:

$$\text{reactants} \rightleftharpoons \text{products} + \text{heat} \tag{10.55}$$

Raising the temperature for an exothermic reaction is similar to increasing the concentrations of the products. As occurs with a chemical change, increasing the amounts of product will move the reaction toward the left, or reactant, side.

For an endothermic reaction, increasing the temperature is equivalent to increasing the concentrations of the reactants. The result is to move the equilibrium toward the product side:

$$\text{heat} + \text{reactants} \rightleftharpoons \text{products} \tag{10.56}$$

and represents an increase in the equilibrium constant.

TABLE 10.3

Effect of Temperature Changes

Temperature Change	Reaction Type	Effect on Reaction	Effect on *K*
Increase	Exothermic	Favors reactants	Decrease
Increase	Endothermic	Favors products	Increase
Decrease	Exothermic	Favors products	Increase
Decrease	Endothermic	Favors reactants	Decrease

IMPORTANT NOTE:

Le Châtelier's principle is important for evaluating what will happen to equilibrium systems. However, it does not explain why. The AP Exam readers want you to use fundamental principles, laws, and theories of chemistry to explain why certain events happen. **Expect to get little, if any, credit for an essay answer based only on Le Châtelier's principle.**

The effects of temperature changes are summarized in Table 10.3. The last column of the table indicates that the actual effect of a change in temperature is a change in the value of the equilibrium constant. In addition to the direction of change, we will see in Chapter 12 on thermodynamics that the amount of increase or decrease in the equilibrium constant is related to the magnitude of the heat of reaction.

SUMMARY

This chapter describes the essence of dynamic equilibrium in chemical systems. The equilibrium law (equation) is described in detail. The reaction quotient, *Q*, can be compared to the equilibrium constant, *K*, to predict in which direction the reaction will go or if it is at equilibrium. In many instances questions concerning equilibrium involve complex accounting for changes in concentration from initial conditions. An equilibrium table is introduced to help keep track of all the variables in a logical manner. This chapter illustrates how equilibrium constants can be determined from simple measurements. There are also calculations that allow us to calculate the composition of a mixture if the value of the equilibrium constant is known along with the initial concentrations. Techniques are described for analytical (exact) solutions to problems as well as estimation or simplification techniques. The concepts of chemical equilibrium are used to illustrate why LeChâtelier's principle works so well.

Important Concepts

Dynamic equilibrium
Equilibrium law
Equilibrium constant and the reaction quotient
Solving equilibrium problems
Le Châtelier's principle

Important Equations

$K = [A]^a[B]^b/[C]^c[D]^d$ for reaction $cC + dD \rightleftharpoons aA + bB$

Practice Exercises

Multiple-Choice

For the first three problems below, one or more of the following responses will apply; each response may be used more than once or not at all in these questions.

    I. K_a
    II. K_{sp}
    III. Q
    IV. K_c
    V. Le Châtelier's principle

1. The effect of temperature on a chemical system is best described using

   (A) I and III
   (B) II
   (C) III and V
   (D) IV
   (E) V

2. This is the correct term(s) to use to determine if a system has come to equilibrium.

   (A) I and III
   (B) II
   (C) III
   (D) IV
   (E) V

3. The term(s) most useful in determining the solubility of a substance is (are)

   (A) I and III
   (B) II
   (C) III and V
   (D) IV
   (E) V

4. A chemical system in equilibrium will

   (A) have the same concentrations of all products and reactants
   (B) form more products if the temperature is increased
   (C) have a specific ratio of product to reactant concentrations
   (D) not have any precipitates
   (E) represent a spontaneous chemical process

5. Chemical equilibrium may be used to describe

   (A) gas phase chemical reactions
   (B) acid and base ionization
   (C) solubility
   (D) A and C
   (E) A, B, and C

6. For the following reaction:

   $$\text{heat} + 2NO_2(g) \rightleftharpoons N_2O_4(g)$$

   which change will not be effective in increasing the amount of $N_2O_4(g)$?

   (A) decreasing the volume of the reaction vessel
   (B) increasing the temperature
   (C) adding N_2 to increase the pressure
   (D) adsorbing the $N_2O_4(g)$ with a solid adsorbant
   (E) adding more $NO_2(g)$ to the reaction vessel

7. The reaction

$$2NO_2(g) \rightleftharpoons N_2O_4(g)$$

has an equilibrium constant of 4.5×10^3 at a certain temperature. What is the equilibrium constant of

$$2N_2O_4(g) \rightleftharpoons 4NO_2(g)?$$

(A) 4.5×10^3
(B) 9.0×10^6
(C) 2.2×10^{-4}
(D) 2.0×10^7
(E) 4.9×10^{-8}

8. The correct form of the solubility product for silver chromate, Ag_2CrO_4, is

(A) $[Ag^+]^2[CrO_4^{2-}]$
(B) $[Ag^+][CrO_4^{2-}]$
(C) $[Ag^+][CrO_4^{2-}]^2$
(D) $[Ag^+]^2[CrO^{2-}]^4$
(E) $[Ag]^2[CrO_4]$

9. For which of the following will $K_p = K_c$?

(A) $MgCO_3(s) + 2HCl(g) \rightleftharpoons MgCl_2(s) + CO_2(g) + H_2O(l)$
(B) $C(s) + O_2(g) \rightleftharpoons CO_2(g)$
(C) $CH_4(g) + 3O_2(g) \rightleftharpoons CO_2(g) + 2H_2O(g)$
(D) $Zn(s) + 2HCl(aq) \rightleftharpoons H_2(g) + ZnCl_2(aq)$
(E) $2NO_2(g) + O_2(g) \rightleftharpoons N_2O_5(g)$

10. Which is an appropriate formulation of the equilibrium law for the reaction

$$MgCO_3(s) + 2HCl(g) \rightleftharpoons MgCl_2(s) + CO_2(g) + H_2O(l)?$$

(A) $\dfrac{[CO_2]}{[HCl]}$

(B) $\dfrac{[MgCl_2][CO_2][H_2O]}{[HCl]^2[MgCO_3]}$

(C) $\dfrac{[HCl]^2[MgCO_3]}{[MgCl_2][CO_2][H_2O]}$

(D) $\dfrac{[CO_2]}{[HCl]^2}$

(E) $\dfrac{[CO_2][H_2O]}{[HCl]^2}$

11. In the reaction

$$2HI(g) \rightleftharpoons H_2(g) + I_2(g)$$

the equilibrium constant is 0.020. If 0.200 mol of HI is placed in a 10.0-L flask, how many moles of $I_2(g)$ will be in the flask when equilibrium is reached?

(A) 0.022
(B) 0.025
(C) 0.0022
(D) 2.2
(E) 0.0025

12. For the reaction

$$2NO_2(g) \rightleftharpoons N_2O_4(g)$$

$K_p = 8.8$ when pressures are measured in atmospheres. Under which of the following conditions will the reaction proceed in the forward direction?

$NO_2(g)$	$N_2O_4(g)$
(A) 0.200 atm	0.352 atm
(B) 250 mm Hg	400 mm Hg
(C) 0.00255 atm	0.000134 atm
(D) 46.5 mm Hg	82.3 mm Hg
(E) 0.138 atm	0.764 atm

13. The solubility product of PbI_2 is 7.9×10^{-9}. What is the molar solubility of PbI_2 in distilled water?

(A) 2.0×10^{-3}
(B) 1.25×10^{-3}
(C) 5.0×10^{-4}
(D) 8.9×10^{-5}
(E) 7.9×10^{-3}

14. The solubility of gold(III) chloride is 1.00×10^{-4} g L^{-1}. What is the solubility product of $AuCl_3$ (molar mass = 303)?

(A) 1.00×10^{-16}
(B) 2.7×10^{-15}
(C) 1.2×10^{-26}
(D) 3.2×10^{-25}
(E) 9.6×10^{-25}

15. If units were used with the equilibrium constant, K_c, for the following reaction:

$$CH_4(g) + 3O_2(g) \rightleftharpoons CO_2(g) + 2H_2O(g)$$

they would be

(A) M^{-2}
(B) M^2
(C) M
(D) M^{-1}
(E) M^3

16. Which of the following CANNOT affect the extent of reaction?

    (A) changing the temperature
    (B) adding a catalyst
    (C) increasing the amounts of reactants
    (D) removing some product
    (E) changing the volume

17. In which of the following cases is the reaction expected to be exothermic?

    (A) Increasing the pressure increases the amount of product formed.
    (B) Increasing the amount of reactants increases the amount of product formed.
    (C) Increasing the temperature increases the amount of product formed.
    (D) Increasing the volume decreases the amount of product formed.
    (E) Increasing the temperature decreases the amount of product formed.

18. A reaction has a very large equilibrium constant of 3.3×10^{13}. Which statement is NOT true about this reaction?

    (A) The reaction is very fast.
    (B) The reaction is essentially complete.
    (C) The reaction is spontaneous.
    (D) The equilibrium constant will change if the temperature is changed.
    (E) The products will react to yield very little reactant.

19. The K_{sp} of AgCl is 1.0×10^{-10}, and the K_{sp} of AgI is 8.3×10^{-17}. A solution is 0.100 M in I^- and Cl^-. What is the molarity of iodide ions when AgCl just starts to precipitate?

    (A) 1.0×10^{-5}
    (B) 9.1×10^{-9}
    (C) 8.3×10^{-7}
    (D) 8.3×10^{-8}
    (E) 1.2×10^{4}

20. One liter of solution contains 2.4×10^{-3} mol of sulfate ions. What is the molar solubility of $BaSO_4$ in this solution? ($K_{sp} = 1.1 \times 10^{-10}$ for $BaSO_4$.)

    (A) 1.05×10^{-5}
    (B) 1.1×10^{-9}
    (C) 2.6×10^{-13}
    (D) 2.2×10^{7}
    (E) 4.6×10^{-8}

21. The equilibrium constant for the reaction

$$H_2(g) + I_2(g) \rightleftharpoons 2HI(g)$$

must be determined. If 1.00 g of HI is placed in a 2.00-L flask, which of the following is LEAST important in determining the equilibrium constant?

    (A) The temperature must remain constant at the desired value.
    (B) Several measurements must be made to assure that the reaction is at equilibrium.
    (C) Only one of the three concentrations needs to be accurately determined.
    (D) All three concentrations must be accurately measured.
    (E) The original mass and volume of the flask must be accurately measured.

22. In an experiment 0.0300 mol each of $SO_3(g)$, $SO_2(g)$, and $O_2(g)$ were placed in a 10.0-L flask at a certain temperature. When the reaction came to equilibrium, the concentration of $SO_2(g)$ in the flask was 3.50×10^{-5} M. What is K_c for the reaction

$$2SO_2(g) + O_2(g) \rightleftharpoons 2SO_3(g)$$

(A) 3.5×10^{-5}
(B) 1.9×10^{7}
(C) 5.2×10^{-8}
(D) 1.2×10^{-9}
(E) 8.2×10^{8}

23. The weak acid H_2A ionizes in two steps with these equilibrium constants:

$$H_2A \rightleftharpoons H^+ + HA^- \qquad K_{a1} = 2.3 \times 10^{-4}$$
$$HA^{-2} \rightleftharpoons H^+ + A^{2-} \qquad K_{a2} = 4.5 \times 10^{-7}$$

What is the equilibrium constant for the reaction:

$$H_2A \rightleftharpoons 2H^+ + A^{2-}$$

(A) 6.8×10^{-11}
(B) 1.0×10^{-10}
(C) 2.3045×10^{-4}
(D) 2.0×10^{-3}
(E) 5.1×10^{2}

ANSWER KEY

1. E	6. C	11. A	16. B	21. D
2. C	7. E	12. B	17. E	22. B
3. B	8. A	13. B	18. A	23. B
4. C	9. B	14. D	19. D	
5. E	10. D	15. D	20. E	

See Appendix 1 for explanations of answers.

Free-Response

Use the principles and techniques of chemical equilibrium to answer the following questions.

(a) The solubility product of HgI_2 is 1.1×10^{-28}.
   (i) What is the molar solubility of HgI_2?
   (ii) What is the molar solubility if HgI_2 is dissolved in 0.00025 molar NaI solution?

(b) At a certain temperature the reaction of hydrogen and chlorine to produce hydrogen chloride, all in the gas phase, has an equilibrium constant of 265. If 25.0 g of HCl is placed in a 150-L vessel and allowed to come to equilibrium, what will the concentrations of all species be?

(c) The gas-phase reaction between oxygen and sulfur dioxide has an equilibrium constant of 8.8×10^{14} at a certain temperature. If 23.4 g of sulfur trioxide is placed in a 20.0-L vessel, calculate the concentration of all species at equilibrium.

ANSWERS

(a) The reaction is $HgI_2(s) \rightleftharpoons Hg^{2+} + 2I^-$, and $K_{sp} = [Hg^{2+}][I^-]^2$.

If the molar solubility of HgI_2 is assigned the variable s, then $[Hg^{2+}] = s$ and $[I^-] = 2s$. Making the appropriate substitutions we get

$$K_{sp} = (s)(2s)^2 = 1.1 \times 10^{-28}$$
$$4s^3 = 1.1 \times 10^{-28}$$
$$s^3 = 2.75 \times 10^{-29}$$
$$s = 3.02 \times 10^{-10} \ M \ HgI_2$$

If the HgI_2 is dissolved in a 0.000250 I^- solution, we write the iodide ion term as $[I^-] = s + 0.000250$. Because we know that the iodide concentration is on the order of 10^{-10} in distilled water, it is safe to assume that $s \ll 0.000250$, so that $[I^-] = 0.000250 \ M$. The $K_{sp} = (s)(0.000250)^2 = 1.1 \times 10^{-28}$. Solving for s, the molar solubility of HgI_2 yields $s = 1.76 \times 10^{-22} \ M$ HgI_2.

(b) We calculate the initial concentration of HCl as

$$mol \ HCl = 25.0 \ g \ HCl \left(\frac{1 \ mol \ HCl}{36.5 \ g \ HCl} \right) = 0.685 \ mol \ HCl$$

molarity HCl = 0.685 mol HCl/150 L = $4.57 \times 10^{-3} \ M$ HCl

The chemical reaction is $H_2(g) + Cl_2(g) \rightleftharpoons 2HCl(g)$.

The equilibrium law is $K_c = 265 = [HCl]^2/[H_2][Cl_2]$.

$[HCl] = (4.57 \times 10^{-3} - 2x)$ and $[H_2] = [Cl_2] = x$. Insert these into the equilibrium law to yield

$$K_c = 265 = (4.57 \times 10^{-3} - 2x)^2/(x)(x)$$

You may want to construct an equilibrium table for these steps.

Now we take the square root of both sides to get

$$16.3 = (4.57 \times 10^{-3} - 2x)/(x)$$

and multiply through by x to get

$$16.3x = 4.57 \times 10^{-3} - 2x$$
$$18.3x = 4.57 \times 10^{-3}$$
$$x = 2.50 \times 10^{-4} \, M = [H_2] = [Cl_2] \text{ and}$$
$$[HCl] = (4.57 \times 10^{-3} - 5.0 \times 10^{-4}) = 4.07 \times 10^{-3} \, M$$

(c) The moles of $SO_3 = 23.4 \, g/(80 \, g/mol) = 0.293 \, mol \, SO_3$, and the molarity of $SO_3 = 0.293 \, mol/20.0 \, L = 0.0146 \, M \, SO_3$.

The reaction is $2SO_2 + O_2 \rightleftarrows 2\, SO_3$, and we write the equilibrium law as $K_c = 8.8 \times 10^{14} = [SO_3]^2/[O_2][SO_2]^2$. If $2x$ moles per liter of SO_3 react, then we can write (use an equilibrium table here if you need to)

$[SO_3] = 0.0146 - 2x$ and $[O_2] = x$ and $[SO_2] = 2x$.

We insert these terms into the equilibrium law to get

$8.8 \times 10^{14} = (0.0146 - 2x)^2/((x)(2x)^2)$. Because the value of the equilibrium constant is very large, we feel safe in saying that $2x \ll 0.0146$, and our equation becomes

$8.8 \times 10^{14} = (0.0146)^2/((x)(2x)^2)$. This equation is easily solved as

$$4x^3 = (0.0146)^2/(8.8 \times 10^{14}) = 2.42 \times 10^{-19}$$
$$x^3 = 6.05 \times 10^{-20} \text{ and } x = 3.93 \times 10^{-7}$$

Calculating the concentrations we get

$[SO_3] = 0.0146 - 0.00000078 = 0.0146$

$[O_2] = x = 3.93 \times 10^{-7}$ and $[SO_2] = 7.86 \times 10^{-7}$.

Kinetics

REACTION RATES

The rate of a chemical reaction is the rate of change in the concentration per unit time. It is expressed as the number of moles per liter that react each second, and the units, in abbreviated form, are always mol L^{-1} s^{-1}. **Reaction rates** are determined by measuring the concentration of one or more of the chemicals involved in the reaction at different times during the course of the reaction. Figure 11.1 illustrates the results of such measurements as a **kinetic curve,** also called a **concentration versus time curve.** The kinetic curve on the left is obtained from five individual measurements of the concentration of a reaction product. These points are connected with a smooth line. The second kinetic curve is one that may be obtained using instrumentation that continuously monitors the concentration of the same product. Figure 11.1 illustrates the increase in product as a reaction occurs. If the concentration of a reactant is measured, a decrease in concentration with time will be recorded as shown in Figure 11.2.

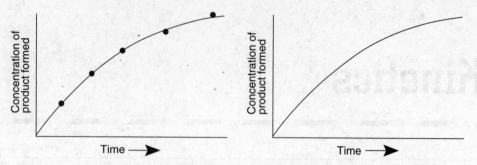

FIGURE 11.1. **Kinetic curves obtained by measuring the formation of a product of a chemical reaction. The first curve is obtained from five individual measurements connected with a smooth line. The second curve is recorded automatically by an instrument designed to monitor the concentration continuously.**

The rate of the production of products (or disappearance of reactants) is obtained from a kinetic curve by determining the slope of the curve at the desired point in time. The slope of a curve is determined in a three-step process:

1. Select the desired point on the curve, and draw a tangent to it.
2. Select two points on the tangent, and determine the concentrations, *c*, and times, *t*, corresponding to these points.
3. Use Equation 11.1 to calculate the slope:

$$\text{Rate} = -\left(\frac{C_2 - C_1}{t_2 - t_1} \right) = -\frac{\Delta C}{\Delta t} \tag{11.1}$$

Figure 11.2 illustrates the construction of the tangents for the determination of the initial rate, *t* = 0, and the rate at 180 seconds, *t* = 180. The small triangles on the tangents connect point 1 and point 2 with a right triangle. The necessary values for ΔC ($C_2 - C_1$) and Δt ($t_2 - t_1$) are the lengths of the sides of these triangles.

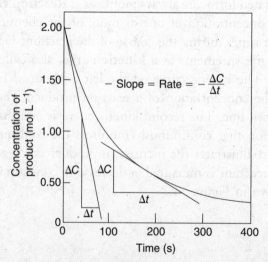

FIGURE 11.2. **Tangents drawn to a kinetic curve, showing how the slope at t = 0 and t = 180 s are determined. This curve decreases with time since it was obtained by measuring the concentration of a reactant.**

Another, less desirable method used to determine the rate of a chemical reaction is to measure the concentration, C, of one reactant or product at only two different times. The rate is then calculated using Equation 11.1. Such a two-point calculation gives the average rate of the reaction over the time interval used.

Since the reactants disappear during a chemical reaction, the rate calculated by measuring a reactant will have a negative sign. If a product is measured, the calculated rate will have a positive sign. By convention, however, chemists always use a positive value for the rate of a reaction, whether it is the positive rate of the appearance of products or the negative rate of disappearance of reactants:

$$\text{Rate} = \frac{\Delta C_{products}}{\Delta t} \tag{11.2}$$

$$\text{Rate} = \frac{-\Delta C_{reactants}}{\Delta t} \tag{11.3}$$

The sign for the rate is implied by calling the rate either the rate for the appearance (implied positive sign) of products or the rate of disappearance (implied negative sign) of reactants. In addition, the rate of appearance (or disappearance) of a substance is divided by its stoichiometric coefficient. The general reaction

$$aA + bB \rightarrow cC + dD$$

has one rate of reaction that is

$$\text{Rate} = \frac{-dA}{a\ dt} = \frac{-dB}{b\ dt} = \frac{+dC}{c\ dt} = \frac{+dD}{d\ dt}$$

Since each rate of formation or disappearance is divided by the corresponding stoichiometric coefficient, we will get the same reaction rate no matter which reactant or product is measured.

The value of the rate of formation or disappearance of a substance can be calculated from the rate of formation or disappearance of any other substance in the reaction.

The value of a reaction rate depends on which reactant or product is measured. After the rate has been measured based on one component of the reaction, the rates of change of the other components may be calculated by a stoichiometric conversion.

Consider the following reaction:

$$2C_2H_6 + 7O_2 \rightarrow 4CO_2 + 6H_2O \tag{11.4}$$

and assume that the rate of reaction was determined by measuring the CO_2 produced and was found to be 2.50 mol CO_2 $L^{-1}s^{-1}$. This can be converted to the rates of appearance or disappearance of all other chemical species by using the stoichiometric relationships in Equation 11.4.

First, the problem is set up in the form of a stoichiometry question. The equation below can be interpreted to read "What will the rate of consumption of C_2H_6 be if the rate of production of CO_2 is 2.50 mol CO_2 L^{-1} s^{-1}?"

$$? \frac{\text{mol } C_2H_6}{L\,s} = \frac{2.50 \text{ mol } CO_2}{L\,s}$$

Since the denominators of both ratios are the same, the problem involves only the conversion of mol CO_2 into mol C_2H_6. The chemical reaction gives us a conversion factor of $\left(\dfrac{2 \text{ mol } C_2H_6}{4 \text{ mol } CO_2} \right)$, which is used to obtain

$$? \frac{\text{mol } C_2H_6}{L\,s} = \frac{2.50 \text{ mol } CO_2}{L\,s} \left(\frac{2 \text{ mol } C_2H_6}{4 \text{ mol } CO_2} \right)$$

Since the units of mol CO_2 cancel, leaving the units desired, the problem is finished, and the answer can be calculated as

$$? \frac{\text{mol } C_2H_6}{L\,s} = \frac{1.25 \text{ mol } C_2H_6}{L\,s}$$

In a similar fashion we can calculate the corresponding rates based on O_2 and H_2O:

$$? \frac{\text{mol } O_2}{L\,s} = \frac{2.50 \text{ mol } CO_2}{L\,s} \left(\frac{7 \text{ mol } O_2}{4 \text{ mol } CO_2} \right)$$

$$= \frac{4.38 \text{ mol } O_2}{L\,s}$$

and

$$? \frac{\text{mol } H_2O}{L\,s} = \frac{2.50 \text{ mol } CO_2}{L\,s} \left(\frac{6 \text{ mol } H_2O}{4 \text{ mol } CO_2} \right)$$

$$= \frac{3.75 \text{ mol } H_2O}{L\,s}$$

$$= 3.75 \text{ mol } H_2O \ L^{-1} \ s^{-1}$$

The intital reaction rates in the following problems all refer to rates as defined in Equation 11.3.

EXERCISE 11.1

Using Figure 11.2, determine the rate, for the reaction illustrated, at 100 s and 200 s after the reaction starts.

Solution

When the tangents are drawn, the slopes are determined to be -6.5×10^{-3} and -2.8×10^{-3} mol L^{-1} s^{-1}. Since this curve represents the disappearance of a reactant, the rate is the negative of the slope, or 6.5×10^{-1} mol L^{-1} s^{-1} at 100 s and 2.8×10^{-3} mol L^{-1} s^{-1} at 200 s.

EXERCISE 11.2

In the reaction

$$N_2(g) + 3H_2(g) \rightarrow 2NH_3(g)$$

the rate of disappearance of $H_2(g)$ is found to be 4.8×10^{-2} mol L^{-1} s^{-1}. What is the rate at which $N_2(g)$ is reacting, and what is the rate at which $NH_3(g)$ is being produced under the same conditions?

Solution

This problem requires the conversion of one rate into another. The initial setup is

$$? \frac{\text{mol } N_2}{L\,s} = \frac{4.8 \times 10^{-2} \text{ mol } H_2}{L\,s}$$

The next step uses the conversion factor for the conversion between $N_2(g)$ and $H_2(g)$ as

$$? \frac{\text{mol } N_2}{L\,s} = \frac{4.8 \times 10^{-2} \text{ mol } H_2}{L\,s} \left(\frac{1 \text{ mol } N_2}{3 \text{ mol } H_2} \right)$$

$$= 1.6 \times 10^{-2} \text{ mol } N_2 \ L^{-1} \ s^{-1}$$

The same steps are used to determine the rate for the production of $NH_3(g)$:

$$? \frac{\text{mol } NH_3}{L\,s} = \frac{4.8 \times 10^{-2} \text{ mol } H_2}{L\,s}$$

$$? \frac{\text{mol } NH_3}{L\,s} = \frac{4.8 \times 10^{-2} \text{ mol } H_2}{L\,s} \left(\frac{2 \text{ mol } NH_3}{3 \text{ mol } H_2} \right)$$

$$= 3.2 \times 10^{-2} \text{ mol } NH_3 \ L^{-1} \ s^{-1}$$

FACTORS THAT AFFECT REACTION RATES

FACTORS THAT AFFECT REACTION RATES

CONCENTRATION:	If a substance is part of the rate law, increasing its concentration (or pressure in the gas phase) will increase the reaction rate.
TEMPERATURE:	The Arrhenius equation shows how the rate constant (and therefore the reaction rate) is exponentially related to the Kelvin temperature.
ABILITY TO MEET:	Reactants in the gas phase or solution react most rapidly because individual molecules of ions react. Solids react more rapidly if very finely divided as in powders or dust.
CATALYST PRESENT:	A catalyst affords an alternate reaction path with a lower activation energy and therefore a larger rate constant (reaction rate).

Concentration, Temperature, and Catalysts

Most people are aware of the factors that increase reaction rates, perhaps without even realizing it. To make a fire burn more fiercely, we add wood to it. A car travels faster if we press on the accelerator pedal to give the engine more gas. In any chemical reaction, the **concentration of reactants** is an important factor in the observed rate.

In addition, to get food to cook faster, we turn up the heat on the stove. To decrease the spoilage of foods, we refrigerate or, even better, we freeze them. Obviously, **temperature** is another important factor in chemical reaction rates. As a consequence, in all rate experiments temperatures must be carefully controlled. It is important to remember that increasing temperature *always* increases reaction rates and decreasing temperature *always* decreases reaction rates.

Finally, a **catalyst** will increase the rate of reaction. A catalyst is a substance that participates in a chemical reaction but does not appear in the balanced equation. Perhaps the most familiar catalysts are those in the catalytic converters in automobiles. The platinum in a catalytic converter provides a surface on which reactants meet and react more efficiently. Although the platinum promotes the effective reaction of two other chemicals, it is not included in the balanced equation.

Other experimental factors, as long as they do not affect the concentration, temperature, or catalysts, will have no effect on the rate of a chemical reaction.

EFFECT OF CONCENTRATION ON REACTION RATES

The effect of concentration on a chemical reaction is expressed in the **rate law.** All rate laws start with the same form

$$\text{Rate} = k[A]^x [B]^y [C]^z \tag{11.5}$$

In this equation k stands for the **rate constant,** and the square brackets indicate the concentrations of the reactants A, B, and C, which have exponents x, y, and z. These exponents are usually small whole numbers. In more complex rate laws, they may be negative numbers or rational fractions. Exponents must be determined from laboratory experiments and have no relationship to the stoichiometric coefficients of the balanced chemical equation.

DETERMINATION OF RATE LAWS

All rate laws must be determined by using the data from a group of well-planned experiments. By changing the concentration of one reactant while holding all other concentrations constant, we can determine whether the change we made has an effect on the rate; and, if it does, we can calculate the exponent for the changed reactant in the rate law. **There is no theoretical way to predict the exponents of a rate law.**

An example of the method used to determine rate laws can be shown using the reaction of peroxydisulfate with iodide ions according to the equation

$$S_2O_8^{2-} + 3I^- \rightarrow 2SO_4^{2-} + I_3^- \tag{11.6}$$

Table 11.1 shows three experiments using this reaction. In experiments 1 and 2 the concentration of iodide ion is the same and the concentrations of peroxydisulfate ions are different. In experiments 2 and 3 the iodide ion concentration is changed while the peroxydisulfate concentration is held constant. The measured initial rate for each experiment is given in the last column.

TABLE 11.1

Kinetic Data for the Peroxydisulfate Reaction at 20.0 °C			
Experiment Number	$[S_2O_8^{2-}]$ (mol L^{-1})	$[I^-]$ (mol L^{-1})	Initial Rate of Reaction (mol L^{-1} s^{-1})
1	0.200	0.200	2.2×10^{-3}
2	0.400	0.200	4.4×10^{-3}
3	0.400	0.400	8.8×10^{-3}

The rate law will have the form

$$\text{Rate} = k[S_2O_8^{2-}]^x [I^-]^y \tag{11.7}$$

where the exponents x and y need to be determined.

To determine the exponent for the peroxydisulfate ion, we focus attention on the change in reaction rate as the concentration of $S_2O_8^{2-}$ is changed in experiments 1 and 2. A rate law can be written for experiment 1 and another for experiment 2 by entering the data from the table into Equation 11.7:

$$\text{Rate}_1 = 2.2 \times 10^{-3} = k(0.200)^x (0.200)^y$$

$$\text{Rate}_2 = 4.4 \times 10^{-3} = k(0.400)^x (0.200)^y$$

The ratio of these two equations is then written (the calculations are usually easier if the larger numbers are used for the numerator) as

$$\frac{\text{Rate}_2}{\text{Rate}_1} = \frac{4.4 \times 10^{-3}}{2.2 \times 10^{-3}} = \frac{k(0.400)^x(0.200)^y}{k(0.200)^x(0.200)^y}$$

The two k's and the two $(0.200)^y$ factors cancel to yield

$$\frac{\text{Rate}_2}{\text{Rate}_1} = \frac{4.4 \times 10^{-3}}{2.2 \times 10^{-3}} = \frac{(0.400)^x}{(0.200)^x}$$

Since both exponents are x, the right-hand term can be rewritten as

$$\frac{\text{Rate}_2}{\text{Rate}_1} = \frac{4.4 \times 10^{-3}}{2.2 \times 10^{-3}} = \left(\frac{0.400}{0.200}\right)^x$$

Solving this gives

$$2.0 = 2.00^x$$
$$= 2.00^1$$

From this result we conclude that exponent $x = 1$. In a similar manner the value of exponent y is determined using experiments 2 and 3, where the concentration of $S_2O_8^{2-}$ is held constant and I^- is varied. The ratio of rate laws for experiments 2 and 3 is

$$\frac{\text{Rate}_3}{\text{Rate}_2} = \frac{8.8 \times 10^{-3}}{4.4 \times 10^{-3}} = \frac{k(0.400)^x(0.400)^y}{k(0.400)^x(0.200)^y}$$

$$2.0 = 2.00^y$$
$$= 2.00^1$$

The value of exponent y is 1. Using the exponents determined in this manner, we write the rate law:

$$\text{Rate} = k[S_2O_8^{2-}]^1[I^-]^1 \quad \text{or} \quad k[S_2O_8^{2-}][I^-] \qquad (11.8)$$

The exponents in the rate law are definitely not the same as the exponents in balanced Equation 11.6.

Once the rate law is known, the rate constant can be calculated. This is done by taking *any* one of the three experiments in Table 11.1, substituting the values into the rate equation, and solving for the rate constant k. The rate constant should be the same whether experiment 1, 2, or 3 is chosen for this calculation. To verify this, we will calculate the rate constant for each experiment and express it in the proper units:

$$\text{Experiment 1: } k = \frac{2.2 \times 10^{-3} \text{ mol L}^{-1} \text{ s}^{-1}}{(0.200 \text{ mol L}^{-1})(0.200 \text{ mol L}^{-1})}$$

$$= 0.055 \text{ L mol}^{-1} \text{ s}^{-1}$$

$$\text{Experiment 2: } k = \frac{4.4 \times 10^{-3} \text{ mol L}^{-1} \text{ s}^{-1}}{(0.400 \text{ mol L}^{-1})(0.200 \text{ mol L}^{-1})}$$

$$= 0.055 \text{ L mol}^{-1} \text{ s}^{-1}$$

$$\text{Experiment 3: } k = \frac{8.8 \times 10^{-3} \text{ mol L}^{-1} \text{ s}^{-1}}{(0.400 \text{ mol L}^{-1})(0.400 \text{ mol L}^{-1})}$$

$$= 0.055 \text{ L mol}^{-1} \text{ s}^{-1}$$

> **REMEMBER THE DIFFERENCE**
>
> REACTION RATE (or simply rate). This varies with the concentration of reactants and time. Rate always has units of mol L^{-1} s^{-1}.
>
> RATE CONSTANT This is a constant value at a fixed temperature for a given reaction. The units of a rate constant depend on the order of the reaction.

We have demonstrated that an easy way to verify the rate law is to calculate the rate constant for each experiment. The rate constants should all agree within experimental error; if they do not, something is wrong with the rate law.

Chemists use the term **order of reaction** to indicate the exponents in the rate law. The reaction we have been considering is said to be **first order** with respect to peroxydisulfate and first order with respect to iodide ions. It is also said to be **second order overall.** Individual reactants have the same orders as their exponents. The order of the overall reaction is the sum of all of the exponents.

To summarize, we can say that in this reaction:

(a) the order with respect to peroxydisulfate is 1 (first order with respect to peroxydisulfate);
(b) the order with respect to iodide ion is one (first order with respect to iodide);
(c) the overall order is $1 + 1 = 2$ (second order overall);
(d) the units for the reaction rate are always mol L^{-1} s^{-1};
(e) the units for this rate constant are (mol L^{-1} s^{-1})/(mol L^{-1})^2 = L mol^{-1} s^{-1}

EXERCISE 11.3

Determine the rate law and value of the rate constant, with its units, for the data in the table below.

Experiment Number	[A] (mol L^{-1})	[B] (mol L^{-1})	Initial Rate of Reaction (mol L^{-1} s^{-1})
1	0.100	0.200	1.1×10^{-6}
2	0.100	0.600	9.9×10^{-6}
3	0.400	0.600	9.9×10^{-6}

Solution

The general form of the rate law will be

$$\text{Rate} = k[\text{A}]^x[\text{B}]^y$$

Taking the ratio of the rate laws for experiments 1 and 2 yields

$$\frac{\text{Rate}_2}{\text{Rate}_1} = \frac{k(0.100)^x (0.600)^y}{k(0.100)^x (0.200)^y}$$

Cancelling the identical terms, k and $(0.100)^x$, on the right and entering the rates on the left yields

$$\frac{9.9 \times 10^{-6}}{1.1 \times 10^{-6}} = \frac{(0.600)^y}{(0.200)^y}$$

$$9 = 3^y$$

Three squared is equal to 9, and therefore $y = 2$.

To determine exponent x, the ratio of the rate laws for experiments 2 and 3 are used:

$$\frac{\text{Rate}_3}{\text{Rate}_2} = \frac{k(0.400)^x (0.600)^y}{k(0.100)^x (0.600)^y}$$

Entering the rates and cancelling as before gives

$$\frac{9.9 \times 10^{-6}}{9.9 \times 10^{-6}} = \frac{(0.400)^x}{(0.100)^x}$$

$$1.0 = 4^x$$

Any number raised to the zero power is equal to 1, and therefore $x = 0$. The rate law is

$$\text{Rate} = k[A]^0 [B]^2 = k[B]^2$$

Since the concentration of A has no effect on the reaction rate, it is not part of the rate law.

The rate constant is determined by taking any of the three experiments, substituting the rate and concentrations into the rate law, and calculating k. Using the data from experiment 1 yields

$$1.1 \times 10^{-6} \text{ mol B L}^{-1} \text{ s}^{-1} = k(0.200 \text{ mol B L}^{-1})^2$$

$$k = \frac{1.1 \times 10^{-6} \text{ mol B L}^{-1} \text{ s}^{-1}}{(0.200 \text{ mol B L}^{-1})^2}$$

$$= 2.75 \times 10^{-5} \text{ L mol}^{-1} \text{ s}^{-1}$$

EXERCISE 11.4

What is the overall order of each of the following rate laws, and what are the units of the rate constant, k, in each of these rate laws?

(a) Rate $= k\,[A]\,[B]\,[C]$ (d) Rate $= k$

(b) Rate $= k\,[X]^2\,[Y]^3$ (e) Rate $= k\,[R]$

(c) Rate $= k\,[M]^2\,[N]$

Solution

(a) Order = 3. Units are $L^2 \text{ mol}^{-2} \text{ s}^{-1}$.

(b) Order = 5. Units are $L^4 \text{ mol}^{-4} \text{ s}^{-1}$.

(c) Order = 3. Units are $L^2 \text{ mol}^{-2} \text{ s}^{-1}$.

(d) Order = 0. Units are $\text{mol } L^{-1} \text{ s}^{-1}$.

(e) Order = 1. Units are s^{-1}.

Effect of Temperature on Reaction Rates

Almost instinctively we add heat when we wish to increase the speed of a reaction. Higher temperatures increase the rates at which foods cook and solids dissolve in water. Cold-blooded animals such as snakes and lizards are almost immobile in cold temperatures, and they sun themselves to warm up in order to hunt for food.

In 1889 Svante Arrhenius developed the equation for the relationship between the rate constant and temperature. It is called the **Arrhenius equation** in his honor.

$$k = Ae^{-E_a/RT} \tag{11.9}$$

This equation includes the rate constant, k; the activation energy, E_a; the universal gas law constant, R, which equals $8.314 \text{ J mol}^{-1} \text{ K}^{-1}$; the Kelvin temperature, T; a proportionality constant, A; and the base of natural logarithms, e. When the natural logarithm of Equation 11.9 is taken, the result is

$$\ln k = \frac{-E_a}{RT} + \ln A \tag{11.10}$$

This equation can be utilized in two ways to eliminate the need to know the value of the constant A. First, a graph of $\ln k$ versus $1/T$ can be constructed after determining the rate constant at a variety of temperatures. This graph, shown in Figure 11.3, is often called an Arrhenius plot. The slope of the line is equal to $-E_a/R$. Arrhenius plots are used to determine the activation energy, E_a, and also the rate constant at any desired temperature.

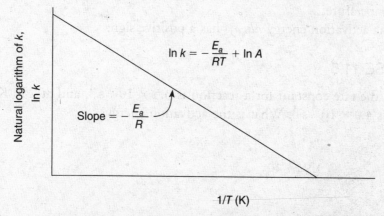

FIGURE 11.3.   An Arrhenius plot of ln k versus 1/T. The activation energy, E_a, is determined from the slope of the straight line.

The second way to apply Equation 11.10 is a two-point approach using only the rate constants determined at two different temperatures. With this method, an

Arrhenius equation is written for the rate constant determined at each temperature, shown by subscripts 1 and 2 in the equations below:

$$\ln k = \frac{-E_a}{RT_1} + \ln A$$

$$\ln k_1 = \frac{-E_a}{RT_2} + \ln A$$

When the second equation is subtracted from the first, the result is

$$\ln k_1 - \ln k_2 = \frac{-E_a}{RT_1} - \frac{-E_a}{RT_2} \tag{11.11}$$

which is commonly rewritten with the ln terms combined on the left and the $-E_a/R$ factored from the right side.

$$\ln\left(\frac{k_1}{k_2}\right) = \frac{-E_a}{R}\left(\frac{1}{T_1} - \frac{1}{T_2}\right) \tag{11.12}$$

This equation has five variables, E_a, k_1, k_2, T_1, and T_2. Given four of these, we can determine the fifth by substitution into the equation. Equation 11.12 may also be written using the ratio of the rates rather than the rate constants:

$$\ln\left(\frac{rate_1}{rate_2}\right) = \frac{-E_a}{R}\left(\frac{1}{T_1} - \frac{1}{T_2}\right) \tag{11.13}$$

TIP

The reaction rate and rate constant are both larger at higher temperatures as long as all other factors are held constant.

The rate and the rate constant are directly proportional to each other as long as the concentrations are held constant.

Often the signs in this equation become confused since the calculation is somewhat involved. It is always possible, however, to test the answer for reasonableness. Two principles must be remembered:

1. The larger rate constant (or rate) will *always* be associated with the higher temperature.
2. The activation energy *always* has a positive sign.

EXERCISE 11.5

At 200 K the rate constant for a reaction is 3.5×10^{-3} s^{-1}, and at 250 K the rate constant is 4.0×10^{-3} s^{-1}. What is the activation energy?

Solution

Using the given data, we can assign 200 K as T_1 and 3.5×10^{-3} as k_1. Similarly, 250 K will be T_2 and 4.0×10^{-3} must be k_2. These are then substituted into the proper places in the equation, along with $R = 8.3145$ J mol^{-1} K^{-1}

$$\ln\left(\frac{3.5 \times 10^{-3}}{4.0 \times 10^{-3}}\right) = \frac{-E_a}{8.3145 \text{ J mol}^{-1} \text{ K}^{-1}}\left(\frac{1}{200 \text{ K}} - \frac{1}{250 \text{ K}}\right)$$

$$-0.1335 = (-1.20 \text{ } 10^{-4} \text{ mol J}^{-1})E_a$$

$$E_a = 1.11 \times 10^3 \text{ J mol}^{-1} = 1.11 \text{ kJ mol}^{-1}$$

EXERCISE 11.6

A common rule of thumb is that, near room temperature, the rate of a reaction will double with each 10 °C rise in temperature. Estimate the activation energy needed for this rule to hold true.

Solution

Choose two temperatures 10 °C apart, such as 290 K and 300 K, which are close to room temperature of 298 K. The rates double, meaning that the rate constants must double. We can assign rate constants of 1.00 to the 290-K temperature and 2.00 to the 300-K temperature. Now the problem is solved as above:

$$\ln\left(\frac{1}{2}\right) = \frac{-E_a}{8.3145 \text{ J mol}^{-1} \text{ K}^{-1}}\left(\frac{1}{200 \text{ K}} - \frac{1}{300 \text{ K}}\right)$$

$$-0.693 = (-1.382 \times 10^{-5} \text{ mol J}^{-1})E_a$$

$$E_a = 5.01 \times 10^4 \text{ J mol}^{-1} = 50.1 \text{ kJ mol}^{-1}$$

Choosing other temperatures near 298 K will result in slightly different answers, but all will be close to 50 kJ mol^{-1}. Many common reactions have activation energies near this value.

EXERCISE 11.7

What is the rate of reaction at 450 °C if the reaction rate is 6.75×10^{-6} mol L^{-1} s^{-1} at 25 °C? The activation energy was previously determined to be 35.5 kJ mol^{-1}.

Solution

Using the equation with rates, we can enter the variables given, after converting kilojoules to joules, to obtain

$$\ln\left(\frac{\text{rate}_1}{6.75 \times 10^{-6}}\right) = \frac{-35,500 \text{ J mol}^{-1}}{8.3145 \text{ J mol}^{-1} \text{ K}^{-1}}\left(\frac{1}{723} - \frac{1}{298}\right)$$

Remembering that the logarithm of a ratio can be written as the logarithm of the numerator minus the logarithm of the denominator, we obtain

$$\ln \text{rate}_1 - \ln (6.75 \times 10^{-6}) = 8.42$$
$$\ln \text{rate}_1 - (-11.91) = 8.42$$
$$\ln (\text{rate}_1) = -3.49$$
$$\text{rate}_1 = 3.05 \times 10^{-2} \text{ mol L}^{-1} \text{ s}^{-1}$$

As expected, the reaction rate is greater at a higher temperature.

APPLICATIONS OF SELECTED RATE LAWS

Zero-Order Reactions

A reaction that is **zero order** has a rate law in which the exponents of all of the reactants are zero. Since any number raised to the zero power is equal to 1 ($x^0 = 1$), the rate law is

$$\text{Rate} = k \tag{11.14}$$

This type of reaction does not depend on the concentration of any reactant. Its rate is equal to the rate constant, and the rate constant has the same units ($\text{mol L}^{-1} \text{ s}^{-1}$) as a rate.

Catalytic reactions are often zero-order reactions. One catalyzed reaction is the decomposition of hydrogen peroxide in the presence of platinum metal:

$$2H_2O_2 \rightarrow 2H_2O + O_2 \tag{11.15}$$

The concentration versus time plot for any zero-order reaction is a straight line, as shown in Figure 11.4.

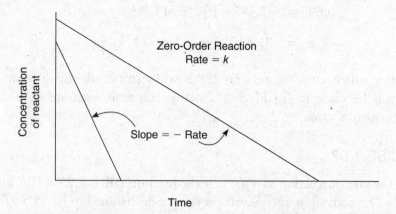

FIGURE 11.4. Kinetic curves of two different zero-order reactions, illustrating the straight-line relationship between concentration and time.

First-Order Reactions

Any rate law in which the sum of the exponents is 1 is a **first-order reaction.** There are a variety of ways in which a reaction can be first order, including the case where the exponents of two reactants are 0.5 each. In the most common case the concentration of one reactant, A, has an exponent of 1 as in the following rate law:

$$\text{Rate} = k[A] \tag{11.16}$$

As a first-order reaction progresses, the concentration of reactant A decreases and the rate decreases. When the concentration of A has decreased to half its original amount, the rate will be half the initial rate. The time required for this to occur is called the **half-life,** $t_{1/2}$. In another half-life the concentration will decrease by half again, to one-fourth of the initial concentration. Figure 11.5 illustrates the shape of the kinetic curve for all first-order reactions.

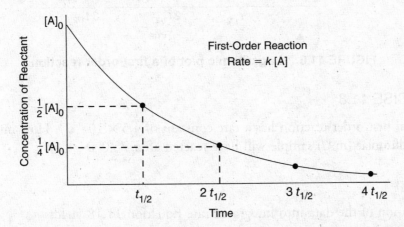

FIGURE 11.5.   Kinetic curve for all first-order reactions. The time at which the concentration is one-half the original concentration is the half-life, $t_{1/2}$.

THE INTEGRATED EQUATION

Using elementary calculus, we can integrate the first-order rate equation to give an equation that allows us to calculate concentrations at any time after the reaction has started:

$$\ln\left(\frac{[A]_0}{[A]_t}\right) = kt \tag{11.17}$$

or

$$\ln[A]_0 - \ln[A]_t = kt \tag{11.18}$$

Each form of the integrated equation contains four variables: the time, t; the rate constant, k; an initial concentration of A, $[A]_0$; and the concentration at some time after the reaction is started, $[A]_t$. Using Equation 11.18, chemists often plot the natural logarithm of the reactant concentration versus time in a graph as shown in Figure 11.6. In this type of graph, only a first-order reaction will result in a straight line, and the slope will be equal to $-k$.

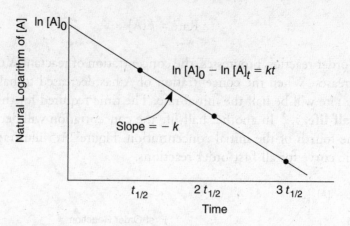

FIGURE 11.6.   Logarithmic plot of a first-order reaction.

EXERCISE 11.8

A certain first-order reaction has a rate constant of 4.5×10^{-3} s^{-1}. How much of a 50.0 millimolar (mM) sample will have reacted after 75.0 s?

Solution

Substitution of the data into integrated rate Equation 11.18 yields

$$\ln(0.0500) - \ln[A]_t = (4.5 \times 10^{-3}\ \text{s}^{-1})(75.0\ \text{s})$$

$$-2.995 - \ln[A]_t = 0.3375$$

$$\ln[A]_t = -3.333$$

$$[A]_t = 0.0356\ M$$

This indicates that the sample is now 35.6 mM. The question asks how much of the sample has reacted; the answer is

$$50.0\ \text{m}M - 35.6\ \text{m}M = 14.4\ \text{m}M$$

Half-Lives

> **TIP**
>
> All radioactive decay obeys first-order kinetics.

As mentioned previously, a first-order reaction has a constant half-life, which is the time required for half of the reactant to be consumed. When half of the reactant is used up, $[A]_t = 0.5\ [A]_0$. Substituting into Equation 11.17 yields

$$\ln\left(\frac{0.5[A]_0}{[A]_0}\right) = -kt_{1/2}$$

$$\ln(0.5) = -kt_{1/2}$$

$$-0.693 = -kt_{1/2}$$

$$t_{1/2} = \frac{0.693}{k} \qquad (11.19)$$

where $t_{1/2}$ is the half-life of the reaction. Half-lives may be used to quickly estimate the fraction of the starting material left after it has been allowed to react for a given number of half-lives.

EXERCISE 11.9

What percentage of a sample has reacted after six half-lives?

Solution

After each half-life, half of the sample present at the start of that half-life will be reacted. After one half-life, $\frac{1}{2}$ of the sample is left; after two half-lives, $\frac{1}{2} \times \frac{1}{2}$ or $\frac{1}{4}$ will remain; after three half-lives, $\frac{1}{2} \times \frac{1}{2} \times \frac{1}{2}$ or $\frac{1}{8}$ remains. After six half-lives $\frac{1}{2} \times \frac{1}{2} \times \frac{1}{2} \times \frac{1}{2} \times \frac{1}{2} \times \frac{1}{2} = \frac{1}{64}$ of the original is left, meaning that 63/64 has reacted. The percentage that has reacted is calculated as

$$\frac{63}{64} \times 100 = 98.4\% \text{ reacted}$$

The most prominent first-order processes in chemistry are those associated with radioactive decay of the elements. The decay of radioactive elements follows first-order kinetics, as shown in Chapter 3.

Second-Order Reactions

Two second-order rate laws are

$$\text{Rate} = k\,[A]^2 \tag{11.20}$$

$$\text{Rate} = k\,[A]\,[B] \tag{11.21}$$

In the special case where [A] = [B] the two rate laws can be integrated to obtain the same equation.

$$\frac{1}{[A]_t} - \frac{1}{[A]_0} = kt \tag{11.22}$$

The half-life of the second-order reactions specified above may be determined by substituting $0.5\,[A]_0$ for $[A]_t$ in Equation 11.21.

$$\frac{1}{0.5[A]_0} - \frac{1}{[A]_0} = kt_{1/2}$$

$$\frac{2}{[A]_0} - \frac{1}{[A]_0} = kt_{1/2}$$

$$\frac{1}{[A]_0} = kt_{1/2} \tag{11.23}$$

$$t_{1/2} = \frac{1}{k\,[A]_0} \tag{11.24}$$

The important information derived from these equations is that the half-life of a second-order reaction depends on the starting concentrations of the reactants. In addition, a straight-line graph will be obtained by plotting 1/[A] versus time. The slope of the line will be equal to *k*, as shown in Figure 11.7.

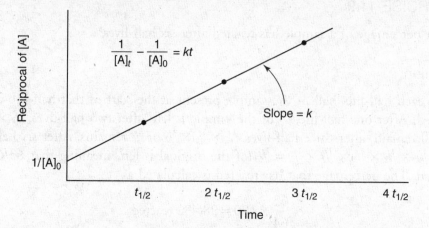

FIGURE 11.7. Plot of 1/[A] versus time, illustrating the straight-line relationship for a second-order reaction.

THEORY OF REACTION RATES

KINETICS THEORIES SUMMARIZED

COLLISION THEORY:
Collision theory states that the rate of a chemical reaction is equal to the collision rate (a very large number) decreased by multiplying by an orientation factor and a minimum energy factor. The actual reaction is visualized as the collision of hard spheres such as billiard balls.

TRANSITION-STATE THEORY:
Transition-state theory looks at the energy changes and geometric changes of molecules as they collide. During the collision process, kinetic energy is converted to potential energy. If this potential energy meets or exceeds the activation energy, the reaction can occur. In addition, there is change in the geometry of the reactants as they become products. The geometry somewhere in the middle of this conversion is called the transition state, and it occurs when the maximum kinetic energy has been converted to potential energy (i.e., at the top of the potential energy profile). The potential energy profile also indicates the heat of reaction.

Collision Theory

The **collision theory** states that the reaction rate is equal to the frequency of effective collisions between reactants. For a collision to be effective, the molecules must collide with sufficient energy and in the proper orientation so that products can form.

The minimum energy needed for a reaction is the activation energy, E_a. If two molecules collide head on, they will stop at some point and all of the kinetic energy will be converted into potential energy. If the molecules strike each other with a glancing blow, however, only part of the kinetic energy will be converted into potential energy. As long as the increase in potential energy is greater than E_a, a reaction is possible. The fraction of all collisions that have the minimum energy needed for reaction can be calculated using the **kinetic molecular theory** of gases discussed in Chapter 7. The fraction of collisions with this minimum energy increases with rising temperature since the average kinetic energy of molecules increases as temperature increases.

In addition to the energy requirement, for an effective collision the molecules must collide in the proper orientation. An example is the reaction of hydrogen iodide molecules with chlorine atoms. In this reaction the chlorine replaces the iodine in the molecule:

$$HI + Cl \rightarrow HCl + I$$

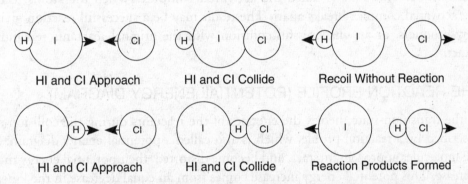

HI and Cl Approach   HI and Cl Collide   Recoil Without Reaction

HI and Cl Approach   HI and Cl Collide   Reaction Products Formed

FIGURE 11.8. Orientation needed for the reaction of hydrogen iodide with chlorine. The top row shows an ineffective collision; the bottom row, an effective collision, forming products.

Figure 11.8 shows hydrogen iodide colliding with a chlorine atom from two directions. When, as in the top row, chlorine collides with the iodide end of HI, the reactants recoil from the collision without a reaction occurring. In the other sequence the chlorine atom collides with the hydrogen end of HI. This collision can cause the iodine atom to be released while the chlorine bonds with the hydrogen. After collision the products HCl and I are present.

The overall reaction rate predicted by the collision theory may be summarized by the equation

$$\text{reaction rate} = N f_e f_o \qquad (11.25)$$

where N represents the number of collisions per second, which depends on the temperature and concentration of the reactants; f_e is the fraction of the collisions with the minimum energy; and f_o is the fraction of collisions with the correct orientation. The fraction f_e will increase as temperature increases, while f_o remains constant for a given reaction.

Transition-State Theory

The **transition-state theory** attempts to describe in detail the molecular configurations and energies as a collision of reactants occurs. This theory recognizes that, as molecules approach on a collision course, they do not act like billiard balls simply bouncing off each other. Instead, as the molecules get closer, their orbitals interact and distort each other. This distortion weakens bonds within the molecules so that at the moment of closest approach some bonds are so weak that they break and new bonds may form.

Using the diagram in Figure 11.8, we can visualize the effective collision in this sequence. First the chlorine approaches the hydrogen end of the HI molecule. As the Cl and HI get closer, the very electronegative Cl starts attracting the electrons that the hydrogen shares with the iodine atom. As a result the H—I bond is weakened and a H—Cl bond starts to form. At the moment of closest approach, the H—I bond is approximately half broken and the H—Cl bond is approximately half formed. This state is called the **activated complex.** When the atoms recoil, the activated complex breaks apart. The result may be a successful reaction giving new products, or an unsuccessful collision with the original reactants remaining intact.

THE REACTION PROFILE (POTENTIAL ENERGY DIAGRAM)

In the transition-state theory, the energies of the reactants during the collision are described by a reaction profile, which is also called a potential energy diagram. As the molecules approach, interact, and become distorted, their potential energy must increase. This potential energy increase comes from an equal decrease in the kinetic energy of the molecules. In other words, as the molecules collide, they slow down and their kinetic energy is converted to potential energy. After they reach the point of closest approach, they recoil and the potential energy is converted back to kinetic energy.

The **reaction profile** plots the increase in potential energy of the reactants as they approach, reaching a maximum at the moment of collision, and then the decrease in potential energy as the products recoil. A reaction profile is shown in Figure 11.9. In a reaction profile the minimum amount of kinetic energy that must be converted into potential energy in order to form products is called the activation energy, E_a. It is often referred to as an energy barrier between the reactants and the products.

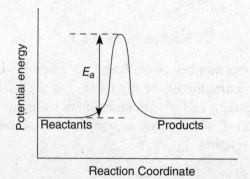

FIGURE 11.9. Reaction profile illustrating the energy barrier between reactants and products.

The transition-state theory is based on the same considerations as the collision theory. First, if reactants collide with enough energy to surmount the energy barrier, a reaction may occur. Second, if the activated complex formed at the moment of collision (top of the energy barrier) has the proper structure, it can proceed to fall apart into products. If it has the wrong structure, however, products cannot form and the molecules recoil as the original reactants. These energy and orientation factors are the same ones that are important to the collision theory.

The major difference in the two theories is that the collision theory views reactions as collisions between hard spheres, similar to the collisions between billiard balls. The transition-state theory, on the other hand, views the collisions as interactions between reactants that are deformed in the collision process. The transition-state theory involves more details about the energy and shapes of the molecules as they collide than the collision theory.

INTERPRETATION OF REACTION PROFILES

Reaction profiles provide a rich source of information about the rates of chemical reactions and are also a graphical view of the conversion of reactants into products. These graphs are often more informative than words alone in describing features of the reaction process.

In a reaction profile, the rate constant, and therefore the rate of a chemical reaction, are inversely related to the height of the energy barrier, E_a. When the activation energy is low, a large proportion of the collisions will have sufficient energy for a reaction to occur. Conversely, a high activation energy indicates that few collisions will have enough energy to convert reactants into products.

Reaction profiles can be used to determine whether a reaction is endothermic or exothermic. This is possible because the potential energy difference between the products and the reactants is equal to the heat of reaction, ΔH:

$$\Delta H = PE_{products} - PE_{reactants} \qquad (11.26)$$

When heat is absorbed from the surroundings, the reaction is endothermic and ΔH has a positive sign. For **endothermic reactions** the potential energy of the products is greater than the potential energy of the reactants. The reverse is true for **exothermic reactions.** The reaction profiles for these two cases are shown in Figure 11.10.

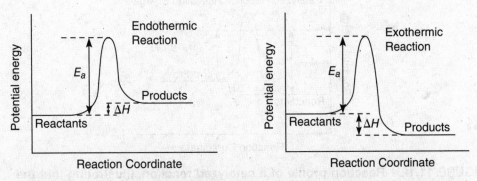

FIGURE 11.10. **Reaction profiles illustrating the difference between an endothermic reaction and an exothermic reaction.**

Another aspect of the reaction profile is that it allows the chemist to explain the reverse as well as the forward chemical reaction. To visualize what occurs in the reverse process, we look at the reaction profile, starting on the product side and proceeding toward the reactant side. When the products are reacting to form reactants, there is a different energy of activation and a different heat of reaction. The activation energy for the reverse reaction is the difference between the potential energy of the products and the maximum energy of the curve. For the heat of reaction, the sign of ΔH will be opposite to that for the forward reaction. Figure 11.11 is the same as Figure 11.10 except that it shows the activation energies and heats of reaction for the reverse reactions.

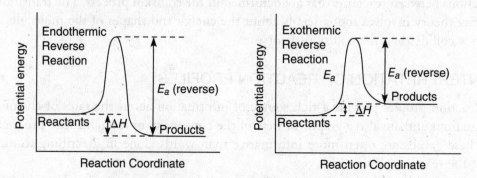

FIGURE 11.11. **Reaction profiles illustrating the activation energy of reverse reactions and the fact that ΔH of the reverse reaction is opposite in sign to ΔH of the forward reaction.**

Reaction profiles also allow us to explain the action of **catalysts** (Figure 11.12). As mentioned previously, a catalyst is a substance that increases the rate of a chemical reaction without itself being reacted. It speeds up a reaction by providing an alternative reaction pathway that has a lower energy barrier in the reaction profile. As a result the energies of activation of both the forward and reverse reactions are simultaneously decreased by the same amount, so that the reaction comes to chemical equilibrium more quickly. It must be kept in mind that a catalyst will not increase the amount of the product formed, nor will it alter the composition of the equilibrium mixture, as indicated by the unchanged potential energy plateaus for the reactants and products.

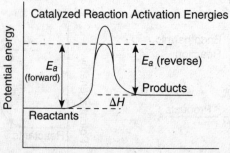

FIGURE 11.12. **Reaction profile of a catalyzed reaction, illustrating that the forward and reverse activation energies are both decreased. The heat of reaction is not affected, nor is the position of equilibrium because the potential energies of the reactants and products are not affected.**

An example of a catalyst is the platinum metal used in the hydrogenation of ethylene. Ethylene has a double bond to which two hydrogen atoms may be added to form ethane:

$$
\underset{\substack{|\\H}}{\overset{\substack{H\\|}}{C}} = \underset{\substack{|\\H}}{\overset{\substack{H\\|}}{C}} + H_2 \rightleftharpoons H - \underset{\substack{|\\H}}{\overset{\substack{H\\|}}{C}} - \underset{\substack{|\\H}}{\overset{\substack{H\\|}}{C}} - H
$$

A mixture of hydrogen and ethylene at room temperature does not show appreciable reaction, but addition of a small amount of finely divided platinum catalyst causes a very rapid reaction. The apparent reason is that the hydrogen molecule is fairly stable, and the energy required to break the hydrogen atoms apart results in a high activation energy. Platinum adsorbs hydrogen readily on its surface and the hydrogen molecule separates into individual hydrogen atoms. These hydrogen atoms then readily react with the ethylene. At the end of the process, the original platinum can be recovered and used again.

The discovery of new, more effective, and more specific catalysts is a major objective of many industrial chemists. Some catalysts are naturally occurring minerals whose properties are often discovered by trial and error. Others are synthetic compounds designed by studying natural catalysts and using chemical methods to improve upon them.

REACTION MECHANISMS

One of the most important uses of chemical kinetics is to decipher the sequence of steps that lead to an observed chemical reaction. Chemists write chemical equations for reactions as a single step. However, most chemical reactions occur in a series of steps called elementary reactions. All of the elementary reactions in a mechanism must add up to the overall balanced equation. The complete sequence of steps is called a **reaction mechanism.**

Elementary Reactions

In a mechanism, the **elementary reactions** usually involve either one molecule breaking apart or the collision of only two reactant molecules. It is extremely rare that three molecules will collide simultaneously, and the simultaneous collision of more than three molecules is so infrequent that such collisions are hardly ever considered. Collisions between just two molecules, however, occur millions of times each second. Therefore, the most probable elementary reaction is one in which only two molecules collide. Another feature of the elementary reaction is that its **coefficients are the exponents** used in the rate law.

A sequence of elementary reactions will always have one step that is slower than all the rest. This slowest step determines the overall rate of reaction and is called the **rate-determining step** or the **rate-limiting step.** Since chemists can measure only the overall reaction rate in the laboratory, they are in fact measuring the rate of the slowest, rate-determining elementary reaction in the mechanism.

Thus the rate law determined for a reaction is directly related to the rate-determining step.

Using these principles, chemists determine the rate law for a chemical reaction. They then postulate a series of elementary reactions based on the fact that one of the steps in the mechanism must obey the experimental rate law. When two or more mechanisms satisfy the rate law requirement, additional experiments must be done to decide which mechanism is correct.

To illustrate the procedure, we consider the reaction

$$H_2 + 2ICl \rightarrow I_2 + 2HCl \tag{11.27}$$

which has rate law

$$Rate = k[H_2][ICl] \tag{11.28}$$

A two-step mechanism that satisfies this rate law is

$$H_2 + ICl \rightarrow HI + HCl \tag{11.29}$$

$$HI + ICl \rightarrow I_2 + HCl \tag{11.30}$$

The two steps in this mechanism, Equations 11.29 and 11.30, add up to the overall reaction. They also involve only two molecules as reactants in each elementary reaction. The rate law describes the first step of the mechanism, which is presumably the slow step. Chemists can easily demonstrate that the second step is a much faster reaction by reacting HI and ICl.

The HI in the above mechanism does not appear in the balanced equation, Equation 11.27. Any chemical species that is part of a mechanism but not part of the balanced equation is called an **intermediate.** In this example, the intermediate HI is also a known compound that made possible the experimental verification of the mechanism. In other cases intermediates are very unstable and often exotic chemical species. Demonstrating the presence of an intermediate provides evidence to support one mechanism over another.

Considering the above example, we see that there is another mechanism that might also be considered. This is a three-step process:

$$H_2 + ICl \rightarrow HI + HCl \tag{11.31}$$

$$H_2 + ICl \rightarrow HI + HCl \tag{11.32}$$

$$2HI \rightarrow H_2 + I_2 \tag{11.33}$$

When added, these three reactions give the same overall reaction. To decide whether this mechanism is possible, scientists determined the rate of decomposition of the intermediate HI to H_2 and I_2 (the last step, Equation 11.33, in the mechanism). This reaction was found to be much slower than the reaction between H_2 and ICl. It was concluded that this is not the correct mechanism since the rate-determining step would give a completely different rate law.

From this description we can see that a kinetic study of a reaction will determine the rate law for the slowest step in the mechanism. Possible mechanisms are proposed, and the correct mechanism must have one step that obeys the observed rate law. If there are still several possible mechanisms, appropriate experiments must be devised to decide which mechanism is correct.

Determining the Rate Laws for Elementary Reactions

The coefficients of the reactants in an elementary reaction are the exponents of the reactant concentrations in the rate law. Therefore the rate law for each step of a mechanism can be predicted directly. One complication is that, after the first step, most of the elementary reactions in a mechanism will have an intermediate as a reactant. To compare an experimental rate law with the rate laws of the elementary reactions, it will be necessary to convert the rate law that has an intermediate into one that has only reactants.

For example, the reaction

$$2NO + O_2 \rightarrow 2NO_2 \tag{11.34}$$

has a possible mechanism consisting of the two elementary reactions

$$NO + O_2 \rightarrow NO_3 \tag{11.35}$$

$$NO_3 + NO \rightarrow 2NO_2 \tag{11.36}$$

The rate law for the first step is

$$\text{Rate} = k[NO][O_2] \tag{11.37}$$

It contains the reactants of the experiment. The rate law for the second step includes the intermediate NO_3:

$$\text{Rate} = k[NO_3][NO] \tag{11.38}$$

To eliminate the NO_3 and to convert it into one of the reactants, we use the **steady-state assumption.** It says that if the second step is the rate-determining step, the first reaction, Equation 11.35, must be relatively fast and reversible. This means that the rate at which NO_3 is formed is equal to the rate at which it disappears:

$$NO_3 \text{ rate formation} = NO_3 \text{ rate disappearance} \tag{11.39}$$

The rate laws governing the forward and reverse reactions in the first step are

$$\text{Rate}_{\text{forward}} = k_f[NO][O_2] \tag{11.40}$$

$$\text{Rate}_{\text{reverse}} = k_r[NO_3] \tag{11.41}$$

Since the forward and reverse rates are equal, we can write

$$k_f[NO][O_2] = k_r[NO_3] \tag{11.42}$$

Solving for $[NO_3]$ gives

$$[NO]_3 = \frac{k_f}{k_r}[NO][O_2] \qquad (11.43)$$

Substituting this result for the intermediate, NO_3, in the rate law gives

$$Rate = k\left(\frac{k_f}{k_r}\right)[NO][O_2][NO] \qquad (11.44)$$

By combining all k, k_f, and k_r rate constants, we can write the rate law for the second step of the mechanism:

$$Rate = k[NO]^2[O_2] \qquad (11.45)$$

EXERCISE 11.10

The kinetics of the following reaction is studied:

$$2NO + O_2 \rightarrow 2NO_2$$

Two possible mechanisms are

$$2NO \rightarrow N_2O_2$$
$$N_2O_2 + O_2 \rightarrow 2NO_2$$

and

$$NO + O_2 \rightarrow NO_3$$
$$NO_3 + NO \rightarrow 2NO_2$$

Describe the method you would use to determine which mechanism is correct.

Solution

The rate laws for each of these elementary reactions can be determined. For the first mechanism they are

$$Rate_1 = k[NO]^2$$

$$Rate_2 = k[N_2O_2][O_2]$$

Using the steady-state assumption to obtain the rate law of the second elementary reaction in terms of measurable reactants, we obtain

$$k_f[NO]^2 = k_r[N_2O_2]$$

and the rate law will be

$$Rate_2 = k\,[NO]^2\,[O_2]$$

For the second mechanism the rate laws are

$$Rate_1 = k\,[NO][O_2]$$

$$Rate_2 = k\,[NO]^2\,[O_2]$$

The second rate law was derived above.

We can see that, if the first step is the slow step, the two mechanisms give two distinctly different rate laws and the decision is clear cut. If the second step is the slow step, however, both mechanisms yield the same rate law. To determine which mechanism is correct, additional experiments must be designed to identify the intermediate, NO_3 or N_2O_2, that is formed during the reaction.

SUMMARY

In order to get to the equilibrium state described in Chapter 10, substances react at a finite rate. As opposed to equilibrium, where the overall concentrations do not change with time, the topic of kinetics in this chapter was all about change. The study of kinetics starts with the determination of reaction rates. Next, well-designed experiments are used to determine rate laws that express how concentrations affect reaction rates. Two special cases, first-order and second-order integrated equations, were discussed In addition to concentration, temperature, the ability of reactants to meet, and the presence of a catalyst also affect the reaction rate. Two theories are used to explain reaction rates. One is the collision theory, and the other is the transition-state theory. These theories are part of virtually every AP Exam.

> **TIP**
>
> Data about reaction kinetics cannot be used to make conclusions about chemical equilibrium. Similarly, the value of an equilibrium constant cannot be used to draw conclusions about reaction rates.

Important Concepts

Reaction rates
Rate laws
Order of reaction
Half-lives
Collision theory
Transition-state theory and reaction profiles
Arrhenius equation

Reaction mechanisms
Elementary processes
Rate-determining step
Intermediate catalyst

Important Equations

$Rate = k\,[A]^x\,[B]^y \cdots$
$t_{1/2} = 0.693/k$

$$\ln\left(\frac{[A_0]}{[A_t]}\right) = kt \text{ for a first-order reaction}$$

$$\ln\left(\frac{k_1}{k_2}\right) = \frac{-E_a}{R}\left(\frac{1}{T_1} - \frac{1}{T_2}\right)$$

Practice Exercises

Multiple-Choice

For the first four problems below, one or more of the following responses applies; each response may be used more than once or not at all in these questions.

I. activation energy
II. orientation
III. potential energy curve
IV. frequency
V. activated complex

1. The heat of a reaction is best deduced from the

(A) I
(B) II
(C) III
(D) IV
(E) V

2. The collision theory involves

(A) I and III
(B) II
(C) II and IV
(D) IV
(E) I, III, and V

3. The transition-state theory involves

(A) I and III
(B) III
(C) III and V
(D) IV
(E) I, III, and V

4. The rate of a chemical reaction is related to

(A) I and III
(B) II
(C) I, II, and IV
(D) IV
(E) I, III, and V

5. The activated complex may be described as

(A) an elementary reaction in a mechanism
(B) the shape of the molecules at the top of the potential energy diagram
(C) the shape of the reaction product
(D) the phase—liquid, solid, or gas—in which a reaction takes place
(E) a coordination state

6. A reaction in which the rate and the rate constant have the same units is

   (A) a radioactive decay
   (B) a second-order reaction
   (C) a reaction with a one-step mechanism
   (D) a first-order reaction
   (E) a zero-order reaction

Questions 7–9 refer to the following diagram:

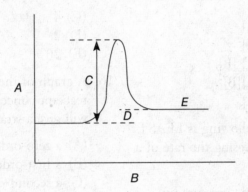

7. In the reaction profile, A, B, and C should be labeled as shown in

	A	B	C
(A)	potential energy	reaction coordinate	activation energy
(B)	heat of reaction	reaction coordinate	potential energy
(C)	potential energy	reaction coordinate	heat of reaction
(D)	heat of reaction	potential energy	activation energy
(E)	activation energy	extent of reaction	heat of reaction

8. In the reaction described by the reaction profile,

   (A) forward E_a > reverse E_a and ΔH is exothermic
   (B) reverse E_a > forward E_a and ΔH is endothermic
   (C) forward E_a < reverse E_a and ΔH is exothermic
   (D) reverse E_a < forward E_a and ΔH is endothermic
   (E) reverse E_a = forward E_a and ΔH is zero

9. Addition of a catalyst to the reaction mixture will affect only

   (A) A
   (B) B
   (C) C
   (D) D
   (E) E

10. A fast reaction should have

    (A) a high activation energy
    (B) a catalyst present
    (C) a large equilibrium constant
    (D) a low activation energy
    (E) an exothermic heat of reaction

11. Which of the following rate laws has a rate constant with units of $L^2 mol^{-2} s^{-1}$?

    (A) Rate = k [A]
    (B) Rate = k [A]^2
    (C) Rate = k [A] [B]
    (D) Rate = k [A][B]^2
    (E) Rate = k [A]^0

12. Which of the following is LEAST effective in increasing the rate of a reaction?

    (A) increasing the pressure by adding an inert gas
    (B) grinding a solid reactant into small particles
    (C) increasing the temperature
    (D) eliminating reverse reactions
    (E) adding a catalyst

13. A first-order reaction has a half-life of 36 min. What is the value of the rate constant?

    (A) $3.2 \times 10^{-4} s^{-1}$
    (B) $1.9 \times 10^{-3} L mol^{-1} s^{-1}$
    (C) $1.2 s^{-1}$
    (D) $0.028 s^{-1}$
    (E) $9.3 \times 10^{-4} L mol^{-1} s^{-1}$

14. If a reactant's concentration is doubled and the reaction rate increases by a factor of 8, the exponent for that reactant in the rate law should be

    (A) 0
    (B) 1
    (C) 2
    (D) 3
    (E) ½

15. In general, if the temperature of a reaction is raised from 300 K to 320 K, the reaction rate will increase by a factor of approximately

    (A) $\dfrac{320 K}{300 K}$
    (B) $\dfrac{22 °C}{2 °C}$
    (C) 4
    (D) 2
    (E) 20

16. A graph of the reciprocal of reactant concentration versus time will give a straight line for

    (A) a zero-order reaction
    (B) a first-order reaction
    (C) a second-order reaction
    (D) both (A) and (C)
    (E) (A), (B), and (C)

17. The rate at which CO_2 is produced in the following reaction:

    $2C_6H_6(g) + 15O_2(g) \rightarrow 12CO_2(g) + 6H_2O(l)$

    is 2.2×10^{-2} mol $L^{-1} s^{-1}$. What is the rate at which O_2 is consumed?

    (A) 2.2×10^{-2} mol $L^{-1} s^{-1}$
    (B) 1.3×10^{-1} mol $L^{-1} s^{-1}$
    (C) 2.8×10^{-2} mol $L^{-1} s^{-1}$
    (D) 1.8×10^{-3} mol $L^{-1} s^{-1}$
    (E) -2.2×10^{-2} mol $L^{-1} s^{-1}$

18. Which of the following will be most helpful in determining the stability or shelf life of a new drug?

    (A) the reaction mechanism for its decomposition
    (B) the rate law for its decomposition
    (C) the Arrhenius plot of the decomposition reaction
    (D) the integrated rate law plot
    (E) the overall chemical reaction

19. An Arrhenius plot is a graph of

    (A) the rate constant versus concentration

    (B) the natural logarithm (ln) of the rate constant versus concentration

    (C) the reciprocal of the rate constant versus ln T

    (D) the rate constant versus ln $(1/T)$

    (E) the natural logarithm (ln) of the rate constant versus $1/T$

20. A first-order reaction has a half-life of 85 s. What fraction of the reactant is left after 255 s?

    (A) $\frac{1}{2}$
    (B) $\frac{1}{4}$
    (C) $\frac{1}{8}$
    (D) $\frac{1}{3}$
    (E) $\frac{7}{8}$

21. A rate law is found to be

    $$\text{Rate} = k[A]^2[B]$$

    The order of the reaction is

    (A) first order
    (B) second order
    (C) third order
    (D) fourth order
    (E) The order of the reaction cannot be determined.

22. A rate law is found to be

    $$\text{Rate} = k[A]^2[B]$$

    Which of the following actions will NOT change the initial reaction rate?

    (A) doubling the concentrations of both A and B

    (B) doubling the concentration of A and halving the concentration of B

    (C) halving the concentration of A and doubling the concentration of B

    (D) halving the concentration of A and quadrupling the concentration of B

    (E) doubling the concentration of A and quadrupling the concentration of B

23. What is the half-life of a reaction that has a first-order rate constant of 2.6×10^{-4} s^{-1}?

    (A) 2.7×10^3 s
    (B) 1.3×10^{-4} s
    (C) 1.2×10^3 s
    (D) 7.7×10^3 s
    (E) 1.3×10^{-4} s

24. Modern automobiles use a catalytic converter to

    (A) increase horsepower by burning more gasoline

    (B) absorb pollutants from the exhaust

    (C) complete the combustion of unburned gases

    (D) cool the exhaust gases

    (E) convert pollutants into water

25. A catalyst will NOT

(A) increase the forward reaction rate

(B) shift the equilibrium to favor the products

(C) alter the reaction pathway

(D) increase the speed at which equilibrium will be achieved

(E) increase the reverse reaction rate

ANSWER KEY

1. C	6. E	11. D	16. C	21. C
2. C	7. A	12. A	17. C	22. D
3. E	8. D	13. A	18. C	23. A
4. C	9. C	14. D	19. E	24. C
5. B	10. D	15. C	20. C	25. B

See Appendix 1 for explanations of answers.

Free-Response

(a) Sketch and label a potential energy diagram (reaction profile) and discuss its implications for chemical reactions. How does the addition of a catalyst alter the diagram? What effect will a change in temperature have?

(b) Describe the collision theory.

(c) Describe the transition-state theory.

(d) How are rate laws and reaction mechanisms related?

(e) Use the table of data below to determine the rate law for the reaction

$$3F + 2H \rightarrow P + 2S$$

Experiment Number	Concentration F (mol/L)	Concentration H (mol/L)	Initial Reaction Rate (mol/L s)
1	0.000345	0.000765	3.24×10^{-8}
2	0.000690	0.000765	3.24×10^{-8}
3	0.000537	0.00765	3.24×10^{-7}

(f) Calculate the rate constant and give its units.

ANSWERS

(a) This figure can be similar to any of the figures from Figure 11.9 to Figure 11.12. The diagrams illustrate how kinetic energy is converted to potential energy during a collision. If enough kinetic energy is converted to potential energy, the reactants can move through the transition state to products. The energy barrier prevents the formation of products if insufficient energy is available. This figure also diagrams an activation energy for the forward reaction and a different activation energy for the reverse reaction. The difference in potential energy from reactants to products is a measure of the heat of reaction.

(b) Collision theory treats reactions as collisions between hard spheres such as billiard balls. Knowing the temperature and concentration, it is possible to calculate the number of collisions gas molecules undergo each second. Very few form products because of two factors. The first is the energy factor where minimum energy is required. The second factor is the orientation factor. Molecules must collide with the correct orientation as well as energy for a reaction to take place.

(c) The transition-state theory starts where the collision theory ends. This theory describes how colliding molecules slow down, converting kinetic energy to potential energy. In a direct collision the molecules eventually stop (the maximum KE is converted to PE) and start to rebound. If the top of the PE diagram has been exceeded, the molecules can recoil as products. If the collision does not exceed the activation energy, the reactants recoil unchanged. The transition-state theory also looks at dynamic changes in shape and bonding as the collision progresses.

(d) The rate law expresses, in terms of the initial reactants, the rate-limiting step of the reaction. A reaction mechanism is a sequence of elementary processes (uni- or bimolecular reactions) that adds up to the overall stoichiometry and also indicates the individual steps that a reaction uses. All reactions have a rate-limiting step, and the rate law must match the proposed mechanism's rate-limiting step and reaction stoichiometry.

(e) In the first two experiments the concentration of H is held constant while the concentration of F is doubled. However, the reaction rate does not change. We must conclude that the exponent is zero and that the reaction is zero order with respect to F. We now know that F does not matter. In Experiment 3 the concentration is ten times the concentration in the first two experiments. At the same time the rate for Experiment 3 is ten times the rates for Experiments 1 and 2. We conclude that the exponent for H is 1 or the reaction is first order with respect to H. The rate law is Rate = k[H].

(f) The rate constant is calculated as k = Rate/[H] for all three experiments, and we get $k = 4.24 \times 10^{-5}$ s^{-1}.

Thermodynamics

CHAPTER 12

- Thermodynamics
- Definition of System
- Definition of Surroundings
- Definition of State Function
- Definition of Standard States
- Potential Energy
- Kinetic Energy
- Specific Heat
- Calorimeters

- First Law of Thermodynamics
- Internal Energy, ΔE
- Heat and Work
- Enthalpy, ΔH
- Heat of Formation
- Entropy, ΔS
- Second Law of Thermodynamics
- Spontaneous Reactions
- Free Energy, ΔG

ESSENTIAL DEFINITIONS

In the discussion of thermodynamics precise terminology is used to avoid confusion and to ensure that experimental results may be compared from one laboratory to another. Some of the important terms defined in this chapter are *system*, *state function*, *standard state*, and the *exo-* and *endo-* prefixes.

System

A **system** is that part of the universe that is under study. Everything else in the universe is called the **surroundings.** When hydrogen and oxygen are placed in a bomb calorimeter to study the formation of water, all of the hydrogen and oxygen atoms and the calorimeter are the system. The surrounding laboratory, the building, the city, and so on are parts of the surroundings.

There are several types of systems. An **open system** can transfer both energy and matter to and from the surroundings. An open bottle of perfume is an example of an open system. A **closed system** is one where energy can be transferred to the surroundings but matter cannot. A well-stoppered bottle of perfume is a closed system. In an **isolated system** there is no transfer of energy or matter to or from the surroundings. A thermos bottle is a close approximation of an isolated system since it minimizes heat transfer to the surroundings. Calorimeters are designed to be as close as possible to a closed system.

State Function

In this chapter we will encounter enthalpy change, ΔH; entropy change, ΔS; free-energy change, ΔG; and energy change, ΔE. The numerical values and mathematical signs of these **state functions** depend only on the difference between the final state and the initial state of the system.

The state of a system is defined by the mass and phase (solid, liquid, or gas) of the matter in the system, as well as the temperature and pressure of the system. For the melting of 1 mole of ice, the initial state is described as 18 g $H_2O(s)$ at 1.00 atm pressure and 273 K, and the final state as 18 g $H_2O(l)$ at 1.00 atm pressure and 273 K. With this precise description all scientists should, within experimental error, obtain identical values of ΔH, ΔS, ΔG, and ΔE for the melting of ice.

Two quantities that are *not* state functions are heat (q) and work (w). The values for these quantities depend on the sequence of steps used to transform matter from the initial state to the final state.

Standard State

The thermodynamic quantities ΔH, ΔS, ΔG, and ΔE are **extensive properties** of matter, meaning that they change as the amount of sample changes. To make these quantities **intensive properties** of matter, we must define precisely the temperature, pressure, mass, and physical state of the substance. A system is in the **standard state** when the pressure is 1 atmosphere, the temperature is 25 °C, and 1 mole of compound is present. When the thermodynamic quantities are determined at standard state, they are intensive properties and a superscript 0 is added to their symbols: as $\Delta H°$, $\Delta S°$, $\Delta G°$ and $\Delta E°$. For a chemical reaction, the standard state involves the number of moles designated by the stoichiometric coefficients in the simplest balanced chemical equation with the smallest possible whole-number coefficients.

Exo- and *Endo-* Prefixes and Sign Conventions

The prefix *exo-*, as in *exothermic*, indicates that energy is being lost from the system to the surroundings. Mathematically, the prefix *exo-* corresponds to a negative sign for numerical thermodynamic quantities. For an exothermic reaction the heat of reaction, ΔH, is a negative number.

The prefix *endo-* indicates that energy is gained from the surroundings. An *endothermic* reaction absorbs heat energy and appears to cool as it progresses. For an endothermic reaction the heat of reaction, ΔH, is a positive number.

TYPES OF ENERGY

Energy takes many forms. Heat and light, along with chemical, nuclear, electrical, and mechanical energy, are some of the common types. Any one of these forms of energy can be converted into any of the other forms. In addition, the **law of conservation of energy** states that energy is never created or destroyed. As a result of these properties, all forms of energy can be converted into heat energy, which we can measure in a calorimeter as described below.

Energy can also be categorized as either kinetic energy, KE, or potential energy, PE. **Kinetic energy** is the energy that matter possesses because of its motion. There is one equation to describe kinetic energy:

$$KE = \frac{1}{2}mv^2 \tag{12.1}$$

When the mass, m, is expressed in kilograms and the velocity, v, in meters per second, the energy units are joules.

Potential energy is stored energy, which may be released under the appropriate conditions, as in a nuclear reaction. There are several forms of potential energy, such as **gravitational energy** and the energy of **electrostatic attraction** between oppositely charged ions. Each form of potential energy is described by its own equation, but these equations are similar to each other. They all have the forms

$$PE_{grav} = K_{grav}\left(\frac{m_1 m_2}{r}\right) \quad \text{and} \quad PE_{elect} = K_{elect}\left(\frac{q_1 q_2}{r}\right) \tag{12.2}$$

The two masses, m, in gravitational attraction and the two charges, q, in electrostatic attraction are separated by a distance, r. K is a proportionality constant that is different for each type of potential energy.

The total energy of a substance is the sum of its kinetic and potential energies.

$$\text{Energy (E)} = \text{potential energy (PE)} + \text{kinetic energy (KE)} \tag{12.3}$$

In chemical substances, the kinetic energy is the motion of the molecules. The potential energy of a chemical is the sum of all attractions, including all the covalent bonds, ionic bonds, or electrostatic attractions in the substance.

MEASUREMENT OF ENERGY

Heat is often called the "lowest form" of energy. In this form it is easily measurable by determining temperature changes caused by the release or absorption of heat in a chemical process.

Specific Heat

Heat energy was originally defined in terms of the calorie, which is the amount of heat needed to raise the temperature of 1 gram of pure water from 14.5 to 15.5 °C. The joule is the metric unit for energy; 1 calorie is equal to exactly 4.184 joules. By virtue of these definitions, 4.184 joules of energy is needed to raise the temperature of 1 gram of water by 1 degree Celsius. This quantity is known as the **specific heat** of water:

$$\text{Specific heat of water} = 4.184 \text{ J g}^{-1}\,°\text{C}^{-1} \tag{12.4}$$

Once the specific heat of water is defined, the specific heat of any other substance can be determined. One method is to immerse a hot object in a known quantity of water and then measure the temperature change that occurs. This method is illustrated in Exercise 12.1.

As an interesting sidelight, in 1818 Pierre Dulong and A.T. Petit discovered that for most metals the specific heat multiplied by the atomic mass of the metal was equal to a constant. The **Dulong and Petit law** is as follows:

$$\text{Specific heat} \times \text{molar mass} \approx 25 \text{ J mol}^{-1}\,°\text{C}^{-1} \tag{12.5}$$

This law helped confirm the molar masses of the elements when disagreements occurred. In addition, specific heat is an intensive physical property of all elements and compounds and can be used to identify substances. In more advanced physical chemistry courses, you will find that this law is predicted by thermodynamics.

Equation 12.6 enables chemists to determine the heat energy of any process by measuring the change in temperature of a known mass of water. The equation is as follows:

$$\text{Heat energy} = (\text{Specific heat})(\text{Mass})(\text{Temperature change})$$
$$q = \text{sp. ht.} \times g \times \Delta T \qquad (12.6)$$

In this equation q, the heat energy, is expressed in joules. The temperature change is always determined as the final temperature minus the initial temperature ($°C_{final} - °C_{initial}$). Equation 12.6 is applicable to any substance, not just water.

There is a distinct difference between heat energy and temperature, as Equation 12.6 indicates. Temperature is a measure of the average kinetic energy of a group of atoms, and a temperature change is a change in the average kinetic energy. Heat energy is produced when both the kinetic energy and the energy of attractions between the atoms in the group change in a chemical or physical process.

EXERCISE 12.1

An insulated cup contains 75.0 g of water at 24.00 °C. A 26.00 g sample of a metal at 85.25 °C is added. The final temperature of the water and metal is 28.34 °C.

(a) What is the specific heat of the metal?
(b) According to the law of Dulong and Petit, what is the approximate molar mass, *MM*, of the metal?
(c) What is the apparent identity of the metal?

Solution

(a) The law of conservation of energy requires that the heat energy gained by the water be exactly equal to the heat energy lost by the metal as it cools in the water. Mathematically this is written as

$$+q_{water} = -q_{metal}$$

Using Equation 12.6, we expand this equation:

$$(\text{sp. heat}_{H_2O})(g_{H_2O})(\Delta T_{water}) = -(\text{sp. heat}_{metal})(g_{metal})(\Delta T_{metal})$$

Entering the data yields

$$4.184\,\text{J}\,\text{g}^{-1}\,°C^{-1}(75.0\,\text{g})(4.34\,°C) = -(\text{sp. heat}_{metal})(26.00\,\text{g})(-56.91\,°C)$$
$$1362\,\text{J} = 1480\,\text{g}\,°C(\text{sp. heat}_{metal})$$
$$\text{sp. heat}_{metal} = 0.920\,\text{J}\,\text{g}^{-1}\,°C^{-1}$$

(b) Using Equation 12.5 for the law of Dulong and Petit we obtain

$$0.920 \text{ J g}^{-1} \text{ °C}^{-1}(\text{molar mass}) = 25 \text{ J mol}^{-1} \text{ °C}^{-1}$$

$$\text{molar mass} = 27 \text{ g mol}^{-1}$$

(c) Aluminum is the only metal with a molar mass of 27 g mol^{-1}.

Calorimeters and Heat Measurements

A **calorimeter** is the device used for measuring the heat energy produced by chemical reactions and physical changes. Figure 12.1 illustrates the design features of a calorimeter. The reaction vessel, usually made of metal, efficiently transfers heat to the rest of the apparatus, which includes a large mass of water in an insulated container that prevents heat from being lost to the surroundings. Also included are a stirrer and a very accurate thermometer. All parts of the calorimeter heat up or cool down as heat is released or absorbed in the chemical process. Water is the major part of the system; however, for the most accurate measurements the specific heats and masses of the reaction vessel, the thermometer, the stirrer, and the container itself must be included in the calculation.

The total **heat capacity**, C, of the calorimeter, sometimes called the calorimeter constant, is defined as the sum of the products of the specific heat and the mass of all components of the calorimeter:

$$C \text{ (J °C}^{-1}) = (\text{sp. ht.})_1 (\text{mass})_1 \quad + \quad (\text{sp. ht.})_2 (\text{mass})_2$$
$$+ \quad (\text{sp. ht.})_3 (\text{mass})_3 \quad + \cdots \quad (12.7)$$

The heat energy produced in the calorimeter is then calculated as

$$q = C \left(°C_{final} - °C_{initial} \right) \quad\quad (12.8)$$

where C replaces the (sp. ht. × g) terms in Equation 12.6.

The most convenient method for determining the heat capacity, C, of a calorimeter is to calibrate the device, using a reaction that will produce a known amount of heat, and then calculate C from the observed temperature change using Equation 12.8.

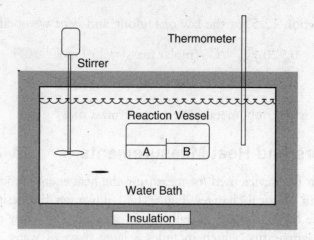

FIGURE 12.1. **Diagram of a calorimeter and its essential components. Reactants A and B may be mixed by rotating the reaction vessel. In a bomb calorimeter the reaction vessel is filled with reactant and O$_2$, which are then ignited with an electric spark.**

EXERCISE 12.2

A calorimeter has a heat capacity of 1265 J °C^{-1}. A reaction causes the temperature of the calorimeter to change from 22.34 °C to 25.12 °C. How many joules of heat were released in this process?

Solution

The heat released in a calorimeter is given in Equation 12.8. We calculate ΔT as 25.12 − 22.34 = +2.78 °C. Entering these data into the equation gives

$$q = 1265 \text{ J °C}^{-1} \text{ (2.78 °C)}$$

$$= 3.52 \times 10^3 \text{ J of heat energy is released}$$

FIRST LAW OF THERMODYNAMICS

The **first law of thermodynamics** states that energy is always conserved. In chemistry this law means that the measurable quantities heat (q) and work (w) must add up to the total energy change in a system:

$$\Delta E = q + w \tag{12.9}$$

The value of q has a positive sign if heat is added to the system. The value of w is positive if work is done on the system. Similarly, q is a negative value if heat is released from the system, and w is a negative value if work is done by the system.

If the system cools, q has a positive sign and the process is said to be endothermic. If the system heats up, it is exothermic and q has a negative sign. As we will see later, work is equal to the pressure times the change in volume, $P \Delta V$. If the volume increases, work is done by the system and w has a negative sign. When work is done on the system, the volume decreases and w has a positive sign.

The change in energy of a system, at constant temperature, is also the difference in potential energy between the final and initial states of the system:

$$\Delta E = PE_{final} - PE_{initial} \tag{12.10}$$

As we will see later, ΔE is mainly heat energy. Therefore a system that increases its potential energy is often said to be endothermic, while a decrease in potential energy indicates an exothermic process.

Work

In Equation 12.9 the energy change is the sum of the heat and the work. Heat is measured using a calorimeter, and work is also easily measured. **Work** is defined as the force applied to an object as it moves a certain distance:

$$\text{Work} = \text{Force} \times \text{Distance moved} \tag{12.11}$$

Force can be defined as the pressure exerted over a given area, so

$$\text{Work} = \text{Pressure} \times \text{Area} \times \text{Distance moved}$$

Multiplying the area by the distance results in volume units or an overall volume change:

$$\text{Work} = \text{Pressure} \times \text{Volume change}$$
$$w = P\Delta V \tag{12.12}$$

Work is the product of the pressure and the change in volume that occurs during a chemical reaction.

EXERCISE 12.3

Demonstrate that work is not a state function by calculating the work involved in expanding a gas from an initial state of 1.00 L and 10.0 atm of pressure to (a) 10.0 L and 1.0 atm pressure, (b) 5.00 L and 2.00 atm and then to 10.0 L and 1.00 atm pressure.

Solution

(a) $w = -P\Delta V = -(1.00 \text{ atm})(10.0 \text{ L} - 5.00 \text{ L})$
$= -9.00 \text{ L atm}$

(b) $w = -P\Delta V = -(2.00 \text{ atm})(5.0 \text{ L} - 1.00 \text{ L}) - (1.00 \text{ atm})(0.0 \text{ L} - 5.00 \text{ L})$
$= -13.0 \text{ L atm}$

In both parts of this exercise the sample starts at the same state (1.00 L and 10.0 atm) and ends in the same state (10.0 L and 1.00 atm). However, the work in units of liter atmospheres is different. This can occur only if w is not a state function.

Use of *R* for the Conversion of Energy Units

We have different energy units—calories, joules, and the newly introduced liter-atmosphere. They are all conveniently related through the universal gas law constant, *R*. To make these conversions let us see how to convert L-atm to joules for the last problem.

$$?\text{joules} = -13.0 \text{ l atm}$$

We will divide by the *R* that has units of L-atm and multiply by the *R* that has units of joules as follows:

$$?\text{joules} = -13.0 \text{ L atm} \left(\frac{\text{mol K}}{0.0821 \text{ L-atm}} \right) \left(\frac{8.31 \text{ J}}{\text{mol K}} \right)$$

After canceling units we end up with the desired joules and -13.0 L-atm becomes -1.32×10^3 J. We use a similar approach to convert to or from other energy units such as electronvolts (EV) or calories.

Definition of q_p, q_v, ΔE, and ΔH

The first law of thermodynamics may be rewritten as

$$\Delta E = q_p - P\Delta V \tag{12.13}$$

The minus sign enters this equation because an increase in volume means that the system does work on the surroundings, and such work has been defined as a negative quantity. The heat term, q, is given the subscript p to indicate that the pressure must be constant.

When the heat energy is measured in a calorimeter that does not allow the volume to change, $P\Delta V$ must be zero. As a result $\Delta E = q_v$, where the subscript v indicates that the volume is held constant. Calorimeters that do not allow the volume to change are called **bomb calorimeters,** a name derived from the heavy stainless steel reaction vessel, which was known to explode if not used correctly.

For most real reactions chemists are interested in the heat generated at constant pressure, q_p. This variable is called the enthalpy change and is given the symbol ΔH. **Enthalpy,** H, is the heat content of a substance, and ΔH is the difference in heat content of the products and reactants:

$$\Delta H = H_{\text{products}} - H_{\text{reactants}} \tag{12.14}$$

Using Equations 12.9 and 12.12, we can write the relationship between ΔE and ΔH as

$$\Delta E = \Delta H - P\Delta V \tag{12.15}$$

For many reactions, the value of ΔH is very large and the value of $P\Delta V$ is relatively small, so that ΔE and ΔH are approximately equal.

Standard Enthalpy Changes and the Standard Heat of Reaction, $\Delta H°$

The heat energy or enthalpy change, ΔH, produced by a chemical reaction is an **extensive property,** since reacting a larger amount of chemicals produces a larger amount of heat. For example, when propane is burned according to the equation

$$CH_3CH_2CH_3(g) + 5O_2(g) \rightarrow 3CO_2(g) + 4H_2O(g) \qquad (12.16)$$

more heat is generated as more propane is burned. To make the heat produced by a reaction an **intensive property,** the amount of chemical that reacts must be specified. The standard heat of a reaction, $\Delta H°$, is generally defined as the heat produced when the number of moles specified in the balanced chemical equation reacts. For the reaction of propane the heat of reaction, $\Delta H°_{react}$, equals −2044 kJ when 1 mole of propane reacts with 5 moles of oxygen as shown in Equation 12.16. The negative sign indicates that a large amount of heat is released in this reaction, making propane an excellent fuel for cooking and heating.

Hess's Law

Hess's law states that, whatever mathematical operations are performed on a chemical equation, the same mathematical operations are applied also to the heat of reaction. Hess's law is summarized as follows:

1. If the coefficients of a chemical equation are all multiplied by a constant, the $\Delta H°_{react}$ is multiplied by that same constant.
2. If two or more equations are added together to obtain an overall reaction, the heats of these equations are also added to give the heat of the overall reaction.

Hess's law allows the chemist to measure $\Delta H°_{react}$ for several reactions and then to combine the equations and their heats to obtain $\Delta H°_{react}$ for a completely different reaction.

For example, we saw above that the burning of propane produces a large amount of heat. The reaction in Equation 12.16 is easy to perform by igniting propane in the presence of oxygen. The reverse reaction for the synthesis of 1 mole of propane from carbon dioxide and water is impossible to perform, but it may be written as

$$3CO_2(g) + 4H_2O(g) \rightarrow CH_3CH_2CH_3(g) + 5O_2(g) \qquad (12.17)$$

Since reversing a reaction is the same as multiplying it by −1, the heat needed for this reaction is +2044 kJ. The change from a negative to positive value is explained on the basis that $\Delta H°$, which is a state function, depends only on the final and initial states of the system. Synthesis of 1 mole of propane has the same final and initial states as the combustion of 1 mole of propane. The only difference is the direction of the process. We must reach the conclusion that $\Delta H°$ has the same magnitude for these two reactions, but they have different signs because they go in opposite directions. Figure 12.2 illustrates this process.

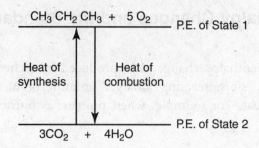

FIGURE 12.2. Diagram illustrating that the heat of a reaction has the same magnitude whether the reaction is run in the forward or reverse direction. The sign of the heat of reaction, however, is positive in one direction and negative in the other.

The general principle that $\Delta H^{\circ}_{\text{forward react}} = -\Delta H^{\circ}_{\text{reverse react}}$ is an essential part of Hess's law.

For the combustion of propane, Hess's first rule (see above) tells us that multiplying the equation by 2 will result in ΔH° also being multiplied by 2:

$$2CH_3CH_2CH_3(g) + 10O_2(g) \rightarrow 6CO_2(g) + 8H_2O(g) \quad \Delta H = 4088 \text{ kJ}$$

If the reaction is multiplied by $^1/_2$, the ΔH° will also be multiplied by $^1/_2$:

$$\tfrac{1}{2}CH_3CH_2CH_3(g) + 2.5O_2(g) \rightarrow 1.5CO_2(g) + 2H_2O(g) \quad \Delta H = 1011 \text{ kJ}$$

If the reaction is multiplied by 6, the ΔH° will also be multiplied by 6:

$$6CH_3CH_2CH_3(g) + 30O_2(g) \rightarrow 18CO_2(g) + 24H_2O(g) \quad \Delta H = -12,264 \text{ kJ}$$

Once ΔH° is multiplied by any coefficient, the system is no longer in standard state and the superscript zero is dropped as shown.

The second of Hess's rules concerns the addition of chemical reactions. To repeat, when chemical reactions are added, all of the reactants are written as reactants of the overall reaction. All of the products are combined as products of the overall reaction. The last step in adding reactions is the cancellation of any identical reactants and products in the overall reaction.

To add the two reactions below, the reactants and products in Equations 12.18 and 12.19 are combined in an overall reaction, Equation 12.20:

$$N_2(g) + O_2(g) \rightarrow 2NO(g) \qquad (12.18)$$

$$2NO(g) + O_2(g) \rightarrow 2NO_2(g) \qquad (12.19)$$

$$N_2(g) + 2NO(g) + 2O_2(g) \rightarrow 2NO(g) + 2NO_2(g) \qquad (12.20)$$

Two molecules of $NO(g)$ are canceled from both the reactant and product sides of Equation 12.20 to obtain the overall reaction:

$$N_2(g) + 2O_2(g) \rightarrow 2NO_2(g) \qquad (12.21)$$

Since the two reactions were added, their heats of reaction are also added:

$$N_2(g) \;+\; O_2(g) \;\rightarrow\; 2NO(g) \qquad \Delta H_1^\circ \;=\; +180.5 \text{ kJ}$$

$$2NO(g) \;+\; O_2(g) \;\rightarrow\; 2NO_2(g) \qquad \Delta H_2^\circ \;=\; -114.1 \text{ kJ}$$

$$\overline{N_2(g) \;+\; 2O_2(g) \;\rightarrow\; 2NO_2(g) \quad \Delta H_{\text{overall}}^\circ \;=\; +66.4 \text{ kJ}}$$

In this example $\Delta H_1^\circ + \Delta H_2^\circ = \Delta H_{\text{overall}}^\circ$.

The heats of many reactions can be determined if the heats of combustion are known for each of the reactants and products. To illustrate this more complex combination of reactions, we can determine the heat of reaction for the synthesis of propane from carbon and hydrogen:

$$3C(s) + 4H_2(g) \rightarrow CH_3CH_2CH_3(g) \qquad (12.22)$$

from the combustion reactions of propane, hydrogen, and carbon. These reactions, along with their ΔH° values, are as follows:

$$(1)\,CH_3CH_2CH_3(g) \;+\; 5O_2(g) \;\rightarrow\; 3CO_2(g) \;+\; 4H_2O(g) \quad \Delta H_1^\circ \;=\; -2044 \text{ kJ}$$

$$(2)\,2H_2(g) \qquad\qquad\quad +\; O_2(g) \;\rightarrow\; 2H_2O(g) \qquad\qquad \Delta H_2^\circ \;=\; -483.6 \text{ kJ}$$

$$(3)\,C(s) \qquad\qquad\qquad\;\; +\; O_2(g) \;\rightarrow\; CO_2(g) \qquad\qquad\quad\;\; \Delta H_3^\circ \;=\; -393.5 \text{ kJ}$$

If it is not immediately obvious how these three equations should be combined, some general principles on how to approach the problem logically are helpful. In the list below, the first three principles tell us how to find which equation to start with. Once Equation 1 is established, the same principles are used to select and manipulate the remaining reactions. If needed, the fourth principle may be used.

1. Focus on the most complex molecules first.
2. Focus only on atoms and molecules that occur in just one reaction.
3. Focus on atoms and molecules that are in the overall equation.
4. Focus on finding atoms and molecules to cancel unneeded ones from already selected equations.

Using principles 1–3, we see that the combustion of propane in Equation 1 should be considered first. This equation contains 1 mole of propane and so does the overall equation, Equation 12.22. However, in the combustion reaction, Equation 1, propane is a reactant, and in Equation 12.22 it is a product. Consequently, Equation 1 must be reversed by multiplying it by −1. This reverses the equation and at the same time changes the sign of ΔH°:

$$3CO_2(g) + 4H_2O(g) \rightarrow CH_3CH_2CH_3(g) + 5O_2(g) \quad \Delta H^\circ = -2044 \text{ kJ} \times (-1)$$

Using principle 3, we now focus on the reaction in Equation 2 and note that it includes $2H_2(g)$ and that we need $4H_2(g)$ in our overall equation. The hydrogens are reactants in both Equation 2 and Equation 12.22, but Equation 2 must be multiplied by 2 so that we have the correct number of H_2 in the final reaction. We will also have to multiply $\Delta H°$ by 2:

$$4H_2(g) + 2O_2(g) \rightarrow 4H_2O(g) \quad \Delta H = -483.6 \text{ kJ} \times 2$$

To demonstrate the use of principle 4, we see that the remaining reaction, Equation 3, contains one $CO_2(g)$ as a product. We need to cancel three $CO_2(g)$ molecules from the already selected equations. Since $CO_2(g)$ is a product in Equation 3 and a reactant in the rearranged Equation 3, the CO_2's will cancel when the reactions are added. However, Equation 3 must be multiplied by 3 so that all three $CO_2(g)$ molecules will cancel:

$$3C(s) + 3O_2(g) \rightarrow 3CO_2(g) \quad \Delta H = 2393.5 \text{ kJ} \times 3$$

Adding the three equations gives us

$$3C(s) + 3O_2(g) + \; + 4H_2(g) + 3O_2(g) + 3CO_2(g) + 4H_2O(g) \rightarrow$$
$$3CO_2(g) + 4H_2O(g) + 5O_2 + CH_3CH_2CH_3(g)$$

After canceling the three $CO_2(g)$, four $H_2O(g)$ and five $O_2(g)$ molecules, the equation becomes

$$3C(s) + 4H_2(g) \rightarrow CH_3CH_2CH_3(g) \tag{12.23}$$

The heat of reaction is the sum of the three $\Delta H°$ values multiplied by the operations performed:

$$
\begin{aligned}
\Delta H°_{\text{overall}} &= \Delta H°_1 \; 3\,(-1) &+& \quad \Delta H°_2 \times 2 &+& \quad \Delta H°_3 \times 3 \\
&= -2044 \text{ kJ} \times (-1) &+& \quad -483.6 \text{ kJ} \times 2 &+& \quad -393.5 \text{ kJ} \times 3 \\
&= +2044 \text{ kJ} &-& \quad 967.2 \text{ kJ} &-& \quad 1180.5 \text{ kJ} \\
&= -103.7 \text{ kJ}
\end{aligned}
$$

The advantage of being able to perform some simple experiments to obtain data for complex, even impossible reactions was recognized very quickly. During the energy crisis of the 1970s, the feasibility of producing alternative fuels was determined from thermochemical calculations.

EXERCISE 12.4

We may wish to synthesize methane from carbon and hydrogen:

$$C(s) + 2H_2(g) \rightarrow CH_4(g)$$

In terms of the heat energy needed for this reaction and the heat of combustion for methane, is this effort worthwhile? The heats of combustion are as follows: $CH_4 = -890.3$ kJ; $H_2 = -571.8$ kJ, and $C = -393.5$ kJ.

Solution

The three combustion reactions are written as

$$CH_4(g) + 2O_2(g) \rightarrow CO_2(g) + 2H_2O(\ell) \quad \Delta H = -890.3 \text{ kJ}$$

$$2H_2(g) + O_2(g) \rightarrow 2H_2O(l) \qquad\qquad \Delta H = -571.8 \text{ kJ}$$

$$C(s) + O_2(g) \rightarrow CO_2(g) \qquad\qquad \Delta H = -393.5 \text{ kJ}$$

To obtain the desired reaction we need to reverse the combustion of methane and add it to the remaining reactions. The heat of this reaction is

$$\Delta H^\circ_{react} = -890.3 \text{ kJ}(-1) + -571.8 \text{ kJ} + -393.5 \text{ kJ}$$
$$= -75.0 \text{ kJ}$$

Ignoring all other factors, we see that the formation of CH_4 produces heat and its combustion produces a much greater amount of heat. On this basis, the synthesis of CH_4 is a feasible process. If, however, the energy required to make $H_2(g)$ from water is included in the calculation, the energy gain is very small and the investment in such a project may be suspect.

In this reaction, methane is formed from the elements carbon and hydrogen. This is known as a formation reaction, and we have determined the heat of formation.

Formation Reactions and Heats of Formation

In the preceding section we saw that the heats of many chemical reactions can be determined if the heats of combustion of all reactants and products are known. For other types of reactions the heat of combustion can also be tabulated. Such a table would be very long and complex, however, and finding the appropriate reactions to combine would be a monumental task. **Heats of formation** allow chemists to tabulate thermochemical data in a short, easy-to-use format.

A **formation reaction** is defined as one in which the reactants are elements in their standard state at 25 °C and 1 atmosphere of pressure, and there is only 1 mole of product. Here are some examples of formation reactions:

$$Fe(s) + \tfrac{1}{2}O_2(g) \rightarrow FeO(s) \tag{12.24}$$

$$2Fe(s) + \tfrac{3}{2}O_2(g) \rightarrow Fe_2O_3(s) \tag{12.25}$$

$$2K(s) + \tfrac{1}{2}H_2(g) + 2O_2(g) + P(s) \rightarrow K_2HPO_4(s) \tag{12.26}$$

Fractional coefficients may be used in formation reactions. Since there is always 1 mole of product, the standard heats of formation, ΔH°_f, are tabulated as the heat produced per mole of product.

Since the reaction can easily be deduced if the product is known, a table of data need contain only the name or formula of the product and its corresponding ΔH°_f. A tabulation of some heats of formation is given in Appendix 3.

Heats of formation of the elements are always zero, whether they are molecules or atoms. The reason is that the formation reaction for an element such as oxygen is defined as

$$O_2(g) \rightarrow O_2(g) \tag{12.27}$$

Since the oxygen is at 25 °C and 1.00 atmosphere pressure as both the product and the reactant, the initial and final states of the oxygen in Equation 12.27 are the same, and their difference must be zero:

$$\Delta H_f^\circ \text{ (any element)} = 0 \tag{12.28}$$

Heats of formation can be calculated from heats of combustion, as was shown in Exercise 12.4.

For the combustion of propane (Equation 12.16):

$$CH_3CH_2CH_3(g) + 5O_2(g) \rightarrow 3CO_2(g) + 4H_2O(g)$$

ΔH° can be calculated using the formation reactions and the tabulated heats of formation. Again, we will need a formation reaction for each of the reactants and products. Elements are excluded, however, since their heats of formation are always zero. In this example, the formation reactions for propane, carbon dioxide, and water are needed:

$$3C(s) \ + \ 4H_2(g) \ \rightarrow \ CH_3CH_2CH_3(g) \qquad \Delta H_f^\circ \ = \ -103.8 \text{ kJ mol}^{-1}$$

$$C(s) \ + \ O_2(g) \ \rightarrow \ CO_2(g) \qquad \Delta H_f^\circ \ = \ -393.5 \text{ kJ mol}^{-1}$$

$$H_2(g) \ + \ {}^{1}\!/_{2}O_2(g) \ \rightarrow \ H_2O(g) \qquad \Delta H_f^\circ \ = \ -241.8 \text{ kJ mol}^{-1}$$

The following operations are performed on these reactions so that they can be combined to yield the combustion reaction:

The propane formation reaction is reversed.

$$CH_3CH_2CH_3(g) \rightarrow 3C(s) + 4H_2(g)$$

$$\Delta H \ = -103.8 \text{ kJ mol}^{-1} \times (-1 \text{ mol})$$

The carbon dioxide formation reaction is multiplied by 3.

$$3C(s) + 3O_2(s) \rightarrow 3CO_2(g)$$

$$\Delta H \ = -393.5 \text{ kJ mol}^{-1} \times 3 \text{ mol}$$

The $H_2O(g)$ formation reaction is multiplied by 4.

$$4H_2(g) + 2O_2(g) \rightarrow 4H_2O(g)$$

$$\Delta H \ = -241.8 \text{ kJ mol}^{-1} \times 4 \text{ mol}$$

After these three operations, adding the three reactions yields the combustion reaction:

$$CH_3CH_2CH_3(g) + 5O_2(g) \rightarrow 3CO_2(g) + 4H_2O(g)$$

The corresponding sum of the heats is the heat of reaction:

$$
\begin{aligned}
\Delta H^\circ &= (-103.8 \text{ kJ mol}^{-1})(-1 \text{ mol}) + (-393.5 \text{ kJ mol}^{-1})(3 \text{ mol}) \\
&\qquad + (-241.8 \text{ kJ mol}^{-1})(4 \text{ mol}) \\
&= -2044 \text{ kJ}
\end{aligned}
$$

After several calculations using heats of formation have been made, a pattern appears. The heat of any reaction will be the sum of $(\Delta H_f^\circ \times \text{mol}_{\text{product}})$ for all the products minus the sum of $(\Delta H_f^\circ \times \text{mol}_{\text{reactant}})$ for all the reactants. The $\text{mol}_{\text{product}}$ and $\text{mol}_{\text{reactant}}$ terms refer to the stoichiometric coefficients in the balanced equation for the reaction:

$$\Delta H^\circ = \sum (\Delta H_f^\circ \times \text{coeff})_{\text{products}} - \sum (\Delta H_f^\circ \times \text{coeff})_{\text{reactants}} \qquad (12.29)$$

> **TIP**
>
> Equation 12.29 is given on the AP Exam.

EXERCISE 12.5

Calculate the heat of combustion of CH_4. The heats of formation are as follows: $\Delta H_f^\circ(CH_4(g)) = -74.8 \text{ kJ mol}^{-1}$, $\Delta H_f^\circ(CO_2(g)) = -110.5 \text{ kJ mol}^{-1}$, and $\Delta H_f^\circ(H_2O(g)) = -241.8 \text{ kJ mol}^{-1}$.

Solution

For the combustion of CH_4 the equation is

$$CH_4(g) + 2O_2(g) \rightarrow CO_2(g) + 2H_2O(g)$$

The heat of this reaction is calculated as

$$
\begin{aligned}
\Delta H_{\text{react}}^\circ &= [(-110.5 \text{ kJ mol}^{-1}) + (-241.8 \text{ kJ mol}^{-1})(2 \text{ mol})] \\
&\qquad - [(-74.8 \text{ kJ mol}^{-1})(1) + (0.00 \text{ kJ mol}^{-1})(2 \text{ mol})] \\
&= -519.3 \text{ kJ}
\end{aligned}
$$

The value of 0.00 kJ mol^{-1} in this calculation represents the heat of formation of $O_2(g)$, which, by definition, is zero.

ENTROPY AND THE SECOND LAW OF THERMODYNAMICS

Entropy is related to the number of different ways in which a system can arrange the particles within the system. Similarly, if we divide the volume of the system into eight units, those two particles can arrange themselves in 56 different ways (8×7). (The first particle can enter any of the eight units and the second can occupy any of the remaining seven, therefore $8 \times 7 = 56$.) If the volume of the system is doubled, we will have 16 volume units the same size as before. The number of different arrangements of our particles is now 240 (16 × 15). That represents an increase in entropy. These examples must be extrapolated to a very large number of infinitesimally small volume units with extremely large numbers of particles (atoms and molecules) and a correspondingly large number of possible arrangements to describe real systems. However, the conclusions are relatively straightforward. An increase in the number of particles increases entropy, an increase in volume increases entropy, changing state (e.g., from liquid to gas) increases entropy, and so on.

We can visualize that water molecules in ice are constrained to the crystal lattice and have virtually no freedom to move and occupy a different volume unit. A water molecule in a liquid can move from one volume unit to another, but it does so rather slowly and is strongly influenced by other water molecules. Finally, in the gas phase the water molecule rapidly moves from one volume unit to another. Therefore we say that water in ice has the least entropy, whereas water in the gas phase has the most entropy. If a chemical reaction produces a gas such as H_2, the increase in possible positions and kinetic energies compared to that in liquids or solids leads to the conclusion that the entropy increases in such a reaction.

Entropy is assigned the symbol, S, and has units of J °C^{-1}. Standard entropy is based on the mole and is written as $S°$ with units of J °C^{-1} mol^{-1}. Values for $S°$ are tabulated in Appendix 3. It should be noted that standard entropy values for elements are not zero as they are for standard heat of formation, $\Delta H_f°$, and standard free energy, $\Delta G_f°$.

Unlike other thermodynamic quantities, such as energy, E, and enthalpy, H, the actual value for the entropy, S, of a substance can be determined. From fundamental principles, a perfect crystal at absolute zero (0 K or −273.16 °C) has zero entropy since there is only one possible arrangement of the atoms. The Boltzmann entropy equation, $S = k \ln w$, where k is the Boltzmann constant and w is the number of microstates, describes the statistical method for determining entropy, S. Since a perfect crystal has one microstate, $k \ln (1) = 0$. Entropy can be determined experimentally since as the temperature of 1 mole of a chemical is increased from absolute zero, the entropy increases and the standard entropy is defined as

$$S° = \frac{q_{\text{rev}}}{T} \tag{12.30}$$

In this equation T represents the temperature in Kelvins, and q_{rev} is the heat added to raise the temperature very slowly from absolute zero up to T. Heat, q, is not a state function, and it seems that S should not be a state function either. However, if heat is always added in a carefully defined manner, the results will always be the same. This carefully defined path is called a **reversible process,** and the heat added is symbolized as q_{rev}. A reversible process is defined as one that occurs in infinitesimally small steps from the initial to the final state.

Entropy changes due to a chemical process are calculated in the same fashion as the heats of reaction. Just as is done for $\Delta H_f°$ values, the tables list entropies for 1 mole of substance, making entropy an intensive physical property. This can be written as

TIP

Equation 12.31 is given on the AP Exam.

$$\Delta S° = \sum (S_f° \times \text{coeff})_{\text{products}} - \sum (S_f° \times \text{coeff})_{\text{reactants}} \tag{12.31}$$

For example, we can calculate the entropy change for the combustion of propane from the data in Appendix 3:

$$CH_3CH_2CH_3(g) + 5O_2(g) \rightarrow 3CO_2(g) + 4H_2O(g)$$

$$\Delta S^\circ = ? \, J \, ^\circ C^{-1}$$

$$\Delta S^\circ = \left[\left(\frac{213.6 \, J}{mol \, K}\right)(3 \, mol) + \left(\frac{188.7 \, J}{mol \, K}\right)(4 \, mol)\right] -$$

$$\left[\left(\frac{205.0 \, J}{mol \, K}\right)(5 \, mol) + \left(\frac{270.2 \, J}{mol \, K}\right)(1 \, mol)\right]$$

$$= 1395.6 \, J \, K^{-1} - 1295.2 \, J \, K^{-1}.$$

$$= +100.4 \, J \, K^{-1} = 100.4 \, J \, ^\circ C^{-1}$$

This result represents an increase in entropy. We might have predicted an increase for this reaction since there are 7 moles of gaseous products and only 6 moles of gaseous reactants. The increase in the number of moles of gas in this reaction is 1 ($\Delta n_g = 1$), indicating that the entropy change is expected to be positive.

In addition to calculating the entropy change from tabulated data, we can estimate the sign and, to some degree, the magnitude of an entropy change for a chemical process. In making such an estimate, the following principles are important:

1. Formation of a gas increases entropy greatly. The greater the position value of Δn_g, the greater the entropy increase. The reverse is true because more negative values of Δn_g represent a decrease in entropy.

2. If $\Delta n_g = 0$, changes from the solid to the liquid phase are the next leading contributors to an increase in entropy. A solid that melts or a solute that dissolves in a solvent both exhibit an increase in entropy. Conversely, the formation of solids results in a decrease in entropy.

3. An increase in temperature increases the entropy of a system, and a temperature decrease lowers the entropy.

GIBBS FREE-ENERGY, ΔG

This thermodynamic quantity was named in honor of J. Willard Gibbs, a preeminent physical chemist who developed the concept. The **free-energy change,** represented by the symbol ΔG°, is the maximum amount of energy available from any chemical reaction. Two forces drive chemical reactions. The first is the enthalpy, ΔH°, which represents the change in the internal potential energy of the atoms. The second is the drive toward an increase in the entropy of the system. If the enthalpy is negative, it means that the internal potential energy of the system is decreased, and this favors a spontaneous reaction. If the entropy increases, a spontaneous reaction is also favored. The combination of these two driving forces is represented as

$$\Delta G^\circ = \Delta H^\circ - T\Delta S^\circ \tag{12.32}$$

TIP

Equation 12.32 is given on the AP Exam.

The Gibbs free-energy equation is derived directly from the **second law of thermodynamics,** which states that any physical or chemical change must result in an increase in the entropy of the universe.

Since $\Delta G°$ is a combination of $\Delta H°$ and $\Delta S°$, we can make some generalizations, shown in Table 12.1, based only on the signs of these quantities.

TABLE 12.1

Relationships of the Algebraic Signs of $\Delta H°$ and $\Delta S°$ to the Sign of $\Delta G°$

$\Delta H°$	$\Delta S°$	$\Delta G°$ as T Increases	Comment
Negative	Positive	Always negative	Always spontaneous
Positive	Negative	Always positive	Always nonspontaneous
Negative	Negative	Becomes positive	Becomes nonspontaneous as *T* increases
Positive	Positive	Becomes negative	Becomes spontaneous as *T* increases

When $\Delta H°$ is negative and $\Delta S°$ is positive, the only value possible for $\Delta G°$ is a negative one. In this case the reaction will be spontaneous at all temperatures. Similarly, when $\Delta H°$ is positive and $\Delta S°$ is negative, $\Delta G°$ must be positive, indicating a nonspontaneous reaction at all temperatures.

When $\Delta H°$ is negative and $\Delta S°$ is negative, $\Delta G°$ may be either negative or positive, depending on the relative magnitudes of $\Delta H°$ and $\Delta S°$. However, the $-T\Delta S°$ term will always be positive. It will have a larger magnitude at high temperatures and a smaller magnitude at low temperatures. This fact suggests that lowering the temperature may make the positive magnitude of $T\Delta S°$ small enough so that, when it is combined with the negative $\Delta H°$, the resulting $\Delta G°$ will be negative and the reaction will be spontaneous.

The reverse is true when $\Delta H°$ is positive and $\Delta S°$ is positive. Again, $\Delta G°$ may be either positive or negative. The same reasoning leads to the conclusion that increasing the temperature will eventually cause the reaction to be spontaneous with a negative $\Delta G°$.

SPONTANEITY

What Does Spontaneous Mean?

When chemists refer to a spontaneous reaction, it is one that occurs without the need for additional energy input after the reactants are mixed and the reaction initiated. Generally, they also mean that $\Delta G°$ for the process has a negative value. This implies that reactions are at standard state. It also implies that reactions are generally considered spontaneous if there are more products than reactants in the equilibrium mixture.

What Does Spontaneous Not Mean?

Spontaneous, or its opposite, nonspontaneous, does not tell us the direction of a reaction. ΔG tells us the direction of a reaction as it approaches equilibrium. These terms do not tell us if reactions that are not at standard state are spontaneous.

As is true of $\Delta H°$ and $\Delta S°$ calculations, we can calculate the value of $\Delta G°$ for a reaction from tabulated values of free energies of formation. These values are listed in a table in Appendix 3. Since temperature is an important variable that affects the value of the free energy, the temperature must be specified. It is most common to list $\Delta G°$ values for room temperature of 25 °C or 298 K. The symbol for free energy incorporates the temperature, as in $\Delta G°_{298}$. In using the free-energy table, we subtract the $\Delta G°_{298}$ of the reactants from the $\Delta G°_{298}$ values of the products in the equation:

$$\Delta G°_{298} = \sum (\Delta G°_{298} \times \text{coeff})_{\text{products}} - \sum (\Delta G°_{298} \times \text{coeff})_{\text{reactants}} \qquad (12.33)$$

For the combustion of propane (Equation 12.16) we obtain

$$\Delta G°_{298} = \left[\left(\frac{-394.4 \text{ kJ}}{\text{mol}} \right)(3 \text{ mol}) + \left(\frac{-228.6 \text{ kJ}}{\text{mol}} \right)(4 \text{ mol}) \right] - \left[\left(\frac{-23.5 \text{ kJ}}{\text{mol}} \right)(1 \text{ mol}) \right]$$

$$= (-1183.2 \text{ kJ} - 914.4 \text{ kJ}) - (-23.5 \text{ kJ})$$

$$= -2074.1 \text{ kJ}$$

> **TIP**
>
> The AP Exam gives you the equation
>
> $\Delta G° = \Sigma \Delta G°_f$ products $- \Sigma \Delta G°_f$ reactants.
>
> Equation 12.33 is the same as this but reminds you to include stoichiometry coefficients.

The negative value indicates that the reaction is spontaneous, as anyone who has used a barbecue grill or propane torch already knows.

We have calculated $\Delta H°$ and $\Delta S°$ for this reaction in preceding sections of this chapter. Using those values, along with a temperature of 298 K, we have a second way to determine the value of $\Delta G°_{298}$:

$$\begin{aligned} \Delta G°_{298} &= \Delta H° & - T\Delta S° \\ &= -2044 \text{ kJ} & - (298 \text{ K})(100.4 \text{ J K}^{-1}) \\ &= -2044 \text{ kJ} & - 29.9 \text{ kJ} \\ &= -2074 \text{ kJ} \end{aligned}$$

Free Energy at Temperatures Other Than 298 K

Using the table of standard free energies of formation in Appendix 3, we can calculate the free-energy change at 298 K. Standard free-energy changes at other temperatures can also be calculated. For this purpose we need to know the standard heat of reaction, $\Delta H°_{\text{react}}$, and the standard entropy change, $\Delta S°_{\text{react}}$, for the reaction. These values are correct for 298 K, but we may assume that they do not change significantly with temperature. Using these values in the free-energy equation with a temperature other than 298 K gives the free-energy change at that different temperature.

EXERCISE 12.6

For a certain reaction, $\Delta H° = +2.98$ kJ and $\Delta S° = +12.3$ J K^{-1}. What is $G°$ at 298 K, 200 K, and 400 K?

Solution

The equation to be solved is

$$\Delta G° = \Delta H° - T\Delta S°$$

Substituting the values given in the problem yields

$$\Delta G^{\circ}_{298} = 2.98\,\text{kJ} - 298\,(12.3\,\text{J K}^{-1})$$
$$= 2.98\,\text{kJ} - 3665\,\text{J}$$

To complete the solution, -3665 J must be converted to -3.67 kJ. Then

$$\Delta G^{\circ} = -0.69\,\text{kJ} \quad \text{(a spontaneous reaction)}$$

At 200 K and 400 K the answers are

$$\Delta G^{\circ}_{200} = 2.98\,\text{kJ} - 200\,\text{K}\,(12.3\,\text{J K}^{-1}) = +0.52\,\text{kJ}$$

and

$$\Delta G^{\circ}_{400} = 2.98\,\text{kJ} - 400\,\text{K}\,(12.3\,\text{J K}^{-1}) = -1.94\,\text{kJ}$$

In this exercise we see that the reaction is not spontaneous at 200 K but is spontaneous at 298 K and 400 K.

EXERCISE 12.7

For a certain reaction, $\Delta H^{\circ} = -13.65$ kJ and a $\Delta S^{\circ} = -75.8$ J K^{-1}. (a) What is ΔG°_{298} at 298 K? (b) Will increasing or decreasing the temperature make the reaction spontaneous? If so, at what temperature will the reaction become spontaneous?

Solution

(a) At 298 K the free energy is

$$\Delta G^{\circ}_{298} = -13.65\,\text{kJ} - 298\,\text{K}\,(-75.8\,\text{J K}^{-1}) = +8.94\,\text{kJ}$$

(b) The reaction is not spontaneous at 298 K. However, since ΔH° and ΔS° both have the same sign, the free energy will change from positive to negative at some temperature. Since the number zero divides the positive numbers from the negative numbers, we may conclude that $\Delta G^{\circ} = 0.00$ kJ is the dividing line between spontaneous and nonspontaneous reactions. Consequently, the free-energy equation is set up as

$$\Delta G^{\circ}_{\text{react}} = \Delta H^{\circ}_{\text{react}} - T\Delta S^{\circ}_{\text{react}}$$

$$0.00\,\text{kJ} = -13.65\,\text{kJ} - T(-75.8\,\text{J K}^{-1})$$

$$T = \frac{13.65\,\text{kJ}}{0.0758\,\text{kJ K}^{-1}} = 180\,\text{K}$$

From this result we predict that the reaction will be spontaneous below 180 K and nonspontaneous above 180 K.

The condensation of a gas and the crystallization of a liquid are two physical processes that are spontaneous at low temperatures and nonspontaneous at higher temperatures.

Free Energy and Equilibrium

When a system is not at standard state, the free-energy change is represented by ΔG, not $\Delta G°$. Equation 12.34 shows the relationship between ΔG and $\Delta G°$:

$$\Delta G = \Delta G° + RT \ln Q \qquad (12.34)$$

TIP

Equation 12.34 is given on the AP Exam.

From Chapter 10 we recall that Q is the reaction quotient. If the value of Q is not equal to the equilibrium constant, further reaction occurs until the system reaches equilibrium.

Using Equation 12.34, we find that when a system is at standard state all concentrations are equal to 1 and $Q = 1$. The natural logarithm of 1 is zero ($\ln 1 = 0$), and consequently $\Delta G = \Delta G°$.

The value of ΔG (without the superscript) tells us whether the reaction will continue and, if so, in which direction it will go. When ΔG is negative, the reaction will proceed in the forward direction. When ΔG is positive, the reaction proceeds in the reverse direction. If ΔG is zero, the reaction is at equilibrium and no further reaction occurs. For the equilibrium condition we find that

$$\Delta G° = -RT \ln K \qquad (12.35)$$

TIP

Equation 12.35 is given on the AP Exam.

by setting $\Delta G = 0$, substituting the equilibrium constant, K, for the reaction quotient, Q, in Equation 12.34, and rearranging.

The relationships between ΔG and $\Delta G°$ are diagrammed in Figure 12.3. The standard free energies, $\Delta G°$, are shown for the reactants on the left side and for the products on the right side of each graph. The difference between the two is $\Delta G°$ for the reaction. In each diagram the curved line connecting the two $G°$ values represents the value of G for the reaction mixture. The slope of the curved line is ΔG, and at the minimum, where the slope = 0, the reaction is in equilibrium.

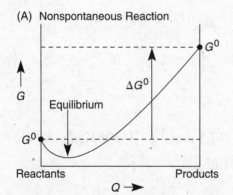
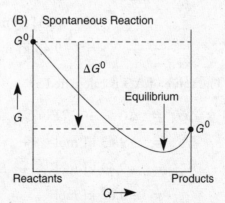

FIGURE 12.3. **Free-energy diagrams for (A) a nonspontaneous reaction and (B) a spontaneous reaction. The difference between the $G°$ points is $\Delta G°$. The slope of the curved line is ΔG, and the equilibrium point is at the minimum of the curve.**

The curves in Figure 12.3 illustrate that in a nonspontaneous reaction a small amount of reactants is converted into products at equilibrium (minimum of curved line). A spontaneous reaction, on the other hand, has most of the reactants converted into products because the minimum is closer toward the product side. The slope of the curved line is negative to the left of the equilibrium point, so that ΔG is negative and the reaction moves toward the products and also toward the minimum. On the right-hand side of the equilibrium point, the slope and ΔG are positive. The reaction proceeds toward the reactants and also toward the minimum.

It is important to note that the designation of a given chemical reaction as spontaneous or nonspontaneous has nothing to do with the direction of the reaction when the chemicals are mixed. Figure 12.3 shows that a mixture of reactants will always proceed in the forward direction (ΔG is negative). However, when the reaction reaches equilibrium, it can be judged as nonspontaneous as in Figure 12.3a or judged as spontaneous as shown in the equilibrium point represented by Figure 12.3b. Similarly, if both products are mixed, some reaction in the reverse direction takes place (ΔG is positive). However, the reaction will be classified depending on the position of equilibrium as spontaneous ($\Delta G°$ is negative) or nonspontaneous ($\Delta G°$ is positive). At standard state, only the temperature can alter the position of equilibrium.

If reactants or products are mixed in nonstandard state proportions, the concentrations, in certain relatively rare situations, may have an effect on whether or not the reaction meets the criteria for being spontaneous.

EXERCISE 12.8

The value of the equilibrium constant is 45 at 298 K. At the same temperature $Q = 35$. Determine the value of $\Delta G°$ for the reaction at 298 K, and show that the value of G indicates the same direction for the reaction as Q predicts.

Solution

The value of $\Delta G°$ is calculated as

$$
\begin{aligned}
\Delta G° &= -RT \ln K \\
&= -(8.314 \text{ J mol}^{-1} \text{ K}^{-1})(298 \text{ K})[\ln (45)] \\
&= -9.43 \text{ kJ mol}^{-1}
\end{aligned}
$$

The value of ΔG is calculated as

$$
\begin{aligned}
\Delta G &= \Delta G° + RT \ln Q \\
&= -9.43 \text{ kJ mol}^{-1} + (8.314 \text{ J mol}^{-1} \text{ K}^{-1})(298 \text{ K})[\ln (35)] \\
&= -9.43 \text{ kJ mol}^{-1} + 8.81 \text{ kJ mol}^{-1} \\
&= -0.62 \text{ kJ mol}^{-1}
\end{aligned}
$$

The value of ΔG predicts that the reaction will proceed in the forward direction. The fact that Q is less than K also indicates that the reaction will proceed in the forward direction.

SUMMARY

Thermodynamics reviewed in this chapter involves the consideration of energy changes in systems at equilibrium. All energy can be considered as either potential energy (energy of position, related to heat content) or kinetic energy (energy of motion, related to temperature). Absolute values for potential or kinetic energy cannot be determined, but changes in these can be measured. State functions $\Delta E°$, $\Delta H°$, $\Delta S°$, and $\Delta G°$ depend only on the initial and final states of the system. The maximum work available from a system is equal to $\Delta G° = \Delta H° - T\Delta S°$. $\Delta E°$ is the change in total internal energy and is measured with a bomb calorimeter. Enthalpy, the heat of a reaction, $\Delta E°$, is measured in a calorimeter at ambient pressure. The standard free energy change is related to the equilibrium constant, and either may be used to determine if a reaction is spontaneous. A spontaneous reaction is generally one where the amount of product is greater than the amount of reactants at equilibrium. This means that $\Delta E°$ is negative and $K_{eq} > 1$.

Important Concepts

Forms of energy
Law of conservation of energy
Hess's law
Standard state
Gibbs free energy, entropy, and enthalpy
Spontaneous and nonspontaneous reactions

Important Equations

$q = (\text{SPECIFIC HEAT}) (\text{MASS}) (\text{TEMPERATURE CHANGE})$

$q = (\text{HEAT CAPACITY}) (\text{TEMPERATURE CHANGE})$

$\Delta E = q - P\Delta V = q_v$

$\Delta H = q_p$

$S = k \ln w = q_{rev}/T$

$\Delta E = q + w$

$\Delta G° = \Delta H° - T\Delta S°$

$\Delta G = \Delta G° - RT \ln Q$

$\Delta G° = - RT \ln K$

Practice Exercises

Multiple-Choice

For the first four problems below, one or more of the following responses will apply; each response may be used more than once or not at all in these questions.

   I. $\Delta G°$
   II. $\Delta S°$
   III. $\Delta H°$
   IV. ΔG
   V. H

1. Which of these must be negative for a reaction to be spontaneous?

   (A) I
   (B) III
   (C) I, II, and IV
   (D) IV
   (E) I, III, and V

2. Which of these cannot be determined?

   (A) I and III
   (B) II
   (C) I, II, and IV
   (D) I, III, and IV
   (E) V

3. Which of these are extensive values?

   (A) I and III
   (B) II
   (C) I, II, and IV
   (D) IV and V
   (E) I, III, and V

4. If _____ is negative and _____ is positive, a reaction will never be spontaneous.

   (A) II, III
   (B) III, IV
   (C) I, II
   (D) IV, III
   (E) I, V

5. When 0.400 g of CH_4 is burned in excess oxygen in a bomb calorimeter that has a heat capacity of 3245 J °C^{-1}, a temperature increase of 6.795 °C is observed. What is the value of q_v?

   (A) 220 kJ mol^{-1}
   (B) −882 kJ
   (C) 477 kJ
   (D) −22.05 kJ
   (E) 8.820 kJ g^{-1}

6. Using the data in question 1, determine $\Delta H°$ for the combustion of methane.

   (A) −22.05 kJ mol^{-1}
   (B) −882 kJ
   (C) +22.05 kJ
   (D) −8.820 kJ g^{-1}
   (E) This value cannot be determined because w is not known.

7. Which of the following describes a system that CANNOT be spontaneous?

   (A) $\Delta H°$ is positive, and $\Delta S°$ is negative.
   (B) $\Delta H°$ is positive, and $\Delta S°$ is positive.
   (C) $\Delta H°$ is negative, and $\Delta S°$ is negative.
   (D) $\Delta H°$ is negative, and $\Delta S°$ is positive.
   (E) $\Delta H°$ is 0.00, and $\Delta S°$ is positive.

8. Which of the following explains the fact that, when KCl is dissolved, water condenses on the outside of the beaker?

(A) $\Delta H°$ is positive, and $\Delta S°$ is negative.
(B) $\Delta H°$ is positive, and $\Delta S°$ is positive.
(C) $\Delta H°$ is negative, and $\Delta S°$ is negative.
(D) $\Delta H°$ is negative, and $\Delta S°$ is positive.
(E) $\Delta H°$ is 0.00, and $\Delta S°$ is negative.

9. The reaction with the greatest expected entropy decrease is

(A) $CH_4(g) + 2O_2(g) \rightarrow CO_2(g) + 2H_2O(g)$
(B) $CH_4(l) + 2O_2(g) \rightarrow CO_2(g) + 2H_2O(g)$
(C) $CH_4(g) + 2O_2(g) \rightarrow CO_2(g) + 2H_2O(l)$
(D) $CH_4(g) + 2O_2(g) \rightarrow CO_2(s) + 2H_2O(g)$
(E) $CH_4(l) + 2O_2(g) \rightarrow CO_2(g) + 2H_2O(l)$

10. Water boils at 100 °C with a molar heat of vaporization of +43.9 kJ. At 100 °C what is the entropy change when water condenses?

$$H_2O(g) \rightarrow H_2O(l)$$

(A) Problem cannot be solved; $\Delta G°$ must also be known.
(B) Problem cannot be solved; this is not a chemical reaction.
(C) -439 J K^{-1}
(D) $+0.439 \text{ J K}^{-1}$
(E) -118 J K^{-1}

11. Which of the following will indicate if a given mixture of chemicals will react in the forward or in the reverse direction?

(A) ΔG
(B) $\Delta G°$
(C) $\Delta S°$
(D) $\Delta H°$
(E) ΔH

12. A gas is allowed to expand from an initial volume of 5.00 L and pressure of 3.00 atm to a volume of 15.0 L and pressure of 1.00 atm. What is the value of w?

(A) +30.0 L atm
(B) +10.0 L atm
(C) −45.0 L atm
(D) +15.0 L atm
(E) −10.0 L atm

13. In question 8 the units of work are given as L atm. To convert L atm to the metric unit of joules, we need to know

(A) Avogadro's constant and Planck's constant
(B) the universal gas law constant in units of $L \text{ atm mol}^{-1} \text{ K}^{-1}$
(C) the universal gas law constant in units of $J \text{ mol}^{-1} \text{ K}^{-1}$
(D) both B and C
(E) A, B, and C

14. Which of the following is the LEAST probable for a combustion reaction?

    (A) $\Delta G°$ is a large negative number.
    (B) $\Delta S°$ is a large negative number.
    (C) $\Delta H°$ is a large negative number.
    (D) K_{eq} is a large positive number.
    (E) Q, the reaction quotient, is a small number.

15. Of the following, which can be precisely determined for a chemical substance?

    (A) entropy, S
    (B) enthalpy, H
    (C) free energy, G
    (D) internal energy, E
    (E) all of these

16. The heat of formation of $CH_3OH(l) = -238.6$ kJ mol^{-1}, of $CO_2(g) = -393.5$ kJ mol^{-1}, and of $H_2O(g) = -241.8$ kJ mol^{-1}. What is $\Delta H°$ for the heat of combustion of methanol to gaseous products?

    (A) -396.7 kJ
    (B) -1277 kJ
    (C) -638.5 kJ
    (D) $+396.7$ kJ
    (E) This value cannot be calculated without the heat of formation for $O_2(g)$.

17. The rate of reaction will be large if

    (A) $\Delta G°$ is a large negative number
    (B) $\Delta S°$ is a large negative number
    (C) $\Delta H°$ is a large negative number
    (D) K_{eq} is a large positive number
    (E) None of the above can be used to estimate reaction rates.

18. Given the following thermochemical data:

$$N_2O_4(g) \rightarrow 2NO_2(g) \quad \Delta H° = +57.93 \text{ kJ}$$
$$2NO(g) + O_2(g) \rightarrow 2NO_2(g) \quad \Delta H° = -113.14 \text{ kJ}$$

determine the heat of the reaction

$$2NO(g) + O_2(g) \rightarrow N_2O_4(g)$$

    (A) 171.07 kJ
    (B) -55.21 kJ
    (C) -171.07 kJ
    (D) $+55.21$ kJ
    (E) -85.54 kJ

19. Which of the following can change the value of $\Delta G°$ for a chemical reaction?

    (A) changes in the total pressure
    (B) changes in the pressures of the reactants
    (C) changes in the concentrations of the reactants
    (D) changes in the temperature in °C
    (E) the presence of a catalyst

20. At what temperature is $K_{eq} = 1.00$ if $\Delta S° = 22.6$ J K^{-1} and $\Delta H° = 15.3$ kJ for a chemical reaction?

    (A) 404 °C
    (B) 677 °C
    (C) 0.67 °C
    (D) 1477 °C
    (E) 1204 °C

21. The standard heat of formation of $SO_3(g)$ is −396 kJ mol^{-1}. The standard entropies of $S(s)$, $O_2(g)$, and $SO_3(g)$ are 31.8, 205.0, and 256 J mol^{-1} K^{-1}, respectively. Calculate the free energy for the decomposition of SO_3 in the reaction

    $$2SO_3(g) \rightarrow 2S(s) + 3O_2(g)$$

    at 25 °C.

    (A) +396 kJ
    (B) −446 kJ
    (C) +346 kJ
    (D) −346 kJ
    (E) +742 kJ

22. The reaction

    $$2C_6H_6(l) + 15O_2(g) \rightarrow 12CO_2(g) + 6H_2O(l)$$

    is expected to have

    (A) a positive ΔH and a negative ΔS
    (B) a negative ΔH and a negative ΔS
    (C) a positive ΔH and a positive ΔS
    (D) a negative ΔH and a negative ΔS
    (E) These predictions cannot be made.

23. The evaporation of any liquid is expected to have

    (A) a positive ΔH and a negative ΔS
    (B) a negative ΔH and a negative ΔS
    (C) a positive ΔH and a positive ΔS
    (D) a positive ΔH and a negative ΔS
    (E) These predictions cannot be made.

24. Which of the following is most likely to be true?

    (A) No products are formed in a nonspontaneous reaction.
    (B) A positive $\Delta G°$ indicates a spontaneous reaction.
    (C) A positive $\Delta S°$ always means that the reaction is spontaneous.
    (D) A spontaneous reaction always goes to completion.
    (E) Combustion of organic compounds has a negative $\Delta H°$.

ANSWER KEY

1. A	6. B	11. A	16. B	21. E
2. E	7. A	12. E	17. E	22. B
3. D	8. B	13. D	18. C	23. C
4. A	9. C	14. B	19. D	24. E
5. D	10. E	15. A	20. A	

See Appendix 1 for explanations of answers.

Free-Response

Answer the following questions using the concepts of thermodynamics and equilibrium and the methods for solving problems.

(a) What parameters define whether or not a given reaction is spontaneous? Based on those parameters, what does it mean to say a reaction is spontaneous? What indicates if a mixture of chemicals will proceed in the forward direction or in the reverse direction?

(b) What is the difference between E, ΔE, and $\Delta E°$?

(c) Tables of thermodynamic data list heat of formation, $\Delta H_f°$, standard free energy, $\Delta G_{298}°$, and entropy, $S°$. Why don't entropy values have a delta symbol, Δ? What other difference does a table of entropy values have?

(d) The value of K_c for the reaction $2NO + O_2 \leftrightarrows N_2O_4$ is 36 at a certain temperature. Calculate K for the following reactions.

   (i) $6NO + 3O_2 \leftrightarrows 3N_2O_4$
   (ii) $NO + \frac{1}{2}O_2 \leftrightarrows \frac{1}{2}N_2O_4$
   (iii) $2N_2O_4 \leftrightarrows 4NO + 2O_2$

FREE-RESPONSE ANSWERS

(a) At this point we have defined a spontaneous reaction as one where K_{eq} is greater than 1.00. This translates into a value of $\Delta G°$ that has a negative sign. (In Chapter 13 we expand this to cases where $E°_{cell}$ has a positive sign.) For the second part of the question, this means that reactions that are symmetrical (equal moles on both sides of the arrow) will be spontaneous as long as more product is formed than reactants left over. For a mixture to proceed in the forward direction, $\Delta G < 0$ or $Q < K$. The opposite conditions are required for the mixture to proceed in the reverse direction.

(b) The difference between the three formats is that E is the total internal energy of a system. This cannot be determined exactly because all motions and attractive forces are not known. ΔE is measurable, as a change in internal energy. However, it refers to the energy change in a system that may be large or small; it is an extensive measure. Last, $\Delta E°$ represents the change in internal energy for a defined system and in an intrinsic property.

(c) Entropies are state functions that can be precisely determined by virtue of the third law of thermodynamics. Therefore we do not need to rely on changes in the state. Another difference is that elements have values of zero for heats of formation and free energies. Elements have nonzero entropies. Finally, entropy has units of J mol^{-1} K^{-1}, whereas the free energy and enthalpy have units of J mol^{-1}.

(d) (i) This reaction has the coefficients multiplied by 3, so that $K = K_c K_c K_c$ $= 4.66 \times 10^4$.

(ii) The coefficients are divided in half, so that $K = (K_c)^{1/2} = 6$.

(iii) In this reaction the reaction is reversed and the coefficients are multiplied by 2, so that $K = (1/(K_c)^2) = 7.7 \times 10^{-4}$.

PART 5

CHEMICAL REACTIONS

Oxidation-Reduction Reactions and Electrochemistry

Chemical reactions in which electrons are transferred from one atom to another are called **oxidation-reduction** reactions. As a group, more reactions may be classified as oxidation-reduction reactions than as acid-base, double-replacement, or complexation reactions combined.

Oxidation is the loss of electrons, and **reduction is the gain of electrons.** When an atom of barium reacts with an atom of sulfur, the barium loses its two valence electrons and is oxidized, while the sulfur gains these two electrons and is reduced.

$$\cdot \ddot{B}a + \cdot \ddot{S} \cdot \rightarrow Ba : \ddot{S} : \qquad (13.1)$$

This reaction may be written as two **half-reactions** that show the individual oxidation and reduction steps:

$$Ba \rightarrow Ba^{2+} + 2e^- \quad \text{(oxidation)} \tag{13.2}$$

$$S + 2e^- \rightarrow S^{2-} \quad \text{(reduction)} \tag{13.3}$$

Although we may write separate half-reactions, they cannot exist without each other. Perhaps the word *redox* was coined to emphasize this point.

In the oxidation-reduction reaction of Equation 13.1, the barium atom loses its electrons to the sulfur atom, causing the sulfur atom to be reduced. Consequently, barium is called the **reducing agent.** The opposite view is that sulfur gains electrons from the barium atom, causing the barium atom to be oxidized. Consequently, sulfur is called the **oxidizing agent.** In any redox reaction, the substance oxidized is the reducing agent and the substance reduced is the oxidizing agent.

OXIDATION NUMBERS

Determining Oxidation Numbers

In more complex reactions it is not always obvious that electrons are transferred. To determine whether a substance has lost or gained electrons, chemists calculate the **oxidation number,** also called the **oxidation state,** of each element. If the oxidation number changes during a reaction, electrons have been transferred.

To determine the oxidation numbers of the elements in a compound or polyatomic ion, we follow a short set of rules:

OXIDATION NUMBER RULE HIERARCHY

1. The oxidation numbers of all atoms add up to the charge on the atom, molecule, or ion.
2. The oxidation number of an alkali metal is +1, for an alkaline earth element it is +2, and for the metals in group IIIA it is +3.
3. The oxidation number of hydrogen is +1, and the oxidation number of fluorine is −1.
4. The oxidation number of oxygen is −2.
5. The oxidation number of a halogen is −1.
6. The oxidation number of nonmetal in group VIA is −2.

The oxidation number rules are a hierarchy, meaning that rule 1 is the most important and rule 6 is the least important. If two rules conflict, the rule closest to the top of the list is obeyed while the one lower on the list is ignored.

Many simple compounds follow the oxidation number rules directly. Since the charge of any element is zero, the oxidation number of any element must be zero, according to rule 1. In compounds such as $CaCl_2$, calcium is +2 according to rule 2, and each chlorine is −1 by rule 5; also, the most important rule, rule 1, is obeyed since +2 and $2 \times (-1)$ add up to zero, the charge on $CaCl_2$. Also, in potassium hydroxide, KOH, potassium is +1 according to rule 2; oxygen is −2 according to rule 4; and hydrogen is +1 according to rule 3. The oxidation numbers all obey rule 1 since $+1 +1 - 2 = 0$, which is the charge on KOH.

Some compounds, however, do not obey one or more of the rules. Since all substances must obey rule 1, we may write rule 1 as an equation in which the total charge is equal to the sum of the elements' oxidation numbers, with each oxidation number multiplied by the number of times the element appears in the formula:

$$\text{Total charge} = (\text{ox. no. 1})(\text{subscript 1}) +$$
$$+ (\text{ox. no. 2})(\text{subscript 2}) + \cdots \qquad (13.4)$$

For example, the oxidation numbers for the elements in perchloric acid, $HClO_4$, are +1 for hydrogen, −1 for chlorine, and −2 for each oxygen if rules 3, 4, and 5 are strictly followed. These assignments violate rule 1, however, since the oxidation numbers add up to −8 and the $HClO_4$ molecule must have a charge of zero. We, therefore, ignore rule 5 in favor of rule 1 and write the equation as

$$
\begin{aligned}
0 \text{ charge} &= (\text{ox. no. H})(1) + (\text{ox. no. Cl})(1) + (\text{ox. no. O})(4) \\
0 &= (+1)(1) + (\text{ox. no. Cl})(1) + (-2)(4) \\
0 &= \text{ox. no. Cl} - 7 \\
\text{ox. no. Cl} &= +7
\end{aligned}
$$

In perchloric acid the oxidation number of the chlorine atom is +7.

Oxidation numbers for polyatomic ions are determined using the same method, being sure that the oxidation numbers add up to the charge on the ion. For example, we might assign all the sulfur and oxygen atoms in the sulfate ion, SO_4^{2-}, oxidation numbers of −2 according to rules 4 and 6. That would disobey rule 1, however, since the sum of these oxidation numbers is $-2 - 8 = -10$, whereas the actual charge of the sulfate ion is −2. Consequently, we abandon rule 6 and work with rules 1 and 4 to write the equation

$$
\begin{aligned}
-2 \text{ charge} &= (\text{ox. no. S})(1) + (\text{ox. no. O})(4) \\
-2 &= \text{ox. no. S} + (-2)(4) \\
-2 &= \text{ox. no. S} - 8 \\
\text{ox. no. S} &= +6
\end{aligned}
$$

Finally, the rules specify oxidation numbers for only 25 of the 111 known elements. However, by using the oxidation number rules and the formulas of the polyatomic ions, chemists can determine the oxidation numbers of most of the other elements. For example, the oxidation numbers for chromium and oxygen in the dichromate ion, $Cr_2O_7^{2-}$, can be determined using rules 1 and 4:

$$
\begin{aligned}
-2 \text{ charge} &= (\text{ox. no. Cr})(2) + (\text{ox. no. O})(7) \\
-2 &= (\text{ox. no. Cr})(2) + (-2)(7) \\
-2 &= (\text{ox. no. Cr})(2) - 14 \\
(\text{ox. no. Cr})(2) &= +12 \\
\text{ox. no. Cr} &= +6
\end{aligned}
$$

Some compounds may contain two elements that are not covered by the oxidation number rules. One such compound is nickel(II) carbonate, $NiCO_3$. In situations like this one we take advantage of knowing that the carbonate ion is CO_3^{2-} and the nickel must be a +2 ion, Ni^{2+}. For the Ni^{2+}, rule 1 applies directly, and the oxidation number is +2. For the carbonate ion we write

$$-2 \text{ charge} = (\text{ox. no. C})(1) + (\text{ox. no. O})(3)$$
$$-2 = (\text{ox. no. C})(1) + (-2)(3)$$
$$-2 = \text{ox. no. C} - 6$$
$$\text{ox. no. C} = +4$$

Using this procedure, we have determined that the oxidation numbers are as follows: +2 for nickel, +4 for carbon, and −2 for each oxygen.

EXERCISE 13.1

Determine the oxidation number of each element in the following formulas:

(a) H_2O_2 (c) K_3PO_4 (e) $Fe_2(C_2O_4)_3$ (g) SO_3^{2-} (i) NH_4^+
(b) $HClO_3$ (d) Na_2CrO_4 (f) MnO_4^- (h) N_2O_5 (j) $FeNH_4(SO_4)_2$

Solution

(a) H = +1; O = −1
(b) H = +1; Cl = +5; O = −2
(c) K = +1; P = +5; O = −2
(d) Na = +1; Cr = +6; O = −2
(e) Fe = +3; C = +3; O = −2
(f) Mn = +7; O = −2

(g) S = +4; O = −2
(h) N = +5; O = −2
(i) N = −3; H = +1
(j) Fe = +2; N = −3; H = +1;
    S = +6; O = −2

Using Oxidation Numbers

Oxidation numbers are used to determine whether or not a substance has been oxidized or reduced. When the permanganate ion reacts in acid solution to form Mn^{2+}, we find that the oxidation number of the manganese is +7 in the permanganate ion and +2 in the Mn^{2+} ion. The change in oxidation number indicates that an oxidation-reduction process has taken place. In addition, the fact that the oxidation number was reduced from +7 to +2 tells us that a reduction of manganese has taken place. Also, the fact that the oxidation number changes by 5 tells us that five electrons have been gained by each manganese atom in the reduction process.

Remember that oxidation numbers are artificial constructs that allow the chemist to draw some conclusions about redox reactions. However, to assign oxidation numbers, we assume that all elements in a compound are ions. This is obviously not true. Therefore, the utility of oxidation numbers is quite limited.

EXERCISE 13.2

Possible reactants and products for oxidation-reduction reactions are given below. In each case, determine the oxidation numbers to tell whether an oxidation or a reduction has occurred and, if so, how many electrons were lost or gained from the starting material.

(a) $Cr_2O_7^{2-}$ to CrO_4^{2-} (d) I^- to IO_3^- (f) Fe to Fe_2O_3

(b) $C_2O_4^{2-}$ to CO_2 (e) $BaCl_2$ to $BaSO_4$ (g) C_2H_4 to C_2H_6

(c) NO_3^- to NO_2

Solution

(a) No change occurs in any oxidation numbers; not a redox process.

(b) C changes from +3 to +4. Two electrons per $C_2O_4^{2-}$ are gained in this oxidation.

(c) N changes from +5 to +4. One electron per NO_3^- is used in this reduction.

(d) I changes from −1 to +5. Six electrons per I^- are used in this oxidation.

(e) Ba does not change; no redox process occurs (Cl_2 and SO_4 are ignored).

(f) Fe changes from 0 to +3. Three electrons per Fe are lost in this oxidation.

(g) C changes from −2 to −3. Two electrons per C_2H_4 are gained in this reduction.

In a redox equation we may now identify what is occurring. For example, Fe^{2+} is often titrated with permanganate ions. The unbalanced reaction is

$$Fe^{2+} + MnO_4^- \rightarrow Mn^{2+} + Fe^{3+} \tag{13.5}$$

From this reaction we see that Fe^{2+} is oxidized to Fe^{3+}. The manganese has an oxidation number of +7 in the permanganate ion and of +2 in the product; therefore the manganese is reduced. We can also say that the permanganate ion is reduced. Because the substance oxidized causes the reduction of the other reactant, this substance is called the reducing agent. Fe^{2+} is oxidized and is also the reducing agent. Similarly, the MnO_4^- ion is reduced and is therefore the oxidizing agent.

Although the manganese in the permanganate ion is the atom actually reduced, chemists never say that the manganese atom is the reducing agent. The terms *oxidizing agent* and *reducing agent* are used with the entire formula unit, which contains the elements that are reduced or oxidized, not the elements themselves.

EXERCISE 13.3

In each of the following unbalanced equations, identify the element that is oxidized, the element that is reduced, and the changes in oxidation number. Also identify the oxidizing and reducing agents.

(a) $O_2 + N_2H_4 \rightarrow H_2O_2 + N_2$

(b) $XeO_3 + I^- \rightarrow Xe + I_2$

(c) $I_2 + OCl^- \rightarrow IO_3^- + Cl^-$

(d) $NO_3^- + Cu \rightarrow NO_2 + Cu^{2+}$

(e) $PbO_2 + Cl^- \rightarrow PbCl_2 + Cl_2$

Solution

The table below lists the required information. The change in oxidation number is indicated after each element in the first two columns.

Element Oxidized	Element Reduced	Oxidizing Agent	Reducing Agent
(a) N (+2)	O (−1)	O_2	N_2H_4
(b) I (+1)	Xe (−6)	XeO_3	I^-
(c) I (+5)	Cl (−2)	OCl^-	I_2
(d) Cu (+2)	N (−1)	NO_3^-	Cu
(e) Cl (+1)	Pb (−2)	PbO_2	Cl^-

BALANCING REDOX REACTIONS

Redox reactions tend to be complex, and attempts to balance them by inspection often fail to achieve an answer within a reasonable length of time. Over the years chemists have devised several methods for balancing redox reactions. The simplest and most versatile is called the **ion-electron method.** To use the ion-electron method, six steps must be performed, in the order given. Steps 1–6 are for reactions that occur in acid (H^+) solution. A seventh step is added if the reaction occurs in basic (OH^-) solution, or if H^+ appears on one side of the equation and OH^- appears on the other side.

ION-ELECTRON METHOD

1. Write two half-reactions, one representing the oxidation, and the other the reduction, that occur in the reaction. It is not necessary to know which is which at this point.
2. In each half-reaction, balance all atoms except hydrogen and oxygen.
3. Balance oxygen atoms in each half-reaction by adding one H_2O molecule for each oxygen atom needed. Never use O_2, OH^-, or any other form of oxygen.
4. Balance the hydrogen atoms by adding hydrogen ions, H^+. Never use H_2, OH^-, or any other form of hydrogen.
5. Balance the charges by adding the proper number of electrons (e^-). If steps 1–4 have been done properly, electrons will be added to the left side of one half-reaction and to the right side of the other.
6. Multiply each half-reaction by the appropriate number so that the two half-reactions have the same number of electrons. Add the half-reactions, and cancel the electrons (**they must cancel**). Also cancel all common ions and molecules. Simplify the coefficients of the equation if possible.
7. Use only if reaction must be balanced in basic solution. Add one OH^- ion for each H^+ ion to both sides of the equation in step 6. Combine the H^+ and OH^- ions on one side of the reaction into H_2O molecules. Cancel H_2O molecules that appear on both sides of the equation, and simplify if possible.

Before using the ion-electron method, however, it is worthwhile to try to balance a reaction by inspection. For example, the unbalanced equation for the reaction of magnesium metal with acid may be written as

$$Mg + H^+ \rightarrow Mg^{2+} + H_2$$

By inspection we see that placing the coefficient 2 in front of H^+ will balance the hydrogen atoms and the charge in this equation:

$$Mg + 2H^+ \rightarrow Mg^{2+} + H_2 \qquad (13.6)$$

There is no need to use the longer ion-electron method.

In the following example, however, the ion-electron method must be used. Standard solutions of iodine are prepared by reacting iodide ions with the iodate ion in acid solution. The skeleton reaction is

$$I^- + IO_3^- \rightarrow I_2$$

The first step of the ion-electron method requires two half-reactions, which are written as

$$I^- \rightarrow I_2$$
$$IO_3^- \rightarrow I_2$$

Even though the skeleton reaction has only one product, there is no reason why both reactants cannot yield the same product, one by oxidation and the other by reduction. Usually two pairs of reactants can be identified to obtain the half-reactions. In addition, H^+, OH^-, or H_2O may be included in the skeleton reaction. These are generally ignored since they are added in later steps.

The second step requires that the iodine atoms be balanced by using the coefficient 2 in both half-reactions:

$$2I^- \rightarrow I_2$$
$$2IO_3^- \rightarrow I_2$$

The third step applies only to the second half-reaction since the first one contains no oxygen. Six water molecules are needed to balance the six oxygens in two iodate ions:

$$2I^- \rightarrow I_2$$
$$2IO_3^- \rightarrow I_2 + \mathbf{6H_2O}$$

The fourth step also involves only the second half-reaction, where 12 H^+ ions are required:

$$2I^- \rightarrow I_2$$
$$\mathbf{12H^+} + 2IO_3^- \rightarrow I_2 + 6H_2O$$

At this point all the atoms are balanced, and we use step 5 to balance the charges with electrons. First we must count the charge on each side of each half-reaction. The first half-reaction has a total charge of −2 on the left and 0 on the right. The second half-reaction has a total charge of +10 on the left and 0 on the right. Two electrons on the right side of the first half-reaction are added to equalize the charges at −2. Ten electrons are added to the left side of the second half-reaction, resulting in a total charge of 0 on both sides:

$$2I^- \rightarrow I_2 + 2e^-$$

$$10e^- + 12H^+ + 2IO_3^- \rightarrow I_2 + 6H_2O$$

In step 6 we can equalize the electrons in the two half-reactions by multiplying the entire first half-reaction by 5:

$$10I^- \rightarrow 5I_2 + 10e^-$$

$$10e^- + 12H^+ + 2IO_3^- \rightarrow I_2 + 6H_2O$$

Adding the equations yields

$$10I^- + 10e^- + 12H^+ + 2IO_3^- \rightarrow$$

$$5I_2 + 10e^- + I_2 + 6H_2O$$

We then cancel 10 electrons from each side, add the iodine molecules on the right to obtain 6 I_2, and divide all of the coefficients by 2 to obtain the final balanced equation:

$$5I^- + 6H^+ + IO_3^- \rightarrow 3I_2 + 3H_2O \qquad (13.7)$$

All atoms must be balanced in the reaction. In addition, in ionic reactions the charges must also balance. In this equation we count a total charge of 0 on both sides of the arrow.

In a third example, the reaction between the permanganate ion and iodide ions occurs in neutral solutions. The unbalanced skeleton reaction is

$$MnO_4^- + I^- \rightarrow I_2 + MnO_2$$

The obvious pairs for the half-reactions are

$$MnO_4^- \rightarrow MnO_2$$

$$I^- \rightarrow I_2$$

In the second half-reaction the iodine atoms need to be balanced:

$$MnO_4^- \rightarrow MnO_2$$

$$2I^- \rightarrow I_2$$

The first half-reaction needs two water molecules to balance the oxygen atoms:

$$MnO_4^- \rightarrow MnO_2 + \mathbf{2H_2O}$$
$$2I^- \rightarrow I_2$$

Four hydrogen ions balance the hydrogen atoms:

$$\mathbf{4H^+} + MnO_4^- \rightarrow MnO_2 + 2H_2O$$
$$2I^- \rightarrow I_2$$

The next step is the balancing of charges. We find that the first half-reaction has a total charge of +3 on the left and 0 on the right, while the second half-reaction has a total charge of −2 on the left and 0 on the right. We add three electrons to the first half-reaction on the left and two electrons to the second half-reaction on the right:

$$\mathbf{3e^-} + 4H^+ + MnO_4^- \rightarrow MnO_2 + 2H_2O$$
$$2I^- \rightarrow I_2 + \mathbf{2e^-}$$

To equalize the electrons, we multiply the first half-reaction by 2 and the second half-reaction by 3:

$$6e^- + 8H^+ + 2MnO_4^- \rightarrow 2MnO_2 + 4H_2O$$
$$6I^- \rightarrow 3I_2 + 6e^-$$

We add the equations:

$$6e^- + 6I^- + 8H^+ + 2MnO_4^- \rightarrow 2MnO_2 + 4H_2O + 3I_2 + 6e^-$$

and cancel the 6 electrons:

$$6I^- + 8H^+ + 2MnO_4^- \rightarrow 2MnO_2 + 4H_2O + 3I_2$$

This reaction is balanced in acid solution. To convert to a basic solution, we use step 7 and add $8OH^-$ to each side:

$$\mathbf{8OH^-} + 6I^- + 8H^+ + 2MnO_4^- \rightarrow 2MnO_2 + 4H_2O + 3I_2 + \mathbf{8OH^-}$$

The $8H^+$ and $8OH^-$ on the left are combined to make $8H_2O$:

$$\mathbf{8H_2O} + 6I^- + 2MnO_4^- \rightarrow 2MnO_2 + 4H_2O + 3I_2 + 8OH^-$$

We then cancel $4H_2O$ from each side to simplify the final equation:

$$4H_2O + 6I^- + 2MnO_4^- \rightarrow 2MnO_2 + 3I_2 + 8OH^- \qquad (13.8)$$

EXERCISE 13.4

Balance each of the following half-reactions in acid solution:

(a) $Fe(s) \rightarrow Fe^{3+}$ (d) $NO_3^- \rightarrow NO_2$
(b) $Cl_2 \rightarrow Cl^-$ (e) $S_2O_3^{2-} \rightarrow SO_4^{2-}$
(c) $Cr^{3+} \rightarrow CrO_4^{2-}$

Solution

(a) $Fe(s) \rightarrow Fe^{3+} + 3e^-$
(b) $2e^- + Cl_2 \rightarrow 2Cl^-$
(c) $4H_2O + Cr^{3+} \rightarrow CrO_4^{2-} + 8H^+ + 3e^-$
(d) $e^- + 2H^+ + NO_3^- \rightarrow NO_2 + H_2O$
(e) $5H_2O + S_2O_3^{2-} \rightarrow 2SO_4^{2-} + 10H^+ + 8e^-$

EXERCISE 13.5

Balance each of the following skeleton redox reactions in the solution indicated:

(a) $Cr_2O_7^{2-} + CH_3CH_2OH \rightarrow Cr^{3+} + CH_3COOH$ (acid solution)
(b) $Cu + NO_3^- \rightarrow NO_2 + Cu^{2+}$ (acid solution)
(c) $MnO_2 + ClO_3^- \rightarrow MnO_4^- + Cl^-$ (basic solution)
(d) $Al + H_2O \rightarrow Al(OH)_4^- + H_2$ (basic solution)

Solution

(a) The balanced half-reactions are

$$6e^- + 14H^+ + Cr_2O_7^{2-} \rightarrow 2Cr^{3+} + 7H_2O$$

$$H_2O + CH_3CH_2OH \rightarrow CH_3COOH + 4H^+ + 4e^-$$

The balanced equation is

$$16H^+ + 2Cr_2O_7^{2-} + 3CH_3CH_2OH \rightarrow 4Cr^{3+} + 3CH_3COOH + 11H_2O$$

> **TIP**
>
> Reduction half-reactions have electrons as reactants.

(b) The balanced half-reactions are

$$Cu \rightarrow Cu^{2+} + 2e^-$$

$$e^- + 2H^+ + NO_3^- \rightarrow NO_2 + H_2O$$

The balanced equation is

$$Cu + 2NO_3^- + 4H^+ \rightarrow 2NO_2 + Cu^{2+}$$

> **TIP**
>
> Oxidation half-reactions have electrons as products.

(c) The balanced half-reactions are

$$6e^- + 6H^+ + ClO_3^- \rightarrow Cl^- + 3H_2O$$

$$2H_2O + MnO_2 \rightarrow MnO_4^- + 4H^+ + 3e^-$$

The balanced equation is

$$2MnO_2 + 2OH^- + ClO_3^- \rightarrow 2MnO_4^- + Cl^- + H_2O$$

(d) The balanced half-reactions are

$$4H_2O + Al \rightarrow Al(OH)_4^- + 4H^+ + 3e^-$$

$$2e^- + 2H^+ + H_2O \rightarrow H_2 + H_2O$$

The balanced equation is

$$2Al + 2OH^- + 6H_2O \rightarrow 2Al(OH)_4^- + 3H_2$$

COMMON OXIDATION-REDUCTION REACTIONS

Single-Replacement (Displacement) Reactions

In **single-replacement** reactions an element replaces an atom in a compound, producing another element and a new compound. For example, the element zinc replaces the hydrogen atom in hydrogen chloride, forming the element H_2 and the new compound zinc chloride:

$$Zn + 2HCl \rightarrow ZnCl_2 + H_2 \tag{13.9}$$

Very active metals, which have the lowest ionization energies, are Li, Na, K, Rb, Cs, Ca, Sr, and Ba. These elements all react with water in single-displacement reactions to form hydrogen. One example is

$$2Na + 2H_2O \rightarrow 2NaOH + H_2 \tag{13.10}$$

Many of these reactions also produce so much heat that the hydrogen ignites.

Active metals do not react with water, but will react with acids in a single-replacement reaction. An example is

$$Mg + 2H^+ \rightarrow Mg^{2+} + H_2 \tag{13.11}$$

The common active metals are Mg, Zn, Pb, Ni, Al, Ti, Cr, Fe, Cd, Sn, and Co.

Inactive metals do not undergo simple single-replacement reactions with either water or acids. The most common inactive metals are Ag, Pt, Au, and Cu. Copper and silver react with concentrated nitric acid in a reaction that produces nitrogen oxides but not hydrogen. Gold reacts with a mixture, called aqua regia, of three parts concentrated HCl and one part concentrated HNO_3. The reaction of very active, active, and inactive metals with water and acid are summarized in Table 13.1.

TABLE 13.1

Summary of Metal Reactions with Water and Acid

Type of Metal	Examples	Comments
Very active metal	Li, Na, K, Rb, Cs, Ca, Sr, Ba	React with H_2O to produce H_2, test gas by igniting a small amount
Active metal	Mg, Zn, Pb, Ni, Al, Ti, Cr, Fe, Cd, Sn, Co	React with acids to form H_2, but not with H_2O
Inactive metal	Ag, Au, Cu, Pt	DO not form H_2 with acids: may react with conc. oxidizing acids HNO_3 and H_2SO_4 or aqua regia

In addition to replacing hydrogen, metals will displace metal ions from their compounds. For instance, copper metal will react with silver nitrate in the single-replacement reaction

$$Cu + 2AgNO_3 \rightarrow Cu(NO_3)_2 + 2Ag \qquad (13.12)$$

The net ionic equation is written as

$$Cu + 2Ag^+ \rightarrow Cu^{2+} + 2Ag \qquad (13.13)$$

In these two reactions, copper is more active than silver and the reaction does occur. The reverse reaction:

$$Cu^{2+} + 2Ag \rightarrow Cu + 2Ag^+ \qquad (13.14)$$

does not occur, however, since silver is less active than copper.

An **activity series** is a listing of metals in the order of their strengths as oxidizing or reducing agents. We can use an activity series to determine whether a certain metal will displace another metal ion from its compounds. Later in this chapter (page 495) we will introduce the concept of standard reduction potentials. A table of standard reduction potentials contains the same information as an activity series. The metal having the lower, or more negative, standard reduction potential is the more active metal.

Only the active and inactive metals will displace each other from compounds. The very active metals listed in Table 13.1 do not displace other metals. Although these metals are certainly reactive enough to result in such displacements, their high activity causes them to react preferentially with water instead.

Nonmetals such as the halogens also participate in single-displacement reactions. The activity series for the halogens, from the strongest to the weakest oxidizing agent, is $F_2 > Cl_2 > Br_2 > I_2$. As a result, adding chlorine to a solution of potassium bromide results in the reaction

$$Cl_2 + 2KBr \rightarrow Br_2 + 2KCl \tag{13.15}$$

The net ionic reaction is

$$Cl_2 + 2Br^- \rightarrow Br_2 + 2Cl^- \tag{13.16}$$

Industrially important single-replacement reactions use carbon to displace oxygen from metal oxides in the refining process. For example, Fe_2O_3 is refined into iron in the reaction

$$2Fe_2O_3 + 3C \rightarrow 4Fe + 3CO_2 \tag{13.17}$$

Reactions of the Permanganate Ion

Permanganate solutions have a deep violet color, and the permanganate ion, MnO_4^-, is a very versatile oxidizing agent. It reacts differently in acidic, neutral, and basic solutions.

In acid solution the half-reaction is

$$MnO_4^- + 8H^+ + 5e^- \rightarrow Mn^{2+} + 4H_2O \tag{13.18}$$

Five electrons are used in this reduction, and the soluble Mn^{2+} ion is almost colorless. In addition, this reaction is slow and the presence of Mn^{2+} ions catalyzes the process.

In neutral or slightly acid solutions the permanganate half-reaction is

$$MnO_4^- + 4H^+ + 3e^- \rightarrow MnO_2 + 2H_2O \tag{13.19}$$

The hydrogen ions in this three-electron half-reaction come from the dissociation of water molecules. The product MnO_2 is an insoluble black precipitate.

In basic solutions the reaction involves a one-electron transfer:

$$MnO_4^- + e^- \rightarrow MnO_4^{2-} \tag{13.20}$$

The reaction mixture changes color from the deep violet of the MnO_4^- ion to a green color for the MnO_4^{2-} ion.

Reactions of Chromium (VI)

When chromium is in the +6 oxidation state, it forms different compounds, depending on the pH of the solution. In neutral or basic solutions the chromate ion, CrO_4^{2-}, predominates. In acid solutions the dichromate ion predominates:

$$2CrO_4^{2-} + 2H^+ \rightleftharpoons Cr_2O_7^{2-} + H_2O \tag{13.21}$$

In very acid solutions the dichromate ion becomes protonated to form chromic acid:

$$Cr_2O_7^{2-} + 2H^+ + H_2O \rightleftharpoons 2H_2CrO_4 \tag{13.22}$$

As the acidity of a solution increases, the strength of chromium(VI) as an oxidizing agent increases. Dichromate salts are dissolved in concentrated sulfuric acid to produce chromic acid, which is a very effective cleaning agent because it oxidizes most organic material. Dichromate salts dissolved in approximately 1 molar acid solutions are used for chemical analysis. For example, the reaction of dichromate with ethyl alcohol is used to test the sobriety of drivers suspected of driving while intoxicated (DWI):

$$8H^+ + Cr_2O_7^{2-} + 3CH_3CH_2OH \rightarrow 2Cr^{3+} + 3CH_3CHO + 7H_2O \tag{13.23}$$

Since the dichromate ion is orange, and the Cr^{3+} ion is green, the change in color from orange to green is a measure of the amount of alcohol present in the breath of a DWI suspect.

Iodine, Hydrogen Peroxide, and Thiosulfate

Iodine is a weak oxidizing agent. A dilute solution of iodine in alcohol, known as tincture of iodine, is an effective antiseptic for minor wounds. Because it is a weak oxidizing agent, iodine may be used in reaction mixtures where chemists want to oxidize only the more active reducing agents that are present.

Hydrogen peroxide, H_2O_2, is another weak oxidizing agent. Its mode of action is to decompose into water and atomic oxygen (O, not O_2). Atomic oxygen normally combines to form molecular O_2. However, in the brief time that it is available, atomic oxygen acts as a very good oxidizing agent if it encounters a suitable reactant. Hydrogen peroxide is used in 3 percent solutions as a household disinfectant and hair bleach. In higher concentrations, 30 percent, it is a very powerful oxidizing agent that must be handled with care. In fact, higher concentrations of H_2O_2 have been used as rocket fuel because of the oxygen this compound supplies.

The thiosulfate ion, $S_2O_3^{2-}$, is a reducing agent, one of the very few that is stable in air because it reacts slowly with O_2. Thiosulfate solutions are used as reducing agents in chemical analysis, often with iodine as the oxidizing agent. Active metals such as zinc and magnesium are also used as reducing agents in chemical reactions.

STOICHIOMETRY OF REDOX REACTIONS

In Chapter 6 stoichiometric calculations that may be used for any chemical reaction are discussed. Stoichiometric calculations for redox reactions are no different from the calculations for other reactions once the balanced equation has been obtained. However, some techniques are more commonly encountered in dealing with redox stoichiometry problems than with other types.

Converting Ionic Equations into Molecular Equations

The ion-electron method for balancing a redox equation results in a net ionic equation. The chemicals on the stockroom shelf are complete compounds, not individual ions. It is necessary, therefore, to translate the ions in a net ionic equation into compounds that can be measured in the laboratory. To do this, we convert a net ionic equation into a molecular equation by adding spectator ions back into the equation. The general principles are as follows:

1. A cation is converted into molecular form by adding negatively charged spectator ions to form a soluble compound. There are many possible choices; unless otherwise specified in the problem, the most convenient spectator ion to add is the chloride ion. If the chloride compound is insoluble, the next most convenient choice is the nitrate ion.

2. An anion is converted into molecular form by adding positively charged spectator ions to form a soluble compound. Again, there are many possible choices; unless otherwise specified in the problem, the most convenient spectator ion to add is the sodium ion. Hydrogen ions are added to anions to make acids if needed.

3. Whatever spectator ions are added to convert anions and cations into molecular forms must also be added to the opposite side of the equation to maintain balance.

To illustrate the process, let us determine the molecular equation for the net ionic equation

$$Ba^{2+} + SO_4^{2-} \rightarrow BaSO_4 \qquad (13.24)$$

Adding two chloride ions to the Ba^{2+} and two sodium ions to the SO_4^{2-} yields the molecular equation. To balance the final equation, 2 NaCl must be added to the product side.

$$BaCl_2 + Na_2SO_4 \rightarrow BaSO_4 + 2NaCl \qquad (13.25)$$

In the reaction used to determine ethyl alcohol:

$$8H^+ + Cr_2O_7^{2-} + 3CH_3CH_2OH \rightarrow 2Cr^{3+} + 3CH_3CHO + 7H_2O \qquad (13.26)$$

a question may be phrased in terms of potassium dichromate: for example, How many grams of potassium dichromate will react with 5.00 grams of ethyl alcohol? It is not necessary to write a complete molecular equation to solve the problem.

We simply add the necessary number of K$^+$ ions to form potassium dichromate on the left and an equal number of potassium ions on the right to obtain

$$8H^+ \ + \ K_2Cr_2O_7 \ + \ 3CH_3CH_2OH \ \rightarrow$$
$$2Cr^{3+} \ + \ 3CH_3CHO \ + \ 7H_2O \ + \ 2K^+ \qquad (13.27)$$

This form of the equation now has the information necessary to solve the problem.

ELECTROCHEMISTRY

The electrons in a balanced half-reaction show the direct relationship between a redox reaction and electricity. A nonspontaneous redox reaction may be forced to occur by adding electric energy in an **electrolytic cell.** A spontaneous redox reaction may be used to create a flow of electrons in a **galvanic cell.**

Electrolysis

TIP
Oxidation occurs at the anode.

TIP
Reduction occurs at the cathode.

In an **electrolysis** experiment a nonspontaneous chemical reaction is forced to occur when two **electrodes** are immersed in an electrically conductive sample, and the electrical voltage applied to the two electrodes is increased until electrons flow. At the electrode supplying the electrons, reduction reactions occur. This electrode is called the **cathode.** At the other electrode, the **anode,** oxidation reactions occur.

One type of sample that may be electrolyzed is a **molten salt.** Salts are composed of ions; when the salt is a solid, the ions are immobile in the crystal lattice. Heating the salt until it melts frees the ions, and the mobility of the ions in the molten salt makes the salt electrically conductive. In the electrolysis of a molten salt that does not contain any polyatomic ions, the cation of the salt will be reduced at the cathode and the anion of the salt will be oxidized at the anode. For example, sodium chloride can be melted, and the electrolysis reactions will be

HELPFUL HINTS
Cathode and *reduction* both start with consonants. *Anode* and *oxidation* both start with vowels.

$$\text{Cathode reaction:} \quad Na^+ + e^- \rightarrow Na \qquad (13.28)$$

$$\text{Anode reaction:} \quad 2Cl^- \rightarrow Cl_2 + 2e^- \qquad (13.29)$$

A diagram of an electrolytic cell is shown is Figure 13.1.

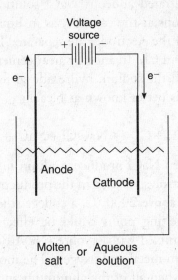

FIGURE 13.1. Setup of an electrolytic cell, described in the text.

Aqueous solutions of salts are also electrically conductive and may be electrolyzed. For solutions, two additional reactions are possible:

$$\text{Cathode: } 2H_2O + 2e^- \rightarrow H_2 + 2OH^- \tag{13.30}$$

$$\text{Anode: } 2H_2O \rightarrow O_2 + 4H^+ + 4e^- \tag{13.31}$$

There are now two possible reactions at each electrode. At the cathode we will have either the reduction of water, as shown in Equation 13.30, or the reduction of a metal ion. At the anode we will have either the oxidation of water or the oxidation of the salt's anion. The following principles may be used to decide which reaction takes place at each electrode:

1. *Cathode:* If the metal ion is a very active metal (Table 13.1), water will be reduced. If the metal ion is an inactive or an active metal, the metal ion will be reduced.

2. *Anode:* If the anion is a polyatomic ion, it generally will not be oxidized. In particular, the sulfate, nitrate, and perchlorate polyatomic anions are not oxidized in aqueous solution. Chloride, bromide, and iodide ions will be oxidized in aqueous solution. If an anion in one salt is oxidized in an aqueous electrolysis, that same anion in any other salt will also be oxidized. For example, since a solution of NaBr results in Br^- being oxidized to Br_2, we may predict that solutions of KBr, $CaBr_2$, NH_4Br, and $AlBr_3$ will all produce Br_2 at the anode.

Electrolysis is important in the industrial production of several chemical materials. Because of the need for electricity, most operations that use electrolysis are located in areas where electricity is inexpensive. Two of these are in the Pacific Northwest and in Niagara Falls close to large hydroelectric generators.

INDUSTRIAL USES FOR ELECTROLYSIS

The electrolysis of concentrated aqueous NaCl solutions, called brine, produces hydrogen and hydroxide ions at the cathode, as in Equation 13.30. At the anode chlorine gas is produced. If the electrodes are separated by a porous membrane, the products are H_2, NaOH, and Cl_2. In another arrangement, if the solution is stirred, the chlorine gas reacts with the sodium hydroxide to form a sodium hypochlorite (NaOCl) solution, which is better known as bleach:

$$2NaOH + Cl_2 \rightarrow NaOCl + NaCl + H_2O \qquad (13.32)$$

The electrolysis of molten NaCl produces sodium metal and chlorine gas.

Another molten-salt electrolysis is used in the production of aluminum. Although Al_2O_3 has a melting point above 2000 °C, a college student, Charles Hall, discovered in 1886 that this melting point could be effectively decreased to around 1000 °C when Al_2O_3 was mixed with cryolite (Na_3AlF_6). When electrolyzed, this mixture produces aluminum metal and oxygen. The more active sodium and fluorine are electrolyzed only when all of the aluminum and oxygen are used up.

Copper, the metal most used for electrical wiring, is refined by electrolysis. Copper ore is often a sulfide of copper that is refined into an impure form of copper by roasting:

$$CuS + O_2 \rightarrow Cu + SO_2 \qquad (13.33)$$

The impure copper is refined by using a bar, an ingot, of crude copper as the anode of an electrolytic cell. The cathode is a small strip of pure copper. During electrolysis the copper is oxidized to Cu^{2+} at the anode and then reduced to copper metal again at the cathode. In the process, impurities such as silver and gold drop to the bottom of the vessel as sludge. The value of the sludge comes close to paying for the electricity used in the refining.

In addition to producing industrial quantities of chemicals, electrolysis is used to electroplate thin layers of a decorative metal on a less expensive metal. Silver and gold are electroplated on iron utensils for decoration. Gold is electroplated on electrical contacts on computer circuit boards to decrease corrosion failure. Chromium is electroplated on automobile parts for decoration and resistance to corrosion.

QUANTITATIVE ELECTROCHEMISTRY

Every balanced half-reaction specifies the number of electrons lost or gained. Consequently, stoichiometric calculations can be used to convert between moles of electrons and moles of all other substances in the half-reaction. If the number of electrons flowing through the electrolysis cell is measured, the quantity of material reacted can be calculated. It was Michael Faraday who discovered the relationship between the mole and electric current. He started with the definition of the coulomb (C), which is the number of electrons that flow past a given point in a wire in exactly 1 second when the current is exactly 1 ampere:

$$1 \text{ coulomb} = 1 \text{ ampere} \times 1 \text{ second} \qquad (13.34)$$

Faraday found that 96,485 coulombs were equal to 1 mole of electrons:

$$1 \text{ mole } e^- = 96,485 \text{ coulombs} \tag{13.35}$$

Faraday's constant, $\mathscr{F}$, is the conversion factor $\left(\dfrac{96,485 \text{ C}}{\text{mol } e^-} \right)$. Using Equations 13.34 and 13.35, we can determine the moles of electrons by measuring the current, I, and the time, t, that the current flows:

$$\text{moles of } e^- = \frac{(I \text{ C/s})(t \text{ s})}{96,485 \text{ C/mole}} = \frac{It}{\mathscr{F}} \tag{13.36}$$

The electric current, I, has units of coulombs per second ($C \text{ s}^{-1}$), time has units of seconds (s), and Faraday's constant is 96,485 coulombs per mole of electrons.

Once the moles of electrons are calculated, the familiar stoichiometric calculations are used to determine other quantities. For example, the half-reaction for the reduction of iodine is

$$I_2 + 2e^- \rightarrow 2I^- \tag{13.37}$$

The moles of I^- produced at an electrode with a current of 0.500 ampere for 90 minutes may be calculated by first determining the moles of electrons:

$$\text{mol } e^- = \frac{(0.500 \text{ C s}^{-1})(90 \text{ min})(60 \text{ s min}^{-1})}{96,485 \text{ C mol}^{-1}}$$

$$= 0.0280 \text{ mol } e^-$$

Using the moles of electrons as the starting point, we can write the stoichiometric calculation:

$$? \text{ mol } I^- = 0.0280 \text{ mol } e^- \left(\frac{2 \text{ mol } I^-}{2 \text{ mol } e^-} \right) = 0.0280 \text{ mol } I^-$$

Chemists often bypass the stoichiometric step by incorporating the stoichiometry into Equation 13.36:

$$\text{moles of } X = \frac{It}{n\mathscr{F}} \tag{13.38}$$

In this equation n is the number of electrons per mole of substance in the balanced half-reaction.

EXERCISE 13.6

A current of 2.34 A is delivered to an electrolytic cell for 85 min. How many grams of (a) Au from $AuCl_3$, (b) Ag from $AgNO_3$, and (c) Cu from $CuCl_2$ will be obtained?

Solution

(a) $Au^{3+} \quad + \quad 3e^- \quad \rightarrow \quad Au$

$$\text{mol Au} = \frac{(2.34 \text{ C s}^{-1})(85 \text{ min})(60 \text{ s min}^{-1})}{(3)(96,485 \text{ C mol}^{-1})}$$

$$= 4.12 \times 10^{-2} \text{ mol Au}$$

$$\text{g Au} = 4.12 \times 10^{-2} \text{ mol Au} \left(\frac{197 \text{ g Au}}{1 \text{ mol Au}} \right) = 8.12 \text{ g Au}$$

(b) $Ag^+ \quad + \quad e^- \quad \rightarrow \quad Ag$

$$\text{mol Ag} = \frac{(2.34 \text{ C s}^{-1})(85 \text{ min})(60 \text{ s min}^{-1})}{(1)(96,485 \text{ C mol}^{-1})} = 0.124 \text{ mol Ag}$$

$$\text{g Ag} = 0.124 \text{ mol Ag} \left(\frac{108 \text{ g Ag}}{1 \text{ mol Ag}} \right) = 13.4 \text{ g Ag}$$

(c) $Cu^{2+} \quad + \quad 2e^- \quad \rightarrow \quad Cu$

$$\text{mol Cu} = \frac{(2.34 \text{ C s}^{-1})(85 \text{ min})(60 \text{ s min}^{-1})}{(2)(96,485 \text{ C mol}^{-1})}$$

$$= 6.18 \times 10^{-2} \text{ mol Cu}$$

$$\text{g Cu} = 6.18 \times 10^{-2} \text{ mol Cu} \left(\frac{63.5 \text{ g Cu}}{1 \text{ mol Cu}} \right) = 3.92 \text{ g Cu}$$

Galvanic Cells

Galvanic cells are used to harness the energy of spontaneous redox reactions. This is done by physically separating the chemicals in the two half-reactions so that the electrons generated by the oxidation half-reaction must flow through an electrical conductor before they can be used in the reduction half-reaction. This flow of electrons can be diverted through cell phones, iPods, laser pointers, and other devices to perform useful work before they reach their destination.

A galvanic cell is constructed as shown in Figure 13.2. All of the reactants in the oxidation half-reaction are placed in the left beaker, and all of the reactants in the reduction half-reaction in the right beaker. If a half-reaction is written with a metal, the metal serves as the electrode for that beaker; otherwise, an inert electrode made of platinum, silver, or gold is used. Electrodes are connected to each other with a metal wire, usually copper, and a device such as a meter (voltmeter or ammeter), motor, or light bulb may be inserted in the electrical circuit. If a voltmeter is used, the positive side of the voltmeter is connected to the cathode and the negative or ground of the voltmeter is connected to the anode to get the correct reading. To complete the circuit, a salt bridge is needed. The flow of charge is carried by electrons in the wires and by ions through the solutions and salt bridge.

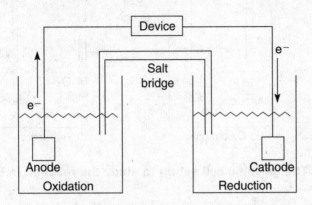

FIGURE 13.2. The galvanic cell.

In setting up a galvanic cell, we start with a balanced redox reaction. For the spontaneous reaction of permanganate with iron(II) in acid solution, the equation is

$$5Fe^{2+} + MnO_4^- + 8H^+ \rightleftharpoons 5Fe^{3+} + Mn^{2+} + 4H_2O \qquad (13.39)$$

To identify where the chemicals should be placed, the half-reactions are written:

$$Fe^{2+} \rightarrow Fe^{3+} + e^- \qquad (13.40)$$

$$5e^- + MnO_4^- + 8H^+ \rightarrow Mn^{2+} + 4H_2O \qquad (13.41)$$

Equation 13.40 is the oxidation half-reaction, and Fe^{2+} and Fe^{3+} are placed in the left beaker. Since neither Fe^{2+} nor Fe^{3+} is a metal, a platinum electrode will be used. MnO_4^-, H^+, and Mn^{2+} must be placed in the right cell since they are the components of the reduction half-reaction (Equation 13.41). Once again, a platinum electrode will be used. A voltmeter will measure the tendency of electrons to flow. The complete cell is shown in Figure 13.3.

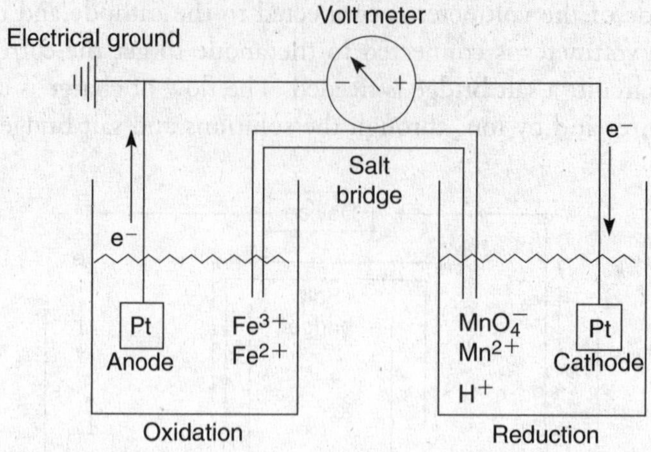

FIGURE 13.3. The galvanic cell set up to study the reaction in Equation 13.39.

To obtain consistent results from a galvanic cell, the variables of temperature, pressure, and concentration must be controlled. For this purpose electrochemists define a standard state for these experiments. Standard state for the galvanic cell is a temperature of 298 K, a pressure of 1.00 atmosphere for all gases, and concentrations of 1.00 molar for all soluble compounds. Concentrations of all solids and pure liquids are also defined as 1.00 molar.

At standard state, the galvanic cell diagram assumes that the reaction of Equation 13.39 is spontaneous and that oxidation and the anode are on the left while reduction and the cathode are on the right. Since this reaction is known to be spontaneous, the cell in Figure 13.3 is correct, and a positive voltmeter reading will be obtained. This voltmeter reading is called the **standard cell voltage,** E°_{cell}, or the **electromotive force,** $\mathscr{EMF}$.

A nonspontaneous reaction may be converted into a spontaneous reaction by reversing the chemical equation. Then the steps outlined above are used to determine the correct setup of the galvanic cell.

In reality we often do not know whether a reaction is spontaneous or nonspontaneous. In such a situation, a chemical equation is written and the galvanic cell is constructed for that equation. When E°_{cell} is measured, it will be either positive or negative. If E°_{cell} is positive, the reaction is spontaneous as written. If E°_{cell} turns out to be negative, the reaction is nonspontaneous as written and should be reversed to construct the galvanic cell.

The chemical reaction and the setup for the galvanic cell are directly related to each other. Knowing one of them allows us to determine the other. The chemicals present in the cell with the anode are written as an oxidation half-reaction; those present in the cell with the cathode, as the reduction half-reaction.

A **cell diagram** is a shorthand method of drawing a galvanic cell. The cell diagram "reads" the galvanic cell from left to right, starting with the anode and ending with the cathode. For the standard-state reaction of permanganate with iron(II) the cell diagram is as follows:

$$\text{Pt} \mid \text{Fe}^{2+}(1 \text{ M}), \text{Fe}^{3+}(1 \text{ M}) \parallel \text{MnO}_4^-(1 \text{ M}), \text{Mn}^{2+}(1 \text{ M}), \text{H}^+(1 \text{ M}) \mid \text{Pt} \qquad (13.42)$$

The single vertical lines represent phase changes between the solid platinum electrodes and the solutions. The double vertical lines represent the salt bridge because it has a phase boundary at either end. Concentrations, if known, are shown in parentheses. The order in which the chemicals are written, within each cell, is not important.

STANDARD REDUCTION POTENTIALS

The standard cell voltage is the difference between the electric potentials (voltages) of the cathode and the anode:

$$E^{\circ}_{cell} = E^{\circ}_{cathode} - E^{\circ}_{anode} \tag{13.43}$$

The $E^{\circ}_{cathode}$ is the standard reduction potential for the reaction occurring at the cathode and represents its tendency to remove electrons from the electrode surface. E°_{anode} is the standard reduction potential for the reaction occurring at the anode and represents its tendency to remove electrons from the anode. It is impossible to independently measure the electromagnetic force values of the cathode and anode. The only measurement possible is the combined E°_{cell}.

If it were possible to measure $E^{\circ}_{cathode}$ or E°_{anode} for any reaction, the standard electrode potentials of all other half-reactions could be determined using Equation 13.43. Since neither $E^{\circ}_{cathode}$ nor E°_{anode} can be independently measured, chemists define the reduction of hydrogen, at standard state, as having a reduction potential of exactly 0.00 volt:

$$H^+(aq) + e^- \rightarrow \tfrac{1}{2}H_2(g) \quad E^{\circ} = 0.00\,\text{volt} \tag{13.44}$$

This definition is used to determine the potentials of other half-reactions in relation to the reduction of hydrogen ions. For example, the reaction of

$$Zn(s) + 2H^+(aq) \rightarrow Zn^{2+}(aq) + H_2(g) \tag{13.45}$$

has a standard cell voltage of $E^{\circ}_{cell} = +0.76$ volt. Since the zinc oxidation occurs at the anode and the reduction of hydrogen ions occurs at the cathode, we may write

$$E^{\circ}_{cell} = +0.76\,\text{V} = E^{\circ}_{H^+} - E^{\circ}_{Zn}$$

The $E^{\circ}_{H^+}$ has been defined as 0.00 volt, and entering that value into the preceding equation gives

$$+0.76\,\text{V} = 0.00\,\text{V} - E^{\circ}_{Zn}$$

Rearranging this equation yields

$$E^{\circ}_{Zn} = -0.76\,\text{V}$$

This is the standard reduction potential for zinc.

Once the standard reduction potential for zinc is known, we may study the reaction

$$Zn(s) + 2Ag^+(aq) \rightarrow Zn^{2+}(aq) + 2Ag(s) \tag{13.46}$$

The standard cell voltage for this reaction is measured as +1.56 volts, so

$$E_{cell}^\circ = +1.56 \text{ V} = E_{Ag^+}^\circ - E_{Zn^{2+}}^\circ$$

Substituting the value for the standard reduction potential of zinc into the above equation yields

$$+1.56 \text{ V} = E_{Ag^+}^\circ - (-0.76 \text{ V})$$

Rearranging and solving yields the standard reduction potential of silver ions:

$$E_{Ag^+}^\circ = +1.56 \text{ V} - 0.76 \text{ V} = +0.80 \text{ V}$$

In a similar manner the standard reduction potentials of all half-reactions are determined. The standard reduction potentials for most half-reactions are known, and some of these are listed in Table 13.2.

With a table of standard reduction potentials, E_{cell}° for any redox reaction can be predicted. If E_{cell}° is positive, the reaction is spontaneous; if it is negative, the reaction is not spontaneous.

EXERCISE 13.7

Write the balanced redox reaction for each of the following, determine E_{cell}°, and state whether or not the reaction is spontaneous:

(a) the single displacement of copper(II) by zinc metal
(b) the single displacement of H^+ by iron, forming Fe^{2+}
(c) the reduction of tin(IV) to tin(II) by Fe(II) forming Fe(III)
(d) the reduction of dichromate ions to chromium(III) ions by Mn^{2+}, forming MnO_2
(e) the oxidation of AsO_3^{3-} to $H_2AsO_4^-$ by the reduction of iodine to iodide ions

Solution

(a) $Cu^{2+} + Zn \rightarrow Zn^{2+} + Cu$
$E_{cell}^\circ = +0.34 \text{ V} - (-0.76 \text{ V}) = +1.10 \text{ V}$
The reaction is spontaneous.

(b) $Fe + 2H^+ \rightarrow Fe^{2+} + H_2$
$E_{cell}^\circ = 0.00 \text{ V} - (-0.44 \text{ V}) = +0.44 \text{ V}$
The reaction is spontaneous.

(c) $Sn^{4+} + 2Fe^{2+} \rightarrow 2Fe^{3+} + Sn^{2+}$
$E_{cell}^\circ = +0.15 \text{ V} - (+0.77 \text{ V}) = -0.62 \text{ V}$
The reaction is not spontaneous.

TABLE 13.2

Standard Reduction Potentials, 25 °C

Half-Reaction					$E°$ (V)
$F_2(g)$	+		$2e^-$	$\rightarrow$ $2F^-$	2.87
Co^{3+}	+		e^-	$\rightarrow$ Co^{2+}	1.82
Au^{3+}	+		$3e^-$	$\rightarrow$ Au	1.50
$Cl_2(g)$	+		$2e^-$	$\rightarrow$ $2Cl^-$	1.36
$O_2(g)$	+ $4H^+$	+	$4e^-$	$\rightarrow$ $2H_2O$	1.23
$Br_2(g)$	+		$2e^-$	$\rightarrow$ $2Br^-$	1.07
$2Hg^{2+}$	+		$2e^-$	$\rightarrow$ Hg_2^{2+}	0.92
Ag^+	+		e^-	$\rightarrow$ Ag	0.80
Hg_2^{2+}	+		$2e^-$	$\rightarrow$ 2Hg	0.79
Fe^{3+}	+		e^-	$\rightarrow$ Fe^{2+}	0.77
I_2	+		$2e^-$	$\rightarrow$ $2I^-$	0.53
Cu^+	+		e^-	$\rightarrow$ Cu	0.52
Cu^{2+}	+		$2e^-$	$\rightarrow$ Cu	0.34
Cu^{2+}	+		e^-	$\rightarrow$ Cu^+	0.15
Sn^{4+}	+		$2e^-$	$\rightarrow$ Sn^{2+}	0.15
S	+ $2H^+$	+	$2e^-$	$\rightarrow$ H_2S	0.14
$2H^+$	+		$2e^-$	$\rightarrow$ H_2	0.00
Pb^{2+}	+		$2e^-$	$\rightarrow$ Pb	−0.13
Sn^{2+}	+		$2e^-$	$\rightarrow$ Sn	−0.14
Ni^{2+}	+		$2e^-$	$\rightarrow$ Ni	−0.25
Co^{2+}	+		$2e^-$	$\rightarrow$ Co	−0.28
Tl^+	+		e^-	$\rightarrow$ Tl	−0.34
Cd^{2+}	+		$2e^-$	$\rightarrow$ Cd	−0.40
Cr^{3+}	+		e^-	$\rightarrow$ Cr^{2+}	−0.41
Fe^{2+}	+		$2e^-$	$\rightarrow$ Fe	−0.44
Cr^{3+}	+		$3e^-$	$\rightarrow$ Cr	−0.74
Zn^{2+}	+		$2e^-$	$\rightarrow$ Zn	−0.76
Mn^{2+}	+		$2e^-$	$\rightarrow$ Mn	−1.18
Al^{3+}	+		$3e^-$	$\rightarrow$ Al	−1.66
Be^{2+}	+		$2e^-$	$\rightarrow$ Be	−1.70
Mg^{2+}	+		$2e^-$	$\rightarrow$ Mg	−2.37
Na^+	+		e^-	$\rightarrow$ Na	−2.71
Ca^{2+}	+		$2e^-$	$\rightarrow$ Ca	−2.87
Sr^{2+}	+		$2e^-$	$\rightarrow$ Sr	−2.89
Ba^{2+}	+		$2e^-$	$\rightarrow$ Ba	−2.90
Rb^+	+		e^-	$\rightarrow$ Rb	−2.92
K^+	+		e^-	$\rightarrow$ K	−2.92
Cs^+	+		e^-	$\rightarrow$ Cs	−2.92
Li^+	+		e^-	$\rightarrow$ Li	−3.05

TIP

The standard reduction potential table on the AP exam may list half-reactions in the reverse order, from the most negative $E°$ to the most positive.

(d) $2H^+ + Cr_2O_7^{2-} + 3Mn^{2+} \rightarrow 3MnO_2 + 2Cr^{3+} + H_2O$

$E^\circ_{cell} = +1.33\ V - (+1.23\ V) = +0.10\ V$

The reaction is spontaneous.

(e) $H_2O + AsO_3^{3-} + I_2 \rightarrow H_2AsO_4^- + 2I^-$

$E^\circ_{cell} = +0.54\ V - (+0.58\ V) = -0.04\ V$

The reaction is not spontaneous.

CELL VOLTAGES

In the preceding section the standard reduction potentials were used to determine the standard cell voltage expected from a galvanic cell. Standard conditions specify the temperature, pressure, and concentrations necessary to obtain these standard voltages, and the voltage measured will differ from the standard voltage if any of these variables are changed. Walter Nernst showed that

$$E_{cell} = E^\circ_{cell} - \frac{RT}{n\mathscr{F}} \ln Q \qquad (13.47)$$

This is called the Nernst equation. E_{cell} is the cell voltage under nonstandard conditions; E°_{cell} is the standard cell voltage; $R = 8.314\ C\ V\ mol^{-1}\ K^{-1}$ is the universal gas constant; T is the Kelvin temperature; n is the number of electrons in the half-reaction; $\mathscr{F} = 96,485\ C\ mol^{-1}$ is Faraday's constant; and Q is the reaction quotient for the chemical reaction. For 298 K, the Nernst equation is often written as

$$E_{cell} = E^\circ_{cell} - \frac{0.0591}{n} \log Q \qquad (13.48)$$

The constants R and $\mathscr{F}$ along with the temperature, 298 K, and a conversion factor from natural logarithms to base 10 logarithms are combined into the constant 0.0591.

Both Equation 13.47 and Equation 13.48 become $E_{cell} = E^\circ_{cell}$ at standard state. At the standard state, all concentrations in the reaction quotient are equal to 1.00, and therefore $Q = 1.00$. The logarithm of 1.00 is zero, and the log term disappears so that $E_{cell} = E^\circ_{cell}$.

For the dichromate reduction half-reaction the Nernst equation is

$$6Fe^{2+} + Cr_2O_7^{2-} + 14H^+ \rightarrow 2Cr^{3+} + 7H_2O + 6Fe^{3+} \qquad (13.49)$$

$$E_{cell} = E^\circ_{cell} - \frac{0.0591}{6} \log \left(\frac{[Cr^{3+}]^2[Fe^{3+}]^6}{[Cr_2O_7^{2-}][H^+]^{14}[Fe^{2+}]^6} \right)$$

In the Nernst equation, Q is always based on the form of the equilibrium law for the complete reaction.

EXERCISE 13.8

The reduction of iron(III) with tin(II) proceeds by the reaction

$$2Fe^{3+} + Sn^{2+} \rightarrow 2Fe^{2+} + Sn^{4+}.$$

In a galvanic cell all of the product concentrations are 0.0355 M and the reactant concentrations are 0.100 M. What is the voltage of the galvanic cell at 25 °C?

$$E^{\circ}_{(Sn^{4+})} = +0.154 \text{ V} \quad \text{and} \quad E^{\circ}_{(Fe^{3+})} = +0.771 \text{ V}$$

Solution

The Nernst equation for the entire reaction is used:

$$E_{cell} = E^{\circ}_{cell} - \frac{0.0591}{n} \log Q$$

where $n = 2$ and Q and E°_{cell} are calculated as

$$Q = \frac{[Fe^{2+}]^2[Sn^{4+}]}{[Fe^{3+}]^2[Sn^{2+}]} = \frac{(0.0355)^2(0.0355)}{(0.100)^2(0.100)} = 0.0447$$

The value for Q and E°_{cell} are substituted into the Nernst equation

$$
\begin{aligned}
E^{\circ}_{cell} &= +0.771 - 0.154 = +0.617 \text{ V} \\
E &= 0.617 - \frac{0.0591}{2} \log (0.0447) \\
&= 0.617 - (-0.040) \\
&= +0.657 \text{ V}
\end{aligned}
$$

STANDARD CELL VOLTAGES AND EQUILIBRIUM

If the two electrodes of a galvanic cell are connected so that electrons flow freely, the reaction will proceed to equilibrium. At equilibrium $E_{cell} = 0$ and the mathematical rearrangement of the two Nernst equations becomes

$$
\begin{aligned}
E^{\circ}_{cell} &= E^{\circ}_{cathode} - E^{\circ}_{anode} \\
&= \frac{RT}{n^{\wedge}} \ln K_{eq} \\
\text{or} &= \frac{0.0591}{n} \log K_{eq} \quad\quad (13.50)
\end{aligned}
$$

where n is the total number of electrons transferred in the redox reaction, and the mathematical combination of the anode reaction quotient and the cathode reaction quotient gives the equilibrium constant, K_{eq}.

The table of standard reduction potentials also provides the information necessary to calculate the equilibrium constant for a redox reaction.

EXERCISE 13.9

Determine the equilibrium constant for each of the following reactions at 298 K:

(a) the single displacement of copper(II) by zinc metal
(b) the single displacement of H^+ by iron, forming Fe^{2+}
(c) the reduction of tin(iv) to tin(II) by the oxidation of Fe(II) being oxidized to Fe(III)
(d) the reduction of dichromate ions to chromium(III) ions Mn^{2+}, forming MnO_2
(e) the oxidation of AsO_3^{3-} to $H_2AsO_4^-$ by the reduction of iodine to iodide ions

Solution

In Exercise 13.7 the reactions and E°_{cell} values were determined. The equation $E^\circ_{cell} = \dfrac{0.0591}{n} \log K_{eq}$ is used in all cases once E°_{cell} and n are determined.

(a) $Cu^{2+} + Zn \rightarrow Zn^{2+} + Cu$
    $E^\circ_{cell} = +0.34\ V - (-0.76\ V) = +1.10\ V$
    Two electrons are transferred, and $K_{eq} = 1.7 \times 10^{37}$.

(b) $Fe + 2H^+ \rightarrow Fe^{2+} + H_2$
    $E^\circ_{cell} = 0.00\ V - (-0.44\ V) = +0.44\ V$
    Two electrons are transferred, and $K_{eq} = 7.8 \times 10^{14}$.

(c) $Sn^{4+} + 2Fe^{2+} \rightarrow 2Fe^{3+} + Sn^{2+}$
    $E^\circ_{cell} = +0.15\ V - (+0.77\ V) = -0.62\ V$
    Two electrons are transferred, and $K_{eq} = 1.0 \times 10^{-21}$.

(d) $2H^+ + Cr_2O_7^{2-} + 3Mn^{2+} \rightarrow 3MnO_2 + 2Cr^{3+} + H_2O$
    $E^\circ_{cell} = +1.33\ V - (+1.23\ V) = +\ 0.10\ V$
    Six electrons are transferred, and $K_{eq} = 1.4 \times 10^{10}$.

(e) $H_2O + AsO_3^{3-} + I_2 \rightarrow H_2AsO_4^- + 2I^-$
    $E^\circ_{cell} = +0.54\ V - (+0.58\ V) = -0.04\ V$
    Two electrons are transferred, and $K_{eq} = 4.4 \times 10^{-2}$.

In these examples we see that fairly small voltages translate into very large or very small equilibrium constants, depending on the sign of E°_{cell}.

FREE-ENERGY CHANGE, ΔG°, AND STANDARD CELL VOLTAGES

As shown by Equation 13.50, the standard cell voltage is related to the equilibrium constant:

$$E^\circ_{cell} = \frac{RT}{n\mathscr{F}} \ln K_{eq}$$

The standard free-energy change, ΔG°, is also related to the equilibrium constant:

$$\Delta G^\circ = -RT \ln K_{eq}$$

The relationship between ΔG° and E°_{cell} is

$$\Delta G^\circ = -n\mathscr{F} E^\circ_{cell} \qquad (13.51)$$

A major method for determining the standard free energy of a reaction is the measurement of the standard cell voltage for the reaction of interest, and then a mathematical conversion using Equation 13.51.

EXERCISE 13.10

Determine the standard free-energy change for each of the following reactions at 298 K:

(a) the single displacement of copper(II) by zinc metal
(b) the single displacement of H^+ by iron, forming Fe^{2+}
(c) the reduction of tin(iv) to tin(II) by the oxidation of Fe(II) to Fe(III)
(d) the reduction of dichromate ions to chromium(III) ions Mn^{2+}, forming MnO_2
(e) the oxidation of AsO_3^{3-} to $H_2AsO_4^-$ by the reduction of iodine to iodide ions

Solution

In Exercise 13.7 the reactions and E°_{cell} were determined. The equation $\Delta G^\circ = -n\mathscr{F} E^\circ_{cell}$ once n and E°_{cell} are known.

(a) $Cu^{2+} + Zn \rightarrow Zn^{2+} + Cu$
$E^\circ_{cell} = +0.34\,V - (-0.76\,V) = +1.10\,V$
Two electrons are transferred and $\Delta G^\circ = -2(96.485)(1.10) = -212$ kJ.
(b) $Fe + 2H^+ \rightarrow Fe^{2+} + H_2$
$E^\circ_{cell} = 0.00\,V - (-0.44\,V) = +0.44\,V$
Two electrons are transferred, and $\Delta G^\circ = -2(96.485)(0.44) = -84.9$ kJ.
(c) $Sn^{4+} + 2Fe^{2+} \rightarrow 2Fe^{3+} + Sn^{2+}$
$E^\circ_{cell} = +0.15\,V - (+0.77\,V) = -0.62\,V$
Two electrons are transferred, and $\Delta G^\circ = -2(96.485)(-0.62) = +120$ kJ.

(d) $2H^+ + Cr_2O_7^{2-} + 3Mn^{2+} \rightarrow 3MnO_2 + 2Cr^{3+} + H_2O$

$E^{\circ}_{cell} = +1.33 \text{ V} - (+1.23 \text{ V}) = +0.10 \text{ V}$

Six electrons are transferred, and $\Delta G^{\circ} = -6(96.485)(0.10) = -57.9$ kJ.

(e) $H_2O + AsO_3^{3-} + I_2 \rightarrow H_2AsO_4^- + 2I^-$

$E^{\circ}_{cell} = +0.54 \text{ V} - (+0.58 \text{ V}) = -0.04 \text{ V}$

Two electrons are transferred, and $\Delta G^{\circ} = -2(96.485)(-0.04) = 7.7$ kJ.

USING GALVANIC CELLS TO DETERMINE CONCENTRATIONS

The Nernst equation suggests a method whereby galvanic cell measurements can be used to determine concentrations. A redox reaction that can be used for this purpose is

$$2Ag^+(aq) + Cu(s) \rightarrow 2Ag(s) + Cu^{2+}(aq) \tag{13.52}$$

The cell diagram for this reaction is

$$Cu \mid Cu^{2+} \parallel Ag^+ \mid Ag \tag{13.53}$$

For this cell, we will use Equation 13.48. We know that $E_{cell} = 0.346$ V, so $E^{\circ}_{cell} = E^{\circ}_{cathode} - E^{\circ}_{anode} = 0.800 \text{ V} - 0.340 \text{ V} = 0.460 \text{ V}$. We can also write Q for this reaction as $\dfrac{[Cu^{2+}]}{[Ag^+]^2}$, leaving out the solid Cu° and Ag°. We can also insert the given concentration for Ag^+ into Q to get $\dfrac{[Cu^{2+}]}{[1.0 \times 10^{-5}]^2}$.

Since $n = 2$, we assemble the Nernst equation as

$$0.346 \text{ V} = 0.460 \text{ V} - \frac{0.0591}{2} \log \frac{[Cu^{2+}]}{[1.0 \times 10^{-5}]^2}$$

$$-0.114 \text{ V} = -\frac{0.0591}{2} \log \frac{[Cu^{2+}]}{[1.0 \times 10^{-5}]^2}$$

$$-0.114 \text{ V} \left(\frac{-2}{0.0591} \right) = \log \frac{[Cu^{2+}]}{[1.0 \times 10^{-5}]^2}$$

$$3.85 = \log \frac{[cu^{2+}]}{[1.0 \times 10^{-5}]^2}$$

$$7.1 \times 10^3 = \frac{[Cu^{2+}]}{[1.0 \times 10^{-5}]^2}$$

$$[Cu^{2+}] = 7.1 \times 10^{-7}$$

The key to using galvanic cell measurements for determining concentrations is that all but one concentration in the Nernst equation must be known and must remain constant during the experiment. In this analysis the silver ion concentration around the cathode was held constant. This constant electrode is also known as the **reference electrode.** The anode that monitored the Cu^{2+} ion concentration is called the **indicator electrode.**

pH MEASUREMENT AND pH ELECTRODES

The concentration of hydrogen ions can be determined by using a galvanic cell in a manner similar to that for the determination of copper(II) ions illustrated above. **pH** is defined as the negative logarithm of the hydrogen ion concentration:

$$pH = -\log [H^+] \qquad (13.54)$$

As in the preceding example, one of the electrodes in the galvanic cell must be a constant reference electrode. Two common reference electrodes for pH measurement are constructed using the half-reactions

$$AgCl(s) + e^- \rightarrow Ag(s) + Cl^- \qquad (13.55)$$

$$Hg_2Cl_2 + 2e^- \rightarrow 2Hg(l) + 2Cl^- \qquad (13.56)$$

The silver-silver chloride reference electrode consists of a silver wire with a coating of insoluble silver chloride in a saturated solution of potassium chloride. The second half-reaction uses the compound calomel, Hg_2Cl_2, and the reference electrode is called a calomel electrode.

The indicator electrode for pH measurements is the **glass electrode,** which consists of a very thin glass membrane. On the inside of the glass electrode is a 1 M solution of hydrochloric acid. The outside of the glass membrane is in contact with the solution to be measured. When the glass membrane has been soaked in water for 24 hours, the glass surface becomes a hydrated gel. The hydrogen ions adsorb to this hydrated glass surface in proportion to the concentration of hydrogen ions in the solution. The 1 M HCl on the inside produces a constant positive charge to the inner membrane, while the solution to be measured produces a smaller positive charge on the outside of the membrane. The difference in charge is reflected as the electrode potential.

A galvanic cell used to measure pH follows the equations

$$E_{cell} = E_{reference} + constant + 0.0591 \log [H^+] \qquad (13.57)$$

and when the constants are combined

$$E_{cell} = C' - 0.0591 \, pH \qquad (13.58)$$

The constant in Equation 13.57 arises because of the variable properties of the glass electrode. It is combined with the potential of the reference electrode (which is also constant) to yield the constant C' in Equation 13.58. In use, a pH electrode must be standardized by immersing it into a buffer solution with a known pH.

Measuring E_{cell} under these conditions allows the calculation of C′. Next the electrode is rinsed with distilled water and immersed in the solution to be measured. The C′ just determined and the E_{cell} value for the unknown solution allow calculation of the pH.

Glass is an excellent insulator and does not allow a large amount of electric current to pass through it. Even the very thin glass membrane of the glass electrode is a good insulator. A special instrument, the pH meter, is needed to make measurements using the glass electrode. When a pH meter is used, the calibration step is done by adjusting the meter reading to the pH of the buffer; the pH of the unknown is then read directly from the meter.

OTHER USES OF GALVANIC CELLS

Galvanic cells can also be used to convert chemical energy into electrical energy. Devices that do this are called **batteries,** and several different batteries are in common use.

The *lead-acid battery* is used to start automobiles. A battery of this type is composed of a lead anode and a lead cathode that is coated with PbO_2. The half-reactions are as follows:

$$PbO_2(s) + 4H^+(aq) + SO_4^{2-}(aq) + 2e^- \rightarrow PbSO_4(s) + -H_2O \text{ (cathode)} \quad (13.59)$$

$$Pb(s) + SO_4^{2-} \rightarrow PbSO_4(s) + 2e^- \quad \text{(anode)} \quad (13.60)$$

As the battery is discharged, it uses up the sulfate ions and the electrodes become coated with lead sulfate. The reactions in Equations 13.62 and 13.63 may be reversed in charging the battery. The reverse reactions regenerate the sulfate ion in solution and reduce the amount of lead sulfate contaminating the surface of the electrodes. Each pair of electrodes produces approximately 2 volts. Six pairs of electrodes are used in a 12-volt car battery.

The *alkaline battery* is the battery used most commonly in flashlights, tape recorders, and TV remote controls. The case of the battery is zinc, which is the anode. The cathode is a graphite rod inserted into a paste made of manganese dioxide, water, and potassium hydroxide. The half-reactions are as follows:

$$Zn(s) + 2OH^-(aq) \rightarrow ZnO(s) + H_2O + 2e^- \quad \text{(anode)} \quad (13.61)$$

$$2MnO_2(s) + H_2O + 2e^- \rightarrow Mn_2O_3(s) + 2OH^-(aq) \text{ (cathode)} \quad (13.62)$$

The total voltage is 1.54 volts. This battery is not rechargeable.

Other batteries of interest are the *nickel-cadmium* or *nicad battery*, the *mercury battery*, and the *silver oxide battery*.

Fuel cells are also used to generate electricity. In a fuel cell, hydrogen is oxidized at an anode and oxygen reduced at a cathode to form water with the production of electricity. A fuel cell is approximately twice as efficient as gas, oil, or coal-powered generators in converting chemical energy into electricity.

IMPORTANT OXIDATION-REDUCTION REACTIONS

Combustion Reactions

Reacting organic compounds with oxygen often results in the production of a large amount of heat along with a flame that is characteristic of the combustion process. The products of **combustion** reactions are usually carbon dioxide and water, as in the combustion of glucose:

$$C_6H_{12}O_6 + 6O_2 \rightarrow 6CO_2 + 6H_2O \tag{13.63}$$

When the amount of oxygen is limited, or there is no oxygen, the products of the reaction may include carbon monoxide or elemental carbon (soot):

$$C_6H_{12}O_6 + 3O_2 \rightarrow 6CO + 6H_2O \text{ (limited } O_2) \tag{13.64}$$

$$C_6H_{12}O_6 \rightarrow 6C + 6H_2O \quad \text{(no } O_2) \tag{13.65}$$

The same products, CO_2 and H_2O, are formed when glucose is metabolized in the body. However, since the process is slow and does not produce a flame, it is not called a combustion reaction. The body uses the heat and energy produced in a more efficient manner in metabolism. Even when the amount of oxygen is limited, CO and C are not produced in metabolic reactions.

Oxidation of Metals

Metallic elements have differing affinities toward oxygen. The very unreactive elements, such as platinum, gold, silver, and copper, do not react with O_2. Silver tarnishes by reacting with small amounts of hydrogen sulfide in the air. Copper reacts with water and carbon dioxide to form a carbonate compound. Some metals, particularly the alkali and alkaline earth metals, however, react readily with oxygen and are completely converted into oxides if exposed long enough.

Magnesium burns with a bright white flame in oxygen. The bright light is used in magnesium flares by the military, in flashbulbs for photography, and in fireworks. The reaction is so energetic that magnesium will continue burning even in an atmosphere of carbon dioxide:

$$2Mg + CO_2 \rightarrow 2MgO + C \tag{13.66}$$

When finely divided into a powder, many metals burn in oxygen. Steel wool burns when placed in a flame, and powdered metals, such as aluminum, are classified as highly combustible.

Aluminum reacts very well with oxygen, as mentioned above. In large sheets or bars of the metal, however, the aluminum oxide formed produces an impervious coating so that complete oxidation does not occur. This oxide layer is only a few molecules thick and is not visible to the eye.

Iron and steel react poorly with gaseous oxygen. The formation of rust requires the presence of water for oxidation to occur; iron does not rust in pure oxygen or in water that contains no oxygen. In this complex electrochemical process, the iron actually acts as the anode of a chemical reaction when moisture is present and as the cathode when it is not. **Corrosion** and its prevention are major concerns of approximately 30 percent of working chemists and chemical engineers. Each year, damage due to corrosion of buildings, bridges, and even computer circuits costs billions of dollars to correct and repair.

SUMMARY

Electrochemistry and oxidation-reduction reactions are combined in this chapter. Defining and identifying oxidation and reduction along with oxidizing agents and reducing agents is described in this chapter. Balancing of the more complex redox reactions is also described in detail. Redox reactions can occur in nature or as part of chemical reactions, including titrations. Chemical energy can be harnessed by using a galvanic or voltaic cell, better known to the consumer as a battery. Standard cell potentials, E°_{cell}, are related to free energies, ΔG°, and equilibrium constants, K_{eq}. Positive values for E°_{cell} indicate a spontaneous reaction. Galvanic cells can be arranged to measure concentrations of ions in solution. One of the most important, the hydrogen ion, is measured in a special galvanic cell with an instrument called a pH meter.

Galvanic cells harness chemical energy to create a flow of electrons in a circuit that can be used to power a flashlight or an MP-3 player. If electrical energy is added to a system, nonspontaneous reactions can be forced to occur in a process called electrolysis. Electrolysis of molten salts produces the redox products of the anions and cations of binary salts. Electrolysis of aqueous solutions produces either the redox products of the salt or hydrogen or oxygen, depending on which substances are more easily oxidized or reduced at the electrode surfaces. Electrolysis is also quantitative, and if the electric current and time are accurately measured, then the amount of substance reacted can be calculated precisely.

Important Concepts

Oxidation and reduction
Oxidation numbers
Ion electron method for balancing equations
Electrolytic and galvanic cells
Standard reduction potentials
Standard cell voltages
Nernst equation

Important Equations

$$E = E^{\circ} - \frac{RT}{n\mathscr{F}} \ln Q$$

$$\text{moles} = \frac{It}{n\mathscr{F}}$$

$$E^{\circ}_{cell} = E^{\circ}_{cathode} - E^{\circ}_{anode}$$

$$E^{\circ}_{cell} = \frac{RT}{n\mathscr{F}} \ln K$$

$$\Delta G^{\circ} = -n\mathscr{F} E^{\circ}_{cell}$$

Practice Exercises

Multiple-Choice

For the first four problems below, one or more of the following responses will apply; each response may be used more than once or not at all in these questions.

    I. Nernst equation
   II. spontaneous
  III. reduction
  IV. oxidation
   V. electrolysis

1. The loss of electrons is a(n) _____ process

  (A) I
  (B) II
  (C) III
  (D) IV
  (E) V

2. In order to produce aluminum, large amounts of electricity are used to perform the _____ of aluminum oxide.

  (A) I
  (B) II
  (C) III
  (D) IV
  (E) V

3. This is used to determine cell voltages when standard state conditions are not present.

  (A) I
  (B) II
  (C) III
  (D) IV
  (E) V

4. _____ cannot occur without _____ also occurring within the same system.

  (A) I, III
  (B) II, V
  (C) III, IV
  (D) IV, V
  (E) V, II

5. Balance the following half-reaction in acid solution:

$$NO_3^- \rightarrow NH_4^+$$

When balanced with the smallest whole-number coefficients, the *sum* of all the coefficients is

  (A) 13
  (B) 26
  (C) 15
  (D) 23
  (E) 21

6. For the reaction between the permanganate ion and iron(II), the reaction is

$$MnO_4^- + 5\ Fe^{2+} + 8\ H^+ \rightarrow 5\ Fe^{3+} + Mn^{2+} + 4\ H_2O$$

Which of the following is the correct form of the Nernst equation for this reaction?

(A) $E = E_{cell}^\circ - \dfrac{RT}{5\mathscr{F}} \ln \dfrac{[Mn^{2+}][Fe^{3+}][H_2O]}{[MnO_4^-][Fe^{2+}][H^+]}$

(B) $E = E_{cell}^\circ - \dfrac{RT}{5\mathscr{F}} \ln \dfrac{[Mn^{2+}][Fe^{3+}]^5[H_2O]}{[MnO_4^-][Fe^{2+}]^5[H^+]}$

(C) $E = E_{cell}^\circ - \dfrac{RT}{5\mathscr{F}} \ln \dfrac{[Mn^{2+}][Fe^{3+}]^5[H_2O]}{[MnO_4^-][Fe^{2+}]^5[H^+]^8}$

(D) $E = E_{cell}^\circ - \dfrac{RT}{8\mathscr{F}} \ln \dfrac{[Mn^{2+}][Fe^{3+}]^5[H_2O]}{[MnO_4^-][Fe^{2+}]^5[H^+]^8}$

(E) $5E = 5E_{cell}^\circ - \dfrac{RT}{8\mathscr{F}} \ln \dfrac{[Mn^{2+}][Fe^{3+}]^5[H_2O]}{[MnO_4^-][Fe^{2+}]^5[H^+]^8}$

7. The following reactions are known to occur spontaneously:

$$Cu\ +\ 2Ag^+\ \rightarrow\ Cu^{2+}\ +\ 2Ag$$
$$Zn\ +\ 2Ag^+\ \rightarrow\ Zn^{2+}\ +\ 2Ag$$
$$Zn\ +\ Cu^{2+}\ \rightarrow\ Zn^{2+}\ +\ Cu$$

The activity series for the three elements from the strongest to the weakest reducing agent is

(A) Cu > Ag > Zn
(B) Zn > Cu > Ag
(C) Ag > Cu > Zn
(D) Ag > Zn > Cu
(E) Zn > Ag > Cu

8. The standard reduction potential for $PbO_2 \rightarrow Pb^{2+}$ is +1.46 V, and the standard reduction potential for $Fe^{3+} \rightarrow Fe^{2+}$ is +0.77 V. What is the standard cell voltage for the reaction

$$4H^+ + PbO_2 + 2Fe^{2+} \rightarrow 2Fe^{3+} + Pb^{2+} + 2H_2O?$$

(A) −0.08 V
(B) +0.69 V
(C) +2.33 V
(D) −0.69 V
(E) −2.33 V

9. Which of the following compounds includes an element with an oxidation number of +5?

(A) ClO_4^-
(B) MnO_4^-
(C) NO_2^-
(D) SO_3^{2-}
(E) NO_3^-

10. Which of the following metals does NOT react with water to produce hydrogen?

(A) Zn
(B) Li
(C) Ca
(D) Na
(E) Rb

11. In the electrolysis of an aqueous solution of $Cu(NO_3)_2$, which of the following is expected to occur?

(A) formation of O_2 at the anode
(B) formation of H_2 at the cathode
(C) deposition of copper metal on the anode
(D) formation of hydroxide ions at the cathode
(E) formation of H^+ at the cathode

12. Sodium metal cannot be electrolyzed from an aqueous Na_2SO_4 solution because

(A) the voltage needed is too high for any available instrument to achieve
(B) water is reduced to O_2 before Na^+
(C) Na^+ has a high over-potential that keeps it from being reduced
(D) H^+ has a more favorable reduction potential than Na^+
(E) Na^+ does electrolyze, but it immediately reacts with water again

13. Which of the following elements has the largest number of possible oxidation states?

(A) Fe
(B) Cl
(C) Ca
(D) Mn
(E) Na

14. For the reaction

$$2I^- + Cl_2 \rightleftharpoons 2Cl^- + I_2$$

the standard cell voltage is $E^\circ_{cell} = +0.82$ V. What is the equilibrium constant for this reaction at 45 °C? ($R = 8.314$ V C mol^{-1} K^{-1}, and $\mathscr{F} = 96{,}485$ C mol^{-1})

(A) 9.8×10^{25}
(B) 1.0×10^{-26}
(C) 1.6×10^5
(D) 6.3×10^{-6}
(E) 2.27

15. Which of the following E°_{cell} values represents a nonspontaneous reaction that produces the greatest amount of product? (Assume the same number of electrons is transferred in each reaction.)

(A) +2.31 V
(B) +0.23 V
(C) −0.12 V
(D) −1.68 V
(E) −1.14 V

16. A metal is electrolyzed from aqueous solution by using an electrical current of 1.23 A for $2\frac{1}{2}$ h, and 3.37 g of metal is deposited. In a separate experiment the number of electrons used for the reduction of the metal is 2. What is the metal?

(A) Al
(B) Ni
(C) Sn
(D) Mg
(E) Au

17. A galvanic cell is set up under nonstandard state conditions, and E_{cell} is measured as -0.16 V. Which of the following is true about this system?

   (A) The galvanic cell is set up incorrectly because of the negative value of E°_{cell}.
   (B) If the contents of the two cells are mixed, the reaction will proceed in the forward direction.
   (C) The reaction is definitely nonspontaneous.
   (D) If the contents of the two cells are mixed, the reaction will proceed in the reverse direction.
   (E) The reaction is definitely spontaneous.

18. A galvanic cell is constructed; and when the temperature of the cell is increased from 20 °C to 30 °C, the cell voltage changes by a factor of 1.034. Which of the following conclusions may be drawn from this observation?

   (A) ΔG° must be positive since the reaction is not spontaneous.
   (B) The increase is due solely to the temperature term in $\Delta G^\circ = -RT \ln K$.
   (C) The information indicates that ΔS° for this reaction must be negative.
   (D) The reaction is spontaneous because ΔG° is negative.
   (E) The voltage change must be due to the change in the equilibrium constant with temperature.

19. Which of the following is NOT commonly produced by electrolysis?

   (A) NaOCl (bleach)
   (B) Al
   (C) Fe
   (D) NaOH
   (E) H_2

20. Which of the following compounds includes an element that has the same oxidation number as the chlorine in sodium chlorate, $NaClO_3$?

   (A) $K_3Fe(CN)_6$
   (B) $KMnO_4$
   (C) $Al(NO_3)_3$
   (D) $(NH_4)_2SO_4$
   (E) $KClO_4$

21. With a current of 1.25 A, how many minutes will be required to deposit 2.00 g of copper on a platinum electrode from a copper(II) nitrate solution? (Faraday's constant = 96,485 C mol^{-1})

   (A) 4859
   (B) 81.0
   (C) 40.5
   (D) 1.35
   (E) 2430

22. What is the minimum number of electrons needed to balance the following half-reaction with whole number coefficients?

$$IO_3^- \rightarrow I_2$$

   (A) 1
   (B) 2
   (C) 5
   (D) 10
   (E) 12

23. Which of the following pairs of constants are NOT mathematically related to each other?

    (A) equilibrium constant and Gibbs free energy
    (B) rate constant and activation energy
    (C) standard cell voltage and equilibrium constant
    (D) standard cell voltage and rate constant
    (E) Gibbs free energy and standard cell voltage

24. Which of the following statements is FALSE?

    (A) Reduction involves a gain of electrons.
    (B) Batteries are galvanic cells.
    (C) A spontaneous reaction always has a positive E°_{cell}.
    (D) Electrolysis reactions always produce a gas at at least one electrode.
    (E) Galvanic cells can be used to determine equilibrium constants.

ANSWER KEY

1. D	6. B	11. A	16. B	21. B
2. E	7. B	12. D	17. D	22. D
3. A	8. B	13. B	18. B	23. D
4. C	9. E	14. A	19. C	24. D
5. D	10. A	15. C	20. C	

See Appendix 1 for explanation of answers.

Free-Response

(a) Balance the following skeleton reaction in acid solution

$$I^- + IO_3^- \rightarrow I_3^-$$

(b) Describe in words what to do to convert the balanced equation in part (a) into a reaction in basic solution.

(c) Write the Nernst equation for the reaction in part (a).

(d) A galvanic (voltaic) cell is set up with a copper wire immersed in a Cu^{2+} solution and a strip of magnesium immersed in a Mg^{2+} solution. If the magnesium ion concentration is 0.00200 M and the cell voltage is measured as 2.15 volts, what is the concentration of the copper ion solution?

(e) A solution containing Cu^{2+} is electrolyzed for 45.0 minutes. The mass of a platinum electrode increases from 38.3421 to 38.4876 grams when the copper is reduced onto it. What was the average electrical current through this electrolytic cell?

ANSWERS

(a) We write two half-reactions, in this case they react to the same product, also balance the iodine atoms.

$$2I^- \rightarrow I_2$$
$$2IO_3^- \rightarrow I_2$$

Next, we balance oxygen by adding one H_2O for every O needed to get

$$2\,I^- \rightarrow I_2$$
$$2IO_3^- \rightarrow I_2 + 6H_2O$$

Now one H^+ is added to balance each H

$$2I^- \rightarrow I_2$$
$$12H^+ + 2IO_3^- \rightarrow I_2 + 6H_2O$$

Now, electrons are added to balance charges:

$$2I^- \rightarrow I_2 + 2e^-$$
$$10e^- + 12H^+ + 2IO_3^- \rightarrow I_2 + 6H_2O$$

Multiply the top half-reaction by 5 to equalize electrons

$$10I^- \rightarrow 5I_2 + 10e^-$$
$$10e^- + 12H^+ + 2IO_3^- \rightarrow I_2 + 6H_2O$$

Add the half-reactions and cancel the electrons to get

$$10I^- + 12H^+ + 2IO_3^- \rightarrow 5I_2 + I_2 + 6H_2O$$

The I_2 terms are added, and all the coefficients are divided by 2 to get

$$5I^- + 6H^+ + IO_3^- \rightarrow 3I_2 + 3H_2O$$

(b) To convert this reaction to a basic solution, add $6OH^-$ to each side of the reaction. On the left combine with the hydrogen ions to make $6H_2O$ while leaving 6 hydroxides on the right. Cancel 3 water molecules from both sides for the simplest equation.

(c) The Nernst equation for the balanced redox reaction is

$$E = E^\circ_{cell} - \frac{0.0591}{n} \log \frac{[I_2]^3}{[I^-]^5[IO_3^-][H^+]^6}$$

(d) $Cu^{2+} + 2e^- \rightarrow Cu^\circ \quad E^\circ = +0.34$
$Mg^{2+} + 2e^- \rightarrow Mg^\circ \quad E^\circ = -2.37$

The spontaneous reaction and standard cell voltage are

$$Mg + Cu^{2+} \rightarrow Mg^{2+} + Cu \quad E^{\circ}_{cell} = +2.71$$

The number of electrons transferred is $n = 2$.

The Nernst equation is $E = 2.71 - (0.0591/2) \log([Mg^{2+}]/[Cu^{2+}])$

$$2.15 = 2.71 - (0.0591/2) \log(0.00200/[Cu^{2+}])$$
$$(2.15 - 2.71)/(-0.0591/2) = \log(0.00200/[Cu^{2+}])$$
$$18.95 = \log(0.00200/[Cu^{2+}])$$

Take the antilog of both sides to get

$$8.93 \times 10^{18} = 0.00200/[Cu^{2+}]$$
$$[Cu^{2+}] = 0.00200/(8.93 \times 10^{18}) = 2.24 \times 10^{-22} \text{ mol L}^{-1}$$

(e) For quantitative electrolysis, the equation

$$\text{moles} = \frac{It}{n\mathscr{F}}$$

is appropriate. For the reduction of copper, $n = 2$ and $t = 2700$ seconds. The mass of copper is 38.4876 g − 38.3421 g = 0.1455 g Cu. This is $\dfrac{1.455 \text{ g}}{63.55 \text{ g/mol}} = 0.00229$ mol Cu. Entering these numbers into the equation above yields

$$0.00229 \text{ moles Cu} = \frac{I \times 2700}{2 \times 96,500}$$

Solving for I yields $I = 0.164$ amperes.

Acids and Bases

CHAPTER 14

- Acid-Base Theories
- Arrhenius Theory
- Brönsted-Lowry Theory
- Lewis Theory
- Naming Acids
- Acid Strengths
- Acid Anhydrides
- Base Anhydrides
- Neutralization Reactions
- Polyprotic Acids
- Conjugate Acid-Base Pairs
- Complexation Reactions
- Ligands

- Coordinate Covalent Bonds
- pH Calculations
- K_w and pK_w
- pH of Weak Acids/Bases
- Hydrolysis Reactions
- pH of Salt Solutions
- Buffers
- Buffer pH Change Calculations
- Preparation of Buffers
- pH of Polyprotic Acid Solutions
- Titration Curves
- Polyprotic Acid Titration Curves
- pH Indicators

ACID-BASE THEORIES

ACID-BASE THEORIES	
THEORY NAME	**BRIEF DESCRIPTION**
Arrhenius	An acid adds hydrogen ions to a solution, and a base adds hydroxide ions to a solution.
Brönsted-Lowry	An acid is a proton donor, and a base is a proton acceptor.
Lewis	An acid is an electron pair acceptor, and a base is an electron pair donor.

Arrhenius Theory

Svante Arrhenius considered an acid to be any substance that increases the concentration of hydrogen ions, H^+, in aqueous solution. When an acid is dissolved in water, the reaction may be written in two different forms. For example, when gaseous hydrogen chloride is dissolved in water, the equation is written as

$$HCl(g) \xrightarrow{\text{H}_2\text{O}} H^+(aq) + Cl^-(aq) \qquad (14.1)$$

> **TIP**
>
> **NOTE:** Most acid-base reactions take place in aqueous solution. The symbol (*aq*) is often omitted for clarity and simplicity.

Modern chemistry recognizes, however, that the hydrogen ion, H^+, is unlikely to exist in aqueous solution. Instead, the hydrogen ion is hydrated or bound to one or more water molecules. H_3O^+ is used to represent the hydration of the hydrogen ion, and the reaction of HCl with water is written as

$$HCl(g) + H_2O(l) \rightarrow H_3O^+(aq) + Cl^-(aq) \qquad (14.2)$$

Common examples of acids are HCl, HBr, H_2SO_4, H_3PO_4, and $HC_2H_3O_2$. The **Arrhenius theory** considers bases to be substances that increase the hydroxide ion, OH^-, concentration when dissolved in water. A typical reaction is as follows:

$$KOH(s) \xrightarrow{H_2O} K^+(aq) + OH^-(aq) \qquad (14.3)$$

Common bases are NaOH, KOH, $Ca(OH)_2$, and $Al(OH)_3$.

Brönsted-Lowry Theory

For a long time it was known that ammonia, NH_3, when dissolved in water increases the hydroxide ion concentration. In fact, ammonia solutions were often called ammonium hydroxide, with the formula NH_4OH. This fit the Arrhenius theory, but not the facts since NH_4OH does not really exist. The **Brönsted-Lowry theory** solves this problem by defining acids as proton donors, as the Arrhenius theory does, but defining bases as proton acceptors. Ammonia, NH_3, is a base since it accepts protons from water molecules in the reaction

$$H_2O(l) + NH_3(g) \rightleftharpoons NH_4^+(aq) + OH^-(aq) \qquad (14.4)$$

Ethylamine, $CH_3CH_2NH_2$, and dimethylamine, $(CH_3)_2NH$, are both bases related to ammonia. In ethylamine the ethyl group, CH_3CH_2-, replaces one hydrogen in ammonia. Two methyl groups, CH_3-, in dimethylamine replace two of the hydrogen atoms in ammonia.

Lewis Theory

G. N. Lewis proposed that acids are substances that accept electron pairs from other atoms, ions, or molecules and bases are substances that donate electron pairs in forming chemical bonds. The reaction between boron trichloride, BCl_3, an electron-deficient compound, and ammonia, NH_3, which has a nonbonding pair of electrons, is an acid-base reaction according to the Lewis definitions. Ammonia is an electron-pair donor and is a base, while boron trichloride is the electron-pair acceptor and is an acid:

$$
\begin{array}{ccccccc}
& H & & \ddot{:}\ddot{C}l\ddot{:} & & & H\!:\!\ddot{C}l\ddot{:} \\
& | & & | & & & | \quad \ddot{} \\
H\!-\!\ddot{N}\!: & + & B\!-\!\ddot{C}l\ddot{:} & \longrightarrow & H\!-\!N\!-\!B\!-\!\ddot{C}l\ddot{:} \\
& | & & | & & & | \quad \ddot{} \\
& H & & \ddot{:}\ddot{C}l\ddot{:} & & & H\!:\!\ddot{C}l\ddot{:}
\end{array}
\qquad (14.5)
$$

The **Lewis theory** is used often to explain the formation of substances called complexes; the Arrhenius and Brönsted-Lowry theories serve to explain the more traditional acid-base reactions.

Modern Concept of Acids

An acid, according to both the Arrhenius and Brönsted-Lowry theories, is any compound having one or more hydrogen atoms that are weakly bound to the rest of the molecule. When dissolved in water, hydrogen ions ionize from the rest of the molecule. Most acids are written with hydrogen as the first element in the formula. Organic acids, however, can be written with hydrogen as the first element or with the —COOH unit that is the functional group for organic acids. The following three formulas all represent ethanoic acid. Ethanoic acid is often called acetic acid.

$$HC_2H_3O_2 \qquad CH_3COOH \qquad \begin{array}{c} OH \\ | \\ CH_3C=O \end{array}$$

ACID-BASE NOMENCLATURE

Binary acids contain hydrogen and one other atom. In aqueous solutions, the name of every binary acid starts with the prefix *hydro-* and ends with the suffix *-ic* on the root of the second atom in the formula. Table 14.1 lists several binary acid names.

TABLE 14.1

Binary Acid Names	
Name	Formula
hydrofluoric acid	HF
hydrochloric acid	HCl
hydrobromic acid	HBr
hydroiodic acid	HI
hydrosulfuric acid	H_2S

Polyatomic anions can be the anions of acids. For these acids, the ending of the polyatomic anion name is changed and the word *acid* is added. If the polyatomic anion name ends in *-ate*, the ending is changed to *-ic*. If the ending is *-ite*, it is changed to *-ous*. Table 14.2 illustrates this principle for several polyatomic anions.

TABLE 14.2

Names of Acids Derived from Polyatomic Anions

Polyatomic Anion	Acid Name	Acid Formula
sulfate	sulfuric acid	H_2SO_4
sulfite	sulfurous acid	H_2SO_3
nitrate	nitric acid	HNO_3
nitrite	nitrous acid	HNO_2
hypochlorite	hypochlorous acid	$HClO$
chlorite	chlorous acid	$HClO_2$
chlorate	chloric acid	$HClO_3$
perchlorate	perchloric acid	$HClO_4$

Organic acids have both common and systematic names. In the systematic name for an organic acid the suffix *-oic* and the word *acid* are added to the root name of the rest of the molecule. Some organic acids and their systematic and common names are listed in Table 14.3.

TABLE 14.3

Organic Acid Names

Systematic Name	Common Name	Formula
methanoic acid	formic acid	$HCOOH$
ethanoic acid	ethanoic (acetic) acid	CH_3COOH
propanoic acid	propanoic acid	CH_3CH_2COOH
butanoic acid	butyric acid	$CH_3CH_2CH_2COOH$
pentanoic acid	valeric acid	$CH_3CH_2CH_2CH_2COOH$
benzoic acid	benzoic acid	C_6H_5COOH

There are two common types of bases, those that have hydroxide ions in their formulas and those that contain nitrogen. The **hydroxide bases** are named by using the name of the metal, with roman numerals if necessary, and then the word *hydroxide*. For example, NaOH, Al(OH)$_3$, and Fe(OH)$_3$ are called sodium hydroxide, aluminum hydroxide, and iron(III) hydroxide, respectively.

Nitrogen bases related to ammonia are amines. Replacing a hydrogen on ammonia with a methyl, —CH$_3$, group produces methylamine. If two methyl groups replace two hydrogen atoms, the compound is called dimethylamine. If a methyl and an ethyl group, —CH$_2$CH$_3$, replace two hydrogen atoms, the compound is called either ethylmethylamine or methylethylamine.

The chloride salt of ammonia is known as ammonium chloride. The chloride salt of methylamine is called methyl ammonium chloride; an alternative name for this salt is methylamine hydrochloride. For chloride salts of nitrogen bases with common names, such as hydrazine, the hydrochloride ending, as in hydrazine hydrochloride, is ordinarily used.

ACIDS

Strong Acids

Strong acids are acids that ionize completely when dissolved in water.

TABLE 14.4

The Six Most Important Strong Acids	
Acid	Ionization Reaction
hydrochloric acid	$HCl \rightarrow H^+ + Cl^-$
hydrobromic acid	$HBr \rightarrow H^+ + Br^-$
hydroiodic acid	$HI \rightarrow H^+ + I^-$
perchloric acid	$HClO_4 \rightarrow H^+ + ClO_4^-$
nitric acid	$HNO_3 \rightarrow H^+ + NO_3^-$
sulfuric acid	$H_2SO_4 \rightarrow H^+ + HSO_4^-$

For sulfuric acid, only the first hydrogen is considered strong; the second hydrogen ion ionizes only slightly. These strong acids are also called mineral acids, as opposed to organic acids.

Weak Acids

All of the remaining acids are weak, meaning that, when they are dissolved in water, only a small percentage of the molecules ionize. Most of the **weak acids** are organic acids. The most common weak mineral acids are HF, H_2CO_3, H_3PO_4, H_3AsO_4, $HClO_3$, $HClO_2$, and $HClO$.

Acid Strengths

The concepts of electronegativity and bonding can be used to compare the strengths of acids based on their chemical formulas. The strength of an acid is inversely proportional to the strength of the bond between the hydrogen and the remainder of the molecule. Strong acids have very weak bonds, and weak acids have stronger bonds. To determine the relative strengths of acids, an estimate of the bond strengths is needed.

As mentioned previously, a **binary acid** is composed of hydrogen and one other atom. Experimental evidence shows that acids get stronger from left to right in a period (row) of the periodic table. Thus, we find that PH_3 is weaker than H_2S, which is weaker than HBr. Across a period the acid strength parallels the electronegativity of the anion. This is reasonable because, as the anion attracts the electrons more strongly, the bond with hydrogen becomes weaker, while the size of the anions remains relatively constant.

Experimental evidence also shows that the strength of binary acids increases from the top of a group to the bottom. Thus HF is weaker than HCl, which is weaker than HBr. Electronegativity cannot explain this sequence of binary acid strengths. Instead, the size of the anion is important. An increase in anion size requires a corresponding increase in bond length. As stated in Chapter 5, longer bond lengths imply weaker bonds. In acids, a weaker bond with hydrogen means a stronger acid.

All mineral acids that are not binary acids are **oxoacids:** acids that contain hydrogen, oxygen, and another element. In all oxoacids, the oxygen atoms are bound to the central atom and the hydrogen atoms are bound to the oxygen atoms. A typical structure is represented by sulfuric acid:

$$
\begin{array}{c}
\text{O} \\
\| \\
\text{H}-\text{O}-\text{S}-\text{O}-\text{H} \\
\| \\
\text{O}
\end{array}
$$

The strength of an oxoacid always depends on the relative strength of the oxygen-hydrogen bond. In an oxoacid the strength of the O—H bond depends on (a) the number of oxygen atoms per hydrogen in the formula and (b) the electronegativity of the central atom in the formula.

Oxoacids that have the same central atom and the same number of hydrogen atoms increase in strength as the number of oxygen atoms increases.

Each extra oxygen withdraws more electron density from the O—H bond, thereby weakening it. The more oxygen atoms per hydrogen, the stronger the acid.

| hypochlorous acid | chlorous acid | chloric acid | perchloric acid |

Weakest ← → Strongest

We can also see that an increase in the number of oxygens helps stabilize the oxoanions by delocalizing the electrons. When there is a very stable anion, it will be less likely to react with H^+ (or hydrolyze water) to form the acid.

Least stable ← → Most stable

The strengths of oxoacids that have the same number of hydrogen and oxygen atoms but a different central atom are affected by the electronegativity of the central atom:

| telluric acid | selenic acid | sulfuric acid |

Weakest ← → Strongest

In the diagram above, the electronegativity of the central atom increases from tellurium to selenium to sulfur, weakening the O—H bond and resulting in stronger acids.

Oxoacids that have the same number of oxygen atoms but different central atoms and different numbers of hydrogens can also be compared. For instance, experiments show that

$$H_3PO_4 \quad < \quad H_2SO_4 \quad < \quad HClO_4$$

In this situation we see that the electronegativity of the central atom increases as the strength of the acid increases. In addition to the electronegativity of the central atom, the oxygen atoms that have no hydrogens attached to them affect the acid strength. H_3PO_4 has only one oxygen that does not have a hydrogen attached to it, and it is attracting electrons from three H—O bonds. In H_2SO_4 there are two oxygens without hydrogens attached, and their effect is concentrated on only two H—O bonds. Finally, in $HClO_4$ three oxygen atoms have no hydrogen attached to them, and their effect is focused on a lone H—O bond. The effect of the electronegativity of the central atom and the effect of the oxygen atoms with no hydrogens attached combine to produce the observed trend in acid strengths. To repeat: when two acids are compared, the acid with more oxygen atoms per hydrogen atom will be the stronger acid.

Every **organic acid** contains the carboxyl group, which is commonly written as

$$-COOH \quad or \quad -C\!\!\stackrel{O}{\underset{O-H}{\diagdown}}$$

TIP

The weaker the O—H bond, the stronger the acid.

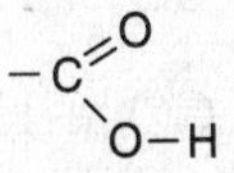

Electronegative atoms such as F, Cl, Br, I, O, and S on nearby carbon atoms will withdraw electron density from the —O—H bond and increase the strength of the acid. The addition of chlorine atoms to ethanoic acid increases the acid strength as shown below.

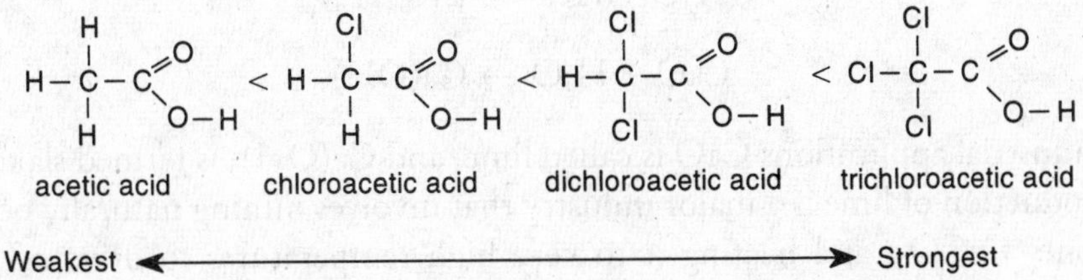

acetic acid < chloroacetic acid < dichloroacetic acid < trichloroacetic acid

Weakest ⟷ Strongest

BASES

Strong Bases

All metal hydroxides are **strong bases.** However, most metal hydroxides are very slightly soluble. Only the hydroxides of group IA metals, strontium and barium, have appreciable solubility; calcium hydroxide is moderately soluble. Soluble hydroxides may cause severe skin burns. Insoluble hydroxides are much less harmful. For instance, $Al(OH)_3$ and $Mg(OH)_2$ can safely be swallowed as an antacid to neutralize excess stomach acid.

Weak Bases

All bases related to ammonia are **weak bases.** Organic compounds related to ammonia are called **amines** and have carbon-containing groups replacing one or more of the hydrogen atoms of ammonia, NH_3. Two such bases, ethylamine and dimethylamine, were described previously.

Base Strengths

The relative strengths of the weak bases may be evaluated based on the electronegativities of the organic functional groups that replace the hydrogen atoms of ammonia. We saw that electronegative substituents such as chlorine increase the strength of organic acids. The reverse is true, however, of organic bases. For example, chloromethylamine is a weaker base than methylamine.

ANHYDRIDES OF ACIDS AND BASES

The word *anhydride* means "without water," and the acidic and basic anhydrides are compounds that, when added to water, become common acids and bases. **Acid anhydrides** are often the oxides of nonmetals. Common acid anhydrides have the following reactions with water:

$$SO_2 + H_2O \rightarrow H_2SO_3 \tag{14.6}$$

$$SO_3 + H_2O \rightarrow H_2SO_4 \tag{14.7}$$

$$CO_2 + H_2O \rightarrow H_2CO_3 \tag{14.8}$$

$$P_2O_5 + 3H_2O \rightarrow 2H_3PO_4 \tag{14.9}$$

Basic anhydrides are the oxides of metals. Some typical reactions with water are these:

$$K_2O + H_2O \rightarrow 2KOH \tag{14.10}$$

$$CaO + H_2O \rightarrow Ca(OH)_2 \tag{14.11}$$

In industrial applications CaO is called lime, and $Ca(OH)_2$ is termed slaked lime. The production of lime is a major industry that involves mining naturally occurring limestone, $CaCO_3$, and heating it to very high temperatures to drive off carbon dioxide in the reaction

$$CaCO_3 \xrightarrow{\text{heat}} CaO + CO_2 \tag{14.12}$$

NEUTRALIZATION REACTIONS

The reactions between acids and bases are called **neutralization reactions.** They are often double-replacement reactions, and the products can be predicted by the methods described in Chapter 4. In most cases, a neutralization reaction can be described as the mixing of an acid with a base to form a salt and water:

$$HBr(aq) + KOH(aq) \rightarrow KBr(aq) + H_2O(l) \qquad (14.13)$$
$$\text{(acid)} \quad \text{(base)} \quad \text{(salt)} \quad \text{(water)}$$

A neutralization reaction may be written as a molecular, ionic, or net ionic equation depending on the type of information the chemist is interested in communicating. For a typical neutralization reaction with soluble acids and bases the three types of reactions would be as follows:

$$HCl + NaOH \qquad \rightarrow \quad NaCl + H_2O \quad \text{(molecular equation)} \quad (14.14)$$

$$H^+ + Cl^- + Na^+ + OH^- \rightarrow \quad Na^+ + Cl^- + H_2O \quad \text{(ionic equation)} \quad (14.15)$$

$$H^+ + OH^- \qquad \rightarrow \quad H_2O \qquad \text{(net ionic equation)} \quad (14.16)$$

If all of the reactants are soluble strong acids and bases, the net ionic equation of a neutralization will *always* be as shown in Equation 14.16.

Acids are often used to dissolve insoluble hydroxides or oxides of metals. For instance, lanthanum ions are used to suppress interferences in atomic absorption spectroscopy. Dissolution of lanthanum oxide, La_2O_3, is achieved by reacting it with either HCl or HNO_3:

$$6HCl \quad + \quad La_2O_3 \quad \rightarrow \quad 2LaCl_3 \quad + \quad 3H_2O \qquad (14.17)$$

$$6HNO_3 \quad + \quad La_2O_3 \quad \rightarrow \quad 2La(NO_3)_3 \quad + \quad 3H_2O \qquad (14.18)$$

These are reactions between an acid and a basic anhydride. In both reactions the chloride and nitrate salts of lanthanum₃ are soluble.

Acid anhydrides such as SO_3 and CO_2 will react with bases as in the following reactions:

$$2NaOH(aq) \quad + \quad SO_3(g) \qquad \rightarrow \quad Na_2SO_4(aq) \quad + \quad H_2O(l) \quad (14.19)$$

$$Ca(OH)_2(aq) \quad + \quad CO_2(g) \quad \rightarrow \quad CaCO_3(s) \quad + \quad H_2O(l) \quad (14.20)$$

Acid and basic anhydrides may react with each other without any water present. For example, lime can react with sulfur trioxide in the reaction

$$SO_3(g) + CaO(s) \rightarrow CaSO_4(s) \qquad (14.21)$$

POLYPROTIC ACIDS

Acids that contain more than one ionizing hydrogen are called **polyprotic acids.** Sulfuric acid, H_2SO_4, and phosphoric acid, H_3PO_4, are two examples. Although ethanoic acid, $HC_2H_3O_2$, has four hydrogen atoms in its formula, it is not a polyprotic acid since it has only one ionizable hydrogen.

All polyprotic acids are weak acids except sulfuric acid, which is unique in that its first proton dissociates completely but the second proton does not. One property of polyprotic acids is that the protons dissociate and react in a stepwise manner; that is, the first proton dissociates or reacts before the second proton. The stepwise dissociation of phosphoric acid is written as shown in Equations 14.22–14.24:

$$H_3PO_4 \rightleftharpoons H_2PO_4^- + H^+ \tag{14.22}$$

$$H_2PO_4^- \rightleftharpoons HPO_4^{2-} + H^+ \tag{14.23}$$

$$HPO_4^{2-} \rightleftharpoons PO_4^{3-} + H^+ \tag{14.24}$$

The product or products formed in the neutralization of a polyprotic acid depend on the amount of base used. When 1 mole of hydroxide ions per mole of phosphoric acid reacts, the equation is

$$H_3PO_4 + OH^- \rightarrow H_2PO_4^- + H_2O \tag{14.25}$$

When 2 moles of hydroxide ions per mole of phosphoric acid react, the equation is

$$H_3PO_4 + 2OH^- \rightarrow HPO_4^{2-} + 2H_2O \tag{14.26}$$

When 3 moles of hydroxide ions per mole of phosphoric acid react, the equation is

$$H_3PO_4 + 3OH^- \rightarrow PO_4^{3-} + 3H_2O \tag{14.27}$$

> **TIP**
>
> Remember that these are combining ratios of H_3PO_4 and OH^-. Therefore, 0.2 mol of H_3PO_4 and 0.4 mol of OH^- follows the stoichiometry written in Equation 14.26.

Salts containing the $H_2PO_4^-$, HPO_4^{2-}, and PO_4^{3-} ions can be prepared by accurately adjusting the amount of base added to the phosphoric acid.

If exactly 1, 2, or 3 moles of hydroxide ions per mole of phosphoric acid are not reacted, a mixture of phosphate salts will form. For example, if 2.25 moles of OH^- are added per mole of phosphoric acid, we may deduce that the first 2 moles of OH^- convert the phosphoric acid to HPO_4^{2-}. The additional 0.25 mole of OH^- converts only some of the HPO_4^{2-} to PO_4^{3-}. Consequently, the final mixture will be a combination of HPO_4^{2-} and PO_4^{3-} salts.

BRÖNSTED-LOWRY THEORY; CONJUGATE ACID-BASE PAIRS

The Brönsted-Lowry theory of acids and bases not only redefined these compounds but also gave us the concept of **conjugate acid-base pairs.** In the equilibrium reaction of ethanoic acid with hydroxide ions:

$$\underset{\text{acid 1}}{HC_2H_3O_2} + \underset{\text{base 2}}{OH^-} \rightleftharpoons \underset{\text{base 1}}{C_2H_3O_2^-} + \underset{\text{acid 2}}{H_2O} \tag{14.28}$$

$HC_2H_3O_2$ is an acid that reacts with the base OH^- in the forward reaction. The products, however, are also acids and bases. $C_2H_3O_2^-$ is a base that can accept an H^+ from the water, while water is an acid since it donates a proton in the reverse reaction. $HC_2H_3O_2$ and $C_2H_3O_2^-$ are called a conjugate acid-base pair. In a similar fashion, H_2O and OH^- are another conjugate acid-base pair in this equation.

Conjugate acid-base pairs always have formulas that differ by only one H^+:

$$\text{Conjugate acid} \rightleftharpoons \text{Conjugate base} + H^+ \qquad (14.29)$$

Relative Strengths of Conjugate Acids and Bases

In a conjugate acid-base pair the relative strengths of the **conjugate acid** and **conjugate base** are determined by the position of the equilibrium in Equation 14.29. If this position results in more products than reactants, the conjugate acid is stronger than the conjugate base. If there are more reactants than products at equilibrium, the conjugate base is the stronger of the pair.

Ethanoic acid dissociates slightly in the reaction

$$HC_2H_3O_2 \rightleftharpoons C_2H_3C_2^- + H^+ \qquad (14.30)$$

and we conclude that ethanoic acid is a weak acid and the ethanoate ion is a stronger base.

The conjugate acid-base pair for ammonia may be written as

$$NH_4^+ \rightleftharpoons NH_3 + H^+ \qquad (14.31)$$

Since an aqueous solution of ammonia has a much higher concentration of ammonia than of the ammonium ion, the equilibrium lies to the right of the double arrow in Equation 14.31. From this fact we conclude that the NH_4^+ ion is a stronger conjugate acid than NH_3 is a conjugate base.

Using similar logic, we can determine the relative strengths of conjugate acids and bases in a complete chemical reaction. For example, when equal numbers of moles of ethanoic acid and base are mixed, this reaction goes virtually to completion, leaving little $HC_2H_3O_2$ and OH^-:

$$HC_2H_3O_2 + OH^- \rightarrow C_2H_3O_2^- + H_2O \qquad (14.32)$$

From the position of the equilibrium, chemists deduce that the OH^- ion is a stronger base than the ethanoate ion, $C_2H_3O_2^-$. At the same time, ethanoic acid is also a stronger acid than the water molecule. Since we already know that ethanoic acid is a weak acid, the water molecule must be a much weaker acid. Similarly, since the ethanoate ion is a relatively strong base, the hydroxide ion is an even stronger base. This information tells us that in the H_2O-OH^- conjugate acid-base pair water is a very weak acid and OH^- is a very strong base.

Another reaction takes place when ethanoic acid is dissolved in distilled water:

$$HC_2H_3O_2 + H_2O \rightleftharpoons C_2H_3O_2^- + H_3O^+ \qquad (14.33)$$

Again, we know that ethanoic acid is a weak acid, and by that definition most of it remains in molecular form. Since the reactants are favored in this equilibrium, we conclude that H_3O^+ is a stronger acid than $HC_2H_3O_2$. Also, the ethanoate ion is a stronger base than water. Knowing the position of equilibrium gives us another method to determine the relative strengths of conjugate acids and bases.

In aqueous solution, the strongest acid is the H_3O^+ ion (often written simply as the H^+ ion) and the strongest base is the OH^- ion. Acids that are stronger than water react with water to produce the H_3O^+ ion:

$$HCl + H_2O \rightarrow Cl^- + H_3O^+ \tag{14.34}$$

Because hydrochloric acid is a very strong acid, this reaction goes all the way to completion, forming the H_3O^+ ion and leaving no HCl molecules. Bases that are stronger than water produce the OH^- ion when dissolved in water. In aqueous solutions all strong acids react completely with the weak base water. All strong bases also react completely, with water acting as a weak acid. This is known as the **leveling effect** of water.

In discussing the strengths of acids and bases, it was mentioned above that the ethanoate ion is a stronger base than water. This fact results in the concept that the anion of a weak acid may be considered as a base. Very strong acids such as HCl and HNO_3 have anions that are extremely weak conjugate bases. On the other hand, very weak acids have anions that are strong conjugate bases. Carbonic acid, H_2CO_3, is a very weak acid, and the carbonate ion, CO_3^{2-}, is a strong conjugate base. Carbonate salts such as sodium carbonate and calcium carbonate (limestone) are rather strong bases and are used in many industrial processes instead of more expensive hydroxide bases such as NaOH and KOH.

Later in this chapter we will see that the strengths of conjugate acids and bases can be expressed numerically as K_a and K_b values. These two values are related to the constant K_w in the equation

$$K_a K_b = K_w \tag{14.35}$$

Equation 14.35 indicates that K_a is inversely proportional to K_b, meaning that strong conjugate acids have weak conjugate bases and vice versa.

Amphiprotic (Amphoteric) Substances

The words *amphiprotic* and *amphoteric* describe the same phenomenon, in which a substance may act as both a conjugate acid and a conjugate base. Amphiprotic salts are anions, and these anions must have at least one proton so that they can act as proton donors. They must also be able to accept a proton to act as bases. The anions of partially neutralized polyprotic acids are always amphiprotic. The common amphiprotic anions are the hydrogen carbonate ion, HCO_3^-; the hydrogen sulfate ion, HSO_4^-; the hydrogen sulfite ion, HSO_3^-; the dihydrogen phosphate ion, $H_2PO_4^-$; and the monohydrogen phosphate ion, HPO_4^{2-}. Each of these ions can accept a proton and act as a base. Each ion can also lose a proton and act as an acid.

Water is an amphiprotic solvent that can act as an acid and as a base as shown in the reaction

$$H_2O + H_2O \rightleftharpoons H_3O^+ + OH^- \tag{14.36}$$

In this reaction one water molecule donates a proton and is an acid, while the other accepts a proton and is a base. In both cases water is an extremely weak acid and an extremely weak base.

LEWIS ACIDS AND BASES; COMPLEXATION REACTIONS

One major advantage of the Lewis theory is to explain **complexation reactions,** which are not covered in the discussions of ionic reactions or covalent reactions. For instance, we know that silver chloride, AgCl, is an insoluble salt. It is not a hydroxide or a basic anhydride, which we know can be dissolved in acids. However, this salt does dissolve in ammonia solutions. An analysis of this process shows that the reaction is

$$AgCl(s) + 2NH_3 \rightarrow Ag(NH_3)_2^+ + Cl^- \tag{14.37}$$

The same reaction occurs, without the Cl^- ions, when silver ions in solution react with ammonia. This reaction is a complexation reaction.

The Lewis theory explains why such ions as $Ag(NH_3)_2^+$ form. In Equation 14.37 the silver ion is a **Lewis acid,** accepting pairs of electrons, while the ammonia donates a pair of electrons to the bond, acting as a **Lewis base.** The Lewis structure shows the nonbonding pair of electrons the ammonia molecule donates:

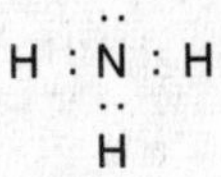

$$\overset{\displaystyle ..}{\underset{\displaystyle H}{H : N : H}}$$

The silver ion started as a silver atom with the electronic configuration [Kr] $5s^1$, $4d^{10}$. In forming the Ag^+ ion, it loses the $5s^1$ electron, resulting in the [Kr] $4d^{10}$ electronic configuration. Therefore it has empty $5s$ and $5p$ orbitals that may accept pairs of electrons. In actual fact, silver ions accept only two pairs of electrons from two ammonia molecules. In a similar fashion we find that all metal ions have available orbitals that may accept electron pairs. We may generalize that metal ions act as Lewis acids.

Ligands

In complexation reactions Lewis bases have many names: **ligands, complexing agents, chelates,** and **sequestering agents.** Most ligands have one pair of electrons to donate, as ammonia does. Some ligands have two pairs of electrons and some have up to six pairs. Ligands that provide more than one electron pair in forming a complex must be large, flexible molecules so that each pair of electrons can be oriented properly to form a bond. The chloride ion has four pairs of electrons but forms only one bond because the remaining six electrons cannot be aligned properly to form additional bonds.

Complexation reactions can be written generally as

$$M^{n+} + xL^{m-} \rightleftharpoons ML_x^{n-xm} \tag{14.38}$$

where M^{n+} is a metal ion with a charge of $+n$ and L^{m-} is a ligand with a charge of $-m$.

Silver tends to accept two electron pairs, and copper accepts four. The other metal ions tend to accept six electron pairs in complexes. With this information we can accurately write most complexation reactions once the number of electron pairs that a ligand can donate has been determined.

Ligands that form only one bond are called **monodentate ligands.** Halide ions are monodentate ligands. Even though they have four available electron pairs, once one electron pair is donated, the other electron pairs are not in the proper positions to make additional bonds. Cyanide ions, CN^-; thiocyanate ions, SCN^-; and anions of weak acids are other common monodentate ligands. Ammonia is a neutral molecule that is a monodentate ligand. Water and carbon monoxide are other monodentate molecular ligands.

Ligands that can form two bonds are called **bidentate ligands.** The most common bidentate ligands are the diamines, such as ethylenediamine, $H_2NCH_2CH_2NH_2$, and the anions of diprotic organic acids, such as the oxalate ion, $^-OOCCOO^-$ or $C_2O_4^{2-}$.

One special ligand is ethylenediaminetetraacetic acid, EDTA:

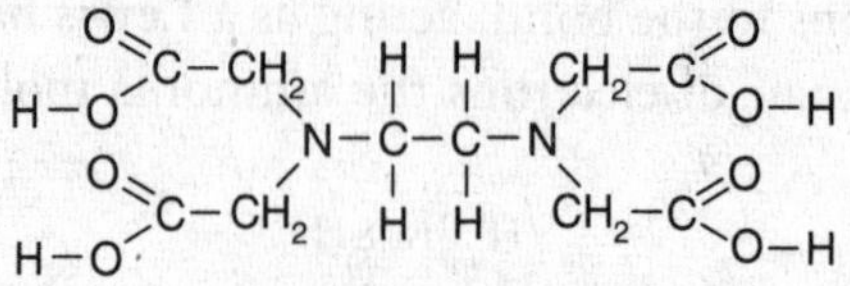

This is a tetraprotic acid. When it loses all four hydrogen ions, forming $EDTA^{4-}$, it has six pairs of electrons to donate. The molecule is flexible enough to allow each of the six pairs to form bonds with a single metal ion. One $EDTA^{4-}$ always reacts with only one metal cation:

$$M^{n+} \quad + \quad EDTA^{4-} \quad \rightarrow \quad M(EDTA)^{n-4}$$

EDTA is an important molecule for chemical analysis of metal ions using simple titration methods. Also, EDTA is found in many consumer products. Cosmetics, drugs, and even foods contain EDTA, which acts as a preservative by forming complexes with metal ions. The same uncomplexed metal ions could act as catalysts to promote oxidation. Complexation reduces or eliminates the catalytic activity and increases the shelf life of consumer products.

Water was mentioned above as a monodentate ligand. In fact, when chemists write an ion with the symbol (*aq*) after it, they are recognizing the fact that all ions are actually complexed to water in aqueous solution. Therefore the $Na^+(aq)$ ion is actually the $Na(H_2O)_6^+$ complex ion. Complexes with other ligands simply replace the water molecules with other electron-pair donors.

The formula of a complex or a complex ion is written as any other chemical formula. For any complex ion the total charge of the ion is the sum of the charges of the ions in the complex. The complex made of Fe^{3+} and six chloride ions, $FeCl_6^{3-}$, has a charge of -3:

$$Fe^{3+} + 6Cl^- \rightleftharpoons FeCl_6^{3-} \tag{14.39}$$

Molecular complexing agents do not affect the charge since they are neutral molecules. Thus Cu^{2+} still has a $+2$ charge when complexed with four molecules of ammonia in $Cu(NH_3)_4^{2+}$:

$$Cu^{2+} + 4NH_3 \rightleftharpoons Cu(NH_3)_4^{2+} \tag{14.40}$$

One interesting application of complexation reactions involves gold mining. Modern gold mining involves spraying dilute cyanide solutions, along with air, on gold-bearing ore. The gold, oxidized to Au^+, complexes with two CN^- ions to form the soluble complex $Au(CN)_2^-$. The liquid containing the gold complex is then treated to recover the concentrated gold.

Coordinate Covalent Bonds

The bonds formed in complexation reactions are covalent bonds. All of the properties of covalent bonds, as well as molecular structure, discussed in Chapter 5 apply also to the bonds in complexes. However, the formation of this covalent bond is unique. Instead of each atom donating one electron to the bond, one atom donates both electrons. Covalent bonds formed in this way are called **coordinate covalent bonds.** Aside from the method of formation, these bonds are true covalent bonds.

QUANTITATIVE ACID-BASE CHEMISTRY

In describing the nature of acids and bases, reference was made to their relative strengths. These strengths can be determined experimentally as the acid and base dissociation constants, K_a and K_b. The qualitative description of acid strengths given above was developed with a knowledge of the experimental values of acid strengths.

pH and pOH: Measurements of Acidity and Basicity

The acidity of a solution is expressed either as the molar hydrogen ion concentration, $[H^+]$, or the **pH** of the solution. Soren Sorensen in 1909 developed the pH system to represent acidity data in a simplified form. The pH and the hydrogen ion concentration are related by the equation

$$pH = -\log[H^+] \tag{14.41}$$

In aqueous solutions of acids and bases the pH ranges from 0 to 14. In very concentrated solutions we may have a negative pH (extremely acidic) or a pH greater than 14 (extremely basic).

In a similar fashion, the molar hydroxide ion concentration is related to the **pOH** by

$$pOH = -\log[OH^-] \tag{14.42}$$

The pH, pOH, $[H^+]$, and $[OH^-]$ are all related to each other. Water dissociates into hydrogen ions and hydroxide ions in the reaction

$$H_2O \rightleftharpoons H^+ + OH^- \tag{14.43}$$

The equilibrium law for Equation 14.43 is

$$K = [H^+][OH^-] = 1.0 \times 10^{-14} \tag{14.44}$$

> **TIP**
>
> K_w, the autoionization constant of water, may be called the ionization constant, the pyrolysis constant, or the ion product for water.

The equilibrium constant in this case is designated as K_w, the **autoionization constant of water.** Taking the negative logarithm of Equation 14.44 gives

$$pK_w = pH + pOH = 14.00 \tag{14.45}$$

Equations 14.44 and 14.45 allow the chemist to determine the pH, pOH, $[H^+]$, or $[OH^-]$ by measuring only one of them and then finding the other three by calculation.

EXERCISE 14.1

Determine $[H^+]$, $[OH^-]$, pH, and pOH, given the following data:

(a) $[H^+] = 2.3 \times 10^{-4}\ M$ (c) pH = 10.67

(b) $[OH^-] = 6.3 \times 10^{-2}\ M$ (d) pOH = 2.34

Solution

Starting with the given information, we use Equations 14.41–14.45 to convert:

	$[H^+]$	$[OH^-]$	pH	pOH
(a)	2.3×10^{-4}	4.3×10^{-11}	3.64	10.36
(b)	1.6×10^{-13}	6.3×10^{-2}	12.80	1.20
(c)	2.1×10^{-11}	4.7×10^{-4}	10.67	3.33
(d)	2.2×10^{-12}	4.6×10^{-3}	11.66	2.34

When $[H^+]$ is greater than $[OH^-]$, the solution is considered to be acidic. A pH less than 7.0 also represents an acid solution. When $[H^+]$ is less than $[OH^-]$, the solution is basic and the pH is greater than 7.0. When $[H^+]$ is equal to the $[OH^-]$, the solution is neutral and the pH is 7.0.

Many of the common substances with which we come into contact are either acidic or basic. Table 14.5 lists some of these substances and their approximate pH values.

TABLE 14.5

Approximate pH Values of Common Substances

Substance	Approximate pH
0.10 *M* HCl	1.00
lemon juice	2.3
vinegar	3.0
wine	2.8–3.8
pickles	3.3
orange juice	3.5
soft drinks	3.0–4.0
beer	4.0–5.0
pure rainwater	5.6
milk	6.3–6.6
blood	7.4
seawater	8.3
milk of magnesia	10.5
ammonia	11.2
lime water	12.4
0.1 *M* NaOH	13.00

> **TIP**
>
> Even distilled water has enough dissolved CO_2 to make it distinctly acidic.

[H⁺] AND pH OF STRONG ACIDS

Except for sulfuric acid, all strong acids are monoprotic and ionize completely when dissolved in water. Consequently, the hydrogen ion concentration of a strong acid solution is equal to the molar concentration of the acid itself:

$$[H^+] = M_{\text{strong acid}} \tag{14.46}$$

Sulfuric acid is the only diprotic strong acid. When this acid is dissolved in water, only one hydrogen ion ionizes, and it acts as a monoprotic acid:

$$H_2SO_4 \rightleftharpoons H^+(aq) + HSO_4^-(aq) \tag{14.47}$$

Also, for sulfuric acid the hydrogen ion concentration is equal to the molar concentration.

The pH of a strong acid solution can be written as

$$pH = -\log[H^+] = -\log M_{\text{strong acid}} \tag{14.48}$$

> **TIP**
>
> The ionization of the first proton of H_2SO_4 makes the solution acidic. This acidic solution suppresses the second ionization step. The result is that H_2SO_4 acts as a strong monoprotic acid when dissolved in water.

[OH⁻], pOH, AND pH OF STRONG BASES

Strong bases dissociate completely when dissolved in water; however, they may have more than one hydroxide ion per formula unit. The concentration of hydroxide ions is equal to the molarity of the base multiplied by the number of hydroxide ions in its chemical formula:

$$[OH^-] = M_{\text{strong base}} \times \text{number of } OH^- \text{ ions per mole} \qquad (14.49)$$

For example, a 0.200 M solution of $Ba(OH)_2$ has a hydroxide concentration of 0.400 M. Once the [OH⁻] is known, the pOH is calculated as

$$pOH = -\log[OH^-]$$

From the pOH the pH is calculated using the relationship

$$pH = 14.00 - pOH$$

EXERCISE 14.2

Determine the pH of each of the following solutions:

(a) 0.020 M HNO_3

(b) 0.00043 M $Ba(OH)_2$

(c) 3.0 g of NaOH dissolved in 250 mL H_2O

(d) 0.00032 mol SO_3 dissolved in 3.4 L H_2O

(e) 0.00098 mol $Ca(OH)_2$ dissolved in 1.3 L H_2O

Solution

> **TIP**
>
> A very quick check of your pH calculations must have acids with pH values below 7 and bases with pH values greater than 7.

(a) [H⁺] = 0.020 M; pH = 1.70

(b) [OH⁻] = (0.00043 M) (2) = 0.00086 M; pOH = 3.07 and pH = 10.93

(c) M_{NaOH} = 0.30; [OH⁻] = 0.30; pOH = 0.52 and pH = 13.48

(d) $SO_3 + H_2O \rightarrow H_2SO_4$; $M_{H_2SO_4} = 9.4 \times 10^{-5}$; [H⁺] = 9.4×10^{-5}; pH = 4.03

(e) $M_{\text{Ca(OH)}_2} = 7.5 \times 10^{-4}$; [OH⁻] = (7.5×10^{-4}) (2) = 1.5×10^{-3}; pOH = 2.82 and pH = 11.18

Calculations for pH can always be checked for reasonableness since acids must have pH values less than 7 and bases must have pH values greater than 7.

pH OF WEAK ACIDS AND WEAK BASES

Hydrofluoric acid is a weak acid that dissociates in water according to the equation

$$HF + H_2O \rightleftharpoons H_3O^+ + F^- \qquad (14.50)$$

Equation 14.50 may also be written in a shorter, more convenient form by eliminating water:

$$HF \rightleftharpoons H^+ + F^- \tag{14.51}$$

Either Equation 14.50 or Equation 14.51 can be used to write the equilibrium law:

$$K_a = \frac{[H^+][F^-]}{[HF]} = \frac{[H_3O^+][F^-]}{[HF]} \tag{14.52}$$

The two forms are identical, and either may be used in the calculations that follow.

The constant K_a is called the **acid dissociation constant;** values for selected weak acids are tabulated in Appendixes 4 and 5. For the example above, since we know the initial concentration of HF and the value of K_a, the hydrogen ion concentration can be calculated. The technique used is similar to the equilibrium calculations in Chapter 10.

EXERCISE 14.3

If 0.100 mol of HF is diluted with distilled water to a volume of 500 mL, what is the pH of the solution?

Solution

To solve this problem, we set up an equilibrium table with the reaction on the first line, and we enter the initial concentration on the INIT. CONC. line. The initial concentration of HF is calculated as

$$\frac{0.100 \text{ mol HF}}{0.500 \text{ L solution}} = 0.200 \; M \text{ HF}$$

REACTION	HF	⇌	H⁺	+	F⁻
INIT. CONC.	0.100/0.500		0.00*		0.00
CHANGE					
EQUILIBRIUM					
SOLUTION					

*The hydrogen ion concentration in distilled water is 1.0×10^{-7}, but in most instances it is considered to be zero.

In the CHANGE row we enter x's to represent the stoichiometric relationships between the reactants and the products. In this table we assign a positive value to the x's under the products because they start at zero and cannot possibly decrease. As a consequence any x's under the reactant(s) must be negative.

REACTION	HF	⇌	H⁺	+	F⁻
INIT. CONC.	0.200 *M*		0.00		0.00
CHANGE	−*x*		+*x*		+*x*
EQUILIBRIUM					
SOLUTION					

The EQUILIBRIUM row is then the sum of the INIT. CONC. and CHANGE rows of the table:

REACTION	HF	⇌	H⁻	1	F⁻
INIT. CONC.	0.200 *M*		0.00		0.00
CHANGE	−*x*		+*x*		+*x*
EQUILIBRIUM	**0.200 − x**		+*x*		+*x*
SOLUTION					

At this point we enter the terms in the EQUILIBRIUM row into the equilibrium law, along with the value of K_a:

$$K_a = \frac{[\text{H}^+][\text{F}^-]}{[\text{HF}]}$$

$$6.6 \times 10^{-4} = \frac{(x)(x)}{0.200 - x}$$

If this equation is solved for x by ordinary means, a quadratic equation results. As before, we can try a simplifying assumption to solve the problem more quickly. The only term that can be simplified is $0.200 - x$. We use the assumption that $x \ll 0.200$ and conclude that $0.200 - x = 0.200$. (Remember that, when we finally calculate x, we must verify that the assumption is true.) Returning to the equation, we find that the assumption yields

$$6.6 \times 10^{-4} = \frac{(x)(x)}{0.200}$$

$$(6.6 \times 10^{-4})(0.200) = x^2$$

$$x = 0.0115$$

Checking the assumption, we find that 0.0115 is less than one-tenth of 0.200 and therefore qualifies as being negligible for this type of calculation. Now we substitute 0.0115 for x in the EQUILIBRIUM row and calculate the SOLUTION row:

REACTION	HF	⇌	H⁺	+	F⁻
INIT. CONC.	0.200		0.00		0.00
CHANGE	−x		+x		+x
EQUILIBRIUM	0.200 − x		+x		+x
SOLUTION	**0.189**		**0.0115**		**0.0115**

With the value for $[H^+] = 0.0115$ M, the pH can be calculated as $-\log(0.0115)$ = 1.94.

Looking at this process a little more closely, we find that, if the assumption is valid, then

$$[H^+] = \sqrt{K_a C_a} \qquad (14.53)$$

where C_a is the initial concentration of the weak acid. Evaluating the conditions under which the assumption holds, we find that it will always be valid as long as $C_a > 100\, K_a$.

Weak bases such as ammonia and methylamine, CH_3NH_2, behave similarly to weak acids. The dissociation reaction for methylamine is

$$CH_3NH_2 + H_2O \rightleftharpoons CH_3NH_3^+ + OH^- \qquad (14.54)$$

Although water cannot be eliminated to simplify the chemical reaction, it does not appear in the equilibrium law since it is the solvent.

$$K_b = \frac{[CH_3NH_3^+][OH^-]}{[CH_3NH_2]} \qquad (14.55)$$

The constant K_b is the **base dissociation constant.** This equilibrium law is similar to the one for HF (Equation 14.52) in that it has two concentration terms in the numerator and one in the denominator. The two laws are mathematically equivalent.

Values of K_b for selected weak bases are given in Appendix 6.

EXERCISE 14.4

What is the pH of a 0.500 M solution of methylamine, $K_b = 4.2 \times 10^{-4}$?

Solution

First we set up the equilibrium table in the same fashion as for the weak acid example in Exercise 14.3.

> **NOTE:**
> We use a factor of 100 here because this will keep the error in x to less than 5% compared with the value of x found when solving using the quadratic equation.

REACTION	CH$_3$NH$_2$	+	H$_2$O	⇌	CH$_3$NH$_2^+$	+	OH$^-$
INIT. CONC.	0.500		55.5		0.00		0.00
CHANGE	−*x*		−*x*		+*x*		+*x*
EQUILIBRIUM	0.500 − *x*		55.5 − *x*		+*x*		+*x*
SOLUTION							

We entered the concentration of water:

$$\frac{1000 \text{ g H}_2\text{O L}^{-1}}{18 \text{ g H}_2\text{O mol}^{-1}} = 55.5 \ M$$

for completeness although it does not appear in the equilibrium law. The terms from the EQUILIBRIUM row are entered into the equilibrium law

$$4.2 \times 10^{-4} = \frac{(x)(x)}{0.500 - x}$$

To simplify, we assume that *x* is very small since 0.500 is more than 100 times larger than K_b. The result is that the 0.500 − *x* term in the denominator becomes 0.500:

$$4.2 \times 10^{-4} = \frac{(x)(x)}{0.500}$$

Solving for *x*, we obtain

$$(4.2 \times 10^{-4})(0.500) = x^2$$

$$x = 0.0145$$

Once *x* is determined, we check the assumption. Another way to judge if our assumption worked is to see if *x* is less than 10% of the initial concentration. We find that it is valid since 0.500 is more than ten times larger than 0.0145. The rest of the equilibrium table is completed using the calculated value of *x*:

REACTION	CH$_3$NH$_2$	+	H$_2$O	⇌	CH$_3$NH$_2^+$	+	OH$^-$
INIT. CONC.	0.500		55.5		0.00		0.00
CHANGE	−*x*		−*x*		+*x*		+*x*
EQUILIBRIUM	0.500 − *x*		55.5 − *x*		+*x*		+*x*
SOLUTION	0.486		55.5		0.0145		0.0145

Using the listed [OH$^-$] we calculate pOH as 1.84, and the pH is 12.16. As with the weak acid, this entire process may be summarized in a simple equation:

$$[OH^-] = \sqrt{C_b K_b} \quad \text{as long as } C_b > 100\, K_b$$

In Exercises 14.3 and 14.4 there are two ways to check the accuracy of the results. First, the answers must be reasonable; a weak acid should have an acidic pH (below 7) and a weak base should have a basic pH (above 7). Second, the calculated values may be entered into the mass action expression, and the result should be close to the given K_a or K_b.

pH OF SALT SOLUTIONS; HYDROLYSIS REACTIONS

When an acid is neutralized with a base, a salt is formed. If the anion of a salt is the conjugate base of a weak acid, it will react with water in a **hydrolysis reaction.** For the fluoride ion the hydrolysis reaction is as follows:

$$F^-(aq) + H_2O(l) \rightleftharpoons HF(aq) + OH^-(aq) \tag{14.56}$$

The conjugate base, F$^-$, of the weak acid HF will produce a basic solution, as shown by the OH$^-$ in Equation 14.56.

If the cation of a salt is the conjugate acid of a weak base, the hydrolysis reaction will result in an acid solution, as shown by the reaction of the ammonium ion:

$$NH_4^+(aq) + H_2O(l) \rightleftharpoons NH_3(aq) + H_3O^+(aq) \tag{14.57}$$

The conjugate acid of a strong base and the conjugate base of a strong acid are extremely weak acids and bases. In fact, they are so weak that they do not hydrolyze in water. For example, Cl$^-$ from the strong acid HCl and Na$^+$ from the strong base NaOH do not hydrolyze in water.

ACIDITY OF SALT SOLUTIONS AND CLASSIFICATION OF SALTS

The cation of a salt will be the conjugate acid of either a strong base or a weak base and the anion of a salt will be the conjugate base of either a strong acid or a weak acid. Table 14.6 describes the different types of salts possible.

The first step in determining the pH of a salt solution is to determine the type of salt. This is done by adding H$^+$ to the anion and OH$^-$ to the cation in the salt to determine the type of acid and base used to prepare the salt. For instance, ammonium chloride is composed of NH$_4^+$ and Cl$^-$ ions. By adding OH$^-$ to the cation and H$^+$ to the anion, NH$_4$OH (NH$_3$ + H$_2$O) and HCl are obtained. From this result we conclude that ammonium chloride is an acid salt prepared from a weak base and a strong acid. We may also quickly conclude that solutions of ammonium chloride will have pH values below 7.

TABLE 14.6

Classification of Salts Based on Acid and Base Used to Prepare Them

Acid Used	Base Use	Salt Type	Salt Solution Will Be
Strong	Strong	Neutral	Close to pH 7*
Strong	Weak	Acidic	Acidic
Weak	Strong	Basic	Basic
Weak	Weak	Mixed	Depends on K_a and K_b

*Remember that all water contains dissolved CO_2 that makes it slightly acidic.

A mixed salt has a cation that is the conjugate acid of a weak base, and an anion that is the conjugate base of a weak acid. The acidity or alkalinity of a mixed salt will be determined by whichever of these conjugates is stronger. Considering the K_a of the weak acid and the K_b of the weak base that form the salt, we conclude that the solution will be acidic if $K_b > K_a$ and basic if $K_a > K_b$.

EXERCISE 14.5

Determine whether the aqueous solutions of each of the following salts will be acid, neutral, or basic:

(a) $CaCl_2$ (c) Na_2SO_3 (e) SrF_2 (g) KNO_2
(b) $NaNO_3$ (d) $KC_2H_3O_2$ (f) NH_4Br (h) Li_2CO_3

Solution

(a) Neutral   (c) Basic   (e) Basic   (g) Basic
(b) Neutral   (d) Basic   (f) Acid   (h) Basic

pH OF SALT SOLUTIONS

The pH of a neutral salt is 7.00. Although we can tell whether a mixed salt solution will be acidic or basic, calculation of the pH is left for higher level courses. We concern ourselves here with calculating the pH value of acidic and basic salts only.

Ammonium chloride was determined in the preceding section to be an acid salt. Since the chloride ion does not hydrolyze, we focus on the hydrolysis of the ammonium ion discussed above to develop the equilibrium law needed to solve problems.

NH_4^+ is the conjugate acid of ammonia, and the hydrolysis reaction is written as either

$$NH_4^+ + H_2O \rightleftharpoons NH_3 + H_3O^+ \tag{14.58}$$

or

$$NH_4^+ \rightleftharpoons NH_3 + H^+ \tag{14.59}$$

The equilibrium law for this reaction is based on either of the two equivalent expressions:

$$K = \frac{[NH_3][H^+]}{[NH_4^+]} = \frac{[H_3O^+][NH_3]}{[NH_4^+]} \tag{14.60}$$

The equilibrium constant K is equal to K_w/K_b, where K_b is the base dissociation constant for ammonia. We can demonstrate this relationship by multiplying the equilibrium law in Equation 14.60 by $\dfrac{[OH^-]}{[OH^-]}$ to obtain

$$K = \left(\frac{[NH_3][H^+]}{[NH_4^+]}\right)\left(\frac{[OH^-]}{[OH^-]}\right) \tag{14.61}$$

In the numerator the product $[H^+][OH^-] = K_w$, which we substitute into Equation 14.61:

$$K = K_w\left(\frac{[NH_3]}{[NH_4^+][OH^-]}\right) \tag{14.62}$$

The term in parentheses is the reciprocal of the equilibrium law for the dissociation of ammonia, and its value is K_b. Performing the substitution yields

$$K = \frac{K_w}{K_b} \tag{14.63}$$

This value is sometimes called the hydrolysis constant for a salt, K_h, but a more accurate name is the acid dissociation constant of the Brönsted-Lowry conjugate acid, symbolized as K_a.

This derivation leads to a universally useful equation and concept. For a conjugate acid-base pair in aqueous solution, the K_a of the conjugate acid multiplied by the K_b of the conjugate base will always be equal to K_w:

$$K_w = K_a K_b \tag{14.64}$$

K_w is the ion product for the dissociation of water:

$$H_2O \rightleftharpoons H^+ + OH^-$$

and

$$K_w = [H^+][OH^-] = 1.00 \times 10^{-14} \tag{14.65}$$

From this derivation we see also that the salt of a weak base acts as a Brönsted-Lowry acid, and we expect its solutions to be acidic.

Once the equilibrium law for acid and basic salts is derived, we can start solving problems. Let us calculate the pH of a 0.100 M solution of ammonium chloride.

Since the equilibrium law has already been described, the next step is to set up the equilibrium table:

REACTION	NH_4^+	$\rightleftharpoons$	NH_3	+	H^+
INIT. CONC.	0.100		0.00		0.00
CHANGE	$-x$		$+x$		$+x$
EQUILIBRIUM	$0.100 - x$		$+x$		$+x$
SOLUTION					

Entering the values of K_w, K_b, and the terms from the EQUILIBRIUM line into the equilibrium law yields

$$\frac{1.0 \times 10^{-14}}{1.8 \times 10^{-5}} = \frac{(x)(x)}{0.100 - x}$$

Assuming that $0.100 \gg x$, we obtain

$$\frac{1.0 \times 10^{-14}}{1.8 \times 10^{-5}} = \frac{(x)(x)}{0.100}$$

which may be solved as

$$(5.6 \times 10^{-10})(0.100) = x^2$$

$$x = 7.4 \times 10^{-6}$$

Since x is much smaller than 0.100, the assumption was valid. We now use the value of x to calculate the values in the SOLUTION row of the table.

REACTION	NH_4^+	$\rightleftharpoons$	NH_3	+	H^+
INIT. CONC.	0.100		0.00		0.00
CHANGE	$-x$		$+x$		$+x$
EQUILIBRIUM	$0.100 - x$		$+x$		$+x$
SOLUTION	0.100		7.4×10^{-6}		7.4×10^{-6}

The pH is then calculated to be

$$pH = -\log(7.4 \times 10^{-6}) = 5.13$$

This result is consistent with the concept that NH_4^+ is a conjugate acid of NH_3 and that the solution must be acidic.

To summarize this calculation in a simpler form, we write

$$[H^+] = \sqrt{\frac{K_w}{K_b} C_s} \quad \text{provided that } C_s > 100 \frac{K_w}{K_b} \tag{14.66}$$

where K_b is the dissociation constant of the weak base used to produce the acidic salt and C_s is the molar concentration of the salt.

For basic salts, calculations are performed in a very similar manner. For example, NaF is a basic salt since it is formed by reacting the strong base NaOH with the weak acid HF. We may ignore Na^+ while writing the hydrolysis equation for F^- as follows:

$$F^- + H_2O \rightleftharpoons HF + OH^- \tag{14.67}$$

The base dissociation constant is calculated by dividing K_w by the acid dissociation constant, K_a, of HF:

$$K_b = \frac{K_w}{K_a} = \frac{[HF][OH^-]}{[F^-]} \tag{14.68}$$

Questions may then be answered by using the procedures outlined before. The general solution can be simplified to

$$[OH^-] = \sqrt{\frac{K_w}{K_a} C_s} \quad \text{provided that } C_s > 100 \frac{K_w}{K_a} \tag{14.69}$$

where C_s is the initial concentration of the salt and K_a is the acid dissociation constant of HF.

The salt KNO_3 does not change the pH of the solution since it is formed from the strong acid HNO_3 and the strong base KOH. The K^+ and NO_3^- ions are such weak Brönsted-Lowry acids and bases that they have no effect on pH.

Two simple equations suffice for calculating the pH values of weak acid, weak base, acid salt, and basic salt solutions:

$$\text{weak acids and acid salts} \quad [H^+] = \sqrt{K_a C_a} \tag{14.70}$$

$$\text{weak bases and basic salts} \quad [OH^-] = \sqrt{K_b C_b} \tag{14.71}$$

For solutions of weak acids or weak bases, K_a and K_b are the acid and base dissociation constants. In salt solutions, K_a and K_b are the conjugate acid and base dissociation constants of the ions in the salt, which are calculated as K_w/K_b and K_w/K_a, respectively. Equations 17.70 and 17.71 work only when the initial concentration, C, is at least 100 times larger than the dissociation constant, K. Otherwise the problem must be solved using the quadratic equation as shown in Chapter 10.

Buffer Solutions

To this point we have considered solutions containing only pure weak acids, weak bases, or their salts dissolved in distilled water. A mixture that contains a conjugate acid-base pair is known as a **buffer solution** because the pH changes by a relatively small amount if a strong acid or base is added to it. Buffer solutions are used to control pH in many biological and chemical reactions. These solutions are governed by the same equilibrium law as are weak acids and weak bases.

For instance, a buffer can be prepared by dissolving 0.200 mole of HF and 0.100 mole of NaF in water to make 1.00 liter of solution. The dissociation reaction of HF is

$$HF(aq) \rightleftharpoons F^-(aq) + H^+(aq) \tag{14.72}$$

and its equilibrium law is

$$K_a = \frac{[F^-][H^+]}{[HF]} = 6.8 \times 10^{-4} \tag{14.73}$$

An equilibrium table is set up using the given information. Since we have 0.100 mole NaF, the concentration of F^- is determined as 0.100 M. The Na^+ in NaF is a spectator ion in the dissociation reaction and is not needed.

REACTION	HF	$\rightleftharpoons$	F⁻	+	H⁺
INIT. CONC.	0.200		0.100		0.00
CHANGE	$-x$		$+x$		$+x$
EQUILIBRIUM	$0.200 - x$		$0.100 + x$		$+x$
SOLUTION					

Substituting the information from the EQUILIBRIUM line into the equilibrium law yields

$$K_a = \frac{(0.100 + x)(x)}{0.200 - x} = 6.8 \times 10^{-4}$$

Assuming that x is small compared to both 0.100 and 0.200, we simplify the equation to

$$K_a = \frac{(0.100)(x)}{0.200} = 6.8 \times 10^{-4}$$

Solving this equation for x yields

$$x = 1.36 \times 10^{-3}$$

This value of x satisfies the assumptions made, and the last line of the equilibrium table can be completed as:

REACTION	HF	⇌	F⁻	+	H⁺
INIT. CONC.	0.200		0.100		0.00
CHANGE	−x		+x		+x
EQUILIBRIUM	0.200 − x		0.100 + x		+x
SOLUTION	**0.199**		**0.101**		**1.36×10^{-3}**

The last column in the table gives us $[H^+] = 1.36 \times 10^{-3}\ M$. The pH of the buffer solution is calculated as 2.87. For almost all buffer solutions the assumptions hold true, and this type of problem is solved by simply entering the given concentrations of the conjugate acid and conjugate base directly into the equilibrium law.

EXERCISE 14.6

Calculate the pH of the following buffer solutions:

(a) 0.250 M ethanoic acid and 0.150 M sodium ethanoate
(b) A solution of 10.0 g each of formic acid and potassium formate dissolved in 1.00 L of H_2O
(c) 0.0345 M ethylamine and 0.0965 M ethyl ammonium chloride
(d) 0.125 M hydrazine and 0.321 M hydrazine hydrochloride

Solution

(a) $K_a = \dfrac{[C_2H_3O_2^-]\,[H^+]}{[HC_2H_3O_2]}$. The given data tell us that $[HC_2H_3O_2] = 0.250\ M$

and $[C_2H_3O_2^-] = 0.150\ M$. Substituting these values into the equation yields

$$1.8 \times 10^{-3} = \frac{(0.150)[H^+]}{0.250}$$

$$[H^+] = 3.0 \times 10^{-3}, \text{ and pH} = 2.52$$

(b) The masses given allow us to calculate concentrations of 0.217 M formic acid and 0.417 M sodium formate. The equilibrium law to use is

$K_a = \dfrac{[CHO_2^-][H^+]}{[HCHO_2]}$. Substituting data gives

$$1.8 \times 10^{-4} = \frac{(0.417)[H^+]}{0.217}$$

$$[H^+] = 9.4 \times 10^{-5}, \text{ and pH} = 4.03$$

(c) The equilibrium law to use is $K_b = \dfrac{[C_2H_5NH_3^+][OH^-]}{[C_2H_5NH_2]}$. We are given

$[C_2H_5NH_2] = 0.0345\ M$ and $[C_2H_5NH_3^+] = 0.0965\ M$. Substituting these values yields

$$4.3 \times 10^{-4} = \dfrac{(0.0965)[OH^-]}{0.0345}$$

$$[OH^-] = 1.5 \times 10^{-4},\ \text{pOH} = 3.82,\ \text{and pH} = 10.18$$

(d) The equilibrium law to use is $K_b = \dfrac{[N_2H_5^+][OH^-]}{[N_2H_4]}$. We are given $[N_2H_4] =$

$0.125\ M$ and $[N_2H_5^+] = 0.321\ M$. Substituting these values yields

$$9.6 \times 10^{-7} = \dfrac{(0.321)[OH^+]}{0.125}$$

$$[OH^-] = 3.7 \times 10^{-7},\ \text{pOH} = 6.43,\ \text{and pH} = 7.57$$

SHORTCUTS IN BUFFER CALCULATIONS

The equilibrium law used for buffers involves a ratio of the concentrations of the conjugate acid and conjugate base. For a ratio, we may use the moles of conjugate acid and moles of conjugate base instead of the acid and base molarities, thereby eliminating a step or two in the calculations.

In addition, if the concentrations of the conjugate acid and conjugate base are equal, their ratio is exactly 1.00. The result is that, with equal molarities or equal numbers of moles of conjugate acid and conjugate base,

$$K_a = [H^+] \quad \text{and} \quad pK_a = pH \tag{14.74}$$

for a buffer made from a weak acid and its conjugate base, and

$$K_b = [OH^-] \quad \text{and} \quad pK_b = pOH \tag{14.75}$$

for a buffer made from a weak base and its conjugate acid.

pH CHANGES IN BUFFERS

Addition of a strong acid to a buffer will decrease the pH slightly, and addition of a strong base will increase the pH slightly. To determine the amount that the pH changes when a strong acid or base is added to a buffer, we must calculate the change in concentration of the conjugate acid and conjugate base in the buffer.

In an ethanoate buffer, ethanoic acid, $HC_2H_3O_2$, is the conjugate acid, and the ethanoate ion, $C_2H_3O_2^-$, is the conjugate base. Addition of a strong acid, represented as H^+, results in the reaction

$$C_2H_3O_2^- + H^+ \rightarrow HC_2H_3O_2 \tag{14.76}$$

Addition of a strong acid increases the concentration of ethanoic acid and decreases the concentration of ethanoate ions. Addition of a strong base to this buffer increases the ethanoate ion concentration and decreases the ethanoic acid concentration as in the reaction

$$HC_2H_3O_2 + OH^- \rightarrow C_2H_3O_2^- \qquad (14.77)$$

To calculate the effect of adding a strong acid or base to a buffer the following steps are followed:

1. Determine the number of moles of the conjugate acid and conjugate base in the original buffer.
2. Determine the moles of strong acid or base added.
3a. If a strong acid is added to the buffer, add the value from step 2 to the conjugate acid and subtract it from the conjugate base.
3b. If a strong base is added to the buffer, add the value from step 2 to the conjugate base and subtract it from the conjugate acid.
4. Substitute the new values for the conjugate acid and conjugate base into the equilibrium law, and calculate the pH as shown in the previous section.

EXERCISE 14.7

An ethanoate buffer is prepared with 0.250 M ethanoic acid and 0.100 M $NaC_2H_3O_2$. If 0.002 mol of solid NaOH is added to 100 mL of this buffer, calculate the change in pH of the buffer due to the addition of the NaOH.

Solution

To calculate the change in pH, the pH of the original buffer is needed, along with the final pH after the base is added. First, the dissociation reaction is written for ethanoic acid:

$$HC_2H_3O_2 \rightleftharpoons H^+ = C_2H_3O_2^-$$

Next the equilibrium law is written:

$$K_a = \frac{[H^+][C_2H_3O_2^-]}{[HC_2H_3O_2]}$$

We calculate the moles of $HC_2H_3O_2$ as 0.100 L $\times$ 0.250 M = 0.0250 mol. The moles of $C_2H_3O_2^-$ ions are 0.100 L $\times$ 0.100 M = 0.0100 mol. The value for K_a and the moles of ethanoate and ethanoic acid we just calculated are substituted into the equilibrium law:

$$1.8 \times 10^{-5} = \frac{[H^+](0.0100)}{0.0250}$$

The hydrogen ion concentration is calculated as $[H^+] = 4.5 \times 10^{-5}$ M, and the pH is 4.35.

To calculate the pH of the buffer after the NaOH is added, we add 0.0020 mol to the $C_2H_3O_2^-$ value and subtract 0.0020 moles from the $HC_2H_3O_2$.

This gives us 0.0120 mol $C_2H_3O_2^-$ and 0.0230 mol $HC_2H_3O_2$. Substituting these new values into the equilibrium law, we have

$$1.8 \times 10^{-5} = \frac{[H^+](0.0120)}{0.0230}$$

$$[H^+] = 3.45 \times 10^{-5} \ M$$

$$pH = 4.46$$

There is an increase of 0.11 pH unit when the NaOH is added. This answer is reasonable since the pH is expected to rise slightly when a base is added.

PREPARATION OF BUFFERS

Preparing a buffer requires several important decisions about the buffer and the system that needs to be buffered. The following steps indicate the general process for preparing a buffer.

1. Decide what pH is required.
2. Decide what final volume is required.
3. Based on Step 1, choose a conjugate acid-base system using tables such as those in Appendices 4, 5, and 6.
4. Determine the moles per liter of acid or base that your reaction will generate.
5. Based on Step 4, the sum of the concentrations of the conjugate acid and conjugate base should be approximately 20 times the value estimated in Step 4.
6. Based on Steps 1, 3, and 5, calculate the separate concentrations of the conjugate acid and base needed.
7. Use Steps 2 and 6 to determine the masses of the conjugate acid and base to use.
8. Measure out the amounts determined in Step 7, dissolve in distilled water, and dilute to the volume determined in Step 2.

There are two alternate approaches to preparing a buffer. In the first approach, based on Steps 2 and 4, assume that all of our buffer material will be the conjugate acid. Then calculate the amount of strong base needed to neutralize enough of the acid to form the conjugate base. Mix the conjugate acid and the strong base to make the buffer.

In the second approach, based on Steps 2 and 4, assume that all of our buffer material will be the conjugate base. Then calculate the amount of strong acid needed to convert part of the conjugate base to the required amount of conjugate acid. Now mix the conjugate base with the desired amount of strong acid to prepare the buffer.

Selection of Buffer Materials

Assume that we need to prepare, at pH 5.00, 100 mL of a buffer that will be capable of buffering 6.0×10^{-5} mole of strong acid for an enzyme experiment. We need to select an appropriate weak acid (Appendix 4) or weak base (Appendix 6) to use for this buffer from the appropriate table. Our table gives only K_a values (or K_b in the case of bases). Instead of converting all of them to pK_a values, we realize that we need a pK_a between 4.00 and 6.00. Converting values in this range back to K_a values shows that for our buffer we require a K_a between 1.0×10^{-4} and 1.0×10^{-6}.

The possible choices are H_3AsO_4, $HCHO_2$, $HC_2H_3O_2$, HF, HNO_2, and HN_3. Since this is a biological experiment, we have to be careful that the buffer has no adverse effects on the biochemicals to be used. Therefore it is wise to avoid H_3AsO_4 because of the arsenic, HF and $HCHO_2$ because of their toxicity, HNO_2 because of its redox properties, and HN_3 because it is generally not readily available. We are left with ethanoic acid, $HC_2H_3O_2$, which is a very popular buffer material.

Next, we need at least 20 times the 6.0×10^{-5} mole of strong acid expected, or 1.2×10^{-3} mole of conjugate acid plus conjugate base. This will be dissolved in 100 mL of solution, so the total molarity will be 1.2×10^{-2} M. Since we already know that 100 mL of buffer is needed, the three methods of preparation can now be considered.

Method 1

As stated above, the first method involves measuring the appropriate amounts of the conjugate acid and base in the desired volume of water. The acid is ethanoic acid, and the conjugate base is the ethanoate ion. Ethanoate ions can be conveniently obtained from the salt sodium ethanoate. It was decided that the total of the conjugate acid and base should be

$$1.2 \times 10^{-2} \ M = [HC_2H_3O_2] + [C_2H_3O_2^-] \tag{14.78}$$

The equilibrium law for the dissociation of ethanoic acid is

$$HC_2H_3O_2 \rightleftharpoons C_2H_3O_2^- + H^+ \tag{14.79}$$

$$K_a = \frac{[C_2H_3O_2^-][H^+]}{[HC_2H_3O_2]} \tag{14.80}$$

Rearranging Equation 14.80 gives

$$\frac{K_a}{[H^+]} = \frac{[C_2H_3O_2^-]}{[HC_2H_3O_2]} \tag{14.81}$$

Since the buffer pH must be 5.00, we enter 1.0×10^{-5} for $[H^+]$.

$$\frac{1.8 \times 10^{-5}}{1.0 \times 10^{-5}} = \frac{[C_2H_3O_2^-]}{[HC_2H_3O_2]} \tag{14.82}$$

$$1.8 = \frac{[C_2H_3O_2^-]}{[HC_2H_3O_2]} \tag{14.83}$$

Equations 14.78 and 14.83 are used to calculate the needed concentrations of ethanoic acid and ethanoate ions:

$$1.8[HC_2H_3O_2] = [C_2H_3O_2^-] \qquad (14.84)$$

Substitute $1.8[HC_2H_3O_2]$ for $[C_2H_3O^-]$ in Equation 14.78 to obtain

$$1.2 \times 10^{-2} = 1.8[HC_2H_3O_2] + [HC_2H_3O_2]$$
$$= 2.8[HC_2H_3O_2]$$
$$4.3 \times 10^{-3} \, M = [HC_2H_3O_2]$$

Then

$$1.8(4.3 \times 10^{-3}) = [C_2H_3O_2^-]$$
$$7.7 \times 10^{-3} \, M = [C_2H_3O_2^-]$$

Since the volume is 100 mL, we can calculate the number of grams of ethanoic acid as

$$? \, g \, HC_2H_3O_2 = 100 \, mL \, HC_2H_3O_2 \left(\frac{4.3 \times 10^{-3} \, mol \, HC_2H_3O_2}{1000 \, mL \, HC_2H_3O_2} \right) \left(\frac{60.0 \, g \, HC_2H_3O_2}{1 \, mol \, HC_2H_3O_2} \right)$$
$$= 0.0258 \, g \, HC_2H_3O_2$$

In a similar calculation the grams of sodium ethanoate can be determined since the molarity of sodium ethanoate is the same as the molarity of the ethanoate ion:

$$? \, g \, NaC_2H_3O_2 = 100 \, mL \, NaC_2H_3O_2 \left(\frac{7.7 \times 10^{-3} \, mol \, NaC_2H_3O_2}{1000 \, mL \, NaC_2H_3O_2} \right) \left(\frac{82.0 \, g \, NaC_2H_3O_2}{1 \, mol \, NaC_2H_3O_2} \right)$$
$$= 0.0631 \, g \, NaC_2H_3O_2$$

To prepare this buffer by Method 1, we need to weigh 0.0258 g of $HC_2H_3O_2$ and 0.0631 g of $NaC_2H_3O_2$ into a 100-mL volumetric flask, add water to the mark, and mix thoroughly.

Method 2

In the second method we have to obtain the same result by measuring only some ethanoic acid and neutralizing part of it with a base such as NaOH. A 0.100 M solution of NaOH can be used for this purpose. We calculate the number of grams of ethanoic acid needed from the 100 mL of solution and the total molarity of 1.2×10^{-2}:

$$? \, g \, HC_2H_3O_2 = 100 \, mL \, HC_2H_3O_2 \left(\frac{12 \times 10^{-2} \, mol \, HC_2H_3O_2}{1000 \, mL \, HC_2H_3O_2} \right) \left(\frac{60.0 \, g \, HC_2H_3O_2}{1 \, mol \, HC_2H_3O_2} \right)$$
$$= 0.0720 \, g \, HCHO$$

The reaction of ethanoic acid with NaOH is

$$NaOH + HC_2H_3O_2 \rightarrow NaC_2H_3O_2 + H_2O$$

We can calculate the volume of NaOH needed to produce 100 mL of 7.7×10^{-3} *M* sodium ethanoate solution:

$$? \text{ mL NaOH} = 100 \text{ mL NaC}_2\text{H}_3\text{O}_2 \left(\frac{7.7 \times 10^{-3} \text{ mol NaC}_2\text{H}_3\text{O}_2}{1000 \text{ mL NaC}_2\text{H}_3\text{O}_2} \right) \left(\frac{1 \text{ mol NaOH}}{1 \text{ mol NaC}_2\text{H}_3\text{O}_2} \right) \left(\frac{1000 \text{ mL NaOH}}{0.100 \text{ mol NaOH}} \right)$$

$$= 7.7 \text{ mL NaOH}$$

In method 2, we need to measure 7.7 mL of 0.100 *M* NaOH and 0.0720 g of $HC_2H_3O_2$ and dissolve in enough water to make 100 mL of solution.

Method 3

In method 3, the grams of $NaC_2H_3O_2$ needed to make a 1.2×10^{-2} *M* solution is calculated:

$$? \text{ mL NaC}_2\text{H}_3\text{O}_2 = 100 \text{ mL NaC}_2\text{H}_3\text{O}_2 \left(\frac{1.2 \times 10^{-2} \text{ mol NaC}_2\text{H}_3\text{O}_2}{1000 \text{ mL NaC}_2\text{H}_3\text{O}_2} \right) \left(\frac{82.0 \text{ g NaC}_2\text{H}_3\text{O}_2}{1 \text{ mol NaC}_2\text{H}_3\text{O}_2} \right)$$

$$= 0.0984 \text{ g NaC}_2\text{H}_3\text{O}_2$$

Then the volume of 0.0500 *M* HCl needed to convert some of this sodium ethanoate into the required amount of ethanoic acid is determined. The reaction of sodium ethanoate with HCl is as follows:

$$HCl + NaC_2H_3O_2 \rightarrow HC_2H_3O_2 + NaCl$$

The volume of HCl needed is calculated as follows:

$$? \text{ mL HCl} = 100 \text{ mL HC}_2\text{H}_3\text{O}_2 \left(\frac{4.3 \times 10^{-3} \text{ mol HC}_2\text{H}_3\text{O}_2}{1000 \text{ mL HC}_2\text{H}_3\text{O}_2} \right) \left(\frac{1 \text{ mol HCl}}{1 \text{ mol HC}_2\text{H}_3\text{O}_2} \right) \left(\frac{1000 \text{ mL HCl}}{0.0500 \text{ mol HCl}} \right)$$

$$= 8.6 \text{ mL HCl}$$

In method 3 we need to weigh 0.0984 g of sodium ethanoate along with 8.6 mL of 0.0500 *M* HCl and dilute the mixture to 100 mL.

In preparing a buffer the chemist uses one of these three methods, usually the easiest one.

pH Values of Polyprotic Acids and Their Salts

Polyprotic acids dissociate in stepwise equilibria, each of which has its own dissociation constant. Each dissociation constant is smaller than the preceding one

$(K_{a1} > K_{a2} > K_{a3})$. When a polyprotic acid such as H_2A dissolves in water, the possible reactions are as follows:

$$H_2A \rightleftharpoons HA^- + H^+ \tag{14.85}$$

$$HA^- \rightleftharpoons A^{2-} + H^+ \tag{14.86}$$

Since K_{a2} is always smaller than K_{a1}, the second dissociation must occur to a lesser extent than the first dissociation. In addition, the first dissociation produces hydrogen ions, and these hydrogen ions further suppress the second dissociation according to Le Châtelier's principle. These two considerations, along with experimental evidence, demonstrate that the first dissociation step is the only important one in determining the $[H^+]$ and pH of a solution of a polyprotic acid.

The equation to use is

$$[H^+] = \sqrt{K_{a1}C_a} \tag{14.87}$$

where C_a is the initial concentration of polyprotic acid.

EXERCISE 14.8

What is the pH value of each of the following solutions?

(a) 0.800 M H_3PO_4
(b) 0.0200 M H_2S
(c) 0.00400 M H_2CO_3

Solution

(a) 1.12
(b) 4.35
(c) 4.37

Similar reasoning is used for the hydrolysis of the fully deprotonated anion of a polyprotic acid, such as A^{2-} in Equation 14.87. The two hydrolysis reactions are these:

$$A^{2-} + H_2O \rightleftharpoons HA^- + OH^- \tag{14.88}$$

$$HA^- + H_2O \rightleftharpoons H_2A + OH^- \tag{14.89}$$

The hydrolysis reaction of Equation 14.88 has the largest equilibrium constant, and the OH^- formed in the first step suppresses the second step according to Le Châtelier's principle. The hydrolysis of a fully deprotonated anion of a polyprotic acid is calculated using the equation

$$[OH^-] = \sqrt{\frac{K_w}{K_{a2}}C_{A^{2-}}} \tag{14.90}$$

EXERCISE 14.9

What is the pH of each of the following solutions?

(a) 3.00 M Na$_3$PO$_4$
(b) 0.0500 M Na$_2$CO$_3$

Solution

(a) 13.41
(b) 11.51

Anions of polyprotic acids, such as HA$^-$ seen recently, are amphiprotic and can act as either conjugate acids or conjugate bases. For salts that contain these anions the hydrogen ion concentration is calculated as the square root of the product of the K_a values for that anion acting as a conjugate acid and as a conjugate base:

$$[H^+] = \sqrt{K_{a1}K_{a2}} \qquad (14.91)$$

EXERCISE 14.10

Calculate the pH of a solution of each of the following ions:

(a) HSO$_3^-$
(b) H$_2$PO$_4^-$
(c) HPO$_4^{2-}$
(d) HCO$_3^-$
(e) HS$^-$

Solution

(a) 4.55
(b) 4.67
(c) 9.77
(d) 8.34
(e) 10.00

Buffer solutions are often made using a conjugate acid and a conjugate base of a polyprotic acid. The pH of such a buffer is governed by the equilibrium law, which includes both the conjugate acid and the conjugate base present in the solution. For phosphoric acid the first dissociation step involves H$_3$PO$_4$ and H$_2$PO$_4^-$, and buffers in the range of pH 1.1–3.1 can be prepared from various mixtures of them. The second dissociation step involves H$_2$PO$_4^-$ and HPO$_4^{2-}$, and salts with these two ions are used to prepare buffers in the range of pH 6.2–8.2. With the HPO$_4^{2-}$ and PO$_4^{3-}$ anions buffers in the pH range 11.3–13.3 can be prepared.

Titration Curves

The titration technique was described in Chapter 6, along with the important calculations that apply to a wide variety of titrations. Acid-base pH calculations are often used to explain what occurs as the experiment is performed. There are four major points of interest during a titration experiment:

1. The start of the titration, where the solution contains only one acid or base.
2. The region where titrant is added up to the equivalence point, and the solution now contains a mixture of unreacted sample and products.
3. The equivalence point, where all of the reactant has been converted into product.
4. The region after the equivalence point, where the solution contains product and excess titrant.

At the start of the titration the sample is a pure acid or base, and the pH is calculated as described in preceding sections. At the equivalence point, the solution is the salt of the acid or base; again, the method of calculating the pH has been described. Between the start and the equivalence point, however, we have a mixture of a conjugate acid and its conjugate base. If a weak acid or base is involved, this is a buffer solution. If only strong acids and bases are used, this region is unbuffered. After the end point, the pH depends on the excess titrant used.

The pH during a titration may be calculated or measured in an experiment. A plot of pH versus volume of titrant added is called a **titration curve.**

The general shapes of strong acid–strong base titration curves are shown in Figure 14.1. There are three important points about these curves. First, there is no buffer region since no weak acids or bases are present. Second, the end-point pH is always 7.0 because the products do not hydrolyze. Third, from the start the pH changes gradually until just before the end point where it changes sharply.

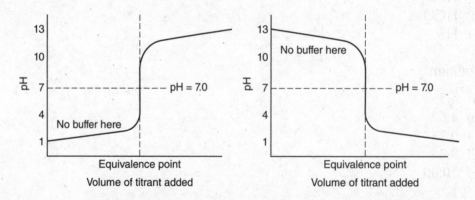

FIGURE 14.1. Titration curves for (left) a strong acid titrated with a strong base and (right) a strong base titrated with a strong acid. The region from the start to the equivalence point is not buffered. The equivalence-point pH is always 7.0.

Comparing Figure 14.1 with Figure 14.2, we find that weak acid and base titrations have a buffer region where a conjugate acid-base pair exists. In the middle of the buffer region, pH = pK_a. A large change in pH occurs just at the start as the buffer mixture is formed, in contrast to strong acid-base titrations. At the equivalence point the salts of weak acids and bases hydrolyze, and the pH is not 7. Salts of weak acids hydrolyze to give basic solutions, and $pH_{ep} > 7$; salts of weak bases will give acidic solutions, $pH_{ep} < 7$.

Titrations of polyprotic acids produce more than one equivalence point because of the sequence of dissociation steps. The titration of phosphoric acid with a strong base is shown in Figure 14.3. Only two end points are observed since the third occurs at a pH too high for observation. At the start, only pure H_3PO_4 is present. At each equivalence point the acid is neutralized to pure $H_2PO_4^-$ and then HPO_4^{2-}; the pH values at the equivalence points are calculated using Equation 14.92. Between the equivalence points there are mixtures of a conjugate acid and its conjugate base, as shown. These are the buffer regions. At the midpoint of each buffer region the pH is equal to the pK_a that governs that particular dissociation step.

Figures 14.2 and 14.3 illustrate why and how a buffer can be made by partial neutralization of a weak acid. The partial neutralization results in a conjugate acid-base mixture, which is a buffer solution.

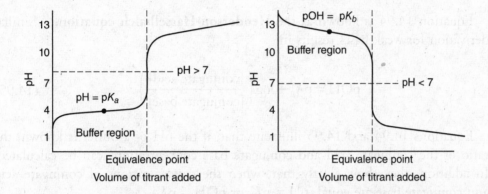

FIGURE 14.2. Titration curves for (left) a weak acid titrated with a strong base and (right) a weak base titrated with a strong acid. Between the start and the equivalence point is a buffer region due to the presence of a conjugate acid-base mixture. Because of the hydrolysis of the products, the equivalence point pH is not 7.0.

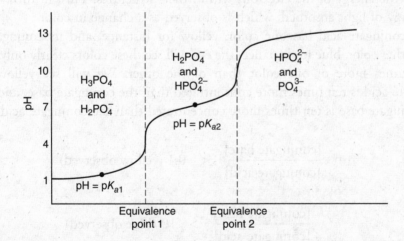

FIGURE 14.3. Titration curve for the titration of phosphoric acid with a strong base.

Henderson-Hasselbalch Equation

The equilibrium law for a weak acid, HA, is written as

$$K_a = \frac{[A^-][H^+]}{[HA]} \qquad (14.92)$$

Taking the logarithm of both sides of Equation 14.92 yields

$$pK_a = pH - \log\left(\frac{[A^-]}{[HA]}\right) \tag{14.93}$$

Rearrangement yields

$$pH = pK_a + \log\left(\frac{[A^-]}{[HA]}\right) = pK_a + \log\left(\frac{[\text{conjugate base}]}{[\text{conjugate acid}]}\right) \tag{14.94}$$

Equation 14.94 is known as the **Henderson-Hasselbalch equation.** A similar derivation for weak bases results in

$$pOH = pK_b + \log\left(\frac{[\text{conjugate acid}]}{[\text{conjugate base}]}\right) \tag{14.95}$$

Equations 14.94 and 14.95 illustrate that, if the pH of a buffer is known, the ratio of the conjugate acid and conjugate base concentrations can be calculated. In addition, they show clearly that, when the concentrations of conjugate acid and conjugate base are equal, $pH = pK_a$ or $pOH = pK_b$.

pH Indicators

Indicators are weak acids and weak bases whose respective conjugate bases and conjugate acids have different colors. The reason is that the loss or gain of a proton changes the energy of the electrons within their structures. This, in turn, changes the energy of light absorbed, which is observed as a change in color.

If a conjugate acid has one color, yellow for instance, and its conjugate base has another color, blue for instance, the eye will see these colors clearly only if there is ten times more of one color than of the other. We will see yellow if the conjugate acid is ten times more concentrated than the conjugate base, and blue if the conjugate base is ten times more concentrated than the conjugate acid:

$$\frac{[\text{conjugate base}]}{[\text{conjugate acid}]} < 0.1 \text{ (yellow observed)} \tag{14.96}$$

$$\frac{[\text{conjugate base}]}{[\text{conjugate acid}]} > 10 \text{ (blue observed)} \tag{14.97}$$

Between these two ratios various shades of green will be observed. Since indicators are weak acids and bases, the significance of this color phenomenon is best understood with the Henderson-Hasselbalch equation, Equation 14.94. Substituting 0.1 into the log term yields

$$pH = pK_a + \log(0.1) = pK_a - 1.0$$

so that the yellow color is observed when the pH of the solution is at least 1 pH unit below the pK_a of the indicator. For the blue color

$$pH = pK_a + \log(10) = pK_a + 1.0$$

and the pH of the solution must be 1 pH unit above the pK_a of the indicator.

This discussion tells us two important properties of indicators used in titrations: (1) the pH at the end point of a titration curve must change by at least 2 pH units very rapidly, and (2) the pK of the indicator must be close to the end-point pH of the titration. From the titration curves we see that the required large change in pH often occurs. Proper selection of an indicator mandates that the pH at the end point and the pK_a of the indicator be close to each other.

SUMMARY

If redox reactions are considered one major class of chemical reactions, the other major group consists of acid-base chemistry. This chapter reviews the concepts of the three major theories: the Arrhenius theory, the Brönsted-Lowry theory, and the Lewis theory. Each theory makes the definition of acids and bases more general, and there will always be questions concerning these theories on the AP Exam. This chapter also reviews the nomenclature applied to acids and bases as well as the relationship between molecular structure, acid-base strengths, and the periodic table.

The quantitative aspects concerning the pH of acid and base solutions are covered in this chapter. The pH of salts, polyprotic acids, and highly charged cations is reviewed. In addition, the concept and quantitative aspects of buffer solutions are covered. The various problems concerning acid-base chemistry are reviewed, and methods for solving problems are presented.

Important Concepts

Arrhenius theory
Brönsted-Lowry theory
Lewis theory
Neutralization
Titration
Buffers
Strong vs weak acids and bases
Predicting relative strengths of acids
Hydrolysis

Important Equations

$$pH = -\log[H^+] \text{ and } pOH = -\log[OH^-]$$

$$pH = pK_a + \log\left(\frac{[\text{conjugate base}]}{[\text{conjugate acid}]}\right)$$

$$K_a = \frac{[H^+][A^-]}{[HA]}$$

$$K_w = [H^+][OH^-]$$

Practice Exercises

Multiple-Choice

For the first four problems below, one or more of the following responses will apply; each response may be used more than once or not at all in these questions.

I. $Cu(NH_3)_4^{2+}$
II. KOH
III. HCO_3^-
IV. CO_3^{2-}
V. SO_3^{2-}

1. Which of the above is the product of a Lewis acid reacting with a Lewis base?

   (A) I and II
   (B) II
   (C) I and III
   (D) IV
   (E) IV and V

2. Which ion is the strongest base?

   (A) I
   (B) II
   (C) III
   (D) IV
   (E) V

3. Which of these imparts a color to a solution?

   (A) I
   (B) II
   (C) III
   (D) IV
   (E) V

4. When an equal number of moles of each is mixed, which of these can be used to prepare a buffer solution?

   (A) I and II
   (B) II and III
   (C) III and IV
   (D) IV and V
   (E) V and I

5. Which of the following has the highest pH?

   (A) 0.100 *M* HCl
   (B) 0.200 *M* $HC_2H_3O_2$
   (C) 0.100 *M* Na_2CO_3
   (D) 0.200 *M* NaCl
   (E) 0.500 *M* $NaC_2H_3O_2$

6. Which of the following CANNOT occur together in solution?

    (A) H_3PO_4 and $H_2PO_4^-$
    (B) HCO_3^- and CO_3^{2-}
    (C) Na^+ and SO_4^{2-}
    (D) $C_2O_4^{2-}$ and $H_2C_2O_4$
    (E) HPO_4^{2-} and PO_4^{3-}

7. When 0.250 mol of NaOH is added to 1.00 L of 0.100 M H_3PO_4, the solution will contain

    (A) HPO_4^{2-}
    (B) $H_2PO_4^-$
    (C) PO_4^{3-}
    (D) A and B
    (E) A and C

8. A buffer with a pH of 10.0 is needed. Which of the following should be used?

    (A) ethanoic acid with a K_a of 1.8×10^{-5}
    (B) ammonia with a K_b of 1.8×10^{-5}
    (C) nitrous acid with a K_a of 7.1×10^{-4}
    (D) $H_2PO_4^-$ and PO_4^{3-} with a K_a of 4.5×10^{-13}
    (E) dimethylamine with a K_b of 1.05×10^{-3}

9. pH is equal to pK_a

    (A) when [conjugate acid] = [conjugate base]
    (B) at the end point of a titration
    (C) in the buffer region
    (D) in the Henderson-Hasselbalch equation
    (E) at equilibrium

10. The pH of a 1.23×10^{-3} M solution of $Al(OH)_3$ aqueous solution is

    (A) 2.91
    (B) 2.43
    (C) 11.09
    (D) 13.52
    (E) 11.57

11. An indicator has a K_a of 6.4×10^{-6}, the conjugate acid is red, and the conjugate base is yellow. At what pH will the solution be red?

    (A) 5.2
    (B) 5.5
    (C) 4.0
    (D) 4.7
    (E) 6.4

12. A buffer has a pH of 4.87. If the buffer is made from a weak acid ($K_a = 3.30 \times 10^{-5}$), and its conjugate base, the $\dfrac{[\text{conjugate base}]}{[\text{weak acid}]}$ ratio is

    (A) 4.87
    (B) 4.48
    (C) 1.00
    (D) 2.45
    (E) 0.41

13. Which of the following statements is correct?

    (A) $HClO_2$ is a stronger acid than $HClO_3$.
    (B) HI is a weaker acid than HCl.
    (C) CH_3COOH is a stronger acid than $CH_2BrCOOH$.
    (D) HNO_3 is a stronger acid than HNO_2.
    (E) H_3PO_4 is a stronger acid than $HClO_4$.

14. What is the pH of a 0.100 M solution of K_2HPO_4? (For H_3PO_4, $pK_1 = 2.15$; $pK_2 = 7.20$; $pK_3 = 12.35$.)

    (A) 1.00
    (B) 13.00
    (C) 9.78
    (D) 6.67
    (E) 4.10

15. Which of the following is the correct method for preparing a buffer solution?

    (A) Mix the correct amounts of a weak acid and its conjugate base.
    (B) Neutralize a weak base partially with strong acid.
    (C) Neutralize a weak acid partially with a strong base.
    (D) Add the appropriate amount of strong acid to an acid salt.
    (E) All of the above methods may be used to prepare buffers.

16. The only acid that is both a strong and a weak acid on dissociation is

    (A) hydrochloric acid
    (B) perchloric acid
    (C) nitric acid
    (D) sulfuric acid
    (E) phosphoric acid

17. Which of the following is the acid anhydride of a monoprotic acid?

    (A) CaO
    (B) SO_3
    (C) FeO
    (D) CO_2
    (E) N_2O_5

18. Which of the following CANNOT be either a Lewis acid or a Lewis base?

    (A) CH_4
    (B) Cu^{2+}
    (C) CO
    (D) Fe^{3+}
    (E) NH_3

19. In the complex ion $Cu(NH_3)_4^{2+}$ the NH_3 is called

    (A) a cation
    (B) a ligand
    (C) a Lewis acid
    (D) an anion
    (E) a conjugate acid

20. The pH of a 0.125 M solution of a weak base is 10.45. What is the pK_b of this base?

    (A) 3.5×10^{-11}
    (B) 6.4×10^{-7}
    (C) 2.8×10^{-4}
    (D) 2.3×10^{-3}
    (E) 1.2×10^{-2}

21. A solution containing HF is titrated with KOH. At the end point of the titration the solution contains

    (A) equal amounts of HF and KOH
    (B) H_2O, H^+, OH^-, K^+, F^-, and HF
    (C) K^+ and F^-
    (D) KF and H_2O
    (E) K^+, F^-, and H_2O

22. A buffer at pH 5.32 is prepared from a weak acid with a pK_a = 5.15. If 100 mL of this buffer is diluted to 200 mL with distilled water, the pH of the dilute solution is

    (A) 5.62
    (B) 5.02
    (C) 5.32
    (D) The identity of the acid is needed to answer the question.
    (E) The concentrations of the acid and the salt are needed to answer the question.

23. If 50.0 mL of a 0.0134 M HCl solution is mixed with 24.0 mL of a 0.0250 M NaOH solution, what is the pH of the final mixture?

    (A) 1.87
    (B) 12.40
    (C) 5.29
    (D) 3.02
    (E) 10.98

24. If 50.0 g of formic acid ($HCHO_2$, $K_a = 1.8 \times 10^{-4}$) and 30.0 g of sodium formate ($NaCHO_2$) are dissolved to make 500 mL of solution, the pH of this solution is

    (A) 4.76
    (B) 3.76
    (C) 3.35
    (D) 4.12
    (E) 3.02

ANSWER KEY

1. C	6. D	11. C	16. D	21. B
2. D	7. E	12. D	17. E	22. C
3. A	8. B	13. D	18. A	23. D
4. C	9. A	14. C	19. B	24. C
5. C	10. E	15. E	20. B	

See Appendix 1 for explanations of answers.

Free-Response

Answer the following questions using the concepts and procedures discussed in this chapter and preceding chapters.

(a) Rank the strengths of phosphoric acid, sulfuric acid, and perchloric acid. Explain your reasoning.

(b) Calculate the pH, pOH, [H⁺], and [OH⁻] of a 0.0035 molar solution of $Ba(OH)_2$.

(c) Sketch and fully label a titration curve for 50.0 mL of 0.0100 *M* ammonia titrated with 0.0200 *M* HBr.

(d) Describe a comprehensive method for preparing a buffer. Be sure to mention all the data you need and calculations that need to be performed.

(e) Write the chemical reaction for propanoic acid (CH_3CH_2COOH) reacting with ammonia. Identify the conjugate acid-base pairs, and if the position of the equilibrium lies toward the product side, indicate their strengths.

ANSWERS

(a) The strengths are ranked in order as $HClO_4 > H_2SO_4 > H_3PO_4$. Stronger acids have weaker bonds between the hydrogen and oxygen in these formulas. The Lewis structures of these acids should be drawn here; some of them are given in Chapter 14. The strengths of these acids depend on two things. First, the electronegativity of the central atom. The more electronegative the central atom, the more electron density is withdrawn from the O–H bond, weakening it and making a stronger acid. Second is the number of oxygen atoms not bonded to hydrogen. These oxygen atoms can attract electrons from the O–H bond, causing the acid to become stronger. We see that $HClO_4$ has three of these oxygen atoms, sulfuric acid has two, and phosphoric acid has only one. Finally, the stability of the anion, due to delocalizing electrons over the oxygens not bonded to H, means that the acid is stronger.

(b) A 0.0035 molar $Ba(OH)_2$ solution is 0.0070 molar in hydroxide ions. Therefore

$[OH^-] = 0.0070$, pOH = 2.15, pH = 14.00 − pOH = 11.85, and $[H^+] = 1.4 \times 10^{-12}$.

(c) This titration curve should look like the second part of Figure 14.2. The end point will be at 25.0 mL and note that halfway to the end point pOH = pK_b.

(d) You should mention that the volume, pH, and buffer capacity (total concentration of conjugate acid and base) need to be decided upon. From the desired pH, an appropriate acid or base with a pK_a within one unit of the desired pH or a pK_b within one unit of the desired pOH should be chosen. Once these are known, the ratio of conjugate acid to conjugate base can be calculated. Finally, the molarities of the conjugate acid and conjugate base are calculated along with the mass or volume of the reagents needed.

(e) $CH_3CH_2COOH + NH_3 \leftrightarrows NH_4^+ + CH_3CH_2COO^-$

The CH_3CH_2COOH and $CH_3CH_2COO^-$ is one conjugate acid-base pair. NH_4^+ and NH_3 is a second conjugate acid-base pair. The side of the reaction favored by the equilibrium contains the weaker acid and base. The NH_4^+ is a weaker acid than CH_3CH_2COOH, and $CH_3CH_2COO^-$ is a weaker base than NH_3. Because the weaker acid and weaker base are on the product side, we predict that equilibrium will lie on the product side.

Organic Chemistry and Polymers

CHAPTER 15

- Carbon
- Carbon Chains
- Structural Isomers
- Stereoisomers
- Cis-Trans Isomers
- Three-Dimensional Structures
- Carbon-Hydrogen Compounds
- Alkanes
- Alkenes
- Alkynes
- Ring Structures
- Functional Groups
- Systematic Nomenclature
- Organic Reactions
- Polymers

Organic chemistry is the study of carbon-containing compounds. Before 1828 the **vital force theory,** which stated that **organic compounds** could be extracted only from living, or once living, organisms, was widely accepted. In 1828, however, the synthesis of urea, $CO(NH_2)_2$, by Friedrich Wöhler demonstrated that inorganic compounds could be converted into organic compounds, and the vital force theory was rapidly abandoned. Today, most known chemical compounds are organic compounds.

In preceding chapters we referred to many organic compounds without considering them as a separate group. This chapter describes the essential features of carbon compounds that should be appreciated.

CARBON

Carbon is the central atom in organic compounds. It exists as an element in three different forms, called **allotropes.** These allotropes are **graphite,** which is the "lead" in pencils; **diamond,** a precious stone; and **buckminsterfullerene.** Graphite consists of layers of sp^2 bonded carbon atoms in six-membered rings which form flat sheets. Each layer is weakly bonded to another with pi bonding. The loosely bound pi electrons allow electrical conduction along two of the three axes of graphite. Movement of electrons from one sp^2 layer to another is difficult, and graphite is considered an insulator if electrons must travel perpendicularly to the layers of six-membered rings. Diamond consists of tetrahedrally bonded carbon atoms in a covalent crystal of extraordinary hardness. Buckminsterfullerene consists of spheres of 60 carbon atoms covalently bonded together.

Carbon has a total of six electrons, four of which are valence electrons. Carbon always forms four bonds, which may be combinations of single, double, and triple bonds. This element never has a nonbonding pair of electrons. Carbon can form four single bonds with a tetrahedral, sp^3, structure. It can bond to three other atoms in a trigonal planar, sp^2, structure, which involves one double bond and two single bonds. Finally, carbon can bond with two other atoms in a linear, sp, structure. These compounds contain either two double bonds, as in CO_2, or one single and one triple bond, as in HCN.

Importance of Carbon

The most important feature of the carbon atom is that it bonds with other carbon atoms to form chains and rings of various sizes and shapes. No other atom in the periodic table can form the variety of structures that carbon can.

Isomers

Two organic compounds may have formulas with exactly the same atoms but different properties. These compounds are called **isomers.** There are three main types of isomers: structural isomers, cis-trans isomers, and stereoisomers.

STRUCTURAL ISOMERS

Structural isomers, also called constitutional isomers, are two or more different compounds that have the same atoms in their formulas. However, the atoms are bonded to each other in different configurations. The hydrocarbon C_5H_{12} has three different structural isomers based on the arrangement of the carbon skeleton, as shown in Figure 15.1. A compound with a single straight chain of carbon atoms is designated as **normal** with the letter n preceding its name, as in n-pentane.

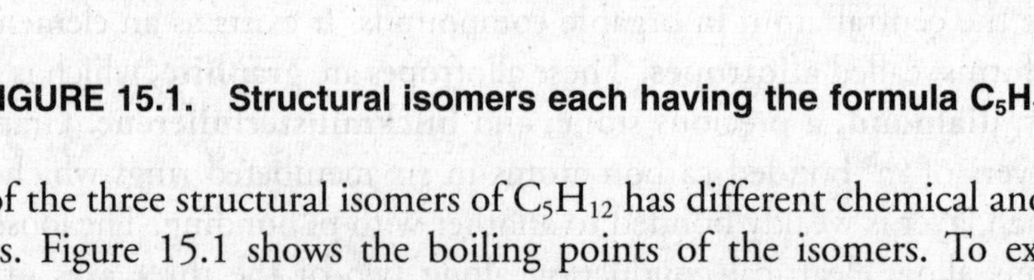

n-pentane
(b.p. = 36.1 °C)

2-methylbutane
(b.p. = 28° C)

2,2-dimethylpropane
(b.p. = −95 °C)

FIGURE 15.1. Structural isomers each having the formula C_5H_{12}.

Each of the three structural isomers of C_5H_{12} has different chemical and physical properties. Figure 15.1 shows the boiling points of the isomers. To explain the difference in boiling points, we see that the n-pentane molecules line up side by side to form many instantaneous dipoles. Because it has the largest London forces, n-pentane has the highest boiling point. The 2,2-dimethylpropane molecules can interact at only a few points, and this compound has the lowest boiling point because of smaller London forces.

Some of the structural isomers having the formula $C_4H_{10}O$ are shown in Figure 15.2.

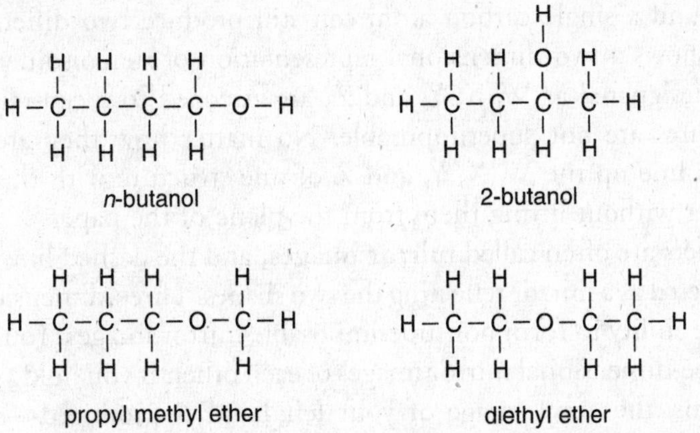

n-butanol

2-butanol

propyl methyl ether

diethyl ether

FIGURE 15.2. Some possible isomers with the formula $C_4H_{10}O$.

Each of these compounds has distinctly different physical and chemical properties. The alcohols, with —OH groups, hydrogen-bond extensively and have high boiling points. The ethers, with C—O—C bonds, have much lower boiling points because the forces of attraction are mainly London forces.

CIS ISOMERS AND TRANS ISOMERS

A carbon-carbon single bond allows the carbon atoms to rotate freely. A C=C double bond is a rigid structure since rotation would require the constant breaking and reforming of the pi bond. This rigidity means that the atoms bound to double-bonded carbon atoms will have fixed orientations. *Cis-* means "on the same side"; *trans-* means "on opposite sides" and refers to groups attached to double-bonded carbon atoms. Figure 15.3 illustrates the structures of *cis-* and *trans-2*-butene. In the cis form the two —CH₃ groups are on the same side. The trans structure has the —CH₃ groups on opposing sides of the double bond.

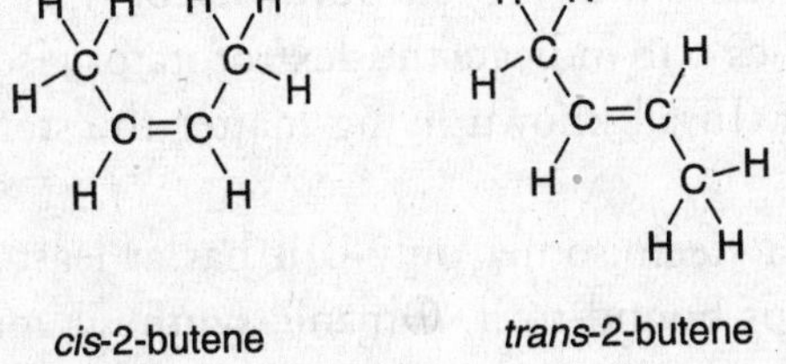

cis-2-butene

trans-2-butene

FIGURE 15.3. Illustration of the difference between the cis and trans isomers of butene.

Since the double-bonded carbon atoms cannot rotate, these two structures are distinctly different. The molecules have different chemical and physical properties as a result of their structures, as shown later in Table 15.2.

cis-trans isomers are stereoisomers, but they are not mirror images. They do not rotate polarized light and are called diasteriomers.

STEREOISOMERS

Stereoisomers have the same formulas, and every atom is bonded to the same atom. Stereoisomers are not structural isomers. The unique feature is that the arrangement of atoms around a single carbon atom can still produce two different molecules. Figure 15.4 shows a two-dimensional representation of stereoisomers. Four different groups, designated as W, X, Y, and Z, are attached to a central carbon atom. These structures are not superimposable. No matter how they are rotated, it is impossible to line up the W, X, Y, and Z of one structure with the W, X, Y, and Z of the other without lifting them from the plane of the paper.

Stereoisomers are often called **mirror images,** and the dashed line in Figure 15.4 can be considered as a mirror reflecting the two shapes. Three-dimensional molecules have the same ability to form nonsuperimposable mirror images. Your left and right hands are three-dimensional mirror images of each other. If you hold your right hand up to a mirror, the exact image of your left hand is produced—try it and see! Stereoisomers that are nonsuperimposable mirror images are called enantiomers.

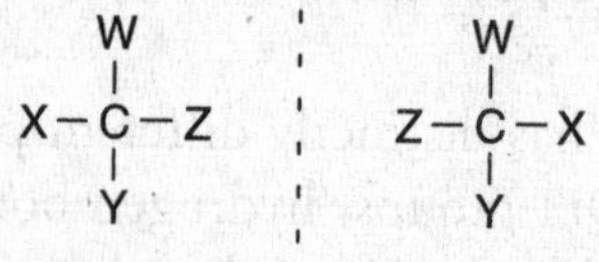

$$
\begin{array}{ccc}
& W & \\
& | & \\
X-C-Z & \vdots & Z-C-X \\
& | & \\
& Y & Y
\end{array}
$$

Stereoisomers or Mirror Images

FIGURE 15.4. Two carbon compounds that have four different groups, W, X, Y, and Z, attached to the central carbon atom. They are not superimposable by rotation. These compounds are reflections of each other, or mirror images.

Stereoisomers, unlike cis and trans isomers, have identical physical properties such as melting and boiling points. The chemical reactivity of two stereoisomers is, for the most part, identical also. However, stereoisomers often have different chemical reactivities in biological systems, where the overall shape of the molecule is important. Most amino acids have two stereoisomers.

Stereoisomers that rotate polarized light are called **optically active isomers** because of this property. One isomer rotates polarized light to the right (**dextrorotatory**) and the other rotates it to the left (**levorotatory**) by an equal amount. One style of nomenclature uses D to indicate the dextrorotatory isomer and L to designate the levorotatory isomer. In all known living matter the stereoisomers are of the L type.

A substance can be a stereoisomer only if it has at least one carbon atom that has four different groups bound to it. Organic synthesis methods may be **stereospecific,** resulting in the exclusive production of one stereoisomer or the other. Other synthetic methods, however, are not stereospecific and result in a 50:50 mixture of D and L isomers. Such a mixture is called a **racemic mixture.** Racemic mixtures do not rotate polarized light because the rotation due to the D isomers is canceled by the equal and opposite rotation of the L isomers.

Three-Dimensional Organic Structures

Chemists and biochemists recognize that the three-dimensional shape of organic compounds is an important consideration in many reactions. Because of the complexity of these compounds, it is advantageous to build models of them rather than drawing them on paper. In all organic compounds the geometry around a carbon bound to four other atoms is tetrahedral (sp^3 hybrid). When a carbon is bound to three other atoms, the shape is trigonal planar (sp^2 hybrid); when the carbon is bound to only two other atoms, the shape is linear (sp hybrid). In addition, oxygen atoms bound to two other atoms have the bent geometry of an sp^3 hybrid. A nitrogen atom bound to three other atoms is tetrahedral, since it has a nonbonding pair of electrons. When bound to two other atoms, a nitrogen atom is trigonal planar (sp^2 hybrid).

CARBON-HYDROGEN COMPOUNDS

Alkanes

Alkanes, are compounds with the general formula C_nH_{2n+2}. The larger, solid alkanes are also called paraffins. These compounds contain all single, or sigma, bonds. Alkanes containing fewer than 5 carbon atoms are gases. Those containing 5–15 carbon atoms are liquids, and those with 16 or more carbon atoms are solids. As a group the alkanes are rather unreactive. Their major use is as a fuel in combustion reactions. Methane, propane, and butane are common gaseous fuels. Gasoline and kerosene are mostly liquid alkanes. Liquid alkanes are used as nonpolar solvents for chemical reactions and for cleaning. The solid alkanes, the paraffins, are often the major components of the wax used in candles.

Table 15.1 lists several alkanes and illustrates the naming system used. The first four alkanes have common names that must be remembered. From pentane on, each name has a prefix (*pent-*, *hex-*, etc.) that indicates the number of carbon atoms in the longest chain. All the names end in -*ane*.

TABLE 15.1

Names and Physical Properties of Selected Alkanes				
Name	Formula: C_nH_{2n+2}	Number of Isomers	Melting Point (°C)	Boiling Point (°C)
methane	CH_4	1	−182.5	−164.0
ethane	C_2H_6	1	−183.3	−88.6
propane	C_3H_8	1	−189.7	−42.1
n-butane	C_4H_{10}	2	−138.3	−0.5
n-pentane	C_5H_{12}	3	−129.7	36.1
n-hexane	C_6H_{14}	5	−95.0	68.9
n-nonane	C_9H_{20}	35	−51.0	150.8
n-dodecane	$C_{12}H_{26}$	355	−9.6	216.3
n-icosane	$C_{20}H_{42}$	366, 319	36.8	343.0

For the normal or straight-chain alkanes the melting and boiling points increase regularly with the number of carbon atoms in the formula. Branched structural isomers have lower melting and boiling points than normal alkanes with the same numbers of carbon atoms. Increased branching decreases melting and boiling points.

Alkenes

Alkenes, also called **olefins,** are compounds with the general formula C_nH_{2n}, where *n* must be 2 or larger. Every alkene has a double bond somewhere in its structure. A compound that contains one or more double or triple bonds is said to be **unsaturated.** A **saturated** compound has only carbon-carbon single bonds. The double bond is a reactive site in alkenes and makes them more reactive than alkanes. Alkenes such as ethylene are the major starting reactants for many chemical syntheses. When the double bond can be located in more than one position in the compound, its location is indicated by a number in front of the name. The number represents the lowest numbered carbon atom with the double bond. Compounds that have two double bonds are called dienes, and those with three double bonds are trienes.

Table 15.2 lists, along with other alkenes, the cis and trans isomers for 2-butene discussed previously and shown in Figure 15.3. The cis isomer has —CH_3 groups on the same side of the double bond, and the trans isomer has the same groups on opposing sides of the double bond. Cis-trans isomers are not optically active; and, as stated earlier, they have different physical and chemical properties.

TABLE 15.2

Names and Properties of Selected Alkenes			
Name	Formula: C_nH_{2n}	Melting Point (°C)	Boiling Point (°C)
ethylene (ethene)*	$CH_2{=}CH_2$	−169	−103.7
propylene (propene)*	$CH_2{=}CHCH_3$	−185.2	−47.4
1-butene	$CH_2{=}CHCH_2CH_3$	−185.3	−6.3
cis-2-butene	$CH_3CH{=}CHCH_3$	−138.9	3.7
trans-2-butene	$CH_3CH{=}CHCH_3$	−105.5	0.9
1,3-butadiene	$CH_2{=}CHCH{=}CH_2$	−108.9	−4.4

*The systematic name is given in parentheses

EXERCISE 15.1

Newspaper reports suggest that saturated fats (molecules with long hydrocarbon chains) and trans unsaturated fats (fats with double bonds) may be equally harmful in causing heart attacks. Suggest why this may be so.

Solution

We know that straight-chain molecules have higher melting points than branched-chain molecules, indicating that it is easier for them to be arranged in an orderly crystalline structure. Trans isomers are essentially straight molecules, while a cis isomer has a definite 120° bend. It seems reasonable that trans isomers could deposit themselves almost as easily in artery walls as straight-chain fats. Deposits of cis fats would be correspondingly more difficult to form because of their bent structure.

Alkynes

Alkynes are compounds with the general formula C_nH_{2n-2}, where n must be 2 or larger. An alkyne has a triple bond somewhere in its structures. Alkynes are very reactive. The most familiar alkyne is acetylene, which is used as a fuel in atomic spectroscopy and welding. A number in front of the name indicates the position of the triple bond.

TABLE 15.3

Names and Properties of Selected Alkynes			
Name	Formula: C_nH_{2n-2}	Melting Point (°C)	Boiling Point (°C)
acetylene (ethyne)	$CH \equiv CH$	−80.8	−84.0 (sublimes)
propyne	$CH \equiv CCH_3$	−101.5	−23.2
1-butyne	$CH \equiv CCH_2CH_3$	−125.7	8.1
2-butyne	$CH_3C \equiv CCH_3$	−32.2	27

Since carbon atoms with triple bonds have only one additional bond, the alkynes do not form cis-trans isomers.

Ring Compounds

In addition to straight chains, carbon atoms can form rings. Rings of six carbon atoms are the most common since the bond angles in these rings are very close to the bond angles in sp^3 or sp^2 hybridized carbon atoms. Five-membered rings are also common, particularly in biochemical molecules.

Cyclohexane is a ring of six carbon atoms with no double bonds. It has two possible structures, shown in Figure 15.5, which can be converted from one to the other.

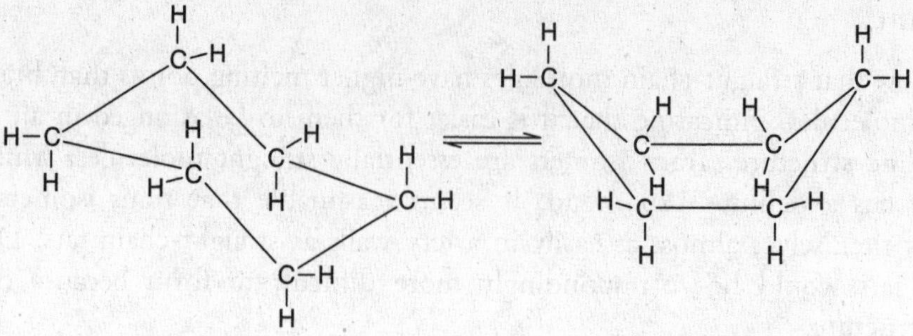

FIGURE 15.5. **The chair and boat forms of cyclohexane.**

The shape of the cyclohexane ring is dictated by the sp^3 hybridization of the carbon atoms, which prefer 109° bond angles. Both the chair and boat forms have these bond angles. Hydrogen atoms on the cyclohexane ring may be replaced by other functional groups to form compounds of added complexity.

Another type of ring is the **benzene** ring. In benzene the carbon atoms have sp^2 hybridization with 120° bond angles. The benzene ring may be visualized as a ring of carbon atoms with alternating double bonds. These are actually resonance structures (as described in Chapter 5), and chemists recognize this fact by drawing a circle in the center of the ring as shown in Figure 15.6.

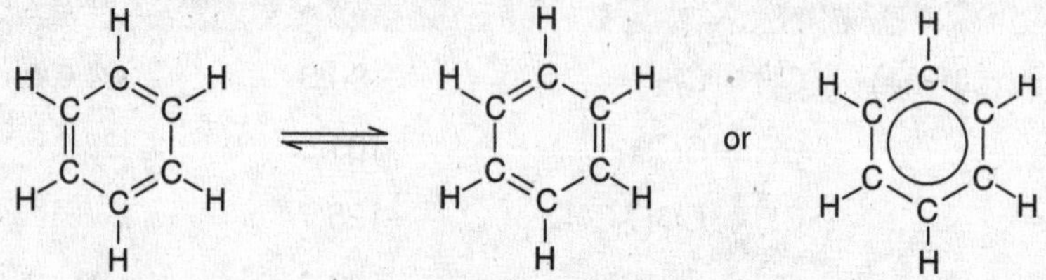

FIGURE 15.6. **Structures of benzene. The two structures on the left are the resonance structures of benzene. The structure on the right is used to symbolize this resonance. Benzene structures are generally drawn without the symbols for carbon and hydrogen.**

The benzene ring is unusually stable because the double bonds are conjugated. In a **conjugated system** every other bond is a double bond. The electrons in the pi orbitals are delocalized, stabilizing the structure. This stability makes it very difficult to break the carbon-carbon bonds in benzene. The typical reactions of benzene involve replacing the hydrogen atoms with other functional groups.

SIDE CHAINS AND FUNCTIONAL GROUPS

Organic molecules can be considered to be structures developed from just a few essential building blocks, called **side chains** or **functional groups.** Some of these building blocks, such as chains and rings, give the molecule its general shape. Other functional groups are responsible for the molecule's characteristic physical properties and chemical reactivity.

Alkyl Side Chains

The alkanes, without a terminal hydrogen, can be considered functional groups. Methane, CH_4, with one hydrogen removed is the methyl group, —CH_3. Ethane without a hydrogen is the ethyl group, —CH_2CH_3. The **alkyl group** is often abbreviated as the symbol R. RH represents an alkane, ROH an alcohol, and R=O an aldehyde.

Aryl Side Chains

Benzene, C_6H_6, and related substances are aromatic compounds. The term *aryl* is derived from the word *aromatic*. The functional **aryl group** for benzene is —C_6H_5, and it is called the phenyl group.

Alcohol Functional Groups—OH

Every **alcohol** has the —OH functional group attached to a carbon, as in ethanol, CH_3CH_2OH. Since the C—O—H bonds are rather strong, the alcohol functional group does not dissociate as the OH⁻ or the H⁺ ion. This functional group does hydrogen-bond to other alcohols and water. A primary alcohol has the —OH group bonded to a carbon that has only one other carbon bonded to it. In a secondary alcohol the —OH group is attached to a carbon that is bonded to two other carbon atoms, and a tertiary alcohol has the —OH bonded to a carbon that is bonded to three other carbon atoms.

Acid Functional Groups —COOH or

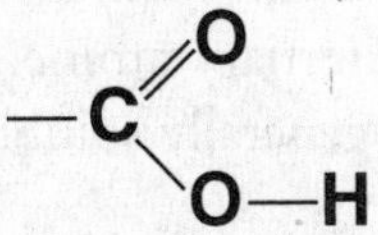

$$-C{\overset{\displaystyle O}{\underset{\displaystyle O-H}{\big\langle}}}$$

Organic acids have the carboxylic acid or carboxyl group. Ethanoic acid has the formula CH_3COOH. The electronegativity of the two oxygen atoms on the same carbon weakens the O—H bond so that hydrogen ions dissociate. Organic acids tend to be weak acids where only a few percent of all molecules dissociate. Carboxyl groups must be located on the terminal carbon of a molecule. Some organic acids have two or even three carboxyl groups.

Aldehyde Functional Groups —C=O

A compound with a terminal —C=O group is called an **aldehyde.** Although this group is polar, only the smaller aldehydes are soluble in water. The aldehydes often have pleasant odors. They are used in perfumes and are found in many beverages such as coffee, tea, beer, and liquor. Formaldehyde, although thought to be carcinogenic, is a major component of urea-formaldehyde foam insulation. It has the structure shown in Figure 15.7.

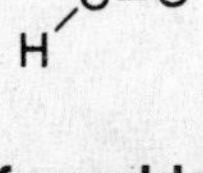

$$\overset{\displaystyle H}{\underset{\displaystyle H}{\big\rangle}}C{=}O$$

FIGURE 15.7.   Structure of formaldehyde, also called methanal.

Ketone Functional Groups

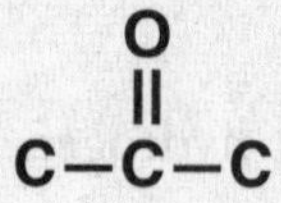

$$\begin{array}{c} O \\ \parallel \\ C-C-C \end{array}$$

Ketones are similar to aldehydes except that the double-bonded oxygen is located on a nonterminal carbon atom. Acetone, a common solvent and nail polish remover, is a ketone; its structure is shown in Figure 15.8. Ketones are found in many natural materials.

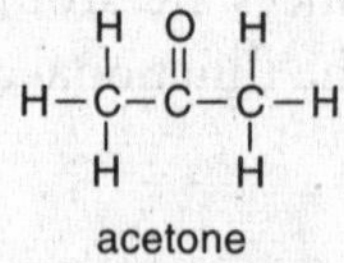

$$\begin{array}{c} \quad\; H \quad O \quad H \\ \quad\; | \quad\;\; \parallel \quad | \\ H-C-C-C-H \\ \quad\; | \qquad\quad | \\ \quad\; H \qquad\quad H \end{array}$$

acetone
(dimethylketone)

FIGURE 15.8. Structure of acetone, a ketone commonly used as a solvent. Its systematic name is propanone.

Amine Functional Groups —NH$_2$

Amines are related to ammonia, NH_3. Replacing one or more hydrogen atoms of ammonia with a carbon chain produces an amine. Replacing one hydrogen with a carbon chain yields a primary amine such as ethylamine, $CH_3CH_2NH_2$. Replacing two hydrogen atoms results in a secondary amine, and replacing three hydrogen atoms yields a tertiary amine. Primary and secondary amines can form hydrogen bonds and are generally soluble in water. Amines are the organic version of bases.

Ether Functional Groups —C—O—C—

Compounds with a carbon-oxygen-carbon unit are known as ethers. Diethyl ether, $CH_3CH_2OCH_2CH_3$, was used as the first anesthetic in medical procedures. With the oxygen in the center of a carbon chain, some of its polarity is diffused in the two separate bonds it forms. As a result ethers are much less polar than ketones and aldehydes of similar size. This lesser polarity translates into lower boiling points and higher vapor pressures than those of aldehydes and ketones. There is a slight polarity, and ethers have higher boiling points than similar alkanes.

Because the oxygen is bonded to two different carbon atoms, ethers are generally less reactive than other oxygen-containing organic compounds. Because of their high vapor pressures, ethers are more flammable than other organic compounds.

Halides —X

The halogens fluorine, chlorine, bromine, and iodine are common organic functional groups. Along with hydrogen, the **halides** are terminal atoms and do not appear in the center of carbon chains. A halide is often represented by the symbol X. Thus the formula CH_3CH_2X can represent fluoroethane, chloroethane, bromoethane, or iodoethane.

SYSTEMATIC NOMENCLATURE FOR ORGANIC COMPOUNDS

The original, or common, names of organic compounds were usually related to the source of the compound. For example, butyric acid was found in butter, caproic acid was isolated from goats, and vinegar was obtained from wine. When the number of known compounds caused this haphazard naming system to become unwieldy, the International Union of Pure and Applied Chemistry, **IUPAC**, developed the systematic nomenclature system. Consequently, many organic compounds have both common and systematic IUPAC names.

The systematic name for an organic compound consists of a prefix, the parent name, and then a suffix. The parent name is based on the longest carbon chain in the molecule. Parent names for ring compounds are based on fundamental ring structures, such as benzene or naphthalene.

Four steps are involved in constructing a name:

1. For the parent name, identify the longest continuous chain of carbon atoms in the structure. The longest chain is not always obvious, as shown in Figure 15.9. In this figure only the carbon backbone is shown. The task is more difficult when all the hydrogen atoms are added to the structure, as in Figure 15.10.

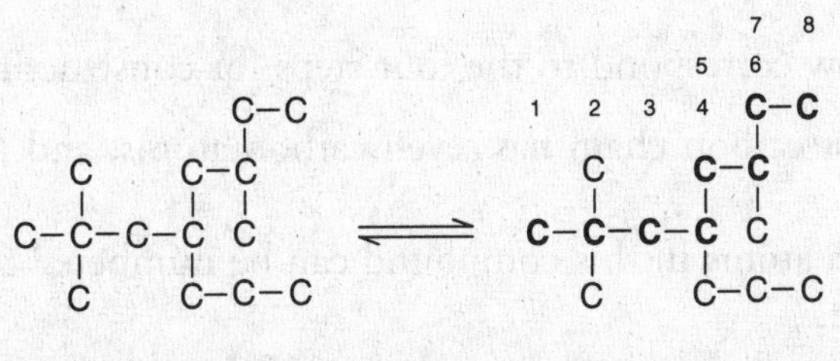

FIGURE 15.9. The longest carbon chain for the structure on the left has eight carbon atoms.

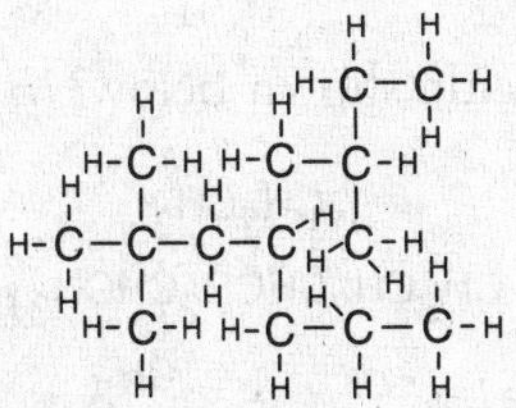

FIGURE 15.10. The compound in Figure 15.9 with all hydrogen atoms attached.

2. Number the carbon atoms in the longest chain. Start at the end closest to a side chain or functional group. In Figure 15.9 the carbon atoms are numbered from left to right.

3. Identify the substituents and the number of the carbon atom to which each is attached. For the structure in Figure 15.9 we have the following:

—CH_3 on carbon 2	2-methyl
—CH_3 on carbon 2	2-methyl
—$CH_2CH_2CH_3$ on carbon 4	4-propyl
—CH_3 on carbon 6	6-methyl

4. For the alcohol, aldehyde, ketone, and acid functional groups, use the terms obtained in step 3 as suffixes; name other functional groups, such as amines and halides, as prefixes. If identical functional groups are found, they are combined and their quantity is indicated by the prefix *di-*, *tri-*, *tetra-*, and so on.

The name of the compound in Figure 15.9 is 2,2,6-trimethyl-4-propyloctane.

EXERCISE 15.2

Name the following compound:

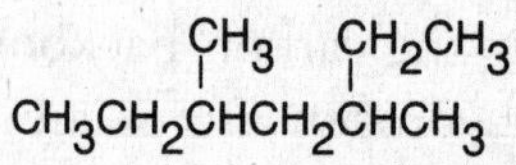

$$\overset{\overset{\displaystyle CH_3}{|}}{CH_3CH_2CH}\overset{}{CH_2}\overset{\overset{\displaystyle CH_2CH_3}{|}}{CH}CH_3$$

Solution

The numbers below correspond to the four steps for constructing a name:

1. The longest carbon chain has seven carbon atoms, and the parent name is heptane.
2. The carbon atoms in this compound can be numbered from either the left or the right.
3. The compound contains a 3-methyl and a 5-methyl group.
4. Combining the methyl groups gives the name 3,5-dimethylheptane.

EXERCISE 15.3

What is the name of the molecule shown below?

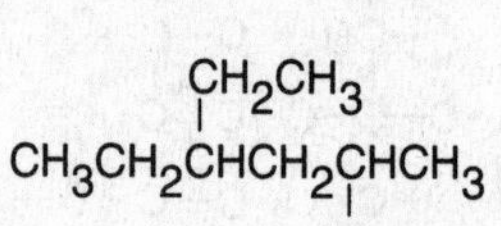

$$\overset{\overset{\displaystyle CH_2CH_3}{|}}{CH_3CH_2CH}CH_2\overset{\underset{\displaystyle CH_3}{|}}{CH}CH_3$$

Solution

The numbers below correspond to the four steps for constructing a name:

1. The longest chain contains six carbon atoms, and the parent name is hexane.
2. The carbon atoms are numbered from right to left.
3. The substituents are 2-methyl and 4-ethyl.
4. The name is 2-methyl-4-ethylhexane.

Systematic nomenclature for the alcohols, ketones, and acids uses suffixes: #-ol for alcohols, #-one for ketones, and #-oic acid for acids. The # sign represents the carbon number of the functional group. To illustrate this principle, Figure 15.11 shows the structure and names of three different alcohols on a pentane carbon chain.

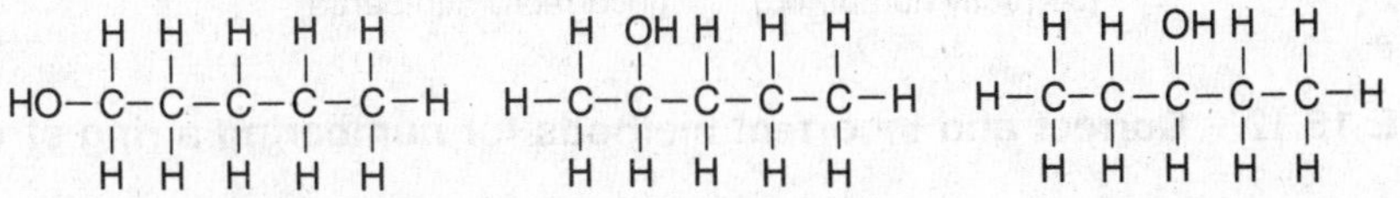

| pentane-1-ol | pentane-2-ol | pentane-3-ol |

FIGURE 15.11. Systematic names for the three alcohols made from *n*-pentane. Alternatively acceptable systematic names are 1-pentanol, 2-pentanol, and 3-pentanol.

The simpler organic compounds are often known by their common names. Some of these are summarized in Table 15.4.

Ring Compounds

Benzene and cyclohexane are the simplest and most common parent structures for ring compounds. Because a ring is circular, the numbering of the carbon atoms follows a different procedure. If there is only one substituent, no number is necessary and the substituent is assumed to be attached to the number 1 carbon atom. Chlorobenzene is a compound with a chlorine atom replacing one of the hydrogen atoms of the benzene ring. When two or more substituents are present on a ring, the carbon atoms are numbered clockwise so that one substituent is on carbon 1 and the remaining substituents are on the lowest numbered carbon atoms possible. In Figure 15.12 the correct and incorrect methods of numbering are shown for two dichlorobenzene molecules.

> **TIP**
>
> The AP Exam does not use common names but may include them as alternate names.

TABLE 15.4

Names of Straight-Chain Organic Compounds with One to Five Carbon Atoms and the Given Functional Groups

Alkane	Acid	Alcohol	Aldehyde	Amine
methane	formic acid (methanoic acid)	methyl alcohol (methanol)	formaldehyde	methylamine
ethane	ethanoic acid (ethanoic acid)	ethyl alcohol (ethanol)	acetaldehyde	ethylamine
propane	propanoic acid	propyl alcohol	propionaldehyde	propylamine
n-butane	butanoic acid	butyl alcohol	butyraldehyde	butylamine
n-pentane	pentanoic acid	pentyl alcohol	pentanaldehyde	1-aminopentane

*Names in parentheses are systematic names. Names with no parenthetical partner are systematic.

1,3-dichlorobenzene
(correctly numbered)

1,5-dichlorobenzene
(incorrectly numbered)

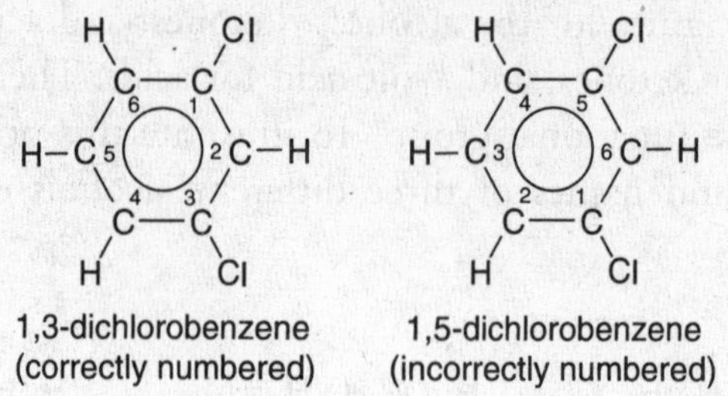

FIGURE 15.12. Correct and incorrect methods for numbering a ring structure.

Once the correct numbering sequence is determined, a ring compound is named in the same way as an alkane.

A frequently used system uses the prefix *ortho-*, *meta-*, or *para-* to indicate the position of a second substituent on a ring. Since the first substituent is always at carbon 1, a substituent in the ortho position is on carbon 2; the meta position is carbon 3; and the para position is carbon 4. 1,3-dichlorobenzene, shown in Figure 15.12, is also known as meta-dichlorobenzene.

ORGANIC REACTIONS

There are literally thousands of different organic reaction types. Several of these are described below.

Combustion

Organic molecules may be burned in excess oxygen to produce carbon dioxide and water. Preceding chapters have shown many of these reactions.

Hydrogenation

Alkenes and alkynes have double and triple bonds. Hydrogen may be added to these bonds in a **hydrogenation** reaction, which is usually catalyzed by platinum. For example, 1-butene is hydrogenated to butane in this reaction:

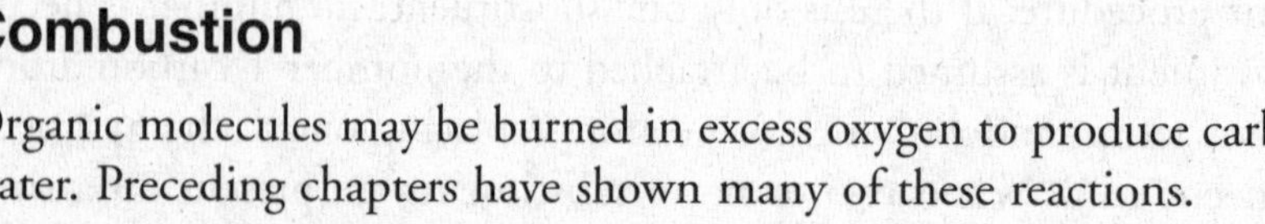

$$\tag{15.1}$$

As stated in a preceding section, compounds containing double or triple bonds are said to be unsaturated. A compound that has no double or triple bonds is termed saturated since it has the maximum number of hydrogen atoms possible.

Halogenation and Hydrohalogenation

Halogenation is a reaction similar to the hydrogenation process. Instead of H_2, a halogen such as Cl_2 or Br_2 may be used. These reactions are generally more vigorous and do not need catalysts.

Hydrohalogenation is the corresponding reaction in which compounds such as HCl and HBr add one hydrogen and one halogen atom to a double bond.

Esterification

In **esterification** an alcohol reacts with an organic acid to produce an **ester** and water. The formation of propyl ethanoate is shown in Equation 15.2.

$$\text{1-propanol} + \text{ethanoic acid} \longrightarrow \text{propyl ethanoate} + H_2O \tag{15.2}$$

In the name of an ester, the name of the alcohol comes first and the name of the acid second. For example, butyl propionate is made from butyl alcohol and propionic acid. Most esters have sweet, fruity aromas. Some of them are used as substitutes for natural products. Some common esters and their aromas are listed in Table 15.5.

TABLE 15.5

Esters and Their Aromas

Name	Aroma
octyl ethanoate	Orange
ethyl methanoate	Rum
methyl butyrate	Apple
ethyl butyrate	Pineapple
isopentyl ethanoate	Banana
pentyl propionate	Apricot
isobutyl methanoate	Raspberry
benzyl ethanoate	Jasmine

Peptide Bond Synthesis

Amino acids contain two functional groups, $-NH_2$ and $-COOH$. A reaction between amino acids forms a bond called the **peptide bond**:

$$\text{glycine} + \text{glycine} \longrightarrow \text{glycylglycine} + H_2O \tag{15.3}$$

The formation of a peptide bond is similar to the formation of an ester. Equation 15.3 shows the formation of one peptide bond, which leaves an NH_2 on one end and a $COOH$ on the other end for the formation of additional peptide bonds. Long chains of amino acids joined by peptide bonds are known as proteins. Only

21 naturally occurring amino acids are found in the thousands of different proteins in the body.

POLYMERS

Polymers are long chains of repeating structural units. One of the simplest polymers is polyethylene. Its structure is shown in Figure 15.13.

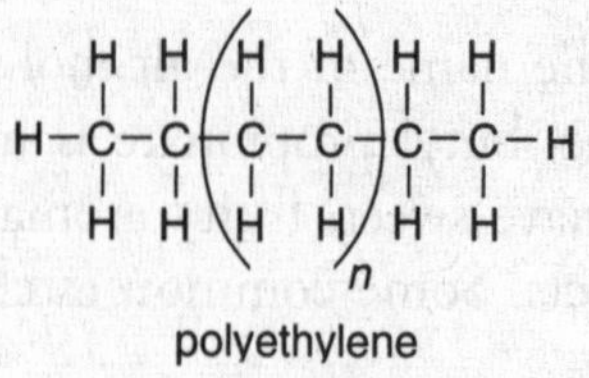

polyethylene

FIGURE 15.13.   General form for drawing the structure of polymer molecules.

The number of repeating ethylene units ($-CH_2CH_2-$) in the complete molecule is n plus the two end groups. The value of n is usually greater than 100 and often exceeds 1000. As the name suggests, the molecule is made from ethylene, $CH_2=CH_2$. Ethylene (ethene) is the **monomer** from which the polyethylene polymer is constructed. Polymers may be linear or branched.

Addition Polymers

A monomer that contains a double bond often reacts in a process called an **addition reaction.** Special catalysts start the reaction by forming free radicals, and the reaction proceeds to form more free radicals in a chain reaction. The net result is that the double bond opens (one bond in the double bond breaks) and reforms with two adjacent molecules. The process is illustrated in Figure 15.14.

Double bond breaks, freeing electrons to form new bonds.

Electrons combine to form new bonds.

FIGURE 15.14.   Elementary steps in the formation of polyethylene.

Some familiar polymers and the monomers from which they are made are listed in Table 15.6.

TABLE 15.6

Common Addition Polymers and Their Uses

Name	Monomer	Typical Uses
polyethylene	$CH_2{=}CH_2$	Bottles, coatings
polypropylene	$CH_2{=}CHCH_3$	Bottles, coatings
polyvinyl chloride	$CH_2{=}CHCl$	Credit cards, pipes
polystyrene	$CH_2{=}CHC_6H_5$	Foamed beads and blocks for packaging
Teflon®	$CF_2{=}CF_2$	Nonstick surfaces

Condensation Polymers

Condensation reactions are similar to the reactions that form esters and peptide bonds. Each reactant must have two functional groups; one is an acid and the other is a base, either —OH or —NH$_2$. We can remember that condensation reactions produce water as a product by recalling that condensation on a cold window or a glass of ice water also produces water.

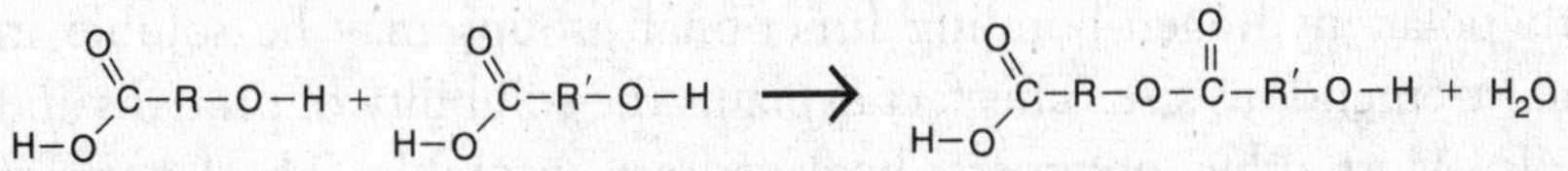

$$(15.4)$$

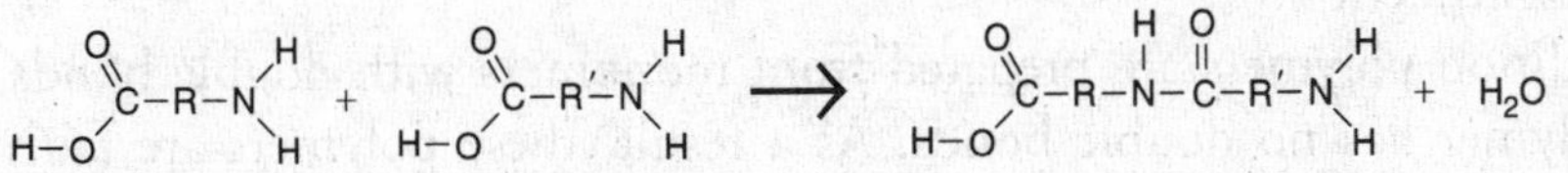

$$(15.5)$$

In the reactions in Equations 15.4 and 15.5, R and R′ represent hydrocarbon groups.

Each monomer may contain an acid and a basic functional group, as shown above, in which case there is only one reactant. Another method involves using one monomer that contains two acid groups and another that contains two basic groups. An equal molar mixture of these two monomers is used to form the polymer:

$$HOOCRCOOH + H_2NR'NH_2 \rightarrow HOOCRCONHR'NH_2 + H_2O \quad (15.6)$$

Starch and cellulose are both condensation polymers of glucose. The different physical and chemical properties of these two polymers are due to the different arrangements of their bonds. Some common condensation polymers and their uses are listed in Table 15.7.

TABLE 15.7

Common Condensation Polymers and Their Uses

Name	Monomer(s)	Typical Uses
Dacron®	terephthalic acid, ethylene glycol	Fibers, fabrics
nylon	1,6-hexanedicarboxylic acid, 1,6-diaminohexane	Fibers, fabrics
proteins	amino acids	Biological reactants, food
disaccharides	Two sugar molecules	Biological energy source
starch	Many glucose molecules	Food source
cellulose	Many glucose molecules	Biological structural material
DNA	nucleotides	Genetic code

Polymer Properties

We may deduce some of the properties of polymers from fundamental chemical principles developed in preceding chapters.

Polymers with polar, hydrogen-bonding functional groups may be soluble in water despite their tremendous size. This fact explains the solubility of proteins and polyvinyl alcohols. Most other polymers, however, are insoluble. The length of these polymers allows them to align with each other, thereby generating substantial London forces of attraction.

Although addition polymers are prepared from monomers with double bonds, the finished polymer has no double bonds. As a result, these polymers are fairly inert. The most inert is **Teflon,** in which the very strong C—F bond cannot be broken easily.

Polymers also have distinct properties related to the absorption of water and gases. Many of these properties depend on the method used to manufacture the polymer. A polymer can be manufactured or treated to increase or decrease the porosity. Branched-chain polymers have different properties than linear polymers. Polymer chemists take advantage of the specific properties of polymers in designing consumer products.

SUMMARY

Organic chemistry is reviewed as an introduction to the topic. The fundamentals of carbon-to-carbon bonding are reviewed with sp, sp^2, and sp^3 geometric structures. These can be assembled into large structures containing many carbons. The concept of isomers in several forms is introduced along with methods for naming simple carbon compounds. The basic functional groups for organic compounds are reviewed and the fundamentals of polymer reactions are covered in this chapter.

Important Concepts

Hybridization of carbon and related shapes
Isomers, structural, cis-trans, and stereo
Alkanes, alkenes, and alkynes
Functional groups
Nomenclature
Polymers

Practice Exercises

Multiple-Choice

For the first four problems below, one or more of the following responses will apply; each response may be used more than once or not at all in these questions.

I. —OH
II. —COOH
III. —NH$_2$
IV. —CHO
V. —CN

1. The amine functional group is

   (A) I
   (B) II
   (C) III
   (D) IV
   (E) V

2. Substances containing functional groups that react to form compounds that usually have pleasant odors.

   (A) I and II
   (B) II and III
   (C) III and IV
   (D) IV and I
   (E) V and III

3. Proteins are made up of many molecules that have at least the functional groups

   (A) I and II
   (B) II and III
   (C) III and IV
   (D) IV and I
   (E) V and II

4. When a methyl group is added to one of these functional groups, a dilute solution of the resulting compound is often used in foods, particularly in salads and pickles.

   (A) I
   (B) II
   (C) III
   (D) IV
   (E) V

5. The aldehyde functional group is

   (A) —C—O—H
   (B) —C=O
   (C) —NH$_2$
   (D) —COOH
   (E) —CH$_3$

6. The *n*-propyl side chain is

   (A) —COOH
   (B) —CH$_3$
   (C) —NH$_2$
   (D) —CH$_2$CH$_2$CH$_2$CH$_3$
   (E) —CH$_2$CH$_2$CH$_3$

7. The peptide bond is formed when an organic acid reacts with a compound containing the functional group

   (A) —C—O—H
   (B) —C=O
   (C) —NH$_2$
   (D) —C≡NC
   (E) —CH$_3$

8. Stereoisomers

   (A) require four different substituents attached to one carbon atom
   (B) have virtually identical physical properties
   (C) often have different biological activities
   (D) are mirror images
   (E) All of the above are true.

9. Under the appropriate reaction conditions propylene will

   (A) form an addition polymer called polypropylene
   (B) react with Br$_2$ to form 1,2-dibromopropane
   (C) in the presence of Pt, react with H$_2$ to form propane
   (D) react with HI
   (E) All of the above reactions occur.

10. The monomer used to form the polymer [—CH$_2$CHCl—]$_n$ is

    (A) CH$_3$CH$_2$Cl
    (B) CH≡CCl
    (C) CH$_n$CHCl$_n$
    (D) CHCH + HCl
    (E) CH$_2$=CHCl

11. The IUPAC systematic name for CH$_3$CH$_2$CH$_2$OH is

    (A) propane-1-ol
    (B) *n*-propanol
    (C) propanol
    (D) 1-methylethane-1-ol
    (E) isopropanol

12. A protein is

    (A) a polysaccharide
    (B) desoxyribonucleic acid
    (C) a polymer of amino acids
    (D) soluble because of the —C=O groups
    (E) a polyester

13. The reaction between a compound containing one —NH$_2$ and a compound containing one —COOH functional group is best described as

    (A) an esterification reaction
    (B) a hydrogenation reaction
    (C) an acid-base reaction
    (D) a hydrolysis reaction
    (E) a combustion reaction

14. To have cis and trans isomers, a compound must have

    (A) *sp* bonded carbon atoms
    (B) *sp^3* bonded carbon atoms
    (C) *sp^2* bonded carbon atoms
    (D) (A) and (B)
    (E) (A) and (C)

15. For compounds with the same number of carbon atoms, the compound with the lowest boiling point is expected to be

    (A) a ketone
    (B) an amine
    (C) an acid
    (D) an alcohol
    (E) an ether

16. Which of the following organic compounds need NOT contain oxygen?

    (A) an alkyne
    (B) an alcohol
    (C) an aldehyde
    (D) a ketone
    (E) an ether

17. An organic base contains the functional group

    (A) —OH
    (B) —COOH
    (C) —NH$_2$
    (D) —C=O
    (E) —Cl

18. At which positions does metadibromobenzene have bromine atoms on the benzene ring?

    (A) 1 and 3
    (B) 1 and 4
    (C) 1 and 2
    (D) 1 and 5
    (E) 1 and 6

19. Of the following molecules which is (are) expected to be planar?

    (A) CH$_4$
    (B) CH$_2$=CH$_2$
    (C) C$_6$H$_6$ (benzene)
    (D) C$_6$H$_{12}$ (cyclohexane)
    (E) B and C

20. The bonding in the benzene molecule, C$_6$H$_6$, in any of its resonance structures, contains

    (A) 6 sigma bonds and 6 pi bonds
    (B) 6 sigma bonds and 12 pi bonds
    (C) 12 sigma bonds and 6 pi bonds
    (D) only sigma bonds
    (E) 12 sigma bonds and 3 pi bonds

21. Which of the following is most likely to form hydrogen bonds?

    (A) an alkyne
    (B) an alcohol
    (C) an aldehyde
    (D) a ketone
    (E) an ether

22. The following compounds have the same number of carbon atoms. Which is expected to have the lowest boiling point?

    (A) an alkyne
    (B) an alcohol
    (C) an aldehyde
    (D) a ketone
    (E) an ether

23. Which of the following always have a constant percentage of carbon in all of their compounds? (Assume that there are no other functional groups on the molecule.)

    (A) alkenes
    (B) alcohols
    (C) aldehydes
    (D) ketones
    (E) ethers

24. CH$_2$O is the empirical formula for

    (A) amino acids
    (B) proteins
    (C) carbohydrates (sugars)
    (D) aldehydes
    (E) DNA and RNA

ANSWER KEY

1. C	6. E	11. A	16. A	21. B
2. A	7. C	12. C	17. C	22. A
3. B	8. E	13. C	18. A	23. A
4. B	9. E	14. C	19. E	24. C
5. B	10. E	15. E	20. E	

See Appendix 1 for explanations of answers.

Free-Response

The following questions focus on the structure and reactions related to organic chemicals.

(a) Expand the following structure and determine the hybridization of each atom except for H.

$$CH_2CHCH_2CH_2OCH_2CCH$$

(b) Write the symbols for the alcohol, amine, carboxyl, and aldehyde functional groups. For each of these functional groups mention a chemical or physical property that can be derived from principles and concepts developed in previous chapters.

(c) Give an example of a polymerization reaction.

(d) Draw and name the structural isomers for hexane.

ANSWERS

(a) Counting carbons from the left, we get $1 = sp^2$, $2 = sp^2$, $3 = sp^3$, $4 = sp^3$, $5 = sp^3$, $6 = sp$, $7 = sp$. The oxygen hybridization is sp^3. Recall that sp is linear, sp^2 is a triangular planar structure, and sp^3 has the shape of a tetrahedron.

(b) Alcohol is R—OH, amine is R—NH$_2$, carboxyl is R—COOH, and aldehyde is R—CHO. The alcohol and amine are both compounds that hydrogen-bond. The carboxyl group is an organic acid that also can hydrogen-bond. The aldehyde group has the fewest interactions. It has a higher vapor pressure and the lowest boiling point for similar sized molecules.

(c) A common polymerization reaction is the formation of polyethylene from ethylene as an addition polymer. Several other polymerization reactions are given in the chapter.

(d) Hexane has six carbon atoms arranged in various linear sequences. The possible different ones are

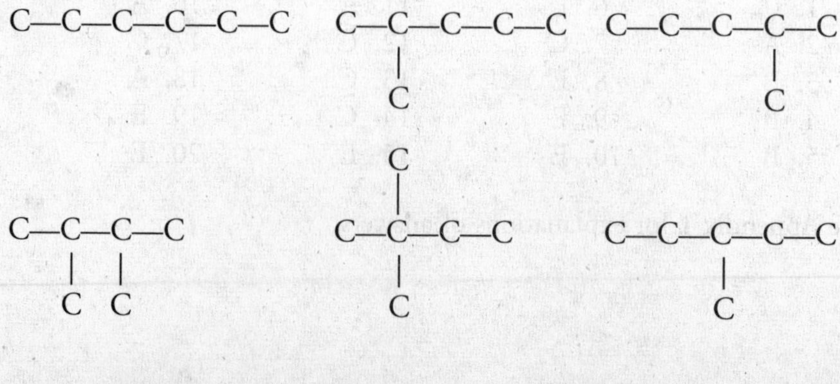

C—C—C—C—C—C   C—C—C—C—C   C—C—C—C—C
                              | |
                              C   C
                              |
                              C

C—C—C   C—C—C—C   C—C—C—C—C
  | | | |
  C   C   C   C

Experimental Chemistry CHAPTER 16

- Data Gathering
- Qualitative/Quantitative
- Accuracy and Precision
- Significant Figures
- Uncertainty
- Rounding
- Graphs
- Independent Variable
- Dependent Variable
- Reading Graduated Scales
- Determining Mass
- Volume Measurement

- Temperature Measurement
- Density
- Specific Heat
- Sample Manipulation
- Separation Techniques
- Instruments
- pH Meters
- Spectrometers
- Reactions
- Qualitative Analysis
- Chemical Hazards
- Safety in the Laboratory

Chemistry is an experimental science. All concepts, laws, and theories are directly supported by experimental evidence. Correctly designed and performed experiments advance knowledge about chemistry, while poor experiments confuse issues and principles and hinder the advancement of science. We need to understand the underlying concepts and techniques that are the basis of good experimental methods.

DATA GATHERING

Information gained in chemical experiments may be **quantitative,** that is, numerical, or it may not involve numbers, in which case the data are **qualitative** observations. If the concentration of an acid is determined to be 0.345 M, a quantitative measurement has been made. When silver nitrate is added to a solution and a white precipitate forms, the result is a qualitative observation. Both types of information are important in any experiment.

All observations and experimental details must be recorded in a **notebook.** The notebook must contain a complete description of the experiment so that any knowledgeable chemist can accurately repeat it. In particular, the description must specify the general idea of the experiment, along with the equipment and chemicals used. Diagrams are often drawn to show how an apparatus is constructed. Detailed information on the mass, volume, and source of every chemical used in the experiment is recorded. When the experiment is performed, careful observations of the reaction are noted as the experiment progresses. Finally, calculations are performed on the data gathered and conclusions are drawn. These conclusions help the chemist design the next experiment.

All information about an experiment is recorded directly in the notebook, which should be a bound, not loose-leaf or spiral, book with numbered pages. Entries are made in ink, and every entry is dated. Data should never be written on scraps of paper and then transcribed later. Erasures or removal of pages is not acceptable. Errors are crossed out in such a way that they are still readable. Although neatness is desirable, it is much more important to have a continuous record of all laboratory activity, whether a particular experiment is successful or not.

CALCULATIONS

Most calculations in chemistry involve simple algebra and two basic approaches. The first is the use of dimensional analysis to convert information from one set of units into another. The second is the use of a memorized equation or law into which data for all variables except one are inserted. The one remaining variable is the unknown for the problem. Correct use of the second method requires that all units be shown to ensure that they cancel properly to obtain the desired units for the numerical answer.

Scientific calculators simplify mathematical operations to the touch of a few keys. Understanding the principles and concepts of chemistry enables us to decide in which order to press those keys. Understanding numbers tells us how to properly interpret the answer that appears on the calculator screen.

Accuracy and Precision

The **accuracy** of a measurement refers to the closeness between the measurement obtained and the true value. Since scientists rarely know the true value, it is generally impossible to determine the accuracy with certainty. One approach to evaluating accuracy is to make a measurement by two completely independent methods. If the results from independent measurements agree, scientists have more confidence in the accuracy of their results.

Accuracy is affected by **determinate errors,** that is, errors due to poor technique or incorrectly calibrated instruments. Careful evaluation of an experiment may eliminate determinate errors.

Precision refers to the closeness of repeated measurements to each other. If the mass of an object is determined as 35.43 grams, 35.41 grams, and 35.44 grams in three measurements, the results may be considered precise. There is no guarantee that they are accurate, however, unless the balance was properly calibrated and the correct methods were used in weighing the object. When proper techniques are used, precise results infer, but do not guarantee, accurate results.

Precision is also a measure of **indeterminate errors,** that is, errors that arise in estimating the last, uncertain, digit of a measurement. Indeterminate errors are random errors and cannot be eliminated. Statistical analysis deals with the theory of random errors.

Significant Figures

Every experimental measurement is made in such a way as to obtain the most information possible from whatever instrument is used. As a result, measurements involve numbers in which the last digit is uncertain to some extent. Scientists characterize a measured number based on the number of **significant figures** it contains.

The number of significant figures in a measurement includes all digits that are not zeros. All zeros to the left of the last nonzero digit are not significant. Imbedded zeros, those between two nonzero digits, are always significant. Trailing zeros are significant only if a decimal point is somewhere in the number. For exponential numbers, the number of significant figures is determined from the digits to the left of the multiplication sign. For example, 8.32×10^3 has three significant figures.

Sometimes there are trailing zeros on the right side of a number. If the number contains a decimal point, trailing zeros are always significant. Trailing zeros that are used to complete a number, however, may or may not be significant. Scientists avoid writing a number such as 12,000 since it does not definitely indicate the number of significant figures. Scientific notation is used instead. Twelve thousand can be written as 1.2×10^4, 1.20×10^4, 1.200×10^4, or 1.2000×10^4, indicating two, three, four, or five significant figures, respectively. It is the responsibility of the experimenter to write numbers in such a way that there is no ambiguity.

EXERCISE 16.1

Determine the number of significant figures in each of the following measured values:

(a) 23.46 mL　　(d) 6.02×10^{23} molecules　　(g) 1.00026×10^{-3} cm
(b) 0.0036 s　　(e) 0.98 mol　　(h) 2.0000 J
(c) 854.236 g　　(f) 0023 m　　(i) 824 mg

Solution

The numbers are repeated with the significant figures in bold type:

(a) **23.46** mL　　(d) **6.02** $\times 10^{23}$ molecules　　(g) **1.00026** $\times 10^{-3}$ cm
(b) 0.**0036** s　　(e) 0.**98** mol　　(h) **2.0000** J
(c) **854.236** g　　(f) 00**23** m　　(i) **824** mg

Some numbers are **exact numbers,** which involve no uncertainty. The number of plates set on a table for dinner may be determined exactly. If five plates are observed and counted, that measurement is exactly 5. In chemistry many defined equalities are exact. For instance, there are exactly 4.184 joules in each calorie. Other exact numbers are stoichiometric coefficients and subscripts in chemical formulas.

The reason for determining the number of significant figures in a measured value that this number tells us how to write the answers to any calculations based on that value. There are two basic rules:

1. The number with the **fewest significant figures** in a multiplication or division problem determines the number of significant figures in the answer. In these calculations an exact number is considered to have an infinite number of significant figures.
2. The number with the **fewest decimal places** in an addition or subtraction problem determines the number of decimal places in the answer. Numbers expressed in scientific notation must all be converted to the same power of 10 before determining which decimal places can be retained.

TIP

Significant Figures

Digits ARE significant if:

1. the digit is not zero.
2. the zero is imbedded.
3. the trailing zero is in a number that has a decimal point.

Digits ARE NOT significant if the zeros are to the left of all nonzero digits.

Uncertainty

There are two types of uncertainty, **absolute uncertainty** and **relative uncertainty.**

The absolute uncertainty is the uncertainty of the last digit of a measurement. For example, 45.47 mL is a measurement of volume, and the last digit is uncertain. The absolute uncertainty is ±0.01 mL. The measurement should be regarded as somewhere between 45.46 and 45.48 mL. The absolute uncertainty should have only one digit. The last digit of the number should be the first digit of the uncertainty; for example, 35.38 ± 0.02 mL.

The relative uncertainty of a number is the absolute uncertainty divided by the number itself. For the above example, the relative uncertainty is

$$\frac{0.01 \text{ mL}}{45.47 \text{ mL}} = 2 \times 10^{-4}$$

The absolute uncertainty governs the principles used for addition and subtraction. The relative uncertainty governs the principles used for multiplication and division.

Rounding

Calculations, especially those done using an electronic calculator, often generate more, and sometimes fewer, significant figures or decimal places than are required by rules 1 and 2 given above. These answers must be rounded to the proper number of significant figures or decimal places. To do this, four steps are followed:

1. The number of digits to be kept in a calculation is determined using rules 1 and/or 2 above.
2. If the digit just after the kept digits is less than 5, the remaining digits are dropped. For example, rounding 6.23499 to three significant figures yields 6.23 because 4 is less than 5.
3. If the digit just after the kept digit is greater than 5, or if it is 5 with additional nonzero digits, 1 is added to the last of the kept digits. For example, rounding 5.5589 to three significant figures yields 5.56. Similarly, 6.34**5002** is rounded to 6.35 because the 5 has a nonzero digit after it.
4. If the digit to be rounded is just 5 or is 5 with all zeros after it, the last digit in the kept digits is rounded to the nearest even number. For example, rounding 2.33**5** to two decimal places yields 2.34, and rounding 6.78**5000** to two decimal places yields 6.78.

EXERCISE 16.2

Perform each of the following calculations, and report the answer with the correct number of significant figures:

(a) 23.456 + 16.0094 + 9.21

(b) 14.98 × 0.00234 × 1.5

(c) $1.46 \times 10^3 - 5.83 \times 10^4$

(d) $(8.236 \times 10^2)(5.55 \times 10^{-3})$

(e) (23.45 − 16.12)/6.233

(f) $(6.02 \times 10^{23})(1.00 \times 10^{-6})/(18.23)$

(g) $(44.23)/(2.33 \times 10^2 - 2.25 \times 10^2)$

Solution

(a) 48.68
(b) 0.053
(c) 5.68×10^4
(d) 4.57
(e) 1.18
(f) 3.30×10^{16}
(g) 6

Significant Figures in Atomic and Molar Masses

All of the atomic masses listed in the periodic table have four or more significant figures. These are measured values and must be included in the determination of the correct number of significant figures in any calculation that uses them. However, most calculations involve fewer than four significant figures, and the significant figures in atomic and molar masses do not affect the number of significant figures in the answers.

As a result, atomic and molar masses are usually rounded to the nearest whole number for convenience. One exception is chlorine, which usually has its atomic mass rounded to 35.5.

EXERCISE 16.3

What percentage error is expected when the atomic mass of each of the following elements is rounded to the nearest whole number?

(a) Na
(b) Ag
(c) Pb
(d) Cl

Solution

The percentage error is calculated as

$$\text{percentage error} = \frac{\text{Measured value} - \text{True value}}{\text{True value}} \times 100$$

In this problem the measured value is the rounded atomic mass, and the true value is the atomic mass.

(a) $\dfrac{0.01}{22.99} \times 100 = +0.04\%$ (c) $\dfrac{-0.2}{207.2} \times 100 = -0.1\%$

(b) $\dfrac{0.132}{107.868} \times 100 = +0.1\%$ (d) $\dfrac{-0.453}{35.453} \times 100 = -1.3\%$

The error for chlorine is ten times that for the other elements. When the atomic mass is rounded to 35.5, the error for chlorine is +0.1%, which is in line with the magnitude of the rounding errors for the other elements.

GRAPHS

A **graph** is used to illustrate the relationship between two variables. Graphs are often a more effective method of communication than tables of data. A graph has two axes: a horizontal axis, usually called the *x*-axis (abscissa), and a vertical axis, called the *y*-axis (ordinate).

It is customary to use the *x*-axis for the **independent variable,** and the *y*-axis for the **dependent variable,** in an experiment. An independent variable is one that the experimenter selects. For instance, concentrations of standard solutions that a chemist prepares are independent variables since any concentrations may be chosen. The dependent variable is a measured property of the independent variable. For instance, the dependent variable may be the amount of light that each of the standard solutions prepared by the chemist absorbs, since the absorbed light is dependent on the concentration. Each data point is an *x,y* pair representing the value of the independent variable and the value of the dependent variable as determined in the experiment.

The first step in constructing a graph is to label the *x*- and *y*-axes to indicate the identity of the independent and dependent variables. Next, the axes are numbered, usually from zero to the largest value expected for each variable. Finally, each data point is plotted by drawing a horizontal line at the value of the dependent variable and a vertical line upward from the value of the independent variable. The intersection of these two lines determines where that data point belongs on the graph.

Most graphs show a linear relationship between two variables. Other graphs, such as those showing kinetic curves, have curved lines. In both cases, data points are plotted on the graph and then the best smooth line is drawn through the points. Lines are never drawn by connecting the data points with straight lines. In very accurate work a statistical analysis called the "method of least squares" is used to determine the best straight line for the data. In most cases, however, the line is drawn by eye, attempting to have all data points as close as possible to the line. The usual result is a line that has the same number of data points above and below it.

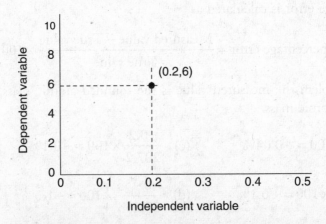

FIGURE 16.1. The positions of the independent and dependent variables in a graph. Dashed lines show the placement of a point representing a value of 0.2 for the independent variable and of 6 for the dependent variable.

Figure 16.2 illustrates that it is incorrect to draw any line beyond the measured data points. The reason is that anything beyond the measured data is unknown. Extending the line implies information that is not verified by experimental data.

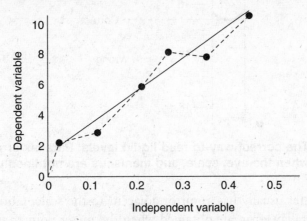

FIGURE 16.2.   The correct way to draw a line using data in a graph. The solid line is the correct line; the dashed line is incorrect.

The slope of a curve or line is often needed as an experimental result. To determine the slope of a line, two points on the line are chosen. The left-hand point has coordinates (x_1, y_1), and the right-hand point has coordinates (x_2, y_2). The values of x and y at these points are determined from the graph, and Equation 16.1 is used to determine the slope:

$$\text{Slope} = \frac{x_2 - x_1}{y_2 - y_1} \qquad (16.1)$$

In a graph with a curved line the slope is determined by drawing a tangent to the curve and then determining the slope of the tangent, as is done for a straight line.

DETERMINATION OF PHYSICAL PROPERTIES

Scale Reading

Many measurements are made by comparing the level of a liquid in a container to a scale etched on the outside of the container or by observing the position of a meter pointer with reference to an adjacent scale. Such a scale is usually a series of lines in which every tenth line is numbered and is usually distinguished also by being longer than the others. The fluid level or meter pointer is never directly in contact with the scale; consequently, incorrect reading techniques can result in **parallax errors.**

The surface tension of a liquid in any container causes the liquid to have a curved surface called a **meniscus.** All glassware is calibrated on the basis that the liquid level corresponds to the bottom of the meniscus. Figure 16.3 demonstrates that, to avoid parallax errors, the eye must be at the same level as the meniscus when measuring liquids.

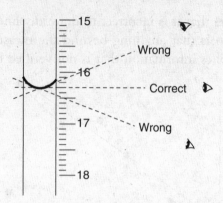

FIGURE 16.3. **The correct way to read liquid levels. Parallax error results when the eye, scale, and meniscus are not lined up horizontally.**

A modern meter usually has a mirror adjacent to the scale. Correct meter readings, with no parallax error, are obtained when the meter pointer and its reflection in the mirror coincide.

Determination of Mass by Weighing

The mass of a chemical substance is obtained by determining the weight of the substance. The weight of a sample is equal to the force of gravity times the mass of the sample:

$$\text{Weight} = (\text{Gravitational constant})(\text{Mass}) \qquad (16.2)$$

If the gravitational constant is known, the mass can be calculated from the measured weight. To avoid this calculation, the mass of a sample may be directly compared to a known mass on a double-pan balance. Modern balances are single-pan balances that are calibrated with standard masses.

Proper use of a balance requires that the balance be calibrated. A sample is always weighed in an appropriate container, never directly on a balance pan. The mass of the empty container is referred to as the tare mass. A sample's mass is the difference between the tare mass and the mass of the sample and the container. A typical notebook entry for a mass determination should look like this:

$$
\begin{array}{r}
24.345 \text{ g total mass} \\
-3.862 \text{ g tare mass} \\
\hline
20.483 \text{ g sample mass}
\end{array}
$$

Liquid Volume Measurement

Graduated cylinders and graduated beakers are used to measure liquid volumes with an accuracy of ±10 percent. More accurate liquid measurements are made with **pipets** or **burets.** Glassware of this type is usually labeled with the letters "**TD,**" which stand for "to deliver" and indicate that the amount of liquid poured or delivered from the pipet or buret is the volume stated. **Volumetric flasks** have the label "**TC,**" which stands for "to contain." Although a 100-mL volumetric flask will contain 100 mL, it will not deliver 100 mL if the liquid is poured out. The difference is the film of solution left on the inner surface of the flask itself.

Pipets come in two types, transfer and measuring. A transfer pipet has a single mark and is used for the most accurate measurements. Measuring pipets are graduated, and any volume may be delivered by stopping the flow of the liquid at the desired point. Before filling, a pipet is rinsed in the solution to be measured; then the pipet is filled above the calibration line with suction from a suction bulb. Quickly placing a finger over the top of the pipet after the suction bulb is removed allows precise control of the flow of liquid. The solution is allowed to drain to the first calibration mark, excess solution is wiped from the tip, and then the solution is allowed to drain by gravity into the receiving flask. When draining is completed, the last drop is removed by momentarily touching the tip of the pipet to the solution surface or the side of the flask. Blowing out the contents of a pipet will deliver the wrong amount of solution.

Burets are long, graduated tubes that hold between 10 and 50 mL of solution. The flow of liquid is controlled by a valve called a **stopcock.** A buret is rinsed with solution, not distilled water, before filling. When filling is completed, the stopcock is opened fully to expel any air from the tip; and, if needed, additional solution is added so that the liquid is near the top graduation mark. In use, the starting volume is recorded, solution is delivered, and, when finished, the final volume is recorded. The difference between the final and initial volumes is the volume delivered. Here is a sample notebook entry:

$$23.86 \text{ mL at the end}$$
$$\underline{-0.23 \text{ mL at the start}}$$
$$23.63 \text{ mL solution delivered}$$

Temperature Measurement

Thermometers are used for temperature measurement. Each thermometer has a line etched near the mercury bulb. This line is the immersion depth. When the thermometer is immersed in a liquid to this etched line, the temperature reading will be the most accurate. The calibration of a thermometer is checked by immersing it in ice water, 0 °C, and then in boiling water, 100 °C.

Broken mercury thermometers must be disposed of carefully because of the hazards posed by elemental mercury.

Determination of Melting and Boiling Points

With sufficient liquid, the boiling point is determined by measuring the temperature as the liquid boils. Accurate determination of the normal boiling point requires that this experiment be done at 1 atmosphere of pressure.

Several instruments are available for determining melting points. For the experiment, a sample is packed into a closed-end capillary tube and inserted in the instrument, which is set to heat the sample slowly. The temperature at which the first crystals start melting is the melting point. The same result can be obtained by attaching the capillary tube to a thermometer so that the sample is next to the mercury bulb. The thermometer and attached sample are immersed in an oil with a high boiling point, and heat is applied until the sample just starts to melt.

Determination of Density

Density is an intrinsic physical property that can be used to identify unknown materials. The general equation for determining density is as follows:

$$\text{Density} = \frac{\text{Mass}}{\text{Volume}} \tag{16.3}$$

Equations 16.4 and 16.5 show the units used for mass and volume for substances in various phases:

$$\text{Density of solids or liquids} = \frac{\text{grams of material}}{\text{cubic centimeters of material}} \tag{16.4}$$

$$\text{Density of gases} = \frac{\text{grams of gas}}{\text{liters of gas}} \tag{16.5}$$

Since density varies somewhat with temperature, we often see the symbol d_{20}, where the subscript indicates the Celsius temperature at which the density was determined.

The density of a solid is determined by measuring first its mass and then its volume. The mass measurement was described in a preceding section. There are two methods to determine the volume of a solid. The first is to measure its dimensions and use trigonometry to calculate the volume. This method works best with a regularly shaped object such as a cylinder or a rectangular solid. The second method is to immerse the solid in a liquid and measure the volume displaced. This is commonly done by placing water in a graduated cylinder and measuring its volume. Then the solid is added and a second volume is determined. The difference in volumes is the volume of the object:

$$\text{Volume of object} = V_{\text{final}} - V_{\text{initial}} \tag{16.6}$$

Dividing the mass by the volume yields the density.

The density of a liquid is determined using a device called a **pycnometer,** which is a small flask with a volume of approximately 5–25 mL. The exact volume is obtained by determining the mass of water needed to completely fill the pyncnometer. Since the density of water is known, the volume of the pyncnometer can be calculated. Next, the mass of unknown liquid needed to fill the pyncnometer is determined. The density is then calculated using Equation 16.4.

The density of a gas is determined by evacuating a large flask with a vacuum pump so that there is virtually no gas inside the flask. The evacuated flask is weighed, and the gas is introduced until its pressure is equal to atmospheric pressure. The flask is then weighed again to obtain the mass of the gas. The volume of the flask is determined from the amount of water it can hold or by some similar technique, and the density is determined using Equation 16.5.

Determination of Specific Heat

The specific heat of a solid, usually a metal, is determined by heating a known mass of the substance to a predetermined temperature and then submerging it in a known quantity of water in an insulated container. The final temperature of the water indicates the temperature increase of the water and the temperature decrease of the metal. The specific heat equation is as follows:

$$q = (\text{Specific heat})(\text{Mass})(\Delta T) \tag{16.7}$$

where ΔT is the temperature change.

Since the heat, q, gained by the water must be equal to the heat lost by the metal, the equality becomes

$$(\text{Specific heat})_{water}(\text{Mass})_{water}(\Delta T)_{water} = -(\text{Specific heat})_{metal}(\text{Mass})_{metal}(\Delta T)_{metal} \tag{16.8}$$

An example of specific heat calculation is given in Chapter 12.

SAMPLE MANIPULATIONS

Heating

Heating can be done in several ways, and the method chosen depends on the equipment on hand and safety factors. Water and aqueous solutions are usually heated with a Bunsen burner. Precautions should be taken to avoid **bumping,** which is a violent burst of boiling that may spatter hot liquid. The best way to avoid bumping is to add boiling chips to the mixture. Liquids heated in test tubes are very likely to bump, and care should be exercised that test tubes are not pointed toward other workers. Burners with open flames should be avoided when any flammable substances are in use in the laboratory.

Bumping occurs because a burner flame superheats one portion of a liquid. Hot-water baths, steam baths, sand baths, and electric heating mantles are effective heating methods that minimize bumping by spreading the applied heat. The fact that the limit for steam and hot water is approximately 100 °C may be a safety feature.

Bumping also occurs when solids, particularly powders, are added to very hot liquids. For this reason solids should be added to cool liquids before heating.

Cooling

Hot solutions or objects are cooled with water or ice. A bath made of crushed ice and water is most effective. Ice by itself is not efficient since much of the flask containing the hot solution is not in contact with the ice. Only heat-resistant laboratory glassware should be used since ordinary glass will shatter with rapid temperature changes.

Lower temperatures (approx. −50 °C) may be obtained using dry ice mixtures, and very low temperatures (−196 °C) with liquid nitrogen. Dry ice and liquid nitrogen can cause injury, however, because of their extremely low temperatures. They also present a smaller, although real, hazard of possible suffocation.

Mixing

Preparation of solutions is the most common mixing operation in chemistry. A solid is dissolved in a solvent by adding the solid slowly, with stirring, to about half of the liquid. When dissolution is complete, the correct amount of liquid is added and mixed well. Grinding the solid to a powder and warming the mixture both speed dissolution.

Preparation of solutions with molar concentration units requires an exact total volume of solution. These solutions are prepared in **volumetric flasks.** A volumetric flask is calibrated with an etched line on its neck to hold a specified volume at a given temperature. If a solution is warmed to speed dissolution, it must be cooled to room temperature before the final volume adjustment.

When a concentrated acid is mixed with water, the acid is always added to the water. Most acids, particularly sulfuric acid, are denser than water and generate a large amount of heat when mixed with water. If water is added to sulfuric acid, it does not mix because of the density of the acid, and the high heat of mixing causes the water to boil and spatter sulfuric acid.

Drying

Drying chemical reagents before use and drying products of reactions are very common operations in the laboratory. Since the drying process involves removing water from a substance, the temperature must be above 100 °C. To avoid decomposition, however, the temperature should also be as low as possible. An oven set between 105° and 110 °C is recommended.

Dilution

A common practice in chemistry laboratories is to prepare a concentrated stock solution of some solute. To make other solutions, the stock solution is diluted with water in the appropriate volume ratio. The dilution law is

$$(C_{initial})(V_{initial}) = (C_{final})(V_{final}) \qquad (16.9)$$

If we have a stock solution of some concentration, $C_{initial}$, we can calculate the volume needed to prepare a given volume at any other concentration.

EXERCISE 16.4

Calculate the volume of a stock 6.00 *M* HCl solution needed to make 1.00 L of a 0.100 *M* HCl solution.

Solution

We can define as follows: $C_{initial} = 6.00\ M$ HCl, $C_{final} = 0.100\ M$ HCl, $V_{final} = 1.00$ L. Entering these values in Equation 16.9 yields

$$(6.00\ M\ \text{HCl})(V_{initial}) = (0.100\ M\ \text{HCl})(1.00\ \text{L})$$

$$V_{initial} = \frac{(0.100\ M\ \text{HCl})(1.00\ \text{L})}{6.00\ M\ \text{HCl}}$$

$$= 0.0166\ \text{L} = 16.6\ \text{mL}$$

To prepare the desired solution, 16.6 mL of the stock solution must be diluted to 1.00 L.

EXERCISE 16.5

How many milliliters of distilled water must be added to 100 mL of 0.250 M KCl to prepare a 0.100 M KCl solution?

Solution

The given values are as follows: $C_{initial}$ 5 0.250 M KCl, $V_{initial} = 100$ mL, $C_{final} = 0.100\ M$ KCl. From these data we calculate V_{final}:

$$(0.250\ M\ \text{KCl})(100\ \text{mL}) = (0.100\ M\ \text{KCl})(V_{final})$$

$$V_{final} = 250\ \text{mL}$$

Since the final volume of solution is the sum of the initial volume and the added distilled water:

$$V_{final} = V_{initial} + V_{water}$$

$$250\ \text{mL} = 100\ \text{mL} + V_{water}$$

$$V_{water} = 150\ \text{mL}$$

Gas Collection

When a gas is generated in a chemical reaction, it may be collected in a **pneumatic trough,** shown in Figure 16.4. A gas-collecting bottle is filled with water, and a tube from the sealed experiment leads under water to the bottle. Any gas evolved displaces the water in the bottle. Details on how to measure the amount of gas collected are described in Chapter 4.

Gases that react with water, however, cannot be collected by displacement of water. Instead, air in a container is displaced by the gas. A gas that has a density greater than the density of air will displace air from the bottom to the top of an upright gas bottle. This process is called upward displacement. When the gas density is less than the density of air, the collecting bottle is inverted and the air is displaced downward as the gas fills the bottle from the top to the bottom. Figure 16.5 illustrates the upward and downward displacement of gas.

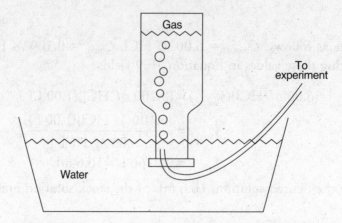

FIGURE 16.4. Diagram of a pneumatic trough.

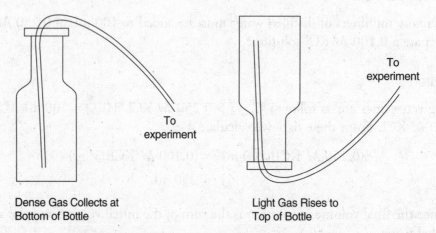

Dense Gas Collects at
Bottom of Bottle

Light Gas Rises to
Top of Bottle

**FIGURE 16.5. Experimental setups for (left) upward displacement of gas and
(right) downward displacement of gas.**

The density of a gas is directly proportional to its molar mass. The average molar mass of air, which is 80 percent nitrogen and 20 percent oxygen, is 29. Gases with molar masses greater than 29 are collected by upward displacement of air; gases with molar masses less than 29, by downward displacement of air.

SEPARATION TECHNIQUES

Precipitation

Solids can be precipitated by chemical reactions such as using Ba^{2+} to **precipitate** sulfate ions as barium sulfate:

$$Ba^{2+}(aq) + SO_4^{2-}(aq) \rightarrow BaSO_4(s) \qquad (16.10)$$

To obtain a pure product the barium solution should be added slowly with rapid stirring. This procedure keeps the barium ion concentration low and the formation of barium sulfate crystals is slowed, thereby preventing the barium sulfate from entrapping impurities during rapid crystal growth. Once precipitation is complete, the mixture is heated to coagulate the small crystals into larger crystals for easy filtration.

Filtration

Filtration is used to separate solid particles from a liquid. Filter paper is folded into a cone and inserted into a funnel. A few drops of water are placed on the filter paper to hold it in place, and then the solution to be filtered is added to the funnel. As liquid flows through the paper under the force of gravity, the solid is left in the filter paper. Precipitates are washed with dilute electrolyte solutions to remove contaminants. Suction or Buchner funnels may be used to speed the filtration process.

Filter paper comes in different grades. Coarse grades allow the liquid to flow faster since the pore size is large. Coarse filter paper cannot be used for fine precipitates, which will pass through the pores. Fine precipitates require filter paper with smaller pores and consequently take longer to filter. Fine precipitates are often heated in a process called digestion in order to form larger crystals, which are more easily filtered.

Centrifugation

For very small amounts of precipitate, **centrifugation** is the preferred method of separation from the solvent. A centrifuge spins samples at high speed, forcing solids to compact at the bottom of a test tube. Each sample-containing test tube inserted in a centrifuge must be balanced by another test tube containing the same amount of liquid. This keeps the centrifuge balanced so that it does not vibrate uncontrollably and perhaps "walk" off the bench.

The clear liquid remaining after centrifugation is called the **supernatant.** When the process is finished, the supernatant is decanted (poured carefully) from the precipitate to separate the liquid from the solid.

Distillation

Distillation is a separation technique whereby a substance with a high vapor pressure (low boiling point) is separated from other substances with lower vapor pressures (higher boiling points). Distillation is performed by boiling a solution and passing the vapor formed through a condenser to recover the vaporized liquid. When a mixture of methyl alcohol (b.p. = 65 °C) and ethyl alcohol (b.p. = 79 °C) is heated to boiling, the vapor formed contains mostly methyl alcohol. When this vapor is condensed, the condensate is enriched in methyl alcohol while the residual in the boiling flask is enriched in ethyl alcohol. Two volatile substances may be purified but not completely separated.

A mixture of a salt in water is an example of a mixture containing a volatile and a nonvolatile substance. This type of mixture allows complete separation of the volatile substance, as in the distillation of seawater to produce pure water.

INSTRUMENTAL TECHNIQUES

pH Determination

A **pH meter** or **pH paper** may be used to determine pH. A pH meter is used when a precise pH value is needed. pH paper serves to estimate the pH of a solution.

A pH meter, with a glass and reference electrode, is standardized with a standard buffer. Then the electrodes are rinsed to remove any buffer and immersed in the sample. The pH is read directly from the meter scale. More accurate pH measurements are made by standardizing the pH meter with two buffers, one with a pH slightly below that of the sample and the other with a pH slightly above that of the sample.

Litmus is a type of pH paper that tells only whether a solution is acid or basic. A clean stirring rod is dipped into the sample, and a drop is transferred to the **litmus paper.** A pink color indicates an acid solution, and a blue color a basic solution. Newer pH papers turn different colors depending on the pH. A drop of sample is placed on the pH paper with a stirring rod, and the resulting color is compared to a color chart to determine the approximate pH.

Spectroscopy

Colored solutions absorb visible light, and this absorption can be measured with a **spectrophotometer.** A spectrophotometer can be used to determine the spectrum of a compound or to determine the concentration of an unknown sample.

The **absorbance** of a sample is the quantity determined in a spectrophotometer. In the first step the wavelength of light to be used for the measurement is selected by adjusting the wavelength dial. The second step is to zero the meter with a **reagent blank,** which contains the same amounts of all reagents and solvents that were used to prepare the sample. In the last step the reagent blank is replaced with the actual sample and the meter is read. More samples can be measured as long as the wavelength is not changed. If the wavelength is changed, the instrument must be zeroed again with the reagent blank.

A spectrum is a graph of the light absorbed by a sample, that is, the absorbance, A, of the sample, versus the wavelength of light. Typically the absorbance of a sample is measured at 10-nm intervals from approximately 400 nm to 700 nm. An absorbance spectrum in the visible-wavelength region is shown in Figure 16.6.

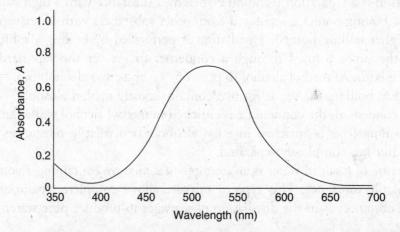

FIGURE 16.6. Absorbance spectrum for a compound that has its maximum absorbance at 515 nm. The sample has a violet color because green light is absorbed and all other wavelengths are observed by eye.

The concentration of a solute is determined at the wavelength of the largest peak in the spectrum. The absorbance of each solution of a series of concentration standards is determined, and a graph called a calibration curve is constructed. After the absorbance of the unknown sample is determined, the corresponding concentration is read from the calibration curve by drawing a horizontal line from the absorbance reading to the curve and then drawing a vertical line from the curve down to the concentration, as shown in Figure 16.7.

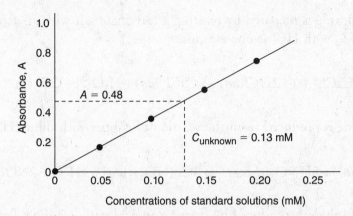

FIGURE 16.7. **Calibration curve used to determine concentrations. Dots represent standard concentrations used to prepare the curve. Dashed line shows that the unknown has an absorbance of 0.48 and that the corresponding concentration is 0.13 millimolar.**

The straight-line graph in Figure 16.7 is a consequence of **Beer's law,** which states that

$$\text{Absorbance (A)} = abc \tag{16.11}$$

In Beer's law a is a constant called the **absorptivity,** which is characteristic of the compound and the wavelength at which the absorbance is measured. The concentration is represented by c, and the thickness of the sample, called the **optical path length,** by b. At a given wavelength, a and b are constant and there is a direct proportionality between the absorbance and the sample concentration. The slope of the line in Figure 16.7 is ab.

EXPERIMENTAL REACTIONS

The laboratory experience is a twofold process. One goal is to become familiar with the techniques and procedures used by chemists. The second is to become familiar with useful chemical reactions.

Synthesis of Gases

Some gases that are synthesized and collected are O_2, H_2, CO_2, NO, H_2S, and NH_3.

Oxygen is prepared by decomposing $KClO_3$ in the presence of MnO_2 as a catalyst:

$$2KClO_3(s) \xrightarrow[\text{MnO}_2]{\text{heat}} 2KCl(s) + 3O_2(g) \tag{16.12}$$

Hydrogen is prepared by reacting an active metal, usually magnesium, with acid:

$$Mg(s) + 2HCl(aq) \rightarrow MgCl_2(aq) + 2H_2(g) \tag{16.13}$$

Carbon dioxide is prepared by reacting a carbonate salt with an acid. The reaction of $CaCO_3$ with HCl is one example:

$$CaCO_3(s) + 2HCl(aq) \rightarrow CaCl_2(aq) + H_2O(l) + CO_2(g) \tag{16.14}$$

Nitric oxide is produced from the reaction of copper with dilute HNO_3:

$$3Cu(s) + 8HNO_3(dil.aq) \rightarrow 3Cu(NO_3)_2(aq) + 4H_2O + 2NO(g) \tag{16.15}$$

Hydrogen sulfide is produced by reacting an acid with a sulfide, for example:

$$FeS(s) + 2HCl(aq) \rightarrow FeCl_2(aq) + H_2S(g) \tag{16.16}$$

Ammonia is produced from the reaction of an ammonium salt with a base:

$$NH_4Cl(aq) + NaOH(aq) \rightarrow NaCl(aq) + H_2O(l) + NH_3(g) \tag{16.17}$$

As stated in a preceding section, gases that do not dissolve in, or react with, water may be collected by displacement of water in a pneumatic trough. Other gases must be collected in the absence of water. Carbon dioxide is 1.5 times as dense as air and can be collected by upward displacement of air from a collecting bottle. Ammonia, which is much lighter than air, can be collected by downward displacement of air from an inverted collecting bottle.

Synthesis of Insoluble Salts

Insoluble salts are often precipitated in double-replacement reactions or in reactions of gases with soluble substances. Some common compounds prepared by double replacement are shown in the following net ionic equations:

$$Ag^+(aq) + Cl^-(aq) \rightarrow AgCl(s) \tag{16.18}$$

$$Ba^{2+}(aq) + SO_4^{2-}(aq) \rightarrow BaSO_4(s) \tag{16.19}$$

$$Fe^{2+}(aq) + SO^{2-}(aq) \rightarrow FeS(s) \tag{16.20}$$

In Equation 16.18 the chloride ion may be replaced by an iodide or a bromide ion. In Equation 16.20 Fe^{2+} can be replaced by almost any metal ion except the alkali metals.

Insoluble carbonates and sulfites are formed by bubbling $CO_2(g)$ and $SO_2(g)$ through solutions containing aqueous metal ions, for example:

$$Ca^{2+}(aq) + CO_2(g) + H_2O(l) \rightarrow CaCO_3(aq) + 2H^+(aq) \qquad (16.21)$$

$$Ni^{2+}(aq) + SO_2(g) + H_2O(l) \rightarrow NiSO_3(s) + 2H^+(aq) \qquad (16.22)$$

Preparation of Soluble Salts

Soluble salts are isolated from aqueous solution by removing water by evaporation with or without applying heat. To obtain a pure sample of salt, the solution must contain only the ions of that salt.

Magnesium nitrate can be prepared by reacting magnesium metal with HNO_3 until no more hydrogen is evolved. HNO_3 is the limiting reactant, and excess magnesium is removed by filtration. The filtrate can be boiled to dryness to recover $Mg(NO_3)_2(s)$:

$$Mg(s, \text{excess}) + HNO_3(aq) \rightarrow Mg(NO_3)_2(aq) + H_2(g) \qquad (16.23)$$

The preparation of calcium chloride uses the same reaction as the preparation of $CO_2(g)$ in Equation 16.14. Care is taken to avoid reacting all of the $CaCO_3(s)$ with HCl, and the reaction mixture is filtered to remove excess $CaCO_3(s)$. The solution that is left contains only $Ca^{2+}(aq)$ and $Cl^-(aq)$ ions, which can be boiled to dryness to recover the $CaCl_2(s)$.

Addition of an acid to an excess of an insoluble base anhydride such as $Fe_2O_3(s)$ can be used to produce soluble salts. Adding H_2SO_4 to an excess of $Fe_2O_3(s)$ will produce $Fe_2(SO_4)_3$:

$$Fe_2O_3(s) + 3H_2SO_4(aq) \rightarrow Fe_2(SO_4)_3(aq) + 3H_2O(l) \qquad (16.24)$$

Reacting a base with an excess of an acid anhydride can also be used to prepare a soluble salt, as in the reaction

$$2NaOH(aq) + O_2(g) \rightarrow Na_2SO_3(aq) + H_2O(l) \qquad (16.25)$$

Soluble salts can be prepared by neutralization reactions as long as neither an excess of acid nor an excess of base is used in the reaction. For instance, LiBr may be prepared by reacting exactly 1 mole of LiOH for every mole of HBr in the reaction

$$LiOH(aq) + HBr(aq) \rightarrow LiBr(aq) + H_2O(l) \qquad (16.26)$$

This method can be used to prepare three different sodium salts from phosphoric acid. NaH_2PO_4 is formed when 1 mole of NaOH is reacted with 1 mole of H_3PO_4. Two moles of NaOH will produce Na_2HPO_4, and 3 moles of NaOH per mole of H_3PO_4 produces Na_3PO_4.

Synthesis of Organic Compounds

A common synthetic organic reaction is the formation of esters by the acid-catalyzed reaction of an alcohol and an organic acid. The reaction of athanoic acid with 1-pentanol produces pentyl ethanoate, which has the odor of bananas:

$$CH_3COOH + CH_3CH_2CH_2CH_2CH_2OH \xrightarrow{H^+}$$
$$CH_3CO_2CH_2CH_2CH_2CH_2CH_3 + H_2O \quad (16.27)$$

Another common reaction is the formation of aspirin, acetylsalicylic acid, from salicylic acid and athanoic acid. Although the anhydride of athanoic acid is used in this reaction, the equation can be written as

QUALITATIVE ANALYSIS OF INORGANIC IONS

Qualitative analysis techniques are used to determine whether or not a sample contains a certain ion. In qualitative analysis an unknown and a reagent are mixed, and the result of the reaction allows us to draw a logical conclusion about the presence or absence of ions in the unknown. Many ions react in a similar manner; and although the addition of one reagent to an unknown may not identify the ion, it limits the possibilities. A sequence of reactions used to analyze a sample is called the qualitative analysis scheme—qual-scheme for short.

The qual-scheme uses selective precipitation to separate the ions in a sample into groups on the basis of their chemical characteristics listed below. When separation is by filtration, the liquid is called the **filtrate.** When the precipitate is separated by centrifugation, the liquid phase is called the **supernatant.**

Table 16.1 lists the main groups in the qual-scheme, starting with the unknown solution. The analysis starts with the sample, called the unknown, and group 1. Precipitates of the ions in each group are removed by filtration or centrifugation before the next group is precipitated.

TABLE 16.1

The Qualitative Analysis Scheme

Group Number	Solution Tested	Precipitating Agent	Precipitated Compounds
1	Unknown	0.1M HCl	$PbCl_2$, Hg_2Cl_2, AgCl
2	Filtrate or supernatant from group 1	H_2S at pH 1	HgS, PbS, CuS, CdS, Bi_2S_3, As_2S_3, Sb_2S_3, SnS_2
3	Filtrate or supernatant from group 2	H_2S at pH 10	MnS, FeS, NiS, CoS, ZnS, $Fe(OH)_3$, $Al(OH)_3$, $Cr(OH)_3$
4	Filtrate or supernatant from group 3	CO_3^{2-} at pH 10	$MgCO_3$, $CaCO_3$, $SrCO_3$, $BaCO_3$
5	Unknown	None	Soluble ions Na^+, K^+, NH_4^+

Separation of chlorides is done first since there are only a few insoluble chlorides, which make a convenient group. The carbonates are separated last since all of the metal ions would precipitate as carbonates, and thus too many ions would precipitate, if this step were done earlier in the scheme. Groups 2 and 3 are mainly sulfides. In acid solution only sulfides with very low K_{sp} values precipitate because the concentration of S^{2-} is low in acid solution. In basic solution the sulfide concentration is much higher, and the more soluble sulfides, with relatively high K_{sp} values, precipitate along with some hydroxides.

After each group has precipitated, additional separations and confirmation tests are run to identify the individual ions. For example, in group 1 the precipitate is washed with hot water to dissolve only $PbCl_2$, which is then confirmed by another precipitation with the chromate ion, CrO_4^{2-}. If a precipitate remains after washing with hot water, ammonia is added to dissolve the silver as the complex ion $Ag(NH_3)_2^+$. Neutralization of the ammonia solution with HCl reprecipitates AgCl and confirms its presence. Any precipitate that is still left when the Ag^+ is dissolved in ammonia is most likely Hg_2Cl_2. In the presence of ammonia Hg_2Cl_2 undergoes a redox reaction to form elemental mercury, causing the precipitate to turn dark gray in the reaction

$$Hg_2Cl_2(s) + 2NH_3(aq) \rightarrow Hg(l) + HgNH_2Cl(s) + NH_4^+(aq) + Cl^-(aq) \quad (16.29)$$

The reactions used and the decision-making process for qualitative analysis are often summarized in a flowchart, as shown in Figure 16.8.

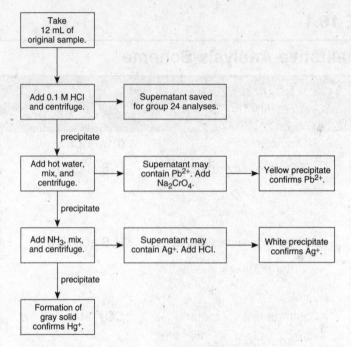

FIGURE 16.8. Flowchart for group 1 qualitative analysis of cations. Hg^+ is a dimer and is often written as Hg_2^{2+}.

Confirmation of the metal sulfides, hydroxides, and carbonates is done in a series of logical steps similar to that for the group 1 ions. A confirmation test is also performed for each metal. Some of these tests are listed in Table 16.2.

TABLE 16.2

Confirmation Spot Tests for Some Metal Ions

Cation	Test Reagent or Method	Observation
Fe^{2+}	$K_3Fe(CN)_6$	Dark blue precipitate
Fe^{3+}	$K_4Fe(CN)_6$	Dark blue precipitate
Cu^{2+}	NH_3	Dark blue solution
Ni^{2+}	Dimethylglyoxime	Red precipitate
Pb^{2+}	CrO_4^{2-}	Orange precipitate
Zn^{2+}	H_2S	White precipitate
NH_4^+	NaOH	Ammonia odor
H^+	Blue litmus paper	Paper turns red.
Na^+	Flame test	Orange flame
K^+	Flame test	Violet flame
Li^+	Flame test	Crimson red flame
Sr^{2+}	Flame test	Bright red flame

Analysis of anions is accomplished by spot tests using selected reagents. Some of these spot tests are listed in Table 16.3.

TABLE 16.3

Spot Tests for Selected Anions

Anion	Test Method	Observation
Carbonate	Add acid	Nonflammable gas evolved
Sulfide	Add acid	Rotten egg odor
Thiocyanate (SCN^-)	Add Fe^{3+}	Blood red complex
Sulfate	Add Ba^{2+}	White precipitate
Chloride	Add Ag^+	White precipitate
Bromide	Add Cl_2, water, and C_6H_{12}	Brown Br_2 in C_6H_{12} layer
Iodide	Add Cl_2, water, and C_6H_{12}	Violet I_2 in C_6H_{12} layer
Hydroxide	Red litmus paper	Paper turns blue.

CHEMICAL HAZARDS

Any human endeavor has some risk associated with it. It is much more likely, however, that you will be injured in an automobile accident or, if you smoke cigarettes, that you will develop cancer than that you will come to harm in a chemistry laboratory. While chemicals can be hazardous, one aim of chemistry education is to train students in the proper handling of these substances. Elimination of all hazards in any field of endeavor is unrealistic. The elimination of all potentially hazardous materials and techniques from the chemistry laboratory would eventually result in a society where no one understood how to handle or use chemicals properly. There are, however, certain hazards of which you should be aware so that you can take sensible precautions.

Highly Flammable Compounds

Organic compounds, particularly low-molar-mass compounds (methane, butane, etc.) and ethers, are highly flammable. They should be kept away from all sparks and flames.

Explosive Compounds

Over time, ethers react to produce explosive peroxides. Empty containers of solvents may contain explosive vapor residues. Dry picric acid is explosive. Nitrogen triiodide and nitroglycerine are shock-sensitive.

Strong Oxidizers

Concentrated perchloric acid, sulfuric acid, nitric acid, and hydrogen peroxide cause almost immediate skin injury. Perchloric acid in contact with organic material causes spontaneous combustion. White phosphorus burns spontaneously in air and emits hazardous fumes.

Compounds Incompatible with Water

The very active metals, Li, Na, K, Rb, Cs, Ca, Ba, and Sr, may react explosively with water.

Compounds with High Heats of Solution

Some substances evolve large amounts of heat when dissolved in water. If unexpected, the heat can cause spattering or can cause you to drop a hot beaker. The very active metals, their soluble oxides and hydroxides, and concentrated sulfuric and nitric acids generate heat. Calcium oxide and phosphorous pentoxide generate large amounts of heat with water. Mixing concentrated bases with concentrated acids is particularly hazardous.

Compounds with Possible Health Hazards

Benzene, chloroform, and carbon tetrachloride are suspected carcinogens and should be used only in a hood. Chlorinated organic compounds, in general, are suspected of being health hazards and should be handled carefully.

SAFETY PRINCIPLES AND EQUIPMENT

The first principle of safe experimentation in a laboratory is as follows: Never design an experiment that risks life or health. An essential part of experimental design is to assess possible safety hazards from all sources. Safety glasses are mandatory at all times; many major chemical corporations will fire a worker on the spot for failing to wear safety glasses. Protective lab clothing is worn when needed.

In experiments with hazardous materials the minimum quantities possible are used, and the work is done in a ventilated hood. When an explosion is even remotely possible, impact-resistant shields are used to protect workers. In a chemistry laboratory fire is always a possibility, and familiarity with exit routes and the location of fire extinguishers, fire blankets, and showers is essential.

SUMMARY

Laboratory work is at the core of chemistry. All the theoretical work must agree with experimental results because theories can be altered, but experimental data cannot. This chapter covers the basics of all experimental chemistry required for this course. You will draw from personal experience from labs performed but the techniques used are described in this chapter. These techniques are divided into data gathering, determination of physical properties, manipulating samples, separating and purifying products, and using simple instrumentation. Also reviewed in this chapter are reactions usually used in the laboratory along with a short description of the qualitative analysis scheme. Many calculations that you have done in the lab are also parts of other chapters in this book. Reviewing your own laboratory notebook to recall general methods, calculations, error analysis, and the purpose of your experiments will be helpful.

Important Concepts

Significant figures and calculations
Graphs
Meter reading
Common experimental reactions
Safety and chemical hazards

Practice Exercises

Multiple-Choice

For the first four problems below, one or more of the following responses will apply; each response may be used more than once or not at all in these questions.

I. sodium ion
II. silver ion
III. bromide ion
IV. sulfide ion
V. ammonium ion

1. This ion gives a bright-orange color in a flame test.

   (A) I
   (B) II
   (C) III
   (D) IV
   (E) V

2. This ion is usually oxidized before it is identified.

   (A) I
   (B) II
   (C) III
   (D) IV
   (E) V

3. Which ions can be identified by a characteristic odor?

   (A) I
   (B) II, and III
   (C) III
   (D) IV and V
   (E) V

4. Which ion is most commonly identified as a white precipitate that dissolves in ammonia?

   (A) I
   (B) II
   (C) III
   (D) IV
   (E) V

5. A 35.25-mL sample is needed. The best piece of glassware to use is

   (A) a buret
   (B) a beaker
   (C) a graduated cylinder
   (D) a volumetric flask
   (E) a pipet

6. Methane is collected by

   (A) upward displacement of air
   (B) displacement of water
   (C) downward displacement of air
   (D) displacement of mercury
   (E) filtration

7. A very fine precipitate is best isolated by

   (A) distillation
   (B) filtration
   (C) vacuum filtration
   (D) centrifugation
   (E) drying

8. The relative uncertainty in the answer to the calculation

   $$\frac{23.00 \times 1.68}{0.1416 \times 0.1332} \text{ is}$$

   (A) 10
   (B) 5×10^{-3}
   (C) 0.01
   (D) 0.0001
   (E) 5×10^{-2}

9. A 1.00 M solution of NaOH is prepared by weighing exactly 40.0 g of NaOH and adding it to exactly 1000 mL of distilled water at room temperature. Which of the following is most likely to be the largest source of error using the above procedure?

   (A) The wrong method is used to prepare the mixture.
   (B) The room temperature is not 25 °C.
   (C) The glassware is incorrectly calibrated.
   (D) NaOH absorbs atmospheric water.
   (E) Carbon dioxide in the water neutralizes some of the NaOH.

10. The vapor pressure of a liquid is measured at several temperatures. When a graph of the data is made,

    (A) temperature is the *x*-axis since it is the dependent variable
    (B) pressure is the *y*-axis since it is the dependent variable
    (C) $1/T$ is the *x*-axis because of Raoult's law
    (D) mole fraction is the *x*-axis and is the independent variable
    (E) the temperature must be in Kelvin units

11. Safety glasses are NOT needed for which of the following?

    (A) weighing samples
    (B) boiling water
    (C) distilling alcohol
    (D) prelab writeups
    (E) titrations

12. Which of the following is the LEAST useful operation when attempting to dissolve a solid?

    (A) adding the solid slowly to the solvent
    (B) grinding the solid to small particles
    (C) drying the solid
    (D) vigorous stirring or shaking
    (E) warming the solution

13. Which of the following dissolves in both acids and bases?

    (A) CaO
    (B) CO_2
    (C) $Al(OH)_3$
    (D) AgCl
    (E) As_2O_3

14. A 25.0-mL sample of a monoprotic acid was titrated to the end point with 20.0 mL of 0.200 M NaOH, and the molarity of the acid was calculated as 0.160 M. After the titration was complete, it was noticed that the buret was not clean, and droplets of solution were seen on the inside of the buret. What can be deduced about the molarity of the unknown?

    (A) The calculation is wrong, and the molarity is really 0.250 M.
    (B) The recorded volume of NaOH is low, and the molarity too high.
    (C) The recorded volume of NaOH is high, and the molarity is too high.
    (D) The recorded volume of NaOH is low, and the molarity is too low.
    (E) The recorded volume of NaOH is high, and the molarity is too low.

15. A sample is brought into a laboratory and mixed with an equal volume of a preservative solution. For analysis a 5.00-mL sample is diluted to 100 mL, and the concentration of chloride ions in the diluted solution is found to be 3.0×10^{-3} M. What is the chloride concentration of the sample?

    (A) 6.0×10^{-2} M
    (B) 6.0×10^{-3} M
    (C) 1.5×10^{-4} M
    (D) 7.5×10^{-5} M
    (E) 1.2×10^{-1} M

16. Qualitative analysis is performed on a colored solution. Addition of HCl results in a white precipitate that dissolves completely in hot water. The solution probably contained

    (A) a transition metal ion and Ag^+
    (B) Pb^{2+} and no other cations
    (C) Pb^{2+} and possibly an alkali metal ion
    (D) Hg_2^{2+} and a transition metal ion
    (E) Pb^{2+} and a transition metal ion

17. A solution is acidified, and a noticeable odor is observed. Which of the following is the most likely source of the odor?

    (A) CH_4
    (B) NH_3
    (C) CO_2
    (D) H_2S
    (E) H_2

18. Flame tests in which a small amount of solution is heated in a flame and the color of the flame is observed are routinely used to confirm the presence of which ion?

    (A) Ca
    (B) K
    (C) Na
    (D) Sr
    (E) all of these

19. The salt $CrCl_3$ may be prepared in pure form by

    (A) reacting an excess of Cr metal with HCl gas
    (B) reacting excess NaCl with Cr_2O_3
    (C) reacting 3 moles of HCl with 1 mole of $Cr(OH)_3$
    (D) reacting 1 mole of Na_2CrO_4 with 3 moles of HCl
    (E) All of the above can be used to make $CrCl_3$.

20. The most common method for determining the molarity of a solution of an acid is

    (A) gravimetric analysis (weighing a precipitate)
    (B) titration with a standard base
    (C) determination of the specific gravity of the acid
    (D) determination of the volume of gas evolved when the solution is reacted with Mg metal
    (E) determination of the pH of the acid

21. Hydrogen sulfide is used to precipitate all of the following EXCEPT

    (A) Cu^{2+}
    (B) Co^{2+}
    (C) Ca^{2+}
    (D) Fe^{3+}
    (E) Cd^{2+}

22. After a buret is filled for a titration, the bubble of air in the tip is not dislodged. What will be the effect?

    (A) No error will result if the bubble does not come out during the titration.
    (B) The volume recorded will be high if the bubble comes out.
    (C) The calculated molarity of the sample will be high if the bubble comes out.
    (D) The mass of sample will be low if the bubble comes out.
    (E) All of the above will be true.

23. If an error is made during an experiment, the appropriate action is to

    (A) stop the experiment and start over again
    (B) make a note of the error in the notebook and finish the experiment
    (C) tear the page(s) out of the notebook and start over again
    (D) adjust the results to correct for the error
    (E) continue the experiment; this is "experimental error"

24. Which of the following is appropriate when constructing a graph of experimental data?

    (A) A line is drawn connecting each data point.
    (B) The independent variable is plotted on the *y*-axis.
    (C) A straight line is always drawn through the data points.
    (D) The line is extended beyond the last data point to the edge of the graph.
    (E) The axes are scaled so that the data fill the graph as completely as possible.

ANSWER KEY

1. A	6. B	11. D	16. E	21. C
2. C	7. D	12. C	17. D	22. E
3. D	8. B	13. C	18. E	23. A
4. B	9. A	14. C	19. C	24. E
5. A	10. B	15. E	20. B	

See Appendix 1 for explanations of answers.

Free-Response

The following questions relate to laboratory work. All of the previous material in ths book may be of use in answering these questions.

(a) Explain whether the desired result will be too high, too low, or unaffected in the following situations. Use appropriate equations to support your conclusions.

    (i) You pipet 10.0 mL of monoprotic acid and titrate with standard base to calculate the molarity of the acid. You mistakenly blow the last drop of solution from the volumetric pipet each time.

    (ii) You determine the molar mass of a gas by determining its density but forget to correct the pressure for the vapor pressure of the water that the gas was collected over.

    (iii) In an experiment you dissolve some lead(II) nitrate but do not shake the volumetric flask well. What error results if you pipet from the top of the flask and what will happen if you pipet from the bottom of the flask?

    (iv) When you prepare a solid compound, you mistakenly wash the sample with hot water instead of ice water.

(b) Separation methods include decanting, filtering, distilling, and chromatography. Give an appropriate example of using each method.

(c) Describe how you can detect and eliminate determinate errors.

(d) Flame tests are used to identify some cations, particularly those that do not tend to form complexes or precipitates. List three elements that can be easily determined using a flame test. Explain why flame tests work based on your knowledge of the structure of the atom.

ANSWERS

(a)

    (i) Because you added a drop or two of extra acid, it will result in a concentration that is higher than it should be.

    (ii) $M = mRT/PV$. Because P is too large (the vapor pressure of water is always subtracted), the result will be smaller than it should be.

    (iii) If not well mixed, the heavy solute will stratify. The results pipeted from the top layers will have lead concentrations lower than expected. Solution pipeted from lower layers will be more concentrated than expected.

    (iv) Hot water generally dissolves substances better than cold water. It is expected that more product will be lost and the results will be low.

(b) Decanting is usually done with a large amount of solid that has large crystals, allowing liquid to be poured off easily without losing solid. Filtering can capture small particles and is preferred in many instances.

Distillation is best used to separate substances that have similar boiling points. If the boiling points are very far apart, other techniques may be better, and if they are too close to each other, chromatography may be needed. Chromatography is excellent for small samples that have very similar, but not identical, physical properties.

(c) Determinate errors are those that can be found and eliminated. Several ways to detect these errors include comparison with standards of known materials and comparison with other workers or labs with different instruments.

(d) Sodium (orange), potassium (violet), and copper (green) are three of the more common flame tests. There are other equally valid methods.

PRACTICE TESTS

On the following pages are three complete practice examinations. Try to duplicate actual test conditions. Follow the time limits and answer all questions as directed. The periodic table may be used for the entire test. The additional tables of data and equations may be used only in Section II, Part A.

Practice Test 1

ANSWER SHEET

1 Ⓐ Ⓑ Ⓒ Ⓓ Ⓔ
2 Ⓐ Ⓑ Ⓒ Ⓓ Ⓔ
3 Ⓐ Ⓑ Ⓒ Ⓓ Ⓔ
4 Ⓐ Ⓑ Ⓒ Ⓓ Ⓔ
5 Ⓐ Ⓑ Ⓒ Ⓓ Ⓔ
6 Ⓐ Ⓑ Ⓒ Ⓓ Ⓔ
7 Ⓐ Ⓑ Ⓒ Ⓓ Ⓔ
8 Ⓐ Ⓑ Ⓒ Ⓓ Ⓔ
9 Ⓐ Ⓑ Ⓒ Ⓓ Ⓔ
10 Ⓐ Ⓑ Ⓒ Ⓓ Ⓔ
11 Ⓐ Ⓑ Ⓒ Ⓓ Ⓔ
12 Ⓐ Ⓑ Ⓒ Ⓓ Ⓔ
13 Ⓐ Ⓑ Ⓒ Ⓓ Ⓔ
14 Ⓐ Ⓑ Ⓒ Ⓓ Ⓔ
15 Ⓐ Ⓑ Ⓒ Ⓓ Ⓔ
16 Ⓐ Ⓑ Ⓒ Ⓓ Ⓔ
17 Ⓐ Ⓑ Ⓒ Ⓓ Ⓔ
18 Ⓐ Ⓑ Ⓒ Ⓓ Ⓔ
19 Ⓐ Ⓑ Ⓒ Ⓓ Ⓔ
20 Ⓐ Ⓑ Ⓒ Ⓓ Ⓔ
21 Ⓐ Ⓑ Ⓒ Ⓓ Ⓔ
22 Ⓐ Ⓑ Ⓒ Ⓓ Ⓔ
23 Ⓐ Ⓑ Ⓒ Ⓓ Ⓔ
24 Ⓐ Ⓑ Ⓒ Ⓓ Ⓔ
25 Ⓐ Ⓑ Ⓒ Ⓓ Ⓔ

26 Ⓐ Ⓑ Ⓒ Ⓓ Ⓔ
27 Ⓐ Ⓑ Ⓒ Ⓓ Ⓔ
28 Ⓐ Ⓑ Ⓒ Ⓓ Ⓔ
29 Ⓐ Ⓑ Ⓒ Ⓓ Ⓔ
30 Ⓐ Ⓑ Ⓒ Ⓓ Ⓔ
31 Ⓐ Ⓑ Ⓒ Ⓓ Ⓔ
32 Ⓐ Ⓑ Ⓒ Ⓓ Ⓔ
33 Ⓐ Ⓑ Ⓒ Ⓓ Ⓔ
34 Ⓐ Ⓑ Ⓒ Ⓓ Ⓔ
35 Ⓐ Ⓑ Ⓒ Ⓓ Ⓔ
36 Ⓐ Ⓑ Ⓒ Ⓓ Ⓔ
37 Ⓐ Ⓑ Ⓒ Ⓓ Ⓔ
38 Ⓐ Ⓑ Ⓒ Ⓓ Ⓔ
39 Ⓐ Ⓑ Ⓒ Ⓓ Ⓔ
40 Ⓐ Ⓑ Ⓒ Ⓓ Ⓔ
41 Ⓐ Ⓑ Ⓒ Ⓓ Ⓔ
42 Ⓐ Ⓑ Ⓒ Ⓓ Ⓔ
43 Ⓐ Ⓑ Ⓒ Ⓓ Ⓔ
44 Ⓐ Ⓑ Ⓒ Ⓓ Ⓔ
45 Ⓐ Ⓑ Ⓒ Ⓓ Ⓔ
46 Ⓐ Ⓑ Ⓒ Ⓓ Ⓔ
47 Ⓐ Ⓑ Ⓒ Ⓓ Ⓔ
48 Ⓐ Ⓑ Ⓒ Ⓓ Ⓔ
49 Ⓐ Ⓑ Ⓒ Ⓓ Ⓔ
50 Ⓐ Ⓑ Ⓒ Ⓓ Ⓔ

51 Ⓐ Ⓑ Ⓒ Ⓓ Ⓔ
52 Ⓐ Ⓑ Ⓒ Ⓓ Ⓔ
53 Ⓐ Ⓑ Ⓒ Ⓓ Ⓔ
54 Ⓐ Ⓑ Ⓒ Ⓓ Ⓔ
55 Ⓐ Ⓑ Ⓒ Ⓓ Ⓔ
56 Ⓐ Ⓑ Ⓒ Ⓓ Ⓔ
57 Ⓐ Ⓑ Ⓒ Ⓓ Ⓔ
58 Ⓐ Ⓑ Ⓒ Ⓓ Ⓔ
59 Ⓐ Ⓑ Ⓒ Ⓓ Ⓔ
60 Ⓐ Ⓑ Ⓒ Ⓓ Ⓔ
61 Ⓐ Ⓑ Ⓒ Ⓓ Ⓔ
62 Ⓐ Ⓑ Ⓒ Ⓓ Ⓔ
63 Ⓐ Ⓑ Ⓒ Ⓓ Ⓔ
64 Ⓐ Ⓑ Ⓒ Ⓓ Ⓔ
65 Ⓐ Ⓑ Ⓒ Ⓓ Ⓔ
66 Ⓐ Ⓑ Ⓒ Ⓓ Ⓔ
67 Ⓐ Ⓑ Ⓒ Ⓓ Ⓔ
68 Ⓐ Ⓑ Ⓒ Ⓓ Ⓔ
69 Ⓐ Ⓑ Ⓒ Ⓓ Ⓔ
70 Ⓐ Ⓑ Ⓒ Ⓓ Ⓔ
71 Ⓐ Ⓑ Ⓒ Ⓓ Ⓔ
72 Ⓐ Ⓑ Ⓒ Ⓓ Ⓔ
73 Ⓐ Ⓑ Ⓒ Ⓓ Ⓔ
74 Ⓐ Ⓑ Ⓒ Ⓓ Ⓔ
75 Ⓐ Ⓑ Ⓒ Ⓓ Ⓔ

The Periodic Table of the Elements

1																	2
H																	**He**
1.0080																	4.0026
3	4											5	6	7	8	9	10
Li	**Be**											**B**	**C**	**N**	**O**	**F**	**Ne**
6.941	9.012											10.810	12.011	14.007	16.00	19.00	20.179
11	12											13	14	15	16	17	18
Na	**Mg**											**Al**	**Si**	**P**	**S**	**Cl**	**Ar**
22.99	24.31											26.98	28.09	30.974	32.06	35.453	39.948
19	20	21	22	23	24	25	26	27	28	29	30	31	32	33	34	35	36
K	**Ca**	**Sc**	**Ti**	**V**	**Cr**	**Mn**	**Fe**	**Co**	**Ni**	**Cu**	**Zn**	**Ga**	**Ge**	**As**	**Se**	**Br**	**Kr**
39.10	40.08	44.96	47.87	50.94	52.00	54.94	55.85	58.93	58.69	63.55	65.41	69.72	72.64	74.92	78.96	79.90	83.80
37	38	39	40	41	42	43	44	45	46	47	48	49	50	51	52	53	54
Rb	**Sr**	**Y**	**Zr**	**Nb**	**Mo**	**Tc**	**Ru**	**Rh**	**Pd**	**Ag**	**Cd**	**In**	**Sn**	**Sb**	**Te**	**I**	**Xe**
85.47	87.62	88.91	91.22	92.91	95.94	[98]	101.07	102.91	106.42	107.87	112.41	114.82	118.71	121.76	127.60	126.90	131.29
55	56	57	72	73	74	75	76	77	78	79	80	81	82	83	84	85	86
Cs	**Ba**	**La**	**Hf**	**Ta**	**W**	**Re**	**Os**	**Ir**	**Pt**	**Au**	**Hg**	**Tl**	**Pb**	**Bi**	**Po**	**At**	**Rn**
132.91	137.33	138.91	178.49	180.95	183.84	186.21	190.23	192.22	195.08	196.97	200.59	204.38	207.2	208.98	[209]	[210]	[222]
87	88	89	104	105	106	107	108	109	110	111	112		114		116		
Fr	**Ra**	**Ac**	**Rf**	**Db**	**Sg**	**Bh**	**Hs**	**Mt**	**Ds**	**Rg**	**Uub**		**Uuq**		**Uuh**		
[223]	226.03	227.03	[267]	[268]	[271]	[272]	[270]	[276]	[281]	[280]	[285]		[289]		[291]		

Lanthanides

58	59	60	61	62	63	64	65	66	67	68	69	70	71
Ce	**Pr**	**Nd**	**Pm**	**Sm**	**Eu**	**Gd**	**Tb**	**Dy**	**Ho**	**Er**	**Tm**	**Yb**	**Lu**
140.12	140.91	144.24	[145]	150.36	151.96	157.25	158.93	162.50	164.93	167.26	168.93	173.04	174.97

Actinides

90	91	92	93	94	95	96	97	98	99	100	101	102	103
Th	**Pa**	**U**	**Np**	**Pu**	**Am**	**Cm**	**Bk**	**Cf**	**Es**	**Fm**	**Md**	**No**	**Lr**
232.04	231.04	238.03	237.05	[244]	[243]	[247]	[247]	[251]	[252]	[257]	[258]	[259]	[262]

Chemistry

SECTION I

TIME: 1 HOUR AND 30 MINUTES

CALCULATORS MAY NOT BE USED WITH SECTION I

REMINDER: For all questions, assume that the temperature is 298 K, the pressure is 1.00 atmospheres of pressure, and solutions are aqueous solutions unless otherwise indicated.

Throughout this practice test the following symbols are used with the definitions provided below, unless otherwise noted in a given problem.

T = temperature

P = pressure

V = volume

S = entropy

M = molarity

m = molal

L, mL = liter(s), milliliter(s)

g = gram(s)

nm = nanometer(s)

H = enthalpy

G = free energy

R = molar gas constant

n = number of moles

atm = atmosphere(s)

J, kJ = joule(s), kilojoule(s)

V = volt(s)

mol = mole(s)

Practice Test 1

SECTION I—PART A

Multiple-Choice Questions

(Calculators May Not Be Used)

75 QUESTIONS

TIME: 90 MINUTES

50% OF TOTAL SCORE

Directions: Each of the lettered choices has a statement immediately following it. Selecting the letter is the same as selecting the statement as your answer. Select the letter that best fits each of the numbered questions that follow. A letter may be used once, more than once or not at all in a set of questions.

Questions 1–4 refer to the following elements.
- (A) Li
- (B) Cu
- (C) Na
- (D) Fe
- (E) Am

1. This Group 1A element gives a bright red color in a flame test.

2. This substance when reacted in most cases results in a 2+ cation. However, its 1+ cation is also well known.

3. When considering trends in the periodic table, this element would be predicted to have the largest atomic radius and also the largest ionic radius in this group.

4. All isotopes of this element are radioactive.

Questions 5–7 refer to the following laws.
- (A) ideal gas law
- (B) law of conservation of matter
- (C) equilibrium law
- (D) rate law
- (E) Hess' law

5. This law must be determined experimentally for each chemical reaction. Once determined, the chemist will know the order of the reaction.

6. If it is necessary to determine the molar mass of a substance, this is the one law that can be used to perform that calculation.

7. Balancing a chemical reaction is an expression of which of these laws?

Questions 8–10 refer to the following compounds.

    (A) sulfur dioxide
    (B) oxalic acid
    (C) phosphoric acid
    (D) carbonic acid
    (E) potassium nitrite

8. This substance is a gas that reacts as an acid. It is implicated in the erosion of marble statues.

9. The salts of this triprotic acid provide some of the buffering in blood and also provide flavor and tartness to cola-type soft drinks.

10. A solution that has a pH greater than 7 will result when this substance is dissolved in water.

Questions 11–14 refer to the following theories.

    (A) VSEPR theory
    (B) Kinetic molecular theory
    (C) Transition-state theory
    (D) Quantum theory
    (E) Atomic theory

11. The fact that electrons in chemical bonds and lone pairs of electrons repel each other is the basis of this theory.

12. Fundamental to this theory is the concept that the behavior of electrons in atoms is best described as waves.

13. Potential energy diagrams that show the activation energy and the enthalpy of reaction are used extensively to illustrate the basics of this theory.

14. This theory includes the concept that the average kinetic energy and temperature are directly related to each other.

SECTION I—PART B

(Calculators May Not Be Used)

Directions: Each of the questions or incomplete statements below is followed by five suggested answers or completions. Select the one that is best in each case and fill in the corresponding oval on the answer sheet.

15. When the volume of a container holding $N_2(g)$ is doubled, the pressure decreases by one-half. The best explanation of this observation is

    (A) the molecules hit the wall less frequently
    (B) Le Châtelier's principle
    (C) the molecules collide less frequently
    (D) molecular collisions become elastic
    (E) molecules are attracted to each other better

16. A certain compound has the name sodium dihydrogen phosphate trihydrate on the bottle. The molecular mass (to three significant figures) of this compound in $g \, mol^{-1}$ is

    (A) 164
    (B) 120
    (C) 110
    (D) 174
    (E) 138

17. When potassium dichromate, $K_2Cr_2O_7$, dissolves in water, it breaks apart into ions that produce an orange-colored solution. The correct representation of these ions is

    (A) no ions are formed because it is a nonelectrolyte
    (B) $2K^+ + 2Cr^{6+} + 7O^{2-}$
    (C) $K_2^{2+} + Cr_2O_7^{2-}$
    (D) $2K^+ + Cr_2O_7^{2-}$
    (E) $2K^+ + Cr_{2+} + O_7^{2-}$

18. Helium is placed in a tube with a pinhole, and it takes 36 minutes for 2.5×10^{-6} mol of the helium to effuse out of the tube. Sulfur dioxide is placed in the same tube under the same conditions. How long will it take for 2.5×10^{-6} mol of the sulfur dioxide to effuse out of the tube?

    (A) 9 minutes
    (B) 4.5 minutes
    (C) 2.4 hours
    (D) 4.8 hours
    (E) 9.6 hours

19. The solubility product of $Fe(OH)_3$ is 1.6×10^{-39}. Which mathematical expression best represents the molar solubility of iron(III) hydroxide?

    (A) $\sqrt{1.6 \times 10^{-39}}$

    (B) $\sqrt[3]{1.6 \times 10^{-39}}$

    (C) $\sqrt[4]{\dfrac{1.6 \times 10^{-39}}{9}}$

    (D) $\sqrt{27 \times 1.6 \times 10^{-39}}$

    (E) $\sqrt[4]{\dfrac{1.6 \times 10^{-39}}{27}}$

20. The geometry of a molecule with d^2sp^3 hybridization and two pairs of nonbonding electrons on the central atom is best described as

    (A) octahedral
    (B) tetrahedral
    (C) square planar
    (D) trigonal bipyramidal
    (E) trigonal planar

21. The pH of 0.01 molar acetic acid ($K_a = 1.8 \times 10^{-5}$) is closest to

    (A) 1
    (B) 2
    (C) 3
    (D) 7
    (E) 11

22. The net ionic equation for the reaction of barium carbonate with hydrochloric acid is

    (A) $Ba^{2+} + 2\ Cl^- \rightarrow BaCl_2$
    (B) $BaCO_3 + 2\ H^+ \rightarrow Ba^{2+} + H_2O + CO_2$
    (C) $BaCO_3 + 2\ H^+ + 2\ Cl^- \rightarrow$
        $BaCl_2 + H_2O + CO_2$
    (D) $Ba^{2+} + CO_3^{2-} + H^+ + Cl^- \rightarrow$
        $BaCl + H_2O + CO_2$
    (E) $BaCO_3 + 2\ HCl \rightarrow$
        $BaCl_2 + H_2O + CO_2$

23. A 0.475 M solution of $FeCl_3$ is prepared for a certain experiment by one group of lab workers. Another group would like to use the same solution but must know the solution's molality. What, if any, additional information is needed to convert the molarity to molality?

    (A) no additional information is needed
    (B) the density of the solution
    (C) the volume of the solution
    (D) the solubility product of KNO_3
    (E) the K_a of nitric acid

24. The geometry of the nitrate ion, NO_3^-, is best described as a

    (A) tetrahedron
    (B) trigonal bipyramid
    (C) linear structure
    (D) triangular planar structure
    (E) square pyramid

25. Which of the following lists the elements in order of decreasing atomic radius (largest radius first)?

    (A) K, Ca, C, F
    (B) F, O, Al, Na
    (C) Rb, Cs, Ba, Ra
    (D) Si, Ge, As, Sb
    (E) Li, Be, Na, C

26. Which of the following can be described as amphiprotic (amphoteric)?

    (A) $HClO$
    (B) H_2SO_4
    (C) PO_4^{3-}
    (D) HSO_3^-
    (E) ClO_4^-

Use the following responses for Questions 27–29.

    I. $HOCl$, $K_a = 3.0 \times 10^{-8}$
    II. $HC_2H_3O_2$, $K_a = 1.8 \times 10^{-5}$
    III. HNO_2, $K_a = 7.1 \times 10^{-4}$

27. A 0.010 M solution of which of the above acids will result in a solution with a pH between 2 and 3?

    (A) I only
    (B) II only
    (C) I and III
    (D) II and III
    (E) III only

28. Which substance(s) can be used to prepare a buffer with a pH of 7.0?

    (A) I and II
    (B) II and III
    (C) III only
    (D) I only
    (E) II only

29. If 0.010 *M* aqueous solutions of the salts of these substances are prepared, which will result in a solution with a pH between 6 and 8?

    (A) I only
    (B) III only
    (C) II only
    (D) I and
    (E) II and III

---

30. The net ionic equation for the precipitation of cobalt(II) nitrate with hydrogen sulfide gas is

    (A) $Co^{2+}(aq) + S^{2-}(aq) \rightarrow CoS(s)$
    (B) $Co(NO_3)_2(s) + H_2S(g) \rightarrow$
        $CoS(s) + 2 HNO_3(aq)$
    (C) $2 Co^{3+}(aq) + 3 S^{2-}(aq) \rightarrow Co_2S_3(s)$
    (D) $H_2S(aq) + Co^{2+}(aq) \rightarrow$
        $CoS(s) + 2 H^+(aq)$
    (E) $H_2S(g) + Co^{2+}(aq) \rightarrow CoS(s) + 2 H^+(aq)$

31. If 0.10 *M* solutions of each of the salts listed below are prepared, which will result in a solution with the highest pH?

    (A) $NaClO$
    (B) $NaClO_3$
    (C) $NaBrO_3$
    (D) $KClO_3$
    (E) $KClO_4$

32. Of the compounds below, which is expected to be the most highly colored?

    (A) $K_2Cr_2O_7$
    (B) KBr
    (C) NH_4NO_3
    (D) Na_2SO_4
    (E) CH_3CH_2COOH

33. Draw the Lewis structures for the following molecules or ions and determine which one is expected to have a bond order of 4/3.

    (A) PCl_5
    (B) NO_2^-
    (C) SO_3
    (D) NH_3
    (E) CO_2

34. A student prepares the following gases and collects them by displacement of water. Which compound will give the lowest percentage yield?

    (A) H_2
    (B) CO_2
    (C) O_2
    (D) NH_3
    (E) Cl_2

35. A radioactive isotope has a half-life, $t_{1/2}$, of 4.5 hours. What is the rate constant for this process?

    (A) $k = \ln(t_{1/2})$
    (B) $k = \dfrac{1}{t_{1/2}} \ln(2)$
    (C) $k = 0.5$
    (D) $k = \ln \dfrac{t_{1/2}}{2}$
    (E) $k = \dfrac{\text{rate}}{t_{1/2}}$

36. Which of the following will not necessarily yield more product in a chemical reaction?

    (A) increasing the temperature
    (B) increasing the amount of reactants
    (C) removing product as it is formed
    (D) decreasing the volume of a reaction where Δn_g is negative
    (E) increasing the volume of a reaction where Δn_g is positive

37. During a titration to determine the molarity of an unknown solution, a chemist has to look upward to read the initial volume of the standard solution in the buret and has to look downward to read the final volume. The result of the experiment

    (A) will be correct since the errors cancel
    (B) will be correct within experimental error
    (C) will be low because the volume measured is too high
    (D) will be low because the volume measured is too low
    (E) will be high because the volume measured is too low

38. Boiling a liquid, A, is represented as $A(l) \rightarrow A(g)$. Which of the following statements describes this process the best?

(A) $\Delta G°$ must be zero
(B) $\Delta S°$ must be positive
(C) $\Delta H°$ must be positive
(D) K_{eq} must be 1.00
(E) $\Delta S°$ must equal $\Delta H°$

39. Which of the following is expected to have the greatest solubility, in grams per 100 mL, in hexane?

(A) CH_3COOH
(B) H_2O
(C) $CH_3CH_2CH_2OH$
(D) $NaCl$
(E) $C_{12}H_{26}$

40. A 0.010 M solution of a weak base has a pH of 9.85. What is the approximate pK_b of this weak base?

(A) 4
(B) 10^{-5}
(C) 10^{-7}
(D) 8
(E) 6

41. Assuming that no ion pairing takes place, what is the expected boiling point of a 2.50 molal solution of Na_2SO_4? The boiling point elevation constant for water is 0.52 °C m^{-1}.

(A) 3.9 °C
(B) 1.3 °C
(C) 2.6 °C
(D) 103.9 °C
(E) 96.1 °C

42. Which of the following measurements cannot be used to determine the molecular mass of a chemical compound?

(A) osmotic pressure
(B) percent composition
(C) freezing-point depression
(D) vapor pressure
(E) gas density

43. A compound has a formula CaX_2. X may be

(A) Br
(B) O
(C) P
(D) N
(E) Ar

44. Using the incorrect name for a compound could be very hazardous in the lab. Which of the following compounds is named correctly?

(A) CaF_2, calcium fluorite
(B) $Mg(C_2H_3O_2)_2$, magnesium ethanoate
(C) $AlCl_3$, aluminum trichloride
(D) FeO, iron(III) oxide
(E) NO_2, nitrogen tetroxide

45. From the following list, the reaction of sodium metal with water is best classified as

(A) a double-replacement reaction
(B) a combustion reaction
(C) a neutralization reaction
(D) an oxidation reaction
(E) a single-replacement reaction

46. When equal masses of the following compounds are dissolved in water, which is expected to conduct electricity the best?

(A) $MgCl_2$
(B) $CH_3CH_2CH_2OH$
(C) SO_3
(D) $KMnO_4$
(E) CH_2O_2

47. Determine the standard cell voltage, $E°_{cell}$, for the reaction

$$Cu^{2+} + Sn^{2+} \rightarrow Cu(s) + Sn^{4+}$$

$$Cu^{2+} + 2\,e^- \rightarrow Cu \quad E° = +0.34 \text{ V}$$

$$Sn^{4+} + 2\,e^- \rightarrow Sn^{2+} \quad E° = +0.15 \text{ V}$$

(A) +0.49 V
(B) +0.19 V
(C) −0.49 V
(D) −0.19 V
(E) +0.051 V

48. An aqueous solution of calcium chloride will react with both reactants in which of the following pairs

    (A) silver nitrate and sodium bromide
    (B) silver sulfate and barium hydroxide
    (C) iodine and potassium permanganate
    (D) potassium nitrate and sodium phosphate
    (E) carbon dioxide and ethanoic acid

49. A colorless solution is formed when which one of the following is dissolved in water?

    (A) potassium chromate
    (B) nickel(II) nitrate
    (C) copper(II) sulfate
    (D) sodium ethanoate
    (E) manganese(II) sulfate

50. Rate the following oxo acids in terms of their relative strengths. Which is the strongest?

    (A) $HClO_2$
    (B) $HBrO_3$
    (C) $HClO_3$
    (D) H_2SO_3
    (E) H_2SeO_3

51. 50.0 mL of a 0.0200 M HCl solution is mixed with 25.0 mL of a 0.0100 M NaOH solution. What is the pH of the final mixture?

    (A) 3.36
    (B) 0.43
    (C) 2.00
    (D) 11.00
    (E) 7.00

52. Barium chloride is a hydrated salt. A 1.25-gram sample of the hydrate is heated in an oven until it does not lose additional mass. The anhydrous $BaCl_2$ has a mass of 1.06 g. What is the formula of the hydrate? The molar mass of $BaCl_2$ = 208.2.

    (A) $BaCl_2 \cdot H_2O$
    (B) $BaCl_2 \cdot 2H_2O$
    (C) $BaCl_2 \cdot 3H_2O$
    (D) $BaCl_2 \cdot 5H_2O$
    (E) $BaCl_2 \cdot 12H_2O$

53. The atomic mass of bromine listed in the periodic table is 79.90. Naturally occurring bromine is a mixture of only two isotopes, bromine-79 and bromine-81. What is the approximate percentage of bromine-81 atoms in natural bromine?

    (A) 10%
    (B) 45%
    (C) 25%
    (D) 65%
    (E) 75%

54. Computer disks, videotapes, and audiotapes use a common chemical, which is

    (A) rust, Fe_2O_3
    (B) Teflon
    (C) sand, SiO_2
    (D) diamond, C
    (E) aluminum, Al

55. Which of the following cannot be a complex ion?

    (A) $Fe(CN)_6^{3-}$
    (B) $Cu(NH_4)_4^{2+}$
    (C) $Co(Cl)_6^{4-}$
    (D) AlF_6^{3-}
    (E) $AgCl_2^-$

56. Below are K_b values for some weak bases. Which should be selected to prepare a buffer at pH 8.9?

    (A) 2.3×10^{-2}
    (B) 5.6×10^{-6}
    (C) 4.3×10^{-4}
    (D) 8.8×10^{-3}
    (E) 8.2×10^{-10}

57. A compound classified as an acid is

    (A) $(CH_3)_2CHCHOHCH_3$
    (B) $CH_3CH{=}O$
    (C) $CH_3OCH_2CH_3$
    (D) $CH_3COCH_2CH_3$
    (E) $CH_3CH_2CH_2COOH$

58. An allotrope of this nonmetallic element is electrically conductive

    (A) sulfur
    (B) oxygen
    (C) carbon
    (D) hydrogen
    (E) chlorine

59. Water is an unusual compound in that it

    (A) forms networks of hydrogen bonds and its solid is less dense than the liquid
    (B) has an unusually large difference between its melting and boiling points
    (C) dissolves a wide variety of other compounds
    (D) has a very high specific heat and heat of vaporization
    (E) all of the above

60. The equilibrium law for the following reaction is

    $$Na_2CO_3(s) + 2\ HCl(g) \rightleftharpoons H_2O(l) + CO_2(g)$$

    (A) $\dfrac{[HCl]^2}{[CO_2]}$

    (B) $\dfrac{[Na_2CO_3][HCl]^2}{[H_2O][CO_2]}$

    (C) $\dfrac{[CO_2]}{[H_2O]}$

    (D) $\dfrac{[H_2O][CO_2]}{[Na_2CO_3][HCl]^2}$

    (E) $\dfrac{[CO_2]}{[HCl]^2}$

61. A certain gas has a volume of 2.50 L and a pressure of 3.00 atmospheres. The pressure on the gas is reduced to 0.500 atmospheres, and the gas is allowed to expand isothermally. Determine the work involved in this process.

    (A) 31.25 liter atmospheres
    (B) 6.25 liter atmospheres
    (C) 37.5 liter atmospheres
    (D) −37.5 liter atmospheres
    (E) −6.25 liter atmospheres

62. The rate law for a reaction is found to be rate = $k[A]^2[B]$. Which of the following is true about this system?

    (A) a plot of log rate versus time will be a straight line
    (B) the units for the rate constant are $mol^2\ L^{-2}\ s^{-1}$
    (C) this reaction is unlikely because it implies the simultaneous collision of two atoms of A and one of B
    (D) it is unlikely that the first step of the mechanism is the rate-limiting step
    (E) all third-order reactions are endothermic

63. The valance bond theory would use which response to describe the interaction of orbitals to form pi bonds?

    (A) the overlap of two s orbitals
    (B) the side-to-side overlap of two p orbitals
    (C) the head-on overlap of two p orbitals
    (D) the combination of two pi orbitals
    (E) the combination of a p and s orbital

64. Which of the following is most probably an optically active organic compound?

    (A) ethanoic acid
    (B) *trans*-2-butene
    (C) *para*-dichlorobenzene
    (D) 2-bromo-2-chlorobutane
    (E) chloroform

65. Identify the compound below with dsp^3 hybridization.

(A) SF_6
(B) PCl_3
(C) $CHBr_3$
(D) I_3^-
(E) BCl_3

66. Determine the electron configuration for the following and determine which pair of ions are isoelectronic with each other.

(A) Ca^{2+} and Mg^{2+}
(B) Br^- and K^+
(C) Fe^{2+} and Cr^{3+}
(D) O^{2-} and F^-
(E) Ar and Ne

67. The periodic table can be used to estimate the relative sizes of the elements and ions. Using that information, predict which pair will have the greatest difference in radius.

(A) K and Na
(B) Cl^- and F^-
(C) O^{2-} and F^-
(D) Na^+ and Br^-
(E) Cl and Br

68. The energy needed to remove one or more electrons from an element can be determined. If we remove electrons one by one from the following elements, which one will most likely have the greatest difference between the second and third ionization energies?

(A) Fe
(B) Na
(C) Sr
(D) Cl
(E) Ar

69. Five grams of an element combines with 19.5 grams of chlorine. It forms an oxide with the formula M_2O_3. What is the element?

(A) Li
(B) Fe
(C) Al
(D) B
(E) Au

70. Phosphoric acid dissociates in three steps with the equilibrium constants

$$K_1 = 7.1 \times 10^{-3}$$

$$K_2 = 6.3 \times 10^{-8}$$

$$K_3 = 4.5 \times 10^{-13}$$

The pH of a 0.100 M solution of K_2HPO_4 is

(A) $-\log K_1$
(B) $-\log \sqrt{(7.1 \times 10^{-3})(6.3 \times 10^{-8})}$
(C) $-\log K_2$
(D) $-\log \sqrt{(K_2)(K_3)}$
(E) 7.00

71. The activation energy for the reverse reaction in the following diagram is

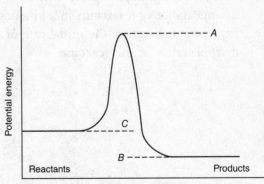

(A) $A - B$
(B) $B - C$
(C) $C - B$
(D) $C - A$
(E) $A - C$

72. Which of the following statements describes the chain of events as the system depicted in the phase diagram is cooled from 150 °C to 25 °C at a constant pressure of 1.00 atmosphere?

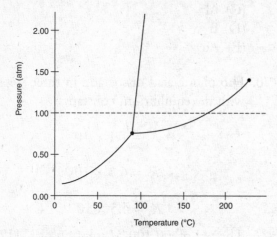

(A) The system starts in the gas phase, condenses to a liquid, and then solidifies.
(B) The system starts as a liquid and then crystallizes to a solid.
(C) The system starts as a gas and crystallizes to a solid.
(D) The system starts as a solid, melts to a liquid, and then vaporizes to a gas.
(E) The system starts as a solid and then sublimes.

73. The following graph represents the disappearance of a reactant in a kinetics experiment. Estimate the initial rate of disappearance of this reactant.

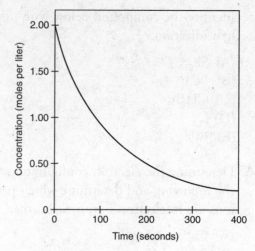

(A) 4×10^{-3} mol L^{-1} s^{-1}
(B) 2.00 mol
(C) 45 mol L^{-1} s^{-1}
(D) 0.022 mol L^{-1} s^{-1}
(E) 90 s

74. The periodic table can be used to get a rough estimate of the ionic nature of a chemical bond. Which of the following is the most ionic based on the positions of the atoms in the periodic table?

(A) Al_2O_3
(B) CCl_4
(C) N_2O
(D) CaF_2
(E) H_2O

75. Determine which of the following compounds contains two pi bonds and nine sigma bonds.

(A) $CH_3CHCHCH_3$
(B) HCN
(C) HCHO
(D) CH_3COOH
(E) $CH_2CHCHCH_2$

STOP

End of Section I.
If you finish before the 90 minutes have elapsed, you may check your work in this section. Do not go on to Section II until you are instructed to do so.

THE FOLLOWING TABLES MAY BE USED TO ANSWER THE QUESTIONS IN SECTION II ONLY.

Standard Reduction Potentials, 25 °C

Half-Reaction	E^0 (volts)	Half-Reaction	E^0 (volts)
$F_2(g) + 2e^- \rightarrow 2F^-$	2.87	$Co^{2+} + 2e^- \rightarrow Co$	−0.28
$Co^{3+} + e^- \rightarrow Co^{2+}$	1.82	$Tl^+ + e^- \rightarrow Tl$	−0.34
$Au^{3+} + 3e^- \rightarrow Au$	1.50	$Cd^{2+} + 2e^- \rightarrow Cd$	−0.40
$Cl_2(g) + 2e^- \rightarrow 2Cl^-$	1.36	$Cr^{3+} + e^- \rightarrow Cr^{2+}$	−0.41
$O_2(g) + 4H^+ + 4e^- \rightarrow 2H_2O$	1.23	$Fe^{2+} + 2e^- \rightarrow Fe$	−0.44
$Br_2(g) + 2e^- \rightarrow 2Br^-$	1.07	$Cr^{3+} + 3e^- \rightarrow Cr$	−0.74
$2Hg^{2+} + 2e^- \rightarrow Hg_2^{2+}$	0.92	$Zn^{2+} + 2e^- \rightarrow Zn$	−0.76
$Ag^+ + e^- \rightarrow Ag$	0.80	$2H_2O(l) + 2e^- \rightarrow H_2(g) + 2OH^-$	−0.83
$Hg_2^{2+} + 2e^- \rightarrow 2Hg$	0.79	$Mn^{2+} + 2e^- \rightarrow Mn$	−1.18
$Fe^{3+} + e^- \rightarrow Fe^{2+}$	0.77	$Al^{3+} + 3e^- \rightarrow Al$	−1.66
$I_2 + 2e^- \rightarrow 2I^-$	0.53	$Be^{2+} + 2e^- \rightarrow Be$	−1.70
$Cu^+ + e^- \rightarrow Cu$	0.52	$Mg^{2+} + 2e^- \rightarrow Mg$	−2.37
$Cu^{2+} + 2e^- \rightarrow Cu$	0.34	$Na^+ + e^- \rightarrow Na$	−2.71
$Cu^{2+} + e^- \rightarrow Cu^+$	0.15	$Ca^{2+} + 2e^- \rightarrow Ca$	−2.87
$Sn^{4+} + 2e^- \rightarrow Sn^{2+}$	0.15	$Sr^{2+} + 2e^- \rightarrow Sr$	−2.89
$S + 2H^+ + 2e^- \rightarrow H_2S$	0.14	$Ba^{2+} + 2e^- \rightarrow Ba$	−2.90
$2H^+ + 2e^- \rightarrow H_2$	0.00	$Rb^+ + e^- \rightarrow Rb$	−2.92
$Pb^{2+} + 2e^- \rightarrow Pb$	−0.13	$K^+ + e^- \rightarrow K$	−2.92
$Sn^{2+} + 2e^- \rightarrow Sn$	−0.14	$Cs^+ + e^- \rightarrow Cs$	−2.92
$Ni^{2+} + 2e^- \rightarrow Ni$	−0.25	$Li^+ + e^- \rightarrow Li$	−3.05

The Periodic Table of the Elements

1																	2
H																	**He**
1.0080																	4.0026
3	4											5	6	7	8	9	10
Li	**Be**											**B**	**C**	**N**	**O**	**F**	**Ne**
6.941	9.012											10.810	12.011	14.007	16.00	19.00	20.179
11	12											13	14	15	16	17	18
Na	**Mg**											**Al**	**Si**	**P**	**S**	**Cl**	**Ar**
22.99	24.31											26.98	28.09	30.974	32.06	35.453	39.948
19	20	21	22	23	24	25	26	27	28	29	30	31	32	33	34	35	36
K	**Ca**	**Sc**	**Ti**	**V**	**Cr**	**Mn**	**Fe**	**Co**	**Ni**	**Cu**	**Zn**	**Ga**	**Ge**	**As**	**Se**	**Br**	**Kr**
39.10	40.08	44.96	47.87	50.94	52.00	54.94	55.85	58.93	58.69	63.55	65.41	69.72	72.64	74.92	78.96	79.90	83.80
37	38	39	40	41	42	43	44	45	46	47	48	49	50	51	52	53	54
Rb	**Sr**	**Y**	**Zr**	**Nb**	**Mo**	**Tc**	**Ru**	**Rh**	**Pd**	**Ag**	**Cd**	**In**	**Sn**	**Sb**	**Te**	**I**	**Xe**
85.47	87.62	88.91	91.22	92.91	95.94	[98]	101.07	102.91	106.42	107.87	112.41	114.82	118.71	121.76	127.60	126.90	131.29
55	56	57	72	73	74	75	76	77	78	79	80	81	82	83	84	85	86
Cs	**Ba**	**La**	**Hf**	**Ta**	**W**	**Re**	**Os**	**Ir**	**Pt**	**Au**	**Hg**	**Tl**	**Pb**	**Bi**	**Po**	**At**	**Rn**
132.91	137.33	138.91	178.49	180.95	183.84	186.21	190.23	192.22	195.08	196.97	200.59	204.38	207.2	208.98	[209]	[210]	[222]
87	88	89	104	105	106	107	108	109	110	111	112		114		116		
Fr	**Ra**	**Ac**	**Rf**	**Db**	**Sg**	**Bh**	**Hs**	**Mt**	**Ds**	**Rg**	**Uub**		**Uuq**		**Uuh**		
[223]	226.03	227.03	[267]	[268]	[271]	[272]	[270]	[276]	[281]	[280]	[285]		[289]		[291]		

Lanthanides

58	59	60	61	62	63	64	65	66	67	68	69	70	71
Ce	**Pr**	**Nd**	**Pm**	**Sm**	**Eu**	**Gd**	**Tb**	**Dy**	**Ho**	**Er**	**Tm**	**Yb**	**Lu**
140.12	140.91	144.24	[145]	150.36	151.96	157.25	158.93	162.50	164.93	167.26	168.93	173.04	174.97

Actinides

90	91	92	93	94	95	96	97	98	99	100	101	102	103
Th	**Pa**	**U**	**Np**	**Pu**	**Am**	**Cm**	**Bk**	**Cf**	**Es**	**Fm**	**Md**	**No**	**Lr**
232.04	231.04	238.03	237.05	[244]	[243]	[247]	[247]	[251]	[252]	[257]	[258]	[259]	[262]

USEFUL INFORMATION FOR SOLVING FREE-RESPONSE PROBLEMS

Atomic Structure

E = energy
ν = frequency
$\lambda = \dfrac{h}{mv}$
$p = mv$
v = velocity
n = principal
$\quad$ quantum number
m = mass
$\Delta E = h\nu$
$c = \lambda\nu$

λ = wavelength
p = momentum
$E_n = \dfrac{-2.178 \times 10^{-18} \text{ joule}}{n^2}$

Speed of light, c $\quad$ = 3.00×10^8 m s^{-1}
Planck's constant, h $\quad$ = 6.63×10^{-34} J s
Boltzmann's constant, k = 1.38×10^{-23} J K^{-1}
Avogadro's number $\quad$ = 6.022×10^{23} mol^{-1}
Electron charge $\quad$ = -1.602×10^{-19} coulomb
1 electron volt/atom $\quad$ = 96.5×10^{23} K J mol^{-1}

Equilibrium and Thermochemistry

$K_a = \dfrac{[H^+][A^-]}{[HA]}$

$K_b = \dfrac{[OH^-][HB^+]}{[B]}$

$K_w = [H^+][OH^-]$
$K_w = 10^{-14}$ at 25 °C
$K_w = K_a K_b$
$\text{pH} = -\log[H^+]$
$\text{pOH} = -\log[OH^-]$
$14.00 = \text{pH} + \text{pOH}$ at 25 °C

$\text{pH} = pK_a + \log \dfrac{[A^-]}{[HA]}$

$\text{pOH} = pK_b + \log \dfrac{[HB^+]}{[B]}$

$pK_a = -\log K_a$
$pK_b = -\log K_b$
$K_p = K_c(RT)^{\Delta n}$
Δn = moles product gas − moles reactant gas
$\Delta S° = \Sigma S°$ products − $\Sigma S°$ reactants
$\Delta H° = \Sigma \Delta H_f°$ products − $\Sigma \Delta H_f°$ reactants
$\Delta G° = \Sigma \Delta G_f°$ products − $\Sigma \Delta G_f°$ reactants
$\Delta G° = \Delta H° - T\Delta S°$
$\quad = -RT \ln K$
$\quad = -n\mathscr{F}E°$

$\Delta G = \Delta G° + RT \ln Q$
$q = mc\,\Delta T$
$C_p = \dfrac{\Delta H}{\Delta T}$

K_a $\quad$ (weak acid)
K_b $\quad$ (weak base)
K_w $\quad$ (water)
K_p $\quad$ (gas pressure)
K_c $\quad$ (concentration)
$S°$ = standard entropy
$H°$ = standard enthalpy
$G°$ = standard free energy
T = temperature
n = moles
m = mass
q = heat
c = specific heat capacity
C_p = molar heat capacity at constant
$\quad$ pressure

1 faraday $\mathscr{F}$ = 96,500 coulombs/mole e^-

Gas constant, R = 8.31 J mol^{-1} K^{-1}
$\quad$ = 0.0821 L atm mol^{-1} K^{-1}
$\quad$ = 8.31 volt coulomb mol^{-1} K^{-1}

Gases, Liquids, Solutions, and Electrochemistry

$$PV = nRT$$

$$\left(P + \frac{n^2 a}{V^2}\right)(V - nb) = nRT$$

$$P_a = P_{total}X_a$$

$$X_a = \frac{mole_a}{total\ moles}$$

$$n = \frac{m}{M}$$

$$K = °C + 273$$

$$\frac{P_1 V_1}{T_1} = \frac{P_2 V_2}{T_2}$$

$$D = \frac{m}{V}$$

$$u_{rms} = \sqrt{\frac{3kT}{m}}$$

$$KE\ per\ molecule = \frac{1}{2}mv^2$$

$$KE\ per\ mole = \frac{3}{2}RTn$$

$$\frac{r_1}{r_2} = \sqrt{\frac{M_2}{M_1}}$$

$$molarity,\ M = \frac{moles\ of\ solute}{liters\ of\ solution}$$

$$molality,\ m = \frac{moles\ of\ solute}{kilograms\ of\ solvent}$$

$$\Delta T_f = iK_f m$$

$$\Delta T_b = iK_b m$$

$$\Pi = \frac{nRTi}{V}$$

$$Q = \frac{[C]^c [D]^d}{[A]^a [B]^b}$$

$$aA + bB \rightleftharpoons cC + dD$$

$$I = \frac{q}{t}$$

$$E_{cell} = E°_{cell} - \frac{RT}{n\mathcal{F}}\ln Q$$

$$\log K = \frac{nE°}{0.0592}$$

P = pressure

V = volume

T = temperature

n = number of moles

D = density

m = mass

v = velocity

u_{rms} = root mean square speed

KE = kinetic energy

r = rate of effusion

M = molar mass

Π = osmotic pressure

i = van't Hoff factor

K_f = molal freezing point depression constant

K_b = molal boiling point elevation

Q = reaction quotient

I = current (amperes)

q = charge (coulombs)

t = time (seconds)

$E°$ = standard reduction potential

K = equilibrium constant

Gas constant, R = 8.31 J mol⁻¹ K⁻¹

= 0.0821 L atm mol⁻¹ K⁻¹

= 8.31 volt coulomb mol⁻¹ K⁻¹

SECTION II
Free-Response Questions

TIME: 1 HOUR AND 35 MINUTES

50% OF TOTAL SCORE

PART A: TIME—55 MINUTES.

Calculators are allowed EXCEPT THOSE WITH QWERTY (TYPEWRITER STYLE) KEYBOARDS. Graphing and programmable calculators may be used. Calculator memories do not have to be cleared of programs and data. You must answer ALL questions; there are no choices.

PART B: TIME—40 MINUTES

USE OF CALCULATORS IS NOT PERMITTED. You must answer ALL questions; there are no choices.

GENERAL INSTRUCTIONS

Times for Parts A and B will be announced separately. You may answer questions within each part in any sequence desired. Divide your time appropriately between each problem.

The periodic table, reduction potentials, and lists containing equations and constants are printed on the preceding pages.

You may answer in pen or pencil. Be sure to write CLEARLY and LEGIBLY. If you make an error, you may save time by crossing it out rather than erasing it. On the actual exam, *write all of your answers in the space provided after each question.* For this test, provide your own paper—approximately 20 sheets.

When instructed, you may begin Part A. When the 55 minutes are up, you must stop work and put away your calculator as directed. DO NOT start work on Part B until instructed to do so. You will be allowed to use the periodic table and the tables of information provided with Part A.

SECTION II

Free-Response Questions

TOTAL TIME = 1 HOUR AND 35 MINUTES

PART A

TIME—55 MINUTES

YOU MAY USE YOUR CALCULATOR ON PART A ONLY

> **Directions:** CLEARLY SHOW THESE STEPS AND METHODS USED IN ARRIVING AT YOUR ANSWERS. It is to your advantage to do this to earn partial credit. Answers without the steps and methods may receive little or no credit. Pay attention to significant figures. On the actual exam, you will need to write all of your answers in the pages of the appropriate booklet. For this test, use a separate sheet of paper.

Answer Question 1 below. (This question is worth 20 percent of Section II.)

1. In reference to the equation below, $SO_3(g)$ is transferred to a 4.00-liter flask at 120 °C and 500 torr to start the experiment.

$$SO_2(g) + O_2(g) \rightleftarrows 2\,SO_3(g)$$

   (a) The flask and its contents are cooled to 25 °C. What is the initial molar concentration of $SO_3(g)$? What is the pressure at 25 °C within the flask?

   (b) The value of $\Delta G°$ for this reaction at 25 °C is −140 kJ. What is equilibrium law? What is the numerical value of the equilibrium constant?

   (c) When the reaction comes to equilibrium, what is the expected partial pressure of $SO_2(g)$?

   (d) After the reaction comes to equilibrium in part (c), an additional 1.00 g of $O_2(g)$ is added to the system. What is the expected partial pressure of $SO_2(g)$ now?

Answer Question 2 below. (This question is worth 20 percent of Section II.)

2. Methane is the main component of natural gas used by many for heating and cooking.

   (a) Write the complete reaction for the combustion of methane when there is a large excess of oxygen available.

   (b) Write a reaction for the incomplete combustion when the amount of oxygen is limited.

   (c) Calculate the volume of oxygen needed to combust each liter of CH_4. Determine the volume of air (20% oxygen) needed to combust a liter of methane.

   (d) Calculate the heat of formation, ΔH_f°, of methane given that the heat of combustion of methane is –885 kJ mol^{-1}. The heats of combustion of hydrogen and carbon are

$$2\ H_2(g) + O_2(g) \rightarrow 2\ H_2O(\ell) \quad \Delta H^\circ = -570.6\ kJ$$

$$C(s) + O_2(g) \rightarrow CO_2(g) \quad \Delta H^\circ = -393.5\ kJ$$

   (e) For the combustion of methane, is entropy expected to increase, decrease, or remain the same? Explain your answer.

Answer Question 3 below. (This question is worth 20 percent of Section II.)

3. An organic substance is synthesized and purified. It is known to contain carbon hydrogen, and oxygen. A 0.500-gram sample is burned in excess oxygen with a copper catalyst to ensure complete combustion. The gases produced pass through a drying tube containing calcium chloride. A second tube contains sodium hydroxide that absorbs the carbon dioxide. The drying tube increases in mass by 0.385 g, and the second tube increases by 1.256 g.

   (a) What is the percentage of carbon in the sample?

   (b) What is the percentage of hydrogen in the sample?

   (c) What is the percentage of oxygen in the sample?

   (d) What is the empirical formula of the compound?

   (e) The conduction of electricity is not increased when the compound is dissolved in water. This compound can be classified as a(n) _____. Explain your answer.

   (f) 15 grams of the compound dissolved in 100 mL of water results in a freezing-point depression of 4.00 °C. What is the molecular formula of the compound?

   (g) There are many ways in which these atoms can be arranged. Draw two of them and show how many sigma and pi bonds are in each structure.

PART B

TIME—40 MINUTES

(Calculators May Not Be Used)

Answer Question 4 below. (This question is worth 5 percent of your total score.)

4. For each of the following three reactions, in part (i) write a BALANCED equation and in part (ii) answer the question about the reaction. In part (i), coefficients should be in terms of lowest whole numbers. Assume that solutions are aqueous unless otherwise indicated. Represent substances in solutions as ions if the substances are extensively ionized. Omit formulas for any ions or molecules that are unchanged by the reaction.

<u>Example:</u>   (i) A strip of magnesium is added to a solution of silver nitrate.

<u>Answer:</u> $Mg + 2 Ag^+ \rightarrow Mg^{2+} + 2 Ag$

   (ii) Which substance is oxidized in the reaction?

<u>Answer:</u>   Magnesium (Mg) metal

(a) (i) A bar of iron metal is immersed in a solution of copper(II) nitrate.
    (ii) What will be observed when this reaction occurs?
(b) (i) Liquid pentane is burned in an excess of oxygen gas.
    (ii) How will the products change if the amount of available oxygen is decreased?
(c) (i) Sulfur trioxide gas is bubbled into limewater, a saturated solution of calcium hydroxide.
    (ii) How does this explain the deterioration of many outdoor statues and monuments?

Directions: TWO essay questions below will count for **15 percent of THE ENTIRE TEST (7.5% EACH)**. You **MUST ANSWER** Questions 5 and 6. Answers to these questions should demonstrate your ability to present logical, coherent, and convincing explanations of chemical facts and observations. Answers are judged on the accuracy of your analysis and on the appropriateness of the details and examples cited. Specific answers and examples are preferred to broad generalizations. Diagrams, illustrations, and equations may be used in the answers.

5. A student performs an experiment to verify the value of R, the gas law constant. The experiment involves the reaction of magnesium metal with hydrochloric acid and collecting a gas.

   (a) Write the balanced chemical equation for the reaction that will be used.

   (b) Draw a diagram of the apparatus and list the equipment needed. Is any tabulated data needed?

   (c) Assuming that the collection device can hold a maximum of 50 mL of gas, set up the calculation for the maximum amount of magnesium metal to be used.

   (d) What considerations go into selecting the volume and molarity of the hydrochloric acid used?

   (e) Give the equation that will be used to analyze the data.

   (f) From your experience with the measurements involved, which measurement will limit the number of significant figures in your answer? Explain.

6. The dichromate ion in acid solution will oxidize ethanol to ethanoic acid. The skeleton reaction is

$$Cr_2O_7^{2-} + CH_3CH_2OH \rightarrow Cr^{3+} + CH_3COOH$$

   (a) Balance the equation.

   (b) Draw and fully label the diagram of a galvanic cell that can be used to study this reaction.

   (c) Describe the accepted standard state for electrochemical measurements.

   (d) If you set up your experiment and make accurate measurements of the cell voltage, what thermodynamic values can be calculated? Use equations to show their relationship to your measurements.

   (e) Do you expect the entropy change for this reaction to be positive, negative, or zero? Explain your answer.

   (f) What will happen if the acid is forgotten and the reaction occurs in a neutral or alkaline solution?

End of Section II.
You may use any remaining time to check your work in this section.

Answer Key
PRACTICE TEST 1

1. A	26. B	51. C
2. B	27. E	52. B
3. E	28. D	53. B
4. E	29. B	54. A
5. D	30. E	55. B
6. A	31. A	56. B
7. B	32. A	57. E
8. A	33. C	58. C
9. C	34. D	59. E
10. E	35. C	60. E
11. A	36. A	61. E
12. D	37. D	62. D
13. C	38. E	63. B
14. B	39. E	64. D
15. C	40. E	65. D
16. D	41. D	66. D
17. D	42. B	67. D
18. C	43. A	68. C
19. E	44. B	69. C
20. C	45. E	70. D
21. C	46. A	71. A
22. B	47. B	72. B
23. B	48. B	73. D
24. D	49. D	74. D
25. A	50. C	75. E

CHAPTER REFERENCES FOR MULTIPLE-CHOICE QUESTIONS

Question	Chapter	Question	Chapter	Question	Chapter
1	1, 16	26	4, 14	51	6, 8, 14
2	2, 4	27	10, 14	52	4, 6
3	2	28	14	53	7, 8, 15
4	2, 3	29	5, 14	54	4, 16
5	11	30	4	55	4, 14
6	7, 16	31	4, 14	56	10, 14
7	1, 6	32	14	57	15
8	4, 5, 7, 14	33	5	58	2, 8
9	10, 14	34	6, 7, 16	59	5, 8
10	4, 14	35	3, 11	60	10, 14
11	5	36	6, 10, 12	61	7, 12
12	1, 2, 5	37	14, 16	62	11
13	7, 11	38	10, 12	63	5
14	7, 11	39	8, 9	64	15
15	7, 8	40	10, 14	65	5
16	2, 4	41	4, 9	66	2, 4
17	4	42	6, 7, 9	67	2, 4
18	5, 7	43	2, 4	68	2
19	4, 10	44	2, 4	69	4, 6
20	5	45	4, 13	70	4, 14
21	14	46	4, 9	71	11, 12
22	4	47	13	72	8
23	4, 8, 9	48	4	73	11
24	5	49	4, 16	74	2, 4
25	2	50	5, 14	75	5, 15

ANSWERS EXPLAINED

Section I—Multiple-Choice

1. **(A)** Lithium will give a red color. Cu is green, and Na is orange.

2. **(B)** Copper is the only one capable of these two oxidation states.

3. **(E)** Americium, Am, is in period 7; the next largest is strontium.

4. **(E)** Americium is the radioactive isotope often used in smoke detectors.

5. **(D)** The exponents determine the order of reaction.

6. **(A)** A measured gas density can be converted into the molar mass.

7. **(B)** The law of conservation of matter is expressed in a balanced chemical equation.

8. **(A)** Sulfur dioxide dissolves in water to form sulfurous acid.

9. **(C)** Phosphoric acid is used for both purposes.

10. **(E)** The remaining substances are acids or acid anhydrides. The nitrite ion hydrolyzes water to form nitrous acid and leaves the OH^- ion.

11. **(A)** Molecular shapes are defined by electron pair repulsion, so they are as far apart as possible.

12. **(D)** Quantum theory is based on Schrödinger's wave mechanics.

13. **(C)** The activated complex is the structure of the reactants as they are converted into products in the transition-state theory.

14. **(B)** It states that the average kinetic energy is proportional to temperature.

15. **(C)** Fewer collisions indicate less interaction between NH_3 molecules. Decreasing the interactions brings the gas closer to the ideal state, defined as having no interactions at all.

16. **(D)** The formula for this compound is $NaH_2PO_4 \cdot 3H_2O$. Adding the atomic masses and rounding to three significant figures gives 174.

17. **(D)** Only potassium, K^+, and dichromate ions are formed.

18. **(C)** SO_2 is 16 times heavier than He, and Graham's law predicts it will take four times as long for SO_2 to effuse. This is 144 minutes or 2.4 hr.

19. **(E)** $K_{sp} = [Fe^{3+}][OH^-]^3 = s(3s)^3 = 27s^4$, where s is the molar solubility. Rearrangement gives response (E).

20. **(C)** The two nonbonding pairs of electrons are located on opposite sides of the central atom, and the remaining atoms are in a square planar shape.

21. **(C)** $H^+ = \sqrt{K_a C_a} = \sqrt{0.01 \times 1.8 \times 10^{-5}} = \sqrt{1.8 \times 10^{-7}}$. This is approximately equal to $10^{-3.5}$, and the pH will be approximately 3.5.

22. **(B)** H_2CO_3 breaks apart into H_2O and CO_2.

23. **(B)** Molality is a temperature-independent concentration unit, and molarity is temperature-dependent. Density is a required conversion factor when switching between these two types of concentration units.

24. **(D)** The nitrate ion is a resonance structure that is trigonal planar.

25. **(A)** The radii decrease from left to right across the table, as well as from the bottom to the top of the table.

26. **(B)** The bisulfite ion is the only one that can donate a proton and also accept a proton in aqueous solution.

27. **(E)** If we multiply each K_a by 0.010 and then take the square root, the result will be the concentration of hydronium (hydrogen) ions. The hydronium ion concentrations are I. 1.7×10^{-5}, II. 4.2×10^{-4}, and III. 2.7×10^{-3}. We also know that the pH of a solution will, at most, be the positive value of the exponent of the hydrogen ion concentration. At its smallest, the pH will be one less than that. Therefore, we can predict that the pH of solution I will range from 4 to 5. The pH of solution II will range from 3 to 4 and that of solution III will range from 2 to 3. Only HNO_2 will result in a solution with a pH between 2 and 3.

28. **(D)** We need to choose the acid with a pK_a within 1 pH unit of the desired pH. Therefore, we need an acid with a pK_a between 6 and 8 or a K_a between 10^{-6} and 10^{-8}. Only HOCl has a pK_a that fits the requirements.

29. **(B)** We calculate the K_b for the salts as K_w/K_b. The three K_b values are I. $K_b = 3.3 \times 10^{-7}$, II. $K_b = 5 \times 10^{-11}$, and III. $K_b = 1.5 \times 10^{-11}$. If we multiply each of these K_b values by the 0.010 M concentrations of each acid and then take the square root, we get the [OH$^-$] concentration. These are I. 5×10^{-5}, II. 2×10^{-6}, and III. 4×10^{-7}. We can estimate the pOH values of the three solutions as I. between 4 and 5, II. between 5 and 6, and III. between 6 and 7. The pH of the solutions then must be I. between 9 and 10, II. between 8 and 9, and III. between 7 and 8. It is apparent that only solution III is within one pH unit of pH 7.

30. **(E)** The equation must represent the conditions given in the problem, and no other response has H_2S as a gas.

31. **(A)** The strongest conjugate base will be obtained from the weakest conjugate acid. Replacing the cations with hydrogens allows us to determine the acid strength and correspondingly the base strength. The acids in order of strength are $HClO_4 \gg HClO_3 > HBrO_3 > HClO$.

32. **(A)** Salts containing transition elements are often colored. Only $K_2Cr_2O_7$ contains a transition element.

33. **(C)** SO_3 has three resonance structures, while NO_2^- has two resonance structures and all the others have only one reasonable structure. The three resonance structures all have three sigma bonds and one pi bond for the three oxygen atoms. This calculates to be 4/3 for the bond order.

34. **(D)** Ammonia is very soluble in water, and very little, if any, gas will be collected. CO_2 is somewhat soluble, and not all CO_2 produced will be collected as a gas.

35. **(C)** The equation is $\ln(2) = kt$. Converting 4.5 hr to 1.62×10^4 s and calculating k results in

$$k = \frac{\ln(2)}{t 1/2} = \frac{\ln(2)}{1.62 \times 10^4 \text{ s}} = 4.3 \times 10^{-5} \text{ s}^{-1}$$

36. **(A)** The effect of temperature depends on whether the reaction is endothermic or exothermic.

37. **(D)** Looking upward, the volume observed is too large, and looking downward, the volume is too small; the net result is a total volume smaller than actually used. This results in a low molarity of the unknown.

38. **(E)** $T\Delta S° = \Delta H°$, not $\Delta S° = \Delta H°$.

39. **(E)** This is the least polar compound and will be most soluble in the nonpolar C_6H_{14}.

40. **(E)** This response is closest to the actual value. For a weak base $[OH^-] = \sqrt{K_b C_b}$. The pOH = 14.00 − 9.85 = 4.15. The $[OH^-] = 10^{-4.15}$. Solving the equation yields (Note that a calculator is not needed.)

$$K_b = \frac{[OH^-]^2}{C_b} = \frac{(10^{-4.15})^2}{0.01} = 10^{-6.30}$$

$$pK_b = 6.30$$

41. **(D)** Na_2SO_4 forms three ions, and the molality of all ions is 7.50. The ΔT is

$$\Delta T = mk_b = 7.50 \ m(0.52 \text{ °C } m^{-1}) = 3.9 \text{ °C}$$

The boiling point is 100 °C + 3.9 °C = 103.9 °C.

42. **(B)** Percent composition can determine empirical formulas but not molar quantities

43. **(A)** A halogen such as Br is the only possible anion in this case.

44. **(B)** Magnesium ethanoate is correctly named.

45. **(E)** $2\ Na + 2\ H_2O \rightarrow 2\ NaOH + H_2(g)$ sodium replaces hydrogen in water; the reaction is a single replacement reaction. D is incorrect because although Na does get oxidized, H^+ is also reduced it is a redox reaction, not just oxidation.

46. **(A)** $MgCl_2$ will produce 3 moles of ions for each mole of compound, resulting in the best conducting solution.

47. **(B)** $E^{\circ}_{cell} = E^{\circ}_{red} - E^{\circ}_{ox} = 0.34 - 0.15 = +0.19$ volt.

48. **(B)** Silver chloride will be insoluble, and calcium hydroxide is insoluble.

49. **(D)** The chromate ion is yellow in its compounds. Soluble salts of the transition metals generally have colored solutions.

50. **(C)** $HClO_3 > HBrO_3$ because of electronegativity of Cl versus Br. $HClO_3 > H_2SO_3 > H_2SeO_3$ because $HClO_3$ has more double-bonded oxygen atoms. $HClO_3 > HClO_2$ for the same reason.

51. **(C)** $(0.050\text{ L HCl})\left(\dfrac{0.0200\text{ mol HCl}}{1\text{ L HCl}}\right) = 0.00100\text{ mol HCl}$

$(0.025\text{ L NaOH})\left(\dfrac{0.0100\text{ mol NaOH}}{1\text{ L NaOH}}\right) = 0.00025\text{ mol NaOH}$

NaOH is the limiting reactant and 0.00075 mol HCl is left over.

$M_{HCl} = \dfrac{0.00075\text{ mol HCl}}{0.075\text{ L HCl}} = 1.00 \times 10^{-2}\ M$ and the pH = 2.00

52. **(B)** ? moles $BaCl_2 = 1.06\text{ g }BaCl_2\left(\dfrac{1\text{ mol }BaCl_2}{208.2\text{ g }BaCl_2}\right)$ = approximately 5×10^{-3} mol $BaCl_2$

? moles $H_2O = (1.25\text{ g} - 1.06\text{ g})\ H_2O\left(\dfrac{1\text{ mol }H_2O}{18\text{ g }H_2O}\right)$ = approximately 10^{-2} mol H_2O

? $\dfrac{\text{mol }H_2O}{\text{mol }BaCl_2} = \dfrac{10^{-2}\text{ mol }H_2O}{5\times10^{-3}\text{ mol }BaCl_2} = 2$ mol H_2O per mole of $BaCl_2$

53. **(B)** We can calculate the percentage of Br-81 in natural bromine as

$$\frac{\%\,^{81}Br}{100}(81) + \frac{100 - \%\,^{81}Br}{100}(79) = 79.90$$

$$\%\,^{81}Br(81) + 7900 - \%\,^{81}Br(79) = 79.90 \times 100$$

$$\%\,^{81}Br(81 - 79) = 7990 - 7900$$

$$\%\,^{81}Br = 90/2 = 45\%$$

54. **(A)** Very pure iron(III) oxide is the magnetic media used in these recording devices.

55. **(B)** This formula makes no sense, the charges do not add up and the ammonium ion cannot be a ligand because all of its valence electrons are used in bonds with hydrogen.

56. **(B)** The pK_b must be within ±1 of the pOH of the desired buffer. The pOH is 14 – 8.9 = 5.1, and we can estimate that the pK of base b is between 5 and 6 and will fulfill the requirements.

57. **(E)** It has the –COOH functional group.

58. **(C)** The allotrope is graphite.

59. **(E)** No other solvent has all of these properties.

60. **(E)** Solids, Na_2CO_3, and pure liquids, H_2O, do not appear in the equilibrium laws.

61. **(E)** The gas expands when pressure is reduced; therefore work is done by the system, and the sign will be negative. Work = $-P\Delta V$ = –(0.5 atm)(15 L – 2.5 L) = – 6.25 L atm.

62. **(D)** If the first step of a mechanism is rate-limiting, it gives the rate law directly. Because the rate law is third order and a three-body collision is highly unlikely, it is therefore unlikely that the first step is rate-limiting.

63. **(B)** The sidewise overlap of p orbitals is used to illustrate the location of electron density in pi bonds.

64. **(D)** This compound has a carbon atom with four different groups bonded to it:

$$CH_3CH_2 - \overset{\overset{\displaystyle Cl}{|}}{\underset{\underset{\displaystyle Br}{|}}{C}} - CH_3$$

65. **(D)** I_3^- has a Lewis structure that requires five pairs of electrons on the central iodine atom.

66. **(D)** Oxygen and fluoride ions are both isoelectronic with argon.

67. **(D)** Sodium cations are about half their atomic radius of the metal, and bromide anions are almost twice the radius of bromine. This results in the largest difference in this group.

68. **(C)** Strontium forms a +2 ion indicating low energy needed to remove the first two electrons and much more energy for removal of the third.

69. **(C)** The chloride must be MCl_3. To estimate the answer we see that 19.5 g is approximately ½ mole of Cl. From the formula we conclude that 5 gram of the metal must be ⅙ mole. Then 1 mole of the metal must be approximately 30 g. Only Al will fit the observed data.

70. **(D)** For an amphiprotic acid ion, the pH of the solution is the average of the two pK values that the ion participates in. The average of pK's is the same as the negative logarithm of the square root of the product of K_2K_3.

71. **(A)** This is the height of the energy barrier when moving from right to left in the diagram

72. **(B)** At 150 °C and 1 atm. pressure, the system is a liquid. At 25 °C and the same pressure, it is a solid.

73. **(D)** A tangent to the start of the reaction should intersect the x-axis at approximately 90 to 120 seconds. The rate is in the range of 2/90 to 2/120 mol/L s. To estimate the answer we note that 2/100 = 0.02 and is within the range of possible values. Response d is the only one close to this that also has the correct units.

74. **(D)** The two elements have the largest difference in electronegativity as estimated by their separation in the periodic table.

75. **(E)** This is 1,3-butadiene, CH_2=CHCH=CH_2. There are six sigma C—H bonds and three sigma C—C bonds. Each double bond contains 1 pi bond for a total of 2.

Section II—Free-Response

Part A

1. In reference to the unbalanced equation below, SO_3 is transferred to a 4.00-liter flask at 120 °C and 500 torr to start the experiment.

$$SO_2(g) + O_2(g) \rightleftarrows 2\,SO_3$$

(a) We use the ideal gas law equation to calculate the moles of sample present in the flask, $PV = nRT$. Initially, 120 °C is converted to 393 K and 500 torr is 500/760 atm or 0.658 atm. Entering these data into the ideal gas law equation yields

$$(0.658\,\text{atm})(4.00\,\text{L}) = n(0.0821\,\text{L atm mol}^{-1}\,\text{K}^{-1})(393\,\text{K})$$
$$n = 0.0816\,\text{mol of SO}_3(g)$$

The initial molar concentration is calculated as:

$$[SO_3] = 0.0816 \text{ mol}/4.00 \text{ L} = 0.0204 \text{ mol/L}$$

Once the number of moles of SO_3 is known, we can calculate the pressure at 25 °C as:

$$P(4.00 \text{ L}) = (0.0816 \text{ mol})(0.0821 \text{ L atm mol}^{-1} \text{ K}^{-1})(298 \text{ K})$$

$$P = 0.499 \text{ atm}$$

(b) For a reaction involving only gases in the equilibrium law $K_p = \dfrac{p_{SO_3}^2}{p_{SO_2}^2 p_{O_2}}$
the value of the free energy is $\Delta G° = -RT \ln K_p$.
To solve this, we write:

$$-140{,}000 \text{ J} = -(8.31 \text{ J mol}^{-1} \text{ K}^{-1})(298 \text{ K}) \ln K_p$$

$$\ln K_p = 56.53$$

$$K_p = 3.6 \times 10^{24}$$

(c) A small amount of SO_3 (we will call it $2x$) will decompose to oxygen and sulfur dioxide. When $2x$ of the SO_3 decomposes, x moles of oxygen and $2x$ moles of SO_2 will result. This can be written as

$$3.6 \times 10^{24} = \frac{\left(p_{SO_3} - 2x\right)^2}{(2x)^2 x} = \frac{(0.499 - 2x)^2}{\left(4x^3\right)}$$

Because the equilibrium constant is so large, it may be assumed that $2x$ is very small compared to 0.499 atm. The result of the approximation is that $0.499 - 2x \cong 0.499$ and

$$3.6 \times 10^{24} = \frac{(0.499)^2}{\left(4x^3\right)}$$

$$4x^3 = 6.9 \times 10^{-26}$$

$$x^3 = 1.7 \times 10^{-26}$$

$$x = 2.6 \times 10^{-9} \text{ atm} = P_{O_2}$$

(d) The partial pressure of the added 1.0 gram (0.0313 mol) of oxygen is

$$P_{O_2}(4.00 \text{ L}) = (0.0313 \text{ mol})(0.0821 \text{ L atm mol}^{-1} \text{ K}^{-1})(298 \text{ K})$$

$$P_{O_2} = 0.191 \text{ atm}$$

The pressure of oxygen from part (c) is vanishingly small compared with 0.191 (i.e., (0.191 atm) + (2.6 × 10⁻⁹) atm = 0.191 atm). Entering this value for the pressure of oxygen into the equations used previously gives us:

$$3.6 \times 10^{24} = \frac{\left(p_{SO_3} - 2x\right)^2}{(2x)^2 x} = \frac{(0.499 - 2x)^2}{(2x^2)(0.191 + x)}$$

Again, x can be assumed to be very small because of the size of the equilibrium constant. So we can write:

$$3.6 \times 10^{24} = \frac{(0.499)^2}{(2x^2)(0.191)}$$

$$2x^2 = 3.6 \times 10^{-25}$$

$$x = 4.255 \times 10^{-13}$$

$$2x = 8.5 \times 10^{-13} \text{ atm}$$

2. Methane is the main component of natural gas used by many for heating and cooking.

(a) $CH_4(g) + 2\,O_2(g) \rightarrow CO_2(g) + 2\,H_2O(l)$

(b) There are two possibilities:

$$2\,CH_4(g) + 3\,O_2(g) \rightarrow 2\,CO(g) + 4\,H_2O(l)$$

and

$$CH_4(g) + O_2(g) \rightarrow C(s) + 2\,H_2O(l)$$

(c) It is not obvious which combustion reaction the problem refers to, so all three will be answered:

For the combustion in excess oxygen there are 2 moles of oxygen for every mole of methane. Avogadro's hypothesis states that in equal volumes of gas, at the same temperature and pressure, there are equal numbers of molecules (also moles). Thus the first reaction will use 1 liter of methane and 2 liters of oxygen the second reaction, producing CO requires 1 liter of methane and 1.5 liters of oxygen. In the last, 1 liter of methane requires 1 liter of oxygen.

When the question is changed to ask how many liters of air are needed, we need to perform that conversion

$$? \text{ L air} = 2 \text{ L O}_2(g) \left(\frac{100 \text{ g air}}{20 \text{ g O}_2} \right) = 10 \text{ L of air is needed when } CO_2 \text{ is the product}$$

$$? \text{ L air} = 1.5 \text{ L O}_2(g) \left(\frac{100 \text{ g air}}{20 \text{ g O}_2} \right) = 7.5 \text{ L of air is needed when } CO \text{ is a product}$$

$$? \text{ L air} = 1 \text{ L O}_2(g) \left(\frac{100 \text{ g air}}{20 \text{ g O}_2} \right) = 5 \text{ L of air is needed when } C \text{ is a product}$$

(d) The formation reaction is $2\ H_2 + C \rightarrow CH_4$

We reverse the reaction for the combustion of methane and then add the combustion reactions for hydrogen and carbon as follows:

$$
\begin{array}{lll}
CO_2(g) + 2\ H_2O(l) \rightarrow CH_4(g) & \Delta H^\circ = +885.0\ kJ \\
2\ H_2(g) + O_2(g) \quad\rightarrow 2\ H_2O(l) & \Delta H^\circ = -570.6\ kJ \\
\underline{C(s) \quad + O_2(g) \quad\rightarrow CO_2(g)} & \underline{\Delta H^\circ = -393.5\ kJ} \\
2\ H_2(g) + C(s) \qquad\rightarrow CH_4 & \Delta H^\circ = -79.1\ kJ
\end{array}
$$

(e) If the water formed is in the gas state, there is little if any increase in entropy. If water condenses to a liquid, there is a large decrease in entropy for the reaction.

3. An organic substance is synthesized and purified. It is known to contain carbon hydrogen, and oxygen. A 0.500-gram sample is burned in excess oxygen with a copper catalyst to ensure complete combustion. The gases produced pass through a drying tube containing calcium chloride. A second tube contains sodium hydroxide that absorbs the carbon dioxide. The drying tube increases in mass by 0.385 g, and the second tube increases by 1.256 g.

(a) All of the carbon in the sample is converted into carbon dioxide that is absorbed on the sodium hydroxide. The mass of CO_2 is 1.256 g. This is converted to the mass of carbon:

$$
?\ g\ C = 1.256\ g\ CO_2 \left(\frac{1\ mol\ CO_2}{44\ g\ CO_2}\right)\left(\frac{1\ mol\ C}{1\ mol\ CO_2}\right)\left(\frac{12\ g\ C}{1\ mol\ C}\right) = 0.3425\ g\ C
$$

What is the percentage of carbon in the sample?

$$
\%\ carbon = \left(\frac{g\ C}{g\ sample}\right) \times 100 = \left(\frac{0.3425\ g\ C}{0.500\ g\ sample}\right) \times 100 = 68.5\%\ C
$$

(b) The hydrogen is completely converted to water. It is absorbed in the drying tube and weighs 0.385 g. This mass is converted to the grams of hydrogen:

$$
?\ g\ H = 0.385\ g\ H_2O \left(\frac{1\ mol\ H_2O}{18\ g\ H_2O}\right)\left(\frac{2\ mol\ H}{1\ mol\ H_2O}\right)\left(\frac{1\ g\ H}{1\ mol\ H}\right) = 0.0428\ g\ H
$$

What is the percentage of hydrogen in the sample?

$$
\%\ hydrogen = \left(\frac{g\ H}{g\ sample}\right) \times 100 = \left(\frac{0.0428\ g\ H}{0.500\ g\ sample}\right) \times 100 = 8.55\%\ H
$$

(c) Since the percentage must add up to 100 for the entire compound, we can calculate $100 - 68.5 - 8.55 = 23.0\%$ oxygen.

(d) Assume that you have a 100-g sample of the compound. Then the percentages also represent masses of each element (this is the quick way to start with gram units that we will need to cancel later). We then have the setup as

$$? \text{ mol C} = 68.5 \text{ g C}$$

$$? \text{ mol H} = 8.55 \text{ g H}$$

$$? \text{ mol O} = 23.0 \text{ g O}$$

We then perform the one-step conversion to moles of each element.

$$? \text{ mol C} = 68.5 \text{ g C} \left(\frac{1 \text{ mol C}}{12 \text{ g C}} \right) = 5.71 \text{ mol C}$$

$$? \text{ mol H} = 8.55 \text{ g H} \left(\frac{1 \text{ mol H}}{1 \text{ g H}} \right) = 8.55 \text{ mol H}$$

$$? \text{ mol O} = 23.0 \text{ g O} \left(\frac{1 \text{ mol O}}{16 \text{ g O}} \right) = 1.44 \text{ mol O}$$

We see that we have the fewest moles of oxygen. We divide the moles of C, H, and O by 1.44 to get 1 mol O, 5.94 mol H, and 3.98 mol C. The moles of hydrogen and carbon are very close to whole numbers, and we can round them off to whole numbers to get an empirical formula of C_4H_6O.

(e) If the compound does not increase the conduction of electricity, it does not break apart into ions upon dissolution. It can be called a nonelectrolyte. Therefore it will be reasonable to use the freezing-point-depression method to determine the molecular mass.

(f) $\Delta T = K_f m$

$$= K_f (\text{moles})/(\text{kg solvent})$$

$$= K_f (\text{g/molar mass})/(\text{kg solvent})$$

$$\text{molar mass} = K_f (\text{g})/(\text{kg solvent})/\Delta T$$

$$= (1.86 \text{ °C mol}^{-1} \text{ kg})(15 \text{ g})/(0.100 \text{ kg})/(4.0 \text{ °C})$$

$$= 69.7 \text{ g mol}^{-1}$$

(g) There are many ways in which these atoms can be arranged. Draw two of them and show how many sigma and pi bonds are in each structure.

Some of the compounds are $CH_2=CHCH_2CHO$, $CH_3CH=CHCHO$, $CH_2=C(CH_3)CHO$, and $CH_3C=CCH_2OH$. Each compound will have two pi bonds as part of two double bonds or one triple bond; there will also be ten sigma bonds.

Part B

4. For these answers, balanced equations are required and only reacting species should be shown. An additional touch would be to add reasonable phases for the substances in the reaction. The instructions indicate that most are reactions in aqueous solution.

(a) (i) A bar of iron metal is immersed in a solution of copper(II) nitrate.

$$Fe(s) + Cu^{2+}(aq) \rightarrow Cu(s) + Fe^{2+}(aq)$$

(ii) What will be observed when this reaction occurs?
Two things will be observed. The blue color of the solution will decrease and eventually disappear. The iron metal will be coated with copper metal with its characteristic color.

(b) (i) Liquid pentane is burned in an excess of oxygen gas.

$$C_5H_{12}(l) + 8O_2(g) \rightarrow 5CO_2(g) + 6H_2O(g)$$

(Note: $6H_2O(l)$ is also acceptable.)

(ii) How will the products change if the amount of available oxygen is decreased?
As oxygen is limited, the carbon dioxide will be replaced with carbon monoxide. Further limitation of oxygen will result in a product of carbon instead of carbon monoxide.

(c) (i) Sulfur trioxide gas is bubbled into limewater, a saturated solution of calcium hydroxide.

$$SO_3(g) + Ca(OH)_2(aq) \rightarrow CaSO_4(aq) + H_2O(l)$$

The net ionic equation is:

$$SO_3(g) + 2OH^-(aq) \rightarrow SO_4^{2-}(aq) + H_2O(l)$$

(ii) How does this explain the deterioration of many outdoor statues and monuments?
Sulfur trioxide is a component of acid rain that reacts with the limestone, CaO, used for many statues.

5. A student performs an experiment to verify the value of R, the gas law constant. The experiment involves the reaction of magnesium metal with hydrochloric acid and collecting a gas.

(a) $Mg(s) + 2\ HCl(aq) \rightarrow Mg^{2+}(aq) + Cl^-(aq) + H_2(g)$

(b) A diagram of the apparatus for collecting a gas is shown below.

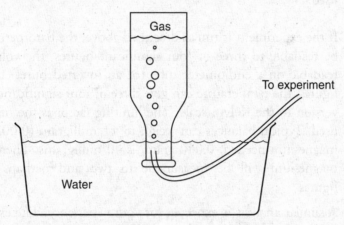

You will need the tabulated values of the vapor pressure of water at various temperatures.

(c) In addition to the 50 mL that the collection tube can hold, we also need to assume that the barometric pressure will be near 1 atm and that the temperature of the lab will be comfortable, somewhere between 19 and 24 °C (let's choose 20 °C). We can use the gas law equation to calculate how many moles of magnesium are needed and then convert that to grams. One final point, if the tube will hold 50 mL, we should use a smaller number just to be on the safe side; 40 mL is a good choice.

$$PV = nRT$$

$$n = \frac{PV}{RT} = \frac{(1\ \text{atm})(0.040\ \text{L})}{(0.0821\ \text{L atm mol}^{-1}\ \text{K}^{-1})(293\ \text{K})}$$

$$n = 1.66 \times 10^{-3}\ \text{mol Mg metal are needed}$$

$$?\ \text{g Mg} = 1.66 \times 10^{-3}\ \text{mol Mg}\left(\frac{24\ \text{g Mg}}{1\ \text{mol Mg}}\right) = 0.040\ \text{g Mg}$$

(d) We must use enough HCl to completely react the magnesium metal. The chemicals' reaction shows that we need 2 moles of HCl for each mole of Mg. That means that $2 \times 1.66 \times 10^{-3}$ mole of HCl is needed. We also need a minimum volume to conveniently set up the pneumatic trough. If that volume is 500 mL, then the molarity of the HCl must be

minimum molarity of HCl =
$(3.32 \times 10^{-3}\ \text{moles})/(0.500\ \text{L}) = 6.64 \times 10^{-3}\ M\ \text{HCl}$

To ensure complete and rapid reaction, the molarity of the HCl can be increased to 0.01 M. This provides a reasonable excess while keeping the acid dilute and safe.

(e) The ideal gas law equation will be used to determine R. It is rearranged as

$$R = \frac{PV}{nT}$$

P will be obtained from the barometric pressure (with a correction for the vapor pressure of water), V is the measured volume of the gas, T is the measured lab temperature, and n is calculated from the mass of magnesium used.

(f) If the experiment is run as described above, the barometric pressure should be readable to three or four significant figures, the volume of the gas is readable on a eudiometer tube (or an inverted buret) to four significant figures, the temperature can give three or four significant figures after conversion to the Kelvin scale. The limiting factor is the mass of magnesium used. Typical balances can weigh to ±1 milligram (0.001 g). The mass of magnesium needed, 0.040 gram, is 40 milligrams. Therefore, the mass of magnesium will be measurable to two and perhaps three significant figures.

A similar analysis is expected for your experience. The exact situation does not need to be replicated.

6. The dichromate ion in acid solution will oxidize ethanol to ethanoic acid. The skeleton reaction is

$$Cr_2O_7^{2-} + CH_3CH_2OH \rightarrow Cr^{3+} + CH_3COOH$$

(a) Write two half-reactions:

$$Cr_2O_7^{2-} \rightarrow Cr^{3+}$$

$$CH_3CH_2OH \rightarrow CH_3COOH$$

Balance all atoms except H and O:

$$Cr_2O_7^{2-} \rightarrow \mathbf{2}\ Cr^{3+}$$

$$CH_3CH_2OH \rightarrow CH_3COOH$$

Balance oxygen atoms with water molecules:

$$Cr_2O_7^{2-} \rightarrow 2\ Cr^{3+} + \mathbf{7\ H_2O}$$

$$\mathbf{H_2O} + CH_3CH_2OH \rightarrow CH_3COOH$$

Balance hydrogens with hydrogen ions:

$$\mathbf{14\ H^+} + Cr_2O_7^{2-} \rightarrow 2\ Cr^{3+} + 7\ H_2O$$

$$H_2O + CH_3CH_2OH \rightarrow CH_3COOH + \mathbf{2\ H^+}$$

Add electrons to balance charge:

$$6\ e^- + 14\ H^+ + Cr_2O_7^{2-} \rightarrow 2\ Cr^{3+} + 7\ H_2O$$

$$H_2O + CH_3CH_2OH \rightarrow CH_3COOH + 2\ H^+ + 2\ e^-$$

Multiply the second reaction by 3 to equalize the electrons:

$$6\ e^- + 14\ H^+ + Cr_2O_7^{2-} \rightarrow 2\ Cr^{3+} + 7\ H_2O$$

$$\mathbf{3\ H_2O + 3\ CH_3CH_2OH \rightarrow 3\ CH_3COOH + 6\ H^+ + 6\ e^-}$$

Add the two equations:

$$6\ e^- + \mathbf{14\ H^+} + Cr_2O_7^{2-} + \mathbf{3\ H_2O} + 3\ CH_3CH_2OH \rightarrow$$

$$3\ CH_3COOH + \mathbf{6\ H^+} + 2\ Cr^{3+} + \mathbf{7\ H_2O} + 6\ e^-$$

Cancel identical items in bold above:

$$\mathbf{8\ H^+} + Cr_2O_7^{2-} + 3\ CH_3CH_2OH \rightarrow 3\ CH_3COOH + 2\ Cr^{3+} + \mathbf{4\ H_2O}$$

(b) The galvanic cell apparatus is shown below.

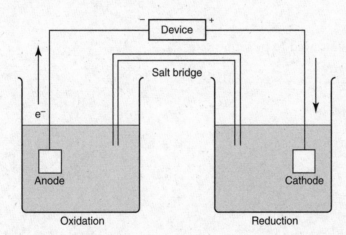

The dichromate is reduced, and it and all the components of its half-reaction go in the right hand beaker. The left beaker contains the ethanol and acetic acid. Both electrodes are platinum because no metals are involved in the half-reactions. The device above is a voltmenter with the positive side connected to the cathode and the negative to the anode.

(c) Standard state for electrochemistry includes a temperature of 25 °C, 1 atm pressure for gases, 1 M solutions for soluble substances, and a saturated solution for slightly soluble substances. In addition, solids and pure liquids are considered to have concentrations of 1 M.

(d) Measurement of E_{cell}° allows us to calculate:

Free energy $\Delta G^{\circ} = -\ nF\ E_{cell}^{\circ}$

Equilibrium constants $E_{cell}^{\circ} = \dfrac{nF}{RT}\ \ln K$

Reaction Spontaneity E_{cell}° if positive indicates a spontaneous reaction

(e) There are no gases involved in this reaction, and therefore the entropy change is not obvious. There are more molecules and ions as reactants than there are products, and this argues for a decrease in entropy as a first approximation.

(f) The dichromate ion, $Cr_2O_7^{2-}$, dissociates into the chromate ion in neutral and alkaline solutions via the following reaction:

$$Cr_2O_7^{2-} + 2\ OH^- \rightarrow 2\ CrO_4^{2-} + H_2O$$

Practice Test 2
ANSWER SHEET

1 (A) (B) (C) (D) (E)
2 (A) (B) (C) (D) (E)
3 (A) (B) (C) (D) (E)
4 (A) (B) (C) (D) (E)
5 (A) (B) (C) (D) (E)
6 (A) (B) (C) (D) (E)
7 (A) (B) (C) (D) (E)
8 (A) (B) (C) (D) (E)
9 (A) (B) (C) (D) (E)
10 (A) (B) (C) (D) (E)
11 (A) (B) (C) (D) (E)
12 (A) (B) (C) (D) (E)
13 (A) (B) (C) (D) (E)
14 (A) (B) (C) (D) (E)
15 (A) (B) (C) (D) (E)
16 (A) (B) (C) (D) (E)
17 (A) (B) (C) (D) (E)
18 (A) (B) (C) (D) (E)
19 (A) (B) (C) (D) (E)
20 (A) (B) (C) (D) (E)
21 (A) (B) (C) (D) (E)
22 (A) (B) (C) (D) (E)
23 (A) (B) (C) (D) (E)
24 (A) (B) (C) (D) (E)
25 (A) (B) (C) (D) (E)

26 (A) (B) (C) (D) (E)
27 (A) (B) (C) (D) (E)
28 (A) (B) (C) (D) (E)
29 (A) (B) (C) (D) (E)
30 (A) (B) (C) (D) (E)
31 (A) (B) (C) (D) (E)
32 (A) (B) (C) (D) (E)
33 (A) (B) (C) (D) (E)
34 (A) (B) (C) (D) (E)
35 (A) (B) (C) (D) (E)
36 (A) (B) (C) (D) (E)
37 (A) (B) (C) (D) (E)
38 (A) (B) (C) (D) (E)
39 (A) (B) (C) (D) (E)
40 (A) (B) (C) (D) (E)
41 (A) (B) (C) (D) (E)
42 (A) (B) (C) (D) (E)
43 (A) (B) (C) (D) (E)
44 (A) (B) (C) (D) (E)
45 (A) (B) (C) (D) (E)
46 (A) (B) (C) (D) (E)
47 (A) (B) (C) (D) (E)
48 (A) (B) (C) (D) (E)
49 (A) (B) (C) (D) (E)
50 (A) (B) (C) (D) (E)

51 (A) (B) (C) (D) (E)
52 (A) (B) (C) (D) (E)
53 (A) (B) (C) (D) (E)
54 (A) (B) (C) (D) (E)
55 (A) (B) (C) (D) (E)
56 (A) (B) (C) (D) (E)
57 (A) (B) (C) (D) (E)
58 (A) (B) (C) (D) (E)
59 (A) (B) (C) (D) (E)
60 (A) (B) (C) (D) (E)
61 (A) (B) (C) (D) (E)
62 (A) (B) (C) (D) (E)
63 (A) (B) (C) (D) (E)
64 (A) (B) (C) (D) (E)
65 (A) (B) (C) (D) (E)
66 (A) (B) (C) (D) (E)
67 (A) (B) (C) (D) (E)
68 (A) (B) (C) (D) (E)
69 (A) (B) (C) (D) (E)
70 (A) (B) (C) (D) (E)
71 (A) (B) (C) (D) (E)
72 (A) (B) (C) (D) (E)
73 (A) (B) (C) (D) (E)
74 (A) (B) (C) (D) (E)
75 (A) (B) (C) (D) (E)

The Periodic Table of the Elements

1																	2
H 1.0080																	**He** 4.0026
3 **Li** 6.941	4 **Be** 9.012											5 **B** 10.810	6 **C** 12.011	7 **N** 14.007	8 **O** 16.00	9 **F** 19.00	10 **Ne** 20.179
11 **Na** 22.99	12 **Mg** 24.31											13 **Al** 26.98	14 **Si** 28.09	15 **P** 30.974	16 **S** 32.06	17 **Cl** 35.453	18 **Ar** 39.948
19 **K** 39.10	20 **Ca** 40.08	21 **Sc** 44.96	22 **Ti** 47.87	23 **V** 50.94	24 **Cr** 52.00	25 **Mn** 54.94	26 **Fe** 55.85	27 **Co** 58.93	28 **Ni** 58.69	29 **Cu** 63.55	30 **Zn** 65.41	31 **Ga** 69.72	32 **Ge** 72.64	33 **As** 74.92	34 **Se** 78.96	35 **Br** 79.90	36 **Kr** 83.80
37 **Rb** 85.47	38 **Sr** 87.62	39 **Y** 88.91	40 **Zr** 91.22	41 **Nb** 92.91	42 **Mo** 95.94	43 **Tc** [98]	44 **Ru** 101.07	45 **Rh** 102.91	46 **Pd** 106.42	47 **Ag** 107.87	48 **Cd** 112.41	49 **In** 114.82	50 **Sn** 118.71	51 **Sb** 121.76	52 **Te** 127.60	53 **I** 126.90	54 **Xe** 131.29
55 **Cs** 132.91	56 **Ba** 137.33	57 **La** 138.91	72 **Hf** 178.49	73 **Ta** 180.95	74 **W** 183.84	75 **Re** 186.21	76 **Os** 190.23	77 **Ir** 192.22	78 **Pt** 195.08	79 **Au** 196.97	80 **Hg** 200.59	81 **Tl** 204.38	82 **Pb** 207.2	83 **Bi** 208.98	84 **Po** [209]	85 **At** [210]	86 **Rn** [222]
87 **Fr** [223]	88 **Ra** 226.03	89 **Ac** 227.03	104 **Rf** [267]	105 **Db** [268]	106 **Sg** [271]	107 **Bh** [272]	108 **Hs** [270]	109 **Mt** [276]	110 **Ds** [281]	111 **Rg** [280]	112 **Uub** [285]		114 **Uuq** [289]		116 **Uuh** [291]		

Lanthanides

58 **Ce** 140.12	59 **Pr** 140.91	60 **Nd** 144.24	61 **Pm** [145]	62 **Sm** 150.36	63 **Eu** 151.96	64 **Gd** 157.25	65 **Tb** 158.93	66 **Dy** 162.50	67 **Ho** 164.93	68 **Er** 167.26	69 **Tm** 168.93	70 **Yb** 173.04	71 **Lu** 174.97

Actinides

90 **Th** 232.04	91 **Pa** 231.04	92 **U** 238.03	93 **Np** 237.05	94 **Pu** [244]	95 **Am** [243]	96 **Cm** [247]	97 **Bk** [247]	98 **Cf** [251]	99 **Es** [252]	100 **Fm** [257]	101 **Md** [258]	102 **No** [259]	103 **Lr** [262]

Chemistry

SECTION I

CALCULATORS MAY NOT BE USED WITH SECTION I

REMINDER: For all questions, assume that the temperature is 298 K, the pressure is 1.00 atmospheres of pressure, and solutions are aqueous solutions unless otherwise indicated.

Throughout this practice test the following symbols are used with the definitions provided below, unless otherwise noted in a given problem.

T = temperature	H = enthalpy
P = pressure	G = free energy
V = volume	R = molar gas constant
S = entropy	n = number of moles
M = molarity	atm = atmosphere(s)
m = molal	J, kJ = joule(s), kilojoule(s)
L, mL = liter(s), milliliter(s)	V = volt(s)
g = gram(s)	mol = mole(s)
nm = nanometer(s)	

Practice Test 2

SECTION I—PART A

Multiple-Choice Questions

(Calculators May Not Be Used)

75 QUESTIONS

TIME: 90 MINUTES

50% OF TOTAL SCORE

Directions: Each of the lettered choices has a statement immediately following it. Selecting the letter is the same as selecting the statement as your answer. Select the letter that best fits each of the numbered questions that follow. A letter may be used once, more than once or not at all in a set of questions.

Questions 1–3 refer to the following compounds.

(A) CH_4
(B) C_6H_6 (benzene)
(C) CH_3CH_2COOH
(D) $CH_3CH_2OCH_2CH_3$
(E) $CH_2{=}CHCH_3$

1. This compound is the monomer used to make polypropylene.

2. This compound is best described using resonance structures.

3. This compound has the lowest boiling point of those listed.

Questions 4–7 refer to the following electronic configurations.

(A) $1s^2 2s^2 2p^6 3s^2 3p^6 4s^2 3d^{10} 4p^6 5s^1 6s^1$
(B) $1s^2 2s^2 2p^6 3s^2 3p^6 4s^2 3d^5$
(C) $1s^2 2s^2 2p^6 3s^2 3p^6 4s^2$
(D) $1s^2 2s^2 2p^6 3s^2 3p^6 4s^2 3d^{10} 4p^6 5s^2 4d^{10} 5p^6$
(E) $1s^2 2s^2 2p^6 3s^2 3p^6 4s^2 3d^{10} 4p^6 5s^2 4d^7$

4. The ground state configuration for an alkaline earth metal.

5. The ground state configuration for the negative ion of a halogen.

6. The ground state configuration for a transition element in period 4.

7. A possible excited state electronic configuration.

Questions 8–10 refer to the following.
    (A) anode
    (B) cathode
    (C) salt bridge
    (D) indicator electrode
    (E) reference electrode

8. Its symbol is the double solid line, // or ||, in a cell diagram.

9. Oxidation occurs here in a galvanic cell.

10. This must have a constant potential so that concentration measurements can be made.

Questions 11–14 refer to the following.
    (A) hydrogen bonding
    (B) molecular crystals
    (C) ionic bonding
    (D) covalent bonding
    (E) London forces (van der Waals forces)

11. The C—H bonds in organic substances involve this.

12. It describes the solid form of CO_2.

13. It is used to explain why nonpolar substances may be condensed to liquids.

14. It is, in general, the strongest intra- and intermolecular attractive force.

SECTION I—PART B

(Calculators May Not Be Used)

> **Directions:** Each of the questions or incomplete statements below is followed by five suggested answers or completions. Select the one that is best in each case and fill in the corresponding oval on the answer sheet.

15. SO_2 and O_2 are mixed in an insulated vessel and sealed so that there is no heat exchange with the surroundings. When this exothermic reaction comes to equilibrium, what best describes what has happened to the system?

    (A) Total energy remains constant and the entropy decreases.
    (B) Total energy increases and the entropy remains constant.
    (C) Total energy remains constant and the temperature of the system increases.
    (D) Total energy decreases and the temperature increases.
    (E) Total energy remains constant and the temperature decreases.

16. Which of the following will have a concentration of chloride ions different from that of the other solutions?

    (A) 0.200 M NaCl
    (B) 0.100 M $AlCl_3$
    (C) 0.100 M $BaCl_2$
    (D) 0.200 M KCl
    (E) 0.0667 M $FeCl_3$

17. When sodium oxalate ($Na_2C_2O_4$) is dissolved in water, it is best represented by

    (A) $Na_2^+ + C_2O_4^{2-}$
    (B) $2\,Na^+ + 2\,C^{3+} + 4\,O^{2-}$
    (C) $2\,Na^+ + C_2O_4^-$
    (D) $Na_2^{2+} + C_2O_4^{2-}$
    (E) $2\,Na^+ + C_2O_4^{2-}$

18. Based upon general rules of electronegativity, which of the following bonds would be expected to have the largest polarity?

    (A) C—N
    (B) C—F
    (C) O—F
    (D) N—O
    (E) C—O

19. Which of the following is expected to have the largest dipole moment?

    (A) CH_4
    (B) CO_2
    (C) SF_6
    (D) HCN
    (E) Cl^-

20. Using the Aufbau (energy) ordering, the correct electronic configuration for the gold(III) ion is

    (A) $1s^2 2s^2 2p^6 3s^2 3p^6 4s^2 3d^{10} 4p^6 5s^2 4d^{10} 5p^6 6s^2 4f^{14} 5d^9$
    (B) $1s^2 2s^2 2p^6 3s^2 3p^6 4s^2 3d^{10} 4p^6 5s^2 4d^{10} 5p^6 6s^1 4f^{14} 5d^{10}$
    (C) $1s^2 2s^2 2p^6 3s^2 3p^6 4s^2 3d^{10} 4p^6 5s^2 4d^{10} 5p^6 6s^2 4f^{14} 5d^6$
    (D) $1s^2 2s^2 2p^6 3s^2 3p^6 4s^2 3d^{10} 4p^6 5s^2 4d^{10} 5p^6 4f^{14} 5d^8$
    (E) $1s^2 2s^2 2p^6 3s^2 3p^6 4s^2 3d^{10} 4p^6 5s^2 4d^{10} 5p^6 6s^2 4f^{14} 5d^{10} 6p^2$

21. A student weighed 2.145 grams of potassium acid phthalate (formula mass of KHP = 204.23), added it to a 100-mL volumetric flask, and added distilled water exactly to the mark. A titration was then performed using 10.0-mL aliquots of the KHP and titration with a standardized 0.1036 molar sodium hydroxide solution. Phenolphthalein was used as an indicator, and each titration was stopped at the first discernible pink color that persisted for 30 seconds. In order, the results obtained were

First titration	9.74 mL
Second titration	9.86 mL
Third titration	9.98 mL
Fourth titration	10.05 mL
Fifth titration	10.18 mL
Sixth titration	10.19 mL
Seventh titration	10.17 mL

What is the most reasonable explanation of the results obtained?

(A) All results are very close and are valid. The average is 10.02 mL.
(B) The variation in results indicates that the glassware was not dry before starting the experiment.
(C) The end point was too faint and should persist until the end of the lab period.
(D) The constant increase suggests that the KHP was not completely dissolved until the fifth titration.
(E) The regular increase suggests that carbon dioxide is being absorbed by the sample.

22. ^{235}U has a half-life of 7.04×10^8 years and may be used to determine the length of time since rocks have solidified. For rocks of the following ages, which can be dated most precisely?

(A) lava from the Mount St. Helens, eruption in 1980
(B) lava from the Mount Etna eruption, 700 BC
(C) lava from lunar craters, 1.5 billion years old
(D) lava from Mount Vesuvius, 5 million years old
(E) lava from Mount Kenya, 20 million years old

23. Photochemical smog in large cities is attributed mostly to emissions from automobiles. The colored gas most associated with this smog is

(A) an oxide of nitrogen
(B) an oxide of sulfur
(C) an oxide of hydrogen
(D) a chloride of sulfur
(E) a nitride of nitrogen

24. The density of a solution must be known if a conversion between which of the following concentration units is possible?

(A) mass fraction to mass (weight) percent
(B) mole fraction to molality
(C) molarity to mole fraction
(D) mass (weight) fraction to molality
(E) mass (weight) fraction to mole fraction

25. A substance has an empirical formula of CH_2. Nitrogen (N_2) effuses through a small hole 2.00 times faster than this compound. What is the formula of the compound?

(A) C_2H_4
(B) C_8H_8
(C) $C_{16}H_8$
(D) C_8H_{16}
(E) C_3H_6

26. Which of the following is correctly named?

    (A) $SbCl_5$, antimony hexachloride
    (B) CH_2CH_2, acetylene
    (C) N_2O_5, nitrogen pentoxide
    (D) CH_3NH_2, ethylamine
    (E) $FeCl_3$, iron(III) chloride

27. Diamond, one of the three allotropes of carbon, is best described as

    (A) an ionic crystal
    (B) a molecular crystal
    (C) a covalent network crystal
    (D) a metallic crystal
    (E) a gaseous crystal

28. The reaction $2 NO_2(g) \rightleftharpoons N_2O_4(g)$ has an equilibrium constant of 234 at a certain temperature. 0.0100 mol of NO_2 and 0.0300 mol of N_2O_4 are placed in a 5.00-liter flask and allowed to come to equilibrium. The reaction that occurs

    (A) increases both NO_2 and N_2O_4 concentrations
    (B) decreases both NO_2 and N_2O_4 concentrations
    (C) increases NO_2 and decreases N_2O_4 concentrations
    (D) decreases NO_2 and increases N_2O_4 concentrations
    (E) the system is at equilibrium and no concentration change is observed

29. The valence electron of potassium can have which of the following sets of quantum numbers that are listed in their customary order?

    (A) 4   1   0   ½
    (B) 4   0   1   ½
    (C) 3   1   1   ½
    (D) 4   0   0   ½
    (E) 5   0   0   -½

30. Water has a freezing-point-depression constant of 1.86 °C per molal. A certain sample has a freezing point of –0.36 °C. What other information is needed to determine the molar mass of a nondissociating solute?

    I.   volume of solution
    II. mass of solute
    III. mass of solvent

    (A) I only
    (B) II only
    (C) III only
    (D) I and II
    (E) II and III

31. What ratio of the mass of Na_2HPO_4 (molar mass = 142) to the mass of NaH_2PO_4 (molar mass = 120) is needed to prepare a buffer with a pH of 7.4? The pK_{a2} for $H_2PO_4^-$ is 7.42.

    (A) 1.58
    (B) 0.63
    (C) 1.87
    (D) 1.18
    (E) 0.53

32. The K_a for hydrofluoric acid is 6.8×10^{-4}. What percentage of HF is dissociated in a 0.080 molar solution where the hydronium ion concentration is $7.4 \times 10^{-3} M$?

    (A) 12.3%
    (B) 4.25%
    (C) 9.2%
    (D) 1.12%
    (E) 23.6%

33. All of the following require the use of the resonance concept to describe their geometry and physical properties except one.

    (A) H_2O
    (B) SO_2
    (C) NO_3^-
    (D) $O-C=O^-$ or HCO_3^-
           | |
           H   O
    (E) O_3

34. A certain chemical reaction is described by a first-order rate law. What is the rate constant, in units of s^{-1}, for the reaction if the half-life is determined to be 228 minutes?

    (A) $\ln(228 \times 60)$

    (B) $\dfrac{\ln(2)}{228 \times 60}$

    (C) $\ln(2)$

    (D) $\ln \dfrac{[A_t]}{[A_0]}$

    (E) $228 \ s^{-1}$

35. A student prepares ten solutions with different concentrations of $KMnO_4$ and measures the absorbance of each using a spectrophotometer. When making a graph of the results, the least desirable thing to do is

    (A) arrange the axes so that the independent variable is on the x-axis and the dependent variable is on the y-axis

    (B) plot each point by finding the coordinates corresponding to the values of the independent and dependent variables

    (C) connect the points in a dot-to-dot fashion

    (D) set the range of values on the axes to include the values found in the experiment

    (E) set the scale of the graph so that it fills the page as completely as possible

36. A gas has a density of 3.79 g L^{-1} at STP. The best choice for the identity of this substance is

    (A) $CH_3CH_2CH_2CH_2CH_2CH_3$
    (B) $CH_3CH_2OCH_2CH_3$
    (C) NaCl
    (D) CH_2Cl_2
    (E) Cl_2

37. There are a variety of ways in which a chemist can judge the strength of a covalent bond. If covalent bond A is stronger than covalent bond B, then

    (A) the vibrational frequency of A is less that of than B

    (B) the bond length of A is generally smaller than that of B

    (C) the electronegativity difference in A is less than that in B

    (D) A dissolves in water better than B

    (E) A absorbs longer wavelength light than B

38. The collision theory and the transition-state theory are the two main theories of reaction rates. Which of the following is (are) part of the collision theory

    I. collision frequency
    II. collision energy
    III. collision orientation

    (A) I and II only
    (B) I and III only
    (C) I only
    (D) I, II, and III
    (E) II and III only

39. A reaction has an equilibrium constant of 3.8×10^3. This constant will change if

    (A) a catalyst is added to the reaction mixture
    (B) additional reactant is added
    (C) the temperature is changed
    (D) the pressure is decreased
    (E) a precipitate is formed

40. Which of the following represents a pair of isotopes of element X?

    (A) $^{21}_{11}X$ and $^{24}_{12}X$

    (B) $^{19}_{9}X$ and $^{39}_{19}X$

    (C) $^{12}_{6}X$ and $^{12}_{6}X$

    (D) $^{14}_{6}X$ and $^{14}_{7}X$

    (E) $^{79}_{35}X$ and $^{81}_{35}X$

41. Which of the following ions may be detected safely by its odor after an appropriate, simple, chemical reaction?

    (A) CO_3^{2-}
    (B) Sr^{2+}
    (C) NO_3^-
    (D) NH_4^+
    (E) MnO_4^-

42. When the skeleton equation below is balanced and all coefficients are reduced to their lowest whole-number values, choose the response that indicates the coefficient for $H^+(aq)$.

    $$Fe_2O_3(s) + C_2O_4^{2-}(aq) + H^+(aq) \rightarrow$$
    $$Fe^{2+}(aq) + CO_2(g) + H_2O(l)$$

    (A) 2
    (B) 3
    (C) 4
    (D) 6
    (E) 9

43. Atoms of an element Y have the ground-state electronic configuration $1s^2, 2s^2, 2p^6, 3s^2, 3p^1$. The element Y will most likely form a compound with oxygen having the following formula:

    (A) YO
    (B) YO_2
    (C) Y_2O_3
    (D) Y_3O_2
    (E) Y_2O

44. Under identical conditions of temperature, volume, and pressure, which of the following gases is expected to deviate the most from ideal gas behavior?

    (A) oxygen
    (B) nitrogen
    (C) methane
    (D) carbon dioxide
    (E) water

45. The reaction $Cu + 2\,Ag^+ \rightarrow 2\,Ag + Cu^{2+}$ is spontaneous. This means that

    (A) $\Delta G°$ for the reaction must be positive
    (B) $E°$ for copper must be positive
    (C) $E°$ for silver must be positive
    (D) $\Delta S°$ for the reaction must be positive
    (E) K_{eq} must be greater than 1.00

46. When the following half-reaction is balanced in an acidic solution with the smallest whole-number coefficients, it will contain

    $$NO_3^- \rightarrow NO$$

    (A) 2 e^- on the right side
    (B) 4 e^- on the left side
    (C) 2 H_2O on the right side
    (D) 2 H^+ on the left side
    (E) a coefficient of 2 for NO

47. When an element emits an alpha particle in a nuclear decay, the original element has its

    (A) atomic number increase by 2 and its atomic mass increase by 4
    (B) atomic number decrease by 1 and the atomic mass stays the same
    (C) atomic mass decrease by 1 and the atomic number remains the same
    (D) atomic mass decrease by 4 and its atomic number decrease by 2
    (E) energy decreased by also emitting a gamma ray

48. Approximately what volume of carbon dioxide, at STP, is needed to precipitate all of the calcium ions in a 100-mL sample of 0.250 M $Ca(NO_3)_2$?

    (A) 560 mL
    (B) 560 L
    (C) 280 mL
    (D) 1.12 L
    (E) 280 L

Practice Test 2

49. Light with the longest wavelength is

    (A) infrared radiation
    (B) visible light
    (C) ultraviolet light
    (D) microwave radiation
    (E) X-rays

50. Which of the following is not commonly used as a catalyst?

    (A) NaCl
    (B) Pt
    (C) Au
    (D) enzymes
    (E) MnO_2

51. Given the following reactions with their equilibrium constants, calculate the equilibrium constant for the overall reaction.

    $Cd^{2+} + 4\ CN^- \rightleftharpoons Cd(CN)_4^{2-}$ $\quad K_f = 7.7 \times 10^{+16}$

    $CdCO_3 \rightleftharpoons Cd^{2+} + CO_3^{2-}$ $\quad K_{sp} = 1.8 \times 10^{-14}$

    $CdCO_3 + 4\ CN^- \rightleftharpoons Cd(CN)_4^{2-} + CO_3^{2-}$ $\quad K_{overall} = ?$

    (A) 1.4×10^3
    (B) $9.5 \times 10^{+2}$
    (C) 4.3×10^{30}
    (D) 2.3×10^{-31}
    (E) $2.2 \times 10^{+6}$

    I. London forces
    II. hydrogen bonds
    III. dipole interactions

52. Intermolecular forces are used to explain many physical properties of liquids and solids. A list of these forces in order from weakest to strongest is

    (A) I, II, and III
    (B) II, I, and III
    (C) I, III, and II
    (D) III, I, and II
    (E) II, III, and I

53. Which of the following reactions is expected to have the largest increase in entropy?

    (A) $2\ NO_2(g) \rightarrow N_2O_4(g)$
    (B) $MgCO_3(s) + 2\ HCl(aq) \rightarrow$
        $CO_2(g) + H_2O(l) + MgCl_2(aq)$
    (C) $C_2H_4(g) + 3\ O_2(g) \rightarrow$
        $2\ CO_2(g) + 2\ H_2O(g)$
    (D) $CuS(s) + O_2(g) \rightarrow Cu(s) + SO_2(g)$
    (E) $KCl(aq) + AgNO_3(aq) \rightarrow$
        $AgCl(s) + KNO_3(aq)$

54. In which chemical system will the moles of each substance in the balanced reaction NOT change with an increase in the volume of the system with no change in the temperature?

    (A) $2\ H_2(g) + O_2(g) \rightarrow 2\ H_2O(g)$
    (B) $N_2(g) + 3\ H_2(g) \rightarrow 2\ NH_3(g)$
    (C) $4\ NO_2(g) + O_2(g) \rightarrow 2\ N_2O_5(g)$
    (D) $CH_4(g) + 2\ O_2(g) \rightarrow CO_2(g) +$
        $2\ H_2O(g)$
    (E) $2\ NO_2(g) \rightarrow N_2O_4(g)$

55. Commercial nitric acid is approximately 15.4 molar. What volume of commercial nitric acid must be used to prepare 5.00 liters of 6.00 molar nitric acid?

    (A) 1.94 mL
    (B) 12.8 mL
    (C) 390 mL
    (D) 1950 mL
    (E) 325 mL

56. How many grams of lead(II) nitrate $(Pb(NO_3)_2)$ must be weighed in order to have exactly 3.00 grams of lead? The molar mass of lead nitrate is 331.

    (A) 4.80 grams
    (B) 0.626 gram
    (C) 130 grams
    (D) 3.90 grams
    (E) 48.0 grams

57. Which of the following does not provide a valid method or enough information to determine the molar mass of an unknown compound?

(A) the density of a gas at STP assuming ideal gas behavior
(B) the osmotic pressure of a solute at a known temperature and concentration in grams per liter
(C) the freezing-point depression of a solvent with a known k_f and the concentration of the solute in grams per kilogram of solvent
(D) the rate of effusion of the unknown gas through a pinhole and the rate of a known gas effusing through the same pinhole under identical conditions
(E) the boiling-point elevation of a solvent with a known k_b and the concentration of the solute in grams per liter of solvent

58. A compound is found to be composed of carbon, hydrogen, and oxygen by qualitative analysis. Quantitative analysis finds that the compound is 38.71 percent carbon and 9.68 percent hydrogen. Which of the following is a possible empirical formula for this compound?

(A) CH
(B) CH_2
(C) CH_3
(D) C_2H_5
(E) CH_3O

59. Two structural isomers, cyclohexane (C_6H_{12}) and 2-hexene (C_6H_{12}), are at the same temperature and pressure. Which physical property is expected to be virtually the same for both compounds?

(A) density of the liquids
(B) boiling point
(C) heat of vaporization
(D) melting point
(E) gaseous density

60. $4 FeS(s) + 7 O_2(g) \rightarrow 2 Fe_2O_3(s) + 4 SO_2(g)$ $\Delta H° = -2432$ kJ mol^{-1}

When the above reaction is at equilibrium for any given pressure, P, and temperature, T, which of the following will shift the position of equilibrium so that more product is formed?

(A) increase the temperature of the system without changing the pressure
(B) add an inert gas to increase the pressure of the system
(C) add a catalyst specific for the forward reaction
(D) remove the Fe_2O_3 as it is formed
(E) remove the SO_2 as it is formed

61. The concentration of H_2 in a flask is 0.023 moles per liter, and the pressure is 380 mm Hg. N_2 is added to this flask to bring the pressure up to 1.00 atmosphere without any change in temperature. What is the concentration of N_2 in the flask?

(A) 0.046 M
(B) 6.6×10^{-5} M
(C) 0.023 M
(D) 1.19 M
(E) 0.012 M

62. In aqueous solution the strongest acid is

(A) HCl
(B) H_3O^+
(C) HBr
(D) HI
(E) all are equally strong

63. The kinetic-molecular theory is best at explaining which of the following?

(A) precipitation reactions
(B) bond vibrations
(C) gas behavior
(D) catalysts
(E) activated complexes

64. A 40.0-milligram sample of pure iron(III) sulfate (molar mass = 400) is dissolved in 1 liter of acidified water. If a base such as sodium hydroxide is added, a precipitate will form. At what pH will iron(III) hydroxide begin to precipitate from this solution? (K_{sp} [Fe(OH)$_3$] = 1.6×10^{-39}.)

(A) 1.6
(B) 2.3
(C) 11.7
(D) 36
(E) 5.8

65. A decrease in pH will be noticed when a small molecule containing one of the following functional groups is added to water. Which of the molecules listed will cause this effect?

(A) $CH_3CH_2CH_2COOH$
(B) $HOCH_2CH_2CH_3$
(C) $CH_3CH_2CH_2NH_2$
(D) $OCHCH_2CH_3$
(E) $HSCH_2CH_2CH_3$

66. In an experiment a student collects hydrogen by displacement of water. The water levels in the bottle and the pneumatic trough are made equal. The barometer in the lab reads 723.2 mm Hg, and the vapor pressure of water at the room temperature of 22 °C is 19.8 mm Hg. What is the pressure of the hydrogen gas collected and what must be measured to determine the mass of hydrogen collected?

(A) 703.4 mm Hg; the volume of H_2 must be determined
(B) 703.4 mm Hg; nothing else is needed
(C) 740.0 mm Hg; the volume of H_2 must be determined
(D) 740.0 mm Hg; the Celsius temperature must be converted to Kelvin
(E) 723.2 mm Hg; the moles of H_2 must be determined

67. Which of the following is not a complex ion?

(A) $Cr_2O_7^{2-}$
(B) $Ag(NH_3)_2^+$
(C) $FeCl_6^{3-}$
(D) $Cd(CN)_4^{2-}$
(E) HgI_4^{2-}

68. Magnesium fluoride has $K_{sp} = 6.6 \times 10^{-9}$. If x represents the molar solubility of MgF_2, what equation should be used to calculate the molar solubility of MgF_2?

(A) $K_{sp} = x^3$
(B) $K_{sp} = x^2$
(C) $K_{sp} = 4x^2$
(D) $K_{sp} = 27x^3$
(E) $K_{sp} = 4x^3$

69. 10.0 grams of a protein is dissolved in distilled water to prepare 500 milliliters of solution at 25 °C, and the osmotic pressure is measured to be 23.2 mm Hg. What mistake will result in the largest error in the results? ($R = 0.0821$ L atm mol^{-1} K^{-1}.)

(A) using 1.0 g rather than the 10.0 g given
(B) not converting 25 °C to 298 K
(C) not converting 23.2 mm Hg into 0.030 atmospheres units
(D) using the wrong value of 8.319 instead of 0.0821 for R
(E) using a 250 mL flask instead of a 500 mL flask for dissolving the sample

70. Which of the following molecules has more than one pi bond?

(A) CO_2
(B) SO_2
(C) NH_3
(D) C_3H_6
(E) CH_2O

71. Which of the following is FALSE about an ideal gas?

    (A) ½ mole will occupy 11.2 liters at STP
    (B) each atom is assumed to have no volume
    (C) attractive forces keep ideal gas molecules in the container
    (D) the average kinetic energy of all ideal gases is the same at STP
    (E) the average velocity of an ideal gas is slightly greater than its most probable velocity

72. An ideal solution of pentane and heptane is prepared, and its vapor pressure is measured as 340 mm Hg at 25 °C. The vapor pressures of pentane and heptane are 440 and 40 mm Hg, respectively, at 25 °C. What is the mole fraction of pentane in this mixture?

    (A) 0.625
    (B) 0.250
    (C) 0.133
    (D) 0.867
    (E) 0.750

73. $Al_2(CO_3)_3 + HCl \rightarrow AlCl_3 + H_2O + CO_2$

    When 5.00 mL of 6.00 molar hydrochloric acid is added to 468 mg of aluminum carbonate (formula mass = 234) according to the above unbalanced reaction, what is the maximum number of millimoles of CO_2 gas that will be evolved?

    (A) 15.
    (B) 5.0
    (C) 23.0
    (D) 6.0
    (E) 1.7

Balance the following chemical equation and use it for the next two questions.

$$SO_2(g) + O_2(g) \rightleftharpoons SO_3(g) \qquad \Delta H = -197 \text{ kJ}$$

74. For the following gas phase reaction

$$2 \, SO_2(g) + O_2(g) \rightarrow 2 \, SO_3(g)$$

under certain conditions that are close to standard state, $\Delta G = 1.38 \text{ kJ mol}^{-1}$ and $\Delta G° = -683 \text{ kJ mol}^{-1}$. For this reaction, under these conditions we can state that

    (A) the reaction will proceed in the forward direction and is considered to be a nonspontaneous reaction
    (B) the reaction will proceed in the reverse direction and is considered to be a nonspontaneous reaction
    (C) the reaction will proceed in the forward direction and is considered to be a spontaneous reaction
    (D) the reaction will proceed in the reverse direction and is considered to be a spontaneous reaction
    (E) the data indicate only that the reaction is not at equilibrium

75. Estimate the mass of $CaCl_2$ (molar mass = 111.0) required to prepare 75.00 mL of a 2.000 molar solution of this salt. Choose the answer below that fits your estimate.

    (A) 150 g
    (B) 16.65 g
    (C) 8.325 g
    (D) 1.65×10^4 g
    (E) 222 g

STOP

End of Section I.
If you finish before the 90 minutes have elapsed, you may check your work in this section. Do not go on to Section II until you are instructed to do so.

THE FOLLOWING TABLES MAY BE USED TO ANSWER THE QUESTIONS IN SECTION II ONLY.

Standard Reduction Potentials, 25 °C

Half-Reaction	E^0 (volts)	Half-Reaction	E^0 (volts)
$F_2(g) + 2e^- \rightarrow 2F^-$	2.87	$Co^{2+} + 2e^- \rightarrow Co$	−0.28
$Co^{3+} + e^- \rightarrow Co^{2+}$	1.82	$Tl^+ + e^- \rightarrow Tl$	−0.34
$Au^{3+} + 3e^- \rightarrow Au$	1.50	$Cd^{2+} + 2e^- \rightarrow Cd$	−0.40
$Cl_2(g) + 2e^- \rightarrow 2Cl^-$	1.36	$Cr^{3+} + e^- \rightarrow Cr^{2+}$	−0.41
$O_2(g) + 4H^+ + 4e^- \rightarrow 2H_2O$	1.23	$Fe^{2+} + 2e^- \rightarrow Fe$	−0.44
$Br_2(g) + 2e^- \rightarrow 2Br^-$	1.07	$Cr^{3+} + 3e^- \rightarrow Cr$	−0.74
$2Hg^{2+} + 2e^- \rightarrow Hg_2^{2+}$	0.92	$Zn^{2+} + 2e^- \rightarrow Zn$	−0.76
$Ag^+ + e^- \rightarrow Ag$	0.80	$2H_2O(l) + 2e^- \rightarrow H_2(g) + 2OH^-$	−0.83
$Hg_2^{2+} + 2e^- \rightarrow 2Hg$	0.79	$Mn^{2+} + 2e^- \rightarrow Mn$	−1.18
$Fe^{3+} + e^- \rightarrow Fe^{2+}$	0.77	$Al^{3+} + 3e^- \rightarrow Al$	−1.66
$I_2 + 2e^- \rightarrow 2I^-$	0.53	$Be^{2+} + 2e^- \rightarrow Be$	−1.70
$Cu^+ + e^- \rightarrow Cu$	0.52	$Mg^{2+} + 2e^- \rightarrow Mg$	−2.37
$Cu^{2+} + 2e^- \rightarrow Cu$	0.34	$Na^+ + e^- \rightarrow Na$	−2.71
$Cu^{2+} + e^- \rightarrow Cu^+$	0.15	$Ca^{2+} + 2e^- \rightarrow Ca$	−2.87
$Sn^{4+} + 2e^- \rightarrow Sn^{2+}$	0.15	$Sr^{2+} + 2e^- \rightarrow Sr$	−2.89
$S + 2H^+ + 2e^- \rightarrow H_2S$	0.14	$Ba^{2+} + 2e^- \rightarrow Ba$	−2.90
$2H^+ + 2e^- \rightarrow H_2$	0.00	$Rb^+ + e^- \rightarrow Rb$	−2.92
$Pb^{2+} + 2e^- \rightarrow Pb$	−0.13	$K^+ + e^- \rightarrow K$	−2.92
$Sn^{2+} + 2e^- \rightarrow Sn$	−0.14	$Cs^+ + e^- \rightarrow Cs$	−2.92
$Ni^{2+} + 2e^- \rightarrow Ni$	−0.25	$Li^+ + e^- \rightarrow Li$	−3.05

The Periodic Table of the Elements

1																	2
H 1.0080																	**He** 4.0026
3 **Li** 6.941	4 **Be** 9.012											5 **B** 10.810	6 **C** 12.011	7 **N** 14.007	8 **O** 16.00	9 **F** 19.00	10 **Ne** 20.179
11 **Na** 22.99	12 **Mg** 24.31											13 **Al** 26.98	14 **Si** 28.09	15 **P** 30.974	16 **S** 32.06	17 **Cl** 35.453	18 **Ar** 39.948
19 **K** 39.10	20 **Ca** 40.08	21 **Sc** 44.96	22 **Ti** 47.87	23 **V** 50.94	24 **Cr** 52.00	25 **Mn** 54.94	26 **Fe** 55.85	27 **Co** 58.93	28 **Ni** 58.69	29 **Cu** 63.55	30 **Zn** 65.41	31 **Ga** 69.72	32 **Ge** 72.64	33 **As** 74.92	34 **Se** 78.96	35 **Br** 79.90	36 **Kr** 83.80
37 **Rb** 85.47	38 **Sr** 87.62	39 **Y** 88.91	40 **Zr** 91.22	41 **Nb** 92.91	42 **Mo** 95.94	43 **Tc** [98]	44 **Ru** 101.07	45 **Rh** 102.91	46 **Pd** 106.42	47 **Ag** 107.87	48 **Cd** 112.41	49 **In** 114.82	50 **Sn** 118.71	51 **Sb** 121.76	52 **Te** 127.60	53 **I** 126.90	54 **Xe** 131.29
55 **Cs** 132.91	56 **Ba** 137.33	57 **La** 138.91	72 **Hf** 178.49	73 **Ta** 180.95	74 **W** 183.84	75 **Re** 186.21	76 **Os** 190.23	77 **Ir** 192.22	78 **Pt** 195.08	79 **Au** 196.97	80 **Hg** 200.59	81 **Tl** 204.38	82 **Pb** 207.2	83 **Bi** 208.98	84 **Po** [209]	85 **At** [210]	86 **Rn** [222]
87 **Fr** [223]	88 **Ra** 226.03	89 **Ac** 227.03	104 **Rf** [267]	105 **Db** [268]	106 **Sg** [271]	107 **Bh** [272]	108 **Hs** [270]	109 **Mt** [276]	110 **Ds** [281]	111 **Rg** [280]	112 **Uub** [285]		114 **Uuq** [289]		116 **Uuh** [291]		

Lanthanides

58 **Ce** 140.12	59 **Pr** 140.91	60 **Nd** 144.24	61 **Pm** [145]	62 **Sm** 150.36	63 **Eu** 151.96	64 **Gd** 157.25	65 **Tb** 158.93	66 **Dy** 162.50	67 **Ho** 164.93	68 **Er** 167.26	69 **Tm** 168.93	70 **Yb** 173.04	71 **Lu** 174.97

Actinides

90 **Th** 232.04	91 **Pa** 231.04	92 **U** 238.03	93 **Np** 237.05	94 **Pu** [244]	95 **Am** [243]	96 **Cm** [247]	97 **Bk** [247]	98 **Cf** [251]	99 **Es** [252]	100 **Fm** [257]	101 **Md** [258]	102 **No** [259]	103 **Lr** [262]

USEFUL INFORMATION FOR SOLVING FREE-RESPONSE PROBLEMS

Atomic Structure

E = energy
v = frequency
$\lambda = \dfrac{h}{mv}$
$p = mv$
v = velocity
n = principal quantum number
m = mass
$\Delta E = hv$
$c = \lambda v$

λ = wavelength
p = momentum
$E_n = \dfrac{-2.178 \times 10^{-18} \text{ joule}}{n^2}$

Speed of light, $c = 3.00 \times 10^8$ m s^{-1}
Planck's constant, $h = 6.63 \times 10^{-34}$ J s
Boltzmann's constant, $k = 1.38 \times 10^{-23}$ J K^{-1}
Avogadro's number = 6.022×10^{23} mol^{-1}
Electron charge = -1.602×10^{-19} coulomb
1 electron volt/atom = 96.5×10^{23} K J mol^{-1}

Equilibrium and Thermochemistry

$K_a = \dfrac{[\text{H}^+][\text{A}^-]}{[\text{HA}]}$

$K_b = \dfrac{[\text{OH}^-][\text{HB}^+]}{[\text{B}]}$

$K_w = [\text{H}^+][\text{OH}^-]$
$K_w = 10^{-14}$ at 25 °C
$K_w = K_a K_b$
pH $= -\log[\text{H}^+]$
pOH $= -\log[\text{OH}^-]$
14.00 $=$ pH + pOH at 25 °C

pH $= pK_a + \log\dfrac{[\text{A}^-]}{[\text{HA}]}$

pOH $= pK_b + \log\dfrac{[\text{HB}^+]}{[\text{B}]}$

$pK_a = -\log K_a$
$pK_b = -\log K_b$
$K_p = K_c(RT)^{\Delta n}$
$\Delta n =$ moles product gas $-$ moles reactant gas
$\Delta S° = \Sigma S°$ products $- \Sigma S°$ reactants
$\Delta H° = \Sigma \Delta H°_f$ products $- \Sigma \Delta H°_f$ reactants
$\Delta G° = \Sigma \Delta G°_f$ products $- \Sigma \Delta G°_f$ reactants
$\Delta G° = \Delta H° - T\Delta S°$
$\quad = -RT \ln K$
$\quad = -n\mathscr{F}E°$

$\Delta G = \Delta G° + RT \ln Q$
$q = mc\,\Delta T$
$C_p = \dfrac{\Delta H}{\Delta T}$

K_a (weak acid)
K_b (weak base)
K_w (water)
K_p (gas pressure)
K_c (concentration)
$S°$ = standard entropy
$H°$ = standard enthalpy
$G°$ = standard free energy
T = temperature
n = moles
m = mass
q = heat
c = specific heat capacity
C_p = molar heat capacity at constant pressure
1 faraday $\mathscr{F} = 96{,}500$ coulombs/mole e^-

Gas constant, R = 8.31 J mol^{-1} K^{-1}
$\quad$ = 0.0821 L atm mol^{-1} K^{-1}
$\quad$ = 8.31 volt coulomb mol^{-1} K^{-1}

Practice Test 2

Gases, Liquids, Solutions, and Electrochemistry

$$PV = nRT$$

$$\left(P + \frac{n^2 a}{V^2}\right)(V - nb) = nRT$$

$$P_a = P_{total}X_a$$

$$X_a = \frac{mole_a}{total\ moles}$$

$$n = \frac{m}{M}$$

$$K = {}^\circ C + 273$$

$$\frac{P_1 V_1}{T_1} = \frac{P_2 V_2}{T_2}$$

$$D = \frac{m}{V}$$

$$u_{rms} = \sqrt{\frac{3kT}{m}}$$

$$KE\ per\ molecule = \frac{1}{2}mv^2$$

$$KE\ per\ mole = \frac{3}{2}RTn$$

$$\frac{r_1}{r_2} = \sqrt{\frac{M_2}{M_1}}$$

$$molarity,\ M = \frac{moles\ of\ solute}{liters\ of\ solution}$$

$$molality,\ m = \frac{moles\ of\ solute}{kilograms\ of\ solvent}$$

$$\Delta T_f = iK_f m$$

$$\Delta T_b = iK_b m$$

$$\Pi = \frac{nRTi}{V}$$

$$Q = \frac{[C]^c [D]^d}{[A]^a [B]^b}$$

$$aA + bB \rightleftharpoons cC + dD$$

$$I = \frac{q}{t}$$

$$E_{cell} = E^\circ_{cell} - \frac{RT}{n\mathscr{F}}\ln Q$$

$$\log K = \frac{nE^\circ}{0.0592}$$

P = pressure

V = volume

T = temperature

n = number of moles

D = density

m = mass

v = velocity

u_{rms} = root mean square speed

KE = kinetic energy

r = rate of effusion

M = molar mass

Π = osmotic pressure

i = van't Hoff factor

K_f = molal freezing point depression constant

K_b = molal boiling point elevation

Q = reaction quotient

I = current (amperes)

q = charge (coulombs)

t = time (seconds)

E° = standard reduction potential

K = equilibrium constant

Gas constant, R = 8.31 J mol^{-1} K^{-1}

= 0.0821 L atm mol^{-1} K^{-1}

= 8.31 volt coulomb mol^{-1} K^{-1}

SECTION II
Free-Response Questions

TIME: 1 HOUR AND 35 MINUTES

50% OF TOTAL SCORE

PART A: TIME—55 MINUTES.

Calculators are allowed EXCEPT THOSE WITH QWERTY (TYPEWRITER STYLE) KEYBOARDS. Graphing and programmable calculators may be used. Calculator memories do not have to be cleared of programs and data. You must answer ALL questions; there are no choices.

PART B: TIME—40 MINUTES

USE OF CALCULATORS IS NOT PERMITTED. You must answer ALL questions; there are no choices.

GENERAL INSTRUCTIONS

Times for Parts A and B will be announced separately. You may answer questions within each part in any sequence desired. Divide your time appropriately between each problem.

The periodic table, reduction potentials, and lists containing equations and constants are printed on the preceding pages.

You may answer in pen or pencil. Be sure to write CLEARLY and LEGIBLY. If you make an error, you may save time by crossing it out rather than erasing it. On the actual exam, *write all of your answers in the space provided after each question.* For this test, provide your own paper—approximately 20 sheets.

When instructed, you may begin Part A. When the 55 minutes are up, you must stop work and put away your calculator as directed. DO NOT start work on Part B until instructed to do so. You will be allowed to use the periodic table and the tables of information provided with Part A.

SECTION II

Free-Response Questions

TOTAL TIME = 1 HOUR AND 35 MINUTES

PART A

TIME—55 MINUTES

YOU MAY USE YOUR CALCULATOR ON PART A ONLY

> **Directions:** CLEARLY SHOW THESE STEPS AND METHODS USED IN ARRIVING AT YOUR ANSWERS. It is to your advantage to do this to earn partial credit. Answers without the steps and methods may receive little or no credit. Pay attention to significant figures. On the actual exam, you will need to write all of your answers in the pages of the appropriate booklet. For this test, use a separate sheet of paper.

Answer Question 1 below. (This question is worth 20 percent of Section II.)

1. At a certain temperature, the following exothermic reaction comes to equilibrium with all reactants and products in the gas phase.

$$H_2(g) + I_2(g) \rightleftarrows 2HI(g)$$

In one experiment at 750 °C, 0.00355 mol of $H_2(g)$ and 0.00435 mol of $HI(g)$ are mixed in a 5.00 L flask. When equilibrium was achieved, the concentration of $I_2(g)$ was found to be 0.000146 mol L^{-1}.

(a) What are the molar concentrations of $H_2(g)$ and $HI(g)$ in this equilibrium mixture?

(b) What is the partial pressure of $I_2(g)$ in the reaction flask?

(c) What is the value of the equilibrium constant K_C?

(d) Show the mathematical relationship between K_C and K_P for this reaction. Explain how you could increase the value of K_C.

(e) What is the mole fraction of $I_2(g)$ in the reaction vessel at equilibrium?

(f) In a new experiment, the previous flask is cleaned out and 0.00100 moles of each substance is introduced. Calculate the concentrations of $H_2(g)$, $I_2(g)$, and $HI(g)$ when the system comes to equilibrium.

Answer Question 2 below. (This question is worth 20 percent of Section II.)

2. When 5.000 grams of an organic compound containing only carbon and hydrogen is burned, the combustion is incomplete, forming CO_2, CO, and H_2O. The water is absorbed by a calcium chloride in a drying tube. The mass of the drying tube increases by 6.432 grams. The carbon dioxide is absorbed by lithium hydroxide, and the increase in mass is found to be 11.125 grams.

(a) How many grams of carbon and hydrogen are in the sample?
(b) What is the empirical formula of the sample?
(d) At 25 °C, 1.000 grams of the compound is introduced into a 1.000-liter flask. The pressure of the gas in the flask is 663 torr. What is the molecular mass of the compound?
(d) What is the molecular formula of the compound?
(e) Draw the structure of this compound and name it.
(f) Describe the geometry of the molecule, list any hybrid bonds, and list the sigma and pi bonding if any.

Answer Question 3 below. (This question is worth 20 percent of Section II.)

3. Two solutions are mixed in equal 100-mL volume portions. The first solution contains 0.150 molar silver nitrate. The second solution contains 0.200 molar calcium chloride. For silver chloride, the $K_{sp} = 1.8 \times 10^{-10}$.

$$2\,AgNO_3(aq) + CaCl_2(aq) \rightarrow 2\,AgCl(s) + Ca(NO_3)_2$$

(a) Write the net ionic equation for this reaction.
(b) What is the limiting reactant?
(c) How many grams of precipitate are formed?
(d) What is the final molarity of the following?
 (i) nitrate ions
 (ii) calcium ions
 (iii) chloride ions
 (iv) silver ions
(e) What is the molar solubility of AgCl in distilled water?
(f) What is the molar solubility of AgCl in a solution that is 0.200 M in $MgCl_2$?

Practice Test 2

PART B

TIME—40 MINUTES

(Calculators May Not Be Used)

Answer Question 4 below. (This question is worth 5 percent of your total score.)

4. For each of the following three reactions, in part (i) write a BALANCED equation and in part (ii) answer the question about the reaction. In part (i), coefficients should be in terms of lowest whole numbers. Assume that solutions are aqueous unless otherwise indicated. Represent substances in solutions as ions if the substances are extensively ionized. Omit formulas for any ions or molecules that are unchanged by the reaction.

Example: (i) A strip of magnesium is added to a solution of silver nitrate.

ANSWER: $Mg + 2 Ag^+ \rightarrow Mg^{2+} + 2 Ag$

  (ii) Which substance is oxidized in the reaction?

ANSWER: Magnesium (Mg) metal

(a) (i) A small piece of potassium is dropped into a large pan of water.
  (ii) Is this reaction exothermic or endothermic?

(b) (i) Sulfuric acid is slowly added to a solution of sodium sulfide.
  (ii) One of the reactants, or products, has a very notable physical property. What is it?

(c) (i) Excess concentrated hydrochloric acid is added to a solution of iron(III) nitrate.
  (ii) What type of acid-base reaction is this?

Practice Test 2

Directions: TWO essay questions below will count for **15 percent of THE ENTIRE TEST (7.5% EACH)**. You **MUST ANSWER** Questions 5 and 6. Answers to these questions should demonstrate your ability to present logical, coherent, and convincing explanations of chemical facts and observations. Answers are judged on the accuracy of your analysis and on the appropriateness of the details and examples cited. Specific answers and examples are preferred to broad generalizations. Diagrams, illustrations, and equations may be used in the answers.

5. The chemical reaction represented below is to be studied electrochemically.

$$Mg^0(s) + 2\ Fe^{3+}(aq) \rightarrow Mg^{2+}(aq) + 2\ Fe^{2+}(aq)$$

(a) Draw a fully labeled diagram of the galvanic cell apparatus showing the appropriate arrangement of all components of the experiment.
(b) What is the E^o_{cell} for this experiment?
(c) What is the equilibrium constant for this reaction?
(d) With your apparatus, you arrange it so that the starting concentrations of the soluble species are all $2.0 \times 10^{-2}\ M$. The volume of all the solutions used is 100 mL. What is the value of E_{cell}?
(e) The two electrodes are short-circuited, and the reaction is allowed to come to equilibrium. What is the value of E_{cell}?
(f) At equilibrium what are the concentrations of all soluble species?

6. The kinetic-molecular theory of gases illustrates why the ideal gas law can be applied to all gases.

(a) Define the properties of an ideal gas as contrasted with a real gas.
(b) What changes need to be made to the ideal gas law equation to account for the properties of real gases.
(c) Under what conditions do all gases most closely approximate ideal gases?
(d) What are the postulates of the kinetic-molecular theory of gases?
(e) How does the kinetic-molecular theory of gases explain each of the following laws?
  (i) Boyle's law
  (ii) Charles' law
  (iii) Gay-Lussac's law

End of Section II.
You may use any remaining time to check your work in this section.

Answer Key
PRACTICE TEST 2

1. E	26. E	51. A
2. B	27. C	52. C
3. A	28. D	53. B
4. C	29. D	54. D
5. D	30. E	55. D
6. B	31. C	56. A
7. A	32. C	57. E
8. C	33. A	58. E
9. A	34. B	59. E
10. E	35. C	60. E
11. D	36. A	61. C
12. B	37. B	62. E
13. E	38. D	63. C
14. C	39. C	64. B
15. C	40. E	65. A
16. B	41. D	66. A
17. E	42. D	67. A
18. B	43. C	68. E
19. D	44. E	69. C
20. D	45. E	70. A
21. D	46. C	71. C
22. C	47. D	72. E
23. A	48. A	73. B
24. C	49. D	74. C
25. D	50. A	75. B

CHAPTER REFERENCES FOR MULTIPLE-CHOICE QUESTIONS

Question	Chapter	Question	Chapter	Question	Chapter
1	5, 15	26	5	51	10, 12
2	5, 15	27	2, 8	52	8, 9
3	5, 8, 15	28	5, 6, 10	53	5, 5, 12
4	1, 2, 4	29	1, 2, 4	54	4, 10, 12
5	1, 2, 4	30	9, 5	55	4, 6, 9, 14
6	1, 2, 4	31	8, 10, 14	56	1, 4, 6
7	1, 4	32	10, 14	57	4, 7, 9
8	13, 16	33	1, 5	58	4, 5, 6
9	13	34	3, 11	59	7, 8, 9
10	13, 14	35	2, 16	60	11, 12
11	5, 15	36	2, 5, 7	61	7, 8
12	5, 8	37	1, 5, 14, 15	62	4, 14
13	5, 8	38	7, 11	63	4, 5, 7, 11
14	4, 5, 8	39	10, 12	64	10, 12, 14
15	12, 16	40	1, 2, 3	65	5, 15, 16
16	4, 6, 9	41	4, 5, 14, 16	66	7, 16
17	4, 9	42	4, 13	67	4, 16
18	5, 9	43	1, 2, 4	68	10, 12
19	5, 9, 14	44	5, 7	69	9, 16
20	1, 2, 4	45	4, 10, 12, 13	70	5, 15
21	6, 14, 16	46	4, 13	71	5, 7
22	3, 11	47	2, 3	72	7, 8
23	5, 7, 15	48	4, 6, 7	73	4, 5, 7
24	6, 8	49	1, 2	74	10, 12
25	5, 7, 11	50	11, 16	75	4, 9, 16

Answers Explained—Practice Test 2

ANSWERS EXPLAINED

Section I—Multiple-Choice

1. **(E)** Propene (propylene) is the monomer for polypropylene.

2. **(B)** Benzene is the only compound that has two or more reasonable resonance structures.

3. **(A)** CH_4 is nonpolar and also is the smallest molecule (fewest London forces per molecule) and is expected to have the lowest boiling point.

4. **(C)** This is the configuration of calcium.

5. **(D)** This is the iodide ion.

6. **(B)** A period 4 transition element has its $3d$ level occupied with one or more electrons and no electrons in the $4p$ or higher levels.

7. **(A)** For the given number of electrons, this configuration should end in $5s^2$. As written, one of these electrons has been excited to the $6s$ level.

8. **(C)** The double line represents the salt bridge.

9. **(A)** Oxidation always occurs at the anode.

10. **(E)** The reference electrode must have a constant voltage.

11. **(D)** Shared electrons, as in the C—H bond, are defined as covalent bonds.

12. **(B)** These include instantaneous and induced dipoles. London forces describe why CO_2 can be solidified.

13. **(E)** London forces, also known as instantaneous dipole or induced dipole attractive forces, provide attractive forces for otherwise nonpolar substances.

14. **(C)** Attractive forces between fully charged ions are the strongest forces, whether in bonds (intramolecular) or between molecules (intermolecular).

15. **(C)** In a closed system the total energy must remain constant. Because the reaction is exothermic, the temperature will increase.

16. **(B)** All others have $[Cl^-] = 0.200$ M.

17. **(E)** The subscript 2 for the sodium indicates that two Na^+ ions form, and the remainder is the oxalate ion.

18. **(B)** Carbon and fluorine are the most widely separated of these pairs in the periodic table and are therefore expected to have the greatest electronegativity difference.

19. **(D)** This is the only unsymmetric molecule, and it is the only one that has any polarity. Dipole moment and polarity are similar concepts.

20. **(D)** Gold will lose the $6s^2$ electrons and one of its outermost d electrons.

21. **(D)** The results show an obvious trend that cannot be ignored, as response (A) suggests. Dry glassware and the intensity of the indicator color should have no effect. The effect of CO_2 would not be as pronounced, and the dissolution of CO_2 is slow.

22. **(C)** Accurate dating of materials by radioactive decay requires that at least 10 percent of the original material decays. In addition, at least 10 percent should be remaining. This translates into a range of approximately 0.33 to 3 half-lives for accurate dating. Only lunar rocks fall in this range for uranium.

23. **(A)** The nitrogen oxides are often brown in color, especially NO_2 and N_2O_4.

24. **(C)** This is the only conversion between temperature-dependent and temperature-independent concentration units.

25. **(D)** The fact that N_2 effuses twice as fast as the unknown compound indicates that the unknown has a mass that is 4 times that of nitrogen: $4 \times 28 = 112$. Dividing the molar mass of 112 by 14, the molar mass of the CH_2, results in the number 8, which indicates there are 8 CH_2 units in the molecular formula. The molecular formula is therefore C_8H_{16}.

26. **(E)** Only iron(III) chloride is correctly named. The correct names, in order, are antimony pentachloride or antimony(V) chloride, ethylene or ethene, dinitrogen pentoxide, methylamine.

27. **(C)** Diamond is a network crystal with each carbon in a tetrahedral network bonded to other carbon atoms throughout the entire crystal.

28. **(D)** Calculate the value of Q:

$$\frac{[N_2O_4]}{[NO_2]^2} = \frac{(0.006)}{(0.002)^2} = \frac{(0.006)}{(0.002)(0.002)} = \frac{(0.003)}{(0.002)} = 1.5$$

Because Q is less than 234, it must increase until equilibrium is reached. This means that N_2O_4 must increase and NO_2 must decrease.

29. **(D)** This is the only set of quantum numbers that can describe a $4s$ electron of potassium

30. **(E)** $\Delta T = K_f \, m = K_f \, \dfrac{\text{g solute/molar mass of solute}}{\text{kilograms of solvent}}$. With the mass of the solvent and the mass of the sample, the only remaining unknown is the molar mass.

31. **(C)** When the pH of the buffer and the pK_a of the salts have the same value, then the ratio of the molarity of the conjugate base to that of the conjugate acid is equal to 1.0. If one mole of each substance is used, the mass ratio will be $\dfrac{142 \text{ g Na}_2\text{HPO}_4}{120 \text{ g NaH}_2\text{PO}_4} = 1.2$

32. **(C)** Given that $[\text{H}^+] = 7.4 \times 10^{-3}$. Divide this by the initial concentration and multiply by 100 to convert to percent. % dissociated $= \dfrac{7.4 \times 10^{-3}}{0.080} \times 100$ $= 9.2\%$

    To estimate, round 0.0074 to 0.008. Then 0.008/0.08 = 0.1, or 10 percent, and we know from the rounding that the actual answer should be slightly less than 10 percent.

33. **(A)** Water has only one possible Lewis structure.

34. **(B)** The equation ln 2 = kt is solved for k. Time must be converted into seconds first.

$$k = \frac{\ln 2}{(228 \text{ min})(60 \text{ s/min})}$$

35. **(C)** A straight line or a smooth curve should be drawn through the data points instead of connecting them dot to dot.

36. **(A)** The problem is set up as $?\ \dfrac{\text{g}}{\text{mol}} = \dfrac{3.79 \text{ g}}{\text{L}}\left(\dfrac{22.4 \text{ L gas}}{1 \text{ mol gas}}\right)$, and the answer can be estimated as $?\ \dfrac{\text{g}}{\text{mol}} = 4 \times 20 = 80$. The molar mass of hexane is 84.

37. **(B)** The stronger the bond, the shorter the bond length. Other properties of strong bonds are high vibrational frequency and absorption of higher-frequency radiation to match.

38. **(D)** All three properties are included in collision theory.

39. **(C)** The ONLY possible way to alter the value of an equilibrium constant is to change the temperature.

40. **(E)** This has the same atomic number and different mass numbers; C represents the same element with identical atomic numbers and mass numbers.

41. **(D)** The addition of base forms ammonia with a characteristic pungent odor.

42. **(D)** The balanced equation is

$$Fe_2O_3(s) + C_2O_4^{2-}(aq) + 6\ H^+(aq) \rightarrow 2\ Fe^{2+}(aq) + 2\ CO_2(g) + 3\ H_2O(l)$$

43. **(C)** The element is identified as aluminum, which only forms a 3+ ion, and Al_2O_3 is the oxide.

44. **(E)** The most polar molecule, water, will deviate from ideal behavior the most.

45. **(E)** When K_{eq} is greater than 1.00, then $\Delta G°$ will be negative $(-\ RT\ \ln K)$ and $E°_{cell}$ will be positive; both are considered indications of spontaneous reactions.

46. **(C)** The balanced half-reaction is

$$3\ e^- + 4\ H^+ + NO_3^- \rightarrow NO + 2\ H_2O$$

47. **(D)** The alpha particle has a mass of 4 and a nuclear charge of 2+.

48. **(A)** The chemical reaction is $CO_2(g) + H_2O(l) + Ca^{2+}(aq) \rightarrow CaCO_3(s) + 2\ H^+(aq)$. The calculation is set up as

$$?\ L\ CO_2 = 0.100\ L \left(\frac{0.250\ mol\ Ca(NO_3)_2}{1\ L\ Ca(NO_3)_2} \right) \left(\frac{1\ mol\ CO_2}{1\ mol\ Ca(NO_3)_2} \right) \left(\frac{22.4\ L\ CO_2}{1\ mol\ CO_2} \right)$$

The answer is estimated as $?\ L\ CO_2 = 0.1 \times 0.25 \times 20 = 0.5\ L = 500\ mL$.

The answer with a calculator is $?\ L\ CO_2 = 0.560\ L = 560\ mL\ CO_2$.

49. **(D)** Microwaves have the longest wavelength.

50. **(A)** Sodium chloride is not a catalyst.

51. **(A)** Because the two reactions are added, the overall equilibrium constant is the product of the two stepwise equilibrium constants:

$$K_{overall} = K_f K_{sp} = (7.7 \times 10^{+16})(1.8 \times 10^{-14}) = 1.4 \times 10^3.$$

To estimate the answer, round to one significant figure:

$$K_{overall} = (8 \times 10^{+16})(2 \times 10^{-14}) = 16 \times 10^2 = 1.6 \times 10^3$$

52. **(C)** The correct sequence is London forces < dipole interactions < hydrogen bonds.

53. **(B)** This reaction has $\Delta n = +1$, meaning that there is one more mole of gas on the product side than on the reactant side of the balanced equation.

54. **(D)** This equation has the same number of moles of gas as products and reactants, $\Delta n = 0$.

55. **(D)** The dilution equation $C_i V_i = C_f V_f$ is solved for V_i because $C_i = 15.4\ M$, $f = 6.00\ M$, and $V_f = 5.00\ L$.

$$V_i = \left(\frac{(6.00\ M)(5.00\ L)}{(15.4\ M)} \right) = 1.95\ L = 1950\ mL$$

The estimated answer is $V_f = \left(\frac{(6.00\ M)(5.00\ L)}{(15\ M)} \right) = \frac{30}{15}\ L = 2\ L$. Note that only the 15.4 M was rounded to 15 to make the problem very simple.

56. **(A)** The problem is set up as

$$?\ g\ Pb(NO_3)_2 = 3.00\ g\ Pb \left(\frac{1\ mol\ Pb}{207\ g\ Pb} \right) \left(\frac{1\ mol\ Pb(NO_3)_2}{1\ mol\ Pb} \right) \left(\frac{331\ g\ Pb(NO_3)_2}{1\ mol\ Pb(NO_3)_2} \right)$$

$$= 4.80\ g\ Pb(NO_3)_2$$

To estimate the answer: $?\ g\ Pb(NO_3)_2 = 3.00 \times \frac{300}{200} = 4.5\ g\ Pb(NO_3)_2$

57. **(E)** For a boiling-point-elevation experiment the concentration must be grams per kilogram of solvent, not grams per liter as given.

58. **(E)** CH_3O is the only possible empirical formula because it has oxygen and the compound is known to contain oxygen. No calculations are needed here. For those interested, the actual empirical formula calculation is

$$?\ mol\ C = 38.71\ g\ C \left(\frac{1\ mol\ C}{12\ g\ C} \right) = 3.226\ mol\ C$$

$$?\ mol\ H = 9.68\ g\ H \left(\frac{1\ mol\ H}{1\ g\ H} \right) = 9.68\ mol\ H$$

$$?\ mol\ O = 51.61\ g\ O \left(\frac{1\ mol\ O}{16\ g\ O} \right) = 3.226\ mol\ O$$

Dividing these three numbers by 3.226 gives 1 C, 3 H, and 1 O, or CH_3O.

59. **(E)** Gas densities depend only on the molar mass, with very slight variations arising from intermolecular attractive forces.

60. **(E)** Only removing SO_2 will have the desired effect according to Le Châtelier's principle. The solid Fe_2O_3 has no effect on equilibrium, it does not appear in the equilibrium law. The heat of reaction indicates that cooling, not heating, will help, and there are no catalysts specific for only one direction in a reaction.

61. **(C)** 380 mm Hg = 0.500 atmosphere. Therefore the addition of N_2 doubles the pressure and correspondingly doubles the concentration of gases in the flask to 0.046 M. Of that concentration, 0.023 is H_2, and the other 0.023 must be N_2.

62. **(E)** All are equal because they are all "leveled" to the strength of H_3O^+.

63. **(C)** Gas behavior is described. This is often confused with the collision theory of reaction rates.

64. **(B)** Precipitation will occur when the concentrations of the ions, multiplied as specified by the K_{sp} expression, exceed the value of K_{sp}.

$$K_{sp} = [Fe^{3+}][OH^-]^3 = 1.6 \times 10^{-39}$$

and

$$[Fe^{3+}] = \frac{40 \times 10^{-3} \text{ gFe}_2(SO_4)_3 \text{ L}^{-1}}{400 \text{ g Fe}_2(SO_4)_3 \text{ mol}^{-1}} \times \frac{2 \text{ mol Fe}^{3+}}{1 \text{ mol Fe}_2(SO_4)_3}$$

$$= 2.0 \times 10^{-4} M$$

Substituting and rearranging, $[OH^-] = \sqrt[3]{\dfrac{1.6 \times 10^{-39}}{2 \times 10^{-4}}}$

$$= \sqrt[3]{8 \times 10^{-36}} = 2 \times 10^{-12}$$

$pOH = -\log(2 \times 10^{-12}) = 11.7$, and the $pH = 14.0 - 11.7 = 2.3$.

Note that if set up properly, the calculations can be done easily without a calculator. To estimate logarithms, we note that the logarithm of a power of 10 is just the exponent itself. This allows the estimation that the pOH is between 11 and 12, and the corresponding pH must be between 2 and 3.

65. **(A)** The —COOH group is indicative of the organic acids and will decrease the pH.

66. **(A)** To calculate moles of gas, the ideal gas law is used: $PV = nRT$. The only unknown besides n is the volume. R can be looked up in a book at any time.

67. **(A)** The dichromate ion is not considered a complex ion; it is a polyatomic anion.

68. **(E)** $K_{sp} = [Mg^{2+}][F^-]^2 = (x)(2x)^2 = 4x^3$, where x is the molar solubility of MgF_2.

69. **(C)** The errors for each response are given in fractional form: (a) 10/1, (b) 298/25, (c) 23/0.03, (d) 8.314/0.0821, (e) 500/250. The largest of these is (c) which is a factor of 700.

70. **(A)** The bonds are O=C=O; each double bond consists of one sigma and one pi bond.

71. **(C)** Ideal gases are assumed to have no attractive or repulsive forces.

72. **(E)** Raoult's law is used: $P_{\text{total}} = X_p(440 \text{ mm Hg}) + (X_h)(40 \text{ mm Hg}) = 340$ mm Hg.

The mole fractions add up to $1.00 = X_p + X_h$. Substitute $(1 - X_p) = X_h$ and solve.

$$340 = 440X_p + 40 - 40X_p$$
$$300 = 400X_p$$
$$X_p = \frac{300}{400} = 0.75$$

73. **(B)** The reaction must first be balanced as

$$Al_2(CO_3)_3 + 6 \text{ HCl} \rightarrow 2 \text{ AlCl}_3 + 3 \text{ H}_2O + 3 \text{ CO}_2$$

Then the correct setup is

? g $Al_2(CO_3)_3$ = 0.005 L HCl
$$\left(\frac{6.00 \text{ mol HCl}}{1 \text{ L HCl}}\right)\left(\frac{1 \text{ mol Al}_2(CO_3)_3}{6 \text{ mol HCl}}\right)\left(\frac{234 \text{ g Al}_2(CO_3)_3}{1 \text{ mol Al}_2(CO_3)_3}\right)$$
$$= 1.17 \text{ g Al}_2(CO_3)_3 = 1170 \text{ mg Al}_2(CO_3)_3$$

To estimate the answer: ? g $Al_2(CO_3)_3$ = 0.005 × 234 = more than 1 gram

Because we have only 400 mg, $Al_2(CO_3)_3$ it is the limiting reactant. Now calculate the moles of CO_2:

? mol CO_2 = 0.400 g $Al_2(CO_3)_3$ $\left(\dfrac{1 \text{ mol Al}_2(CO_3)_3}{234 \text{ g Al}_2(CO_3)_3}\right)\left(\dfrac{3 \text{ mol CO}_2}{1 \text{ mol Al}_2(CO_3)_3}\right)$
$$= 0.0051 \text{ mol CO}_2$$

To estimate the answer: ? mol CO_2 = 0.4 × 3/200 = 1.2/200 = 0.006 mol CO_2. Note that because only the denominator was rounded to a smaller number, we also know that the real answer is slightly less than 0.006 and can choose the correct response.

74. **(C)** The ΔG value is negative. That tells us that the reaction will proceed in the forward direction. The $\Delta G°$ tells us the position of equilibrium. Since it is negative and the concentrations are near standard conditions, we conclude that the equilibrium mixture will contain more products than reactants.

75. **(B)** The problem is correctly set up as

? g $CaCl_2$ = 0.0750 L $CaCl_2$ $\left(\dfrac{2.00 \text{ mol CaCl}_2}{1 \text{ L CaCl}_2}\right)\left(\dfrac{111 \text{ g CaCl}_2}{1 \text{ mol CaCl}_2}\right) = 16.65$ g $CaCl_2$

The answer is estimated as ? g $CaCl_2$ = 0.075 × 2 × 100 = 15 g $CaCl_2$. (Note that 111 was rounded to 100.)

Section II—Free-Response

Part A

1. (a) We can set up an ICE table, but the reaction is simple enough that it is not needed. The first thing to do is to calculate the initial concentrations of the two given substances, H_2 and HI, by dividing by the volume of the flask.

$$[H_2]_{initial} = (0.00355 \text{ mol } H_2/5.00 \text{ L}) = 0.000710 \text{ mol } H_2 \text{ L}^{-1}$$

$$[HI]_{initial} = (0.00435 \text{ mol HI}/5.00 \text{ L}) = 0.000870 \text{ mol HI L}^{-1}$$

$$H_2(g) + I_2(g) \rightleftarrows 2HI(g)$$
$$+x \quad +x \quad -2x$$

Since there is no I_2 present initially, it must have come from the HI. As we can see from above, the 0.000146 mol I_2 L^{-1} is designated as x. Since $2x$ of the HI reacted, this means that 0.000292 mol HI L^{-1} must have decomposed. The remaining HI is $0.000870 - 0.000292 = 0.000578$ mol HI L^{-1}. The concentration of H_2 must increase by x, and we calculate $0.000710 + 0.000146 = 0.000856$ mol H_2 L^{-1}.

(b) We have the information to solve the ideal gas law, $PV = nRT$. In 1.00 liter, we have 0.000146 moles of I_2. The temperature is $273 + 750 = 1023$ K. The value of R to use is 0.0821 L atm mol^{-1} K^{-1}. We will solve for the partial pressure in units of atmospheres.

$$P(1.00 \text{ L}) = (0.000146 \text{ mol})(0.0821 \text{ L atm mol}^{-1} \text{ K}^{-1})(1023 \text{ K})$$
$$P = 0.0123 \text{ atm}$$

(c) The equilibrium law is

$$K_C = \frac{[HI]^2}{[H_2][I_2]}$$

We take the information from part (a) and calculate:

$$K_C = (0.000578)^2/(0.000856)(0.000146) = 2.67$$

(d) We know, or can derive, the relationship between K_C and K_P, which is

$$K_P = K_C (RT)^{\Delta n_g}$$

Recall that the term Δn_g is the difference between the moles of gaseous products and reactants. We see that $\Delta n_g = 0$ and therefore $K_C = K_P$.

(e) We can calculate:

$$\text{mol } I_2 = (0.000146 \, M)(5.00 \text{ L}) = 0.000730 \text{ mol } I_2$$
$$\text{mol } H_2 = (0.000856 \, M)(5.00 \text{ L}) = 0.00428 \text{ mol } H_2$$
$$\text{mol HI} = (0.000578 \, M)(5.00 \text{ L}) = 0.00289 \text{ mol HI}$$

The total moles in the system is 0.00790 mol.

The mol fraction is $X = (0.000730)/0.00790 = 0.0924$

(f) We will set up an ICE table to solve this problem:

Reaction	$H_2(g)$	+	$I_2(g)$	$\rightleftarrows$	$2HI(g)$
INITIAL CONCENTRATION	0.000200		0.000200		0.000200
CHANGE	$+x$		$+x$		$-2x$
EQUILIBRIUM	$0.000200 + x$		$0.000200 + x$		$0.000200 - 2x$

We then write the equilibrium law as:

$$K_C = 2.67 = \frac{(0.000200 - 2x)^2}{(0.000200 + x)(0.000200 + x)}$$

We take the square root of both sides to get:

$$K_C = 2.67 = \frac{(0.000200 - 2x)^2}{(0.000200 + x)(0.000200 + x)}$$

$$K_C = 1.63 = \frac{(0.000200 - 2x)}{(0.000200 + x)}$$

$$x = -0.000039 \text{ (notice the negative sign!)}$$

By entering our value of x in the above equilibrium line of our ICE table, we get:

$$[H_2] = [I_2] = 0.000169 \, M$$
$$[HI] = 0.000278 \, M$$

2. (a) We can calculate the mass of hydrogen in the water as

$$? \text{ g H} = 6.432 \text{ g } H_2O \left(\frac{1 \text{ mol } H_2O}{18 \text{ g } H_2O}\right)\left(\frac{2 \text{ mol H}}{1 \text{ mol } H_2O}\right)\left(\frac{1 \text{ g H}}{1 \text{ mol H}}\right) = 0.715 \text{ g}$$

$? \text{ g C} = 5.00 \text{ g compound} - 0.715 \text{ g H} = 4.285 \text{ g C}$

(b) The simplest ratio of moles of carbon and hydrogen is calculated as

$$? \text{ mol H} = 0.715 \text{ g H} \left(\frac{1 \text{ mol H}}{1 \text{ g H}} \right) = 0.715 \text{ mol H}$$

$$? \text{ mol C} = 4.285 \text{ g C} \left(\frac{1 \text{ mol C}}{12 \text{ g C}} \right) = 0.357 \text{ mol C}$$

To normalize the data above to small whole numbers, the smallest value is divided into both results:

$$(0.357/0.357) \text{ mol C} = 1 \text{ mol C}$$

$$(0.715/0.357) \text{ mol H} = 2 \text{ mol H}$$

The empirical formula is CH_2.

(c) To determine the molecular mass we use the ideal gas law $PV = nRT$. We substitute $n = $ g/molar mass to get

$$PV = \left(\frac{\text{g } RT}{\text{molar mass}} \right), \text{ and this is rearranged to molar mass} = \left(\frac{\text{g } RT}{PV} \right)$$

Substituting the data from the problem yields

$$\text{molar mass} = \left(\frac{(1.00 \text{ g}) \ (0.0821 \text{ L atm mol}^{-1} \text{ K}^{-1}) \ (298 \text{ K})}{(663 \text{ torr}/760 \text{ torr atm}^{-1}) \ (1.00 \text{ L})} \right)$$

$$= 28.0 \text{ g mol}^{-1}$$

(d) The number of empirical formula units per molecule is determined by dividing the molecular mass by the mass of the empirical formula unit:

no. empirical formula units $= 28.0$ g mol^{-1}/14.0 g emp-form-unit$^{-1} = 2$

Therefore the molecular formula is twice the empirical formula, or C_2H_4.

(e) The only reasonable structure is

$$\begin{array}{ccc} \text{H} & & \text{H} \\ \diagdown & & \diagup \\ & \text{C} = \text{C} & \\ \diagup & & \diagdown \\ \text{H} & & \text{H} \end{array}$$

It is named ethene (ethylene).

(f) This compound contains five sigma bonds and one pi bond. The pi bond is part of the double bond. Each carbon atom is sp^2-hybridized. Because sp^2 hybrids are planar, the molecule is totally planar.

3. (a) The net ionic equation is:

$$Ag^+(aq) + Cl^-(aq) \rightarrow AgCl(s)$$

(b) and (c) To solve parts (b) and (c) together, calculate the mass of AgCl produced from each reactant. The limiting reactant will produce the smallest amount of product, thus answering both parts (a) and (b).

? g AgCl =

$$100 \text{ mL AgNO}_3 \left(\frac{0.150 \text{ mol AgNO}_3}{1000 \text{ mL AgNO}_3} \right) \left(\frac{2 \text{ mol AgCl}}{2 \text{ mol AgNO}_3} \right) \left(\frac{143 \text{ g AgCl}}{1 \text{ mol AgCl}} \right)$$

= 2.145 g AgCl = 2.14 g AgCl when properly rounded

? g AgCl =

$$100 \text{ mL CaCl}_2 \left(\frac{0.200 \text{ mol CaCl}_2}{1000 \text{ mL CaCl}_2} \right) \left(\frac{2 \text{ mol AgCl}}{1 \text{ mol CaCl}_2} \right) \left(\frac{143 \text{ g AgCl}}{1 \text{ mol AgCl}} \right)$$

= 5.72 g AgCl

The limiting reactant is $AgNO_3$ because it produces the least product (it limits the amount of product); 2.14 grams of AgCl is produced.

(d) For the spectator ions, we can use the dilution equation, so that for the nitrate ion:

(i) $C_i V_i = C_f V_f$

$(0.150 \ M)(100 \text{ mL}) = (x \ M)(200 \text{ mL})$

$x \ M = 0.075 \ M \text{ NO}_3$

(ii) $C_i V_i = C_f V_f$

$(0.200 \ M)(100 \text{ mL}) = (x \ M)(200 \text{ mL})$

$x \ M = 0.100 \ M \text{ Ca}^{2+}$

(iii) The original chloride ion concentration was $0.200 \times 2 = 0.400 \ M$. It reacted with all of the 0.150 M, silver ions leaving $0.400 - 0.150 = 0.250$

M chloride. The dilution is then

$C_i V_i = C_f V_f$

$(0.250 \ M)(100 \text{ mL}) = (x \ M)(200 \text{ mL})$

$x \ M = 0.125 \ M \text{ Cl}^-$

(iv) $K_{sp} = 1.8 \times 10^{-10} = [\text{Ag}^+][\text{Cl}^-] = [\text{Ag}^+] \times 0.125$

$[\text{Ag}^+] = 1.44 \times 10^{-9}$

(e) We write the solubility product as $K_{sp} = [\text{Ag}^+][\text{Cl}^-] = 1.8 \times 10^{-10}$.

If s moles per liter of AgCl dissolves, then s moles per liter of Ag^+ will be present along with s moles per liter of Cl^- ions. We enter s for both concentrations to get

$$s \times s = s^2 = 1.8 \times 10^{-10}$$

Solving for s we get $s = 1.3 \times 10^{-5}$ moles per liter as the molar solubility of AgCl.

(f) In this part of the problem, we consider dissolving AgCl in a solution that already contains a common ion, the chloride ion. We will use the format above, but we will add the chloride ions from the $MgCl_2$ to the chloride ions, s, from the AgCl.

First we calculate the chloride ions in the solution of $MgCl_2$ as:

$$\frac{?\,\text{mol Cl}^-}{\text{L}_{\text{solution}}} = \frac{0.200 \text{ mol MgCl}_2}{\text{L}_{\text{solution}}}\left(\frac{2 \text{ mol Cl}^-}{1 \text{ mol MgCl}_2}\right) = 0.400 \, M \text{ Cl}^-$$

We then write the equilibrium law as:

$$1.8\times10^{-10} = \left[\text{Ag}^+\right]\left[\text{Cl}^-\right] = (s)(0.400+s)$$

(At this point we assume that $\ll 0.400$.)

$$1.8\times10^{-10} = (s)(0.400)$$

$$s = 1.8\times10^{-10}/0.400 = 4.5\times10^{-10}$$

We note that that $4.5 \times 10^{-10} \ll 0.400$, so the assumption is true.

We can be confident that the molar solubility of AgCl under these conditions is $4.5 \times 10^{-10} \, M$.

Part B

4. For these answers, balanced equations are required and only reacting species should be shown. An additional touch would be to add reasonable phases for the substances in the reaction. The instructions indicate that most are reactions in aqueous solution.

(a) (i) A small piece of potassium is dropped into a large pan of water.

$$\text{K}(s) + 2\text{H}_2\text{O}(l) + \rightarrow 2\text{K}^+(aq) + 2\text{OH}^-(aq) + \text{H}_2(g)$$

(ii) Is this reaction exothermic or endothermic?

We observe a very vigorous reaction and the ignition of the hydrogen in some cases. This indicates an exothermic process.

(b) (i) Sulfuric acid is slowly added to a solution of sodium sulfide.

$$\text{S}^{2-}(aq) + 2\text{H}^+(aq) \rightarrow \text{H}_2\text{S}(g)$$

(ii) One of the reactants, or products, has a very notable physical property. What is it?

The hydrogen sulfide gas has the very distinctive odor of rotten eggs.

(c) (i) Excess concentrated hydrochloric acid is added to a solution of iron(III) nitrate.

$$Fe^{3+}(aq) + 6\,Cl^-(aq) \rightarrow FeCl_6^{3-}(aq)$$

(ii) What type of acid-base reaction is this?

A complex ion is formed. The reaction is a Lewis acid (the Fe^{3+}) reacting with the chloride ion, which is a Lewis base.

5. (a) The complete apparatus should resemble the following. It may be reversed, but the polarity of the meter must also be reversed.

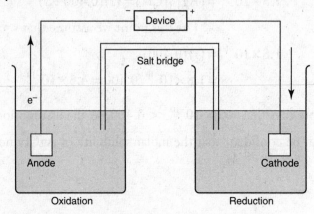

The two half-reactions are $Mg \rightarrow Mg^{2+} + 2\,e^-$ (oxidation)

and $\qquad\qquad Fe^{3+} + e^- \rightarrow Fe^{2+}$ (reduction)

The left cell should have a magnesium electrode and Mg^{2+} in solution, The right-hand cell should have an inert electrode such as Pt and Fe^{2+} and Fe^{3+} in solution. The device indicated should be a voltmeter. The negative side should be connected to the anode and the positive to the cathode to get the polarity correct.

(b) $= E^\circ_{\text{reduction}} - E^\circ_{\text{oxidation}}$

$\qquad = 0.77 - (-2.37)$

$\qquad = 3.14$ volts

(c) $E^\circ_{\text{cell}} = \dfrac{RT}{n\mathscr{F}} \ln K_{eq} = \dfrac{0.0592}{n} \log K_{eq}$

$\qquad 3.14 = (0.0592/2) \log K_{eq}$

$\qquad \log K_{eq} = 106$

$\qquad\quad K_{eq} = 1 \times 10^{106}$ [Note that K_{eq} is so large that a calculator cannot be used.]

(d) $E_{cell} = E_{cell}^{\circ} - \dfrac{BT}{n\mathscr{F}}\ln Q = E_{cell}^{\circ} - \dfrac{0.0592}{n}\log Q$

$\quad = 3.14 - (0.0592/2)\log\left(\dfrac{[Mg^{2+}][Fe^{2+}]^2}{[Fe^{3+}]^2}\right)$

$\quad = 3.14 - (0.0592/2)\log\left(\dfrac{[2.0\times10^{-2}][2.0\times10^{-2}]^2}{[2.0\times10^{-2}]^2}\right)$

$\quad = 3.14 - (0.0592/2)\log(2.0\times10^{-2})$

$\quad = 3.14 - (0.0592/2)(-1.69)$

$\quad = 3.14 + 0.050$

$\quad = 3.19$ volts

(e) After the cell comes to equilibrium, there is no flow of electrons in either direction. Therefore the cell voltage is zero.

(f) With the size of the equilibrium constant, all of the reactants can be considered as converted to products. Therefore the 2×10^{-2} M Fe^{3+} will increase the concentration of Fe^{2+} to 4×10^{-2} M, and the concentration of Mg^{2+} becomes 3×10^{-2} M. From this point, x moles per liter of Mg^{2+} will react in the reverse direction with $2x$ moles of Fe^{2+} to form $2x$ moles of Fe^3. This is summarized in the equilibrium table below.

Reaction	Mg	+	2 Fe^{3+}	→	Mg^{2+}	+	2 Fe^{2+}
Initial	Solid		0		3×10^{-2}		4×10^{-2}
Change	+x		+$2x$		$-x$		$-2x$
Equilibrium	1		$2x$		$3\times10^{-2} - x$		$4\times10^{-2} - 2x$
Answer			7.0×10^{-56}		3×10^{-2}		4×10^{-2}

The equilibrium law may be written as

$$K_{eq} = \dfrac{[Mg^{2+}][Fe^{2+}]^2}{[Fe^{3+}]^2} = \dfrac{(3\times10^{-2} - x)(4\times10^{-2} - 2x)^2}{(2x)^2}$$

Because of the size of K_{eq}, we may assume that x is very small compared to the concentrations. The assumptions simplify the numerator to

$$10^{106} = \dfrac{(3\times10^{-2})(4\times10^{-2})^2}{(2x)^2} = \dfrac{(4.8\times10^{-5})}{4x^2}$$

$$10^{106}\,x2 = 1.2\times10^{-5}$$

$$x2 = 1.2\times10^{-111} \quad 12\times10^{-112}$$

$$x = 3.5\times10^{-56}\,M = [Fe^{3+}]$$

$$2x = 7.0\times10^{-56} = [Fe^{3+}]$$

$$[Fe^{2+}] = 4\times10^{-2}\,M$$

$$[Mg^{2+}] = 3\times10^{-2}\,M$$

6. The kinetic-molecular theory of gases illustrates why the ideal gas law can be applied to all gases.

(a) An ideal gas is defined as one where the molecules have no volume and the molecules exhibit no attractive or repulsive forces toward each other.

(b) The ideal gas law is $PV = nRT$. For real gases the equation is $\left(P + \dfrac{n^2 a}{V^2}\right)(V - nb) = nRT$

The *a* in the first parentheses is related to the attractive forces of the real molecule. Polar molecules have large values for *a*, and nonpolar molecules have relatively small values for *a*. The term *b* in the second parentheses corrects for the real volume of the molecule in a real gas. Larger molecules have larger values for *b*.

(c) Gases are least ideal as they approach the point of condensing, and they are most like an ideal gas the farther they are from condensing. Because condensation is aided by low temperatures and high pressures, the reverse, high temperatures and low pressures, are the conditions where most gases obey the ideal gas law.

(d) The kinetic-molecular theory says that gases are composed of many small particles moving very rapidly. The molecules are far apart and rarely collide. When collisions do occur, they are completely elastic. Finally, the theory postulates that the average kinetic energy of the particles is related to the temperature of the gas.

(e) How does the kinetic-molecular theory of gases explain the following?

   (i) Boyle's law relates pressure and volume (PV = constant). When the volume increases, there are fewer collisions of the particles with the container walls. These collisions are what cause the effect called pressure. Therefore a larger volume causes a decrease in pressure. The reverse argument is also true.

   (ii) Charles's law relates temperature and volume (T/V = constant). To maintain pressure as the volume is increased, the number of collisions with the walls must increase. This is done with a corresponding increase in temperature that is an increase in kinetic energy and a corresponding increase in molecular velocity. The faster the molecules move, the more collisions with the walls occur each second.

   (iii) Gay-Lussac's law is the relationship between pressure and temperature (P/T = constant). This predicts an increase in temperature if the pressure increases with no change in volume. This agrees with the theory because an increase in temperature increases the kinetic energy and this results in more collisions per second with the container walls that is observed as a pressure increase.

Practice Test 3

ANSWER SHEET

1 Ⓐ Ⓑ Ⓒ Ⓓ Ⓔ
2 Ⓐ Ⓑ Ⓒ Ⓓ Ⓔ
3 Ⓐ Ⓑ Ⓒ Ⓓ Ⓔ
4 Ⓐ Ⓑ Ⓒ Ⓓ Ⓔ
5 Ⓐ Ⓑ Ⓒ Ⓓ Ⓔ
6 Ⓐ Ⓑ Ⓒ Ⓓ Ⓔ
7 Ⓐ Ⓑ Ⓒ Ⓓ Ⓔ
8 Ⓐ Ⓑ Ⓒ Ⓓ Ⓔ
9 Ⓐ Ⓑ Ⓒ Ⓓ Ⓔ
10 Ⓐ Ⓑ Ⓒ Ⓓ Ⓔ
11 Ⓐ Ⓑ Ⓒ Ⓓ Ⓔ
12 Ⓐ Ⓑ Ⓒ Ⓓ Ⓔ
13 Ⓐ Ⓑ Ⓒ Ⓓ Ⓔ
14 Ⓐ Ⓑ Ⓒ Ⓓ Ⓔ
15 Ⓐ Ⓑ Ⓒ Ⓓ Ⓔ
16 Ⓐ Ⓑ Ⓒ Ⓓ Ⓔ
17 Ⓐ Ⓑ Ⓒ Ⓓ Ⓔ
18 Ⓐ Ⓑ Ⓒ Ⓓ Ⓔ
19 Ⓐ Ⓑ Ⓒ Ⓓ Ⓔ
20 Ⓐ Ⓑ Ⓒ Ⓓ Ⓔ
21 Ⓐ Ⓑ Ⓒ Ⓓ Ⓔ
22 Ⓐ Ⓑ Ⓒ Ⓓ Ⓔ
23 Ⓐ Ⓑ Ⓒ Ⓓ Ⓔ
24 Ⓐ Ⓑ Ⓒ Ⓓ Ⓔ
25 Ⓐ Ⓑ Ⓒ Ⓓ Ⓔ

26 Ⓐ Ⓑ Ⓒ Ⓓ Ⓔ
27 Ⓐ Ⓑ Ⓒ Ⓓ Ⓔ
28 Ⓐ Ⓑ Ⓒ Ⓓ Ⓔ
29 Ⓐ Ⓑ Ⓒ Ⓓ Ⓔ
30 Ⓐ Ⓑ Ⓒ Ⓓ Ⓔ
31 Ⓐ Ⓑ Ⓒ Ⓓ Ⓔ
32 Ⓐ Ⓑ Ⓒ Ⓓ Ⓔ
33 Ⓐ Ⓑ Ⓒ Ⓓ Ⓔ
34 Ⓐ Ⓑ Ⓒ Ⓓ Ⓔ
35 Ⓐ Ⓑ Ⓒ Ⓓ Ⓔ
36 Ⓐ Ⓑ Ⓒ Ⓓ Ⓔ
37 Ⓐ Ⓑ Ⓒ Ⓓ Ⓔ
38 Ⓐ Ⓑ Ⓒ Ⓓ Ⓔ
39 Ⓐ Ⓑ Ⓒ Ⓓ Ⓔ
40 Ⓐ Ⓑ Ⓒ Ⓓ Ⓔ
41 Ⓐ Ⓑ Ⓒ Ⓓ Ⓔ
42 Ⓐ Ⓑ Ⓒ Ⓓ Ⓔ
43 Ⓐ Ⓑ Ⓒ Ⓓ Ⓔ
44 Ⓐ Ⓑ Ⓒ Ⓓ Ⓔ
45 Ⓐ Ⓑ Ⓒ Ⓓ Ⓔ
46 Ⓐ Ⓑ Ⓒ Ⓓ Ⓔ
47 Ⓐ Ⓑ Ⓒ Ⓓ Ⓔ
48 Ⓐ Ⓑ Ⓒ Ⓓ Ⓔ
49 Ⓐ Ⓑ Ⓒ Ⓓ Ⓔ
50 Ⓐ Ⓑ Ⓒ Ⓓ Ⓔ

51 Ⓐ Ⓑ Ⓒ Ⓓ Ⓔ
52 Ⓐ Ⓑ Ⓒ Ⓓ Ⓔ
53 Ⓐ Ⓑ Ⓒ Ⓓ Ⓔ
54 Ⓐ Ⓑ Ⓒ Ⓓ Ⓔ
55 Ⓐ Ⓑ Ⓒ Ⓓ Ⓔ
56 Ⓐ Ⓑ Ⓒ Ⓓ Ⓔ
57 Ⓐ Ⓑ Ⓒ Ⓓ Ⓔ
58 Ⓐ Ⓑ Ⓒ Ⓓ Ⓔ
59 Ⓐ Ⓑ Ⓒ Ⓓ Ⓔ
60 Ⓐ Ⓑ Ⓒ Ⓓ Ⓔ
61 Ⓐ Ⓑ Ⓒ Ⓓ Ⓔ
62 Ⓐ Ⓑ Ⓒ Ⓓ Ⓔ
63 Ⓐ Ⓑ Ⓒ Ⓓ Ⓔ
64 Ⓐ Ⓑ Ⓒ Ⓓ Ⓔ
65 Ⓐ Ⓑ Ⓒ Ⓓ Ⓔ
66 Ⓐ Ⓑ Ⓒ Ⓓ Ⓔ
67 Ⓐ Ⓑ Ⓒ Ⓓ Ⓔ
68 Ⓐ Ⓑ Ⓒ Ⓓ Ⓔ
69 Ⓐ Ⓑ Ⓒ Ⓓ Ⓔ
70 Ⓐ Ⓑ Ⓒ Ⓓ Ⓔ
71 Ⓐ Ⓑ Ⓒ Ⓓ Ⓔ
72 Ⓐ Ⓑ Ⓒ Ⓓ Ⓔ
73 Ⓐ Ⓑ Ⓒ Ⓓ Ⓔ
74 Ⓐ Ⓑ Ⓒ Ⓓ Ⓔ
75 Ⓐ Ⓑ Ⓒ Ⓓ Ⓔ

The Periodic Table of the Elements

1																		2
1 **H** 1.0080																		**He** 4.0026
3 **Li** 6.941	4 **Be** 9.012											5 **B** 10.810	6 **C** 12.011	7 **N** 14.007	8 **O** 16.00	9 **F** 19.00	10 **Ne** 20.179
11 **Na** 22.99	12 **Mg** 24.31											13 **Al** 26.98	14 **Si** 28.09	15 **P** 30.974	16 **S** 32.06	17 **Cl** 35.453	18 **Ar** 39.948
19 **K** 39.10	20 **Ca** 40.08	21 **Sc** 44.96	22 **Ti** 47.87	23 **V** 50.94	24 **Cr** 52.00	25 **Mn** 54.94	26 **Fe** 55.85	27 **Co** 58.93	28 **Ni** 58.69	29 **Cu** 63.55	30 **Zn** 65.41	31 **Ga** 69.72	32 **Ge** 72.64	33 **As** 74.92	34 **Se** 78.96	35 **Br** 79.90	36 **Kr** 83.80
37 **Rb** 85.47	38 **Sr** 87.62	39 **Y** 88.91	40 **Zr** 91.22	41 **Nb** 92.91	42 **Mo** 95.94	43 **Tc** [98]	44 **Ru** 101.07	45 **Rh** 102.91	46 **Pd** 106.42	47 **Ag** 107.87	48 **Cd** 112.41	49 **In** 114.82	50 **Sn** 118.71	51 **Sb** 121.76	52 **Te** 127.60	53 **I** 126.90	54 **Xe** 131.29
55 **Cs** 132.91	56 **Ba** 137.33	57 **La** 138.91	72 **Hf** 178.49	73 **Ta** 180.95	74 **W** 183.84	75 **Re** 186.21	76 **Os** 190.23	77 **Ir** 192.22	78 **Pt** 195.08	79 **Au** 196.97	80 **Hg** 200.59	81 **Tl** 204.38	82 **Pb** 207.2	83 **Bi** 208.98	84 **Po** [209]	85 **At** [210]	86 **Rn** [222]
87 **Fr** [223]	88 **Ra** 226.03	89 **Ac** 227.03	104 **Rf** [267]	105 **Db** [268]	106 **Sg** [271]	107 **Bh** [272]	108 **Hs** [270]	109 **Mt** [276]	110 **Ds** [281]	111 **Rg** [280]	112 **Uub** [285]		114 **Uuq** [289]		116 **Uuh** [291]		

Lanthanides

58 **Ce** 140.12	59 **Pr** 140.91	60 **Nd** 144.24	61 **Pm** [145]	62 **Sm** 150.36	63 **Eu** 151.96	64 **Gd** 157.25	65 **Tb** 158.93	66 **Dy** 162.50	67 **Ho** 164.93	68 **Er** 167.26	69 **Tm** 168.93	70 **Yb** 173.04	71 **Lu** 174.97

Actinides

90 **Th** 232.04	91 **Pa** 231.04	92 **U** 238.03	93 **Np** 237.05	94 **Pu** [244]	95 **Am** [243]	96 **Cm** [247]	97 **Bk** [247]	98 **Cf** [251]	99 **Es** [252]	100 **Fm** [257]	101 **Md** [258]	102 **No** [259]	103 **Lr** [262]

Chemistry

SECTION I

TIME: 1 HOUR AND 30 MINUTES

CALCULATORS MAY NOT BE USED WITH SECTION I

REMINDER: For all questions, assume that the temperature is 298 K, the pressure is 1.00 atmospheres of pressure, and solutions are aqueous solutions unless otherwise indicated.

Throughout this practice test the following symbols are used with the definitions provided below, unless otherwise noted in a given problem.

T = temperature

P = pressure

V = volume

S = entropy

M = molarity

m = molal

L, mL = liter(s), milliliter(s)

g = gram(s)

nm = nanometer(s)

H = enthalpy

G = free energy

R = molar gas constant

n = number of moles

atm = atmosphere(s)

J, kJ = joule(s), kilojoule(s)

V = volt(s)

mol = mole(s)

Practice Test 3

SECTION I—PART A

Multiple-Choice Questions

(Calculators May Not Be Used)

75 QUESTIONS

TIME: 90 MINUTES

50% OF TOTAL SCORE

Directions: Each of the lettered choices has a statement immediately following it. Selecting the letter is the same as selecting the statement as your answer. Select the letter that best fits each of the numbered questions that follow. A letter may be used once, more than once or not at all in a set of questions.

Questions 1–4 refer to the following elements.

(A) Ca
(B) S
(C) Fe
(D) N
(E) Cs

1. One of the above elements is a transition metal.

2. One of the above elements forms a polyatomic anion where it has an oxidation number of +6

3. Which of the above has the smallest atomic radius?

4. Which of the above is the least electronegative element?

Questions 5–7 refer to the following compounds:

(A) phosphates
(B) carbonates
(C) sulfites
(D) oxides
(E) sulfates

5. Compounds in this category often are reducing agents.

6. Compounds in this category that also contain a metal atom are basic anhydrides.

7. Compounds in this category are commonly used in fertilizers to promote root growth.

Questions 8–11 refer to the following elements.

    (A) Na
    (B) Li
    (C) Si
    (D) S
    (E) Ar

8. One of these elements is used to vulcanize rubber.

9. This element has the lowest ionization energy.

10. This element is the basis of semiconductor technology.

11. Of the elements listed here, it is an element in a strong acid.

Questions 12–14 refer to the following compounds.

    (A) $KMnO_4$
    (B) NH_3
    (C) $H_2C_2O_4$
    (D) K_2HPO_4
    (E) SiO_2

12. This compound is a common oxidizing agent.

13. There are several complexing agents listed. Which one is a base?

14. This compound is a common primary standard for acid-base titrations.

SECTION I—PART B

(Calculators May Not Be Used)

> **Directions:** Each of the questions or incomplete statements below is followed by five suggested answers or completions. Select the one that is best in each case and fill in the corresponding oval on the answer sheet.

15. The ideal gas law may be used to derive all of the following except

    (A) Graham's law of gas effusion
    (B) Boyle's law
    (C) Charles's law
    (D) Avogadro's principle
    (E) Gay-Lussac's law

16. It takes 40.0 mL of 0.100 M NaOH to titrate 488 mg of a solid monoprotic acid to the phenolphthalein end point. What is the molecular mass of the acid?

    (A) 221
    (B) 122
    (C) 68
    (D) 1.2×10^5
    (E) 1.2×10^{-1}

17. What is the molar solubility of silver chromate? (K_{sp} [Ag_2CrO_4] = 1.2×10^{-12}.)

    (A) 1.2×10^{-12}
    (B) $\sqrt{1.2 \times 10^{-12}}$
    (C) $\sqrt[3]{1.2 \times 10^{-12}}$
    (D) $\sqrt[3]{\dfrac{1.2 \times 10^{-12}}{4}}$
    (E) $\sqrt[3]{\dfrac{1.2 \times 10^{-12}}{8}}$

18. A substance that contains both ionic and covalent bonds is

    (A) KBr
    (B) $HClO_4$
    (C) Na_2SO_4
    (D) $AlCl_3$
    (E) $CH_3CH_2CH_2CH_2Cl$

19. Which of the following has the highest pH?

    (A) the end point of a strong acid titrated with a strong base
    (B) the end point of a weak acid titrated with a strong base
    (C) the end point of a weak base titrated with a strong acid
    (D) the end point of a strong base titrated with a strong acid
    (E) the end point of a weak acid titrated with a weaker base

20. Determine which of the following elements is expected to be diamagnetic.

    (A) Ca
    (B) Fe
    (C) Cr
    (D) Mn
    (E) Co

21. A hygroscopic substance requires special care when handling and when it is stored storing it in a laboratory. Why does it need this special care?

    (A) it reacts with atmospheric oxygen
    (B) it spontaneously combusts
    (C) it spontaneously combusts when in contact with organic matter
    (D) it absorbs water from the atmosphere
    (E) it oxidizes readily

22. Very fine precipitates are most easily separated by

    (A) distillation
    (B) filtration
    (C) centrifugation
    (D) evaporation
    (E) vacuum filtration

23. Which of the following cannot be used to predict if a reaction is spontaneous or not at a given temperature?

    (A) $\Delta G°$
    (B) K_c
    (C) $E°_{cell}$
    (D) $\Delta H°$
    (E) $\Delta S°$ and $\Delta H°$

24. Which of the following cannot be a Lewis acid?

    (A) Fe^{2+}
    (B) Fe^{3+}
    (C) NH_4^+
    (D) BCl_3
    (E) H^+

25. Which of the following is not a conjugate acid-base pair?

    (A) H_2SO_4 and SO_4^{2-}
    (B) HCl and Cl^-
    (C) NH_3 and NH_2^-
    (D) HPO_4^{2-} and PO_4^{3-}
    (E) H_2S and HS^-

26. The CH_4 molecule has a tetrahedral geometry. This shape is best explained using

    (A) hybrid orbitals
    (B) Hund's rule
    (C) the transition-state theory
    (D) Lewis structures
    (E) electron configurations

27. A reaction has the rate law: rate = $k[A][B]^2$. The units for the rate constant are

    (A) s^{-1}
    (B) $mol\ L^{-1}\ s^{-1}$
    (C) $L^2\ mol^{-2}\ s^{-1}$
    (D) $L\ mol^{-1}\ s^{-1}$
    (E) $mol\ s^{-1}$

28. A concentration-versus-time plot will be curved for all of the following except

    I. a zero-order reaction
    II. a first-order reaction
    III. a second-order reaction

    (A) I only
    (B) II only
    (C) III only
    (D) I and II
    (E) I and III

29. A certain gas has a molar solubility in CCl_4 of $2.7 \times 10^{-4}\ M$ when its pressure is 330 mm Hg. What is its solubility, at the same temperature, if the pressure of the gas is decreased to 220 mm Hg?

    (A) $2.8 \times 10^{-4}\ M$
    (B) $1.8 \times 10^{-4}\ M$
    (C) $1.25 \times 10^{-4}\ M$
    (D) $7.37 \times 10^{-5}\ M$
    (E) $4.0 \times 10^{-4}\ M$

30. Which of the following radioactive isotopes is most likely to decay by emitting a beta particle?

    (A) ^{235}U
    (B) ^{40}Cl
    (C) ^{35}Ar
    (D) ^{246}Cf
    (E) ^{21}Na

31. Which of the following is produced commercially by electrolysis?

    (A) sodium hydroxide
    (B) aluminum
    (C) hydrogen
    (D) chlorine
    (E) all of these

32. Which pair of substances cannot be the major components of (coexist in) an aqueous solution?

    (A) OH^- and H^+
    (B) $H_2PO_4^-$ and HPO_4^{2-}
    (C) $HOCl$ and OCl^-
    (D) SO_4^{2-} and SO_3^{2-}
    (E) H_2CO_3 and CO_3^{2-}

Use the following list of ionization constants for Problems 33 and 34.

Acid	Acid Dissociation Constant
I. $HC_2H_3O_2$	1.8×10^{-5}
II. HCN	6.2×10^{-10}
III. HNO_2	7.1×10^{-4}
IV. $HCHO_2$	1.8×10^{-4}
V. $HOBr$	2.1×10^{-9}

33. When each of these acids is titrated, the one with the highest pH at its end point will be

    (A) I
    (B) II
    (C) III
    (D) IV
    (E) V

34. A buffer with a pH of 4.00 is required for a certain experiment. Which acid(s), along with their salts, can be used to prepare this buffer?

    (A) I
    (B) II
    (C) III and IV
    (D) II and V
    (E) I, III, and IV

35. Which of the following substances cannot be a reducing agent?

    (A) Ag
    (B) I^-
    (C) Fe^{3+}
    (D) Cr^{3+}
    (E) Cl^-

36. A 10.0-gram sample of ethane (C_2H_6) is burned in an enclosed vessel with 40.0 grams of oxygen. Which of the following best describes the mixture when the reaction is complete?

    (A) the limiting reactant is C_2H_6
    (B) the theoretical yield of CO_2 is 29.3 grams
    (C) 2.67 grams of oxygen is left over when the reaction is complete
    (D) when finished, the total mass of the system is 50.0 grams
    (E) all of the above

37. What is the molar mass of a nonelectrolyte that causes water to boil at 101.0 °C when 23.2 grams is dissolved in 100 grams of distilled water? (k_b for $H_2O = 0.51$ °C molal^{-1}.)

    (A) 8.6
    (B) 78.9
    (C) 12.6
    (D) 45.1
    (E) 116

38. A pure hydrocarbon is burned in air, and 66 grams of CO_2 along with 27 grams of water are collected. What is the empirical formula of the hydrocarbon?

    (A) CH_2
    (B) CH
    (C) C_2H_3
    (D) C_2H_4
    (E) C_3H_4

39. All of the following sets of quantum numbers are correct **except**

(A) 4	3	1	$+\frac{1}{2}$
(B) 3	2	-2	$-\frac{1}{2}$
(C) 5	0	0	$-\frac{1}{2}$
(D) 2	2	-1	$+\frac{1}{2}$
(E) 3	2	0	$+\frac{1}{2}$

40. When measuring the moles of gas in a certain experiment, the volume of the flask was 43 mL, the temperature was 24.3 °C, and the pressure was 745 torr. The value of R is 0.8205 L atm mol^{-1} K^{-1}. The correct value to report for the number of moles of gas in this experiment is

    (A) 2×10^{-3} mol
    (B) 1.7263×10^{-3} mol
    (C) 1.726×10^{-3} mol
    (D) 1.73×10^{-3} mol
    (E) 1.7×10^{-3} mol

41. Which structure(s) have 90° bond angles in the molecule?

    I. tetrahedron
    II. trigonal bipyramid
    III. octahedron
    IV. trigonal planar molecule

    (A) I
    (B) I and II
    (C) II and III
    (D) IV
    (E) I, II, and III

42. The boiling point of bromine is −7.2 °C, and its heat of vaporization is +15 kJ mol^{-1}. What is the entropy change when one mole of $Br_2(l)$ is converted to $Br_2(g)$ at −7.2 °C?

    (A) −2.08 kJ K^{-1}
    (B) −2,080 J K^{-1}
    (C) 56.4 kJ K^{-1}
    (D) 56.4 J K^{-1}
    (E) 17.7 J K^{-1}

43. What is the mole fraction of sodium chloride in a solution that is 2.00 molal in NaCl? The molar mass of H_2O = 18.0, and NaCl = 58.5.

    (A) 0.500
    (B) 0.0348
    (C) 0.965
    (D) 0.117
    (E) 0.883

44. A 25.0-milliliter sample of 0.080 molar silver nitrate is added to excess sodium chloride. After coagulating, filtering, and drying, the precipitate of AgCl (molar mass = 143) weighed exactly 212 milligrams. What was the percentage yield of this experiment?

    (A) 25%
    (B) 56%
    (C) 74%
    (D) 32.8%
    (E) 100%

45. A galvanic cell is constructed to study the reaction

    $$Zn(s) + Cu^{2+}(aq) \rightarrow Cu(s) + Zn^{2+}(aq)$$

    What is the voltage of the cell if $[Cu^{2+}] = 0.200\ M$ and $[Zn^{2+}] = 0.0200\ M$. ($E^{\circ}_{cell} = +1.10$ V.)

    (A) 1.07 V
    (B) 1.10 V
    (C) 1.13 V
    (D) 1.02 V
    (E) 1.40 V

46. The net ionic equation for the reaction of sodium sulfate with iron(II) chloride is

    (A) $SO_4^{2-}(aq) + Fe^{2+}(aq) \rightarrow FeSO_4(s)$
    (B) $Cl^-(aq) + Si^+(aq) \rightarrow SiCl(s)$
    (C) $SO_3^{2-}(aq) + Fe^{2+}(aq) \rightarrow FeSO_3(s)$
    (D) $Na_2SO_4(s) + Fe^{2+}(aq) \rightarrow$
    $FeSO_4(s) + 2\ Na^+(aq)$
    (E) $SO_3^{2-}(aq) + FeCl_3(s) \rightarrow$
    $Fe_2(SO_3)_2(aq) + 3\ Cl^-(aq)$

47. 3.00 moles of neon in a flask has a pressure of 1.50 atm. The pressure rises to 4.50 atmospheres when 1.00 mole of hydrogen and some oxygen are added. How many moles of oxygen are added?

    (A) 1.50
    (B) 3.00
    (C) 5.00
    (D) 2.50
    (E) 4.50

48. Write the Lewis structure for each of the following molecules and then indicate which of the following molecules will have a Lewis structure that contains a double bond?

    (A) NH_3
    (B) NO_3^-
    (C) CH_4
    (D) Cl_2
    (E) BF_3

49. Electromagnetic radiation is usually divided into wavelength regions with names that are somewhat descriptive of their nature. Each region has a different energy range. If the wavelength regions are listed from the lowest energy to the highest energy, the sequence would be

    (A) ultraviolet < visible < infrared
    (B) visible < infrared < microwave
    (C) X-rays < infrared < ultraviolet
    (D) microwaves < visible < infrared
    (E) visible < ultraviolet < X-rays

50. A 50.0-mL sample of hydrochloric acid with an unknown concentration is titrated with 0.125 *M* sodium hydroxide. Which of the following is true concerning this experiment?

    (A) the volume of NaOH used will be less than 50.0 mL
    (B) the end point will be at a pH greater than 7
    (C) the color change of the indicator will be from colorless to pink
    (D) the reaction must be standardized by adding KHP
    (E) the equivalence point will have a pH of exactly 7

Questions 51–52

Balance the following half-reaction for an acid solution using the smallest possible whole number coefficients and entering the proper number of electrons where needed. Use your result for the next two questions.

$$MnO_4^{2-} \rightarrow Mn^{2+}$$

51. In the balanced half-reaction

    (A) there should be 4 electrons on the left side
    (B) there should be 3 electrons on the right side
    (C) this is not a half reaction, no electrons are needed
    (D) there should be 2 electrons on the left side
    (E) there should be 2 electrons on the right side

52. The hydrogen ions in the half-reaction appear as

    (A) 2 on the right side
    (B) 4 on the left side
    (C) 3 on the left side
    (D) 4 on the right side
    (E) 8 on the left side

53. An oxide of nitrogen is found to contain 25.9 percent nitrogen. The empirical formula for this substance is

    (A) NO
    (B) NO_2
    (C) N_2O
    (D) N_2O_3
    (E) N_2O_5

54. Using the trends in the periodic table, determine which ion is expected to have the largest radius.

    (A) K^+
    (B) I^-
    (C) Cl^-
    (D) Na^+
    (E) Ba^{2+}

55. The trend for atomic radii to decrease from left to right across a period of the periodic table is ascribed to

    (A) electrons that attract each other to compress the outer orbitals
    (B) increasing nuclear mass has a gravitational effect contracting the atomic size
    (C) increasing effective nuclear charge more than offsets the repulsion of added electrons in the same shell
    (D) the question is wrong, atomic radii increase across each period
    (E) Hund's rule, the Pauli exclusion principle, and the Heisenberg uncertainty principle all combine to explain this phenomenon.

56. Use the following balanced redox equation to determine which of the statements listed below is true.

$$Cr_2O_7^{2-}(aq) + 3\ Sn^{2+}(aq) + 14\ H^+(aq) \rightarrow$$
$$2\ Cr^{3+}(aq) + 3\ Sn^{4+}(aq) + 7\ H_2O(l)$$

    (A) dichromate ions are oxidized by Sn(II) ions.
    (B) hydrogen ions are reduced to H_2O
    (C) chromium is reduced from the +6 to the +3 oxidation state
    (D) the oxidation state of chromium does not change
    (E) the large coefficient for H^+ forces the reaction toward completion

57. $IO_3^-(aq) + H^+(aq) \rightarrow I_2(aq) + H_2O(l)$

    When the above half-reaction is balanced by entering the electrons where needed and using the smallest whole-number coefficients possible, there will be
    (A) 5 electrons on the left-hand side
    (B) no electrons are needed
    (C) 3 electrons on the right-hand side
    (D) 10 electrons on the left-hand side
    (E) 8 electrons on the right-hand side

58. The reaction $2\ HgCl_2 + C_2O_4^{2-} \rightarrow$

    $2\ Cl^- + 2\ CO_2 + Hg_2Cl_2(s)$ has a rate law of

$$rate = k[HgCl_2][C_2O_4^{2-}]^2$$

    Overall, this is a _____ order reaction
    (A) first
    (B) second
    (C) zero
    (D) third
    (E) fourth

59. In the reaction in the previous problem which of the following statements is true?

    (A) the exponents in the rate law come from the coefficients in the balanced reaction
    (B) the magnitude of the rate constant will indicate the how far the reaction goes toward completion
    (C) chloride ions will have no effect on the reaction rate
    (D) the rate constant varies with the square of the $C_2O_4^{2-}$ concentration
    (E) the half-life of this reaction is $0.693/k$

60. Which of the following best represents a phase diagram?

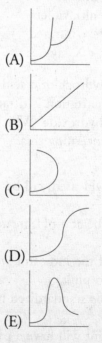

    (A)

    (B)

    (C)

    (D)

    (E)

61. The critical pressure is the

    (A) end of the liquid-vapor equilibrium line on a phase diagram
    (B) pressure above which no increase in temperature will result in the liquid vaporizing to a gas
    (C) pressure required for boiling to occur at any given temperature
    (D) pressure where solid, liquid, and gas are all at equilibrium
    (E) pressure where the compound is thermodynamically unstable

62. $2 Fe^0 + 3 Cu^{2+} \rightarrow 2 Fe^{3+} + 3 Cu^0$

    If this reaction is spontaneous, then the standard cell voltage (potential), $E°_{cell}$, and the standard free energy, $\Delta G°$,
    (A) are both positive
    (B) are both negative
    (C) will have $E°_{cell}$ positive and $\Delta G°$ negative
    (D) will have $E°_{cell}$ negative and $\Delta G°$ positive
    (E) will both equal zero

63. Radioactive decay often results in the emission of alpha particles, beta particles, positrons, or neutrons. If in the radioactive decay of $^{210}_{83}Bi$ a single particle is observed to be emitted, which of the following is the most unlikely product?

    (A) $^{209}_{83}Bi$
    (B) $^{208}_{79}Au$
    (C) $^{210}_{82}Pb$
    (D) $^{206}_{81}Tl$
    (E) $^{210}_{84}Po$

64. In the electrolysis of water at 26 °C the total pressure of the gas collected is 740 mm Hg. At that same temperature the vapor pressure of water is tabulated as 25 mm Hg. What is the partial pressure of the oxygen?

    (A) 740 mm Hg
    (B) 1.00 atm
    (C) 715 mm Hg
    (D) 25 mm Hg
    (E) 765 mg Hg

65. Which of the following has physical properties that are best explained in the context of a covalent network crystal?

    (A) $CaCl_2$
    (B) SiO_2
    (C) Sn
    (D) Pt
    (E) $[—CF_2—]_n$

66. The most polar covalent bond is found between which two elements based upon the electronegativity of the elements?

    (A) C—N
    (B) P—F
    (C) S—O
    (D) Si—C
    (E) O—P

67. Which of the following does not form hydrogen bonds?

    (A) HCN
    (B) HF
    (C) CH_3NH_2
    (D) CH_3COOH
    (E) $ClCH_2OH$

68. When 1 mole of H_3PO_4 reacts with 2 moles of KOH, one of the products is water. The other is

    (A) KH_2PO_4
    (B) K_2HPO_4
    (C) K_3PO_4
    (D) a mixture of K_2HPO_4 and K_3PO_4
    (E) a mixture of KH_2PO_4 and K_2HPO_4

69. At constant temperature and pressure, what is the maximum number of liters of O_2 that can react with 1.50 liters of SO_2 to form SO_3?

    (A) 1.50 L
    (B) 1.00 L
    (C) 3.00 L
    (D) 4.50 L
    (E) 0.75 L

70. The compound that contains 10.4 percent oxygen is

    (A) NaOH
    (B) CaO
    (C) Al_2O_3
    (D) BaO
    (E) $Ca(OH)_2$

71. Write the ionization reaction for HCN and then write the correct equilibrium law for the dissociation of the weak acid HCN.

    (A) $K_a = \dfrac{[CN^-][H^+]}{[HCN]}$

    (B) $K_a = \dfrac{[CN^-][H_3O^+]}{[HCN][H_2O]}$

    (C) $K_a = \dfrac{[HCN][H^+]}{[CN^-]}$

    (D) $K_a = \dfrac{[HCN]}{[CN^-][H^+]}$

    (E) $K_a = \dfrac{[HCN][H_2O]}{[CN^-][H_3O^+]}$

72. Which of the following represents a reaction that can occur at the anode of an electrolytic cell?

    (A) $Al^{3+} + 3\ e^- \rightarrow Al(s)$
    (B) $Br_2 + 2\ e^- \rightarrow 2\ Br^-$
    (C) $2\ H^+ + 2\ e^- \rightarrow H_2(g)$
    (D) $2\ I^- \rightarrow I_2 + 2\ e^-$
    (E) $Zn^{2+} + 2\ e^- \rightarrow Zn(s)$

73. A cathode ray tube (Crooke's tube) was used to determine the

    (A) *e/m* ratio of the proton
    (B) numerical charge of the electron
    (C) existence of neutrons
    (D) numerical charge of the proton
    (E) *e/m* ratio of the electron

74. A halogen, X, and an alkaline earth metal, M, will form a compound with the formula

    (A) MX
    (B) MX_2
    (C) MX_3
    (D) M_2X_3
    (E) M_3X_2

75. The electronic configuration $1s^2 2s^2 2p^6 3s^2 3p^6$ represents all of the following except

    (A) Ti^{4+}
    (B) Cr^{6+}
    (C) S^{2-}
    (D) V^{3+}
    (E) Ca^{2+}

End of Section 1.
If you finish before the 90 minutes have elapsed, you may check your work in this section. Do not go on to Section II until you are instructed to do so.

THE FOLLOWING TABLES MAY BE USED TO ANSWER THE QUESTIONS IN SECTION II ONLY.

Standard Reduction Potentials, 25 °C

Half-Reaction	E^0 (volts)	Half-Reaction	E^0 (volts)
$F_2(g) + 2e^- \rightarrow 2F^-$	2.87	$Co^{2+} + 2e^- \rightarrow Co$	−0.28
$Co^{3+} + e^- \rightarrow Co^{2+}$	1.82	$Tl^+ + e^- \rightarrow Tl$	−0.34
$Au^{3+} + 3e^- \rightarrow Au$	1.50	$Cd^{2+} + 2e^- \rightarrow Cd$	−0.40
$Cl_2(g) + 2e^- \rightarrow 2Cl^-$	1.36	$Cr^{3+} + e^- \rightarrow Cr^{2+}$	−0.41
$O_2(g) + 4H^+ + 4e^- \rightarrow 2H_2O$	1.23	$Fe^{2+} + 2e^- \rightarrow Fe$	−0.44
$Br_2(g) + 2e^- \rightarrow 2Br^-$	1.07	$Cr^{3+} + 3e^- \rightarrow Cr$	−0.74
$2Hg^{2+} + 2e^- \rightarrow Hg_2^{2+}$	0.92	$Zn^{2+} + 2e^- \rightarrow Zn$	−0.76
$Ag^+ + e^- \rightarrow Ag$	0.80	$2H_2O(l) + 2e^- \rightarrow H_2(g) + 2OH^-$	−0.83
$Hg_2^{2+} + 2e^- \rightarrow 2Hg$	0.79	$Mn^{2+} + 2e^- \rightarrow Mn$	−1.18
$Fe^{3+} + e^- \rightarrow Fe^{2+}$	0.77	$Al^{3+} + 3e^- \rightarrow Al$	−1.66
$I_2 + 2e^- \rightarrow 2I^-$	0.53	$Be^{2+} + 2e^- \rightarrow Be$	−1.70
$Cu^+ + e^- \rightarrow Cu$	0.52	$Mg^{2+} + 2e^- \rightarrow Mg$	−2.37
$Cu^{2+} + 2e^- \rightarrow Cu$	0.34	$Na^+ + e^- \rightarrow Na$	−2.71
$Cu^{2+} + e^- \rightarrow Cu^+$	0.15	$Ca^{2+} + 2e^- \rightarrow Ca$	−2.87
$Sn^{4+} + 2e^- \rightarrow Sn^{2+}$	0.15	$Sr^{2+} + 2e^- \rightarrow Sr$	−2.89
$S + 2H^+ + 2e^- \rightarrow H_2S$	0.14	$Ba^{2+} + 2e^- \rightarrow Ba$	−2.90
$2H^+ + 2e^- \rightarrow H_2$	0.00	$Rb^+ + e^- \rightarrow Rb$	−2.92
$Pb^{2+} + 2e^- \rightarrow Pb$	−0.13	$K^+ + e^- \rightarrow K$	−2.92
$Sn^{2+} + 2e^- \rightarrow Sn$	−0.14	$Cs^+ + e^- \rightarrow Cs$	−2.92
$Ni^{2+} + 2e^- \rightarrow Ni$	−0.25	$Li^+ + e^- \rightarrow Li$	−3.05

The Periodic Table of the Elements

1																	2
H 1.0080																	**He** 4.0026
3 **Li** 6.941	4 **Be** 9.012											5 **B** 10.810	6 **C** 12.011	7 **N** 14.007	8 **O** 16.00	9 **F** 19.00	10 **Ne** 20.179
11 **Na** 22.99	12 **Mg** 24.31											13 **Al** 26.98	14 **Si** 28.09	15 **P** 30.974	16 **S** 32.06	17 **Cl** 35.453	18 **Ar** 39.948
19 **K** 39.10	20 **Ca** 40.08	21 **Sc** 44.96	22 **Ti** 47.87	23 **V** 50.94	24 **Cr** 52.00	25 **Mn** 54.94	26 **Fe** 55.85	27 **Co** 58.93	28 **Ni** 58.69	29 **Cu** 63.55	30 **Zn** 65.41	31 **Ga** 69.72	32 **Ge** 72.64	33 **As** 74.92	34 **Se** 78.96	35 **Br** 79.90	36 **Kr** 83.80
37 **Rb** 85.47	38 **Sr** 87.62	39 **Y** 88.91	40 **Zr** 91.22	41 **Nb** 92.91	42 **Mo** 95.94	43 **Tc** [98]	44 **Ru** 101.07	45 **Rh** 102.91	46 **Pd** 106.42	47 **Ag** 107.87	48 **Cd** 112.41	49 **In** 114.82	50 **Sn** 118.71	51 **Sb** 121.76	52 **Te** 127.60	53 **I** 126.90	54 **Xe** 131.29
55 **Cs** 132.91	56 **Ba** 137.33	57 **La** 138.91	72 **Hf** 178.49	73 **Ta** 180.95	74 **W** 183.84	75 **Re** 186.21	76 **Os** 190.23	77 **Ir** 192.22	78 **Pt** 195.08	79 **Au** 196.97	80 **Hg** 200.59	81 **Tl** 204.38	82 **Pb** 207.2	83 **Bi** 208.98	84 **Po** [209]	85 **At** [210]	86 **Rn** [222]
87 **Fr** [223]	88 **Ra** 226.03	89 **Ac** 227.03	104 **Rf** [267]	105 **Db** [268]	106 **Sg** [271]	107 **Bh** [272]	108 **Hs** [270]	109 **Mt** [276]	110 **Ds** [281]	111 **Rg** [280]	112 **Uub** [285]		114 **Uuq** [289]		116 **Uuh** [291]		

Lanthanides

58 **Ce** 140.12	59 **Pr** 140.91	60 **Nd** 144.24	61 **Pm** [145]	62 **Sm** 150.36	63 **Eu** 151.96	64 **Gd** 157.25	65 **Tb** 158.93	66 **Dy** 162.50	67 **Ho** 164.93	68 **Er** 167.26	69 **Tm** 168.93	70 **Yb** 173.04	71 **Lu** 174.97

Actinides

90 **Th** 232.04	91 **Pa** 231.04	92 **U** 238.03	93 **Np** 237.05	94 **Pu** [244]	95 **Am** [243]	96 **Cm** [247]	97 **Bk** [247]	98 **Cf** [251]	99 **Es** [252]	100 **Fm** [257]	101 **Md** [258]	102 **No** [259]	103 **Lr** [262]

USEFUL INFORMATION FOR SOLVING FREE-RESPONSE PROBLEMS

Atomic Structure

E = energy

ν = frequency

$\lambda = \dfrac{h}{mv}$

$p = mv$

v = velocity

n = principal quantum number

m = mass

$\Delta E = h\nu$

$c = \lambda\nu$

λ = wavelength

p = momentum

$E_n = \dfrac{-2.178\times10^{-18}\ joule}{n^2}$

Speed of light, $c = 3.00 \times 10^8$ m s^{-1}

Planck's constant, $h = 6.63 \times 10^{-34}$ J s

Boltzmann's constant, $k = 1.38 \times 10^{-23}$ J K^{-1}

Avogadro's number = 6.022×10^{23} mol^{-1}

Electron charge = -1.602×10^{-19} coulomb

1 electron volt/atom = 96.5×10^{23} K J mol^{-1}

Equilibrium and Thermochemistry

$K_a = \dfrac{[H^+][A^-]}{[HA]}$

$K_b = \dfrac{[OH^-][HB^+]}{[B]}$

$K_w = [H^+][OH^-]$

$K_w = 10^{-14}$ at 25 °C

$K_w = K_a K_b$

$pH = -\log[H^+]$

$pOH = -\log[OH^-]$

$14.00 = pH + pOH$ at 25 °C

$pH = pK_a + \log\dfrac{[A^-]}{[HA]}$

$pOH = pK_b + \log\dfrac{[HB^+]}{[B]}$

$pK_a = -\log K_a$

$pK_b = -\log K_b$

$K_p = K_c(RT)^{\Delta n}$

Δn = moles product gas − moles reactant gas

$\Delta S° = \Sigma S°$ products − $\Sigma S°$ reactants

$\Delta H° = \Sigma \Delta H°_f$ products − $\Sigma \Delta H°_f$ reactants

$\Delta G° = \Sigma \Delta G°_f$ products − $\Sigma \Delta G°_f$ reactants

$\Delta G° = \Delta H° - T\Delta S°$

$\quad = -RT \ln K$

$\quad = -n\mathscr{F}E°$

$\Delta G = \Delta G° + RT \ln Q$

$q = mc\,\Delta T$

$C_p = \dfrac{\Delta H}{\Delta T}$

K_a (weak acid)

K_b (weak base)

K_w (water)

K_p (gas pressure)

K_c (concentration)

$S°$ = standard entropy

$H°$ = standard enthalpy

$G°$ = standard free energy

T = temperature

n = moles

m = mass

q = heat

c = specific heat capacity

C_p = molar heat capacity at constant pressure

1 faraday $\mathscr{F}$ = 96,500 coulombs/mole e^-

Gas constant, R = 8.31 J mol^{-1} K^{-1}

$\quad$ = 0.0821 L atm mol^{-1} K^{-1}

$\quad$ = 8.31 volt coulomb mol^{-1} K^{-1}

Gases, Liquids, Solutions, and Electrochemistry

$$PV = nRT$$

$$\left(P + \frac{n^2 a}{V^2}\right)(V - nb) = nRT$$

$$P_a = P_{\text{total}} X_a$$

$$X_a = \frac{\text{mole}_a}{\text{total moles}}$$

$$n = \frac{m}{M}$$

$$K = {}^\circ C + 273$$

$$\frac{P_1 V_1}{T_1} = \frac{P_2 V_2}{T_2}$$

$$D = \frac{m}{V}$$

$$u_{\text{rms}} = \sqrt{\frac{3kT}{m}}$$

$$\text{KE per molecule} = \frac{1}{2}mv^2$$

$$\text{KE per mole} = \frac{3}{2}RTn$$

$$\frac{r_1}{r_2} = \sqrt{\frac{M_2}{M_1}}$$

$$\text{molarity, } M = \frac{\text{moles of solute}}{\text{liters of solution}}$$

$$\text{molality, } m = \frac{\text{moles of solute}}{\text{kilograms of solvent}}$$

$$\Delta T_f = iK_f m$$

$$\Delta T_b = iK_b m$$

$$\Pi = \frac{nRTi}{V}$$

$$Q = \frac{[C]^c [D]^d}{[A]^a [B]^b}$$

$$aA + bB \rightleftharpoons cC + dD$$

$$I = \frac{q}{t}$$

$$E_{\text{cell}} = E_{\text{cell}}^\circ - \frac{RT}{n\mathscr{F}} \ln Q$$

$$\log K = \frac{nE^\circ}{0.0592}$$

P = pressure

V = volume

T = temperature

n = number of moles

D = density

m = mass

v = velocity

u_{rms} = root mean square speed

KE = kinetic energy

r = rate of effusion

M = molar mass

Π = osmotic pressure

i = van't Hoff factor

K_f = molal freezing point depression constant

K_b = molal boiling point elevation

Q = reaction quotient

I = current (amperes)

q = charge (coulombs)

t = time (seconds)

E° = standard reduction potential

K = equilibrium constant

Gas constant, R = 8.31 J mol^{-1} K^{-1}

= 0.0821 L atm mol^{-1} K^{-1}

= 8.31 volt coulomb mol^{-1} K^{-1}

SECTION II
Free-Response Questions

TIME: 1 HOUR AND 35 MINUTES

50% OF TOTAL SCORE

PART A: TIME—55 MINUTES.

Calculators are allowed EXCEPT THOSE WITH QWERTY (TYPEWRITER STYLE) KEYBOARDS. Graphing and programmable calculators may be used. Calculator memories do not have to be cleared of programs and data. You must answer ALL questions; there are no choices.

PART B: TIME—40 MINUTES

USE OF CALCULATORS IS NOT PERMITTED. You must answer ALL questions; there are no choices.

GENERAL INSTRUCTIONS

Times for Parts A and B will be announced separately. You may answer questions within each part in any sequence desired. Divide your time appropriately between each problem.

The periodic table, reduction potentials, and lists containing equations and constants are printed on the preceding pages.

You may answer in pen or pencil. Be sure to write CLEARLY and LEGIBLY. If you make an error, you may save time by crossing it out rather than erasing it. On the actual exam, *write all of your answers in the space provided after each question.* For this test, provide your own paper—approximately 20 sheets.

When instructed, you may begin Part A. When the 55 minutes are up, you must stop work and put away your calculator as directed. DO NOT start work on Part B until instructed to do so. You will be allowed to use the periodic table and the tables of information provided with Part A.

SECTION II

Free-Response Questions

TOTAL TIME = 1 HOUR AND 35 MINUTES

PART A

TIME—55 MINUTES

YOU MAY USE YOUR CALCULATOR ON PART A ONLY

> **Directions:** CLEARLY SHOW THESE STEPS AND METHODS USED IN ARRIVING AT YOUR ANSWERS. It is to your advantage to do this to earn partial credit. Answers without the steps and methods may receive little or no credit. Pay attention to significant figures. On the actual exam, you will need to write all of your answers in the pages of the appropriate booklet. For this test, use a separate sheet of paper.

Answer Question 1 below. (This question is worth 20 percent of Section II.)

1. An acidic solution contains 2.0 grams per liter each of V^{3+} and Ni^{2+}. The solubility products of their hydroxides are 4.0×10^{-35} and 6×10^{-16}, respectively.

   (a) Write the dissolution reactions and K_{sp} expressions for both the V^{3+} and Ni^{2+} ions.

   (b) Calculate the molar solubility of each of these hydroxides in distilled water.

   (c) Sodium hydroxide solution is added until a slight cloudiness indicates that something is starting to precipitate. Show by calculations what is precipitating.

   (d) What is the pH of the solution at this point?

   (e) Balances are generally accurate to ±1 milligram. What will the [OH⁻] be when all but the last milligram of the first precipitant has been precipitated?

   (f) Will the second element precipitate before or after the first element has been reduced to 1 milligram? Can these two ions be quantitatively separated using their hydroxide precipitates? Explain.

Answer Question 2 below. (This question is worth 20 percent of Section II.)

2. One of the breakthroughs in understanding atomic structure was the empirical and then the theoretical explanation of atomic spectra.

   (a) Succinctly describe the empirical and theoretical background of the hydrogen spectrum. Include the following: a distinction between empirical and theoretical explanations of nature, how spectra are produced, and how a spectrum is measured. How are energy, wavelength and frequency related to each other?

   (b) Calculate the energy and determine the wavelength of light associated with an electron's transition from the fourth energy level to the second energy level in the hydrogen atom.

   (c) Is this light emitted or absorbed? Explain. Is this light in the visible region of the spectrum? If so, what color is it?

   (d) A certain chemical bond has an energy of 360 kJ mol^{-1}. What wavelength of electromagnetic radiation is needed to break this bond? What spectral region is this wavelength in?

Answer Question 3 below. (This question is worth 20 percent of Section II.)

3. 1.04 grams of an unknown monoprotic weak acid is dissolved in 400 mL of distilled water.

   (a) Write the general dissociation reaction for the weak acid.

   (b) A pH electrode and a reference electrode are placed in the solution, and it is titrated with 0.100 M NaOH. Data for the titration curve are gathered and plotted. The inflection point of the plot is at 34.6 mL. How many moles of NaOH were used to neutralize the weak acid and how many moles of weak acid were present in the sample? What is the molecular mass of this weak acid?

   (c) Assuming that the pH meter is accurately standardized, which of the following three measurements will give the more reliable value for the K_a of this weak acid? (1) at the start of the titration, (2) at the midpoint of the titration, or (3) at the end point of the titration. Explain your answer.

   (d) The pH of the mixture was 3.78 when 12.0 mL of NaOH was added. What is the K_a of the weak acid?

PART B

TIME—40 MINUTES

(Calculators May Not Be Used)

Answer Question 4 below. (This question is worth 5 percent of your total score.)

4. For each of the following three reactions, in part (i) write a BALANCED equation and in part (ii) answer the question about the reaction. In part (i), coefficients should be in terms of lowest whole numbers. Assume that solutions are aqueous unless otherwise indicated. Represent substances in solutions as ions if the substances are extensively ionized. Omit formulas for any ions or molecules that are unchanged by the reaction.

<u>Example:</u>   (i) A strip of magnesium is added to a solution of silver nitrate.

<u>Answer:</u> $Mg + 2\ Ag^+ \rightarrow Mg^{2+} + 2\ Ag$

(ii) Which substance is oxidized in the reaction?

<u>Answer:</u>   Magnesium (Mg) metal

(a) (i) Finely divided manganese dioxide is added to a dilute solution of hydrogen peroxide.

(ii) What is the purpose of manganese dioxide in this reaction?

(b) (i) A dilute solution of chlorine is added to a solution of potassium bromide.

(ii) Identify the oxidizing agent, and state why it was selected.

(c) (i) Equal volumes of equal molar solutions of phosphoric acid and barium hydroxide are mixed.

(ii) One of the substances in this reaction is amphiprotic. Identify the substance, and explain why it was chosen.

Directions: TWO essay questions below will count for **15 percent of THE ENTIRE TEST (7.5% EACH)**. You **MUST ANSWER** Questions 5 and 6. Answers to these questions should demonstrate your ability to present logical, coherent, and convincing explanation of chemical facts and observations. Answers are judged on the accuracy of your analysis and on the appropriateness of the details and examples cited. Specific answers and examples are preferred to broad generalizations. Diagrams, illustrations, and equations may be used in the answers.

5. Gravimetric analysis is the quantitative measurement of a precipitate that can be related to the amount of a substance in an unknown sample. A classic gravimetric experiment is the determination of sulfate in a sample by precipitating $BaSO_4$. All gravimetric procedures have similar concepts even if the samples vary.

   (a) Describe the steps of a gravimetric experiment starting with a solid sample.
   (b) What critical measurements are made in this process (a critical measurement is one that enters into the calculations)
   (c) What experimental errors can occur? Note whether or not the error will increase or decrease the result.
   (d) Give a general equation for converting the mass of a precipitate of $BaSO_4$ to grams of SO_4 and another equation for converting to grams of sulfur.
   (e) What compound contains the highest percentage of sulfate by mass? What solid contains the highest percentage of sulfate by mass? Why would we want to know this information?
   (f) A 0.500-gram sample of an unknown results in 0.500 grams of precipitate. What is the percentage of sulfate in the unknown? Is this a reasonable percentage? What is the percentage of sulfur?

6. The following data are generated for a certain chemical reaction that has the stoichiometry

$$2 A + 3 B \rightarrow 2 P + S$$

Experiment	Concentration of A (moles per liter)	Concentration of B (moles per liter)	Initial Reaction Rate (moles S L^{-1} s^{-1})
1	2.3×10^{-3}	6.1×10^{-4}	2.61×10^{-3}
2	4.1×10^{-3}	5.5×10^{-4}	7.49×10^{-3}
3	6.6×10^{-4}	1.2×10^{-3}	4.23×10^{-4}
4	4.1×10^{-3}	5.8×10^{-3}	7.90×10^{-2}

(a) What is the rate law for this reaction? (Be sure to include the units.)

(b) What is the overall order of the reaction? What is the order with respect to A? With respect to B?

(c) What is the numerical value of the rate constant?

(d) The $\Delta H°$ of this reaction is –65 kJ. At 25 °C, if 1 liter of 1.00 molar A and 1.00 liter of 1.50 molar B are rapidly mixed, will the solution boil? (Assume no heat loss and use a specific heat capacity of 4.184 J g^{-1} °C^{-1} for the solution and a heat of vaporization of water of 43.9 kJ mol^{-1}.)

STOP

End of Section II.
You may use any remaining time to check your work
in this section.

Answer Key

PRACTICE TEST 3

1. C	26. A	51. A
2. B	27. C	52. E
3. D	28. A	53. E
4. E	29. B	54. B
5. C	30. B	55. C
6. D	31. E	56. C
7. A	32. E	57. D
8. C	33. B	58. D
9. A	34. E	59. C
10. C	35. C	60. A
11. D	36. E	61. A
12. A	37. E	62. C
13. B	38. A	63. B
14. D	39. D	64. C
15. A	40. E	65. B
16. B	41. C	66. B
17. D	42. D	67. A
18. C	43. B	68. B
19. B	44. A	69. E
20. A	45. C	70. D
21. D	46. C	71. A
22. C	47. C	72. D
23. D	48. B	73. E
24. C	49. E	74. B
25. A	50. E	75. D

CHAPTER REFERENCES FOR MULTIPLE-CHOICE QUESTIONS

Question	Chapter	Question	Chapter	Question	Chapter
1	1, 2	26	1, 2, 4	51	4, 13
2	1, 4, 13	27	2, 5	52	4, 13
3	2, 4	28	11	53	4, 6
4	2, 4, 5	29	7, 9	54	2, 4
5	2, 4, 13	30	1, 3	55	1, 2, 4
6	2, 4, 14	31	2, 13	56	4, 13
7	2, 4, 16	32	14, 16	57	13
8	2, 15, 16	33	4, 14	58	11
9	2, 4	34	13, 14	59	11
10	2, 4	35	2, 4, 13	60	8, 9
11	2, 4, 14	36	4, 5, 6, 15	61	8, 9
12	4, 13	37	4, 5, 9	62	10, 12
13	4, 5, 10	38	2, 4, 5, 6	63	2, 3
14	15, 16	39	1, 2	64	7, 13
15	7	40	7, 16	65	4, 8
16	4, 6, 14	41	5	66	2, 5
17	10, 12	42	7, 8, 12	67	5, 8, 9, 14
18	4, 5	43	4, 8, 9	68	4, 5, 14
19	5, 14	44	4, 6	69	5, 6, 7
20	5, 14	45	10, 12	70	4, 6
21	1, 2, 4	46	4	71	10, 12, 14
22	10, 16	47	4, 7	72	4, 13
23	10, 12	48	5	73	1, 2
24	10, 14	49	1, 2	74	2, 4
25	10, 14	50	6, 14	75	2, 4

ANSWERS EXPLAINED

Section I—Multiple-Choice

1. **(C)** Iron is the only transition element on the list.

2. **(B)** Sulfur is the only element that can have a +6 oxidation state, in the sulfate ion, for example. Nitrogen is known to be +5 in the nitrate ion.

3. **(D)** Nitrogen is the smallest; sulfur is the next smallest.

4. **(E)** Cesium is the least electronegative, being closest to the lower left corner of the periodic table.

5. **(C)** They are oxidized to sulfates.

6. **(D)** Metal oxides are basic anhydrides.

7. **(A)** Phosphorus is an essential nutrient along with nitrogen and potassium.

8. **(C)** Sulfur is used for vulcanizing rubber.

9. **(A)** Sodium has the lowest ionization energy.

10. **(C)** Silicon is the basis of most semiconductor devices, transistors, and integrated circuits.

11. **(D)** Sulfur is found in the strong acid H_2SO_4.

12. **(A)** Potassium permanganate is the only oxidizing agent on the list.

13. **(B)** Ammonia is a base. $H_2C_2O_4$ is the other complexing agent listed, but it is an acid.

14. **(D)** EDTA is a preservative because it complexes metal ions that often catalyze reactions that cause spoilage.

15. **(A)** Graham's law of effusion can be derived from the kinetic molecular theory and the equation for kinetic energy.

16. **(B)** The calculation is molar mass $= \dfrac{488 \text{ mg acid}}{40 \times 0.1 \text{ mmol acid}} = 122$ mg/mmol, or 122 g/mol. Note that the molarity times the mL of base is the mmol of base, and for a monoprotic acid it is also the mmol of the unknown acid.

17. **(D)** $K_{sp} = [Ag^+]^2[CrO_4^{2-}] = (2x)^2(x) = 4x^3$. Therefore we divide the K_{sp} by 4 and then take the cube root as indicated by response (D).

18. **(C)** The sodium ion and sulfate ion constitute the ionic bond. The bonding between the sulfur and the oxygen in the sulfate ion is covalent.

19. **(B)** At the end point all that exists is the salt of the acid and base. A salt of a weak acid and a strong base will hydrolyze water to produce the most basic solution with the highest pH.

20. **(A)** Diamagnetic elements have all electron spins paired. This occurs only for Ca. Fe is ferromagnetic, and the rest are paramagnetic.

21. **(D)** All hygroscopic materials must be sealed tightly to keep them from absorbing water from the air.

22. **(C)** This is the easiest method without loss of precipitate. Extremely fine solids such as colloids may require an ultracentrifuge.

23. **(D)** The heat of reaction must be combined with the entropy to determine the free energy change.

24. **(C)** The ammonium ion cannot accept electron pairs.

25. **(A)** This pair differs by more than one H^+.

26. **(A)** p orbitals are oriented $90°$ apart, and s orbitals have no directionality. The only way to explain the four identical bonds with a tetrahedral structure with $109°$ angles is by using hybrid orbitals.

27. **(C)** Concentration units of $mol\ L^{-1}$ for each concentration are divided into the rate units of $mol\ L^{-1}\ s^{-1}$.

28. **(A)** The rate law for a zero-order reaction is rate $= k$, and the slope of the kinetic curve must be the same at all times; consequently, it is a straight line.

29. **(B)** Gas solubility is directly proportional to the partial pressure of the gas (Henry's law):

    $[gas] = k\,P_{gas}$. When the pressure is reduced to two-thirds of its original value, then the solubility will be two-thirds of the original value.

30. **(B)** Beta particles increase the atomic number when emitted. Elements with low atomic numbers emit beta particles when their mass is greater than the average mass in the periodic table. This brings the proton/electron ratio closer to unity, which is the stable ratio. Heavy elements such as U and Cf usually emit alpha particles rather than beta particles.

31. **(E)** The NaOH remains in solution after H_2 is produced in the electrolysis of brine solutions.

32. **(E)** The H_2CO_3 will transfer a proton to CO_3^{2-} to give the bicarbonate ion, HCO_3^-. Either H_2CO_3 or CO_3^{2-} will be left over as the excess reactant.

33. **(B)** The weakest conjugate acid will have the strongest conjugate base.

34. **(E)** To be used for a buffer, the pK_a of a weak acid must be within ±1 of the desired pH of the buffer. The pK_a of the acid must be between 3.0 and 5.0. The three acids—methanoic, ethanoic, and nitrous—fit those requirements.

35. **(C)** Fe^{3+} cannot be oxidized any further and therefore cannot cause reduction nor be a reducing agent.

36. **(E)** We solve this as a limiting reactant problem. First, the balanced chemical equation is written as:

$$2\,C_2H_6 + 7\,O_2 \rightarrow 4\,CO_2 + 6\,H_2O$$

Then we calculate the mass of CO_2 formed from each of the reactant masses as

$$?g\,CO_2 = 10\,g\,C_2H_6\left(\frac{1\,mol\,C_2H_6}{30\,g\,C_2H_6}\right)\left(\frac{4\,mol\,CO_2}{2\,mol\,C_2H_6}\right)\left(\frac{44\,g\,CO_2}{1\,mol\,CO_2}\right)$$
$$= 29.3\,g\,CO_2$$

$$?g\,CO_2 = 40\,g\,O_2\left(\frac{1\,mol\,O_2}{32\,g\,O_2}\right)\left(\frac{4\,mol\,CO_2}{7\,mol\,O_2}\right)\left(\frac{44\,g\,CO_2}{1\,mol\,CO_2}\right)$$
$$= 31.4\,g\,CO_2$$

These calculations tell us that the theoretical yield of CO_2 is 29.3 g CO_2. It also tells us that the limiting reactant is C_2H_6. Finally, the law of conservation of matter tells us that the total mass of the system cannot be more or less than the initial 50 g of the reactants given. Three of the four possibilities are true; it is likely that the fourth is also true. We can prove it with the following calculation.

$$?g\,O_2 = 10\,g\,C_2H_6\left(\frac{1\,mol\,C_2H_6}{30\,g\,C_2H_6}\right)\left(\frac{7\,mol\,O_2}{2\,mol\,C_2H_6}\right)\left(\frac{32\,g\,O_2}{1\,mol\,O_2}\right)$$
$$= 37.3\,g\,O_2\,react$$

Since 37.3 g of O_2 react, $40.0 - 37.3 = 2.7$ g O_2 is left.

37. **(E)** The equation $\Delta T = k_b\,m$ is readily solved to find that the solution is 2 molal, or 2 moles in 1000 grams of solvent. The given solution contains 232 grams in 1000 grams of solvent. Therefore we conclude that 2 moles weigh 232 g and 1 mole has a mass of 116 g.

38. **(A)** We can calculate that 66 grams of CO_2 is 1.5 moles of CO_2 and also 1.5 moles of carbon. Similarly, 27 grams of water is 1.5 moles of water or 3 moles of hydrogen. In whole numbers there is 1 mole of carbon combined with every 2 moles of hydrogen for an empirical formula of CH_2.

39. **(D)** These numbers are listed traditionally in the order n, ℓ, m_ℓ, m_s. Because ℓ can be no greater than $n-1$, response (D) represents the quantum numbers that are incorrect.

Answers Explained—Practice Test 3

40. **(E)** This question is asking you to determine the correct number of significant figures to report in your answer. We use the ideal gas law for this calculation, $PV = nRT$, to calculate the number of moles, n, present. The calculation is

$$n = \left(\frac{PV}{RT}\right) = \frac{(745 \text{ torr}/(760 \text{ torr/atm}))(43 \text{ mL}/(1000 \text{ mL/L}))}{(0.08205 \text{ L atm mol}^{-1} \text{ K}^{-1})(297.4 \text{ K})}$$

$$n = \frac{(0.980 \text{ atm})(0.043 \text{ L})}{(0.08205 \text{ L atm mol}^{-1} \text{ K}^{-1})(297.4 \text{ K})}$$

$$n = 0.0017263 \text{ moles}$$

This is rounded to 0.0017 moles.

41. **(C)** Trigonal bipyramid and octahedral structures have 90° bond angles.

42. **(D)** At a phase change $\Delta G° = 0 = \Delta H° - T\Delta S°$.

Therefore $\Delta S° = \dfrac{\Delta H°}{T} = \dfrac{15000 \text{ J}}{265.8 \text{ K}} = 56.4 \text{ J K}^{-1}$.

This result can be estimated as $\dfrac{15000 \text{ J}}{250 \text{ K}}$ = approx. 60 J K^{-1}. Be careful to use the correct units.

43. **(B)** A 2.00 molal solution contains 2.00 moles of solute and 1000 g H_2O = 55.5 mol H_2O.

$$X = \frac{2.00 \text{ mol NaCl}}{2.00 \text{ mol NaCl} + 55.5 \text{ mol H}_2\text{O}} = \frac{2}{57.5} = 0.0348.$$

Approximating the result to be between 0.03 and 0.04 also allows the correct answer to be chosen.

44. **(A)** We estimate the milligrams of AgCl that can be made with the given silver nitrate:

$$? \text{ g AgCl} = 25.00 \text{ mL AgNO}_3 \left(\frac{0.080 \text{ mol AgNO}_3}{\text{L AgNO}_3}\right)\left(\frac{1 \text{ mol AgCl}}{1 \text{ mol AgNO}_3}\right)$$

$$\left(\frac{143.3 \text{ g AgCl}}{1 \text{ mol AgCl}}\right)$$

= 287 mg AgCl (Note how the milli units did not cancel.)

$$\text{percentage yield} = \frac{\text{actual yield}}{\text{theoretical yield}} \times 100$$

$$= \frac{212 \text{ mg AgCl}}{287 \text{ mg AgCl}} \times 100$$

$$= 74\%$$

To estimate the answer, set up the problem as above and then round the 143.3 to 150 to get

$? \text{ g AgCl} = 25 \times 0.08 \times 150 = 300.$ Then $\dfrac{212}{300} \times 100$ = approx. 70%

45. **(C)** $E_{cell} = E^{\circ}_{cell} - \dfrac{0.0591}{n} \log Q$

$Q = \dfrac{[Zn^{2+}]}{[Cu^{2+}]}$ and $n = 2$.

$E_{cell} = 1.10 \text{ V} - \dfrac{0.0591}{2} \log \dfrac{0.0200 \, M}{0.200 \, M}$

$\qquad = 1.10 \text{ V} - 0.03 \log(0.1)$

$\qquad = 1.10 - 0.03 \, (-1)$

$\qquad = 1.13 \text{ V}$

46. **(C)** This is the correct equation because sodium sulfite is soluble but iron(II) sulfite is not.

47. **(C)** If 1.5 atmospheres represents 3 moles, then 4.5 atmospheres must mean that 9 moles are in the flask. Of the 9 moles, 3 moles are neon and 1 mole is hydrogen. Therefore the remaining 5 moles must be the oxygen.

48. **(B)** The nitrate ion has a double bond.

49. **(E)** This is the correct order from least energetic visible light to most energetic X-rays. The remaining responses are not necessarily true.

50. **(E)** This is a strong acid titrated with a strong base, and the equivalence point is at pH 7.

51. **(A)** The balanced half-reaction is $8 \, H^+ + MnO_4^{2-} + 4 \, e^- \rightarrow Mn^{2+} + 4 \, H_2O$.

52. **(E)** The balanced half-reaction is $8 \, H^+ + MnO_4^{2-} + 4 \, e^- \rightarrow Mn^{2+} + 4 \, H_2O$.

53. **(E)** The compound must contain $100 - 25.9 = 74.1$ percent oxygen. We may estimate that in 100 grams of the compound, the 25.9 g of nitrogen is slightly less than 2 moles. Similarly, the 74.1 grams of oxygen is slightly less than 5 moles. A reasonable estimate is N_2O_5. The actual calculations agree.

$? \text{ mol N} = 25.9 \text{ g N} \left(\dfrac{1 \text{ mol N}}{14 \text{ g N}} \right) = 1.85 \text{ mol N}$

$? \text{ mol O} = 74.1 \text{ g O} \left(\dfrac{1 \text{ mol O}}{16 \text{ g O}} \right) = 4.63 \text{ mol O}$

Dividing both by 1.85 yields 1 mole N and 2.5 mol O. The simplest ratio is N_2O_5.

54. **(B)** Barium and iodide ions are isoelectronic with xenon. Barium ions are much smaller because barium has more protons than electrons. Iodide ions are larger because iodide has more electrons than protons.

55. **(C)** More protons attracting electrons from approximately the same distance cause the decrease in size.

56. **(C)** These are the correct oxidation numbers for chromium.

57. **(D)** The balanced equation is $2IO_3^-(aq) + 10\ H^+(aq) + 10\ e^- \rightarrow I_2(aq) + 6\ H_2O(l)$.

58. **(D)** The sum of the exponents in the rate law is 3.

59. **(C)** Chloride ions do not appear as a factor in the rate law and therefore have no effect in this case.

60. **(A)** This has the general form of the phase diagram, with a triple point clearly shown.

61. **(A)** The critical point is at the end of the liquid/vapor equilibrium line.

62. **(C)** These conditions indicate spontaneous reactions.

63. **(B)** This product shows a loss of mass of 2 amu and a loss of nuclear charge of 4; none of the particles has these properties.

64. **(C)** 740 mm Hg – 25 mm Hg = 715 mm Hg

65. **(B)** Sand and quartz are tetrahedral arrays of silicon atoms connected to each other by oxygen atoms.

66. **(B)** All other choices are adjacent to each other in the periodic table and have lower electronegativity differences.

67. **(A)** Hydrogen must be bonded to F, O, or N to form hydrogen bonds.

68. **(B)** Two hydrogen ions are replaced by two potassium ions based on the mole ratio given.

69. **(E)** The balanced reaction is $2\ SO_2 + O_2 \rightarrow 2\ SO_3$. This reaction shows that every liter of O_2 reacts with 2 liters of SO_2. From this it is readily estimated that $1.5/2 = 0.75$ L O_2 is needed. The factor-label calculation is just as simple.

$$?\ L\ O_2 = 1.5\ L\ SO_2 \left(\frac{1\ mol\ SO_2}{22.4\ L\ SO_2}\right)\left(\frac{1\ mol\ O_2}{2\ mol\ SO_2}\right)\left(\frac{22.4\ L\ O_2}{1\ mol\ O_2}\right) = 0.75\ L\ O_2$$

70. **(D)** Percent composition = $\dfrac{\text{atomic mass element} \times n}{\text{molar mass of compound}} \times 100$. n is the number of atoms of the element in the compound's formula. Because oxygen has a mass of 16, the compound must have a mass of $160 \times n$ to be 10 percent nitrogen. All the compounds are too light except for BaO.

71. **(A)** The dissociation reaction is $HCN \rightleftharpoons H^+ + CN^-$. Water is considered a pure liquid and does not appear in the equilibrium law. Written as $HCN + H_2O \rightleftharpoons H_3O^+ + CN^-$, the same equilibrium law results with H_3O^+ replacing H^+.

72. **(D)** Oxidation occurs at the anode, and the only oxidation half-reaction is (D).

73. **(E)** The main determination was the *e/m* ratio of the electron.

74. **(B)** Halogens always are 1– charge ions, and the alkaline earth metals are always 2+ charged ions.

75. **(D)** All of these are isoelectronic with argon except V^{3+}.

Section II—Free-Response

Part A

1. A solution contains 2.0 grams per liter each of V^{3+} and Ni^2. The solubility products of their hydroxides are 4.0×10^{-35} and 6×10^{-16}, respectively.

   (a) $V(OH)_3 \rightleftharpoons V^{3+} + 3\ OH^-$ $K_{sp} = [V^{3+}]\ [OH^-]^3$
   $Ni(OH)_2 \rightleftharpoons Ni^{2+} + 2\ OH^-$ $K_{sp} = [Ni^{2+}]\ [OH^-]^2$

   (b) If *s* moles per liter of each salt dissolves, then the concentration of the metal ions will be *s* moles per liter. For the vanadium compound, the concentration of hydroxide ions will be $3s$, and for the nickel compound, the hydroxide concentration will be $2s$. The solubility product equations become

   $$K_{sp}\ (V(OH)_3) = (s)\ (3s)^3 = 27s^4$$
   $$4.0 \times 10^{-35} = 27s^4$$
   $$s = 1.1 \times 10^{-9}\ M\ V(OH)_3$$
   $$K_{sp}\ (Ni(OH)_2) = (s)\ (2s)^2 = 4s^3$$
   $$6 \times 10^{-16} = 4s^3$$
   $$s = 5.31 \times 10^{-6}\ M\ Ni(OH)_2$$

   (c) We need to calculate the molar concentrations of the metal ions.

   $$?\ \frac{mol\ V^{3+}}{L\ solution} = \frac{2\ g\ V^{3+}}{L\ solution} \left(\frac{1\ mol\ V^{3+}}{50.94\ g\ V^{3+}} \right) = 3.9 \times 10^{-2}\ M$$

   $$?\ \frac{mol\ Ni^{2+}}{L\ solution} = \frac{2\ g\ Ni^{2+}}{L\ solution} \left(\frac{1\ mol\ Ni^{2+}}{58.69\ g\ Ni^{2+}} \right) = 3.4 \times 10^{-2}\ M$$

then

$$K_{sp} = [V^{3+}] [OH^-]^3$$
$$4.0 \times 10^{-35} = 3.9 \times 10^{-2} [OH^-]^3$$
$$[OH^-]^3 = 1.03 \times 10^{-33}$$
$$[OH^-] = 1.00 \times 10^{-11}$$

$$K_{sp} = [Ni^{2+}] [OH^-]^2$$
$$6 \times 10^{-16} = 3.4 \times 10^{-2} [OH^-]^2$$
$$[OH^-]^2 = 1.76 \times 10^{-14}$$
$$[OH^-] = 1.32 \times 10^{-7}$$

The vanadium(III) hydroxide starts precipitating (K_{sp} is reached when the concentration of hydroxide is four orders of magnitude less than that of the nickel(II) hydroxide). Therefore the vanadium will precipitate first.

(d) The pH is calculated as pH = $- \log [H^+] = - \log \dfrac{K_w}{[OH^-]}$

Using the latter expression, pH = $- \log \left(\dfrac{1.0 \times 10^{-14}}{1.0 \times 10^{-11}} \right)$
$$= - \log 1.0 \times 10^{-3}$$
$$= 3.00$$

(e) One milligram of V^{3+} per liter is 2.0×10^{-5} molar V^{3+}.

$$K_{sp} = [V^{3+}] [OH^-]^3$$
$$4.0 \times 10^{-35} = 2.0 \times 10^{-5} [OH^-]^3$$
$$[OH^-]^3 = 2.0 \times 10^{-30}$$
$$[OH^-] = 1.3 \times 10^{-10}$$
$$pH = 4.11$$

(f) We can calculate the pH at which the Ni^{2+} will start precipitating based on the calculations in part (c). The result was that $[OH^-] = 1.32 \times 10^{-7}$ *M* for the $Ni(OH)_2$. The pH of this solution is 7.12.

We can conclude that less than 1 milligram of V^{3+} will remain in solution at pH 4.11 and that the Ni^{2+} will not start precipitating until the pH rises above 7.12. Therefore, separation of the two ions based on their hydroxides is possible.

2. One of the breakthroughs in understanding atomic structure was the empirical and then the theoretical explanation of atomic spectra.

(a) An empirical explanation is one where a mathematical equation is developed to fit the observations but the reasons why the equation fits the data are not fully understood. An example is the Rydberg equation. A theoretical explanation is one where the mathematical description of observations can be related to fundamental constants and variables, as in the equation describing energy levels within the atom.

An atomic spectrum is a plot of light intensity as a function of either wavelength or frequency. Wavelength, λ, and frequency, ν, are inversely proportional to each other. This relationship is best expressed in the equation $\lambda \nu = c$. The symbol c stands for the speed of light. In addition, the frequency and wavelength are related to the energy of a photon of light so that:

$$E = h\nu = hc/\lambda$$

In this equation, the symbol E stands for energy in joules and h is Planck's constant. Atomic spectra are produced by adding energy, usually heat or electrical energy, to an element in its ground state. The element with the added energy is said to be in an excited state. It releases this energy in the form of light and then returns to its ground state. The spectrum is recorded with electronic detectors, or it may be recorded on photographic film.

(b) The energy of the nth energy level in the hydrogen atom is given by the equation in the table of equations given: $E_n = \dfrac{-2.178 \times 10^{-18}}{n^2}$ joule . We can calculate the energy of each level and take the difference. Because the electron passes from the fourth to the second energy level, it is already known that energy is emitted (the mathematical sign will be negative) and that this energy will be in the form of light.

For $n = 2$, $E_2 = \dfrac{-2.178 \times 10^{-18}}{2^2}$ joule $= -5.445 \times 10^{-19}$ joule

For $n = 4$, $E_4 = \dfrac{-2.178 \times 10^{-18}}{4^2}$ joule $= -1.3669 \times 19^{-19}$ joule

$\Delta E = E_{\text{final}} - E_{\text{initial}} = (-5.445 \times 10^{-19} \text{ joule}) - (-1.3669 \times 19^{-19} \text{ joule})$

$= -4.078 \times 10^{-19}$ joule energy is given off (light emitted).

To calculate the wavelength of a photon we combine the relationships $E = h\nu$ and $\lambda \nu = c$ to get $E = hc/\lambda$.

$E = 4.078 \times 10^{-19}$ joule $= (6.63 \times 10^{-34} \text{ J s})(3.00 \times 10^8 \text{ m s}^{-1})/ \lambda$

$\lambda = 4.877 \times 10^{-7}$ m $= 487.7$ nm (this is green light in the visible spectrum)

(c) First we need to calculate the energy per bond, not the moles of bonds as given.

? J/bond $= s\left(\dfrac{360 \text{ kJ}}{1 \text{ mole bonds}}\right)\left(\dfrac{1 \text{ mol bonds}}{6.02 \times 10^{23} \text{ bonds}}\right) = 5.98 \times 10^{-19}$ J/bond

Using the equations above for calculating the wavelength,

$E = 5.98 \times 10^{-19}$ joule $= (6.63 \times 10^{-34} \text{ J s})(3.00 \times 10^8 \text{ m s}^{-1})/ \lambda$

$\lambda = 3.32 \times 10^{-7}$ m $= 332$ nm (this is in the ultraviolet region of the spectrum).

3. 1.00 grams of an unknown monoprotic weak acid is dissolved in 400 mL of distilled water.

(a) $HA \rightarrow H^+ + A^-$

or $HA + H_2O \rightarrow H_3O^+ + A^-$

The reaction with NaOH is

$HA + NaOH \rightarrow H_2O + Na^+ + A^-$

(b) We start by calculating the moles of the weak acid:

$$M \times L = \text{millimoles}$$

$$(0.100\ M)(34.6\ \text{mL}) = 3.46\ \text{millimoles HCl}$$

Based on the equation above, this is also 3.46 millimoles of HA.

molecular mass = 6.00 grams/3.46×10^{-3} moles = 300.5 g/mol

(c) The pH is changing rapidly at both the start of the titration and at the end point. It is difficult to precisely determine the pH at those points. Midway to the end point, the solution is buffered, and the slope of the titration curve is at its lowest. This is the most reliable point at which to determine the K_a.

(d) The Henderson-Hasselbalch equation states:

$$pH = pK_a + \log \frac{[A^-]}{[HA]}$$

For our purposes, 12.0/34.6 is the fraction that is converted to A^-, and 22.6/34.6 is the fraction that remains as the weak acid HA. Substituting these values, we obtain

$$3.78 = pK_a + \log \frac{(12.0/34.6)}{(22.6/34.6)}$$

The 34.6 cancels in the fraction, and taking the logarithm of the ratio 12.0/22.6 yields –0.275.

$$3.78 = pK_a + (-0.275)$$

$$pK_a = 3.78 + 0.275 = 4.06$$

Part B

4. For these answers, balanced equations are required and only reacting species should be shown. An additional touch would be to add reasonable phases for the substances in the reaction. The instructions indicate that most are reactions in aqueous solution.

(a) (i) Finely divided manganese dioxide is added to a dilute solution of hydrogen peroxide.

$$H_2O_2(aq) \rightarrow 2H_2O(l) + O_2(g)$$

(ii) What is the purpose of manganese dioxide in this reaction?

The manganese dioxide, MnO_2, acts as a catalyst to speed up this reaction.

(b) (i) A dilute solution of chlorine is added to a solution of potassium bromide.

$$2Br^-(aq) + Cl_2(aq) \rightarrow 2Cl^-(aq) + Br_2(aq)$$

(ii) Identify the oxidizing agent, and state why it was selected.

The oxidizing agent is $Cl_2(aq)$. An oxidizing agent causes oxidation and is itself reduced. The $Cl_2(aq)$ causes the bromide ion to be oxidized, and the chlorine is reduced to chloride ions. It fits the definition of an oxidizing agent.

(c) (i) Equal volumes of equal molar solutions of phosphoric acid and barium hydroxide are mixed.

$$H_3PO_4(aq) + 2OH^-(aq) \rightarrow HPO_4^{2-}(aq) + 2H_2O(l)$$

(ii) One of the substances in this reaction is amphiprotic. Identify the substance, and explain why it was chosen.

The $HPO_4^{2-}(aq)$ is amphiprotic since it can act as an acid and donate a proton to become $PO_4^{3-}(aq)$. It can also act as a base to react with a proton to become $H_2PO_4^-(aq)$.

5. Gravimetric analysis is the quantitative measurement of a precipitate that can be related to the amount of a substance in an unknown sample. A classic gravimetric experiment is the determination of sulfate in a sample by precipitating $BaSO_4$. All gravimetric procedures have similar concepts even if the samples vary.

(a) Steps that are important are

(i)   Dry sample.

(ii) Weigh sample.

(iii) Dissolve sample.

(iv) Adjust pH if needed.

(v)   Add precipitating agent and test for complete precipitation.

(vi) Filter and wash precipitate.

(vii) Dry and weigh precipitate.

(viii) Calculate results and report.

(b) The two weighing steps are critical to the answer.

(c) Each of the steps listed above can have an effect on the result. These are briefly noted.

(i)   Dry sample (incomplete drying will result in a low result).

(ii)   Weigh sample (errors in weighing can be positive or negative).

(iii) Dissolve sample (incomplete dissolution may add to the mass of the precipitate).

(iv)   Adjust pH if needed (the wrong pH may make the sample more soluble, thus giving a low result).

(v)   Add precipitating agent and test for complete precipitation (if precipitation was not complete, result will be low).

(vi)   Filter and wash precipitate (if filter lets the precipitate through or if washing redissolves the precipitate, the results will be low)

(vii) Dry and weigh precipitate (incomplete drying will give a high result).

(viii) Calculate results and report (calculation errors can be either positive or negative).

(d) $? \text{ g } SO_4^{2-} = \text{weighed g } BaSO_4 \left(\dfrac{1 \text{ mol } BaSO_4}{233 \text{g } BaSO_4} \right) \left(\dfrac{1 \text{ mol } SO_4^{2-}}{1 \text{ mol } BaSO_4} \right)$

$\left(\dfrac{98 \text{ g } SO_4^{2-}}{1 \text{ mol } SO_4^{2-}} \right)$

$? \text{ g } S = \text{weighed g } BaSO_4 \left(\dfrac{1 \text{ mol } BaSO_4}{233 \text{g } BaSO_4} \right) \left(\dfrac{1 \text{ mol } S}{1 \text{ mol } BaSO_4} \right) \left(\dfrac{32 \text{ g } S}{1 \text{ mol } S} \right)$

(e) H_2SO_4 has the highest percentage of sulfate per mole (98%). The next element, He, does not form compounds with sulfate ions. Lithium after that will give us Li_2SO_4 (87%). It has a very high percentage, but the rare $BeSO_4$ (91%) has a higher percentage of sulfate. The utility of this information is that it places an upper limit on the possible percentage of sulfate in a real sample.

(f) Use the equations above to determine the mass of sulfate and sulfur.

$? \text{ g } SO_4^{2-} = 0.500 \text{ g } BaSO_4 \left(\dfrac{1 \text{ mol } BaSO_4}{233 \text{g } BaSO_4} \right) \left(\dfrac{1 \text{ mol } SO_4^{2-}}{1 \text{ mol } BaSO_4} \right)$

$\left(\dfrac{98 \text{ g } SO_4^{2-}}{1 \text{ mol } SO_4^{2-}} \right)$

$= 0.210 \text{ g } SO_4^{2-}$

$? \text{ g } S = 0.500 \text{ g } BaSO_4 \left(\dfrac{1 \text{ mol } BaSO_4}{233 \text{ g } BaSO_4} \right) \left(\dfrac{1 \text{ mol } S}{1 \text{ mol } BaSO_4} \right) \left(\dfrac{32 \text{ g } S}{1 \text{ mol } S} \right)$

$= 0.0687 \text{ g } S$

The percentages are

$$\% \; SO_4^{2-} = \left(\frac{g \; SO_4^{2-}}{g \; sample} \right) \times 100 = \left(\frac{0.210 \; g \; SO_4^{2-}}{0.500 \; g \; sample} \right) \times 100 = 42.0\%$$

$$\% \; S = \left(\frac{g \; S}{g \; sample} \right) \times 100 = \left(\frac{0.0687 \; g \; S}{0.500 \; g \; sample} \right) \times 100 = 13.7\%$$

6. The following data are generated for a certain chemical reaction that has the stoichiometry

$$2 \; A + 3 \; B \rightarrow 2 \; P + S$$

Experiment	Concentration of A (moles per liter)	Concentration of B (moles per liter)	Initial Reaction Rate (moles S L^{-1} s^{-1})
1	2.3×10^{-3}	6.1×10^{-4}	2.61×10^{-3}
2	4.1×10^{-3}	5.5×10^{-4}	7.49×10^{-3}
3	6.6×10^{-4}	1.2×10^{-3}	4.23×10^{-4}
4	4.1×10^{-3}	5.8×10^{-3}	7.90×10^{-2}

(a) To determine the exponents for a rate law, we run two experiments where all reactant concentrations are kept the same except for one that is varied. We then write separate rate laws for the experiment and take a ratio of them as follows:

$$rate(1) = k[A]^x[B]^y \quad and \quad rate(2) = k[A]^x[B]^y$$

In the table above, Experiments 2 and 4 have the same concentration of A, whereas B varies. Entering the concentrations, we get

$$7.49 \times 10^{-3} = k(4.1 \times 10^{-3})^x(5.5 \times 10^{-4})^y$$
and
$$7.90 \times 10^{-2} = k(4.1 \times 10^{-3})^x(5.8 \times 10^{-3})^y$$

We make a ratio of the two equations, usually with the equation with the largest rate as the numerator (for convenience)

$$\frac{7.90 \times 10^{-2}}{7.49 \times 10^{-3}} = \frac{k(4.1 \times 10^{-3})^x(5.8 \times 10^{-3})^y}{k(4.1 \times 10^{-3})^x(5.5 \times 10^{-4})^y}$$

The k and the $(4.1 \times 10^{-3})^x$ terms cancel, leaving

$$\frac{7.90 \times 10^{-2}}{7.49 \times 10^{-3}} = \frac{(5.8 \times 10^{-3})^y}{(5.5 \times 10^{-4})^y} = \left(\frac{(5.8 \times 10^{-3})}{(5.5 \times 10^{-4})} \right)^y$$

$10.54 = (10.54)^y$. The only value of y that will maintain the equality is 1.

To obtain the exponent x for compound A we set up another ratio. Any other two experiments will do because we already know the exponent $y = 1$. We will try the remaining two experiments.

Experiment	Concentration of A (moles per liter)	Concentration of B (moles per liter)	Initial Reaction Rate (moles S L^{-1} s^{-1})
1	2.3×10^{-3}	6.1×10^{-4}	2.61×10^{-3}
2	4.1×10^{-3}	5.5×10^{-4}	7.49×10^{-3}
3	6.6×10^{-4}	1.2×10^{-3}	4.23×10^{-4}
4	4.1×10^{-3}	5.8×10^{-3}	7.90×10^{-2}

$$\frac{2.61 \times 10^{-3}}{4.23 \times 10^{-4}} = \frac{k(2.3 \times 10^{-3})^x (6.1 \times 10^{-4})^1}{k(6.6 \times 10^{-4})^x (1.2 \times 10^{-3})^1} = \frac{(2.3 \times 10^{-3})^x}{(6.6 \times 10^{-4})^x}(0.508)$$

$$\frac{6.17}{0.508} = \left(\frac{(2.3 \times 10^{-3})}{(6.6 \times 10^{-4})}\right)^x = (3.48)^x$$

$12.15 = (3.48)^x$. By trial and error we can determine that $x = 2$.

The rate law is rate = $k[A]^2[B]$.

All concentrations have units of mol/L, and the rate has units of mol L^{-1} s^{-1}. The rate constant has units of L^2 mol^{-2} s^{-1}.

(b) This is a third-order reaction overall, and it is second order with respect to [A] and first order with respect to [B].

(c) Take any experiment's data and substitute into the rate law. For example

$$rate = k[A]^2[B]$$

$$2.61 \times 10^{-3} \text{ mol } L^{-1} s^{-1} = k(2.3 \times 10^{-3})^2(6.1 \times 10^{-4})$$

$$k = 1.86 \times 10^3 \text{ } L^2 \text{ mol}^{-2} \text{ s}^{-1}$$

(d) The $\Delta H°$ of –65 kJ is the amount of heat generated if 2 moles of A react with 3 moles of B (the stoichiometry of the reaction). The conditions given react with half that amount. Therefore we expect –32.5 kJ of energy. We can calculate the final temperature of the mixture by using the equation

$$\text{Heat}(q) = (\text{specific heat})(\text{mass})(T_f - T_i)$$

$$32.5 \text{ kJ} = (4.184 \text{ J g}^{-1} °C^{-1})(2000 \text{ g } H_2O)(T_f - 25 °C)$$

$$32,500 \text{ J} = 8368 T_f - 8368 (25 °C)$$

$$T_f = 28.8 °C$$

The energy released does not bring the mixture anywhere near boiling.

Appendix 1

Answer Explanations for End-of-Chapter Questions

CHAPTER 1

1. **(B)** The Heisenberg uncertainty principle says that the more precisely we know the position, the less precisely we know its momentum, and vice versa.

2. **(C)** Hund's rule says that all orbitals at a given energy level are filled with one electron each before pairing occurs. The Pauli exclusion principle means that no two electrons can have all four quantum numbers the same or two unpaired electrons cannot share an orbital.

3. **(A)** The Rydberg equation, if the larger value of n is set to an infinitely large number (indicating complete removal of the electron) can easily provide this answer.

4. **(B)** The s and f are symbols used for orbitals, while the other symbols represent quantum numbers.

5. **(E)** only s and f represent orbitals. Calcium can have s orbitals filled while in the ground state. It has no f orbitals and the only way this could have an electron is in excited state.

6. **(C)** The quantum number m_ℓ indicates the shape of the orbital.

7. **(E)** The order of energies is as follows: microwave < infrared < visible < ultraviolet < X rays.

8. **(C)** Solve the equation $\lambda \nu = c$, where $c = 3.00 \times 10^8$ m s^{-1} and $\nu = 4.00 \times 10^{14}$ s^{-1}.

$$\lambda = \frac{3.00 \times 10^8 \text{ m s}^{-1}}{4.00 \times 10^{14} \text{ s}^{-1}} = 7.5 \times 10^{-7} \text{ m} = 750 \text{ nm}$$

9. **(E)** Arsenic has a total of 15 p electrons: 6 in period 2, 6 in period 3, and 3 in period 4.

10. **(B)** These quantum numbers represent a $3d$ electron (the second quantum number is 2) in the third energy level (the first quantum number is 3). Only Fe has an incompletely filled d sublevel.

11. **(E)** When $\ell = 0$ it is an s orbital; $\ell = 1$, a p orbital; $\ell = 2$, a d orbital.

12. **(D)** $A = 31$ and $Z = 15$. In a neutral atom the electrons and protons are each equal to Z. The differences between A and Z is the number of neutrons = 16.

13. **(D)** An element that has an initial d^4 or d^9 electron configuration is usually going to promote an s electron to give a d^5 or d^{10} configuration, which is more stable. Ag is one of those elements.

14. **(C)** The electronic configuration of a noble gas always ends with np^6. These gases also have an ns^2 in their structure, where n is the highest principal level.

15. **(A)** Energy is needed to increase the value of n, and energy is released when n decreases. The more levels by which the electron increases, the greater the energy needed.

16. **(D)** Phosphorous is in Group VA and has five valence electrons.

17. **(C)** The Millikan oil-drop experiment determined the charge of the electron independent of the mass.

18. **(C)** The law of multiple proportions is a consequence of the atomic theory, not part of it.

19. **(E)** Chromium has an electron configuration of [Cr] $3d^5 4s^1$, giving it six unpaired electrons.

20. **(B)** The ℓ or second quantum number defines orbital shape.

21. **(D)** 2.36×10^{-34} is the largest common divisor of all five measurements.

22. **(A)** The $4d$ electrons appear in the fifth period of the periodic table.

23. **(B)** The f orbitals in the sixth and seventh periods hold 14 electrons.

24. **(C)** Valence electrons are the electrons in the outermost energy level of the atom. Only the s and p electrons are in the same energy level as the period in the periodic table.

25. **(D)** $E = h\nu$ is the equation for determining the energy of a photon.

26. **(E)** Bohr's model of the atom demonstrated that the Rydberg equation could be used for all of the listed purposes.

CHAPTER 2

1. **(E)** Bismuth is a metal, and nitrogen is a nonmetal.

2. **(D)** Bromine is a halogen.

3. **(B)** Strontium has 38 protons.

4. **(D)** Bromine is a liquid under normal conditions.

5. **(C)** Uranium at the bottom of the periodic table is the largest.

6. **(A)** A differentiating electron is the electron present in one atom but not in the atom just before it in the periodic table. For transition elements, these are *d* electrons.

7. **(B)** Boiling points vary regularly within groups (columns), and Ni and Pt are just above and below the Pd atom.

8. **(D)** Boiling points decrease for metals and increase for nonmetals from the top to the bottom of a group.

9. **(A)** In general, an element close to F has the higher electronegativity of any pair of atoms. Except for response A, the element closest to F is listed second.

10. **(E)** An element with only two valence electrons must be in the second group from the left of the table. These are the alkaline earth elements.

11. **(D)** Both the number of electrons and the number of protons are equal to the atomic number of Ar, 18. The number of neutrons is the difference between the atomic mass and the atomic number ($40 - 18 = 22$).

12. **(C)** Only H is not mentioned in the chapter as having common allotropes.

13. **(A)** Since Be is in the second group, it is expected to lose two electrons easily while the third ionization is very difficult. The other atoms can lose three or more electrons with relative ease and have lower-third ionization energies.

14. **(D)** Li and Mg are diagonally related in size and many physical properties stemming from atomic and ionic size. Although Si and C are next to each other in the fourth group, Si is a metalloid and semiconductor, and C is not. There is a larger difference between Si and C than between Mg and Li.

15. **(D)** Most elements are metals.

16. **(E)** We can determine the number of neutrons only if a specific isotope is selected.

17. **(A)** Electrons are the first parts of the atoms to encounter each other in a collision between atoms.

18. **(C)** Sn is the symbol for tin.

19. **(A)** In a period, the largest atom is closest to the left side of the periodic table. In a group, the atoms increase in size from top to bottom. Therefore K is a larger atom than Ca.

20. **(B)** Hg is the furthest from the line dividing the metals from the nonmetals and therefore is least likely to be a metalloid.

21. **(D)** The effective (unshielded by core electrons) nuclear charge increases, contracting a shell of e^-.

$$\ln\left(\frac{N_0}{N_t}\right)$$

$$\frac{\ln(2)}{t_{1/2}} = k$$

CHAPTER 3

1. **(C)** The electron is called a beta particle by nuclear scientists.

2. **(A)** Heavy radioisotopes tend to disintegrate by loss of an alpha particle.

3. **(D)** The gamma ray is a form of electromagnetic radiation that has no mass or charge.

4. **(E)** The difference in the species is 3 mass units but no charge units. Only three neutrons can account for this change.

$$_{92}^{235}\text{U} \rightarrow \,_{36}^{94}\text{Kr} + \,_{56}^{139}\text{Ba} + 3\,_{0}^{1}\text{N}$$

5. **(A)** Alpha particles have the least penetrating ability, and neutrons have very great penetrating ability.

6. **(A)** The unknown particle needs a mass of 4 and an atomic number of 2. This is an alpha particle or a helium nucleus.

7. **(C)** To estimate the answer we note that 70% decomposition is slightly less than 75% decomposition. We know that 2 half-lives will leave 25% of the original amount and 75% will have decomposed. Therefore the time involved must be a little less than 2 half-lives. Since 2 half-lives is 11,460 years, we can judge that, of the given possibilities, 9950 is the most reasonable choice.

To calculate an exact answer we start with $\ln\left(\dfrac{N_0}{N_t}\right) = kt$, and for the half-life $\dfrac{\ln(2)}{t_{1/2}} = k$. Given the half-life of 5730 years for ^{14}C, the value of k is calculated as

$$k = \frac{\ln(2)}{5730 \text{ yr}} = 1.21 \times 10^{-4} \text{ yr}^{-1}$$

Using the first equation and setting $N_0 = 100$, and since N_t is the amount that is left, we obtain

$$N_t = 100 - 70 = 30.$$

Solving the first equation for t gives

$$t = \frac{\ln(100/30)}{1.21 \times 10^{-4} \text{ yr}^{-1}} = 9950 \text{ yr}$$

8. **(B)** $\ln(2) = kt$. Solving for k gives

$$k = \frac{\ln(2)}{(16.3 \text{ min})(60 \text{ s/min})} = 7.1 \times 10^{-4} \text{ s}^{-1}$$

9. **(D)** The reaction is

$$^{1}_{0}n + {}^{235}_{92}U \rightarrow {}^{139}_{56}Ba + 3{}^{1}_{0}n + {}^{94}_{36}Kr$$

10. **(B)** The atomic mass of natural copper is given as 65.55 in the periodic table. Since Cu comes before Bi in the periodic table, it will probably emit either a beta particle (electron) or a positron (positive-charge electron). Emitting a beta particle will reduce the n/p ratio to a value closer to the ratio in Cu-65.

11. **(D)** One week has 7 days × 24 h/day = 168 h. If we divide 168 h by the half-life of 37.5 h, we get 4.7 half-lives of ^{82}Br in one week. Four half-lives leaves $\dfrac{1}{16}$ or 144 mg of ^{82}Br. Five half-lives leaves $\dfrac{1}{32}$ or 72 mg of the original amount of ^{82}Br. The only answer between these two is 88.1 mg. Alternatively, k can be calculated from $\ln(2) = kt_{1/2}$, and then $\ln\dfrac{N_0}{N_t} = kt$ is used to calculate N_t.

12. **(D)** Select the isotope where the n/p ratio is closest to the ratio determined from the average atomic mass in the periodic table. The zinc isotope comes closes to this ratio.

13. **(E)** $\ln\left(\dfrac{N_0}{N_t}\right) = kt$. We let $N_0 = 100$, $N_t = 75$, and $t = 3.24$ h. After entering the data, we solve for k in units of hours.

$$\ln\left(\frac{100}{75}\right) = k(3.24 \text{ h})$$

$$k = \left(\frac{0.288}{3.25 \text{ hr}}\right) = 0.0888 \text{ h}^{-1}$$

$$= 0.089 \text{ h}^{-1}$$

14. **(A)** $\ln(2) = kt$. Solve for k:

$$k = \frac{\ln(2)}{(32.5 \text{ days})(24 \text{ hr/day})(60 \text{ min/hr})(60 \text{ s/min})} = 2.47 \times 10^{-7} \text{ s}^{-1}$$

$$\text{Rate} = kN_0 = (2.47 \times 10^{-7} \text{ s}^{-1})(8.00 \times 10^7 \text{ nuclei})$$

$$= 19.8 \text{ nuclei s}^{-1} \text{ decay}$$

CHAPTER 4

1. **(C)** The bromide ion is Br^-.

2. **(E)** Of the possibilities listed, only calcium and phosphate have the required opposite charges to form an ionic compound: $Ca_3(PO_4)_2$.

3. **(D)** There are six possible compounds: NH_4Br, $(NH_4)_3PO_4$, $CaBr_2$, $Ca_3(PO_4)_2$, $FeBr_3$, and $FePO_4$.

4. **(E)** Of the six compounds, only the iron(III) and calcium phosphates are insoluble. Only iron(III) phosphate is one of the responses.

5. **(D)** All potassium salts are soluble. See solubility rules.

6. **(B)** The sulfides are insoluble except for those that have alkali metals or the ammonium ion as the cation. See solubility rules.

7. **(E)** MnO_4^- is defined as the permanganate ion.

8. **(C)** Two ammonium ions and one oxalate ion are obtained. The anion does not decompose into other species, and the ammonium cation is enclosed in parentheses with a subscript of 2 indicating that two NH_4^+ ions must result.

9. **(E)** Iron is a transition element with more than one possible charge and requires the use of the Stock system for its name. Because it is a +3 ion, it is called iron(III). The NO_3 identifies the compound as a nitrate.

10. **(C)** The balanced reaction is

$$2C_6H_6 + 15O_2 \rightarrow 12CO_2 + 6H_2O$$

The sum of these coefficients is 35.

11. **(C)** Potassium loses one electron, and the preceding noble gas in the periodic table is Ar.

12. **(C)** Na and F have electronic configurations of their ions that are identical to that of Ne. Each atom in the other pairs is isoelectric with different noble gases.

13. **(D)** Aluminum always forms a +3 ion and needs three Cl ions with a -1 charge each in order to form a neutral molecule.

14. **(E)** This is the electronic configuration of the Ar atom. Each of these ions is isoelectronic with Ar.

15. **(D)** The correct name for this ion is the chlorite ion.

16. **(B)** Calcium ions are +2 and oxide ions -2, so the correct formula is CaO.

17. **(E)** The dihydrogen phosphate ion, $H_2PO_4^-$, needs one K^+ for a correct formula. In the other formulas there is an excess of a positive or negative charge.

18. **(C)** The products of this reaction are hydrogen gas and aluminum bromide.

19. **(D)** The solubilities must be known to write correct net ionic equations; these equations do not predict solubility.

20. **(D)** This is the only reaction that has the same number of each atom as reactants and products.

21. **(C)** PbCl is the only insoluble substance in this equation.

CHAPTER 5

1. **(A)** CBr_4 is a tetrahedron. NH_3 is a triangular pyramid but is related to the basic tetrahedron.

2. **(E)** Only $H-C \equiv N$ is a linear structure, with a 180° bond angle.

3. **(E)** HCN has two pi bonds.

4. **(E)** HCN, being linear, must lie in one plane. The SO_3 is trigonal planar and also is flat.

5. **(B)** PF_5 has ten electrons surrounding the phosphorus atom.

6. **(A)** The elements Ba, Zn, C, and Cl are arranged in order from the lower left (Ba) to the upper right corner of the periodic table (Cl).

7. **(C)** All the choices have a C atom. Of the second atom in each bond, the O atom is the most electronegative.

8. **(D)** In SF_2Cl_4 the two F atoms may be arranged opposite each other for a nonpolar molecule, or next to each other for a polar molecule.

9. **(C)** The carbonate ion has only one double bond in each resonance structure and therefore one pi bond. Although the entire ion has a charge of –2, it is nonpolar.

10. **(C)** In SF_5^- there are five atoms bound to the sulfur and one nonbonding pair of electrons. This is an octahedral structure and a sp^3d^2 hybrid.

11. **(D)** H_2S is similar in structure to water. With two nonbonding electron pairs, the molecule has a bent structure.

12. **(A)** The cyanide ion, CN^-, is electronically identical to N_2.

13. **(C)** The octahedron is sp^3d^2 not sp^3d.

14. **(C)** The Lewis structure for SO_3 involves one O atom with a double bond and two O atoms with single bonds. The double bond can be placed in three positions, resulting in three resonance structures.

15. **(B)** Ammonia has a nonbonding pair of electrons.

16. **(E)** All of these statements are true.

17. **(C)** The sulfite ion is SO_3^{2-}. There are six valence electrons on each oxygen atom and on the sulfur atom, totaling 24 electrons. Two electrons are added for the –2 charge, resulting in a total of 26 electrons.

18. **(B)** Two nitrogen atoms give the prefix *di-* in dinitrogen and three oxygen atoms give the *tri-* in trioxide.

19. **(A)** A 60° angle does not occur in any hybrid structure considered in this text.

20. **(A)** The *sp* hybrid requires that two atoms be bound to the central atom with no nonbonding electron pairs. None of the structures named in the question fulfills these requirements. SO_2 is sp^2, SF_6 is sp^3d^2, SCl_4 is sp^3d, and SCl_2 is sp^3.

21. **(D)** The direction, or orientation, of a bond in space has nothing to do with bond strength.

CHAPTER 6

1. **(D)** The formula of an ionic compound is an empirical formula, We can add the masses of each element to obtain the molar mass, and we can calculate the percentage of iron and chlorine in the compound using the mass data.

2. **(A)** The molarity is defined as moles of solute dissolved in a liter of solution. The volume of almost everything changes with temperature, and therefore the molarity changes with temperature.

3. **(D)** We can determine the empirical formula if we know the percentage composition as shown in this chapter.

4. **(C)** We can determine the molecular formula if we know the empirical formula and the molar mass.

5. **(B)** $? \text{ g KClO}_3 = 0.200 \text{ L KClO}_3 \left(\dfrac{0.150 \text{ mol KClO}_3}{1 \text{ L KClO}_3} \right) \left(\dfrac{122.5 \text{ g KClO}_3}{1 \text{ mol KClO}_3} \right)$

    $= 3.68 \text{ g KClO}_3$

6. **(E)** The reaction is

    $$\text{NaOH} + \text{HNO}_3 \rightarrow \text{NaNO}_3 + \text{H}_2\text{O}$$

    $? \text{ mL NaOH} = 35.0 \text{ mL HNO}_3 \left(\dfrac{0.345 \text{ mol HNO}_3}{1000 \text{ mL HNO}_3} \right) \left(\dfrac{1 \text{ mol NaOH}}{1 \text{ mol HNO}_3} \right)$

    $\times \left(\dfrac{1000 \text{ mL NaOH}}{0.130 \text{ mol NaOH}} \right)$

    $= 92.9 \text{ mL NaOH}$

7. **(A)** $? \text{ g CaCO}_3 = 3.00 \text{ L CO}_2 \left(\dfrac{1 \text{ mol CO}_2}{22.4 \text{ L CO}_2} \right) \left(\dfrac{1 \text{ mol CaCO}_3}{1 \text{ mol CO}_2} \right) \left(\dfrac{100 \text{ g CaCO}_3}{1 \text{ mol CaCO}_3} \right)$

    $= 13.4 \text{ g CaCO}_3$

8. **(D)** $? \text{ mol H} = 14.3 \text{ g H} \left(\dfrac{1 \text{ mol H}}{1.0 \text{ g H}} \right) = 14.3 \text{ mol H}$

    $? \text{ mol C} = 85.7 \text{ g C} \left(\dfrac{1 \text{ mol C}}{12.0 \text{ g C}} \right) = 7.14 \text{ mol C}$

    $\dfrac{14.3 \text{ mol H}}{7.14} = 2 \text{ mol H}$ and $\dfrac{7.14 \text{ mol C}}{7.14} = 1 \text{ mol C.}$

    The empirical formula is CH_2.

9. **(C)** One Al atom, 3 N atoms, and 9 O atoms add up to a mass of 213.

10. **(D)** $? \text{ mg Na}_2\text{SO}_4 = 0.100 \text{ L Na}^+ \left(\dfrac{0.00100 \text{ mol Na}^+}{1 \text{ L Na}^+} \right) \left(\dfrac{1 \text{ mol Na}_2\text{SO}_4}{2 \text{ mol Na}^+} \right)$

$$\times \left(\dfrac{142 \text{ g Na}_2\text{SO}_4}{1 \text{ mol Na}_2\text{SO}_4} \right)$$

$$= 7.1 \times 10^{-3} \text{ g Na}_2\text{SO}_4 = 7.1 \text{ mg Na}_2\text{SO}_4$$

11. **(A)** $? \text{ cm} = 400 \text{ nm} \left(\dfrac{10^{-9} \text{ m}}{1 \text{ nm}} \right) \left(\dfrac{1 \text{cm}}{10^{-2} \text{ m}} \right) = 4.00 \times 10^{-5} \text{ cm}$

12. **(D)** $? \text{ mol Al} = 1 \text{ mol Fe} \left(\dfrac{8 \text{ mol Al}}{9 \text{ mol Fe}} \right) = \dfrac{8}{9} \text{ mol Al}$

13. **(C)** The reaction is $2\text{KOH} + \text{H}_2\text{SO}_4 \rightarrow \text{K}_2\text{SO}_4 + 2\text{H}_2\text{O}$

$$\dfrac{\text{mol H}_2\text{SO}_4}{1 \text{ L H}_2\text{SO}_4} = \dfrac{0.125 \text{ mol KOH}}{1 \text{ L KOH}} \left(\dfrac{1 \text{ mol H}_2\text{SO}_4}{2 \text{ mol KOH}} \right) \left(\dfrac{35.4 \text{ mL KOH}}{50.0 \text{ mL H}_2\text{SO}_4} \right)$$

$$= 0.0443 \text{ M H}_2\text{SO}_4$$

14. **(E)** The mass of the empirical formula unit is 14 g unit^{-1}.

$$\dfrac{83.5 \text{ g mol}^{-1}}{14 \text{ g unit}^{-1}} = 5.96 \text{ unit mol}^{-1}$$

This rounds to 6 empirical formula units per mole and a molecular formula of C_6H_{12}.

15. **(D)** $? \text{ g Fe} = 1 \text{ atom Fe} \left(\dfrac{1 \text{ mol Fe}}{6.02 \times 10^{23} \text{ atoms Fe}} \right) \left(\dfrac{55.85 \text{ g Fe}}{1 \text{ mol Fe}} \right)$

$$= 9.28 \times 10^{-23} \text{ g Fe}$$

16. **(B)** $? \text{ g SO}_2 = 4.00 \text{ L SO}_2 \left(\dfrac{1 \text{ mol SO}_2}{22.4 \text{ L SO}_2} \right) \left(\dfrac{64 \text{ g SO}_2}{1 \text{ mol SO}_2} \right) = 11.4 \text{ g SO}_2$

17. **(D)** Molar mass of K_3PO_4 is 212. Mass of potassium in one mole of K_3PO_4 is $3 \times 39 = 117$.

$$\%\text{K} = \dfrac{117 \text{ g K}}{212 \text{ g K}_3\text{PO}_4} \times 100 = 55.2\% \text{ K}$$

18. **(C)** $? \text{ g C} = 0.0357 \text{ g CO}_2 \left(\dfrac{1 \text{ mol CO}_2}{44 \text{ g CO}_2} \right) \left(\dfrac{1 \text{ mol C}}{1 \text{ mol CO}_2} \right) \left(\dfrac{12 \text{ g C}}{1 \text{ mol C}} \right)$

$= 0.097 \text{ g C}$

$\%\text{C} = \dfrac{0.097 \text{ g C}}{0.200 \text{ g sample}} \times 100 = 48.7\% \text{ C}$

19. **(B)** It is the kilogram that is an S.I. unit, not the gram.

20. **(A)** Since the question asks for the amount of product, we do the calculations using each reactant and choose the smaller answer.

$? \text{ g AgCl} = 20.0 \text{ g AgNO}_3 \left(\dfrac{1 \text{ mol AgNO}_3}{170 \text{ g AgNO}_3} \right) \left(\dfrac{2 \text{ mol AgCl}}{2 \text{ mol AgNO}_3} \right) \left(\dfrac{143.5 \text{ g AgCl}}{1 \text{ mol AgCl}} \right)$

$= 16.9 \text{ AgCl}$

$= 15.0 \text{ g CaCl}_2 \left(\dfrac{1 \text{ mol CaCl}_2}{111 \text{ g CaCl}_2} \right) \left(\dfrac{2 \text{ mol AgCl}}{1 \text{ mol CaCl}_2} \right) \left(\dfrac{143.5 \text{ g AgCl}}{1 \text{ mol AgCl}} \right)$

$= 38.8 \text{ g AgCl}$

16.9 g AgCl is the correct answer and also defines $AgNO_3$ as the limiting reactant.

21. **(D)** Determine the limiting reactant (this was done in the preceding example, but another method is shown here):

$? \text{ g AgNO}_3 = 15.0 \text{ g CaCl}_2 \left(\dfrac{1 \text{ mol CaCl}_2}{111 \text{ g CaCl}_2} \right) \left(\dfrac{2 \text{ mol AgNO}_3}{1 \text{ mol CaCl}_2} \right) \left(\dfrac{170 \text{ g AgNO}_3}{1 \text{ mol AgNO}_3} \right)$

$= 45.9 \text{ g AgNO}_3$

This calculation shows that we need 45.9 g $AgNO_3$ to react all of the $CaCl_2$. The problem only gives us 20.0 g. Therefore $AgNO_3$ is used up first and is the limiting reactant. Now use the given amount of the limiting reactant to calculate the number of grams of $CaCl_2$ that react.

$? \text{ g CaCl}_2 = 20.0 \text{ g AgNO}_3 \left(\dfrac{1 \text{ mol AgNO}_3}{170 \text{ AgNO}_3} \right) \left(\dfrac{1 \text{ mol CaCl}_2}{2 \text{ mol AgNO}_3} \right) \left(\dfrac{111 \text{ g CaCl}_2}{1 \text{ mol CaCl}_2} \right)$

$= 6.53 \text{ g CaCl}_2$

Since 6.53 g $CaCl_2$ react, $15.0 - 6.53 = 8.47$ g $CaCl_2$ must be left.

22. **(B)** $? \text{ g Cl} = 5.86 \text{ g AgCl} \left(\dfrac{1 \text{ mol AgCl}}{143.5 \text{ g AgCl}} \right) \left(\dfrac{1 \text{ mol Cl}}{1 \text{ AgCl}} \right) \left(\dfrac{35.5 \text{ g Cl}}{1 \text{ mol Cl}} \right)$

$= 1.45 \text{ g Cl}$

$? \text{ \%Cl} = \dfrac{1.45 \text{ g Cl}}{50.0 \text{ g sample}} \times 100 = 2.90\%$

23. **(D)** $? \text{ L air} = 1 \text{ mol CH}_4 \left(\dfrac{2 \text{ mol O}_2}{1 \text{ mol CH}_4} \right) \left(\dfrac{22.4 \text{ L O}_2}{1 \text{ mol O}_2} \right) \left(\dfrac{1 \text{ L air}}{0.2 \text{ L O}_2} \right)$

$= 224 \text{ L air}$

24. **(C)** Assume a 100-g sample of the compound. Then 25% H = 25 g H and 75% C = 75 g C.

Calculate moles of each:

$? \text{ mol H} = 25 \text{ g H} \left(\dfrac{1 \text{ mol H}}{1 \text{ g H}} \right) = 25 \text{ mol H}$

$? \text{ mol C} = 75 \text{ g C} \left(\dfrac{1 \text{ mol C}}{12 \text{ g C}} \right) = 6.25 \text{ mol C}$

$$\dfrac{6.25 \text{ mol C}}{6.25} = 1 \text{ mol C}$$

Also,

$$\dfrac{25 \text{ mol H}}{6.25} = 4 \text{ mol H}$$

The formula is CH_4.

25. **(B)** We calculate:

$? \text{ molecules C}_{12}\text{H}_{22}\text{O}_{11} = 2.5 \times 10^{-6} \text{ g C}_{12}\text{H}_{22}\text{O}_{11} \left(\dfrac{1 \text{ mol C}_{12}\text{H}_{22}\text{O}_{11}}{342 \text{ g C}_{12}\text{H}_{22}\text{O}_{11}} \right)$

$\left(\dfrac{6.02 \times 10^{23} \text{ molecules C}_{12}\text{H}_{22}\text{O}_{11}}{1 \text{ mol C}_{12}\text{H}_{22}\text{O}_{11}} \right)$

$= 4.4 \times 10^{15} \text{ molecules of C}_{12}\text{H}_{22}\text{O}_{11}$

CHAPTER 7

1. **(A)** Boyle's law says that $P \times V =$ constant. Therefore as the pressure decreases the volume must increase.

2. **(D)** The effusion rate will increase if the molecular mass is decreased, as it does going from ethane to methane.

3. **(B)** The relationship between temperature and average kinetic energy is a fundamental principle.

4. **(A)** Volume and pressure change in opposite directions because they are inversely proportional. All other pairs are directly proportional to each other.

5. **(A)** The more volume per molecule keeps them widely separated, therefore reducing interactions. Increased velocity due to an increased temperature also reduces attractions between gas molecules.

6. **(A)** This problem gives all of the information needed to solve the ideal gas law equation, $PV = nRT$, for volume:

$$V = \frac{nRT}{P} = \frac{(2.50 \text{ mol})(0.0821 \text{ L atm mol}^{-1} \text{ K}^{-1})(318 \text{ K})}{1.50 \text{ atm}}$$

$$= 43.5 \text{ L}$$

All of the values given in the problem are used directly in this equation except that the temperature of 45 °C must be converted to 318 K. The other choices represent incorrect combinations of the given data.

7. **(C)** This is another calculation using the ideal gas law equation:

$$n = \frac{PV}{RT} = \frac{(800 \text{ mm Hg}/760 \text{ mm Hg atm}^{-1})(6.45 \text{ L})}{(0.0821 \text{ L atm mol}^{-1} \text{ K}^{-1})(297 \text{ K})}$$

$$= 0.278 \text{ mol}$$

In this example the pressure in millimeters of mercury had to be converted to atmospheres, and the Celsius temperature to Kelvin units. As in the preceding example, the incorrect responses involve simple errors in application of the ideal gas law equation.

8. **(D)** At STP 1 mol of an ideal gas occupies 22.4 L. The mass of 22.4 L of gas is the molar mass of the substance. The density is then

$$\text{density} = \frac{\text{mass}}{\text{liters}} = \frac{20.18 \text{ g mol}^{-1}}{22.4 \text{ L mol}^{-1}}$$

$$= 0.901 \text{ g L}^{-1}$$

Equation 4.12 could also be solved to obtain the density:

$$\frac{g}{v} = \frac{(\text{molar mass}) P}{RT} = \frac{(20.18 \text{ g mol}^{-1})(1.00 \text{ atm})}{(0.0821 \text{ L atm mol}^{-1} \text{ K}^{-1})(273)} = 0.901 \text{ g L}^{-1}$$

9. **(D)** To solve the problem, take the ratio of two ideal gas law equations as follows:

$$\frac{P_i V_i}{P_f V_f} = \frac{n_i R T_i}{n_f R T_f}$$

and cancel variables that are kept constant, n, R, and T. Rearrange the remaining variables.

$$P_f = \frac{P_i V_i}{V_f}$$

$$\frac{(58 \text{ mm Hg})(155 \text{ mL})}{1000 \text{ mL}} = 8.99 \text{ mm Hg}$$

Note that the units of volume must be the same for V_i and V_f, and so 1.00 L was converted to 1000 mL.

10. **(B)** The pressure is directly proportional to the number of moles of gas, and the error in the pressure will be directly reflected in the moles of hydrogen reported. By Dalton's law of partial pressures

$$P_{\text{total}} = P_{H_2} + P_{H_2O}$$

The vapor pressure of water at 30 °C is given as 31.82 mm Hg. The correct pressure of hydrogen is

$$P_{H_2} = P_{\text{total}^-} + P_{H_2O}$$

$$= 745 \text{ mm Hg} - 31.82 \text{ mm Hg}$$

$$= 713.2 \text{ mm Hg}$$

The error is

$$\text{Error} = \frac{\text{measured} - \text{true}}{\text{true}} \times 100$$

$$= \frac{745 - 713.2}{713.2}$$

$$= +4.5\%$$

11. **(B)** This answer is obtained by using the ideal gas law equation $PV = nRT$ to calculate the pressure of each gas:

$$P_{CO_2} = \frac{nRT}{V} = \frac{(0.016 \text{ mol } CO_2)(0.0821 \text{ L atm mol}^{-1} \text{ K}^{-1})(298 \text{ K})}{2.50 \text{ L}}$$

$$= 0.157 \text{ atm}$$

$$P_{CH_4} = \frac{nRT}{V} = \frac{(0.035 \text{ mol } CH_4)(0.0821 \text{ L atm mol}^{-1} \text{ K}^{-1})(298 \text{ K})}{2.50 \text{ L}}$$

$$= 0.343 \text{ atm}$$

Dalton's law of partial pressures gives

$$P_{\text{total}} = 0.157 \text{ atm} + 0.343 \text{ atm} = 0.500 \text{ atm}$$

Conversion of 0.500 atm to mm Hg yields 380 mm Hg.

Another approach to solving this problem would be to add the moles of the two gases and then to use the ideal gas law to calculate the total pressure directly.

$$P = \frac{nRT}{V} = \frac{(0.035 \text{ mol CH}_4 + 0.016 \text{ mol CO}_2)(0.0821 \text{ L atm mol}^{-1} \text{ K}^{-1})(298 \text{ K})}{2.50 \text{ L}}$$

$$= 0.500 \text{ atm}$$

This yields a pressure of 380 mm Hg.

12. **(B)** Carbon dioxide is soluble to some extent in water. To obtain the amount dissolved in the water, the amount actually obtained is subtracted from the theoretical yield. The chemical reaction is

$$2C_2H_6 + 7O_2 \rightarrow 4CO_2 + 6H_2O$$

The theoretical yield is

$$? \text{ g CO}_2 = 1.50 \text{ g C}_2\text{H}_6 \left(\frac{1 \text{ mol C}_2\text{H}_6}{30 \text{ g C}_2\text{H}_6} \right)\left(\frac{4 \text{ mol CO}_2}{2 \text{ mol C}_2\text{H}_6} \right)\left(\frac{44 \text{ g CO}_2}{1 \text{ mol CO}_2} \right)$$

$$= 4.40 \text{ g CO}_2$$

From the gas collection data:

$$n = \frac{PV}{RT} = \frac{(746 \text{ mm Hg}/760 \text{ mm Hg/atm})(2.00 \text{ L})}{(0.0821 \text{ L atm mol}^{-1} \text{ K}^{-1})(298 \text{ K})}$$

$$= 0.0802 \text{ mol CO}_2$$

$$\text{g CO}_2 = 0.0802 \text{ mol CO}_2 \left(\frac{44 \text{ CO}_2}{1 \text{ mol CO}_2} \right) = 3.53 \text{ g CO}_2$$

The difference between the theoretical yield and actual yield is

$$4.40 \text{ g} - 3.53 \text{ g} = 0.87 \text{ g CO}_2$$

This amount is apparently dissolved in the water of the pneumatic trough.

13. **(B)** Heating a gas increases the pressure, while increasing the volume decreases the pressure. From the information given, it is impossible to tell whether the temperature increase will dominate and cause an increase in pressure or whether the volume increase will dominate and cause an overall pressure decrease. All of the other options will definitely cause either an increase or a decrease in pressure of the gas.

14. **(D)** On the average, heavier molecules move more slowly.

15. **(E)** Both an increase in the force of the collisions with the container walls and an increase in the frequency of collisions lead to the increase in pressure with increased temperature.

16. **(D)** Ideal gases have no volume or attractive forces.

17. **(D)** The further a gas is from its condensation point (boiling point), the more it behaves as an ideal gas. Condensation can be achieved by cooling a gas and/or by increasing its pressure. Therefore low pressure and high temperature will cause a gas to be as far as possible from its condensation point and therefore to behave most like an ideal gas.

18. **(E)** Since a real gas has a volume associated with it, its volume must be greater than that of the corresponding ideal gas:

$$V_{ideal} = V_{real} - nb$$

Similarly, the pressure of a real gas is less than that of the ideal gas since attractions cause the trajectories of the molecules to be curved rather than linear. This curved trajectory decreases the frequency of collisions with the walls and therefore the pressure:

$$P_{ideal} = P_{real} + \frac{an^2}{v^2}$$

19. **(D)** The rate of effusion is inversely proportional to the square root of the molecular masses:

$$\sqrt{\frac{molar\ mass_1}{molar\ mass_2}} = \frac{v_2}{v_1}$$

Entering the given data we have

$$\sqrt{\frac{154\ g\ CCl_4\ mol^{-1}}{44\ g\ CO_2\ mol^{-1}}} = \frac{6.3 \times 10^{-2}\ mol\ CO_2\ s^{-1}}{v_1} = 1.87$$

$$v_1 = \frac{6.3 \times 10^{-2}}{1.87} = 3.4 \times 10^{-2}\ mol\ CCl_4\ s^{-1}$$

20. **(D)** At STP, 22.4 L of a gas is equivalent to 1 mol. The mass of 22.4 L of this gas is equivalent to

$$22.4\ L\ mol^{-1} \times 3.48\ g\ L^{-1} = 78\ g\ mol^{-1}$$

Of the responses, only C_6H_6 and CaF_2 have molar masses of 78. However, CaF_2 is an ionic solid and highly unlikely to be in the gaseous state; therefore C_6H_6, benzene, is the most appropriate answer.

21. **(A)** Rearrange the ideal gas equation to read

$$n = \frac{PV}{RT} = \frac{(0.450\ atm)(2.50\ L)}{(0.0821\ L\ atm\ mol^{-1}\ K^{-1})(300\ K)} = 0.0457\ mol$$

22. **(A)**

$$\frac{P_1V_1}{P_2V_2} = \frac{n_1RT_1}{n_2RT_2}$$

R, and T and V are constants and cancel, so that $P_1/P_2 = n_1/n_2$. Entering the data from the experiment.

$$\frac{0.800 \text{ atm}}{1.10 \text{ atm}} = \frac{n_1}{n_1 + 0.200}$$

and solving for n_1 yields $n_1 = 0.532$ mol. Two-thirds of n_1 is oxygen, and one-third is nitrogen.

Therefore there is 0.355 mol of O_2 in the flask.

23. **(D)**

$$\frac{P_1V_1}{P_2V_2} = \frac{n_1RT_1}{n_2RT_2}$$

In this problem n, R, and T cancel, and $P_1V_1 = P_2V_2$. Entering data from the problem yields

$$(245 \text{ mm Hg})(1.50 \text{ L}) = (P_2)(0.350 \text{ L}).$$

Solving for P_2 gives 1050 mm Hg.

24. **(D)** The mass of air in the flask = $5.00 \text{ L}\left(\frac{1.290 \text{ g}}{L}\right)$ = 6.45 g. The flask

must weigh 543.251 – 6.45 = 536.80 g. The mass of the new gas is 566.11 – 536.80 = 29.31 g. The density of the gas is

$$d = \frac{29.31 \text{ g}}{5.00 \text{ L}} = 5.86 \text{ g L}^{-1}.$$

$$? \frac{\text{g}}{\text{mol}} = \frac{5.86 \text{ g}}{L}\left(\frac{22.4 \text{ L}}{1 \text{ mol}}\right) = 131 \text{ g mol}^{-1}, \text{ which is the atomic mass of xenon.}$$

25. **(A)** $? \text{L H}_2 = 0.100 \text{ g Mg}\left(\frac{1 \text{ mol Mg}}{24.31 \text{ g Mg}}\right)\left(\frac{1 \text{ mol H}_2}{1 \text{ mol Mg}}\right)\left(\frac{22.4 \text{ L H}_2}{1 \text{ mol H}_2}\right)$

$$= 0.0921 \text{ L H}_2$$
$$= 92.1 \text{ mL H}_2$$

CHAPTER 8

1. **(B)** London forces are the weakest and hydrogen bonds the strongest.

2. **(B)** Salts with ionic bonds have the highest melting points.

3. **(D)** Sublimation is the transformation of a solid directly to a gas.

4. **(A)** Iron is in period 3 and has the largest electron cloud, which is expected to be most polarizable.

5. **(A)** CF_4 is the only compound that does not have an N—H, O—H, or F—H bond in its structure.

6. **(D)** C_6H_{14} does not hydrogen-bond and has fewer London forces than C_8H_{18}.

7. **(A)** The larger the heat of vaporization, the larger the change in boiling point when the pressure is changed.

8. **(C)** Aluminum is a soft metal that will not even scratch glass.

9. **(C)** Although the instantaneous dipoles are very weak attractive forces, in a large molecule many of them can exceed the strength of a hydrogen bond. In this case, they cause $C_{30}H_{62}$ to be a solid at room temperature.

10. **(E)** The unit cell and its dimensions can be used, with Avogardro's number, to determine the density.

11. **(A)** The lengths of the horizontal plateaus represent the heats of fusion and vaporization. Although not identical, they are closest to being the same.

12. **(E)** The slope of the curve for the solid, liquid, or gas has units of degree Celsius per joule. Therefore, $\dfrac{1}{slope \times mass}$ is the specific heat with the required units of $J\ g^{-1}\ °C^{-1}$.

13. **(D)** The lower left line is the equilibrium line between the solid and gas phases. It ends at the triple point at approximately 95 °C.

14. **(D)** The point at which the liquid-gas equilibrium line intersects the dotted line representing 1 atm of pressure is the normal boiling point, approximately 180 °C.

15. **(A)** The coordinates of these conditions fall in the liquid region of the phase diagram.

16. **(E)** Only 250 °C and 2.00 atm pressure are clearly beyond the critical point.

17. **(D)** The rapid evaporation leaves molecules with lower average KE, and this is felt as coolness since the liquid absorbs heat from the skin. Water evaporates more slowly, and the effect is less pronounced.

18. **(A)** The bubbles must be ethanol. Ethanol does not decompose to hydrogen and oxygen at these temperatures. These bubbles cannot be steam (water vapor) or air because they are not present in pure ethanol.

19. **(A)** All atoms in a diamond are covalently bonded, producing a very large structure.

20. **(A)** Viscosity, boiling point, and heat of vaporization are all directly related to the strength of intermolecular attractions. In a liquid substance with low intermolecular attractions, the values for other properties will also be low.

CHAPTER 9

1. **(E)** The sudden reduction of CO_2 pressure when the bottle is uncapped causes the solubility of the CO_2 to decrease dramatically, causing the fizzing.

2. **(A)** Osmotic pressure measurements can easily determine molar masses that are extremely large. Other methods mentioned in this chapter will give too small a response to be detected.

3. **(C)** Vapor pressure and Raoult's law both consider vapor pressure.

4. **(B)** Ion pairing is most easily measured using the freezing-point-depression technique of those mentioned.

5. **(C)** $\dfrac{\text{mol CdCl}_2}{\text{L CdCl}_2} = \dfrac{140 \text{ g CdCl}_2}{100 \text{ mL CdCl}_2}\left(\dfrac{1 \text{ mol CdCl}_2}{183 \text{ g CdCl}_2}\right)\left(\dfrac{1 \text{ mL}}{10^{-3} \text{ L}}\right) = 7.65 \ M$

6. **(E)**
$$P = P^0 \, X_{\text{solvent}}$$

From this we get

$$X_{\text{solvent}} = \frac{P}{P^0} = \frac{456 \text{ mm Hg}}{832 \text{ mm Hg}} = 0.548$$

Mole fraction of solute is

$$X_{\text{solute}} = 1.000 - X_{\text{solvent}} = 1.000 - 0.548 = 0.452$$

7. **(B)** Surface tension changes depending on the type of molecule dissolved and its intermolecular attractive forces, rather than the amount of solute particles present. Surface tension is not a colligative property.

8. **(C)** The least polar substance, C_6H_6 will be most soluble in the nonpolar hexane.

9. **(E)** The osmotic pressure calculation uses molarity units; the others use molality or mole fraction units.

10. **(D)** Vapor pressure measurements may deviate significantly from Raoult's law, resulting in errors if the solution is not an ideal solution.

11. **(A)** The density of the solution must be measured. The other values are either not needed or readily available in tables.

12. **(A)** The sucrose will produce the largest molality of soluble particles resulting in the largest $\Delta T = k_b$ m.

13. **(E)** A release of heat energy when substances are mixed indicates stronger attractive forces in the solution than in the pure solvents.

14. **(B)** Enzymes have limited solubility and high molar masses. Consequently, the osmotic pressure method is the best choice.

15. **(B)** $25 \text{ ppb Pb}^{2+} = \dfrac{25 \text{ g Pb}^{2+}}{10^9 \text{ g solution}} \left(\dfrac{1 \text{ mol Pb}^{2+}}{207 \text{ g Pb}^{2+}} \right) \left(\dfrac{10^3 \text{ g solution}}{1 \text{ L solution}} \right)$

$$= 1.2 \times 10^{-7} \text{ M Pb}^{2+}$$

16. **(C)** Increasing temperature always decreases the solubility of a gas such as O_2.

17. **(B)** It is most reasonable that the vapor pressure and boiling point of any solution will lie between the vapor pressures and boiling points of the pure liquids. Only the pressure of 0.750 atm fulfills these conditions.

18. **(B)** $\Delta T = mk_f$ and $m = \dfrac{k_f}{\Delta T} = \dfrac{-10.0 \,°C}{-1.86 \,°C} = 5.4$ molal

Of the listed possibilities, only KCl will result in a solution that is 5.4 molal in ions (both K^+ and Cl^-).

19. **(B)** CH_3OH is an organic alcohol, methanol, and is a nonelectrolyte.

20. **(D)** $\dfrac{7.5 \,°C}{0.52 \,°C \cdot m^{-1}} = 14.4$ molal

$$\Delta T = (1.86 \,°C \cdot m^{-1})(14.4 \text{ m}) = 26.8 \,°C$$

The freezing point is 0 °C – 26.8 °C = – 26.8 °C.

21. **(E)** Glucose is a nonelectrolyte; the first three choices are ionic and the acid CH_3CH_2COOH ionizes slightly. Therefore glucose produces the lowest molality of particles (molecules or ions), resulting in the lowest boiling-point increase.

22. **(B)** The solubility of a gas in a liquid is directly proportional to the vapor pressure above the liquid. This is Henry's law, and the proportionality constant is called the Henry's law constant.

$$C = k_H P$$

The units of the Henry's law constant depend on the units of concentration and pressure used. In this problem we calculate k_H as $(0.975\ \text{g L}^{-1})/(1.00\ \text{atm}) = 0.975\ \text{g L}^{-1}\ \text{atm}^{-1}$. With this constant we can now calculate the new concentration as

$$C = 0.975\ \text{g L}^{-1}\ \text{atm}^{-1}\ (0.212\ \text{atm})$$

$$= 0.207\ \text{g L}^{-1}$$

23. **(E)** Two forces for dissolution are the energy, which is unfavorable since the solution cools, and the entropy change. The entropy change must be great enough to overcome the energy deficit.

24. **(A)**
$$X = \frac{\text{mol}_{\text{ethanol}}}{\text{mol}_{\text{ethanol}} + \text{mol}_{\text{water}}}$$

$$= \frac{20.0\ \text{g}/46\ \text{g mol}^{-1}}{20.0\ \text{g}/46\ \text{g mol}^{-1} + 30.0\ \text{g}/18\ \text{g mol}^{-1}}$$

$$= \frac{0.435}{0.435 + 1.667} = 0.207$$

25. **(C)** We solve the osmotic pressure equation. First convert the 0.12% solution to 0.12 g in 0.10 liters. Next, 4 °C is 277 K and the pressure (Π) is $(2.25/760)$ atm. The equation to solve is

$$\Pi V = nRT$$

We rearrange this to read:

$$\text{molar mass} = \frac{gRT}{PV}$$

$$\text{molar mass} = \frac{(0.12\ \text{g})(0.0821\ \text{L atm mol}^{-1}\ \text{K}^{-1})(277\ \text{K})}{(2.25/760\ \text{atm})(0.10\ \text{L})}$$

$$\text{molar mass} = 9.3 \times 10^3\ \text{g mol}^{-1}$$

CHAPTER 10

1. **(E)** Le Châtelier's principle addresses this topic the best of the items listed.

2. **(C)** We calculate the value of Q and if it matches the value of the equilibrium constant, then the system is at equilibrium. If Q does not match the value of the equilibrium constant, the system is not at equilibrium.

3. **(B)** The K_{sp} is the solubility product and is specific for systems that are slightly soluble.

4. **(C)** A system in equilibrium is governed by the equilibrium law, which specifies a specific ratio of product to reactant concentrations.

5. **(E)** All of these are governed by equilibrium laws.

6. **(C)** Increasing the pressure by adding an inert gas does not change the concentrations of the reactants or products.

7. **(E)** The reaction has been reversed, causing the equilibrium constant to be inverted. In addition, the coefficients are doubled and the equilibrium constant is squared. The new equilibrium constant is

$$K = \frac{1}{(4.5 \times 10^{+3})^2} = 4.9 \times 10^{-8}$$

8. **(A)** The reaction is

$$Ag_2CrO_4(s) \rightleftharpoons 2\ Ag^+(aq) + CrO_4^{2-}(aq)$$

and $K_{sp} = [Ag^+]^2[CrO_4^{2-}]$.

9. **(B)** $K_p = K_c$ when $\Delta n_g = 0$. Only the combustion of carbon has these properties.

10. **(D)** Only HCl and CO_2 should appear in the equilibrium law. Solids and pure liquids have constant concentrations and are included in the equilibrium constant. Response A does not have the correct exponent for HCl.

11. **(A)** $K_c = \frac{[H_2]\ [I_2]}{[HI]^2} = 0.020$

$$\text{Initial concentration of HI} = \frac{0.200\ \text{mol HI}}{10.0\ \text{L}} = 0.0200\ M$$

REACTION	2HI	$\rightleftharpoons$	H$_2$	+	I$_2$
INITIAL CONDITIONS	0.200 *M*		0 *M*		0 *M*
CHANGE	−2x		+x		+x
EQUILIBRIUM	0.0200 − 2x		+x		+x
ANSWER					

$$\frac{[H^2][I^2]}{[HI]^2} = 0.020 = \frac{(x)(x)}{(0.0200 - 2x)^2}$$

Take the square root of both sides to obtain

$$0.1414 = \frac{x}{0.0200 - 2x}$$

$$(0.1414)(0.0200 - 2x) = x$$

$$0.002828 - 0.2828x = x$$

$$0.002828 = x + 0.2828x$$

$$0.002828 = 1.2828x$$

$$\frac{0.002828}{1.2828} = x$$

$$x = 2.2 \times 10^{-3}$$

REACTION	2HI	⇌	H₂	+	I₂
INITIAL CONDITIONS	**0.200 *M***		**0 *M***		**0 *M***
CHANGE	−2x		+x		+x
EQUILIBRIUM	0.0200 − 2x		+x		+x
ANSWER	0.01559 M		2.2 × 10⁻³ M		2.2 × 10⁻³ M

Since this is a 10.0-L flask, it will contain 2.2×10^{-2} or 0.022 mol $I_2(g)$.

12. **(B)** For this reaction: $\quad Q = \dfrac{P_{N_2O_4}}{P_{NO_2}^2}$

The reaction will go in the forward direction if Q is less than K_p. Q must be calculated from pressures in atmospheres. Only choice B results in a value of Q less than K_p.

13. **(B)** $\quad PbI_2(s) \rightleftharpoons Pb^{2+}(aq) + 2I^-(aq)$ and $K_{sp} = [Pb^{2+}][I^-]^2$

REACTION	PbI₂	⇌	Pb²⁺	+	2I⁻
INITIAL CONDITIONS	**Solid**		**0 *M***		**0 *M***
CHANGE	−x		+x		+2x
EQUILIBRIUM	Solid		+x		+2x
ANSWER					

$$7.9 \times 10^{-9} = (x)(2x)^2$$

$$= 4x^3$$

$$1.98 \times 10^{-9} = x^3$$

$$1.25 \times 10^{-3} = x \text{ (This is also the molar solubility.)}$$

14. **(D)** Molar solubility $= \dfrac{1.00 \times 10^{-4} \text{ g AuCl}_3 / 303 \text{ g AuCl}_3 \text{ mol}^{-1}}{1 \text{ L}}$

$$= 3.3 \times 10^{-7} \ M$$

REACTION	$AuCl_2$	$\rightleftharpoons$	Au^{3+}	$+$	Cl^-
INITIAL CONDITIONS	**Solid**		**0 *M***		**0 *M***
CHANGE	$-x$		$+x$		$+3x$
EQUILIBRIUM	Solid		$+x$		$+3x$
ANSWER					

$$K_{sp} = [Au^{3+}][Cl^-]^3 = (x)(3x)^3$$

Since $x = 3.3 \times 10^{-7}$,

$$K_{sp} = (3.3 \times 10^{-7})(3 \times 3.3 \times 10^{-7})^3$$

$$= 3.2 \times 10^{-25}$$

15. **(D)**

$$K_c = \frac{[CO_2][H_2O]^2}{[CH_4][O_2]^3} = \frac{(M)(M)^2}{(M)(M)^3} = \frac{1}{M} = M^{-1}$$

16. **(B)** A catalyst affects only the rate at which equilibrium is reached, not the amount of product formed.

17. **(E)** In an exothermic reaction the amount of product decreases with increasing temperature.

$$\text{Reactants} \rightleftharpoons \text{Products} + \text{Heat}$$

which is the general equation for an exothermic reaction, illustrates that adding heat by increasing the temperature forces the reaction toward the reactant side.

18. **(A)** The reaction rate has nothing to do with the size of the equilibrium constant.

19. **(D)**

$$[Ag^+] = \frac{K_{sp}}{[Cl^-]} = \frac{1.0 \times 10^{-10}}{0.100} = 1.0 \times 10^{-9} \ M$$

when AgCl starts to precipitate.

$$[I^-] = \frac{K_{sp}}{[Ag^+]} = \frac{8.3 \times 10^{-17}}{1.0 \times 1.0^{-9}} = 8.3 \times 10^{-8} \ M$$

20. **(E)** $K_{sp} = 1.1 \times 10^{-10} = [Ba^{2+}][SO_4^{2-}] = [x]([x] + 2.4 \times 10^3)$

Assume $2.4 \times 10^{-3} >> x$ so that

$$1.1 \times 10^{-10} = (x)(2.4 \times 10^{-3})$$

and

$$x = \frac{1.1 \times 10^{-10}}{2.4 \times 10^{-3}} = 4.6 \times 10^{-8} \ M$$

Here x is the concentration of Ba^{2+} and the molar solubility of $BaSO_4$.

21. **(D)** When the initial amount of HI is known, only one: I_2, H_2, or HI needs to be measured to determine the equilibrium constant. Therefore, it is *not* important to measure all three concentrations.

22. **(B)**

REACTION	$2SO_2$	$\rightleftharpoons$	O_2	+	$2SO_3$
INITIAL CONDITIONS	0.00300 *M*		0.00300 *M*		0.00300 *M*
CHANGE	2x		x		–2x
EQUILIBRIUM	0.00300 + 2x		0.00300 + x		0.00300 – 2x
ANSWER	3.50×10^{-5} *M*				

From the last two lines,

$$3.50 \times 10^{-5} = 0.00300 + 2x$$

Solve for x to get $x = 0.00148$, which yields

$$[O_2] = 0.00152 \ M \quad \text{and} \quad [SO_3] = 0.00596 \ M$$

on the answer line.

Then

$$K_{eq} = \frac{[SO_3]^2}{[SO_2]^2[O_2]} = \frac{(0.00596)^2}{(3.50 \times 10^{-5})^2(0.00152)} = 1.9 \times 10^7$$

23. **(B)** Since the two reactions are added to obtain the overall reaction.

$$K_{overall} = K_{a1}K_{a2} = (2.3 \times 10^{-4})(4.5 \times 10^{-7}) = 1.0 \times 10^{-10}$$

CHAPTER 11

1. **(C)** The potential energy curve illustrated the heat of reaction as the difference in the system's potential energy between the reactants and products.

2. **(C)** The orientation for an effective collision and the frequency of collisions are all part of the collision theory.

3. **(E)** The activation energy, potential energy curves, and activated complexes are all used to describe the transition-state theory concepts.

4. **(C)** These items all affect the rate of a reaction. The others describe other features of reaction rate theory.

5. **(B)** The activated complex may be described as the structure of the colliding reactants.

6. **(E)** A zero-order reaction, Rate = k, satisfies the condition stated in the question.

7. **(A)** The x-axis (B) is the reaction coordinate, the y-axis (A) is the potential energy, and C represents the activation energy.

8. **(D)** This is an endothermic reaction, with the activation energy larger for the forward reaction than for the reverse reaction.

9. **(C)** Catalysts have an effect on the activation energy.

10. **(D)** The lower the activation energy, the faster the reaction.

11. **(D)** These are the units for a third-order rate constant.

12. **(A)** Adding an inert gas has no effect on the reaction rate.

13. **(A)** $\ln(2) = kt_{1/2}$

$$k = \frac{\ln(2)}{t_{1/2}} = \frac{0.693}{(36 \text{ min})(60 \text{ s min}^{-1})} = 3.2 \times 10^{-4} \text{ s}^{-1}$$

14. **(D)**

$$\frac{\text{Rate}_1}{\text{Rate}_2} = \frac{(\text{Conc}_1)^x}{(\text{Conc}_2)}$$

where x is the exponent in the rate law. Consequently, $8 = 2^x$, based on the information in the question, and x must be 3.

15. **(C)** The reaction rate is an exponential function given by the Arrhenius equation. Since the activation energy is not given in this problem, we rely on the rule of thumb that the reaction rate doubles for each 10 °C increase in temperature. A 20 °C increase in temperature will increase the rate by a factor of approximately 4.

16. **(C)** Integrated rate equations for second-order reactions involve $\dfrac{1}{[A]}$ terms.

17. **(C)**

$$? \frac{\text{mol O}_2}{\text{L s}} = \frac{2.2 \times 10^{-2} \text{ mol CO}_2}{\text{L s}} \left(\frac{15 \text{ mol O}_2}{12 \text{ mol CO}_2} \right)$$

$$= 2.8 \times 10^{-2} \text{ mol O}_2 \text{ L}^{-1}\text{s}^{-1}$$

18. **(C)** The Arrhenius plot is most useful in determining the stability or shelf-life of consumer products since it allows the chemist to predict the rate of reaction at any temperature.

19. **(E)** The Arrhenius equation is

$$\ln k = \frac{-E_a}{RT} + \text{const}$$

Consequently $\ln k$ is plotted versus $\frac{1}{T}$.

20. **(C)**

$$\frac{255 \text{ s}}{85 \text{s half-life}^{-1}} = 3 \text{ half-lives}$$

$$\left(\frac{1}{2} \right)^3 = \frac{1}{8}$$

21. **(C)** The sum of the exponents of all concentrations is the order of the reaction.

22. **(D)**

$$k[\text{A}]^2[\text{B}] = k \left[\frac{\text{A}}{2} \right]^2 [4\text{B}]$$

23. **(A)** The half-life and rate constant are related by the expression:

$$\ln(2) = kt_{1/2}$$

The half-life will be

$$t_{1/2} = \ln(2)/k = 0.692/(2.6 \times 10^{-4} \text{ s}^{-1})$$
$$t_{1/2} = 2.7 \times 10^3 \text{ s}$$

24. **(C)** The main purpose of the catalytic converter is to reduce hydrocarbon emissions by combusting any unburned or partially burned fuel.

25. **(B)** The catalyst will not alter K_{eg} and therefore will not alter the position of equilibrium. All other effects will be observed.

CHAPTER 12

1. **(A)** A negative value for $\Delta G°$ always indicates a spontaneous reaction.

2. **(E)** The enthalpy content H cannot be determined exactly. Only the difference in enthalpy between two states is readily measurable.

3. **(D)** Only those items with the superscript "o" are intensive in nature. Each incorporates the "per mole" term in their units. H and ΔG depend on how large the system is.

4. **(A)** This combination of entropy and enthalpy will always yield a positive value for $\Delta G°$.

5. **(D)** $q_v = -$ (heat capacity)$(\Delta T) = - (3245 \text{ J } °C^{-1})(6.795 °C) = - 22.05 \text{ kJ}$

   The sign is negative since heat is released in the reaction.

6. **(B)** $$\Delta H = q_p$$

   In question 1, q_v was determined. To convert between q_v and q_p requires calculation of the work or $P \Delta V$. However, $CH_4 + 2O_2 \rightarrow CO_2 + 2H_2O$ and $\Delta n_g = 0$ so no volume change is expected and $q_p = q_v$.

   $$\Delta H° = \frac{q_p}{\text{mol CH}_4} = \frac{22.05 \text{ kJ}}{0.4 \text{ g}/16 \text{ g mol}^{-1}} = -822 \text{ kJ}$$

7. **(A)** A system where $\Delta G°$ is always positive will always be nonspontaneous. For this to occur, $\Delta H°$ must be positive and $\Delta S°$ must be negative.

8. **(B)** The condensation of water indicates that the solution becomes cold and the solution process is endothermic. Consequently, $\Delta H°$ must be positive; and if solution occurs, $\Delta G°$ must be negative, requiring that $\Delta S°$ be positive.

9. **(C)** When CH_4 is burned, the greatest decrease in the moles of gas $(\Delta n_g = -2)$ is obtained if liquid water forms.

10. **(E)** During a phase change $\Delta G° = 0$ and $T\Delta S° = \Delta H°$.

    Calculate

    $$\Delta S° = \frac{\Delta H°}{T} = \frac{+43900 \text{ J}}{373 \text{ K}} = +118 \text{ J K}^{-1}$$

11. **(A)** ΔG tells us the direction of the reaction. $\Delta G°$ indicates the position of equilibrium at standard state. Neither ΔS nor ΔH alone indicate the direction of the reaction or the position of equilibrium.

12. **(E)** $w = - P\Delta V = - (1.00 \text{ atm})(10.0 \text{ L}) = - 10.0 \text{ L atm}$.

13. **(D)** Both forms of the universal gas constant are needed. The ratio of the two forms of this constant has units of $\frac{\text{L atm}}{\text{J}}$ is used as the factor label for the conversion.

14. **(B)** Except for small molecules, the entropy change is a large positive value.

15. **(A)** Entropy can be experimentally determined. Only changes in the other thermodynamic quantities can be measured.

16. **(B)** The reaction is

$$2CH_3OH(l) + 3O_2(g) \rightarrow 2CO_2(g) + 4H_2O(g)$$

$$\Delta H° = [(2 \text{ mol})(- 393.5 \text{ kJ mol}^{-1}) + (4 \text{ mol})(- 241.8 \text{ kJ mol}^{-1})]$$
$$- [(2 \text{ mol})(- 238.6 \text{ kJ mol}^{-1})]$$

$$= - 1277 \text{ kJ}$$

17. **(E)** The rate of a reaction cannot be determined from thermodynamic quantities.

18. **(C)** The $\Delta H°$ value for first reaction is subtracted from the value for second reaction.

$$\Delta H° = \Delta H° \text{ (second react)} - \Delta H° \text{ (first react)}$$

$$= - 113.14 \text{ kJ} - 57.93 \text{ kJ}$$

$$= - 171.07 \text{ kJ}$$

19. **(D)** Only changes in temperature change $\Delta G°$. The fact that the temperature change is measured in Celsius degrees is not significant. Thermodynamic calculations involving temperature require conversion to the Kelvin scale.

20. **(A)**
$$\Delta G° = - RT \ln K_{eq} = 0.00$$

since the natural log of 1.00 is zero

$$\Delta G° = \Delta H° - T\Delta S°$$

and since $\Delta G° = 0$, we find

$$\Delta H° = T\Delta S°$$

$$T = \frac{\Delta H°}{\Delta S°} = \frac{5300 \text{ J mol}^{-1}}{22.6 \text{ J mol}^{-1} \text{ K}^{-1}} = 677 \text{ K}$$

$$C = 677 \text{ K} - 273 \text{ K} = 404 \text{ C}$$

21. **(E)** $\Delta H° = - (2 \text{ mol})(- 396 \text{ kJ mol}^{-1})$

$$= +792 \text{ kJ (heats of formation for elements} = 0)$$

$$\Delta S° = (3 \text{ mol O}_2) (205 \text{ J mol}^{-1} \text{ K}^{-1}) + (2 \text{ mol S})(31.8 \text{ J mol}^{-1} \text{ K}^{-1})$$
$$- (2 \text{ mol SO}_3)(256 \text{ J mol}^{-1} \text{ K}^{-1})$$

$$= 167 \text{ J K}^{-1}$$

$$\Delta G° = 792 \text{ kJ} - (298 \text{ K})(0.167 \text{ kJ K}^{-1}) = 742 \text{ kJ}$$

22. **(B)** We expect the combustion of organic compounds to produce heat and be exothermic. $\Delta H°$ negative. Since 15 mol of reactant gas produces 12 mol of product gas, we expect a decrease in entropy, $\Delta S°$ negative.

23. **(C)** Heat must be added to evaporate a liquid, and therefore ΔH° is positive. Since the liquid becomes a gas, ΔS° is also positive.

24. **(E)** Combustion of organic compounds produces heat; therefore ΔH° is negative.

CHAPTER 13

1. **(D)** Oxidation is often defined as a loss of electrons.

2. **(E)** Aluminum is produced by electrolysis.

3. **(A)** The Nernst equation relates the standard state and nonstandard state conditions of a galvanic cell.

4. **(C)** Reduction and oxidation must always occur together.

5. **(D)** The balanced half-reaction is

$$8e^- + 10H^+ + NO_3^- \rightarrow NH_4^+ + 3H_2O$$

6. **(B)** The half-reaction is

$$5e^- + 8H^+ + MnO_4^- \rightarrow Mn^{2+} + 4H_2O$$

and B is the correct form.

7. **(B)** The first two reactions show that Cu and Zn metals are more effective reducing agents than Ag. The last reaction shows that Zn is a better reducing agent than Cu. Consequently, Zn is the strongest reducing agent, and Ag the weakest, of the three.

8. **(B)**
$$E^\circ_{cell} = E^\circ_{reduction} - E^\circ_{oxidation}$$

Since Fe is oxidized and Pb is reduced,

$$E^\circ_{cell} = + 1.46 - (+0.77) = +0.69 \text{ V}$$

9. **(E)** The nitrogen in the nitrate ion has an oxidation number of +5.

10. **(A)** Zinc reacts with acids. The others are very active and react with water.

11. **(A)**
$$2H_2O \rightarrow O_2 + 4H^+ + 4e^-$$

occurs at the anode.

$$Cu^{2+} + 2e^- \rightarrow Cu$$

occurs at the cathode.

12. **(D)** H ions are reduced before Na ions.

13. **(B)** Chlorine can have −1, +1, +3, +5, and +7 for its oxidation states.

14. **(A)**
$$\Delta G^\circ = - n \mathscr{F} E^\circ_{cell}$$

and

$$\Delta G^\circ = - RT \ln K$$

Therefore

$$-n \mathscr{F} E^\circ_{cell} = - RT \ln K$$

Calculate

$$\ln K = -\frac{n \hat{\ } E^\circ_{cell}}{-RT} = \frac{(2)(96,485 \text{ C mol}^{-1})(0.82 \text{ V})}{(8.314 \text{ V C mol}^{-1} \text{ K}^{-1})(318 \text{ K})} = 59.85$$
$$K = 9.8 \times 10^{25}$$

15. **(C)** A nonspontaneous reaction must have a negative value for E°_{cell}. The value closest to zero, here −0.12 V, will produce the greatest amount of product.

16. **(B)**
$$\text{mol} = \frac{It}{n\mathscr{F}} = \frac{(1.23 \text{ A})(2.5 \text{ h})(60 \text{ min h}^{-1})(60 \text{ s min}^{-1})}{(2)(96,485 \text{ C mol}^{-1})} = 0.0574 \text{ mol}$$

$$\text{Molar mass} = \frac{\text{g compound}}{\text{mol compound}} = \frac{3.37 \text{ g}}{0.0574 \text{ mol}} = 58.7 \text{ g mol}^{-1}$$

Nickel has a molar mass of 58.7.

17. **(D)** Since the reaction is not at standard state, we cannot tell whether or not it would be spontaneous under standard conditions. The chemicals in the cell will react in the reverse direction when mixed.

18. **(B)** The ratio of Kelvin temperatures is
$$\frac{303}{293} = 1.034$$

Thus the change is due simply to the temperature change. This indicates also that for this reaction K is the same at these two temperatures and $\Delta H^\circ = 0.0$ kJ.

19. **(C)** Iron is produced in high-temperature kilns by reduction with carbon; the others are produced by electrolysis.

20. **(C)** Chlorine has an oxidation number of +5. The nitrogen in the nitrate ion NO_3^- is also +5.

21. **(B)**
$$Cu^{2+} + 2e^- \rightarrow Cu^0$$
$$\text{mol} = \frac{\text{g Cu}}{\text{molar mass}} = \frac{It}{nF}$$

Rearranging the equation yields

$$t = \frac{(g\ Cu)nF}{(molar\ mass)I} = \frac{(2.00\ g)(2)(96,485\ C\ mol^{-1})}{(63.5\ g\ mol^{-1})(1.25\ C\ s^{-1})} = 4862\ s = 81\ min$$

22. **(D)** The balanced half-reaction is

$$10e^- + 12H^+ + 2IO_3^- \rightarrow I_2 + 6H_2O$$

23. **(D)** The standard cell voltage is a measure of the equilibrium constant or Gibbs free energy, not a rate constant.

24. **(D)** Electrolysis does not necessarily produce a gas at either the anode or the cathode.

CHAPTER 14

1. **(C)** Only these two species actually exists in solution KOH might be considered as a Lewis acid (K^+) and a Lewis base (OH^-). However, KOH really does not exist to any extent in solution.

2. **(D)** The copper-ammonia complex has negligible basicity. The carbonate ion will be more basic than the bicarbonate ion. Also, since sulfurous acid is stronger than carbonic acid, their conjugate bases have their strengths reversed and the carbonate ion will be a stronger base than the sulfite ion.

3. **(A)** The copper ammonia complex is deep blue.

4. **(C)** Of the given compounds we could neutralize some bicarbonate with the KOH to make a buffer. However, if we add equal moles of KOH and bicarbonate, all the bicarbonate is converted to carbonate and no buffer exists. Equal moles of bicarbonate and carbonate will produce a buffer.

5. **(C)** Carbonic acid is a weaker acid than ethanoic acid, and its salt will be a stronger base, giving a solution with a higher pH. NaCl has no effect on the pH, and A and B are acids with pH values less than 7.

6. **(D)** In solution only conjugate acid-base pairs exist. The oxalate ion and oxalic acid differ by more than one H^+. The ions of sodium sulfate (choice C) can exist in solution together.

7. **(E)** The solution contains 0.100 mol of H_3PO_4, and the addition of 0.250 mol of NaOH represents 2.5 mol base for each mole of phosphoric acid. The first 2 mol of base convert the H_3PO_4 to HPO_4^{2-}. The next 0.5 mol of base converts only half of the HPO_4^{2-} to PO_4^{3-}, and the resulting solution is a mixture of these two ions.

8. **(B)** We need a weak acid with a pK_a within 1 pH unit of 10.00 or a weak base with a pK_b within 1 pH unit of 4.00 (pOH). Ammonia comes closest to these requirements.

9. **(A)** When the concentrations of the conjugate acid and base are equal, $pH = pK_a$.

10. **(E)**

$$[OH^-] = 3 \times 1.23 \times 10^{-3} = 3.69 \times 10^{-3} \, M$$

$$pOH = -\log(3.69 \times 10^{-3}) = 2.43$$

$$pH = 14.00 - 2.43 = 11.57$$

11. **(C)** $pK_a = 5.19$. At pH values below 4.19 the indicator will be red, and at pH values above 6.19 the indicator will be yellow. From 4.19 to 6.19 it will be various shades of orange, the combination of red and yellow.

12. **(D)** Using the Henderson-Hasselbach equation gives

$$pH = pK_a + \log\left(\frac{[\text{conj. base}]}{[\text{conj. acid}]}\right)$$

Solving this yields

$$4.87 = 4.48 + \log\left(\frac{[\text{conj. base}]}{[\text{conj. acid}]}\right)$$

$$0.39 = \log\left(\frac{[\text{conj. base}]}{[\text{conj. acid}]}\right)$$

$$2.45 = \left(\frac{[\text{conj. base}]}{[\text{conj. acid}]}\right)$$

13. **(D)** The greater number of oxygen atoms on HNO_3 weakens the bond with hydrogen, causing it to be a strong acid. Nitrous acid, HNO_2, is a weak acid.

14. **(C)** The anion is amphiprotic and can act as both an acid and a base.

$$pH = \frac{pK_2 + pK_3}{2} = \frac{7.20 + 12.35}{2} = 9.78$$

15. **(E)** Each of these methods results in a solution containing a conjugate acid and its conjugate base in significant amounts, therefore resulting in a buffer solution.

16. **(D)** Sulfuric acid, H_2SO_4, dissociates its first proton completely and is a strong acid. The second proton does not dissociate completely and is weak.

17. **(E)** The reaction is $N_2O_5 + H_2O \rightarrow 2HNO_3$.

18. **(A)** CH_4 lacks the capacity either to accept or to donate electron pairs.

19. **(B)** In this ion, NH_3, is the ligand, which is another name for a Lewis base.

20. **(B)**
$$pOH = 14.00 - 10.45 = 3.55$$

and

$$[OH^-] = 2.82 \times 10^{-4} \, M$$

For a weak base

$$[OH^-] = \sqrt{K_b C_b}$$

Then

$$K_b = \frac{[OH^-]^2}{C_b} = \frac{7.95 \times 10^{-8}}{0.125} = 6.4 \times 10^{-7}$$

21. **(B)** An aqueous solution always contains H^+ and OH^- ions. In this solution the major solutes are K^+ and F^-, while a small amount of F^- hydrolyzes to form HF.

22. **(C)** Addition of distilled water to a buffer does not change the pH unless a very large amount of water is added to a small volume of buffer.

23. **(D)** The equation is

$$NaOH + HCl \rightarrow NaCl + H_2O$$

$$mmol \; HCl = (0.0134 \, M)(50.0 \, mL) = 0.670 \; mmol \; HCl$$

$$mmol \; NaOH = (0.0250 \, M)(24.0 \, mL) = 0.600 \; mmol \; NaOH$$

There is an excess of 0.070 mmol HCl, and the total volume is

$$50.0 + 24.0 = 74.0 \; mL$$

$$M_{acid} = \frac{0.070}{74} = 9.46 \times 10^{-4} \, M$$

$$pH = -\log(9.46 \times 10^{-4}) = 3.02$$

24. **(C)** This is a buffer solution:

$$[HCHO_2] = \frac{50.0 \, g/46 \, g \, mol^{-1}}{0.500 \, L} = 2.17$$

$$[NaCHO_2] = \frac{30.0 \, g/68 \, g \, mol^{-1}}{0.500 \, L} = 0.882$$

$$[H^+] = \frac{K_a[HCHO_2]}{[NaCHO_2]} = \frac{1.8 \times 10^{-4}(2.17)}{0.882} = 4.4 \times 10^{-4}$$

$$pH = -\log(4.4 \times 10^{-4}) = 3.35$$

CHAPTER 15

1. **(C)** The amine functional group is related to ammonia. The simplest is a primary amine —NH_2.

2. **(A)** Alcohols and acids (—COOH) react to form esters. Most esters have pleasant or fruity odors.

3. **(B)** Amino acids are the building blocks of proteins. As their name suggest, they contain the amine functional group (—NH_2) and the acid functional group (—COOH).

4. **(B)** Adding a methyl group to the acid functional group produces ethanoic acid. Dilute (5%) solutions of ethanoic acid are known as vinegar.

5. **(B)** A terminal —C=O is an aldehyde functional group.

6. **(E)** Propane contains three carbons, and propyl groups also contain three carbon atoms.

7. **(C)** Condensation reactions between organic acids and amines form peptide bonds.

8. **(E)** All of these are properties of stereoisomers.

9. **(E)** All of these reactions occur.

10. **(E)** Since [—CH_2CHCl—]$_n$ does not contain a peptide or ester bond, it must be an addition polymer formed from a monomer that contains a double bond.

11. **(A)** Propane-l-ol is the correct IUPAC name.

12. **(C)** Proteins are condensation polymers of the 21 natural amino acids.

13. **(C)** These are the functional groups for an organic acid and an organic base, and they participate in a neutralization or acid-base reaction. They could also form a peptide bond, but that is not one of the choices.

14. **(C)** Cis-trans isomers involve side groups attached to double-bonded carbon atoms only. In carbon compounds sp^2 bonding represents a double bond.

15. **(E)** An ether has the lowest polarity of all choices and should have the lowest boiling point.

16. **(A)** An alkyne contains triple bonds but does not contain oxygen.

17. **(C)** —NH_2 is the organic base functional group.

18. **(A)** Meta positioning represents one group on the 1 carbon and another on the 3 carbon in a benzene ring.

19. **(E)** Both benzene and ethylene are flat molecules because of the double bonds, which are sp^2 (trigonal planar) in geometry.

20. **(E)** There are six C—C sigma bonds, six C—H sigma bonds, and three pi bonds in the benzene ring.

21. **(B)** The alcohol is the only compound with an —OH group, needed for hydrogen bonding.

22. **(A)** An alkyne has virtually no polarity. Every other choice has an oxygen that is polar to some extent, resulting in a higher boiling point.

23. **(A)** The general formula for alkenes is C_nH_{2n}, and the percentage of carbon will always be 85.7%.

24. **(C)** The name carbohydrate is derived from the empirical formula for these compounds, CH_2O.

CHAPTER 16

1. **(A)** The sodium ion gives an orange flame test.

2. **(C)** The bromide ion is oxidized to bromine and then extracted into an organic solvent, producing a distinctive brown color.

3. **(D)** The sulfide ion and ammonium ion are converted to gases that have odors. Acid is added to sulfides to produce $H_2S(g)$ that smells like rotten eggs. Adding base to an ammonium salt produces ammonia that has a characteristic sharp odor.

4. **(B)** Silver ions are precipitated with chloride. It can be confirmed by dissolving the silver chloride in ammonia, forming a soluble $Ag(NH_3)_2^+$ complex ion.

5. **(A)** A buret will measure this volume most accurately.

6. **(B)** Methane does not dissolve in water, so displacement of water is used to collect this compound.

7. **(D)** Centrifugation is best. Fine precipitates clog filters. Drying and distillation are very time-consuming and do not remove any soluble ions present.

8. **(B)** The answer to the calculation must have three significant figures and is 2.05×10^3. The relative uncertainty is

$$\frac{0.01 \times 10^3}{2.05 \times 10^3} = 0.005 \text{ or } 5 \times 10^{-3}$$

9. **(A)** To prepare a molar solution, the solvent is added to the solute to achieve the total volume of solution desired. By adding the solute to 1 L of water the final volume will not be 1 L.

10. **(B)** Pressure is the dependent variable and is the y-axis of the graph.

11. **(D)** Prelab write-ups are done at home or in the library before entering the lab. Safety glasses are always required for any work in the lab.

12. **(C)** Drying the solid allows an accurate determination of the compound measured but has little effect on the dissolution process.

13. **(C)** $Al(OH)_3$ is amphiprotic. An acid will neutralize the hydroxide ions, and a base will form a hydroxy complex, which is soluble.

14. **(C)** For a monoprotic acid titrated with a monobasic base, $M_a V_a = M_b V_b$. The molarity of the acid is calculated as

$$M_a = \frac{M_b V_b}{V_a}$$

Droplets left in the buret mean that the volume of base reported is larger than the volume that the buret actually delivered. Since the V_b reported is larger than it should be, the M_a calculated using the above equation is too high.

15. **(E)** $C_i V_i = C_f V_f$. When the analysis is started,

$$C_i(5.00 \text{ mL}) = (3.0 \times 10^{-3} M)(100 \text{ mL})$$

$$= 6.0 \times 10^{-2} M$$

The preservative step adds an equal volume of preservative to the sample. Choose any two volumes so that the final volume is twice the initial sample volume. Then

$$C_i(50 \text{ mL}) = (6.0 \times 10^{-2} M)(100 \text{ mL})$$

$$= 1.2 \times 10^{-1} M$$

16. **(E)** Unlike $PbCl_2$, the other chloride precipitates, $AgCl$ and Hg_2Cl_2, are not soluble in hot water. Since all the precipitate dissolves, they are not present. Also, since the unknown was colored, we may deduce that a transition metal ion is present.

17. **(D)** CO_2 and H_2S are the only gases produced when a solution is acidified. CO_2 has no odor, but H_2S smells like rotten eggs. NH_3 is formed when a solution is made alkaline.

18. **(E)** All of these ions are confirmed with flame tests.

19. **(C)** $CrCl_3$ is a soluble salt. Reacting precisely 3 moles of HCl with 1 mole of $Cr(OH)_3$ will produce a solution containing only Cr^{3+} and Cl^- ions, which can be dried to obtain pure $CrCl_3$. The other mixtures contain other ions in addition to Cr^{3+} and Cl^-. This results either in an impure product or in reactions that will not form $CrCl_3$.

20. **(B)** Titration is the most common method. Choices (A) and (D) are possible, but not widely used, methods; (E) and (C) are imprecise methods.

21. **(C)** Calcium sulfide is relatively soluble and is not precipitated in the qual-scheme.

22. **(E)** This is the only choice left because (A), (B), (C), and (D) are all correct.

23. **(A)** The correct approach is to abandon the experiment and start again.

24. **(E)** The data for the experiment should fill most of the graph.

Appendix 2

Electronic Configurations of the Elements

H	$1s^1$				
He	$1s^2$				
Li	$1s^2$	$2s^1$			
Be	$1s^2$	$2s^2$			
B	$1s^2$	$2s^2 2p^1$			
C	$1s^2$	$2s^2 2p^2$			
N	$1s^2$	$2s^2 2p^3$			
O	$1s^2$	$2s^2 2p^4$			
F	$1s^2$	$2s^2 2p^5$			
Ne	$1s^2$	$2s^2 2p^6$			
Na	$1s^2$	$2s^2 2p^6$	$3s^2$		
Mg	$1s^2$	$2s^2 2p^6$	$3s^2$		
Al	$1s^2$	$2s^2 2p^6$	$3s^2 3p^1$		
Si	$1s^2$	$2s^2 2p^6$	$3s^2 3p^2$		
P	$1s^2$	$2s^2 2p^6$	$3s^2 3p^3$		
S	$1s^2$	$2s^2 2p^6$	$3s^2 3p^4$		
Cl	$1s^2$	$2s^2 2p^6$	$3s^2 3p^5$		
Ar	$1s^2$	$2s^2 2p^6$	$3s^2 3p^6$		
K	$1s^2$	$2s^2 2p^6$	$3s^2 3p^6$	$4s^1$	
Ca	$1s^2$	$2s^2 2p^6$	$3s^2 3p^6$	$4s^2$	
Sc	$1s^2$	$2s^2 2p^6$	$3s^2 3p^6$	$4s^2 3d^1$	
Ti	$1s^2$	$2s^2 2p^6$	$3s^2 3p^6$	$4s^2 3d^2$	
V	$1s^2$	$2s^2 2p^6$	$3s^2 3p^6$	$4s^2 3d^3$	
Cr	$1s^2$	$2s^2 2p^6$	$3s^2 3p^6$	$4s^1 3d^5$	
Mn	$1s^2$	$2s^2 2p^6$	$3s^2 3p^6$	$4s^2 3d^5$	
Fe	$1s^2$	$2s^2 2p^6$	$3s^2 3p^6$	$4s^2 3d^6$	
Co	$1s^2$	$2s^2 2p^6$	$3s^2 3p^6$	$4s^2 3d^7$	
Ni	$1s^2$	$2s^2 2p^6$	$3s^2 3p^6$	$4s^2 3d^8$	
Cu	$1s^2$	$2s^2 2p^6$	$3s^2 3p^6$	$4s^1 3d^{10}$	
Zn	$1s^2$	$2s^2 2p^6$	$3s^2 3p^6$	$4s^2 3d^{10}$	
Ga	$1s^2$	$2s^2 2p^6$	$3s^2 3p^6$	$4s^2 3d^{10} 4p^1$	
Ge	$1s^2$	$2s^2 2p^6$	$3s^2 3p^6$	$4s^2 3d^{10} 4p^2$	
As	$1s^2$	$2s^2 2p^6$	$3s^2 3p^6$	$4s^2 3d^{10} 4p^3$	
Se	$1s^2$	$2s^2 2p^6$	$3s^2 3p^6$	$4s^2 3d^{10} 4p^4$	
Br	$1s^2$	$2s^2 2p^6$	$3s^2 3p^6$	$4s^2 3d^{10} 4p^5$	
Kr	$1s^2$	$2s^2 2p^6$	$3s^2 3p^6$	$4s^2 3d^{10} 4p^6$	
Rb	$1s^2$	$2s^2 2p^6$	$3s^2 3p^6$	$4s^2 3d^{10} 4p^6$	$5s^1$
Sr	$1s^2$	$2s^2 2p^6$	$3s^2 3p^6$	$4s^2 3d^{10} 4p^6$	$5s^2$
Y	$1s^2$	$2s^2 2p^6$	$3s^2 3p^6$	$4s^2 3d^{10} 4p^6$	$5s^2 4d^1$

Zr	$1s^2$	$2s^2 2p^6$	$3s^2 3p^6$	$4s^2 3d^{10} 4p^6$	$5s^2 4d^2$
Nb	$1s^2$	$2s^2 2p^6$	$3s^2 3p^6$	$4s^2 3d^{10} 4p^6$	$5s^1 4d^4$
Mo	$1s^2$	$2s^2 2p^6$	$3s^2 3p^6$	$4s^2 3d^{10} 4p^6$	$5s^1 4d^5$
Tc	$1s^2$	$2s^2 2p^6$	$3s^2 3p^6$	$4s^2 3d^{10} 4p^6$	$5s^2 4d^5$
Ru	$1s^2$	$2s^2 2p^6$	$3s^2 3p^6$	$4s^2 3d^{10} 4p^6$	$5s^1 4d^7$
Rh	$1s^2$	$2s^2 2p^6$	$3s^2 3p^6$	$4s^2 3d^{10} 4p^6$	$5s^1 4d^8$
Pd	$1s^2$	$2s^2 2p^6$	$3s^2 3p^6$	$4s^2 3d^{10} 4p^6$	$5s^1 4d^{10}$
Ag	$1s^2$	$2s^2 2p^6$	$3s^2 3p^6$	$4s^2 3d^{10} 4p^6$	$5s^1 4d^{10}$
Cd	$1s^2$	$2s^2 2p^6$	$3s^2 3p^6$	$4s^2 3d^{10} 4p^6$	$5s^2 4d^{10}$
In	$1s^2$	$2s^2 2p^6$	$3s^2 3p^6$	$4s^2 3d^{10} 4p^6$	$5s^2 4d^{10} 5p^1$
Sn	$1s^2$	$2s^2 2p^6$	$3s^2 3p^6$	$4s^2 3d^{10} 4p^6$	$5s^2 4d^{10} 5p^2$
Sb	$1s^2$	$2s^2 2p^6$	$3s^2 3p^6$	$4s^2 3d^{10} 4p^6$	$5s^2 4d^{10} 5p^3$
Te	$1s^2$	$2s^2 2p^6$	$3s^2 3p^6$	$4s^2 3d^{10} 4p^6$	$5s^2 4d^{10} 5p^4$
I	$1s^2$	$2s^2 2p^6$	$3s^2 3p^6$	$4s^2 3d^{10} 4p^6$	$5s^2 4d^{10} 5p^5$
Xe	$1s^2$	$2s^2 2p^6$	$3s^2 3p^6$	$4s^2 3d^{10} 4p^6$	$5s^2 4d^{10} 5p^6$

Cs	$1s^2$	$2s^2 2p^6$	$3s^2 3p^6$	$4s^2 3d^{10} 4p^6$	$5s^2 4d^{10} 5p^6$	$6s^2$
Ba	$1s^2$	$2s^2 2p^6$	$3s^2 3p^6$	$4s^2 3d^{10} 4p^6$	$5s^2 4d^{10} 5p^6$	$6s^2$
La	$1s^2$	$2s^2 2p^6$	$3s^2 3p^6$	$4s^2 3d^{10} 4p^6$	$5s^2 4d^{10} 5p^6$	$6s^2 4f^1$
Ce	$1s^2$	$2s^2 2p^6$	$3s^2 3p^6$	$4s^2 3d^{10} 4p^6$	$5s^2 4d^{10} 5p^6$	$6s^2 4f^2$
Pr	$1s^2$	$2s^2 2p^6$	$3s^2 3p^6$	$4s^2 3d^{10} 4p^6$	$5s^2 4d^{10} 5p^6$	$6s^2 4f^3$
Nd	$1s^2$	$2s^2 2p^6$	$3s^2 3p^6$	$4s^2 3d^{10} 4p^6$	$5s^2 4d^{10} 5p^6$	$6s^2 4f^4$
Pm	$1s^2$	$2s^2 2p^6$	$3s^2 3p^6$	$4s^2 3d^{10} 4p^6$	$5s^2 4d^{10} 5p^6$	$6s^2 4f^5$
Sm	$1s^2$	$2s^2 2p^6$	$3s^2 3p^6$	$4s^2 3d^{10} 4p^6$	$5s^2 4d^{10} 5p^6$	$6s^2 4f^6$
Eu	$1s^2$	$2s^2 2p^6$	$3s^2 3p^6$	$4s^2 3d^{10} 4p^6$	$5s^2 4d^{10} 5p^6$	$6s^2 4f^7$
Gd	$1s^2$	$2s^2 2p^6$	$3s^2 3p^6$	$4s^2 3d^{10} 4p^6$	$5s^2 4d^{10} 5p^6$	$6s^2 4f^7 5d^1$
Tb	$1s^2$	$2s^2 2p^6$	$3s^2 3p^6$	$4s^2 3d^{10} 4p^6$	$5s^2 4d^{10} 5p^6$	$6s^2 4f^9$
Dy	$1s^2$	$2s^2 2p^6$	$3s^2 3p^6$	$4s^2 3d^{10} 4p^6$	$5s^2 4d^{10} 5p^6$	$6s^2 4f^{10}$
Ho	$1s^2$	$2s^2 2p^6$	$3s^2 3p^6$	$4s^2 3d^{10} 4p^6$	$5s^2 4d^{10} 5p^6$	$6s^2 4f^{11}$
Er	$1s^2$	$2s^2 2p^6$	$3s^2 3p^6$	$4s^2 3d^{10} 4p^6$	$5s^2 4d^{10} 5p^6$	$6s^2 4f^{12}$
Tm	$1s^2$	$2s^2 2p^6$	$3s^2 3p^6$	$4s^2 3d^{10} 4p^6$	$5s^2 4d^{10} 5p^6$	$6s^2 4f^{13}$
Yb	$1s^2$	$2s^2 2p^6$	$3s^2 3p^6$	$4s^2 3d^{10} 4p^6$	$5s^2 4d^{10} 5p^6$	$6s^2 4f^{14}$
Lu	$1s^2$	$2s^2 2p^6$	$3s^2 3p^6$	$4s^2 3d^{10} 4p^6$	$5s^2 4d^{10} 5p^6$	$6s^2 4f^{14} 5d^1$
Hf	$1s^2$	$2s^2 2p^6$	$3s^2 3p^6$	$4s^2 3d^{10} 4p^6$	$5s^2 4d^{10} 5p^6$	$6s^2 4f^{14} 5d^2$
Ta	$1s^2$	$2s^2 2p^6$	$3s^2 3p^6$	$4s^2 3d^{10} 4p^6$	$5s^2 4d^{10} 5p^6$	$6s^2 4f^{14} 5d^3$
W	$1s^2$	$2s^2 2p^6$	$3s^2 3p^6$	$4s^2 3d^{10} 4p^6$	$5s^2 4d^{10} 5p^6$	$6s^2 4f^{14} 5d^4$
Re	$1s^2$	$2s^2 2p^6$	$3s^2 3p^6$	$4s^2 3d^{10} 4p^6$	$5s^2 4d^{10} 5p^6$	$6s^2 4f^{14} 5d^5$
Os	$1s^2$	$2s^2 2p^6$	$3s^2 3p^6$	$4s^2 3d^{10} 4p^6$	$5s^2 4d^{10} 5p^6$	$6s^2 4f^{14} 5d^6$
Ir	$1s^2$	$2s^2 2p^6$	$3s^2 3p^6$	$4s^2 3d^{10} 4p^6$	$5s^2 4d^{10} 5p^6$	$6s^2 4f^{14} 5d^7$
Pt	$1s^2$	$2s^2 2p^6$	$3s^2 3p^6$	$4s^2 3d^{10} 4p^6$	$5s^2 4d^{10} 5p^6$	$6s^1 4f^{14} 5d^9$
Au	$1s^2$	$2s^2 2p^6$	$3s^2 3p^6$	$4s^2 3d^{10} 4p^6$	$5s^2 4d^{10} 5p^6$	$6s^1 4f^{14} 5d^{10}$
Hg	$1s^2$	$2s^2 2p^6$	$3s^2 3p^6$	$4s^2 3d^{10} 4p^6$	$5s^2 4d^{10} 5p^6$	$6s^2 4f^{14} 5d^{10}$
Tl	$1s^2$	$2s^2 2p^6$	$3s^2 3p^6$	$4s^2 3d^{10} 4p^6$	$5s^2 4d^{10} 5p^6$	$6s^2 4f^{14} 5d^{10} 6p^1$
Pb	$1s^2$	$2s^2 2p^6$	$3s^2 3p^6$	$4s^2 3d^{10} 4p^6$	$5s^2 4d^{10} 5p^6$	$6s^2 4f^{14} 5d^{10} 6p^2$
Bi	$1s^2$	$2s^2 2p^6$	$3s^2 3p^6$	$4s^2 3d^{10} 4p^6$	$5s^2 4d^{10} 5p^6$	$6s^2 4f^{14} 5d^{10} 6p^3$
Po	$1s^2$	$2s^2 2p^6$	$3s^2 3p^6$	$4s^2 3d^{10} 4p^6$	$5s^2 4d^{10} 5p^6$	$6s^2 4f^{14} 5d^{10} 6p^4$
At	$1s^2$	$2s^2 2p^6$	$3s^2 3p^6$	$4s^2 3d^{10} 4p^6$	$5s^2 4d^{10} 5p^6$	$6s^2 4f^{14} 5d^{10} 6p^5$
Rn	$1s^2$	$2s^2 2p^6$	$3s^2 3p^6$	$4s^2 3d^{10} 4p^6$	$5s^2 4d^{10} 5p^6$	$6s^2 4f^{14} 5d^{10} 6p^6$

Fr	$1s^2$	$2s^2 2p^6$	$3s^2 3p^6$	$4s^2 3d^{10} 4p^6$	$5s^2 4d^{10} 5p^6$	$6s^2 4f^{14} 5d^{10} 6p^6$	$7s^1$
Ra	$1s^2$	$2s^2 2p^6$	$3s^2 3p^6$	$4s^2 3d^{10} 4p^6$	$5s^2 4d^{10} 5p^6$	$6s^2 4f^{14} 5d^{10} 6p^6$	$7s^2$
Ac	$1s^2$	$2s^2 2p^6$	$3s^2 3p^6$	$4s^2 3d^{10} 4p^6$	$5s^2 4d^{10} 5p^6$	$6s^2 4f^{14} 5d^{10} 6p^6$	$7s^2 6d^1$
Th	$1s^2$	$2s^2 2p^6$	$3s^2 3p^6$	$4s^2 3d^{10} 4p^6$	$5s^2 4d^{10} 5p^6$	$6s^2 4f^{14} 5d^{10} 6p^6$	$7s^2 6d^2$

Pa	$1s^2$	$2s^2 2p^6$	$3s^2 3p^6$	$4s^2 3d^{10} 4p^6$	$5s^2 4d^{10} 5p^6$	$6s^2 4f^{14} 5d^{10} 6p^6$	$7s^2 5f^2 6d^1$
U	$1s^2$	$2s^2 2p^6$	$3s^2 3p^6$	$4s^2 3d^{10} 4p^6$	$5s^2 4d^{10} 5p^6$	$6s^2 4f^{14} 5d^{10} 6p^6$	$7s^2 5f^3 6d^1$
Np	$1s^2$	$2s^2 2p^6$	$3s^2 3p^6$	$4s^2 3d^{10} 4p^6$	$5s^2 4d^{10} 5p^6$	$6s^2 4f^{14} 5d^{10} 6p^6$	$7s^2 5f^4 6d^1$
Pu	$1s^2$	$2s^2 2p^6$	$3s^2 3p^6$	$4s^2 3d^{10} 4p^6$	$5s^2 4d^{10} 5p^6$	$6s^2 4f^{14} 5d^{10} 6p^6$	$7s^2 5f^6$
Am	$1s^2$	$2s^2 2p^6$	$3s^2 3p^6$	$4s^2 3d^{10} 4p^6$	$5s^2 4d^{10} 5p^6$	$6s^2 4f^{14} 5d^{10} 6p^6$	$7s^2 5f^7$
Cm	$1s^2$	$2s^2 2p^6$	$3s^2 3p^6$	$4s^2 3d^{10} 4p^6$	$5s^2 4d^{10} 5p^6$	$6s^2 4f^{14} 5d^{10} 6p^6$	$7s^2 5f^7 6d^1$
Bk	$1s^2$	$2s^2 2p^6$	$3s^2 3p^6$	$4s^2 3d^{10} 4p^6$	$5s^2 4d^{10} 5p^6$	$6s^2 4f^{14} 5d^{10} 6p^6$	$7s^2 5f^9$
Cf	$1s^2$	$2s^2 2p^6$	$3s^2 3p^6$	$4s^2 3d^{10} 4p^6$	$5s^2 4d^{10} 5p^6$	$6s^2 4f^{14} 5d^{10} 6p^6$	$7s^2 5f^{10}$
Es	$1s^2$	$2s^2 2p^6$	$3s^2 3p^6$	$4s^2 3d^{10} 4p^6$	$5s^2 4d^{10} 5p^6$	$6s^2 4f^{14} 5d^{10} 6p^6$	$7s^2 5f^{11}$
Fm	$1s^2$	$2s^2 2p^6$	$3s^2 3p^6$	$4s^2 3d^{10} 4p^6$	$5s^2 4d^{10} 5p^6$	$6s^2 4f^{14} 5d^{10} 6p^6$	$7s^2 5f^{12}$
Md	$1s^2$	$2s^2 2p^6$	$3s^2 3p^6$	$4s^2 3d^{10} 4p^6$	$5s^2 4d^{10} 5p^6$	$6s^2 4f^{14} 5d^{10} 6p^6$	$7s^2 5f^{13}$
No	$1s^2$	$2s^2 2p^6$	$3s^2 3p^6$	$4s^2 3d^{10} 4p^6$	$5s^2 4d^{10} 5p^6$	$6s^2 4f^{14} 5d^{10} 6p^6$	$7s^2 5f^{14}$
Lr	$1s^2$	$2s^2 2p^6$	$3s^2 3p^6$	$4s^2 3d^{10} 4p^6$	$5s^2 4d^{10} 5p^6$	$6s^2 4f^{14} 5d^{10} 6p^6$	$7s^2 5f^{14} 6d^1$

Appendix 3

Thermodynamic Data for Selected Elements, Compounds, and Ions (25 °C)

Substance	ΔH_f° (kJ/mol^{-1})	S° (J mol^{-1} K^{-1})	ΔG_f° (kJ/mol^{-1})
aluminum			
Al(s)	0	28.3	0
AlCl$_3$(s)	−704	110.7	−629
Al$_2$O$_3$(s)	−1676	51.0	−1576.4
Al$_2$(SO$_4$)$_3$(s)	−3441	239	−3100
barium			
Ba(s)	0	66.9	0
BaCO$_3$(s)	−1219	112	−1139
BaCl$_2$(s)	−860.2	125	−810.8
Ba(OH)$_2$(s)	−998.22	−8	−875.3
Ba(NO$_3$)$_2$(s)	−992	214	−795
BaSO$_4$(s)	−1465	132	−1353
bromine			
Br$_2$(l)	0	152.2	0
Br$_2$(g)	+30.9	245.4	3.11
HBr(g)	−36	198.5	53.1
calcium			
Ca(s)	0	41.4	0
CaCO$_3$(s)	−1207	92.9	−1128.8
CaF$_2$(s)	−741	80.3	−1166
CaCl$_2$(s)	−795.8	114	−750.2
CaO(s)	−635.5	40	−604.2
Ca(OH)$_2$(s)	−986.6	76.1	896.76
CaSO$_4$(s)	−1433	107	−1320.3
carbon			
C(s, graphite)	0	5.69	0
C(s, diamond)	+1.88	2.4	+2.9
CCL$_4$(l)	−134	214.4	−65.3

Substance	ΔH_f° (kJ/mol^{-1})	S° (J mol^{-1} K^{-1})	ΔG_f° (kJ/mol^{-1})
$CO(g)$	−110	197.9	−137.3
$CO_2(g)$	−394	213.6	−394.4
$CO_2(aq)$	−413.8	117.6	−385.98
$H_2CO_3(aq)$	−699.65	187.4	−623.08
$HCO_3^-(aq)$	−691.99	91.2	−586.77
$CO_3^{2-}(aq)$	−677.14	−56.9	527.81
$HCN(g)$	+135.1	201.7	+124.7
$CN^-(aq)$	+150.6	94.1	+172.4
$CH_4(g)$	−74.9	186.2	−50.79
$C_2H_2(g)$	+227	200.8	+209
$C_2H_4(g)$	+51.9	219.8	+68.12
$C_2H_6(g)$	−84.5	229.5	−32.9
$C_2H_8(g)$	−104	269.9	−23
$C_2H_{10}(g)$	−126	310.2	−17.0
$C_6H_6(l)$	+49.0	173.3	+124.3
$CH_3OH(l)$	−238	126.8	−166.2
$C_2H_5OH(l)$	−278	161	−174.8
$HCHO_2(g)$	−363	251	+335
$HC_3H_3O_2(l)$	−487.0	160	−392.5
$HCHO(g)$	−108.6	218.8	−102.5
$CH_3CHO(g)$	−167	250	−129
$(CH_3)_3CO(l)$	−248.1	200.4	−155.4
$C_6H_5CO_2H(s)$	−385.1	167.6	−245.3
chlorine			
$Cl_2(g)$	0	223.0	0
$HCl(g)$	−92.5	186.7	−95.27
$HCl(aq)$	−167.2	56.5	−131.2
$HClO(aq)$	−131.3	106.8	−80.21
chromium			
$Cr(s)$	0	23.8	0
$CrCl_2(s)$	−326	115	−282
$CrCl_3(s)$	−563.2	126	−493.7
copper			
$Cu(s)$	0	33.15	0
$CuCl(s)$	−137.2	86.2	−119.87
$CuCl_2(s)$	−172	119	−131
$CuSo_4(s)$	−771.4	109	−661.8
fluorine			
$F_2(g)$	0	202.7	0
$F^-(aq)$	−332.6	13.8	−278.8
$HF(g)$	−271	173.5	−273
hydrogen			
$H_2(g)$	0	130.6	0
$H_2O(l)$	−286	69.96	−237.2
$H_2O(g)$	−242	188.7	−228.6

Substance	ΔH_f° (kJ/mol^{-1})	S° (J mol^{-1} K^{-1})	ΔG_f° (kJ/mol^{-1})
iron			
Fe(s)	0	27	0
Fe$_2$O$_3$(s)	−822.2	90.0	−741.0
Fe$_3$O$_4$(s)	−1118.4	146.4	−1015.4
lead			
Pb(s)	0	64.8	0
PbCl$_2$(s)	−359.4	136	−314.1
PbS(s)	−100	91.2	−98.7
PbSO$_4$(s)	−920.1	149	−811.3
lithium			
Li(s)	0	28.4	0
LiF(s)	−611.7	35.7	−583.3
LiCl(s)	−408	59.29	−383.7
magnesium			
Mg(s)	0	32.5	0
MgCO$_3$(s)	−1113	65.7	−1029
MgF$_2$(s)	−1113	79.9	−1056
MgCl$_2$(s)	−641.8	89.5	−592.5
MgO(s)	−601.7	26.9	−569.4
Mg(OH)$_2$(s)	−924.7	63.1	−833.9
manganese			
Mn(s)	0	32.0	0
MnO$_4^-$(aq)	−542.7	191	−449.4
KMnO$_4$(s)	−813.4	171.71	−713.8
MnO$_2$(s)	−520.9	53.1	−466.1
nitrogen			
N$_2$(g)	0	191.5	0
NH$_3$(g)	−46.0	192.5	−16.7
NH$_4$Cl(s)	−314.4	94.6	−203.9
NO(g)	+90.4	210.6	+86.69
NO$_2$(g)	+34	240.5	+51.84
N$_2$O(g)	+81.5	220.0	+103.6
HNO$_3$(l)	−174.1	155.6	−79.9
oxygen			
O$_2$(g)	0	205.0	0
O$_3$(g)	+143	238.8	+163
OH$^-$(aq)	−230.0	−10.75	−157.24
potassium			
K(s)	0	64.18	0
KF(s)	−567.3	66.6	−537.8
KCl(s)	−436.8	82.59	−408.3
KOH(s)	−424.8	78.9	−379.1
K$_2$SO$_4$(s)	−1433.7	176	−1316.4

Substance	ΔH_f° (kJ/mol^{-1})	S° (J mol^{-1} K^{-1})	ΔG_f° (kJ/mol^{-1})
silver			
Ag(s)	0	42.55	0
AgCl(s)	−127.1	96.2	−109.8
AgNO$_3$(s)	−124	141	−32
sodium			
Na(s)	0	51.0	0
NaF(s)	−571	51.5	−545
NaCl(s)	−413	72.38	−384.0
NaOH(s)	−426.8	64.18	−382
Na$_2$SO$_4$(s)	−1384.49	149.49	−1266.83
sulfur			
S(s, rhombic)	0	31.8	0
SO$_2$(g)	−297	248	−300
SO$_3$(g)	−396	256	−370
H$_2$SO$_4$(aq)	−909.3	20.1	−744.5
SF$_6$(g)	−1209	292	−1105
tin			
Sn(s, white)	0	51.6	0
SnCl$_4$(l)	−511.3	258.6	−440.2
zinc			
Zn(s)	0	41.6	0
ZnCl$_2$(s)	−415.1	111	−369.4

Appendix 4

Ionization Constants of Weak Acids

Monoprotic Acid	Name	K_a
HIO_3	iodic acid	1.69×10^{-1}
HNO_2	nitrous acid	7.1×10^{-4}
HF	hydrofluoric acid	6.8×10^{-4}
$HCHO_2$	formic acid	1.8×10^{-4}
$HC_3H_5O_3$	lactic acid	1.38×10^{-4}
$HC_7H_5O_2$	benzoic acid	6.28×10^{-5}
$HC_4H_7O_2$	butanoic acid	1.52×10^{-5}
HN_3	hydrazoic acid	1.8×10^{-5}
$HC_2H_3O_2$	ethanoic acid	1.8×10^{-5}
$HC_3H_5O_2$	propanoic acid	1.34×10^{-5}
$HOCl$	hypochlorous acid	3.0×10^{-8}
HCN	hydrocyanic acid	6.2×10^{-10}
HC_6H_5O	phenol	1.3×10^{-10}
HOI	hypoiodous acid	2.3×10^{-11}
H_2O_2	hydrogen peroxide	1.8×10^{-12}

Appendix 5

Ionization Constants of Polyprotic Acids

Polyprotic Acid	Name	K_{a1}	K_{a1}	K_{a3}
H_2SO_4	sulfuric acid	Large	1.0×10^{-2}	
H_2CrO_4	chromic acid	5.0	1.5×10^{-6}	
$H_2C_2O_4$	oxalic acid	5.6×10^{-2}	5.4×10^{-5}	
H_3PO_3	phosphorous acid	3×10^{-2}	1.6×10^{-7}	
H_2SO_3	sulfurous acid	1.2×10^{-2}	6.6×10^{-8}	
H_2SeO_3	selenous acid	4.5×10^{-3}	1.1×10^{-8}	
$H_2C_3H_2O_4$	malonic acid	1.4×10^{-3}	2.0×10^{-6}	
$H_2C_8H_4O_4$	phthalic acid	1.1×10^{-3}	3.9×10^{-6}	
$H_2C_4H_4O_6$	tartaric acid	9.2×10^{-4}	4.3×10^{-5}	
H_2CO_3	carbonic acid	4.5×10^{-7}	4.7×10^{-11}	
H_3PO_4	phosphoric acid	7.1×10^{-3}	6.3×10^{-8}	4.5×10^{-13}
H_3AsO_4	arsenic acid	5.6×10^{-3}	1.7×10^{-7}	4.0×10^{-12}
$H_3C_6H_5O_7$	citric acid	7.1×10^{-4}	1.7×10^{-5}	6.3×10^{-6}

Appendix 6

Ionization Constants of Weak Bases

Weak Base	Name	K_b
$(CH_3)_2NH$	dimethylamine	9.6×10^{-4}
CH_3NH_2	methylamine	4.4×10^{-4}
$CH_3CH_2NH_2$	ethylamine	4.3×10^{-4}
$(CH_3)_3N$	trimethylamine	7.4×10^{-5}
NH_3	ammonia	1.8×10^{-5}
N_2H_4	hydrazine	9.6×10^{-7}
C_5H_5N	pyridine	1.5×10^{-9}
$C_6H_5NH_2$	aniline	4.1×10^{-10}

Appendix 7

Solubility Product Constants

Salt	Dissolution Reaction	K_{sp}
fluorides		
MgF_2	$MgF_2(s) \rightleftharpoons Mg^{2+}(aq) + 2F^-(aq)$	6.6×10^{-9}
CaF_2	$CaF_2(s) \rightleftharpoons Ca^{2+}(aq) + 2F^-(aq)$	3.9×10^{-11}
BaF_2	$BaF_2(s) \rightleftharpoons Ba^{2+}(aq) + 2F^-(aq)$	1.7×10^{-6}
PbF_2	$PbF_2(s) \rightleftharpoons Pb^{2+}(aq) + 2F^-(aq)$	3.6×10^{-8}
chlorides		
$CuCl$	$CuCl(s) \rightleftharpoons Cu^+(aq) + Cl^-(aq)$	1.9×10^{-7}
$AgCl$	$AgCl(s) \rightleftharpoons Ag^+(aq) + Cl^-(aq)$	1.8×10^{-10}
$PbCl_2$	$PbCL_2 \rightleftharpoons Pb^{2+}(aq) + 2CL^-(aq)$	1.7×10^{-5}
bromides		
$CuBr$	$CuBr(s) \rightleftharpoons Cu^+(aq) + Br^-(aq)$	5×10^{-9}
$AgBr$	$AgBr(s) \rightleftharpoons Ag^+(aq) + Br^-(aq)$	5.0×10^{-13}
$PbBR_2$	$PbBr_2(s) \rightleftharpoons Pb^{2+}(aq) + 2Br^-(aq)$	2.1×10^{-6}
iodides		
CuI	$CuI(s) \rightleftharpoons Cu^+(aq) + I^-(aq)$	1×10^{-12}
AgI	$AgI(s) \rightleftharpoons Ag^+(aq) + I^-(aq)$	8.3×10^{-17}
PbI_2	$PbI_2(s) \rightleftharpoons Pb^{2+}(aq) + 2I^-(aq)$	7.9×10^{-9}
hydroxides		
$Mg(OH)_2$	$Mg(OH)_2(s) \rightleftharpoons Mg^{2+}(aq) + 2OH^-(aq)$	7.1×10^{-12}
$Ca(OH)_2$	$Ca(OH)_2(s) \rightleftharpoons Ca^{2+}(aq) + 2OH^-(aq)$	6.5×10^{-6}
$Mn(OH)_2$	$MN(OH)_2(s) \rightleftharpoons Mn^{2+}(aq) + 2OH^-(aq)$	1.6×10^{-13}
$Fe(OH)_2$	$Fe(OH)_2(s) \rightleftharpoons Fe^{2+}(aq) + 2OH^-(aq)$	7.9×10^{-16}
$Fe(OH)_3$	$Fe(OH)_3(s) \rightleftharpoons Fe^{3+}(aq) + 3OH^-(aq)$	1.6×10^{-39}
$Ni(OH)_2$	$Ni(OH)_2(s) \rightleftharpoons Ni^{2+}(aq) + 2OH^-(aq)$	6×10^{-16}
$Cu(OH)_2$	$Cu(OH)_2(s) \rightleftharpoons Cu^{2+}(aq) + 2OH^-(aq)$	4.8×10^{-20}
$Cr(OH)_3$	$Cr(OH)_3(s) \rightleftharpoons Cr^{3+}(aq) + 3OH^-(aq)$	2×10^{-30}
$Zn(OH)_2$	$Zn(OH)_2(s) \rightleftharpoons Zn^{2+}(aq) + 2OH^-(aq)$	3.0×10^{-16}
$Cd(OH)_2$	$Cd(OH)_2(s) \rightleftharpoons Cd^{2+}(aq) + 2OH^-(aq)$	5.0×10^{-15}

Salt	Dissolution Reaction	K_{sp}
sulfites		
$CaSO_3$	$CaSO_3(s) \rightleftharpoons Ca^{2+}(aq) + SO_3^{2-}(aq)$	3×10^{-7}
$BaSO_3$	$BaSO_4(s) \rightleftharpoons Ba^{2+}(aq) + SO_3^{2-}(aq)$	8×10^{-7}
sulfates		
$CaSO_4$	$CaSO_4(s) \rightleftharpoons Ca^{2+}(aq) + SO_4^{2-}(aq)$	2.4×10^{-5}
$BaSO_4$	$BaSO_4(s) \rightleftharpoons Ba^{2+}(aq) + SO_4^{2-}(aq)$	1.1×10^{-10}
Ag_2SO_4	$Ag_2SO_4(s) \rightleftharpoons 2Ag^+(aq) + SO_4^{2-}(aq)$	1.5×10^{-5}
$PbSO_4$	$PbSO_4(s) \rightleftharpoons Pb^{2+}(aq) + SO_4^{2-}(aq)$	6.3×10^{-7}
chromates		
Ag_2CrO_4	$Ag_2CrO_4(s) \rightleftharpoons 2Ag^+(aq) + CrO_4^{2-}(aq)$	1.2×10^{-12}
Hg_2CrO_4	$Hg_2CrO_4(s) \rightleftharpoons Hg_2^{2+}(aq) + CrO_4^{2-}(aq)$	2.0×10^{-9}
$PbCrO_4$	$PbCrO_4(s) \rightleftharpoons Pb^{2-}(aq) + CrO_4^{2-}(aq)$	1.8×10^{-14}
carbonates		
$MgCO_3$	$MgCO_3(s) \rightleftharpoons Mg^{2+}(aq) + CO_3^{2-}(aq)$	3.5×10^{-8}
$CaCO_3$	$CaCO_3(s) \rightleftharpoons Ca^{2+}(aq) + CO_3^{2-}(aq)$	9.3×10^{-10}
$BaCO_3$	$BaCO_3(s) \rightleftharpoons Ba^{2+}(aq) + CO_3^{2-}(aq)$	5.0×10^{-9}
$MnCO_3$	$MnCO_3(s) \rightleftharpoons Mn^{2+}(aq) + CO_3^{2-}(aq)$	5.0×10^{-10}
$CuCO_3$	$CuCO_3(s) \rightleftharpoons Cu^{2+}(aq) + CO_3^{2-}(aq)$	2.3×10^{-10}
Ag_2CO_3	$Ag_2CO_3(s) \rightleftharpoons 2Ag^{2+}(aq) + CO_3^{2-}(aq)$	8.1×10^{-12}
Hg_2CO_3	$Hg_2CO_3(s) \rightleftharpoons Hg_2^{2+}(aq) + CO_3^{2-}(aq)$	8.9×10^{-17}
$ZnCO_3$	$ZnCO_3(s) \rightleftharpoons Zn^{2+}(aq) + CO_3^{2-}(aq)$	1.0×10^{-10}
$PbCO_3$	$PbCO_3(s) \rightleftharpoons Pb^{2+}(aq) + CO_3^{2-}(aq)$	7.4×10^{-14}
sulfides		
MnS	$MnS(s) \rightleftharpoons Mn^{2+}(aq) + S^{2-}(aq)$	3.0×10^{-11}
FeS	$FeS(s) \rightleftharpoons Fe^{2+}(aq) + S^{2-}(aq)$	8.0×10^{-19}
CoS	$CoS(s) \rightleftharpoons Co^{2+}(aq) + S^{2-}(aq)$	5.0×10^{-22}
NiS	$NiS(s) \rightleftharpoons Ni^{2+}(aq) + S^{2-}(aq)$	4.0×10^{-20}
CuS	$CuS(s) \rightleftharpoons Cu^{2+}(aq) + S^{2-}(aq)$	8.0×10^{-37}
Cu_2S	$Cu_2S(s) \rightleftharpoons 2Cu^{2+}(aq) + S^{2-}(aq)$	3.0×10^{-49}
Ag_2S	$Ag_2S(s) \rightleftharpoons 2Ag^+(aq) + S^{2-}(aq)$	8.0×10^{-51}
Tl_2S	$Tl_2S(s) \rightleftharpoons 2Tl^+(aq) + S^{2-}(aq)$	6.0×10^{-22}
ZnS	$ZnS(s) \rightleftharpoons Zn^{2+}(aq) + S^{2-}(aq)$	2.0×10^{-25}
CdS	$CdS(s) \rightleftharpoons Cd^{2+}(aq) + S^{2-}(aq)$	1.0×10^{-27}
HgS	$HgS(s) \rightleftharpoons HG^{2+}(aq) + S^{2-}(aq)$	2.0×10^{-53}
SnS	$SnS(s) \rightleftharpoons Sn^{2+}(aq) + S^{2-}(aq)$	1.3×10^{-26}
PbS	$PbS(s) \rightleftharpoons Pb^{2+}(aq) + S^{2-}(aq)$	3.0×10^{-28}
In_2S_3	$In_2S_3(s) \rightleftharpoons 2In^{3+}(aq) + 3S^{2-}(aq)$	4.0×10^{-70}

Glossary

A

absolute uncertainty   The uncertainty of ±1 in the last digit of a measurement. If this uncertainty is different from ±1, it is written as part of the number; for example, 23.45 ± 0.05 indicates an uncertainty of ±5 in the last digit.

absolute zero   The lowest possible temperature, 0.0 K or – 273.16 °C.

absorbance, *A*   A measure of the amount of light absorbed by a chemical.

absorptivity, *a*   A constant the value of which depends on the sample and the wavelength at which the measurement is made in spectroscopy.

accuracy   The degree of closeness between a measured value and the true value.

acid   Any substance that donates protons, or as Lewis acids are electron-pair acceptors.

acid anhydride   The oxide of a nonmetal that forms an acid when dissolved in water.

acid dissociation constant, K_a   The value of the equilibrium law for the dissociation of a weak acid.

activated complex   The structures of colliding molecules at the moment of collision, generally thought to be intermediate between the structures of the products and of the reactants.

activation energy   The increase in potential energy, due to a molecular collision, necessary to convert a reactant into a product.

activity series   A listing of elemental substances in the order of their ability to be oxidized or reduced. This listing makes it possible to predict whether an element will cause the oxidation or the reduction of an ion of another element.

addition reaction   The reaction in which a double bond opens to form two additional single bonds.

adhesive force   The attractive force between two dissimilar substances.

alcohol   An organic compound with an —OH group.

aldehyde   An organic compound with a terminal —CHO group.

alkali metals   The extremely reactive elements in the first group (column) of the periodic table. They all have ns^1 electrons as valence electrons.

alkaline earth metals   The very reactive elements in the second group (column) of the periodic table. They all have ns^2 electrons as valence electrons.

alkanes   Organic compounds with the general formula C_nH_{2n+2}.

alkenes   Organic compounds with double bonds in their structures.

alkyl group   A functional group that is alkane in nature.

alkynes   Organic compounds with triple bonds in their structures.

allotrope(s)   One or more distinct forms of an element; classification as an allotope is based on structure and physical properties. For example, diamond and graphite are two allotropes of carbon.

alpha particle   A helium nucleus.

amines   Compounds, related to ammonia, in which one or more of the hydrogen atoms in ammonia have been replaced by organic functional groups.

amino acid   An organic acid that contains both an amine and an acid functional group on adjacent carbon atoms.

amorphous   A term meaning "without structure."

amphiprotic (amphoteric)   A term designating a substance that can act as both a conjugate acid and a conjugate base.

amphoteric   *See* **amphiprotic**.

anhydride   The oxide of a metal or nonmetal that reacts with water to form an acid or a base, respectively.

anion   An ion with a negative charge.

aqueous   A term designating a system that involves water or a chemical mixture or solution having water as the solvent.

Arrhenius equation   The equation that relates temperature to rate constant. $K = Ae^{-Ea/RT}$.

Arrhenius theory   The theory that an acid increases hydrogen ion concentration when dissolved and that a base increases hydroxide ion concentration when dissolved.

aryl group   A functional group that is aromatic in nature.

atom   A fundamental particle of chemistry. At present, 109 atoms are known and are arranged in an orderly manner in the periodic table.

atomic mass, A   The relative mass of an element as compared to the mass of the isotope C-12, which is defined as exactly 12.

atomic number, Z   The number of an element in the periodic table; also, a number representing the number of protons in the nucleus of an atom.

atomic orbital   The orbital structure of an element; also, an orbital within an element.

atomic symbol   A one- or two-letter abbreviation of an element's name. Some symbols (e.g., Pb for lead) are derived from Latin names of the elements.

autopyrolysis constant of water, K_w   The value of the equilibrium law for the dissociation of water into H^+ and OH^-.

Avogadro's number   A quantity equal to 6.02×10^{23}.

Avogadro's principle   A statement of the direct relationship between the moles of a gas and the volume of that gas.

axial atom   A term used to describe the position of an atom in a covalent molecule of the AX_5 or AX_6 basic structure. The axial atoms are on the vertical axis of the molecule in positions similar to the Earth's North and South Poles.

azimuthal quantum number, ℓ   The quantum number that specifies the sublevel in which an electron is located; ℓ may be any number from zero up to $n - 1$.

B

balanced reaction   A chemical equation that has the smallest whole-number coefficients for the reactants and products so that there is the same number of atoms of each element on both sides of the arrow.

barometer   A closed-end manometer used for measuring atmospheric pressure.

base   Any compound that increases the hydroxide concentration of a solution or is a proton acceptor. Lewis bases are electron-pair donors.

base anhydride   The oxide of a metal that forms a base when dissolved in water.

base dissociation constant, K_b   The value of the equilibrium law for the dissociation of a weak base.

basic structure   One of five basic geometrics—linear, triangular planar, tetrahedral, trigonal planar, or octahedral—that a molecule may take.

battery A galvanic cell used to produce electricity for consumer items such as flashlights, portable radios, and heart pacemakers.

Beer's law The statements that the absorbance of a sample is the product of the absorptivity, optical path length, and sample concentration; $A = abc$.

beta particle, β An electron.

bidentate ligand A Lewis base that donates two pairs of electrons.

binary acid An acid that contains hydrogen and one other element in its formula.

body-centered cubic (bcc) A cubic structure in which one atom is at each of the eight corners and one atom is in the center of the unit cell.

Bohr atom The model of the atom developed by Niels Bohr. This model views electrons as circling the nucleus like a miniature solar system.

boiling point, normal The temperature at which the vapor pressure of a liquid is equal to 760 millimeters of mercury (1.00 atmospheres); also, the temperature at which a gas condenses. Also called condensation point.

boiling-point-elevation constant, k_b The temperature increase of the boiling point per molal noun of solute particles.

bonding electron pair A pair of electrons that participate in a covalent bond.

bond order The average number of bonds per atom covalently bonded to a central atom.

Boyle's law The law that expresses the inverse relationship between the volume and the pressure of a gas; $PV = $ constant.

Bragg equation The equation that relates the atomic dimensions in a crystal to the angles at which monochromatic X rays will undergo constructive reinforcement.

Brönsted-Lowry theory The theory that acids are proton donors and bases are proton acceptors.

buckminsterfullerene The allotrope of carbon that has the formula C_{60}.

buffer capacity The moles of strong acid or strong base required to change the pH of a liter of buffer by 1 pH unit.

buffer solution An aqueous solution containing a conjugate acid and its conjugate base in a molar ratio greater than 0.1 and less than 10.0.

bumping The violent boiling that occurs when a solution becomes superheated.

buret A tube approximately 1 centimeter in diameter that is used for measuring liquid volumes of 10–100 milliliters.

C

calorimeter An instrument used to determine heat energy.

catalyst A substance that speeds up the rate of a chemical reaction by providing an alternative reaction pathway with a lower activation energy.

cation An ion with a positive charge.

cell voltage, E The voltage of a galvanic cell under nonstandard state conditions.

centrifugation The process of separating a solid from a liquid by spinning it rapidly to artificially increase the gravitational force.

chain reaction A nuclear reaction that produces more neutrons than were needed to initiate it, therefore causing more reactions than occurred in the preceding step.

Charles's law The law that expresses the direct relationship between the temperature and volume of a gas: $P/V = $ constant.

chelate A Lewis base that usually has more than one pair of electrons to donate.

***cis* isomer** An isomer with substituents on the same side of the double bond.

Clausius-Clapeyron equation The equation that relates vapor pressure to heat of vaporization.

closed system A system in which mass cannot be lost to or gained from the surroundings.

coefficient   A number placed in front of a chemical formula to represent the number of molecules of that substance that are included in the equation. This number multiplies the number of atoms in the formula unit.

cohesive force   The sum of all the attractive forces in a pure substance.

colligative property   Any one of several physical properties of a solution that change depending on the amount of solute particles present in the solution.

collision theory   The theory of kinetics, which relates reaction rates to the frequency, energy, and orientation of molecules in collisions.

complex   *See* **complex ion.**

complexation reaction   A reaction between a Lewis acid and a Lewis base.

complexing agent   *See* **Lewis base.**

complex ion   A combination of one or more compounds or anions with a metal ion by coordinate covalent bonding. Also called a complex.

compound   A combination of two or more elements into a distinct substance with definite physical properties.

concentrated   A qualitative term indicating a large amount of solute in a given amount of solvent.

concentration   An expression of the amount of solute mixed with a solvent.

concentration vs time curve   A graph of reactant or product concentration as a function of time.

condensation   The conversion of a gas into a liquid.

condensation point   *See* **boiling point.**

condensation reaction   A polymerization reaction between an acid and either an alcohol or an amine.

conjugate acid   Any substance that has a proton that may be donated to a base.

conjugate acid-base pair   A pair of substances whose formulas differ by one H^+ ion.

conjugated double bonds   A series of two or more double bonds, each separated by only a single bond in a molecule.

cooling curve   A graph showing the changes that occur while a substance is cooling.

coordinate covalent bond   The covalent bond between two atoms that is formed when one substance donates both electrons.

covalent bond   The bond between two atoms that arises from the sharing of a pair of electrons.

covalent compound   A chemical compound in which the atoms are held together with covalent bonds.

covalent crystal   A crystal that consists of only one molecule. All atoms are joined to others with covalent bonds. Also called network crystal.

critical mass   The minimum mass of a fissile (fissionable) material needed to sustain a nuclear chain reaction.

critical point   The temperature and pressure above which a gas cannot be condensed to a liquid.

crystal lattice   The arrangement of atoms, ions, or molecules in a crystal structure.

crystallization point   *See* **melting point.**

cyclotron   A machine for accelerating charged particles, which are usually used to bombard targets in an effort to generate nuclear reactions.

D

Dalton's law of partial pressures   The law that the total pressure of a gas is the sum of the individual pressures of all the gases in the mixture; $P_{total} = p_1 + p_2 + \ldots$.

decay   The spontaneous emission of a particle in a radioactive event.

ΔE The energy change due to a chemical reaction. It is equal to the heat and work of the reaction or to the change in potential energy of the products as compared to that of the reactants.

$\Delta G°$ The standard free-energy change for a reaction. The temperature is 298 K unless otherwise indicated.

$\Delta G_f°$ The standard free-energy change of formation corresponding to the formation of 1 mole of product from its elements at 298 K.

ΔH The enthalpy change occurring in a chemical process. Without the super-script zero it indicates an extensive property. Often called the heat of reaction.

$\Delta H°$ The standard enthalpy change occurring in a reaction; refers to the heat produced or absorbed when the moles of reactants specified in the chemical reaction react at standard state.

$\Delta H_f°$ The standard heat of formation, which is the heat produced or absorbed when 1 mole of a product is formed.

ΔS The change in entropy between the final state and the initial state in a chemical process. Without the superscript zero it indicates an extensive property.

$\Delta S°$ The standard entropy change of a chemical process; an intensive property based on the moles of substance in the balanced chemical reaction.

dependent variable   The variable that an experiment measures; its value depends on the value of the independent variable.

derived structure   A molecular structure that is derived from one of the five basic structures.

detergent   A chemical substance that has both polar and nonpolar properties and is soluble in both polar and nonpolar solvents.

determinant error   An error associated with faulty instruments, calibrations, or techniques.

dextrorotatory   A term describing an optical isomer that rotates polarized light to the right.

diagonal relationship   A term indicating that two elements in periods 2A and 3A and two groups, x and $x + 1$, respectively, have similar properties; for example, Li and Mg or B and Si.

diagonal trend   A term describing the fact that some properties of atoms vary regularly from the lower left corner to the upper right corner of the periodic table. Electronegativity, ionization energy, and electron affinity are some of these diagonal relationships.

diamond   An allotrope of carbon in which all carbon atoms have sp^3 hybridization.

diatomic   A term describing a molecule that contains only two atoms (e.g., HCl, H_2).

differentiating electron   The electron that differentiates one element from an adjacent element in the periodic table.

diffraction   The combination of light waves that results in either constructive or destructive reinforcement.

diffusion   The movement of molecules from one place to another by random motion.

dilute   A qualitative term indicating a small amount of solute in a given amount of solvent.

dimer   A substance composed of two identical molecular or ionic units.

dipole   A polar molecule. The term *dipole* reminds us that a polar molecule has only two poles, one positive and one negative.

direct relationship   A relationship between two variables whereby one must increase when the other increases.

dispersion forces   *See* **London forces.**

dissociation   The breakup of an ionic compound into ions.

double-replacement reaction   A chemical equation in which the cation of one substance replaces the cation of a second substance. At the same time, the cation of the second substance replaces the cation of the first substance.

***dsp*³ hybrid orbital**   An orbital formed from one s, three p, and one d orbital. The electrons in these orbitals are all equal in energy. Structures are all related to the trigonal bipyramid.

d^2sp^3 hybrid orbital   An orbital formed from one *s*, three *p*, and two *d* orbitals. The electrons in the hybrid all have the same energy. Structures are all related to the octahedron.

ductile   A term describing the property of being able to be drawn into wire forms.

Dulong and Petit's law   The law stating that the specific heat of a metal multiplied by its molar mass is equal to a constant of approximately 25 J mol^{-1} °C^{-1}.

dynamic equilibrium   The state in which a chemical process is going in the forward direction at the same rate that it is going in the reverse direction and the concentrations of reactants and products remain constant. Equilibrium follows the kinetic period, in which reaction occurs and the concentrations of reactants and products change.

E

EDTA   Ethylenediamine tetraacetic acid, a very useful complexing agent.

effusion   The movement of molecules through a small hole from one container to another.

electrode   A metal placed in a liquid to transfer electrons in a galvanic or electrolytic cell.

electrolysis   The process of using electric current to reduce a chemical substance at the cathode and oxidize a chemical substance at the anode.

electrolyte   A substance that dissociates or ionizes completely into ions in solution.

electrolytic cell   An arrangement of electrodes and a power source used to force nonspontaneous redox reactions to occur.

electron affinity   The energy released or absorbed in adding an electron to an atom.

electron deficient   A term describing a Lewis structure that has fewer than an octet of electrons around one or more of its atoms, except hydrogen.

electron   1. The unit of negative charge in the atom. The diffuse electron cloud surrounds the dense nucleus. 2. One of the three particles, along with the proton and neutron, that make up an atom. An electron has a negative charge and virtually no mass in comparison to the neutron and proton. Electrons are arranged in an orderly fashion around the nucleus in a diffuse electron cloud. There are as many electrons as protons in an element.

electronic configuration   A listing of the electrons within an atom, based on the sublevels that are filled and the relative energies of these sublevels. For example, the electronic configuration for silicon is $1s^22s^22p^63s^23p^2$.

electronegativity   A measure of an atom's tendency to attract electrons. Fluorine has the highest, and francium the lowest, electronegativity in the periodic table.

electroneutrality, law of   A statement of the fact that no chemical compound has a net charge. In addition, an element has no net charge.

electrostatic potential energy   The energy of attraction of two oppositely charged particles, or the energy of repulsion of two like-charged particles.

element   Any one of the 117 distinct particles, known as atoms, that are currently known. Each has distinct chemical and physical properties.

elementary reaction   One reaction in a mechanism. It is usually bimolecular or unimolecular, and its coefficients are the exponents in the rate law.

empirical formula   The formula that gives only the simplest whole-number ratio of the atoms that make up a compound. *See also* **molecular formula; structural formula.**

endothermic   A term describing any process that absorbs heat from the surroundings. Endothermic processes cool the system.

end point   The point, where neither reactant is in excess, that marks the end of a titration experiment.

enthalpy   The heat content of a chemical substance. Enthalpy is related to the internal potential energy of the substance.

entropy   Entropy of a system is proportional to the number of microstates in that system. This is a more precise definition than the randomness of a chemical system.

enzyme One of many naturally occurring catalysts found in biological materials.

enzyme-substrate complex The activated complex formed in an enzyme-catalyzed reaction.

equatorial atom A term used to describe the position of an atom in a covalent molecule of the AX_5 or AX_6 basic structure. The equatorial atoms are around the center of the molecule in positions similar to the Earth's equator.

equilibrium constant, K The numerical value of the equilibrium law. The only variable that has an effect on the equilibrium constant is temperature.

equilibrium law The basic equation governing chemical equilibrium. Each balanced chemical equation has its own equilibrium law.

equilibrium table A table of data used to summarize the numerical data and stoichiometric relationships of an equilibrium system.

equivalent A value that is determined for a substance by dividing its mass by the equivalent weight.

equivalent weight The mass of a compound that loses or gains 1 mole of electrons in an oxidation-reduction reaction. In acid-base reactions it is the mass that furnishes or reacts with 1 mole of H^+ ions.

escape energy The minimum kinetic energy of a liquid molecule that is needed for transformation into a gas.

ether An organic compound containing the —C—O—C— functional group.

eudiometer A closed-end manometer.

evaporation The transformation of a liquid into a gas at a temperature below the boiling point.

exact number A number that has no uncertainty. Exact numbers include stoichiometric coefficients, formula subscripts, and most defined quantities.

excess reactant Any reactant that is not completely consumed in a chemical reaction.

exothermic A term describing any process that gives off heat to the surroundings. Exothermic processes heat the system.

extrinsic property A physical or chemical property that varies in proportion to the amount of matter.

F

face-centered cubic (fcc) A cubic structure in which atoms are at the corners and an atom is on each cube face of the unit cell.

factor label A ratio used to convert a number with one set of units into the equivalent number with different units.

Faraday's constant, $\mathscr{F}$ The relationship between the coulomb and moles of electrons; 96,485 C = 1 mol e^-.

filtrate The liquid that passes through a filter.

first law of thermodynamics The law that states that in any chemical or physical process all energy is conserved.

first-order reaction A reaction with a rate law having exponents that add up to 1. The rate law is Rate = $k[A]$.

fissile A term describing a nucleus that is capable of undergoing nuclear fission.

fluorescence A property of some atoms and molecules that allows them to absorb photons of light and reemit them very rapidly, but with a different energy. The emitted light always has a lower energy and longer wavelength than the absorbed light.

formal charge The charge on an atom in a covalent compound, calculated by assuring that all bonding electrons are equally shared.

formation reaction A chemical reaction in which the reactants are elements at standard state and the product is 1 mole of one compound.

formula   The representation of a chemical substance using chemical symbols and appropriate subscripts for the numbers of atoms and superscripts to represent charges if the substance is an ion.

free radical   A molecule that contains an unpaired electron in its Lewis structure.

freezing-point depression constant, k_f   The temperature decrease in the freezing point per molal °C m^{-1} of solute particles.

functional group   A group of atoms on an organic compound that represent a characteristic chemical entity.

G

galvanic cell   The experimental setup used to convert chemical energy into electric energy. All batteries are galvanic cells.

gamma ray   A high-energy photon often emitted in a nuclear reaction.

gas   A state of matter characterized by the ability to flow and fill any container completely without regard to the amount of gas in the container.

Gay-Lussac's law   The law that expresses the direct relationship between the temperature and pressure of a gas; P/T = constant.

Gibbs free energy   The maximum energy from any chemical reaction.

Graham's law of effusion   The law that relates the rate at which gases pass through a small hole to the mass of the molecule; $\sqrt{m_1/m_2} = \bar{v}_2/\bar{v}_1$.

gram-atomic mass   *See* **atomic mass.**

gram-molar mass   *See* **molar mass.**

graph   A pictorial method of presenting and evaluating experimental data.

graphite   An allotrope of carbon in which carbon has sp^2 hybridization.

gravitational potential energy   The attractive energy of any two masses toward each other.

group   A column in the periodic table.

H

half-life   The time required for half of the reactants to be consumed in a first- order chemical reaction or a radioactive decay.

halide   An organic compound with a halogen (e.g., $CH_3 CH_2 Cl$) in its structure.

halogen   An element in the next to last group of the periodic table. Halogens are reactive elements with ns^2, np^5 valence electron structures.

halogenation   The addition of a halogen to a double or triple bond.

heat capacity   The amount of heat energy that a system needs in order for its temperature to change by 1 °C.

heating curve   A graph showing the changes that occur while a substance is heated.

heat of combustion   The heat released when 1 mole of sample, usually an organic compound, is completely burned in oxygen to form CO_2 and H_2O.

heat of fusion   The heat energy needed to convert a solid into a liquid. The units are either joules per gram or joules per mole.

heat of reaction   See ΔH.

heat of vaporization   The heat energy needed to convert a liquid into a gas. The units are either joules per gram or joules per mole.

Henderson-Hasselbach equation   A derivation of the equilibrium law obtained by taking the negative logarithm of the equilibrium law; pH = pK_a + log([conjugate base]/[conjugate acid]).

Henry's law   The law that expresses the relationship between the solubility of a gas and its partial pressure; $c = kp$.

Hess's law   The law that states that heats of reaction are additive when chemical reactions are added.

Hund's rule   The rule that every orbital in a sublevel must fill with one electron before a second electron of opposite spin can be added to any orbital in that sublevel.

hybrid orbital   An orbital constructed by combining electrons, usually froms s and p orbitals, into a new orbital where all the electrons have the same properties. These orbitals are designated as sp, sp^2, sp^3, dsp^3, and d^2sp^3.

hydrate   A substance that contains a fixed number of water molecules. The water molecules are written separately from the formula itself and connected to it with a dot in the center of the line between the chemical formula and the water molecules (e.g., $CuSO_4 \cdot 5\ H_2O$).

hydrogenation   The addition of H_2 to a double or triple bond.

hydrogen bonding   An extra-strength dipole-dipole attractive force due to a large electronegativity difference between hydrogen and nitrogen, oxygen, or fluorine.

hydrohalogenation   The addition of a hydrogen and a halogen to a double or triple bond by using a binary acid such as HF, HCl, HBr, or HI.

hydrolysis reaction   The reaction of a substance, usually a conjugate acid or base, with water.

I

ideal gas   A gas that obeys the ideal gas law; conceptually, a gas molecule with no volume and no attractive forces with other molecules.

ideal gas law   The law that relates temperature, pressure, volume, and moles of gas; $PV = nRT$.

independent variable   The variable in an experiment that is under the control of the experimenter.

indeterminate error   An error in estimating the uncertain digit in a measurement. Also called random error.

indicator   A chemical added to a titration experiment that changes color at the end point.

indicator electrode   An electrode placed in a sample in order to measure the concentration of an ion in the sample.

induced dipoles   A dipole formed by the interaction of a nonpolar substance and either a polar substance or an instantaneous dipole.

initial conditions   A quantitative description of a chemical system at the start of a reaction.

inspection method   A method for balancing chemical equations. The inspection involves counting the number of each atom present in the equation and then balancing by adding appropriate coefficients to the reactants or products.

instantaneous dipole   A distortion of the electron cloud around an atom or molecule that gives the atom or molecule momentary polarity.

intermediate   A substance that appears in the elementary reactions of a mechanism but not in the overall balanced equation.

intermolecular forces   The attractive forces—dipole-dipole attractions, London forces, and hydrogen bonding—between molecules and atoms that allows them to condense into liquids and solidify into solids.

internuclear axis   An imaginary line connecting the nuclei of two atoms.

intrinsic property   A physical or chemical property that does not change with the amount of matter.

inverse relationship   A relationship between two variables whereby one must increase if the other decreases.

ion   An element that has lost or gained one or more electrons. *See also* **polyatomic ion.**

ion-electron method   A method for balancing more complex oxidation-reduction equations. It involves a logical sequence of steps described in Chapter 13.

ionic bond   The attraction of a negative anion for a positive cation.

ionic compound   A chemical compound composed of negatively charged anions and positively charged cations. The unit is held together by the attraction of the positive charges toward the negative charges.

ionic crystal   A crystal formed from cations and anions where the main attractive force is the attraction of positive charges toward negative charges.

ionization   The removal or addition of an electron from an atom or molecule. Also the formation of ions when a molecular substance dissolves in water.

ionization energy   The energy required to remove an electron completely from an atom.

isoelectronic   A term describing any two atoms that have identical electronic configurations. These atoms may be ions or elements.

isolated system   A system in which neither mass nor energy is transferred to or from the surroundings.

isomers   Distinctly different compounds that have the same elemental composition.

isotope   A form of an element with a specified number of protons, neutrons, and electrons.

IUPAC   International Union of Pure and Applied Chemistry, which sets nomenclature standards.

K

K The symbol for the equilibrium constant, often written as K_{eq}.

K_a The acid dissociation constant, a special term denoting the equilibrium of a weak acid. A weak acid dissociation is always in the form

$$HA + H_2O \rightleftharpoons A^- + H_3O^+$$

K_b The base dissociation constant, a special term denoting the equilibrium of a weak base. A weak base dissociation is always in the form

$$B + H_2O \rightleftharpoons BH^+ + OH^-$$

K_c The equilibrium constant used when the reactants and products are specified as concentrations.

K_d The dissociation constant, a special term used mainly to describe the dissociation of complex ions.

K_{eq} See K.

K_f The formation constant, a special term used to describe the formation of complex ions.

K_p The equilibrium constant used when the reactants and products are specified in terms of partial pressures.

K_{sp} The solubility product, a special term denoting the fact that the equilibrium is between a solid and its solution products.

K_w The autopyrolysis constant of water, equal to 1.0×10^{-14} at 25 °C.

ketone   An organic compound with a nonterminal —C=O group.

kinetic curve   A graph of reactant or product concentration as a function of time.

kinetic energy   The energy that matter possesses because of its motion; $KE = \frac{1}{2}mv^2$.

kinetic molecular theory   The theory of the motion of molecules in the gas phase, which explains gas pressure, effusion and diffusion rates, and the effect of temperature on the behavior of gases.

L

leveling effect   An expression of the fact that the strongest acid in water is the H^+ (H_3O^+) ion and the strongest base is the hydroxide ion, OH^-.

levorotatory   A term describing an optical isomer that rotates polarized light to the left.

Lewis acid   Any substance that can accept electron pairs.

Lewis base   Any substance that can donate electron pairs. Also called ligand; complexing agent; sequestering agent.

Lewis structure   A molecular structure based on the concept that all atoms try to achieve the noble gas electronic configuration by sharing electrons.

Lewis theory   The theory that acids are electron-pair acceptors and bases are electron-pair donors.

ligand   *See* **Lewis base.**

limiting reactant   The reactant that is completely consumed in a reaction, causing it to stop. Also called limiting reagent.

limiting reagent   *See* **limiting reactant.**

linear   A term referring to atoms aligned in a straight line; a three-atom arrangement with a $180°$ bond angle.

liquid   A state of matter characterized by its ability to flow in order to fill any container from the bottom up.

litmus paper   A type of pH paper using the indicator litmus, which is pink in acid and blue in base.

lock and key   The analogy used to depict how an enzyme recognizes reactants on the basis of physical geometry as well as chemical characteristics.

London forces   The attractive forces from instantaneous dipoles. These forces are due to the possibility that the electron clouds around atoms and molecules are not perfectly symmetrical at all times. Also called dispersion forces.

lone pairs   Electron pairs in Lewis structures that are not used for bonding.

M

magnetic quantum number, m_ℓ   The quantum number that specifies the orbital in which an electron is located and the orientation of the orbital in space; m_ℓ may be any number from -1 to $+1$, including zero.

malleable   A term describing the property of being able to be hammered into new shapes.

manometer   A device used for measuring gas pressures.

mass   A quantity of matter.

mass fraction (wt/wt)   A concentration unit defined as the mass of one solute divided by the total mass of the solution.

mass-volume fraction (wt/vol)   A concentration unit defined as the mass of one solute in a given total volume of solution.

melting point, normal   The temperature at which a solid melts at 1.00 atmosphere of pressure; also, the temperature at which a liquid becomes a solid. Also called crystallization point.

meniscus   The curved surface of a liquid in a tube or container.

metal   A substance with characteristic properties of high electrical conductivity, malleability, and a metallic silver or yellow luster.

metallic crystal   A crystal formed from a metal in the periodic table. Metallic crystals are malleable and ductile and conduct electricity.

metalloid   An element that has properties of both metals and nonmetals.

metastable   A term describing a physical situation in which a material is stable unless disturbed.

metric base unit   One of seven basic units of measurement in the metric system. More complex units are combinations of base units.

metric prefix   A prefix (e.g., *milli-*, *pico-*) used with a metric base unit to represent a specific exponential value.

Michaelis-Menton equation   The rate equation that applies to enzyme-catalyzed reactions.

mirror image   A description of stereoisomers in which the structure of one isomer is the reflection of the other in a mirror.

molality (*m*)   A concentration unit defined as the number of moles of solute dissolved in 1 kilogram of solvent.

molarity (*M*)   A concentration unit defined as the number of moles of solute in 1 liter of solution.

molar mass   The sum of the gram-atomic masses of all atoms in a chemical formula. Also called molecular mass.

molar volume   The volume of 1 mole of gas, usually at standard temperature and pressure (STP).

mole (mol)   The quantity of any substance that contains 6.02×10^{23} units of that substance.

mole fraction (χ)   A concentration unit defined as the number of moles of solute divided by the total number of moles in a solution.

molecular crystal   A crystal formed from a molecule. The attractive forces that hold molecular crystals together are London forces, dipole-dipole attractions, hydrogen bonding, or a combination of these.

molecular formula   The formula for a molecular or covalent substance, showing all of the atoms that comprise the molecular unit. A molecular formula may be simplified to an empirical formula if all of the subscripts can be divided by a common number. *See also* **empirical formula; structural formula.**

molecular mass   *See* **molar mass.**

molecular orbital   An orbital created by the pairing of electrons from different atoms. This orbital encircles the atoms that are bonded together.

molecule   A group of atoms bound together by covalent bonds with zero total charge.

molten salt   A solid salt that has been heated to a temperature where it becomes a liquid. Also called a fused salt.

monochromatic   A term describing light that has a single wavelength.

monodentate ligand   A Lewis base that donates one pair of electrons.

monomer   One of the individual repeating units of a polymer.

N

natural abundance   The percentage of an isotope of an element found in nature.

network crystals   *See* **covalent crystal.**

neutralization reactions   A chemical reaction of an acid with a base.

neutron   One of three particles, along with the electron and proton, that make up an atom. This particle has no charge, but has a mass approximately equal to the proton mass. Neutrons and protons make up the bulk of the mass of the atom.

noble gas   An element in the last group in the periodic table. Noble gases are unusually stable elements and all have ns^2, np^6 valence electrons.

nonbending electron pair   A pair of electrons in a Lewis structure that is not shared with any other atoms.

nonelectrolyte   A substance that does not dissociate at all in solution.

nonmetal   An element that is not metallic. Nonmetals do not conduct electricity well and do not have a shiny metallic luster. They are located in the upper right portion of the periodic table.

nonpolar   A term describing a bond or molecule that has its charge distributed evenly. Only diatomic elements form truly nonpolar bonds. Symmetrical molecules are nonpolar.

normal   A term describing an organic compound in which all carbon atoms are arranged in one straight chain.

normality   The number of equivalents of a substance dissolved in 1 liter of solution.

nuclear charge   The number of positive charges in the nucleus. This is the same as the number of protons in the nucleus (Z) and is also the atomic number.

nuclear fission   A radioactive decay process initiated by the absorption of a neutron; it results in a large nucleus dividing roughly in half.

nuclear fusion   The combination of two nuclei to form a new atom.

nuclear mass   The total mass of the nucleus. This is the sum of the masses of the protons and neutrons in the nucleus. Since electrons have virtually no mass, it is also the atomic mass (A) of the isotope.

nuclear reactor   A device that uses a nuclear reaction to create heat energy for the purpose of generating electricity.

nucleon   Either a proton or a neutron, both of which are fundamental particles of the nucleus.

nucleus   The center of an atom, which contains the protons and neutrons. The nucleus is extremely dense and comprises a very small fraction of the atom's volume; the rest of the atom is empty space.

O

octahedron   A geometric structure of six atoms covalently bound to a central atom. Each atom is 90° from any other.

octet rule   A simple but effective rule stating that covalent molecules tend to have octets of electrons around each atom in their structures. These octets simulate the electron configurations of the noble gases.

open system   A system in which matter and energy can be exchanged with the surroundings.

optical isomer   A stereoisomer that rotates polarized light.

optical path length, b   The thickness of a sample in a spectroscopic experiment.

orbital   A region of space that may be occupied by a maximum of two electrons. The shape of an orbital is defined by the sublevel it is in. The orientation of the orbital depends on its assigned quantum number, m_ℓ. Every orbital in a given sublevel must be filled with one electron before a second electron may fill the orbital.

orbital diagram   A diagram in which boxes represent individual valence orbitals. Electrons are represented as arrows to show the spins of the electrons in each orbital.

order of reaction   The sum of the exponents in a rate law.

organic acid   An acid containing carbon and the —COOH functional group.

organic compound   A compound composed of carbon and usually hydrogen.

osmotic pressure   The pressure needed to stop the migration of solute through a semipermeable membrane.

oxidation   The loss of electrons; also, the increase in oxidation number.

oxidizing agent   A substance that causes another to be oxidized; also, a substance that itself is reduced.

oxoacids   An acid that contains hydrogen, oxygen, and another element in its formula, excluding organic acids.

P

partial pressure   The pressure of a single gas in a mixture of gases.

parts per billion (ppb)   A unit of measurement similar in concept to percent, obtained by multiplying a fraction by one billion (10^9).

parts per million (ppm) A unit of measurement similar in concept to percent, obtained by multiplying a fraction by one million (10^6).

Pauli exclusion principle The requirement that no two electrons in an atom have the same set of four quantum numbers, n, ℓ, m_ℓ, and m_s.

percent (%) A unit of measurement meaning parts per hundred, obtained by multiplying a fraction by 100.

period A row in the periodic table.

periodic table The table in which the elements are arranged in an orderly fashion that shows the relationships of their chemical and physical characteristics.

pH The negative logarithm of the hydrogen ion concentration in a solution; pH = –log H^+.

phase diagram A graph showing the relationship between temperature and pressure and the conversion of matter among three states: solid, liquid, and gas.

pH indicator A weak acid or a weak base whose conjugate acid and conjugate base have different colors. An indicator changes color indicating the end point of a titration.

pH meter An electronic device used to measure the pH values of solutions.

pH paper Paper with a pH indicator absorbed on it so that it changes color depending on the pH of the solution; pH paper is used to estimate pH. *See also* **litmus paper.**

phosphorescence A property of some atoms and molecules that allows them to absorb photons of light and reemit them seconds to hours later. The emitted light always has a lower energy, longer wavelength than the absorbed light.

pi bond A bond made from the sideways overlap of two p orbitals. The electron density of a pi bond lies outside the internuclear axis.

pipet A narrow tube calibrated for precise measurement of small volumes of liquids.

pK_w The negative logarithm of the autopyrolysis constant of water, equal to 14.00.

planar triangle A geometric structure of four atoms, three bonded to a central atom with 120° angles between the atoms, which are all in the same plane.

Planck's constant (h) The constant relating the energy of a photon to its frequency.

pneumatic trough An experimental setup for collecting gases by the displacement of water.

pOH The negative logarithm of the hydroxide ion concentration in a solution; pOH= – log OH^-.

polar A term describing the property of a covalent bond or molecule of having one end more positive than the other.

polarizability The tendency for an atom's electron cloud to be deformed so that polarity is created.

polyatomic ion An ion composed of more than one atom covalently bonded together. A polyatomic ion acts as a unit in most chemical reactions.

polymer A long-chain organic molecule with repeating units.

polyprotic acid An acid with two or more ionizable hydrogen atoms in its formula.

positron A positive electron.

potential energy The energy of matter that may, under appropriate conditions, be converted into work.

precipitate 1. (v.) To cause the formation of a solid by a chemical reaction. 2. (n.) The solid formed as a result of a chemical reaction.

precision The degree of closeness of a group of repeated measurements to each other.

pressure The force per unit area; gas molecules exert this force by collisions with the container walls.

principal quantum number, n The quantum number that specifies the energy level of the atom in which an electron is located; n may be any integer from 1 to infinity.

product The result of a chemical reaction. Products are placed on the right-hand side of a chemical equation.

proton One of three particles, along with the electron and neutron, that make up an atom. The proton has a positive charge, equal in magnitude (but with the opposite sign) to the charge of the electron. The number of protons is equal to the atomic number, Z, of an element. Protons and neutrons make up the bulk of the mass of an atom.

Q

Q *See* **reaction quotient.**

qualitative A term referring to a description of a physical or chemical property without the use of numbers or equations.

qualitative analysis A logical sequence of experiments and observations used to determine the composition of a sample.

quantitative A term referring to a description of a physical or chemical property using numbers or equations.

quantum number One of four numbers (n, ℓ, m_ℓ, m_s) used in the wave- mechanical model of the atom to describe an electron in an atom.

R

racemic mixture An equal molar mixture of L and D optical isomers.

radioactive disintegration series A sequence of radioactive disintegrations from a heavy isotope to a lighter, stable isotope in more than one step.

radioactivity The property that some unstable nuclei have of decaying spontaneously with the emission of a small particle and/or energy.

radioisotope A radioactive isotope of an element.

random error *See* **indeterminate error.**

randomness A qualitative description of the disorder of the molecules in any sample.

Raoult's law The law that expresses the relationship between the vapor pressure of a solution and the mole fraction of solute in that solution.

rate constant, *k* A constant in the direct relationship between the amount of reacting substance and the rate of the reaction.

rate-determining step The slowest reaction in a mechanism, which limits the overall rate of reaction. Also called rate-limiting step.

rate law The mathematical relationship between reactant concentrations and reaction rate. The general form of a rate law is Rate = $k[A]^x[B]^y[C]^z$.

rate-limiting step *See* **rate-determining step.**

reactant One of the starting materials in a chemical reaction. The reactants are placed on the left side of a chemical equation.

reaction mechanism The detailed series of elementary reactions that add up to the overall reaction. *See also* **elementary reaction.**

reaction profile A plot of the potential energy of molecules as they collide in a reaction, illustrating the nature of the activation energy.

reaction quotient, *Q* The value of the ratio of the equilibrium law when a chemical system is not in a state of equilibrium. The value of Q in comparison to that of K indicates the direction of the reaction.

reaction rate The velocity, in moles per liter per second, at which reactants are converted into products in a chemical reaction.

reagent blank A solution used to set the zero point of a spectrophotometer.

real gas A gas in which there are attractive forces between the molecules and the molecules have a finite volume.

redox A word coined to combine the terms *reduction* and *oxidation*. It indicates that reduction and oxidation always occur together.

reducing agent A substance that causes another substance to be reduced; also, a substance that is oxidized.

reduction The gain of electrons; also, the decrease in oxidation number.

reference electrode An electrode in a galvanic cell, which has a constant potential since the concentrations of all reactants are kept constant.

relative mass A term describing the fact that the masses of atoms in the periodic table are relative, without units, as compared to the mass of the carbon-12 isotope.

relative uncertainty The absolute uncertainty of a measurement divided by the value of the measurement.

resonance structure A Lewis structure that can be drawn in more than one equally probable way. The actual structure is a mixture of all possible resonance structures.

reversible process A chemical or physical process that can be changed from one state to another and then back to the original state. A reversible process takes place in infinitesimally small steps.

S

$S°$ The standard entropy of 1 mole of a substance.

saturated A term describing an organic compound in which all carbon-carbon bonds are single bonds. Saturated compounds have the maximum number of hydrogen atoms; that is, they are saturated with hydrogen.

saturated solution A solution that has the maximum amount of solute dissolved in it.

scientific notation A method of writing numbers in which the significant figures are numbers from 1 to 10 and they are multiplied by 10 raised to the appropriate power to indicate the position of the decimal point.

second law of thermodynamics The law that states that in all physical and chemical processes the overall entropy of the universe must increase.

second-order reaction A reaction with a rate law having exponents that add up to 2. The rate law is Rate = $k[A][B]$ or Rate = $[A]^2$.

semipermeable membrane A thin, solid material through which certain molecules can diffuse while others cannot; may be visualized as a barrier with small holes that allow only molecules of a certain size to pass through.

sequestering agent *See* **Lewis base.**

shell The old term for the principal energy level of an electron; the region in space where electrons are located around the nucleus of the atom. Energy levels are numbered starting with the energy level closest to the nucleus. The number of the principal energy level is also known as the principal quantum number.

sigma bond A covalent bond formed by the sharing of a pair of electrons. The electron pair is located along the internuclear axis between the two atoms that share it.

significant figures All the digits in a measurement except preceding zeros.

simple cubic A term describing a cubic structure with one atom in each of the eight corners of a unit cell.

solid A state of matter characterized by a rigid structure that retains its shape without a container.

solubility A property of a solute that refers to the maximum amount of that solute that can be dissolved in a solvent. This term can be a qualitative or quantitative description of a solute.

solute The substance—gas, liquid, or solid—dissolved in the solvent.

solution A uniform mixture of chemicals. In a solution it is impossible to distinguish separate solute and solvent particles.

solvent   Typically, the liquid phase in which a gas, another liquid, or a solid is dissolved. In a mixture of two or more liquids the solvent is the liquid present in the largest amount.

sp **hybrid orbital**   An orbital constructed from an *s* and a *p* orbital into one where both have equal energies. The resulting structures related to this orbital are linear.

sp^2 **hybrid orbital**   An orbital constructed from an *s* and two *p* orbitals. The resulting orbitals all have the same energy. Structures related to this orbital are triangular planar.

sp^3 **hybrid orbital**   An orbital constructed from an *s* and three *p* orbitals. The resulting orbitals all have the same energy. Structures related to this orbital are tetrahedral.

specific heat   An intrinsic property of matter that describes the quantity of heat energy needed to raise the temperature of 1 gram of substance by 1 °C.

spectrophotometer   An instrument for determining the amount of light absorbed by a sample.

spin quantum number, m_s The quantum number that specifies the spin of an electron as either $+\frac{1}{2}$ or $-\frac{1}{2}$. Two electrons in the same orbital must have opposite spins.

spontaneous reaction   Any reaction that occurs without outside assistance; quantitatively, any reaction that has an equilibrium constant greater than 1.

standard cell voltage, E°_{cell} The voltage of a galvanic cell when the system is at standard state; also, the combination of two standard reduction potentials as $E^\circ_{cell} = E^\circ_{cathode} - E^\circ_{anode}$.

standard reduction potential, E The potential (voltage) of a reduction half- reaction at standard state.

standard state   Defined temperature, pressure, and concentrations. In electrochemistry the standard state is 1.00 atmosphere pressure, 298 K or 25 °C, and 1.00 molar concentrations for all soluble compounds. Solids and pure liquids are also defined as 1.00 molar.

standard temperature and pressure   A state defined as having a temperature of 0 °C and 1 atmosphere of pressure.

state function   A variable whose value depends only on the initial and final states of the system. State functions are ΔH, ΔS, ΔG, and ΔE.

steady-state assumption   The assumption that, in evaluating rate constants for elementary reactions, the concentrations of intermediates may be mathematically eliminated by assuming that all prior fast steps are in chemical equilibrium.

stereoisomers   Compounds that have the same formula and same bonding but differ in the geometric arrangement of the atoms.

stereospecific   A term describing a chemical reaction that produces only one stereoisomer.

stoichiometry   Mathematical relationships between chemical substances in a chemical equation.

stopcock   The valve on the end of a buret.

STP   *See* **standard temperature and pressure.**

strong acid   An acid that dissociates completely when dissolved in water.

strong base   A base that dissociates completely when dissolved in water.

structural formula   A formula that shows the actual arrangement of atoms within the molecule and the bonds between the atoms. *See also* **empirical formula; molecular formula.**

structural isomers   Compounds with the same formula but with the atoms bonded in different arrangements.

sublevel   A subdivision of an energy level. Electrons in each principal energy level are localized in sublevels. Each sublevel has a distinct shape associated with it. Sublevels are numbered from zero up to one less than the number of the principal energy level. These sublevel numbers are the azimuthal quantum numbers, ℓ. Sublevels are also designated by the letters *s, p, d, f.*

subscript   A number placed to the right of, and slightly below, the symbol for an element to represent the number of times that atom is present in the formula unit.

substrate(s)   The reactant(s) in an enzyme-catalyzed reaction.

supercritical fluid   A gas at a temperature and pressure above the critical point. Such a fluid has properties of both a gas and a liquid.

supersaturated solution   A metastable solution that has more than the maximum amount of solute dissolved in it.

supercooling   The property of some materials capable of being cooled to temperatures below their melting points without solidifying. Supercooled solutions are also supersaturated and are metastable.

supernatant   The liquid remaining above a solid after centrifugation.

surface tension   The added attractive force per molecule at the surface of a liquid. Surface tension causes liquids to assume shapes that minimize surface area.

surfactant   A substance that lowers the surface tension of liquids.

surroundings   All parts of the universe not included in the system being studied.

symmetrical   A term describing a geometrical property whereby a structure may be rotated by some angle less than 360° and after rotation the molecule has the same configuration as before.

system   The portion of the universe that is under study.

T

TC ("to contain")   A label on glassware indicating that the item is calibrated to contain the indicated volume.

TD ("to deliver")   A label on glassware indicating that the calibration is based on the volume delivered.

Teflon   The addition polymer of $CF_2 = CF_2$ with extraordinary nonstick properties.

tetrahedron   A geometric structure with four atoms bound to a central atom by covalent bonds. Each bond is equidistant from any other with a bond angle of 109°.

thermodynamics   The study of energy changes in chemical and physical processes.

titration   An experimental procedure for reacting two solutions in order to determine the quantity or concentration of one of the solutions.

titration curve   A plot of pH versus the volume of titrant added to a sample.

tracer   A radioactive element used to detect the movement of materials in a complex system.

trans **isomer**   An isomer with substituents on opposite sides of a double bond.

transition element   An element having a *d* electron as the differentiating electron in its electronic configuration.

transition-state theory   The reaction-rate theory that details the events and energy changes that occur as two molecules collide.

transuranium element   Any of the 17 elements from atomic number 93 to 109.

triangular bipyramid   A geometric structure with five atoms covalently bound to a central atom. Three atoms in the equatorial position are 120° from each other. Two additional atoms in the axial positions are 90° from the equatorial atoms.

triple point   The temperature and pressure at which all three states of matter—solid, liquid, and gas—are in equilibrium.

U

unit cell   The fundamental building block of crystals. An entire crystal is formed by repetitive stacking of the unit cells.

universal gas constant, *R*   The constant needed to relate the temperature, pressure, volume, and moles of gas in the ideal gas law, $PV = nRT$.

universe   The entirety of all matter and space that exist.

unsaturated A term describing an organic compound that contains one or more double or triple bonds in its structure.

unsaturated solution A solution in which the solute concentration is less than the maximum amount possible.

V

vacuum distillation The laboratory technique of vaporizing and condensing a liquid for the purpose of purification. Vacuum distillation is used to reduce the boiling points of heat-sensitive compounds.

valence electrons The outermost *s* and *p* electrons in an atom. The number and the arrangement of valence electrons define chemical and physical properties.

valence shell electron-pair repulsion (VSEPR) theory A method of evaluating molecular structure by relating the number of bonding and nonbonding electron pairs on an atom to its geometrical structure.

van der Waals forces *See* **London forces.**

vapor pressure The pressure developed by a liquid or solid in a closed container at a constant temperature.

viscosity The ability of a fluid to flow. The more easily a fluid flows, the lower is its viscosity.

vital force theory The theory, now discredited, that all organic molecules must be formed in living matter.

volumetric flask A flask calibrated to contain a precise volume of liquid.

volume-volume fraction (vol/vol) A concentration unit defined as the volume of one liquid solute divided by the total volumes of the liquids mixed to prepare a solution.

W

weak acid An acid that dissociates slightly when dissolved in water.

weak electrolyte A substance that partially dissociates into ions in solution.

weight The force developed by the gravitational attraction of two masses.

weighted average An average that depends on the abundance of the objects being averaged.

wetting The spreading of a liquid on a surface that occurs because the adhesive forces overcome the cohesive forces in the liquid.

X

X ray The high-energy electromagnetic radiation emitted in nuclear decay events or when certain metals are bombarded with energetic electrons.

Z

zero-order reaction A reaction in which the rate is independent of reactant concentration. The rate law is Rate = *k*.

Index

How to Use the CD-ROM

The software is not installed on your computer; it runs directly from the CD-ROM. Barron's CD-ROM includes an "autorun" feature that automatically launches the application when the CD is inserted into the CD-ROM drive. In the unlikely event that the autorun feature is disabled, follow the manual launching instructions below.

Windows®
Insert the CD-ROM and the program should launch automatically. If the software does not launch automatically, follow the steps below.
1. Click on the Start button and choose "My Computer."
2. Double-click on the CD-ROM drive, which will be named *AP_Chemistry.exe*
3. Double-click *AP_Chemistry.exe* application to launch the program.

Macintosh®
1. Insert the CD-ROM.
2. Double-click the CD-ROM icon.
3. Double-click the *AP_Chemistry* icon to start the program.

SYSTEM REQUIREMENTS

The program will run on a PC with:
Window® Intel® Pentium II 450 MHz
or faster, 128MB of RAM
1024 × 768 display resolution
Windows 2000, XP, Vista
CD-ROM Player

The program will run on a Macintosh® with:
PowerPC® G3 500 MHz
or faster, 128MB of RAM
1024 × 768 display resolution
Mac OS X v.10.1 through 10.4
CD-ROM Player